The Water-Food-Energy Nexus

Green Chemistry and Chemical Engineering

Series Editor:
Sunggyu Lee
Ohio University, Athens, Ohio, USA

The Water-Food-Energy Nexus
Processes, Technologies, and Challenges

I. M. Mujtaba

R. Srinivasan

N. O. Elbashir

CRC Press
Taylor & Francis Group
Boca Raton London New York

CRC Press is an imprint of the
Taylor & Francis Group, an **informa** business

CRC Press
Taylor & Francis Group
6000 Broken Sound Parkway NW, Suite 300
Boca Raton, FL 33487-2742

© 2018 by Taylor & Francis Group, LLC
CRC Press is an imprint of Taylor & Francis Group, an Informa business

No claim to original U.S. Government works

Printed on acid-free paper

International Standard Book Number-13: 978-1-4987-6083-6 (Hardback)
                                   13: 978-1-138-74607-7 (Paperback)

To our families and children

Contents

SECTION I Water

SECTION II   Food

SECTION III   Energy

SECTION IV Sustainable Future

Preface

Water, food, energy, and quality of life go hand in hand. The food we eat, the house we live in, the transports we use, and the things we cannot do without 24/7/365 determine our quality of life and require sustainable and steady supplies of water, food, and energy. Exponential growth in population and the fundamental right to have basic food and standards of living require increasing amounts of water and energy. The quantity of available freshwater and energy sources that directly affect the cost of production (irrigation and energy) and the transportation (energy) of food are diminishing. In addition, there is increased water pollution due to industrial uses of water. The direct use of such water for human consumption as well as irrigation for food production is prohibitive and requires technological solutions. Securing sustainable water, food, and energy supplies are more important challenges today for scientists and engineers than ever before.

With the above in mind, Professors Mujtaba and Elbashir organized workshops in Qatar and in India in 2015. The Qatar workshop was on energy and water security and was coordinated by Professors Mujtaba and Elbashir and funded by the British Council (UK) and Texas A&M University (USA). Thirteen participants from the UK and 15 from Qatar (academics and industrialists) presented stimulating and state-of-the-art research and knowledge transfer ideas in energy and water over 3 days. The Indian workshop was on water, food, and energy nexus and was coordinated by professors Mujtaba and Srinivasan and funded by the Royal Society (UK) and the Department of Science and Technology (India). Three participants from the UK and 15 from India (academics and industrialists) presented stimulating and state-of-the-art research and knowledge transfer ideas in water, food, and energy over 3 days. A total of 40 presentations were made and both events received a great deal of national press coverage.

The developments in energy-efficient water production, management, wastewater treatment, and energy-efficient processes for food and essential commodities were widely discussed at these workshops. This book presents those technical discussions for wider public benefit around the globe.

The book has 37 contributions (most from the two workshops mentioned earlier) and is divided into four sections:

- Section I: Water
- Section II: Food
- Section III: Energy
- Section IV: Sustainable Future

Section I includes 10 contributions on water desalination, water management, and wastewater treatment.

Water desalination covers the state of the art in mode-based research in desalination together with the global water–energy challenge in desalination and forward osmosis-based desalination for agricultural irrigation.

Water management covers topics on sustainable water management in industrial cities, water network synthesis, and water quality monitoring.

Wastewater treatment includes four contributions on the removal of endocrine, water conservation, life cycle assessment into the synthesis of wastewater treatment plants, and appropriate technologies for supplying safe drinking water.

Section II includes five contributions on food. The contributions cover advances in cereal processing, clean technology for sustainable food security, bioenergy in food production, water and energy consumption in food processing, and a mathematical model for food cooking undergoing phase changes.

Section III includes 16 contributions on fossil fuel, biofuel, synthetic fuel, and renewable energy, and carbon capture.

Fossil fuel includes two contributions on energy-efficient crude oil transport and the process industry economics of crude oil and petroleum derivatives.

Biofuel has two contributions: biodiesel production from renewable sources and synthesis of biodiesel from used cooking oil.

Synthetic fuel and renewable energy includes five contributions on gas-to-liquid (GTL)-derived synthetic fuel, the role of alternative aviation fuel, a modeling approach for the GTL Fischer–Tropsch reactor and carbon footprint, a distributed renewable energy system and management, and demand for and generation of a smart grid.

Carbon capture contains seven contributions on the rotating packed bed for carbon capture, integration of natural gas combined cycle power generation and chemical absorption based carbon capture, postcombustion carbon capture, integration of supercritical coal-fired power plant and carbon capture, experimental and theoretical modeling of carbon capture and sequestration chain, and the performance of organic polymers for carbon capture.

Section IV includes six contributions on a sustainable future. The topics cover the role of molecular thermodynamics in developing processes and products for a sustainable future, green engineering in process systems, the fundamental aspect of petrochemical water splitting, petrochemical approaches to solar hydrogen generation, a design and operation strategy of energy-efficient process, and the sustainability of process, supply chain, and enterprise.

MATLAB® is a registered trademark of The MathWorks, Inc. For product information, please contact:

The MathWorks, Inc.
3 Apple Hill Drive
Natick, MA 01760-2098 USA
Tel: 508-647-7000
Fax: 508-647-7001
E-mail: info@mathworks.com
Web: www.mathworks.com

Editors

I. M. Mujtaba is a professor of computational process engineering and currently the head of the School of Engineering at the University of Bradford. He earned BSc Eng and MSc Eng degrees in chemical engineering at the Bangladesh University of Engineering and Technology in 1983 and 1984, respectively, and earned a PhD at the Imperial College London in 1989. He is a fellow of the IChemE, a chartered chemical engineer, and the current chair of IChemE's Computer Aided Process Engineering Subject Group. He was the chair of the European Committee for Computers in Chemical Engineering Education from 2010 to 2013.

Professor Mujtaba leads research into dynamic modeling, simulation, optimization, and control of batch and continuous chemical processes with specific interests in distillation, industrial reactors, refinery processes, desalination, and crude oil hydrotreating focusing on energy and water. He has managed several research collaborations and consultancy projects with industry and academic institutions in the United Kingdom, Italy, Hungary, Malaysia, Thailand, and Saudi Arabia. He has published more than 245 technical papers and has delivered more than 60 invited lectures, seminars, or short courses around the world. He has supervised 26 PhD students to completion and is currently supervising 8 PhD students. He is the author of *Batch Distillation: Design and Operation* (Imperial College Press, 2004), which is based on his 18 years of research in batch distillation. Professor Mujtaba has edited *Application of Neural Network and Other Learning Technologies in Process Engineering* (Imperial College Press, 2001) and *Composite Materials Technology: Neural Network Applications* (CRC Press, 2009).

R. Srinivasan is a professor of chemical engineering and holder of the Institute chair at the Indian Institute of Technology Gandhinagar. Previously he was at the National University of Singapore (NUS) and AStar's Institute of Chemical and Engineering Sciences. Raj earned a BTech at the Indian Institute of Technology Madras in 1993 and a PhD at Purdue University in 1998, both in chemical engineering. He was a research associate at the Honeywell Technology Center, Minneapolis, before joining NUS.

Raj's research program is targeted at developing artificial intelligence–inspired systems engineering approaches for the design and operation of complex systems. His research has resulted in over 400 peer-reviewed journal papers and conference presentations. He is an editor of *Process Safety and Environmental Protection* journal and the *Journal of Frugal Innovation,* and on the editorial board of several other journals. His research has been recognized with several best paper awards.

N. O. Elbashir holds a joint appointment as a professor in the Chemical Engineering Program and the Petroleum Engineering Program at Texas A&M University at Qatar, and he is the director of Texas A&M's Gas and Fuels Research Center, a major research center that involves 27 faculty members from both the Qatar and College Station campuses of Texas A&M University (http://gfrc.tamu.edu/). He has extensive research and teaching experience from four different countries around the world, including his previous position as researcher at the BASF R&D Catalysts Center in Iselin, New Jersey. His research activities focus on the design of advanced reactors, catalysts, and conversion processes for natural gas, coal, and CO_2 to ultraclean fuels, and value-added chemicals. He has established several unique global research collaboration models between academia and industry with research funds exceeding 12 million dollars during the past 6 years. He holds several US and European patents and a large number of scientific publications in peer-reviewed journals, conference papers, technical industry reports, and invited and conference presentations. The scholarship of his research activities has received awards from the Qatar Foundation, BASF Corp., the Gordon Research Conferences, Texas A&M University Qatar, the American Institute of Chemical Engineers, Shell, and the Third and Fourth International Gas Processing Conference.

Contributors

Ayman Abdelaziz
College of Engineering
Swansea University
Swansea, United Kingdom

Sumaiya Zainal Abidin
Faculty of Chemical Engineering and Natural
    Resources Universiti Malaysia Pahang
and
Center of Excellence for Advanced Research
    in Fluid Flow
Universiti Malaysia Pahang
Gambang, Pahang Malaysia

Arief Adhitya
Process Science and Modelling
Institute of Chemical and Engineering
    Sciences
Singapore

Wajdi Ahmed
Texas A&M University at Qatar
Doha, Qatar

Nasser Al-Habsi
Department of Food Science and Nutrition
College of Agricultural and Marine
    Sciences
Sultan Qaboos University
Muscat, Oman

Sabla Alnouri
Department of Chemical Engineering
Indian Institute of Technology Delhi
Hauz Khas, New Delhi, India

Mudhar A. Al-Obaidi
Chemical Engineering Division
University of Bradford
Bradford, West Yorkshire, United Kingdom

Salih M. Alsadaie
Chemical Engineering Division
University of Bradford
Bradford, West Yorkshire, United Kingdom

Ali Altaee
Qatar Energy and Environment Research
    Institute
The Qatar Foundation
Doha, Qatar

Maryam Aryafar
Department of Chemical Engineering
Surrey University
Guildford, United Kingdom

Mert Atilhan
Department of Chemical Engineering
Texas A&M University at Qatar
Doha, Qatar

Santanu Bandyopadhyay
Department of Energy Science and Engineering
Indian Institute of Technology Bombay
Powai, Mumbai, Maharashtra, India

Rajasekhar Batchu
Department of Electrical Engineering
Indian Institute of Technology Gandhinagar
Gandhinagar, Gujarat, India

Dilip Kumar Behara
Department of Chemical Engineering
Indian Institute of Technology Kanpur
Kanpur, Uttar Pradesh, India

Chitta Ranjan Behera
Department of Chemical Engineering
Indian Institute of Technology Gandhinagar
Gandhinagar, Gujarat, India

A. Beigzadeh
Canada Centre for Mineral and Energy Technology
Natural Resources Canada
Ottowa, Canada

J. Bensabat
Environmental and Water Resources
    Engineering Ltd
Haifa, Israel

G. C. Boulougouris
Molecular Thermodynamics and Modelling of
    Materials Laboratory
Institute of Nanoscience and
    Nanotechnology
National Center for Scientific Research
    "Demokritos"
Agia Paraskevi, Attica, Greece

S. Brown
Department of Chemical and Biological
    Engineering
University of Sheffield
Sheffield, United Kingdom

Richard Butterfield
College of Engineering
Swansea University
Swansea, United Kingdom

A. Ceroni
Institut national de l'environnement industriel et
    des risques
Parc Technologique ALATA
Verneuil-en-Halatte, France

Benoît Chachuat
Department of Chemical Engineering
Imperial College London
London, United Kingdom

S. Y. Chen
School of Chemical Engineering
Dalian University of Technology
Dalian, People's Republic of China

Hanif A. Choudhury
Texas A&M University at Qatar
Doha, Qatar

A. Collard
Department of Chemical Engineering
University College London
London, United Kingdom

S. Cooreman
ArcelorMittal Global R&D Gent-OCAS NV
Zelzate, Belgium

Nishith B. Desai
Department of Energy Science and
    Engineering
Indian Institute of Technology Bombay
Powai, Mumbai, Maharashtra, India

Valentina Depetri
PSE-Lab, Process Systems Engineering
    Laboratory
CMIC Department
Polytechnic University of Milan
Milan, Italy

Emilio Diaz-Bejarano
Department of Chemical Engineering
Imperial College London
London, United Kingdom

Ioannis G. Economou
Chemical Engineering
Texas A&M University at Qatar
Doha, Qatar

and

Molecular Thermodynamics and Modelling of
    Materials Laboratory
Institute of Nanoscience and Nanotechnology
National Center for Scientific Research
    "Demokritos"
Agia Paraskevi, Attica, Greece

N. O. Elbashir
Department of Chemical Engineering
Texas A&M University at Qatar
Doha, Qatar

Mahmoud El-Halwagi
Department of Chemical Engineering
Indian Institute of Technology Delhi
Hauz Khas, New Delhi, India

M. Fairweather
School of Chemical and Process Engineering
University of Leeds
Leeds, United Kingdom

S. A. E. G. Falle
School of Mathematics
University of Leeds
Leeds, United Kingdom

R. Farret
Institut national de l'environnement industriel
    et des risques
Parc Technologique ALATA
Verneuil-en-Halatte, France

Y. Flauw
Institut national de l'environnement industriel
    et des risques
Parc Technologique ALATA
Verneuil-en-Halatte, France

Mohammed Ghouri
Department of Chemical Engineering
Texas A&M University at Qatar
Doha, Qatar

Sina Gilassi
Department of Chemical Engineering
Petronas University of Technology
Seri Iskandar, Malaysia

Ravindra Gudi
Department of Chemical Engineering
Indian Institute of Technology Bombay
Powai, Mumbai, Maharashtra, India

Miao Guo
Centre for Process Systems Engineering
Department of Chemical Engineering
Imperial College London
London, United Kingdom

Iskandar Halim
Process Science and Modelling Institute of
    Chemical and Engineering Sciences
Singapore

Malak Hamdan
Department of Chemical Engineering
Surrey University
Guildford, United Kingdom

J. Hébrard
Institut national de l'environnement industriel
    et des risques
Parc Technologique ALATA
Verneuil-en-Halatte, France

R. Hojjati Talemi
ArcelorMittal Global R&D
    Gent-OCAS NV
Zelzate, Belgium

D. Jamois
Institut national de l'environnement industriel
    et des risques
Parc Technologique ALATA
Verneuil-en-Halatte, France

Lakshmi E. Jayachandran
Agricultural and Food Engineering
    Department
Indian Institute of Technology Kharagpur
Kharagpur, West Bengal, India

Aprajeeta Jha
Agricultural and Food Engineering
    Department
Indian Institute of Technology Kharagpur
Kharagpur, West Bengal, India

Atuman S. Joel
Process and Energy Systems Engineering Group
University of Hull
Hull, United Kingdom

Kalpesh Joshi
Department of Electrical Engineering
Indian Institute of Technology Gandhinagar
Gandhinagar, Gujarat, India

Nitin Kaistha
Department of Chemical Engineering
Indian Institute of Technology Kanpur
Kanpur, Uttar Pradesh, India

Kumaran Kannaiyan
Micro Scale Thermo-Fluids Laboratory
Texas A&M University at Qatar
Doha, Qatar

Subhankar Karmakar
Centre for Environmental Science and
    Engineering
and
Interdisciplinary Program in Climate Studies
and
Centre for Urban Science and Engineering
Indian Institute of Technology Bombay
Powai, Mumbai, Maharashtra, India

Amir Khan
Faculty of Engineering and Informatics
School of Engineering
University of Bradford
Bradford, West Yorkshire, United Kingdom

DoYeon Kim
Centre for Process Systems Engineering
Department of Chemical Engineering
Imperial College London
London, United Kingdom

C. Kolster
Centre for Process Systems Engineering
Department of Chemical Engineering
Imperial College London
London, United Kingdom

Panagiotis Krokidas
Department of Chemical Engineering
Texas A&M University at Qatar
Doha, Qatar

Adekola Lawal
Process Systems Enterprise Ltd
London, United Kingdom

Jonathan G. M. Lee
School of Chemical Engineering and Advanced
    Materials
Merz Court
Newcastle University
Newcastle upon Tyne, United Kingdom

Patrick Linke
Department of Chemical Engineering
Indian Institute of Technology Delhi
Hauz Khas, New Delhi, India

Xiaobo Luo
Process and Energy Systems Engineering Group
Department of Chemical and Biological
    Engineering
University of Sheffield
Sheffield, United Kingdom

Lin Ma
Energy 2050
University of Sheffield
Sheffield, United Kingdom

Sandro Macchietto
Department of Chemical Engineering
Imperial College London
London, United Kingdom

N. Mac Dowell
Centre for Process Systems Engineering
Department of Chemical Engineering
Imperial College London
London, United Kingdom

Naila Mahdi
Texas A&M University at Qatar
Doha, Qatar

H. Mahgerefteh
Department of Chemical Engineering
University College London
London, United Kingdom

Davide Manca
PSE-Lab, Process Systems Engineering
     Laboratory
CMIC Department
Polytechnic University of Milan
Milan, Italy

Flavio Manenti
Department of Chemistry, Materials and
     Chemical Engineering "Giulio Natta" (CMIC)
Polytechnic University of Milan
Milan, Italy

S. Martynov
Department of Chemical Engineering
University College London
London, United Kingdom

Elisabetta Mercuri
PSE-Lab, Process Systems Engineering
     Laboratory
CMIC Department
Polytechnic University of Milan
Milan, Italy

Vasileios K. Michalis
Department of Chemical Engineering
Texas A&M University at Qatar
Doha, Qatar

Othonas A. Moultos
Engineering Thermodynamics, Process and
     Energy Department
Delft University of Technology
Delft, The Netherlands

I. M. Mujtaba
Chemical Engineering Division
University of Bradford
Bradford, West Yorkshire, United Kingdom

Bhallamudi S. Murty
Department of Civil Engineering
Indian Institute of Technology Madras
Chennai, Tamil Nadu, India

Nasr Mohammad Nasr
Texas A&M University at Qatar
Doha, Qatar

Laial Bani Nassr
Texas A&M University at Qatar

A. Niemi
Department of Earth Sciences
Uppsala University
Uppsala, Sweden

I. K. Nikolaidis
Molecular Thermodynamics and Modelling
     of Materials Laboratory
Institute of Nanoscience and Nanotechnology
National Center for Scientific Research
     "Demokritos"
Agia Paraskevi, Attica, Greece

Olajumoke Ololade Odejimi
College of Engineering
Swansea University
Swansea, United Kingdom

Ojasvi
Department of Chemical Engineering
Indian Institute of Technology Kanpur
Kanpur, Uttar Pradesh, India

Eni Oko
Process and Energy Systems Engineering Group
Department of Chemical and Biological
     Engineering
University of Sheffield
Sheffield, United Kingdom

Akeem K. Olaleye
Process and Energy Systems Research Group
University of Hull
Hull, United Kingdom

Raj Ganesh S. Pala
Department of Chemical Engineering
Indian Institute of Technology Kanpur
Kanpur, Uttar Pradesh, India

Davide Papasidero
Department of Chemistry, Materials and
     Chemical Engineering "Giulio Natta" (CMIC)
Polytechnic University of Milan

Raj Patel
Chemical Engineering Division
University of Bradford
Bradford, West Yorkshire, United Kingdom

Ligy Philip
Department of Civil Engineering
Indian Institute of Technology Madras
Chennai, Tamil Nadu, India

Laura Piazza
Food Science and Technology
University of Milan
Milan, Italy

Sauro Pierucci
Department of Chemical Engineering
CMIC Department, Polytechnic University of
Milan
Milan, Italy

Naran M. Pindoriya
Department of Electrical Engineering
Indian Institute of Technology Gandhinagar
Gandhinagar, Gujarat, India

Andrey V. Porsin
Boreskov Institute of Catalysis
and
UNICAT Ltd.
Novosibirsk, Russia

R. T. J. Porter
Department of Chemical Engineering
University College London
London, United Kingdom

Mohamed Pourkashanian
Energy 2050
University of Sheffield
Sheffield, United Kingdom

C. Proust
Institut national de l'environnement industriel
et des risques
Parc Technologique ALATA
Verneuil-en-Halatte, France

Channarong Puchongkawarin
Centre for Process Systems Engineering
Department of Chemical Engineering
Imperial College London
London, United Kingdom

Soumya Ranjan Purohit
Agricultural and Food Engineering Department
Indian Institute of Technology Kharagpur
Kharagpur, West Bengal, India

Mohammad Shafiur Rahman
Department of Food Science and Nutrition
College of Agricultural and Marine Sciences
Sultan Qaboos University
Muscat, Oman

Nejat Rahmanian
School of Engineering
University of Bradford
Bradford, United Kingdom

Haile-Selassie Rajamani
Sustainable Energy and Smart Grid Group
Faculty of Engineering and Information Sciences
University of Wollongong Dubai
Dubai, United Arab Emirates

Colin Ramshaw
Process and Energy Systems Engineering Group
Department of Chemical and Biological
Engineering
University of Sheffield
Sheffield, United Kingdom

D. Rebscher
Federal Institute for Geosciences and Natural
Resources
Hannover, Germany

Reza Sadr
Micro Scale Thermo-Fluids Laboratory
Mechanical Engineering Program
Texas A&M University at Qatar
Doha, Qatar

Basudeb Saha
School of Engineering
London South Bank University
London, United Kingdom

C. Salvador
Canada Centre for Mineral and Energy
Technology
Natural Resources Canada
Ottawa, Canada

Sudipta Sarkar
Department of Civil Engineering
Indian Institute of Technology Roorkee
Roorkee, India

Chintan Savla
Department of Chemical Engineering
Indian Institute of Technology Bombay
Powai, Mumbai, Maharashtra, India

R. Segev
Environmental and Water Resources
    Engineering Ltd
Haifa, Israel

Arup K. SenGupta
Department of Civil and Environmental
    Engineering
Lehigh University
Bethlehem, Pennsylvania

Nilay Shah
Centre for Process Systems Engineering
Department of Chemical Engineering
Imperial College London
London, United Kingdom

Munawar A. Shaik
Department of Chemical Engineering
Indian Institute of Technology Delhi
Hauz Khas, New Delhi, India

Adel O. Sharif
Qatar Energy and Environment Research
    Institute
The Qatar Foundation
Doha, Qatar

Gyan Prakash Sharma
Department of Chemical Engineering
Indian Institute of Technology Kanpur
Kanpur, Uttar Pradesh, India

Sri Sivakumar
Department of Chemical Engineering
Indian Institute of Technology Kanpur
Kanpur, Uttar Pradesh, India

M. T. Sowgath
Chemical Engineering Department
Bangladesh University of Engineering and
    Technology
Dhaka, Bangladesh

Babji Srinivasan
Department of Chemical Engineering
Indian Institute of Technology Gandhinagar
Gandhinagar, Gujarat, India

R. Srinivasan
Department of Chemical Engineering
Indian Institute of Technology Gandhinagar
Gandhinagar, Gujarat, India

P. Srinivasa Rao
Agricultural and Food Engineering Department
Indian Institute of Technology Kharagpur
Kharagpur, West Bengal, India

David C. Stuckey
Department of Chemical Engineering
Imperial College London
London, United Kingdom

Sarojini Tiwari
Department of Chemical Engineering
Indian Institute of Technology Gandhinagar
Gandhinagar, Gujarat, India

Chedly Tizaoui
College of Engineering
Swansea University
Swansea, United Kingdom

P. P. Tripathy
Agricultural and Food Engineering Department
Indian Institute of Technology Kharagpur
Kharagpur, West Bengal, India

D. M. Tsangaris
Molecular Thermodynamics and Modelling of
    Materials Laboratory
Institute of Nanoscience and Nanotechnology
National Center for Scientific Research
    "Demokritos"
Agia Paraskevi, Attica, Greece

Ioannis N. Tsimpanogiannis
Environmental Research Laboratory
National Center for Scientific Research
    "Demokritos"
Agia Paraskevi, Attica, Greece

Ruh Ullah
Department of Chemical Engineering
Qatar University
Doha, Qatar

Arun Prakash Upadhyay
Department of Chemical Engineering
Indian Institute of Technology Kanpur
Kanpur, Uttar Pradesh, India

D. Van Hoecke
ArcelorMittal Global R&D Gent-OCAS NV
Zelzate, Belgium

Vikas Varekar
Centre for Environmental Science and
    Engineering
Indian Institute of Technology Bombay
Powai, Mumbai, Maharashtra, India

Yannic Vaupel
AVT Process Systems Engineering
RWTH Aachen University
Aachen, Germany

Niki Vergadou
Molecular Thermodynamics and Modelling of
    Materials Laboratory
Institute of Nanoscience and Nanotechnology
National Center for Scientific Research
    "Demokritos"
Agia Paraskevi, Attica, Greece

Saiyed R. Wahadj
Qatar Energy and Environment Research
    Institute
The Qatar Foundation
Doha, Qatar

Meihong Wang
Process and Energy Systems Engineering Group
Department of Chemical and Biological
    Engineering
University of Sheffield
Sheffield, United Kingdom

J. L. Wolf
Federal Institute for Geosciences and Natural
    Resources
Hannover, Germany

R. M. Woolley
School of Chemical and Process Engineering
University of Leeds
Leeds, United Kingdom

KeJun Wu
School of Chemical Engineering and
    Advanced Materials
Merz Court
Newcastle University
Newcastle upon Tyne, United Kingdom

Cafer T. Yavuz
Graduate School of EEWS
Korea Advanced Institute of Science and
    Technology
Daejeon, Republic of Korea

J. L. Yu
School of Chemical Engineering
Dalian University of Technology
Dalian, People's Republic of China

K. E. Zanganeh
Canada Centre for Mineral and Energy
    Technology
Natural Resources Canada
Ottowa, Canada

Guillermo Zaragoza
Center for Energy, Environmental and
    Technological Research
Almería Solar Platform
Tabernas, Almería, Spain

Y. Zhang
School of Chemical Engineering
Dalian University of Technology
Dalian, People's Republic of China

I

Water

1

I. M. Mujtaba,
Salih M. Alsadaie,
Mudhar A.
Al-Obaidi, and
Raj Patel
University of Bradford

M. T. Sowgath
*Bangladesh University of
Engineering and Technology*

Davide Manca
*Polytechnic University
of Milan*

Sudipta Sarkar
*Indian Institute of
Technology Roorkee*

Arup K. SenGupta
Lehigh University

Ali Altaee, Saiyed
R. Wahadj, and
Adel O. Sharif
The Qatar Foundation

Guillermo Zaragoza
Almería Solar Platform

Malak Hamdan and
Maryam Aryafar
Surrey University

Desalination

1.1 Model-Based Techniques in Desalination Processes: A Review

*I. M. Mujtaba, Salih M. Alsadaie, Mudhar A. Al-Obaidi,
Raj Patel, M. T. Sowgath, and Davide Manca*

1.1.1 Introduction

After air, water is the most essential commodity for all living beings on the earth. Quality water and quality life go hand in hand. The food we eat, the houses we live in, the transports we use, and the things we cannot do without in 24/7/365 determine our quality of life and require sustainable and steady water supplies (IChemE Technical Roadmap, 2007).

Exponential growth in population (Figure 1.1) and improved standards of living (together with water pollution due to industrial use of water) are increasing the freshwater demand and are putting a serious strain on the quantity of naturally available freshwater. By the year 2030, the global needs of water will be 6900 billion m³/day compared to 4500 billion m³/day required in 2009 (2030 Water Resources Group, 2009). With most of the accessible water around us being saline (94% of the world's water), desalination technology is vital for our sustainability.

Desalination markets grew significantly in the last decades. Desalination Market (2005) reported that the total world desalination capacity was about 30 million m³/day in 2005 and forecasted the capacity in 2010 and 2015 to be about 44 and 62 million m³/day, respectively. Due to low cost of fossil fuels (until 2005), thermal desalination was the preferred option for the Gulf Region with the greatest installed capacity. Although a smaller expansion was forecasted in Asia until 2015, Asia will become the fastest growing market in the long run due to its enormous population and economic growth leading to a water demand that cannot be fulfilled with conventional water sources. The Republic of Korea aimed to have its first large seawater desalination plant up and running by August 2012 with a capacity of 45,000 m³/day using reverse osmosis (RO) technology (The International Desalination and Water Reuse Quarterly industry website, http://www.desalination.biz). In 2011, the Tamil Nadu state government in India announced to set up a third desalination plant with a capacity of 200,000 m³/day to meet the growing water needs of the city of Chennai (Source: The International Desalination and Water Reuse Quarterly industry website, http://www.desalination.biz).

The commonly used industrial desalination processes can be classified broadly into two groups: (1) thermal processes and (2) membrane processes (Figure 1.2). Although thermal process [mainly multistage flash (MSF)] is the oldest and still the dominating process for large-scale production of freshwater, RO process has been continuously increasing its market share (Global Water Intelligence, 2009), and in 2010, RO desalination accounted for 60% of desalination plants' capacity, while MSF represented only about 26% (Global Water Intelligence, 2011). Figure 1.3 shows a dramatic increase in the desalination capacity of new RO worldwide in the period from 1980 to 2016 compared to thermal desalination (Desalination Market, 2010). The RO desalination capacity, as shown in Figure 1.3, has increased approximately 20 times in this period, while thermal processes capacity remained at its initial level. This dramatic increase in the capacity of RO has contributed to the advancement of the RO membrane technology and design of efficient high-pressure centrifugal pumps. The new high-productivity membrane elements consisting of higher surface area, enhanced permeability, and denser membrane packing, will yield more quantity of freshwater per membrane element (Singh, 2008). The enhancement of membrane technologies is due to the fact that membranes are used in many different separation processes rather than only in water desalination.

Typical cost of freshwater production in 2004–2005 was around 0.7 US$/m³ for MSF-based thermal process and around 0.6 US$/m³ for RO process. The cost of freshwater produced by membrane treatment has shown a dramatic reduction trend over several years. This remarkable progress has been made mainly through two aspects: huge improvements in membrane material and incorporation of the energy recovery devices in RO systems (Greenlee et al., 2009), which significantly reduced the energy requirements. Khawaji et al. (2008) reported that the unit energy consumption for seawater desalination has been reduced to as low as 2 kWh/m³ compared to 4 kWh/m³ consumed in a thermal process such as MSF

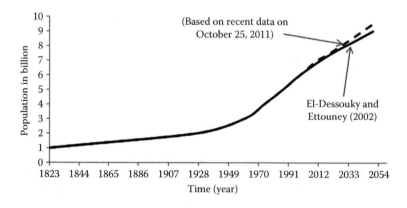

FIGURE 1.1   World population from 1823 to 2050. (Reprinted from *Fundamentals of Salt Water Desalination*, El-Dessouky, H.T., Ettouney, H.M., Copyright 2002, Elsevier; Dotted line: http://www.telegraph.co.uk; October 25, 2011.)

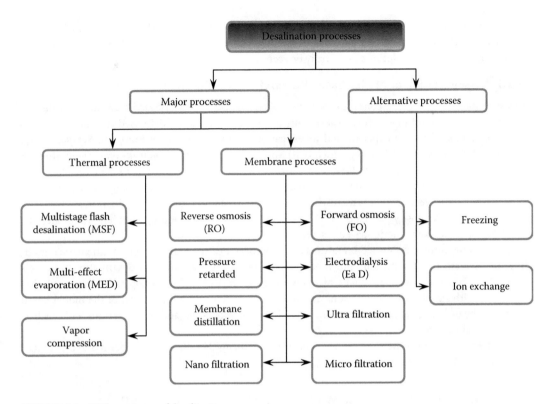

FIGURE 1.2 Different types of desalination processes.

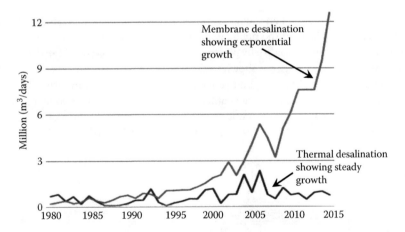

FIGURE 1.3 Global desalination capacity increase by technology, 1980–2016. (Adapted from Global Water Intelligence, *Desalination Market 2010: Global Forecast and Analysis*, Media Analytics Ltd, Oxford, UK, 2010; Misdan, N. et al., *Desalination*, 287, 228–237, 2012.)

distillation. However, it should be noted that, thermal desalination is more energy intensive and costly compared to membrane based desalination, it can better deal with high-salinity feed water and delivers even higher permeate quality in terms of freshwater salinity (Fritzmann et al., 2007; Misdan et al., 2012).

Numerous investigations have been conducted in the past decades to develop more sustainable technological solutions that would meet the increasing water demand (Greenlee et al., 2009; Misdan et al., 2012). The ongoing objective is still the improvement in design, operation, and control of desalination processes to ensure quality water (in terms of salinity and other dissolved solids such as boron) at cheaper price with lower environmental impact. However, exploitation of the full potential of model-based techniques in such solutions can hardly be seen.

1.1.1.1 Dissemination of Model-Based Research in Desalination

Process Systems Engineering community makes extensive use of model-based techniques in design, operation, control, and in designing experiments due to the fact that model-based techniques are less expensive compared to any experimental investigation. The yearly event of *European Symposium on Computer-Aided Process Engineering* (since 1991) and the 3-year event of *International Symposium on Process Systems Engineering* (since 1982) and the *Computers and Chemical Engineering Journal* (published by Elsevier since 1979) cover design, operation, control, and process integration of many processes except desalination (very limited). Interestingly, most reported literatures on desalination are based on experiments and are mostly published in *Desalination Journal* (since 1966 by Elsevier), although periodically it publishes some model-based research works.

There are only limited published literatures that have been dealing with rigorous mathematical modeling, mathematical optimization, and model-based control of MSF and RO desalination processes (since 1957). However, the main focus of many of these works was to develop tailor-made solution algorithms with simplifying assumptions and was not to exploit the full potential of the model-based techniques by utilizing many available commercial process modeling (and optimization) tools such as ASPEN and gPROMS.

This chapter highlights the state of the art of future challenges in the use of model-based techniques in desalination for making freshwater.

1.1.2 State of the Art: Use of Model-Based Techniques in MSF Desalination Process

Mujtaba (2008, 2011, 2012) reflected on the use of model-based techniques in MSF desalination process, which includes studies available in the public domain until 2011. The work presented in those papers, together with recent developments, is briefly presented in this section. A typical MSF desalination process and detailed features of a flash stage are shown in Figures 1.4 and 1.5. The process consists of essentially a brine heater and a number of flashing and condensing stages connected in series. The seawater (W_s) flows through the condenser tubes and enters at temperature T_s and is heated up to a temperature T_1 in the rejection section by condensation of product water vapor gained by flashing of seawater. Part of the leaving seawater is rejected to the sea (C_W), and the other part is used as make-up (F) to be mixed with the

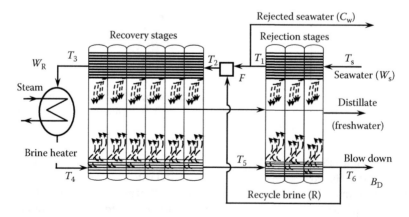

FIGURE 1.4 A typical MSF desalination process. (Adapted from Hawaidi, E.A.M., Mujtaba, I.M., *Chem. Eng. J.*, 165, 545–553, 2010.)

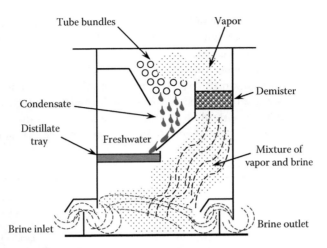

FIGURE 1.5 Details of consecutive stages of MSF.

recirculating brine (R) at T_6. The mixture at T_2 is then fed to the heat recovery section where it is heated to a temperature T_3. In the brine heater, the preheated seawater is further heated to the highest possible temperature T_4, which is called the top brine temperature (TBT). From there it passes through the flashing stages at a lower pressure, partly flashes into vapor, which is condensed on the condenser tubes and is collected in the distillate tray across the stages. The brine then leaves the recovery section at temperature T_5 and rejection section at T_6 where a part of the brine is blown down (B_D) and the rest is recirculated (R) (Soliman, 1981). It should be noted that many alternative configurations of the MSF process can be generated depending on the way the seawater is fed and brine is recycled (El-Dessouky and Ettouney, 2002).

To date, for a given *fixed or seasonal freshwater demand profile, seawater composition, seasonal seawater temperature,* and *time-dependent fouling profiles,* the main issues in an MSF process have been to determine the following:

Design parameters: These include TBT, fouling factor, number of stages, width and height of the stages, heat transfer area (number of tubes in the condensers), materials of construction, vent line orifice (air and noncondensables), demister size and materials, interstage brine transfer device, brine heater area, and size of freshwater storage tank (El-Dessouky and Ettouney, 2002; Hawaidi and Mujtaba, 2010; Said et al., 2010; Alsadaie and Mujtaba, 2014).

Operation parameters: These include seawater flow and temperature, steam flow and temperature, brine recycle flow and level, seawater rejection, scale formation and antiscale dosage, and maintenance schedule (Hawaidi and Mujtaba, 2010; Said et al., 2013).

Cost: The process involves capital, operating (utilities, cleaning), pretreatment, and posttreatment (chemicals) costs. Majority of the experimental- or model-based studies of the past and recent are focused on maximizing the profitability of operation, or maximizing the recovery ratio, or maximizing the plant performance (GOR—gained output ratio) ratio, or minimizing the cost, or minimizing the external energy input, by optimizing design and operating parameters (Rosso et al., 1997; Mussati et al., 2001; Tanvir and Mujtaba, 2008; Hawaidi and Mujtaba, 2010, 2011).

1.1.2.1 Steady-State Modeling of MSF Desalination Process and Application

Although flash distillation has been around us since the beginning of the 20th century and the Office of Saline Water, was created in 1952 to develop economical flash distillation-based desalination processes (Cadwallader, 1967), the model-based technique in MSF desalination was first applied in 1970 (Mandil and Ghafour, 1970). Referring to Figures 1.4 and 1.5, an MSF process model should include an accurate description (via mathematical equations) of (1) mass/material balance, (2) energy balance, (3) thermal efficiency, (4) physical properties (such as heat capacity, density, boiling point temperature elevation due to salinity, heat of vaporization), (5) heat transfer coefficients (reflecting effect of fouling and noncondensable gases), (6) pressure and temperature drop, (7) geometry of brine heater, demister, condenser, stages, and vents, (8) interstage flow (orifice), (9) thermodynamic losses including the nonequilibrium allowance and demister losses, and (10) kinetic model for salt deposition and corrosion (Mujtaba, 2010).

Table 1.1 describes the evolution of steady-state MSF process models over the last half century (only the major developments published in international journals are included). Helal et al. (1986) set the

TABLE 1.1 Steady-State Models for MSF since 1970

Author (Year)	Description of the Model	Limitation
Mandil and Ghafour (1970)	Approximate lumped parameter model, constant thermo-physical properties	Constant value for the physical properties make the model's equations linearized
Coleman (1971)	Stage-to-stage model, linear and simplified TE (boiling point temperature elevation) correlation for different temperature range, no fouling/scaling	No fouling and the model equations are linearized
Helal et al. (1986)	Detailed stage-to-stage model, nonlinear TE correlation and other physical properties as function of (temperature, seawater composition), temperature loss due to demister included, heat transfer coefficient (HTC) via polynomial fit (fouling included)	Linearization and decomposition of the equations is a very complicated process and requires many mathematical manipulation steps. No consideration of fouling
Rosso et al. (1997)	Model similar to that of Helal et al. (1986) and carried out different simulation studies. Model validation with plant data	Noncondensable gases not included, no heat, constant fouling value
El-Dessouky et al. (1995) and El-Dessouky and Ettouney (2002)	Model based on Helal et al., (1986) but includes heat losses to the surroundings, constant inside/outside tube fouling factors, pressure drop across demister, number of tubes in the condenser and tube material, constant nonequilibrium allowance	Noncondensable gases are not included. Empirical correlations are not accurate for wide range of operating conditions
Mussati et al. (2001)	Detailed stage-to-stage model but with constant thermo-physical properties	Constant value for the physical properties make the model less accurate
Sommariva et al. (2004)	The model connects plant efficiency and environmental impact through energy balance around the system boundary	The paper neglects the environmental impact from salinity
Tanvir and Mujtaba (2008)	Model based on Helal et al. (1986), but included NN (Neural Network)-based correlation for TE calculation	No noncondensable gases, fixed fouling, focus only on TE
Aminian (2010)	NN (RBF Neural Network)-based correlation for TE calculation in MSF desalination	Used a model from literature to develop TE correlation
Hawaidi and Mujtaba (2010, 2011)	Model based on Helal et al. (1986), but includes dynamic brine heater fouling and dynamic seawater temperature profile. Also included dynamic intermediate storage tank to enhance flexibility in operation	No heat losses, no NCGs
Said et al. (2010, 2013)	Model based on Helal et al. (1986), but includes effect of NCGs and fouling factors on overall HTC. Considers regular (variable) and irregular (variable) water demand	Effect of venting system design affecting NCGs buildup is neglected, which can affect HTC
Al-Fulaij et al. (2010, 2011) and Al-Fulaij (2011)	Rigorous modeling, CFD-based demister modeling	No heat losses, the effect of NCGs on overall heat transfer coefficient is neglected
Alsadaie and Mujtaba (2014)	Model based on Helal et al. (1986), but includes effect of venting system design	Improved performance by controlling NCGs buildup and thus overall HTC

scene for serious model-based activities in MSF desalination. An accurate estimation of temperature elevation (TE) due to salinity is important in developing a reliable process model. Several correlations for estimating the TE exist in the literature (El-Dessouky and Ettouney, 2002; Tanvir and Mujtaba, 2006; Aminian, 2010).

As highlighted in Mujtaba (2008, 2010, 2012), at high temperature, water with soluble salts allows deposits to form scale that can reduce the heat transfer rate and can increase specific energy consumption and operating costs. Baig et al. (2011) showed that an increase of the fouling resistance from 0 to 0.001 m²K/W in the brine heater can result in 400% decrease in the overall heat transfer coefficient. This can cause frequent shutdowns of the plant for cleaning. Although a good amount of studies were carried out on the experimental study of scaling and corrosion, only a handful of such studies focused on the modeling (or attempt to modeling) of scale formation in MSF process (Wangnick, 1995; Mubarak, 1998; Al-Anezi and Hilal, 2007; Hawaidi and Mujtaba, 2010; Said et al., 2013; Al-Rawajfeh et al., 2014). Most of these models, except for those of Hawaidi and Mujtaba (2010) (Figure 1.6) and Said et al. (2013), have been developed and studied on their own but have not been a part of the MSF process models referred to in Table 1.1.

The presence of noncondensable gases such as carbon dioxide, nitrogen, and oxygen, resulting from air leakage to stages and the release of dissolved gases in brine are of great concern for the performance of the MSF process. Seifert and Genthner (1991) and Genthner and Seifert (1991) developed an analytical model to estimate the amount of noncondensable gases (NCGs) in the MSF chambers and their variation from stage to stage depending on the venting points. Recently, Said et al. (2010) and Alsadaie and Mujtaba (2014) studied the effect of NCGs on the performance of MSF process. Different number and location of venting system were varied to obtain the optimum number and location of venting points with respect to the permissible level of NCGs concentration in the stages and performance ratio (Alsadaie and Mujtaba, 2014).

Mussati et al. (2004, 2005) applied a rigorous desalination model in optimizing design and operation (synthesis) of a dual-purpose (energy and water) seawater desalination plant using the MINLP techniques. Most recently, model-based techniques have been employed to study the effect of time-varying fouling (Figure 1.6), seasonal variation of seawater temperature, daily and/or seasonal variation of freshwater demand (Figure 1.7) on the operation (Table 1.2), design and cost of production (Figure 1.8 and Table 1.2) (Hawaidi and Mujtaba, 2010; 2011; Hawaidi, 2012; Said et al., 2012, 2013), and to identify flexible maintenance opportunities (see Section 1.1.2.3).

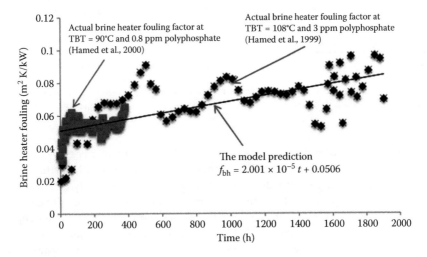

FIGURE 1.6 Brine heater fouling f_{bh} model. (Adapted from Hawaidi, E.A.M., Mujtaba, I.M., *Chem. Eng. J.*, 165, 545–553, 2010.)

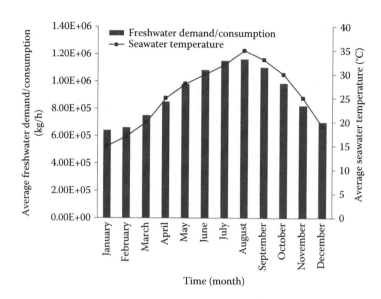

FIGURE 1.7   Average monthly seawater temperature and freshwater demand/consumption profiles during a year.

TABLE 1.2   Summary of Optimization Results

Month	f_{bh} (m² K/kW)	N (Opt)	$F \times 10^6$ (kg/h)	$R \times 10^6$ (kg/h)	T_s (°C)	W_s (kg/h)	TOC ($/M)	TMC ($/M)
January	0.065	10	2.4	3.48	93.63	119386	346837	856056
February	0.078	10	3.25	3.06	94.67	140774	395139	870652
March	0.093	10	3.75	3.61	93.57	144784	413188	922358
April	0.108	11	4.39	4.64	96.63	157034	470029	1011793
May	0.121	12	5.55	5.48	98.14	175697	538310	1111598
June	0.135	14	6.40	6.41	98.93	178609	570792	1204499
July	0.150	15	7.30	7.35	100.29	192269	624094	1286866
August	0.164	19	7.55	7.55	98.93	158491	551316	1324165
September	0.178	15	7.04	7.04	100.54	181176	593126	1255898
October	0.192	13	5.78	5.59	99.68	165034	526707	1130611
November	0.206	11	4.46	4.29	98.84	151604	466344	1008108
December	0.222	10	3.49	3.33	97.69	133263	400363	909583

Source: Hawaidi, E.A.M., Simulation, optimisation and flexible scheduling of MSF desalination process under fouling, PhD, University of Bradford, 2013.

N, number of recovery stages; *F*, make up seawater feed to the recovery stages (Figure 1.4); *R*, recycled brine; T_s, steam temperature; W_s, steam flow rate; TOC, total operating cost; TMC, total monthly cost; *N*, *F*, and *R* were optimized while TMC was minimized.

MSF desalination is considered very energy intensive and faces many challenges to cut the cost off and improve the market shares (profitability). While raw material costs are competitive in the global economy, the only way to achieve the target is by reducing the operational cost (labor, utility, etc.). It should be noted that the operating cost is affected by the maintenance and cleaning cost. Therefore, designing the plant to extend the period of running without shutting down the plant can heavily reduce the annual operating cost.

In recent years, Sowgath and Mujtaba (2008), Hawaidi and Mujtaba (2010, 2011), Hawaidi (2012), and Said et al. (2013) observed that, for a fixed or variable freshwater demand, seawater temperature dictates the optimum number of flashing chambers (winter season requiring less number of stages than the summer, Table 1.2 and Figure 1.8). Table 1.2 shows that the optimum number of stages in winter season

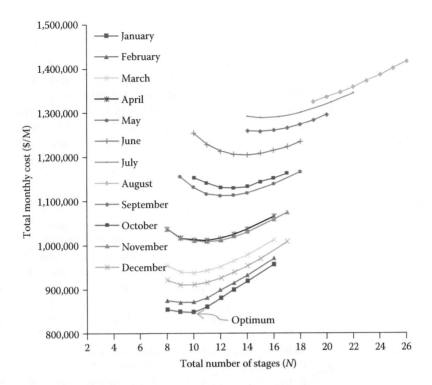

FIGURE 1.8 The variation of total monthly cost with total number of stages during a year.

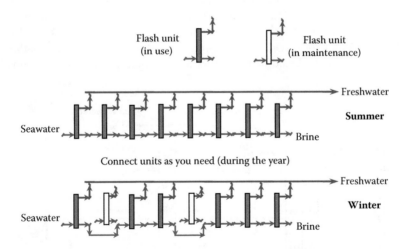

FIGURE 1.9 Flexible design and maintenance opportunity in MSF. (Adapted from Sowgath, T., Mujtaba, I. M., *Chem. Eng.*, 804, 28–29, 2008.)

is 10 while the maximum number of stages in summer season is 19. Therefore, if a plant is designed based on the optimum number of stages required in summer then a new way of designing MSF processes is required, which opens up huge opportunities for flexible maintenance and operation schedule throughout the year (without shutting down the plant fully for maintenance as other seasons require a lower number of stages). Figure 1.9 explains the flexibility of moving some stages for scheduled maintenance without shutting the plant down.

1.1.2.2 Dynamic Modeling of MSF Desalination Process

Dynamic models are used for troubleshooting, fault detection, reliability, start-up and shutdown conditions, and to implement advanced control (Gambler and Badreddin, 2004). Earlier attempts to develop a dynamic model were made by Glueck and Bradshaw (1970), Delene and Ball (1971), and Yokoyama et al. (1977); however, due to the lack of information and powerful computer tools, significant deviations in their results occurred. Based on the steady-state model developed by Helal et al. (1986), more dynamic models were developed later by Husain et al. (1993), Reddy et al. (1995), Aly and Marwan (1995), and Maniar and Deshpande (1996). A complete model for dynamic simulation was developed by Thomas et al. (1998) and Mazzotti et al. (2000). Shivayyanamath and Tewari (2003) developed a model for the analysis of start-up characteristics, and Gambler and Badreddin (2004) developed a dynamic model for control purpose. Mass holdup, concentration, and temperature were used as state (dynamic) variables in these models. However, more recently, Sowgath (2007) used a NN-based correlation to estimate TE in their dynamic model. Hawaidi and Mujtaba (2011) developed a NN-based correlation to predict dynamic freshwater demand throughout the day in any season. They coupled a steady-state MSF process model with a dynamic storage tank model (Figure 1.10) and developed an optimization framework to optimize the plant operation subject to dynamic seawater temperature and dynamic (throughout the day and throughout all seasons) freshwater demand. Figure 1.11 shows a typical temperature profile in a particular day in autumn. Figure 1.12 shows the optimum freshwater production rate from the MSF process, the dynamic tank level (not violating upper and lower tank level), and the dynamic consumption profile.

Said et al. (2013) carried out a similar study, but the process model included a dynamic fouling model.

Al-Fulaij et al. (2011) developed a dynamic model that includes demister losses and noncondensable gases. Among these studies, a very limited number of publications considered the effect of NCGs in their model. Earlier studies (Alasfour and Abdulrahim, 2009; Said et al., 2010) focused on the effect of NCGs on the heat transfer rate, and a fixed value for the amount of NCGs was considered in every stage of the MSF process. In practice, the existence of a venting system is crucial to enhance the heat transfer rate by removing the NCGs to the atmosphere or to the evacuating system. The evacuating system can be in series for some stages and in parallel for other stages. This allows the variation of NCGs from stage to stage. It should be noted that in most MSF plants, venting points in the first

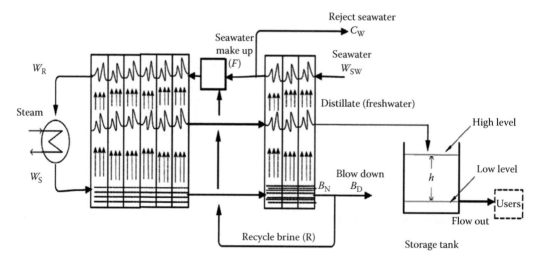

FIGURE 1.10 A typical MSF desalination process with storage tank. (Adapted from Hawaidi, E.A.M., Mujtaba, I.M., *Ind. Eng. Chem. Res.*, 50, 10604–10614, 2011.)

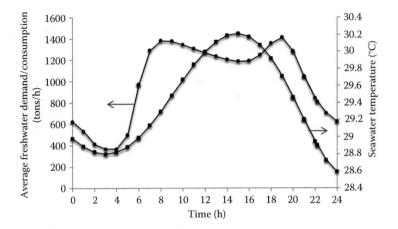

FIGURE 1.11 Dynamic seawater temperature and freshwater demand profiles.

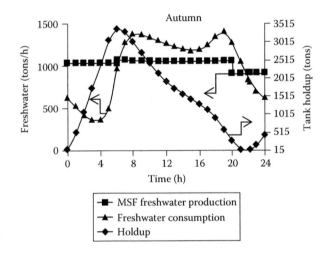

FIGURE 1.12 Variations of freshwater production of MSF, freshwater consumption, and freshwater holdup during a day in autumn. (Adapted from Hawaidi, E.A.M., Mujtaba, I.M., *Chem. Eng. J.*, 165, 545–553, 2010.)

and last stages are very important due to the large release of NCGs in the first stage and to vent the accumulated NCGs from the remaining stages. Increasing the number of venting points more than necessary will result in undue vapor losses through the vent and thus a higher energy consumption by the process.

More recently, Alsadaie and Mujtaba (2014) studied the effect of venting system design on the overall performance of MSF process using a detailed dynamic process model based on dynamic models of Reddy et al. (1995) and Al-Fulaij et al. (2011). The model includes the mass flow rate of NCGs in the material balance equations with supporting correlations for physical properties calculations.

Recently, Sowgath and Mujtaba (2015) presented a dynamic model to conduct a dynamic optimization for the MSF process. For a fixed freshwater demand and a maximum performance ratio, a steam temperature profile was optimized with regard to the variation of the seawater temperature. They assumed that during a particular day of a particular season, the seawater temperature changes dynamically from 33°C to 39°C. For simplicity, the dynamic temperature profile was replaced by a discrete temperature profile as shown in Figure 1.13. They aimed to obtain optimum steam temperature at those discrete points, which are needed to off-set the change in seawater temperature while ensuring constant water demand and maximum performance ratio. Table 1.3 summarizes the results.

Stage temperature was treated as a dynamic variable and Figure 1.14 shows the dynamic temperature profile of the first stage (TBT). They assumed that the plant operates at steady state right up to the point of change of seawater temperature (e.g., until 9 am at which point seawater temperature changes from 33°C to 35°C). Figure 1.14 shows the dynamic profile. As can be seen, the process reaches the next steady state after 8 s of the step change in the seawater temperature. Therefore, Figure 1.14 only includes 16 s operation prior and after each step change of which the second 8 s operation is dynamic.

The interest of using a computational fluid dynamic model (CFD) has recently received some attention in the area of MSF modeling. As a part of some efforts to simulate the MSF process using CFD, a 2D simulation model for flash chamber was developed to study the vapor flow through the demister (Khamis Mansour et al., 2011) and address the optimum location of jumping plate (weir) and its number inside the MSF chamber (Khamis Mansour and Fath, 2013a,b). Janajreh et al. (2013) and Al-Fulaij et al. (2014) presented a 2D simulation model to study the vapor flow and pressure drop through the MSF demister. Although these studies were good attempts to apply CFD models in MSF, they are still far away from simulating the whole MSF plant using CFD. Table 1.4 lists some important studies featuring dynamic models.

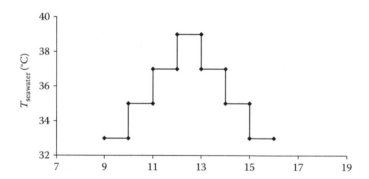

FIGURE 1.13   Discrete seawater temperature profile.

TABLE 1.3   Summary of the Optimization Results

Run	$T_{seawater}$ (°C)	T_{steam} (°C)	PR
1	33	93.96	11.45
	35	95.76	
2	35	95.76	11.54
	37	97.55	
3	37	97.55	11.63
	39	99.35	
4	39	99.35	11.54
	37	97.51	
5	37	97.51	11.45
	35	95.72	
6	35	95.72	11.36
	33	93.94	

Note: The freshwater demand for all cases = 9×10^5 (kg/h).

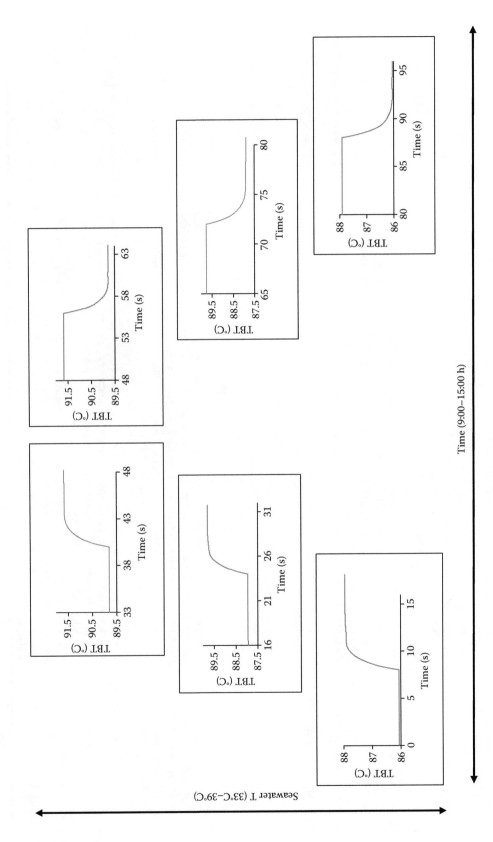

FIGURE 1.14 Dynamic profile of the first stage temperature (TBT profile) at discrete seawater temperature.

TABLE 1.4 Dynamic Models for MSF since 1970

Authors (Year)	Type/Feature of Model	Limitations
Glueck and Bradshaw (1970)	Simulation. The flash stage is divided into four compartments, streams are interacting materially and thermally among themselves	Combination of vapor space and distillate in differential energy balance results in over specified model
Delene and Ball (1971)	Simulation. Design of a digital code to simulate MSF by considering the MSF process consisting of two well-mixed tanks, holdup of cooling brine is included	Simple and no fouling or noncondensable gases
Yokoyama et al. (1977)	Simulation. Model using Runge–Kutta–Gill method to predict start-up characteristics such as flashing brine and coolant temperature	Estimated value of orifice coefficient is questionable
Husain et al. (1993)	Simulation. Dynamic model with flashing and cooling brine using SPEEDUP package and tridiagonal matrix method formulation	No venting system for noncondensable gases and no fouling
Reddy et al. (1995)	Simulation. Improvement of Hussain's model by including brine recycle, a holdup, and interstage flow.	No venting system for noncondensable gases and no fouling
Aly and Marwan (1995)	Simulation. Dynamic model solved by combination of Newton–Raphson and Runge–Kutta methods	No venting system for noncondensable gases and no fouling
Maniar and Deshpande (1996)	Control. SPEEDUP package for dynamic model, empirical corrections for the evaporation rates, some controlled variables are investigated	No venting system for noncondensable gases and no fouling
Thomas et al. (1998)	Simulation code written in C and implemented in a UNIX system. The flashing stage is divided into four compartments	Difference between the actual and model results due to noninclusion of the controller
Mazzott et al. (2000)	Simulation. Dynamic model including stage geometry, variation of the physical properties as a function of temperature and concentration	Not compared with the previous results due to the lack of detailed information in the literature
Tarifa and Scenna (2001)	Dynamic modeling for fault diagnosis. Developed numerical solution algorithm	No venting system for noncondensable gases and no fouling. TBT is very high (120°C) and can lead to severe fouling
Shivayyanamath and Tewari (2003)	Simulation. Use of FORTRAN 95 and Runge–Kutta method, stage-to-stage calculations, variation of the physical properties with temperature and salinity	No venting system for noncondensable gases, fouling and distillate holdup are neglected
Gambler and Badreddin (2004)	MATLAB®/Simulink®-based dynamic model for analysis and control design	Lack of real time data
Sowgath (2007) and Sowgath and Mujtaba (2015)	Optimization. gPROMS-based dynamic model considering nonequilibrium effects, demister pressure drop, and the brine and distillate holdup equations. NN correlation used for boiling point elevation. Variable seawater temperature	No venting system for NCGs and no fouling
Hawaidi and Mujtaba (2011)	Optimization using gPROMS. Dynamic model for storage tank linked to freshwater line. Variable freshwater demand and variable seawater temperature	Dynamic model for the storage tank but not for the MSF process. Fouling is not included
Al-Fulaij et al. (2011)	Dynamic model that includes the demister losses, distillate flashing, and NCGs. The model is solved with gPROMS	No heat, the effect of noncondensable gases is neglected
Said et al. (2013)	Optimization using gPROMS. Dynamic model for storage tank linked to freshwater line. Variable freshwater demand, variable seawater temperature, dynamic fouling	Dynamic model for the storage tank but not for the MSF process
Alsadaie and Mujtaba (2016)	Optimization and control. gPROMS is used to apply generic model control and solve optimization problem	Only two variables are optimized and controlled

1.1.3 State of the Art: Use of Model-Based Techniques in RO Desalination Process

RO is a type of membrane process commonly used for seawater and brackish water desalination. In industrial applications, a typical RO water desalination system consists of the following stages as shown in Figure 1.15: (1) Physical–chemical pretreatment, (2) Pumping unit, (3) Feed water supply unit, (4) Membrane element assembly unit, (5) Energy recovery system, (6) Posttreatment, and (7) Control system (not shown in the figure).

Since the 1960s (Loeb and Sourirajan, 1963), membrane processes have been rapidly developing and are now surpassing thermal desalination processes (in new plant installations; Misdan et al., 2012; Figure 1.3). Fritzmann et al. (2007) reviewed the state of the art of reverse osmosis (RO) desalination and quoted 108 references with the terms "modeling" and "simulation" appearing only once, and "optimization" twice. However, none of these references dealt with model-based RO process simulation or optimization. Misdan et al. (2012) reviewed the development, challenges, and future prospects of seawater RO desalination by thin film composite membrane and quoted 107 references with the terms "modeling" and "optimization" appearing only once. Again, interestingly none of these references dealt with model-based RO process simulation or optimization. In a water research journal, Greenlee et al. (2009) presented today's challenges, water sources and technology in RO desalination. In more than 200 references, there was only one mention of the model-based RO optimization of Vince et al. (2008). Kim et al. (2009) presented an overview of systems engineering approaches for a large-scale seawater desalination plant with a RO network and observed that since 1965 there were only 30 research articles which had used model-based techniques (to some extent) for simulation and optimization of RO processes. However, only about half of these refer to seawater or brackish water desalination application.

1.1.3.1 Steady-State Modeling of RO Desalination Process

1.1.3.1.1 Membrane Transport Models

There are mainly three types of RO membrane transport models:

a. *Nonporous or homogeneous membrane models* (such as the solution-diffusion, solution-diffusion-imperfection models, and extended solution-diffusion models)

   Solution-diffusion model—was originally developed by Lonsdale et al. (1965). It assumes that the solvent and solute are independent of each other and dissolve in the nonporous and homogeneous surface layers of the membrane. Then, both of them diffuse across the membrane as a consequence of its chemical potential gradient such as the applied pressure and the

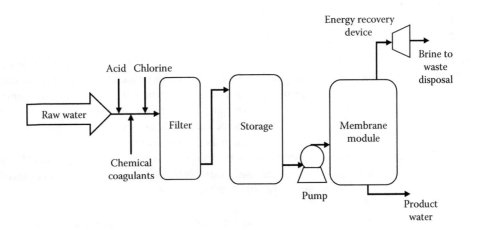

FIGURE 1.15 A typical RO desalination process.

concentration difference across the membrane for water and salt, respectively. In this model, the differences in solubility and diffusivity are considered to influence strongly the fluxes through the membrane.

Solution-diffusion-imperfection model—was developed by Sherwood et al. (1965) and is a modification of the solution-diffusion model that includes pore flow and diffusion of solute and solvent through the membrane.

b. *Pore models* (such as the finely porous, preferential sorption-capillary flow and surface force-pore flow models)

Preferential sorption-capillary flow model—was developed by Sourirajan (1970). It assumes that the separation is due to both surface phenomena and fluid transport through the pores. In this model, the membrane is assumed to be microporous, and its barrier layer has chemical properties (preferential sorption for the solvent or preferential repulsion for the solutes). As a result, a layer of an almost pure solvent is preferentially sorbed on the surface and in the pores of the membrane. Solvent is forced through the membrane capillary pores under pressure from that layer.

Surface force-pore flow model—was developed by Matsuura and Sourirajan (1981) and it is a modification of the preferential sorption-capillary flow model which allows characterization and specification of a membrane precisely as a function of pore size distribution along with a quantitative measure of the surface forces that appear between solute–solvent and the membrane wall inside the transport corridor.

c. *Irreversible thermodynamics models* (such as Kedem–Katchalsky and Spiegler–Kedem models)

Kedem and Katchalsky model—Kedem and Katchalsky (1958) proposed the first model based on irreversible thermodynamics. This approach considers the membrane as a black box. It assumes that the membrane is very near to equilibrium and the transport through the membrane is slow. The main principle of this model is that there is coupling between the fluxes of the components that move through the membrane. They introduced a new parameter named: Staverman reflection coefficient that links both the solute and solvent fluxes. This model describes the solvent and solute fluxes in terms of pressure and osmotic differences for solvent flux and osmotic variation and average concentration for solute.

Spiegler and Kedem model—Spiegler and Kedem (1966) modified the Kedem and Katchalsky model. They assumed that the solute flux is a combination of diffusion and convection. Also, the model considered the convective coupling aspects of the solute transport. According to them, the average concentration does not represent the solute and solvent fluxes; therefore, the flux equations are developed in terms of local coefficients and then integrated for the entire membrane layer.

1.1.3.1.2 Concentration Polarization (CP) Models

During the passage of some components through a semi-permeable membrane, accumulation of the rejected solutes on the front of the membrane surface occurs and a concentration gradient is formed with higher concentrations directly at the membrane surface (Figure 1.16). This results in a physical equilibrium between the convective transport (permeate flux due to the pressure) and diffusion (from the membrane wall to the bulk feed solution due to the concentration gradient). This phenomenon is known as CP (Fritzmann et al., 2007). It is crucial to determine the concentration of solutes at the surface of the membrane and use it in the estimation of the transport parameters for the rigours transport model instead of the bulk concentration (often used in simple model). The main operational variables that dictate CP are (1) pressure difference across the membrane, (2) solute concentration in the feed, and (3) hydrodynamics (turbulence).

Several models were proposed to estimate CP and determine the solute concentration at the surface of the membrane such as analytical film theory (FT) model (Michaels, 1968) and retained solute (RS)

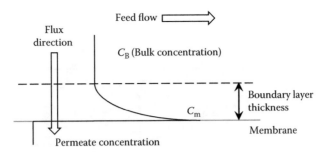

FIGURE 1.16 Schematic diagram of CP.

model (Song and Yu, 1999). Kim and Hoek (2005) compared analytical CP models with a rigorous CP model and experimental CP data. They found that FT model accurately predicted experimental permeate flux and salt rejection data.

1.1.3.1.3 Fouling Models

Fouling and scaling have significant impact on the design of RO process, and lead to dynamic adjustment of operation parameters if certain freshwater demand is to be met. The most critical obstacle that restricts further growth and wider application of membrane separation processes is fouling. It is caused by the adsorption of solutes on the membrane surface, which block the pores and form a gel layer (Madaeni et al., 2011). Fouling affects the operational reliability and increases the production cost, therefore many studies (mainly experimental) were focused on this subject (Baker and Dudley, 1998; Al-Bastaki and Abbas, 2004; Oh et al., 2009).

Oh et al. (2009) included discrete fouling factors in their simplified model based on the solution-diffusion model to measure the performance of RO systems that were used later for optimization purposes. Zhu et al. (1997) proposed a simple mathematical representation for decay in water flux due to fouling, which was also used by See et al. (1999). Al-Bastaki and Abbas (2004) proposed a more detailed analytical expression to represent decay in water flux due to membrane fouling. However, these correlations are applicable for the specific membrane type only with specific values of pressure and feed salinity (Table 1.5). As the salt concentration in seawater around the world varies markedly, these correlations offered limited use (Barello et al., 2014). Figure 1.17 shows the calculated profiles of water permeability constant given by the two correlations.

Sassi and Mujtaba (2011b) used the fouling model of Al-Bastaki and Abbas (2004) in the RO optimization studies. It should be noted that water flux may also affect the rate of fouling build up on the

TABLE 1.5 Correlations for the Evaluation of the Dynamic Water Permeability Constant

Correlation 1 (experimental based): Al-Bastaki and Abbas (2004):

$$K_w = 0.68 K_{w_0} e^{\frac{79}{t+201}}$$

Membrane type: Spiral wound, Filmtec BW30-400; Feed concentration: 25.40 (g/L); Pressure: 12 (bar); Operating time: 1500 (days)

Correlation 2 (theoretical): Zhu et al. (1997):

$$K_w = 0.68 K_{w_0} e^{\frac{-t}{328}}$$

Used in Simulation with Hollow fiber membrane, Dupont B-10; Feed concentration: 34.80 (g/L) Pressure: 62–70 (atm); Operating time: 370 (days).

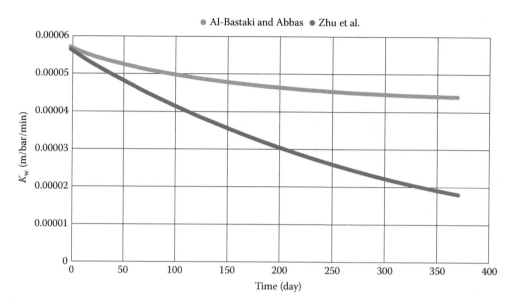

FIGURE 1.17   Dynamic water permeability trends. (Adapted from Barello, M. et al., *Desalination*, 345, 101–111, 2014.)

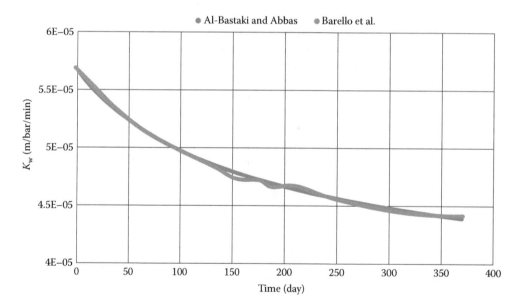

FIGURE 1.18   Dynamic water permeability constant profiles by Al-Bastaki and Abbas correlation, and Barello et al. correlation. Membrane type and operating conditions same as Al-Bastaki and Abbas. (Adapted from Barello, M. et al., *Desalination*, 345, 101–111, 2014.)

membrane surface. However, to understand the effect of flux on the fouling one requires a detailed hydrodynamic model of the membrane channels (Wardeh and Morvan, 2008; Vrouwenvelder et al., 2009).

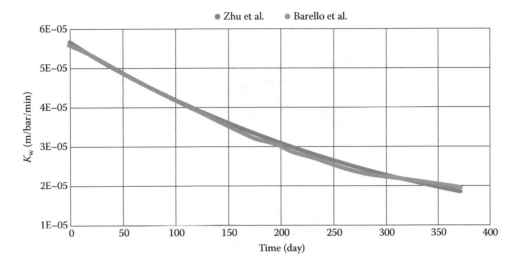

FIGURE 1.19 Dynamic water permeability constant profiles by Zhu et al. correlation and Barello et al. correlation. Membrane type and operating conditions same as Zhu et al. (1997). (Adapted from Barello, M. et al., *Desalination*, 345, 101–111, 2014.)

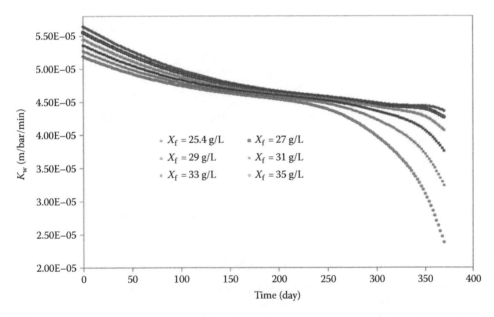

FIGURE 1.20 Dynamic water permeability constant for different feed salinity (pressure same as Al-Bastaki and Abbas, 2004).

To overcome the limitations of the earlier correlations, Barello et al. (2014) developed an Artificial Neural Network (ANN)-based correlation to predict a dynamic decay in water permeability constants due to fouling of the RO membranes. Figures 1.18 and 1.19 show that the ANN-based correlations can predict the dynamic permeability constants very closely to those obtained by the existing correlations for the same membrane type, operating pressure range, and feed salinity. However, the novel feature of the ANN-based correlation is that it is able to predict dynamic permeability constant profile for any of

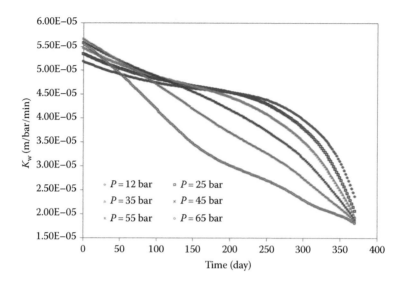

FIGURE 1.21   Dynamic water permeability constant for different pressure (feed salinity same as Zhu et al., 1997).

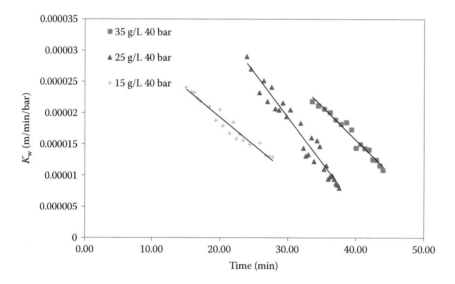

FIGURE 1.22   Dynamic water permability constant profiles at different feed salinity.

the two membrane types (used in earlier correlations) and for any operating pressure and a wide range of feed salinity values (which were not featured in the earlier correlations).

Figures 1.20 and 1.21 demonstrate the robustness of Barello et al.'s correlation as it can predict the constant profiles of the dynamic water permeability at various process conditions (feed salinity and pressure).

Barello et al.'s correlation highlighted for the first time a very interesting observation that water permeability constant is not only a function of pressure as always thought to be the case (Figure 1.21) but is also a function of feed salinity (Figure 1.20). Barello et al. (2015) verified this theoretical observation experimentally for tubular membranes which is represented in Figure 1.22. Figure 1.22 shows the dynamic water permeability constant profiles for three different initial feed salinity at 40 bar.

TABLE 1.6 Models for RO Membranes since 1958

Authors (Year)	Type/Description of Model
Kedem and Katchalsky (1958)	Model based on irreversible thermodynamics approach and describes the solvent and solute fluxes in terms of pressure and osmotic differences for solvent flux and osmotic variation and average concentration for solute
Spiegler and Kedem (1966)	Modified Kedem and Katchalsky model and assumed that the solute flux is a combination of diffusion and convection.
Lonsdale et al. (1965)	Model assumes that the solvent and solute dissolve in the nonporous and homogeneous surface layers of the membrane
Sherwood et al. (1965)	Modified Lonsdale et al. model and includes pore flow and diffusion of solute and solvent through the membrane
Michaels (1968)	Analytical FT model to estimate CP (due to buildup of solute along the membrane surface)
Sourirajan (1970)	Preferential sorption-capillary flow model assumes that the separation is due to both surface phenomena (preferential sorption for solvent) and fluid transport through the pores
Burghoff et al. (1980)	Improved the solution-diffusion model by incorporating the pressure dependence to the solute chemical potential
Matsuura and Sourirajan (1981)	Modified Sourirajan model and allows characterization and specification of a membrane as a function of pore size distribution along with surface forces
Mehdizadeh and Dickson (1989)	Invented the modified surface force-pore flow model by modifying the imperfect water material balance of the surface force-pore flow model within axial and radial positions
Song and Yu (1999)	RS model to estimate CP
Zhu et al. (1997)	Simple model for decay in water flux due to fouling
Al-Bastaki and Abbas (2004)	Detailed analytical expression to represent decay in water flux due to membrane fouling
Jain and Gupta (2004)	Merged the CP theory into the modified surface force-pore flow model
Abbas and Al-Bastaki (2005)	Neural Network-based modeling of an RO water desalination process
Ahmad et al. (2005) and Ahmad et al. (2007)	Developed Spiegler and Kedem model to consider solute–solute interactions in a multicomponent system
Gupta et al. (2007)	Developed Spiegler and Kedem model by counting the boundary layer thickness and investigates the contribution rate of the convection flux
Lee and Elimelech (2007)	Developed the finely porous model which considered viscous flow in addition to diffusion via micro-porous membranes
Wardeh and Morvan (2008, 2011)	Developed CFD model for evaluating fluid flow and CPs through RO membrane channels
Barello et al. (2014)	Developed neural network-based correlation for estimating water permeability constant in RO desalination process under membrane fouling
Wang et al. (2014)	Investigated the extended solution-diffusion transport mode that considered semi-quantitative representations of the free volume of the membrane in a given water chemistry under realistic filtration conditions
Attarde et al. (2015)	Developed a mathematical model based on solution-diffusion model for pressure retarded osmosis (PRO) and forward osmosis (FO)
Barello et al. (2015)	Developed experimental-based correlations showing the dependence of water and salt permeability constants on feed salinity for the first time

In summary, the solution-diffusion model was used in most of the membrane applications. However, Villafafila and Mujtaba (2003) considered simulation and optimization of RO process using the irreversible thermodynamics-based transport model.

Table 1.6 highlights the evolution of RO process models (for different components of the process) over the last half century.

1.1.3.2 Dynamic Modeling of RO Desalination Process

Most of the dynamic RO models were obtained by system identification using real data so that they are only valid for a particular plant, working at the selected operating point (Alatiqi et al., 1989;

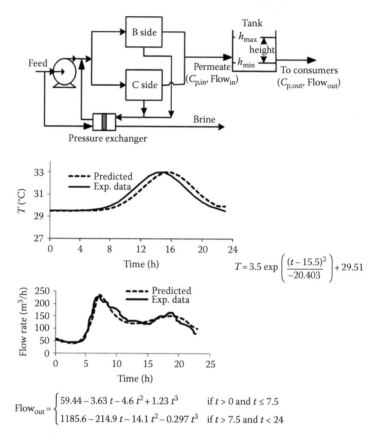

$$T = 3.5 \exp\left(\frac{(t-15.5)^2}{-20.403}\right) + 29.51$$

$$\text{Flow}_{out} = \begin{cases} 59.44 - 3.63\,t - 4.6\,t^2 + 1.23\,t^3 & \text{if } t > 0 \text{ and } t \le 7.5 \\ 1185.6 - 214.9\,t - 14.1\,t^2 - 0.297\,t^3 & \text{if } t > 7.5 \text{ and } t < 24 \end{cases}$$

FIGURE 1.23 Two-stage RO system with storage tank and dynamic temperature and freshwater demand profile. (Adapted from Sassi, K.M., Mujtaba, I.M, *Comput. Chem. Eng.*, 59, 101–110, 2013a.)

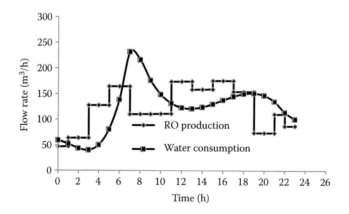

FIGURE 1.24 Dynamic freshwater production and demand profiles. (Adapted from Sassi, K.M., Mujtaba, I.M, *Comput. Chem. Eng.*, 59, 101–110, 2013a.)

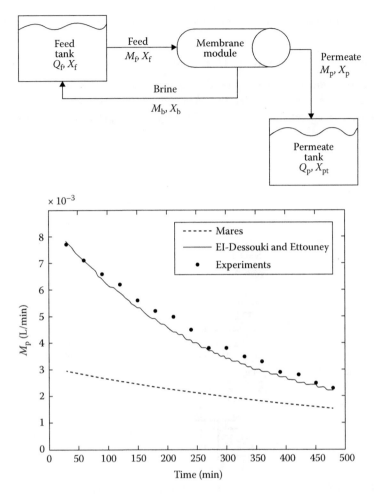

FIGURE 1.25 Dynamic batch RO system and models validation for permeate flow rate. (Adapted from Barello, M. et al., *Comput. Chem. Eng.*, 83, 139–156, 2015.)

Robertson et al., 1996; Abbas, 2006). There are only a few dynamic models based on mass and energy balances, and solution-diffusion models available in the literature (Gambier et al., 2007; Bartman et al., 2009).

Sassi and Mujtaba (2013a) coupled steady-state RO process model with a dynamic storage tank model and developed an optimization framework to optimize the plant operation subject to dynamic seawater temperature in summer and dynamic (throughout the day) freshwater demand (Figure 1.23). The violation of tank lower and upper levels is represented by a mathematical equation and is added as path constraint in the optimization framework. Figure 1.24 shows the optimum freshwater production rate from the RO process, dynamic tank level (not violating upper and lower tank level), and the dynamic consumption profile. Membrane replacement scheduling and maintenance is an essential area. Use of two-side (B-side and C-side) RO stage allows an effective scheduling and maintenance option.

Barello et al. (2015) considered two different models for batch RO system that is inherently dynamic. The first model was based on a solution-diffusion model (El-Dessouky and Ettouney, 2002), and the second one was based on an irreversible thermodynamics model (Meares, 1976) and validated the models against the experimental data (dynamic permeate flow rate) of the batch RO system with tubular membrane (Figure 1.25). Clearly, El-Dessouky and Ettouney's (2002) model shows a closer fit to the experimental data.

1.1.3.3 Network Optimization in RO Desalination Process

Generally, the most common arrangements of the membrane modules in Figure 1.12 are shown in Figure 1.26. These are (1) series, (2) parallel, and (3) tapered arrays. Although applied in RO waste treatment process (not in seawater desalination), El-Halwagi (1992) presented the most comprehensive

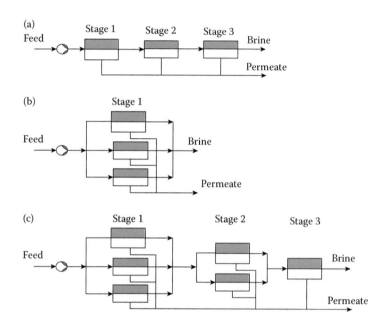

FIGURE 1.26   RO membrane configurations. (a) Series, (b) parallel, and (c) tapered.

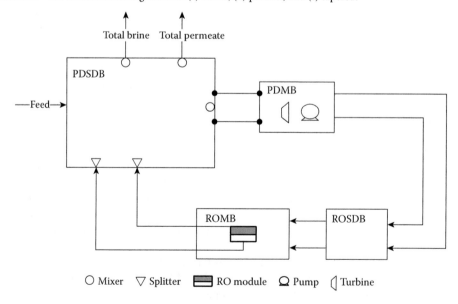

FIGURE 1.27   Superstructure of RO configuration.

model-based RO network synthesis problem formulation and solution by considering a superstructure configuration (Figure 1.27). The RO network were described using four boxes:

- Pressurization/depressurization stream distribution box (PDSDB)
- Pressurization/depressurization matching box (PDMB)
- RO stream-distribution box (ROSDB)
- RO matching box (ROMB)

The functions of the distribution boxes were to represent all possible grouping of stream splitting, mixing, bypassing, and recycling. The matching boxes determine all possible streams matching to units. With this formulation, all possible structure arrangements could be represented (such as those presented in Figure 1.23).

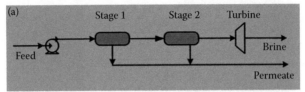

Salinity (15000–45000 ppm), seawater temperature (15°C–25°C)

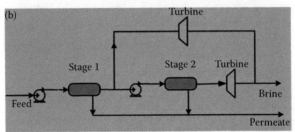

Salinity (35000–45000 ppm), seawater temperature (30°C–40°C)

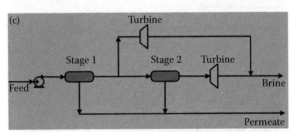

Salinity (50000 ppm), seawater temperature (15°C–25°C)

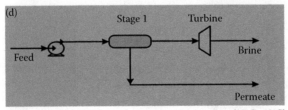

Salinity (45000–50000 ppm), seawater temperature (35°C–40°C)

FIGURE 1.28 Optimum RO networks at different feed salinity and seawater temperature. (Adapted from Sassi and Mujtaba 2011a; Sassi, and Mujtaba, 2012.)

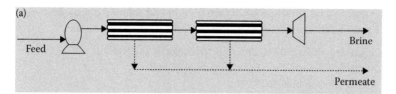

Fouling in the first stage (50%–60%)

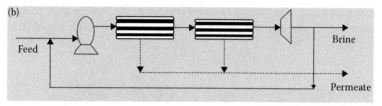

Fouling in the first stage (70%)

FIGURE 1.29   Optimum RO network under different fouling distributions.

Several limitations of earlier works (Fan et al., 1968; Bansal and Wiley, 1973; Evangelista, 1989) on the optimization of multistage RO network have been addressed in El-Halwagi's work. Zhu et al. (1997)

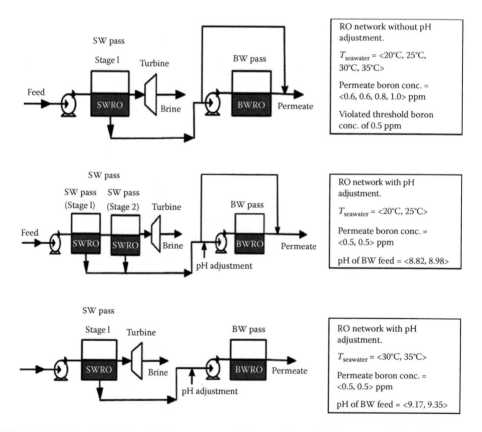

FIGURE 1.30   Optimal RO network for boron removal with or without pH adjustment. (Adapted from Sassi and Mujtaba, 2013.)

extended El-Halwagi's work by including scheduling aspect in the network synthesis and considered membrane fouling and seawater desalination application. See et al. (1999) extended the work of Zhu et al.'s and included membrane cleaning and regeneration.

More recently, Sassi and Mujtaba (2011a, 2012) and Sassi (2012) considered RO network synthesis problems (for RO desalination processes with hollow fiber membrane without fouling) for a wide range of salinity and seawater temperature resulting in a different optimum network for different conditions (Figure 1.28). Also, Sassi and Mujtaba (2011b) studied in detail the effect of different fouling distributions (utilizing Al-Bastaki and Abbas's fouling model) in stages on the overall network design and operation (Figure 1.29) in RO desalination processes with spiral wound membrane. In all synthesis exercises, total cost of freshwater production was minimized and an MINLP-based optimization technique was used within gPROMS software.

Boron plays a significant role in the growth of creatures, but uncontrolled exposure brings harmful effects to living organisms including plants. The average boron concentration in Mediterranean Sea is about 6.5 ppm and the pH of Mediterranean Sea is 7.95. Different countries have different threshold boron concentrations for drinking water (<0.5 ppm being the most common). Boron removal by RO process depends mainly on the temperature, pressure, and the boric acid/borate ion ratio, the latter being mainly dependent on seawater pH. Increasing pH and operating pressure of the feed water enhances the boron rejection while increasing seawater temperature reduces the boron rejection (Patroklou et al., 2013; Sassi and Mujtaba, 2013a; Patroklou and Mujtaba 2014). For a given freshwater demand with specified salt concentration, Sassi and Mujtaba (2013b) have developed an MINLP-based optimization framework for RO network design for controlling boron concentration in the permeate. For different seawater temperatures (20°C–35°C), Figure 1.30 shows an optimal RO network with and without pH adjustment.

Saif and Almansoori (2015) have developed MINLP-based optimization to optimize the process layout and operation conditions for seawater RO process with controlled boron concentration and minimum capital and operational costs.

1.1.4 State of the Art: Hybrid MSF/RO Desalination Process

Integration of a seawater RO unit with an MSF distiller provides the opportunity to blend the products of the two processes. Such arrangement allows operating the RO unit with relatively high TDS (total dissolved solids) and thus reduces the replacement rate of the membranes (Hamed, 2005). In addition, this integration, instead of operating in parallel, can improve the performance of MSF and reduce the cost of desalted water (Calì et al., 2008). Although the discussions on hybrid MSF/RO date back to the 1980s, Helal et al. (2003, 2004a,b) and Marcovecchio et al. (2005) presented the detailed model-based feasibility studies of a hybrid MSF/RO desalination process. Marcovecchio et al. (2009) presented a model to study the alternative and feasible arrangement of a hybrid RO/MSF system. Their objective was to determine optimal equipment sizes and operating conditions while minimizing the freshwater production cost. Skiborowski et al. (2012) developed a superstructure mathematical model using an MINLP technique to optimize the hybrid RO/MED plant.

Most of the previous studies considered several ways of connecting the RO with MSF process. According to Helal et al. (2003), these arrangements could be as follows:

- Independent two-stage RO and brine recycle MSF plants with common intake-outfall facilities.
- Heated MSF cooling water reject being fed to the single-stage RO plant.
- The two hot parts of the MSF blowdown and cooling reject being mixed to form the feed to the single-stage RO plant.
- The two plants being fully integrated with part of the RO reject forming a portion of the MSF make-up.
- Once-through MSF-RO hybrid plant with the MSF unit upstream of the single-stage RO unit.

- Once-through MSF-RO hybrid plant fully integrated with the single-stage RO unit upstream of the MSF unit.
- The rejected stream from MSF rejection section is fed to single-stage RO plant and all the reject from the RO is used as make-up for the MSF plant.

Helal et al. (2004b) concluded that the cost of freshwater from MSF plant could be reduced by up to 24% through hybridization with RO technology. Marcovecchio et al. (2009), on the other hand, concluded that the optimal arrangement of a hybrid system depends strongly on the seawater conditions (salinity and temperature) together with the freshwater demand.

1.1.5 State of the Art: Hybrid Power/MSF Plants (Dual Purpose)

A cogeneration plant, often called a dual-purpose plant, is the one that supplies heat for a thermal desalination unit and produces electricity for distribution to the electrical grid. Most of the MSF distillation plants, especially in Arabian Gulf countries, are paired with power plants in a cogeneration configuration (Al-Mutaz and Al-Namlah, 2004). This type of combination is considered to be more thermodynamically efficient and economically feasible than single-purpose power generation and water production plants (Hamed et al., 2006) and reduces the energy needed for thermal desalination by one-third to one-half (Winter et al., 2002).

Because of their dual function to provide heat and power, cogenerating plants are often complex facilities to design and operate. Therefore, the use of mathematical models and optimization software is playing an important role in design and operation by providing a very detailed analysis. A number of different studies have been reported in the literature for cost analysis of dual purpose plants (Kamal and Sims, 1997; Safi and Korchani, 1999; Uche et al., 2001; Hajeeh et al., 2003; Mussati et al., 2003, 2004, 2005; Cardona and Piacentino, 2004; Kamal, 2005; Hamed et al., 2006; El-Nashar, 2008; Ferreira et al., 2010; Marcovecchio et al., 2010; Luo et al., 2011; Ghelichzadeh et al., 2012; Manesh et al., 2012; Sanaye and Asgari, 2013; Xianli et al., 2014).

Most of the previous optimization studies share a common goal: *reduce the operating cost and increase profit*. Other researchers attempted to increase the performance ratio (PR) of the desalination plant by using some utilities from power stations. The majority of the current and planned dual-purpose desalination plants use either fossil fuels or nuclear power as their source of energy. However, triple hybrid plants [Power/Desalination/renewable energy sources (RES)] open the opportunity for future challenge.

1.1.6 State of the Art: Use of Renewable Energy in MSF and RO Process

Coupling RES such as solar, wind, and geothermal energy with traditional fossil fuel-based energy systems in water desalination can lead to energy-efficient processes with significantly reduced environmental impact in terms of greenhouse gas emissions and carbon footprint. The solar technologies produce a large amount of heat, which typically suits the thermal desalination. The wind energy-based electricity, on the other hand, is often combined with membrane desalination (IEA-ETSAP and IRENA, 2012). Tokui et al. (2014) reported that the sustainable operation for MSF and RO can be achieved if about 40%–50% of the total power consumption is supplied by RES. Although, in 2008, 10% of the generated electricity worldwide was produced by RES (Shatat et al., 2013), current information on desalination shows that only 1% of total desalinated water is based on energy from renewable sources (IEA-ETSAP and IRENA, 2012). Interestingly, the countries suffering from shortage of drinking water sources have significant amounts of RES. The limited use of renewable energy can be attributed to more than one problem such as lack of information, education, training, and research. Moreover, the use of expensive energy storage systems, due to the stochastic nature of RES, usually limits the exploitation of RES.

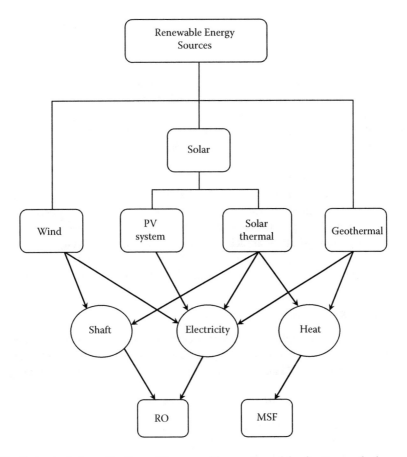

FIGURE 1.31 Technological combinations of the renewable energies and desalination methods.

Various methods (experimental to simple calculations) to evaluate the economic viability of coupling RES with desalination plants have been published; however, serious model-based techniques have not been considered in this area yet.

The development of a hybrid desalination/renewable energy model is very complex because of uncertain renewable energy supplies, load demands, and large number of parameters. Normally, renewable energy and desalination process design software was developed separately (Bourouni et al., 2011). However, the need for such a model is crucial to provide extensive analysis for combining desalination systems with RES.

Connolly et al. (2010) presented a review of 37 different computer tools that can be used to analyze the integration of renewable energy. However, none of them were used for desalination processes. Rizzuti et al. (2007) edited a book comprising 26 research articles from around the world on desalination coupled to renewable energies (solar and wind); however, only two articles considered a model-based MSF process (Bogle et al., 2006) and a hybrid MSF/RO process (Fois et al., 2007). Mathioulakis et al. (2007) showed possible combinations of renewable energies to be used in MSF and RO processes (Figure 1.31) but no model-based techniques were discussed. Model-based simulation of renewable energy-driven desalination systems was provided by Koroneos et al. (2007). Bourouni et al (2011) presented a model based on Genetic Algorithms to study possible arrangements for coupling small RO units to RES (i.e., photovoltaic and wind). The objective function was to minimize water production cost.

1.1.7 State of the Art: Environmental Impact Modeling

Due to increasing environmental legislation (EL), the activities in the area of assessment and quantification of environmental impact (EI) are gaining importance in desalination processes (Hoepner, 1999; El Din et al., 2000; Hoepner and Lattemann, 2003). It is well known that the two main impacts on the environment are from the brine discharges (salinity and temperature) and from the greenhouse gas emissions due to high-energy consumptions.

Sommariva et al. (2004), for the first time, attempted to establish model-based relations between the improvement in plant efficiency (PE) and EI in thermal desalination systems. However, the impact of the salinity of brine discharge was neglected. Vince et al. (2008a) provided a simple model-based LCA (Life Cycle Analysis) tool to provide help at the decision-making stage for designing, operating, and choosing an appropriate process to minimize EI. Mujtaba (2010) summarized some of the major work since 1999 on quantifying EI in MSF. In most cases, EI issues are dealt in a reactive mode where EI from an existing process is assessed and, based on the current EL, the operations are adjusted. The preventive mode requires that new design and operations are achieved based on a set/desired EI target (Vince et al., 2008a). While trial and error based on experimental studies is time-consuming and expensive, studying these via model-based techniques is less time-consuming and inexpensive and remains a future challenge.

1.1.8 Conclusions

Discussions presented in earlier sections captured the current state of the art of using model-based techniques in desalination, specially in MSF and RO processes. Figure 1.32 summarizes what has been done over the last few decades. Figure 1.32 also summarizes the area of opportunities to be engaged in the use

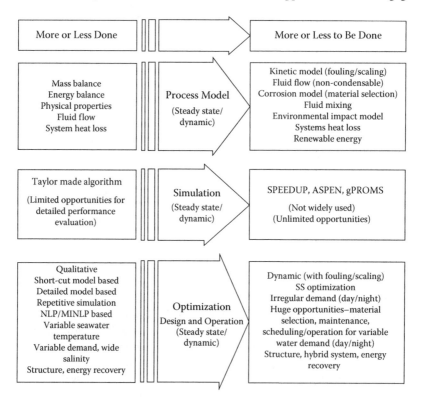

FIGURE 1.32 State of the art and future opportunities in model-based studies in desalination.

of model-based techniques in desalination for making freshwater. In process modeling, steady-state- and dynamic model-based techniques are well known in the literature. However, few studies considered the scale formation (kinetic model) and the effect of noncondensable gases. Environmental impact model and integration of RES with desalination systems are other new areas that require more attention due to the new environmental regulations.

Commercial software such as SPEEDUP, ASPEN, and gPROMS have been employed for modeling desalination processes. However, their capabilities have not been fully explored. In terms of optimization, several studies have been carried out on the optimization of design and operation parameters of MSF desalination processes based on fixed water demand and fixed water temperature. However, studies with variable freshwater demand and seawater temperature have been very limited. The possibility of the variation of recovery stages with seasonal changes opens huge opportunities to explore the feasibility of carrying out maintenance and schedule operations without shutting the plant down.

In summary, development of economically and environmentally efficient desalination processes is the future challenge. This can be achieved by better design, operation, and control and model-based techniques.

References

2030 Water Resources Group. 2009. Charting our water future—Economic frameworks to inform decision-making.

Abbas, A. 2006. Model predictive control of a reverse osmosis desalination unit. *Desalination*, 194, 268–280.

Abbas, A. and Al-Bastaki, N. 2005. Modeling of an RO water desalination unit using neural networks. *Chemical Engineering Journal*, 114, 139–143.

Ahmad, A. L., Chong, M. F. and Bhatia, S. 2005. Mathematical modeling and simulation of the multiple solutes system for nanofiltration process. *Journal of Membrane Science*, 253, 103–115.

Ahmad, A. L., Chong, M. F. and Bhatia, S. 2007. Mathematical modeling of multiple solutes system for reverse osmosis process in palm oil mill effluent (POME) treatment. *Chemical Engineering Journal*, 132, 183–193.

Al-Anezi, K. and Hilal, N. 2007. Scale formation in desalination plants: Effect of carbon dioxide solubility. *Desalination*, 204, 385–402.

Alasfour, F. N. and Abdulrahim, H. K. 2009. Rigorous steady state modeling of MSF-BR desalination system. *Desalination and Water Treatment*, 1, 259–276.

Alatiqi, I. M., Ghabris, A. H. and Ebrahim, S. 1989. System identification and control of reverse osmosis desalination. *Desalination*, 75, 119–140.

Al-Bastaki, N. and Abbas, A. 2004. Long-term performance of an industrial water desalination plant. *Chemical Engineering and Processing: Process Intensification*, 43, 555–558.

Al-Fulaij, H. F. 2011. Dynamic modeling of multi stage flash (MSF) desalination plant. PhD, University College London.

Al-Fulaij, H., Cipollina, A., Bogle, D. and Ettouney, H. 2010. Once through multistage flash desalination: gPROMS dynamic and steady state modeling. *Desalination and Water Treatment*, 18, 46–60.

Al-Fulaij, H., Cipollina, A., Ettouney, H. and Bogle, D. 2011. Simulation of stability and dynamics of multistage flash desalination. *Desalination*, 281, 404–412.

Al-Fulaij, H., Cipollina, A., Micale, G., Ettouney, H. and Bogle, D. 2014. Eulerian-Eulerian modelling and computational fluid dynamics simulation of wire mesh demisters in MSF plants. *Engineering Computations*, 31, 1242–1260.

Al-Mutaz, I. S. and Al-Namlah, A. M. 2004. Characteristics of dual purpose MSF desalination plants. *Desalination*, 166, 287–294.

Al-Rawajfeh, A. E., Ihm, S., Varshney, H. and Mabrouk, A. N. 2014. Scale formation model for high top brine temperature multi-stage flash (MSF) desalination plants. *Desalination*, 350, 53–60.

Alsadaie, S. M. and Mujtaba, I. M. 2016. Modelling and simulation of MSF desalination plant: The effect of venting system design for non-condensable gases. *Chemical Engineering Transactions*, 39, 1615–1620.

Aly, N. H. and Marwan, M. A. 1995. Dynamic behavior of MSF desalination plants. *Desalination*, 101, 287–293.

Aminian, A. 2010. Prediction of temperature elevation for seawater in multi-stage flash desalination plants using radial basis function neural network. *Chemical Engineering Journal*, 162, 552–556.

Attarde, D., Jain, M., Chaudhary, K. and Gupta, S. K. 2015. Osmotically driven membrane processes by using a spiral wound module—Modeling, experimentation# and numerical parameter estimation. *Desalination*, 361, 81–94.

Baig, H., Antar, M. A. and Zubair, S. M. 2011. Performance evaluation of a once-through multi-stage flash distillation system: Impact of brine heater fouling. *Energy Conversion and Management*, 52, 1414–1425.

Baker, J. S. and Dudley, L. Y. 1998. Biofouling in membrane systems—A review. *Desalination*, 118, 81–89.

Bansal, I. K. and Wiley, A. J. 1973. A mathematical model for optimizing the design of reverse osmosis systems.

Barello, M., Manca, D., Patel, R. and Mujtaba, I. M. 2014. Neural network based correlation for estimating water permeability constant in RO desalination process under fouling. *Desalination*, 345, 101–111

Barello, M., Manca, D., Patel, R. and Mujtaba, I. M. 2015. Operation and modeling of RO desalination process in batch mode. *Computers and Chemical Engineering*, 83, 139–156.

Bartman, A. R., Christofides, P. D. and Cohen, Y. 2009. Nonlinear model-based control of an experimental reverse-osmosis water desalination system. *Industrial and Engineering Chemistry Research*, 48, 6126–6136.

Bogle, I. D. L., Cipollina, A. and Micale, G. 2006. Dynamic modeling tools for solar powered desalination process during transient operations. In *Solar Desalination for the 21st Century*, Rizzuti, L. et al. (Editors), Spinger.

Bourouni, K., M'Barek, T. B. and Al Taee, A. 2011. Design and optimization of desalination reverse osmosis plants driven by renewable energies using genetic algorithms. *Renewable Energy*, 36, 936–950.

Burghoff, H. G., Lee, K. L. and Pusch, W. 1980. Characterization of transport across cellulose acetate membranes in the presence of strong solute–membrane interactions. *Journal of Applied Polymer Science*, 25, 323–347.

Cadwallader, E. A. 1967. Carbon dioxide—The key to economical desalination. *Industrial and Engineering Chemistry*, 59, 42–47.

Calì, G., Fois, E., Lallai, A. and Mura, G. 2008. Optimal design of a hybrid RO/MSF desalination system in a non-OPEC country. *Desalination*, 228, 114–127.

Cardona, E. and Piacentino, A. 2004. Optimal design of cogeneration plants for seawater desalination. *Desalination*, 166, 411–426.

Coleman, A. K. 1971. Optimization of a single effect, multi-stage flash distillation desalination system. *Desalination*, 9, 315–331.

Connolly, D., Lund, H., Mathiesen, B. V. and Leahy, M. 2010. A review of computer tools for analysing the integration of renewable energy into various energy systems. *Applied Energy*, 87, 1059–1082.

Delene, J. G. and Ball, S. J. 1971. A digital computer code for simulating large MSF evaporator desalting plant dynamics. Oak Ridge National Laboratory Report# ORNL-TM-2933.

Desalination Market. 2005. A global assessment & forecast. *Global Water Intelligence*. http://www.globalwaterintel.com/.

El-Dessouky, H. T. and Ettouney, H. M. 2002. *Fundamentals of Salt Water Desalination*, Elsevier.

El-Dessouky, H., Shaban, H. I. and Al-Ramadan, H. 1995. Steady-state analysis of multi-stage flash desalination process. *Desalination*, 103, 271–287.

El Din, A. M. S., Arain, R. A. and Hammoud, A. A. 2000. On the chlorination of seawater. *Desalination*, 129, 53–62.

El-Halwagi, M. M. 1992. Synthesis of reverse-osmosis networks for waste reduction. *AIChE Journal*, 38, 1185–1198.

El-Nashar, A. M. 2008. Optimal design of a cogeneration plant for power and desalination taking equipment reliability into consideration. *Desalination*, 229, 21–32.

Evangelista, F. 1989. Design and performance of two-stage reverse osmosis plant. *Chemical Engineering and Processing: Process Intensification*, 25, 119–125.

Fan, L. T., Cheng, C. Y., Ho, L. Y. S., Hwang, C. L. and Erickson, L. E. 1968. Analysis and optimization of a reverse osmosis water purification system Part I. Process analysis and simulation. *Desalination*, 5, 237–265.

Ferreira, E. M., Balestieri, J. A. P. and Zanardi, M. A. 2010. Optimization analysis of dual-purpose systems. *Desalination*, 250, 936–944.

Fois, E., Lallai, A. and Mura, G. 2007. Desalted water from a hybrid RO/MSF plant with rdf combustion: Modelling and economics. In *Solar Desalination for the 21st Century*, Springer.

Fritzmann, C., Löwenberg, J., Wintgens, T. and Melin, T. 2007. State-of-the-art of reverse osmosis desalination. *Desalination*, 216, 1–76.

Gambier, A., Krasnik, A. and Badreddin, E. 2007. Dynamic modeling of a simple reverse osmosis desalination plant for advanced control purposes. In *American Control Conference, 2007*. IEEE, 4854–4859.

Gambler, A. and Badreddin, E. 2004. Dynamic modelling of MSF plants for automatic control and simulation purposes: A survey. *Desalination*, 166, 191–204.

Genthner, K. and Seifert, A. 1991. A calculation method for condensers in multi-stage evaporators with non-condensable gases. *Desalination*, 81, 349–366.

Ghelichzadeh, J., Derakhshan, R. and Asadi, A. 2012. Application of alternative configuration of cogeneration plant in order to meet power and water demand. *Chemical Engineering Transactions*, 29, 775–780.

Global Water Intelligence. 2009. *IDA Desalination Yearbook 2008–2009*. International Desalination Association.

Global Water Intelligence. 2010. *Desalination Market 2010: Global Forecast and Analysis*. Media Analytics Ltd, Oxford, UK.

Global Water Intelligence. 2011. *IDA Desalination Yearbook 2010–2011*. International Desalination Association.

Glueck, A. R. and Bradshaw, R. W. 1970. A mathematical model for a multistage flash distillation plant. In *Proceedings of the 3rd International Symposium on Fresh Water from the Sea*, 95–108.

Greenlee, L. F., Lawler, D. F., Freeman, B. D., Marrot, B. and Moulin, P. 2009. Reverse osmosis desalination: Water sources, technology, and today's challenges. *Water Research*, 43, 2317–2348.

Gupta, V. K., Hwang, S.-T., Krantz, W. B. and Greenberg, A. R. 2007. Characterization of nanofiltration and reverse osmosis membrane performance for aqueous salt solutions using irreversible thermodynamics. *Desalination*, 208, 1–18.

Hajeeh, M., Mohammad, O., Behbahani, W. and Dashti, B. 2003. A mathematical model for a dual-purpose power and desalination plant. *Desalination*, 159, 61–68.

Hamed, O. A. 2005. Overview of hybrid desalination systems—Current status and future prospects. *Desalination*, 186, 207–214.

Hamed, O. A., AL-Sofi, M. A. K., Mustafa, G. M. and Dalvi, A. G. 1999. Performance of different anti-scalants in multi-stage flash distillers. *Desalination*, 123, 185–194.

Hamed, O. A., AL-Sofi, M. A. K., Imam, M., Ba-Mardouf, K., AL-Mobayed, A. S. and Ehsan, A. 2000. Evaluation of polyphosphonate anti-scalant at low dose rate in the ALJubail PhaseII MSF plant, Saudi Arabia. *Desalination*, 128, 275–280.

Hamed, O. A., Al-Washmi, H. A. and Al-Otaibi, H. A. 2006. Thermoeconomic analysis of a power/water cogeneration plant. *Energy*, 31, 2699–2709.

Hawaidi, E. A. M. 2013. Simulation, optimisation and flexible scheduling of MSF desalination process under fouling. PhD, University of Bradford.

Hawaidi, E. A. M. and Mujtaba, I. M. 2010. Simulation and optimization of MSF desalination process for fixed freshwater demand: Impact of brine heater fouling. *Chemical Engineering Journal*, 165, 545–553.

Hawaidi, E. A. M. and Mujtaba, I. M. 2011. Meeting variable freshwater demand by flexible design and operation of the multistage flash desalination process. *Industrial and Engineering Chemistry Research*, 50, 10604–10614.

Helal, A. M., El-Nashar, A. M., Al-Katheeri, E. and Al-Malek, S. 2003. Optimal design of hybrid RO/MSF desalination plants Part I: Modeling and algorithms. *Desalination*, 154, 43–66.

Helal, A. M., El-Nashar, A. M., Al-Katheeri, E. S. and Al-Malek, S. A. 2004a. Optimal design of hybrid RO/MSF desalination plants Part II: Results and discussion. *Desalination*, 160, 13–27.

Helal, A. M., El-Nashar, A. M., Al-Katheeri, E. S. and Al-Maler, S. A. 2004b. Optimal design of hybrid RO/MSF desalination plants Part III: Sensitivity analysis. *Desalination*, 169, 43–60.

Helal, A. M., Medani, M. S., Soliman, M. A. and Flower, J. R. 1986. A tridiagonal matrix model for multistage flash desalination plants. *Computers and Chemical Engineering*, 10, 327–342.

Hoepner, T. 1999. A procedure for environmental impact assessments (EIA) for seawater desalination plants. *Desalination*, 124, 1–12.

Hoepner, T. and Lattemann, S. 2003. Chemical impacts from seawater desalination plants—A case study of the northern Red Sea. *Desalination*, 152, 133–140.

Husain, A., Hassan, A., Al-Gobaisi, D. M. K., Al-Radif, A., Woldai, A. and Sommariva, C. 1993. Modelling, simulation, optimization and control of multistage flashing (MSF) desalination plants Part I: Modelling and simulation. *Desalination*, 92, 21–41.

IChemE. 2007. Technical roadmap: A roadmap for 21st century chemical engineering.

IEA-ETSAP and IRENA. 2012. Water desalination using renewable energy.

Jain, S. and Gupta, S. K. 2004. Analysis of modified surface force pore flow model with concentration polarization and comparison with Spiegler–Kedem model in reverse osmosis systems. *Journal of Membrane Science*, 232, 45–62.

Janajreh, I., Hasania, A. and Fath, H. 2013. Numerical simulation of vapor flow and pressure drop across the demister of MSF desalination plant. *Energy Conversion and Management*, 65, 793–800.

Kamal, I. 2005. Integration of seawater desalination with power generation. *Desalination*, 180, 217–229.

Kamal, I. and Sims, G. V. 1997. Thermal cycle and financial modeling for the optimization of dual-purpose power-cum-desalination plants. *Desalination*, 109, 1–13.

Kedem, O. 1989. Commentary on 'Thermodynamic analysis of the permeability of biological membranes to non-electrolytes' by O. Kedem and A. Katchalsky Biochim. Biophys. Acta 27 (1958) 229–246. *Biochimica et Biophysica Acta (BBA)—General Subjects*, 1000, 411–430.

Khamis Mansour, M. and Fath, H. E. S. 2013a. Numerical simulation of flashing process in MSF flash chamber. *Desalination and Water Treatment*, 51, 2231–2243.

Khamis Mansour, M. and Fath, H. E. S. 2013b. Comparative study for different demister locations in multistage flash (MSF) flash chamber (FC). *Desalination and Water Treatment*, 51, 7379–7393.

Khamis Mansour, M., Fath, H. E. S. and El-Samni, O. 2011. Computational fluid dynamics study of MSF flash chambers Su-components; i-vapor flow through demister. In *The 15th International Water Technology Conference, (IWTC15)*, Alexandria, Egypt.

Kim, S. and Hoek, E. M. V. 2005. Modeling concentration polarization in reverse osmosis processes. *Desalination*, 186, 111–128.

Kim, Y. M., Kim, S. J., Kim, Y. S., Lee, S., Kim, I. S. and Kim, J. H. 2009. Overview of systems engineering approaches for a large-scale seawater desalination plant with a reverse osmosis network. *Desalination*, 238, 312–332.

Koroneos, C., Dompros, A. and Roumbas, G. 2007. Renewable energy driven desalination systems modelling. *Journal of Cleaner Production*, 15, 449–464.

Lee, S. and Elimelech, M. 2007. Salt cleaning of organic-fouled reverse osmosis membranes. *Water Research*, 41, 1134–1142.

Loeb, S. and Sourirajan, S. 1963. Saline water conversion-II. *Advances in Chemistry Series*, 38, 117.

Lonsdale, H. K., Merten, U. and Riley, R. L. 1965. Transport properties of cellulose acetate osmotic membranes. *Journal of Applied Polymer Science*, 9, 1341–1362.

Luo, C., Zhang, N., Lior, N. and Lin, H. 2011. Proposal and analysis of a dual-purpose system integrating a chemically recuperated gas turbine cycle with thermal seawater desalination. *Energy*, 36, 3791–3803.

Madaeni, S. S., Sasanihoma, A. and Zereshki, S. 2011. Chemical cleaning of reverse osmosis membrane fouled by apple juice. *Journal of Food Process Engineering*, 34, 1535–1557.

Mandil, M. A. and Ghafour, E. E. A. 1970. Optimization of multi-stage flash evaporation plants. *Chemical Engineering Science*, 25, 611–621.

Manesh, M. H. K., Ghalami, H., Amidpour, M. and Hamedi, M. H. 2012. A new targeting method for combined heat, power and desalinated water production in total site. *Desalination*, 307, 51–60.

Maniar, V. M. and Deshpande, P. B. 1996. Advanced controls for multi-stage flash (MSF) desalination plant optimization. *Journal of Process Control*, 6, 49–66.

Marcovecchio, M. G., Mussati, S. F., Aguirre, P. A. and Nicolás, J. 2005. Optimization of hybrid desalination processes including multi stage flash and reverse osmosis systems. *Desalination*, 182, 111–122.

Marcovecchio, M., Mussati, S., Scenna, N. and Aguirre, P. 2009. Hybrid desalination systems: Alternative designs of thermal and membrane processes. *Computer Aided Chemical Engineering*, 27, 1011–1016.

Marcovecchio, M., Scenna, N., Aguirre, P. and Mussati, S. 2010. Global optimal design of electricity and fresh water plants. In *2nd International Conference on Engineering Optimization*, Lisbon, Portugal.

Mathioulakis, E., Belessiotis, V. and Delyannis, E. 2007. Desalination by using alternative energy: Review and state-of-the-art. *Desalination*, 203, 346–365.

Matsuura, T. and Sourirajan, S. 1981. Reverse osmosis transport through capillary pores under the influence of surface forces. *Industrial and Engineering Chemistry Process Design and Development*, 20, 273–282.

Mazzotti, M., Rosso, M., Beltramini, A. and Morbidelli, M. 2000. Dynamic modeling of multistage flash desalination plants. *Desalination*, 127, 207–218.

Meares, P. 1976. *Membrane Separation Process*, Elsevier Scientific, Oxford.

Mehdizadeh, H. and Dickson, J. M. 1989. Theoretical modification of the surface force-pore flow model for reverse osmosis transport. *Journal of Membrane Science*, 42, 119–145.

Michaels, A. S. 1968. New separation technique for CPI. *Chemical Engineering Progress*, 64, 31.

Misdan, N., Lau, W. J. and Ismail, A. F. 2012. Seawater reverse osmosis (SWRO) desalination by thin-film composite membrane—Current development, challenges and future prospects. *Desalination*, 287, 228–237.

Mubarak, A. 1998. A kinetic model for scale formation in MSF desalination plants. Effect of antiscalants. *Desalination*, 120, 33–39.

Mujtaba, I. M. 2008. Keynote lecture. CAPE FORM 2008, Thessaloniki, Greece.

Mujtaba, I. M. 2010. Status and future development. In *Process Systems Engineering: Volume 7: Dynamic Process Modeling*, Georgiadis, M. C., Banga, J. R. and Pistikopoulos, E. N. (Editors), John Wiley and Sons.

Mujtaba, I. M. 2011. Modelling multistage flash desalination process – Current status and future development. In *Dynamic Process Modelling*, Pistokopoulos, E. N. et al. (Editors), Vol 7, Chapter 9, Wiley-VCH, Germany, 287–317.

Mujtaba, I. M. 2012. The role of PSE community in meeting sustainable freshwater demand of tomorrow's world via desalination. *Computer Aided Chemical Engineering*, 31, 91–98.

Mussati, S., Aguirre, P. and Scenna, N. J. 2001. Optimal MSF plant design. *Desalination*, 138, 341–347.

Mussati, S., Aguirre, P. and Scenna, N. 2003. Dual-purpose desalination plants. Part I. Optimal design. *Desalination*, 153, 179–184.

Mussati, S. F., Aguirre, P. A. and Scenna, N. J. 2004. A rigorous, mixed-integer, nonlineal programming model (MINLP) for synthesis and optimal operation of cogeneration seawater desalination plants. *Desalination*, 166, 339–345.

Mussati, S. F., Aguirre, P. A. and Scenna, N. J. 2005. Optimization of alternative structures of integrated power and desalination plants. *Desalination*, 182, 123–129.

Oh, H.-J., Hwang, T.-M. and Lee, S. 2009. A simplified simulation model of RO systems for seawater desalination. *Desalination*, 238, 128–139.

Patroklou, G. and Mujtaba, I. M. 2014. Economic optimisation of seawater reverse osmosis desalination with boron rejection. In *24th European Symposium on Computer Aided Process Engineering, Volume 33*, Klemeš, J., Varbanov, P. and Liew, P. Y. (Editors), Elsevier, 1381–1386

Patroklou, G., Sassi, K. M. and Mujtaba, I. M. 2013. Simulation of boron rejection by seawater reverse osmosis desalination. *Chemical Engineering Transactions*, 32, 1873–1878.

Reddy, K. V., Husain, A., Woldai, A. and Al-Gopaisi, D. M. K. 1995. Dynamic modelling of the MSF desalination process. In *Proceedings of the IDA and WRPC World Conference on Desalination and Water Treatment*, 227–242.

Rizzuti, L., Ettouney, H. M. and Cipollina, A. 2007. *Solar Desalination for the 21st Century: A Review of Modern Technologies and Researches on Desalination Coupled to Renewable Energies*, Springer, Dordrecht.

Robertson, M. W., Watters, J. C., Desphande, P. B., Assef, J. Z. and Alatiqi, I. M. 1996. Model based control for reverse osmosis desalination processes. *Desalination*, 104, 59–68.

Rosso, M., Beltramini, A., Mazzotti, M. and Morbidelli, M. 1997. Modeling multistage flash desalination plants. *Desalination*, 108, 365–374.

Safi, M. J. and Korchani, A. 1999. Cogeneration applied to water desalination: Simulation of different technologies. *Desalination*, 125, 223–229.

Said, S. A., Emtir, M. and Mujtaba, I. M. 2013. Flexible design and operation of multi-stage flash (MSF) desalination process subject to variable fouling and variable freshwater demand. *Processes*, 1, 279–295.

Said, S. A., Mujtaba, I. M. and Emtir, M. 2010. Modelling and simulation of the effect of non-condensable gases on heat transfer in the MSF desalination plants using gPROMS software. *Computer Aided Chemical Engineering*, 28, 25–30.

Said, S. A., Mujtaba, I. M. and Emtir, M. 2012. Flexible design and operation of MSF desalination process: Coping with uniform and irregular freshwater demand. In *Proceedings of the 9th International conference on Computational Management*, April 18–20, London.

Saif, Y. and Almansoori, A. 2015. Synthesis of reverse osmosis desalination network under boron specifications. *Desalination*, 371, 26–36.

Sanaye, S. and Asgari, S. 2013. Four E analysis and multi-objective optimization of combined cycle power plants integrated with multi-stage flash (MSF) desalination unit. *Desalination*, 320, 105–117.

Sassi, K. M. 2012. Optimal scheduling, design, operation and control of reverse osmosis based desalination. PhD, University of Bradford.

Sassi, K. M. and Mujtaba, I. M. 2011a. Optimal design of reverse osmosis based desalination process with seasonal variation of feed temperature. *Chemical Engineering Transactions*, 25, 1055–1060.

Sassi, K. M. and Mujtaba, I. M. 2011b. Optimal design and operation of reverse osmosis desalination process with membrane fouling. *Chemical Engineering Journal*, 171, 582–593.

Sassi, K. M. and Mujtaba, I. M. 2012. Effective design of reverse osmosis based desalination process considering wide range of salinity and seawater temperature. *Desalination*, 306, 8–16.

Sassi, K. M. and Mujtaba, I. M. 2013a. Optimal operation of RO system with daily variation of freshwater demand and seawater temperature. *Computers and Chemical Engineering*, 59, 101–110.

Sassi, K. M. and Mujtaba, I. M. 2013b. MINLP based superstructure optimization for boron removal during desalination by reverse osmosis. *Journal of Membrane Science*, 440, 29–39.

See, H. J., Vassiliadis, V. S. and Wilson, D. I. 1999. Optimisation of membrane regeneration scheduling in reverse osmosis networks for seawater desalination. *Desalination*, 125, 37–54.

Seifert, A. and Genthner, K. 1991. A model for stagewise calculation of non-condensable gases in multi-stage evaporators. *Desalination*, 81, 333–347.

Shatat, M., Worall, M. and Riffat, S. 2013. Opportunities for solar water desalination worldwide: Review. *Sustainable Cities and Society*, 9, 67–80.

Sherwood, T. K., Brian, P. L. T., Fisher, R. E. and Dresner, L. 1965. Salt concentration at phase boundaries in desalination by reverse osmosis. *Industrial and Engineering Chemistry Fundamentals*, 4, 113–118.

Shivayyanamath, S. and Tewari, P. K. 2003. Simulation of start-up characteristics of multi-stage flash desalination plants. *Desalination*, 155, 277–286.

Singh, R. 2008. Sustainable fuel cell integrated membrane desalination systems. *Desalination*, 227, 14–33.

Skiborowski, M., Mhamdi, A., Kraemer, K. and Marquardt, W. 2012. Model-based structural optimization of seawater desalination plants. *Desalination*, 292, 30–44.

Soliman, M. A. 1981. A mathematical-model for multistage flash desalination plants. *Journal of Engineering Sciences*, 7, 143–150.

Sommariva, C., Hogg, H. and Callister, K. 2004. Environmental impact of seawater desalination: Relations between improvement in efficiency and environmental impact. *Desalination*, 167, 439–444.

Song, L. and Yu, S. 1999. Concentration polarization in cross-flow reverse osmosis. *AIChE Journal*, 45, 921.

Sourirajan, S. 1970. *Reverse Osmosis*, Logos Press Ltd, London.

Sowgath, M. T. 2007. Neural network based hybrid modelling and MINLP based optimisation of MSF desalination process within gPROMS: Development of neural network based correlations for estimating temperature elevation due to salinity, hybrid modelling and MINLP based optimisation of design and operation parameters of MSF desalination process within gPROMS. Dissertation, University of Bradford.

Sowgath, T. and Mujtaba, I. 2008. Less of the foul play. *Chemical Engineer*, 804, 28–29.

Sowgath, T. M. and Mujtaba, I. 2015. Meeting the fixed water demand of MSF desalination using scheduling in gPROMS. *Chemical Engineering Transactions*, 45, 451–456.

Spiegler, K. S. and Kedem, O. 1966. Thermodynamics of hyperfiltration (reverse osmosis): Criteria for efficient membranes. *Desalination*, 1, 311–326.

Tanvir, M. S. and Mujtaba, I. M. 2006. Neural network based correlations for estimating temperature elevation for seawater in MSF desalination process. *Desalination*, 195, 251–272.

Tanvir, M. S. and Mujtaba, I. M. 2008. Optimisation of design and operation of MSF desalination process using MINLP technique in gPROMS. *Desalination*, 222, 419–430.

Tarifa, E. E. and Scenna, N. J. 2001. A dynamic simulator for MSF plants. *Desalination*, 138, 349–364.

Thomas, P. J., Bhattacharyya, S., Patra, A. and Rao, G. P. 1998. Steady state and dynamic simulation of multi-stage flash desalination plants: A case study. *Computers and Chemical Engineering*, 22, 1515–1529.

Tokui, Y., Moriguchi, H. and Nishi, Y. 2014. Comprehensive environmental assessment of seawater desalination plants: Multistage flash distillation and reverse osmosis membrane types in Saudi Arabia. *Desalination*, 351, 145–150.

Uche, J., Serra, L. and Valero, A. 2001. Thermoeconomic optimization of a dual-purpose power and desalination plant. *Desalination*, 136, 147–158.

Villafafila, A. and Mujtaba, I. M. 2003. Fresh water by reverse osmosis based desalination: Simulation and optimisation. *Desalination*, 155, 1–13.

Vince, F., Aoustin, E., Bréant, P. and Marechal, F. 2008a. LCA tool for the environmental evaluation of potable water production. *Desalination*, 220, 37–56.

Vince, F., Marechal, F., Aoustin, E. and Breant, P. 2008b. Multi-objective optimization of RO desalination plants. *Desalination*, 222, 96–118.

Vrouwenvelder, J. S., Van Paassen, J. A. M., Kruithof, J. C. and Van Loosdrecht, M. C. M. 2009. Sensitive pressure drop measurements of individual lead membrane elements for accurate early biofouling detection. *Journal of Membrane Science*, 338, 92–99.

Wang, J., Mo, Y., Mahendra, S. and Hoek, E. M. V. 2014. Effects of water chemistry on structure and performance of polyamide composite membranes. *Journal of Membrane Science*, 452, 415–425.

Wangnick, K. 1995. How incorrectly determined physical and constructional properties in the seawater and brine regimes influence the design and size of an MSF desalination plant–stimulus for further thoughts. In *Proceedings of the IDA World Congress on Desalination and Water Science*, Abu Dhabi, United Arab Emirates, 201–218.

Wardeh, S. and Morvan, H. P. 2008. CFD simulations of flow and concentration polarization in spacer-filled channels for application to water desalination. *Chemical Engineering Research and Design*, 86, 1107–1116.

Wardeh, S. and Morvan, H. P. 2011. CFD modelling of reverse osmosis channels with potential applications to the desalination industry. *Chemical Product and Process Modeling*, 6(2). doi: 10.2202/1934-2659.1599.

Winter, T., Pannell, D. J. and McCann, L. M. J. 2002. The economics of desalination and it's potential application to Australia. Australian Agricultural and Resource Economics Society.

Xianli, W. U., Yangdong, H. U., Lianying, W. U. and Hong, L. I. 2014. Model and design of cogeneration system for different demands of desalination water, heat and power production. *Chinese Journal of Chemical Engineering*, 22, 330–338.

Yokoyama, K., Ikenaga, Y., Inoue, S. and Yamamoto, T. 1977. Analysis of start-up characteristics of commercial MSF plant. *Desalination*, 22, 395–401.

Zhu, M., El-Halwagi, M. M. and Al-Ahmad, M. 1997. Optimal design and scheduling of flexible reverse osmosis networks. *Journal of Membrane Science*, 129, 161–174.

1.2 Addressing the Global Water–Energy Challenge through Energy-Efficient Desalination

Sudipta Sarkar and Arup K. SenGupta

1.2.1 Introduction

The dwindling availability of freshwater globally is a matter of concern at the present time. It necessitates the conversion of huge quantity of saline water available in oceans and other brackish water sources to freshwater through a process widely known as desalination process. The present state-of-the-art technologies for desalination are not energy efficient so that they can be universally used for the conversion of saline water to freshwater [1,2]. Therefore, the need of the hour is to develop next-generation low-energy desalination facilities for both brackish and seawater. The presently available desalination technologies can be classified into two major categories: thermal multistage flash distillation and membrane processes. The membrane-based reverse osmosis (RO) processes have made significant advances during the last few decades and are being mostly used around the world [3–10]. Desalination of sea or brackish water using membrane is a physical process of separation of salt and water phases across a semipermeable membrane. Separation of the phases takes place under the influence of chemical potential gradient created either by the application of pressure (RO, nanofiltration, etc.), concentration gradient (forward osmosis), electrical potential (electrodialysis), or combinations of them. Although widely used, the cost-effectiveness of RO processes is adversely affected by two major shortcomings:

1. High-energy consumption per unit of freshwater produced.
2. Lack of durability of the RO membrane and its susceptibility to fouling.

While the current research emphasis is on the improvement of the membrane material for robustness and higher permeate flux [8,10,11–16], recent progresses have been made toward the development of forward osmosis (FO) and membrane distillation processes as well [17–21]. All membrane-based processes have two common features: (1) Membrane at the saline and pure water interface is the most important part of the process. (2) The energy efficiency of the process is directly related to the performance of the membrane. Continuous improvements in the membrane-based processes, especially the RO process, involved developing better membranes in terms of permeability and resilience against fouling. Undoubtedly, RO is by far the most cost-efficient and popular method of desalination at present. However, the process still suffers from major drawbacks that tend to limit its wide-scale applications for the production of freshwater from brackish or sea water sources.

In order to attain energy efficiency, a membrane-based desalination process is targeted to have a maximum possible recovery of product water from sea or brackish water. Such an effort always produces in the feed side a reject stream that is more concentrated than the original feed sea or brackish water. Inside a boundary layer formed at the surface of the membrane in the feed side, the effect of the increase

TABLE 1.7   Typical Seawater Composition

Ion	Valence	Concentration (M)
Na^+	1	0.448
Cl^-	1	0.545
SO_4^{2-}	2	0.0281
Ca^{2+}	2	0.0124
Ba^{2+}	2	1.53×10^{-7}
Sr^{2+}	2	0.000148
Mg^{2+}	2	0.05327

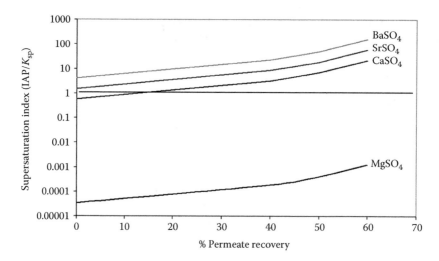

FIGURE 1.33 Calculated supersaturation indices for sulfate salts of different divalent cations at different permeate water recovery for RO membrane desalination of seawater.

in the concentration is much more pronounced, causing the membrane to effectively work against a high concentration differential between the feed and the product. Such an effect is known as concentration polarization. At the elevated concentrations inside the boundary layer, some of the ions present in the seawater tend to form precipitates of their respective salts (scales) when their concentrations exceed the corresponding solubility product constant (K_{sp}) values. Inorganic fouling or scaling of the membranes caused by salt precipitates tend to decrease the product water flux. It also necessitates frequent maintenance of the fouled membranes. Frequent maintenance results in significant reduction of the plant availability. The scales formed by the precipitation of carbonate salts can theoretically be reversed by acid-cleaning process. However, scales produced by sulfate salts such as $CaSO_4$ or $BaSO_4$ are irreversible; pH does not have any impact on their solubility. Antiscalants, which are normally salts of poly(acrylic acid) or poly(phosphonic acid), are routinely administered into the feed water to inhibit scaling of sulfate salts. However, use of antiscaling chemicals can only retard the process of scaling by binding to the metal ions but does not prevent its formation in the long run. The antiscalants are expensive chemicals and arguably can cause detrimental effects when the concentrate rich with antiscalants are discharged into the environment.

Achieving higher product water recovery for a desalination plant means better energy efficiency. In many cases, it offers better options for concentrate management, especially for those plants located inland or using brackish water. However, any attempt to increase the product water recovery from the membrane module makes the reject stream more concentrated. Increase in the concentration of seawater is equivalent to increase of ion activity product of the scale-forming ions that may lead to enhanced precipitation leading to scale formation. Therefore, any attempt to increase the energy efficiency of a

TABLE 1.8 Solubility Product Values of Different Sulfate Salts under Ideal Condition at 25°C

Salt	Chemical Formula	Solubility Product (K_{sp})
Calcium sulfate	$CaSO_4$	6.3×10^{-5}
Barium sulfate	$BaSO_4$	1.08×10^{-10}
Strontium sulfate	$SrSO_4$	2.82×10^{-7}
Magnesium sulfate	$MgSO_4$	4.67

Source: Norwich, N.Y.: Knovel, 2008 *Critical Tables*, 2nd Edition, 2008.

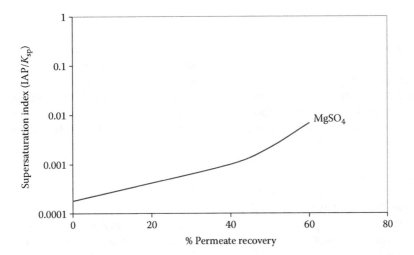

FIGURE 1.34 Supersaturation index of magnesium sulfate at various permeate recovery considering that all cations in seawater have been replaced by magnesium.

desalination process by increasing the water recovery is limited by the scale-forming potential of salts at the membrane surfaces. Since the extent of scaling caused by precipitation of $CaSO_4$ is more prevalent than other sparingly soluble inorganic salts, it is the primary focus of the present study. The methodology proposed here is amenable to application in case of other sparingly soluble salts including calcium phosphate or barium sulfate.

1.2.1.1 Solubility of Different Sulfate Salts and Scaling Potential

The supersaturation index, which expresses the level of saturation of a solution with respect to various mineral salts, is given by

$$SI_i = \frac{IAP}{K_{sp,i}}, \tag{1.1}$$

where IAP is the ion activity product and $K_{sp,i}$ is the solubility product for the salt i. A saturation index value greater than unity means that the solubility limit for the salt has been exceeded, thus indicating a potential scaling problem.

Control of scaling is necessary for the improvement of a membrane-based desalination process. Because of the relative abundance of sulfate ions in seawater, the sulfate salts of calcium, barium, and strontium are significant as potential scale-formers. While the precipitates of other salts of calcium,

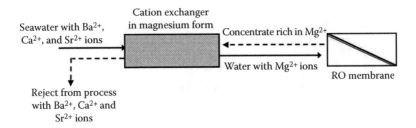

FIGURE 1.35 Elimination of scale-forming ions by magnesium ions in hybrid RIX-MEM process.

barium, and strontium, such as carbonate salts, can however be controlled and cleaned by a swing in pH, the scales of sulfate salts cannot be cleaned or controlled by change in pH.

Table 1.7 shows a typical sea water composition collected from the Atlantic Ocean and Figure 1.33 represents the variation in the theoretical supersaturation indices of sulfate salts of different ions with changes in the percentage permeate water recovery. The nominal salt rejection has been considered to be 99%. The indices have been calculated based on the composition of different key ions present in a typical seawater sample as indicated in Table 1.7. It may be observed from Figure 1.33 that even with modest 30% permeate recovery, there is a huge potential for scaling due to the precipitation of the sulfate salts of calcium, barium, and strontium.

It is interesting to note that all the scale-forming ions belong to Gr. 2A of the periodic table. Also, magnesium ions, another Gr. 2A element, although present in higher concentration compared to the other elements in the same group, did not show any precipitation potential even at a very high water recovery ratio. The reason behind such a difference in the scaling potential lies in the fact that magnesium salts of sulfate have very high solubility product value (K_{sp}) compared to the sulfate salts of other cations belonging to the same group. The solubility product values of different sulfate salts are listed in Table 1.8. The supersaturation index value for $MgSO_4$ is more than two orders of magnitude lower than unity and thus, it does not pose any threat with respect to scaling/membrane fouling at higher recovery.

Figure 1.34 shows the modified scaling potential diagram of the seawater for a hypothetical case where all the cations including sodium ions of seawater are replaced by equivalent concentration of magnesium ions. It is observed that even with complete replacement of all the cations, the system can run at high permeate water recovery ratio without any potential risk of formation of precipitates of sulfate salts. This analysis is independent of any specific type of RO membrane to be used. Thus, it is possible to avoid the sulfate-induced scaling if the scale-forming cations are replaced by magnesium ions. Such replacement of scale-forming cations by magnesium ions is possible through incorporation of a cation exchange step ahead of the membrane desalination step.

Here, we propose a novel reversible ion exchange-membrane (RIX-MEM) process for desalination that alters the chemistry of the RO feed solution in such a way as to eliminate the potential for scale formation by cations, without requiring any continuous need of extra chemicals or energy.

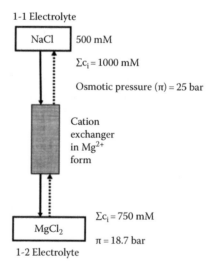

FIGURE 1.36 Reduction in theoretical osmotic pressure of sodium chloride following passage through ion exchangers presaturated with divalent magnesium ion.

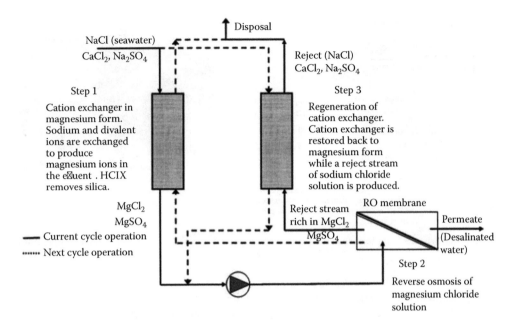

FIGURE 1.37 Schematic of the RIX-MEM desalination process describing three major operational steps.

1.2.1.2 RIX-MEM: Process Fundamentals

The central element of the RIX-MEM process is that a process of reversible ion exchange precedes membrane-based desalination processes, such as FO, RO, and nanofiltration, to alter the water chemistry of membrane-feed water in such a way that the scale-forming divalent cations, such as calcium, barium, strontium, etc., are replaced by nonscale-forming magnesium ions. The underlying scientific bases of the RIX-MEM process are as follows:

1. Magnesium salts of sulfate, carbonate, and fluoride ions are more soluble compared to the sulfate, carbonate, and fluoride salts of calcium, barium, or strontium.
2. For strong acid cation exchangers, scale-forming calcium, barium, and strontium ions are preferred to magnesium ions; magnesium is thus preferentially displaced by calcium, barium, or strontium in a fixed-bed column containing cation exchanger presaturated in magnesium form.
3. Ion exchange processes work on the exchange of equivalents of ions; osmotic pressure of an aqueous solution is governed by molar concentrations of the ions/solutes.

Figure 1.35 shows the concept of the hybrid process where the scale-forming cations of barium, calcium, and strontium shall be replaced by magnesium ions in the cation exchange step prior to the RO process. The dashed lines show that the process can be reversed by regenerating the cation exchanger back to magnesium form by using the concentrate from the RO process.

Considering Na^+ and Cl^- ions to be the primary constituents in seawater, it is possible that the osmotic pressure of the seawater can be greatly reduced by using suitably engineered ion exchange process ahead of the membrane desalination step, as illustrated in Figure 1.36. When Na^+ ions are exchanged in equivalent concentrations for divalent magnesium ions, there is a resultant decrease in the molar concentrations of ions in the solution. This drop actually helps to reduce the osmotic pressure of the resulting solution. Thus, for a RIX-MEM process, it is possible to achieve greater energy economy compared to the conventional RO process. Also, the ion exchange process is reversible; therefore, sodium chloride can be reproduced from $MgCl_2$ by reversing the flow direction as shown by the dashed lines in Figure 1.36.

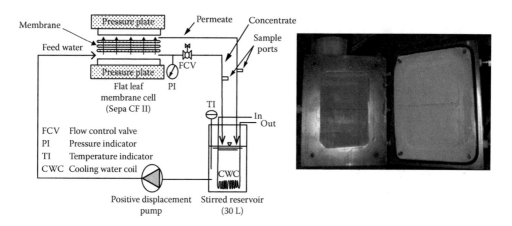

FIGURE 1.38 Schematic of the experimental setup and photograph of GE Osmonics flat-leaf test apparatus with a RO membrane.

Process details and description of different process steps are described hereafter. Two trains of cation exchange columns and one RO membrane are needed and the schematic of the RIX-MEM desalination process is illustrated in Figure 1.37.

Step 1 Seawater is passed through a cation exchanger in magnesium form leading to the following exchange reactions:

$$\overline{(RSO_3^-)_2 Mg^{2+}} + 2Na^+ \Leftrightarrow 2\overline{(RSO_3^-)Na^+} + Mg^{2+} \tag{1.2}$$

$$\overline{(RSO_3^-)_2 Mg^{2+}} + Ca^{2+} \text{or } Ba^{2+} \Leftrightarrow \overline{(RSO_3^-)_2 Ca^{2+} \text{ or } Ba^{2+}} + Mg^{2+}. \tag{1.3}$$

Overbar denotes the solid exchanger phase, while RSO_3^- represents the sulfonic acid fixed functional group of the ion exchanger. The resulting $MgCl_2$ (2-1 electrolyte) from the cation exchange step has 30% lower osmotic pressure but the equivalent electrolyte concentration remains the same.

Step 2 The resultant solution from Step 1 is subjected to RO. Since the osmotic pressure of the solution is now lower than the original sea or brackish water and fouling potential is nearly eliminated as it contains only magnesium as cation along with other anions, higher product water recovery is attainable with lower membrane area requirement. The product water recovery being higher, the energy consumption per unit volume of product water is lower than the conventional RO process.

Step 3 The reject stream from the membrane, now rich in magnesium, passes through the previously exhausted cation exchange column from Step 1. The column is transformed back into magnesium form and the resulting effluent mostly contains NaCl (1-1 electrolyte) along with other cations like calcium, magnesium, etc. and anions like sulfate, bicarbonate, etc.

$$2\overline{(RSO_3^-)Na^+} + Mg^{2+} \Leftrightarrow \overline{(RSO_3^-)_2 Mg^{2+}} + 2Na^+ \tag{1.4}$$

$$\overline{(RSO_3^-)_2 \ Ca^{2+} \text{ or } Ba^{2+}} + Mg^{2+} \Leftrightarrow \overline{(RSO_3^-)_2 Mg^{2+}} + Ca^{2+} \text{ or } Ba^{2+}. \tag{1.5}$$

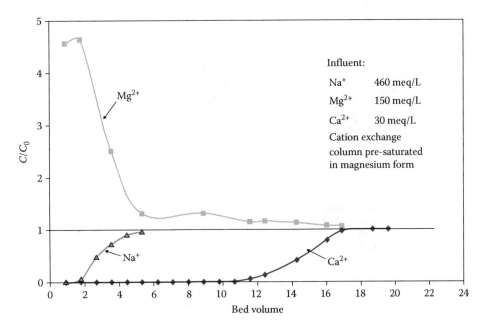

FIGURE 1.39 Breakthrough profile of different ions for a synthetic seawater solution passed through cation exchange column presaturated in magnesium form. (C_0 and C are the influent and effluent concentration, respectively, of a component at any point of operation).

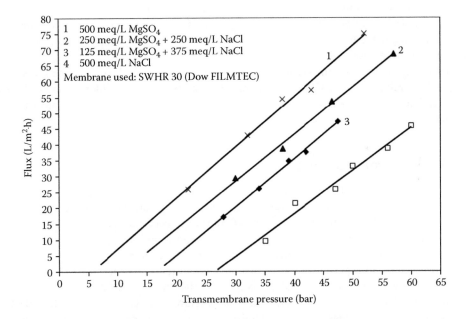

FIGURE 1.40 Flux obtained for a RO membrane at different transmembrane pressures for solutions of same equivalent concentrations, but different proportions of magnesium and sodium salts.

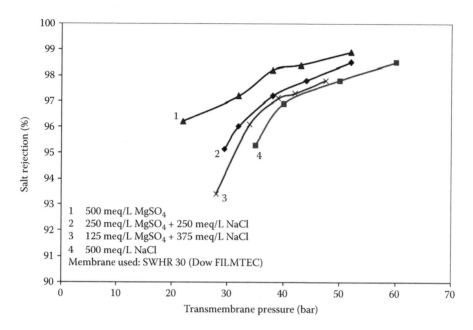

FIGURE 1.41 Salt rejection profile by RO membrane when subjected to different transmembrane pressures for solutions with different proportions of magnesium and sodium salts, but same equivalent concentration of electrolyte.

Ideally no external regenerant is required and the ion exchange unit is now ready for operation in Step 1. In principle, in RIX-MEM process, cation exchangers can switch back and forth between magnesium and sodium forms without needing any external regenerant and desalination is accomplished with higher permeate water recovery.

In this chapter, we focus on the experimental validation of the general basis of the process to show that the scale-forming potential of seawater can be significantly reduced by tailored combination of ion exchange and RO process. We also report the study results on the sustainability of the process run over many without needing any external chemical.

1.2.2 Materials and Methods

A flat-leaf test cell (SEPA CF II, GE Osmonics) with an effective membrane area of 140 cm^2 was used to perform RO experiments. The unit was operated at different transmembrane pressures ranging from 2 to 60 bar transmembrane pressure using SWHR 30 reverse osmosis membrane (DOW Chemicals Co., Midland, MI). As per the manufacturer, SWHR 30 is a polyamide thin film composite membrane with nominal water flux of 38.28 L/(m^2·h) at 32,000 ppm NaCl concentration. A flow sheet and photograph of the closed-loop experimental setup is shown in Figure 1.38. The flat-leaf unit, the high pressure pump, and the tubings were all made of SS-316 to limit corrosion. The membranes were initially conditioned by subjecting to 30 bar transmembrane pressure for at least 2 h under deionized water. During the test, the system was allowed to stabilize for at least 1 h under a particular pressure before samples were taken and flow rate was recorded. The reservoir of about 30 L capacity was completely mixed during test runs. The temperature was maintained at 22°C–25°C during the test runs. For every transmembrane pressure, three samples were collected from the permeate side under steady-state conditions at an interval of 15 min.

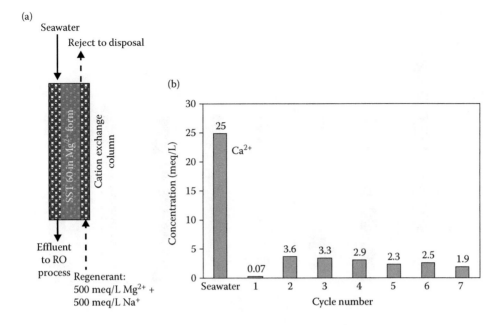

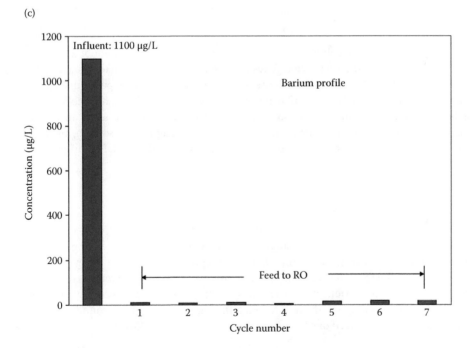

FIGURE 1.42 (a) Schematic of experimental column run, (b) calcium concentration, and (c) barium concentration in synthetic seawater and effluent over number of cycles.

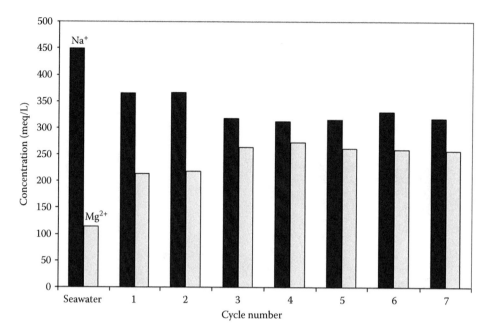

FIGURE 1.43   Sodium and magnesium concentrations in the influent and effluent obtained after number of cycles of operation.

All fixed-bed ion exchange column runs were carried out in a glass column of 2.54×10^{-2} m (1 in.) diameter using a strong acid cation exchange resin SST 60 (Purolite Co., Philadelphia, PA). Capacity of the resin was 3.6 eq/kg in Na^+ form. The column was fitted with a peristaltic pump that fed the electrolyte solution at a constant rate and the effluent was collected at an ELDEX fraction collector.

Conductivity of each sample was measured immediately after collection using an ORION (model 120) conductivity meter. Concentrations of the cations, namely magnesium, sodium, and calcium were measured using a flame-type atomic absorption spectrometer (AAnalyst 200, PerkinElmer).

1.2.3 Results

1.2.3.1 Magnesium Ion Exchange

A synthetic seawater solution containing 460 meq/L sodium, 30 meq/L calcium, 150 meq/L magnesium ions as cations was processed through a column containing cation exchange resins (Purolite SST 60) presaturated in magnesium form. Figure 1.39 represents the breakthrough profile of different cations in the effluent of the column. Magnesium ions in the ion exchange resins were replaced by other cations, and as a result, its concentration in the effluent was much higher than its influent. Understandably, sodium was the first to break through the column at around two bed volumes. Calcium ions, as it is a preferred one by the ion exchanger, had a breakthrough much later, at around 12 bed volumes.

1.2.3.2 Runs with RO Membrane

Solutions containing different relative proportions of sodium chloride and magnesium sulfate, but with constant total equivalent concentration of electrolyte similar to seawater were subjected to pressure-driven RO process using the flat-leaf-type membrane cell test apparatus. Figure 1.40 shows plots of the permeate flux data recorded at different transmembrane pressures for different feed solutions. It is noted that the incremental replacement of sodium and chloride ions in the feed solution by equivalent concentration of magnesium and sulfate ions, respectively, helped to achieve higher flux of permeate at same transmembrane pressures. Figure 1.41 represents the salt rejection profile of the RO membrane for the experiment. It may be noted that higher proportion of magnesium in the feed solution resulted in better salt rejection by the membrane.

1.2.3.3 Reversibility of the RIX-MEM Process

To validate the reversibility of the RIX-MEM process in the laboratory, we performed cyclical runs according to the schematic in Figure 1.42a. The synthetic seawater solution containing 450 meq/L sodium ions, 100 meq/L magnesium, 25 meq/L calcium, and 1100 μg/L barium ions of volume equivalent to four bed volumes (BV) was fed to a column filled with cation exchange resin (SST 60, Purolite Co., PA) in magnesium form (forward run). The used column was regenerated using a solution of 2BV containing 500 meq/L sodium and 500 meq/L magnesium ions (reverse run). Figures 1.42b and c show calcium and barium concentrations, respectively, at the exit to the column during the forward run. Concentrations of both these ions approached a steady state with a value much lower than that of the synthetic seawater. It may be noted that in the RIX-MEM process, the effluent from the forward run of the ion exchange column is used as the feed to the RO membrane.

Figure 1.43 shows the sodium and magnesium ion concentrations observed at the column exit for the forward run in the cyclic process. Sodium concentration in the effluent (feed solution to the RO process in the RIX-MEM process) has increased with the progression of cycles, but overall, the composition of cations in the effluent approaches a steady state after the fifth cycle.

1.2.4 Discussion

1.2.4.1 Modification in Feed Water Composition Improves Energy Efficiency of the RO Process

For a RO process, water flux (J_W) and transmembrane pressure (Δp) are interrelated through the following relationship:

$$J_W = K_W(\Delta p - \Delta \pi), \tag{1.6}$$

where $\Delta \pi$ is the osmotic pressure difference between the feed and the permeate solution and K_w is the water permeability coefficient, an intrinsic property of the membrane.

The permeate side solution should, ideally for an RO, have low concentration of ions and its osmotic pressure can be considered as negligible. Therefore, osmotic pressure of the feed solution influences the RO process. It is observed from Figure 1.40 that the gradual replacement of sodium and chloride ions by equivalent concentration of magnesium and sulfate ions, respectively, reduced the osmotic pressure of the feed solution. As a result, it was possible to achieve higher water flux (J_w) across the membrane at the same transmembrane pressure. Hence, alteration of feed water chemistry by incorporation of an appropriate ion exchange step ahead of the RO process should help attain better energy efficiency for the overall desalination process. RIX-MEM process includes a cation exchange step ahead of RO process where magnesium-sodium exchange coupled with RO process helps attain better energy efficiency than the conventional RO process. The pretreatment step should also reduce the scaling potential of the RO feed water as the scale-forming ions are suitably eliminated as shown in Figure 1.42.

1.2.4.2 Sustainability of the Energy Efficiency over Multiple Cycle Run

The sustainability of the process depends on the differential affinity of cations toward a cation exchanger. The affinity depends on many factors that include valence, hydrated ionic radii, etc., of the ions in consideration. In general, for dilute solutions, the affinity sequence of the cations for common strong acid cation exchangers (such as SST-60 or C-100 resins) is as follows:

$$Ba^{2+} > Pb^{2+} > Sr^{2+} > Ca^{2+} > Ni^{2+} > Cd^{2+} > Cu^{2+} > Co^{2+} > Zn^{2+} > Mg^{2+}$$

$$> Ag^+ > Cs^+ > K^+ > NH_4^+ > Na^+ > Li^+. \tag{1.7}$$

It is evident that breakthrough of ions from a cation exchange column shall follow the reverse sequence. Breakthrough pattern of different ions as in Figure 1.39 is in agreement with the above affinity sequence. The general cation exchange reactions as mentioned in Equations 1.2 and 1.3 take place when synthetic solution of seawater is passed through a column containing cation exchanger in magnesium form. Considering ideality, the selectivity coefficient for a heterovalent exchange reaction as depicted in Equation 1.2 can be presented as follows:

$$K_{Mg/Na} = \frac{y_{Mg} x_{Na}^2}{x_{Mg} y_{Na}^2} \times \frac{C_T}{Q}. \tag{1.8}$$

And for homovalent exchange reaction 1.3, the selectivity coefficient is defined as follows:

$$K_{Ca/Mg} = \frac{y_{Ca} x_{Mg}}{x_{Ca} y_{Mg}}, \tag{1.9}$$

where x and y represent the fractional concentration of the ions at aqueous and resin phase, respectively, while C_T is the total equivalent cation concentration in the aqueous phase and Q is the exchanging capacity of the cation exchange resin.

In an ion exchange column, relative preference of one cation over the other for the ion exchange resin is represented by the separation factor that determines the breakthrough behaviors of the ions. Magnesium/sodium and calcium/magnesium separation factors are represented by $\alpha_{Mg/Na}$ and $\alpha_{Ca/Na}$, respectively. They are defined as follows:

$$\alpha_{Mg/Na} = \frac{y_{Mg} x_{Na}}{x_{Mg} y_{Na}} \tag{1.10}$$

and

$$\alpha_{Ca/Na} = \frac{y_{Ca} x_{Na}}{x_{Ca} y_{Na}}. \tag{1.11}$$

An analysis of Equations 1.8 through 1.11 reveals that the separation factor is not a constant for heterovalent exchange, like those involving sodium/calcium or sodium/magnesium exchanges. For a constant ion-exchange capacity of the resin, the separation factor value changes depending on the electrolyte concentration in the aqueous phase. Thus, the ion exchanger's preference for a divalent ion (e.g., Mg^{2+} ion) over a monovalent ion (e.g., Na^+ ion) rapidly decreases as the electrolyte concentration increases. Beyond a particular electrolyte concentration, the selectivity reverses making monovalent ions preferred over divalent ions. The electrolyte concentration at which this reversal occurs depends on the relative affinities of the exchanging ions. For homovalent exchange, however, there is no effect of aqueous phase electrolyte concentration on the separation factor. Therefore, relative preference of calcium over magnesium remains the same regardless of the electrolyte concentration.

An analysis of the results obtained during the column run as shown in Figure 1.39 provided the following values of the different separation factors:

$$\alpha_{Ca/Mg} = 16.2, \ \alpha_{Mg/Na} = 0.309 \text{ and } \alpha_{Ca/Na} = 5.$$

The synthetic seawater influent to the column had a total electrolyte concentration of 620 meq/L. According to the affinity sequence in Equation 1.7, for dilute solutions the separation factor for Mg/Na exchange is expected to be >1. However, the separation factor of <1 as observed suggests selectivity reversal for Mg/Na exchange at total electrolyte concentration equivalent to seawater. Following points may be noted: (1) There is magnesium/sodium selectivity reversal for concentrated electrolyte solution such as seawater; (2) calcium/sodium separation factor value is significantly >1; selectivity reversal is absent even at concentration equivalent to seawater; (3) affinity for barium and strontium ions are much higher than calcium; hence, logically there shall be no reversal for barium/sodium or strontium/sodium selectivity for seawater; (4) barium, strontium, and calcium are preferred by the ion exchanger over magnesium even for concentrated solution such as seawater, as it is a homovalent exchange of ions.

Reversal of magnesium–sodium selectivity due to high electrolyte concentration has a favorable impact during the forward run resulting in elution of magnesium ions from the column. Thus, the osmotic pressure of the solution should reduce, making it possible to attain high water recovery at the RO step. However, higher water recovery produces more concentrated return water from the RO step. In the proposed process, the concentrate is used for regeneration. Therefore, during regeneration there will be a further drop in the magnesium/sodium separation factor. So, instead of magnesium ion getting back into ion exchanger phase as it was hypothesized in the hybrid process, magnesium is eluted out into the reject stream causing an overall leakage of magnesium from the system. To sustain the process, there is a need to inject magnesium ions in to the system so as to compensate for the loss of magnesium from the system. Seawater itself contains a significant concentration of magnesium ions and can replenish for the loss to some extent. Therefore, the system is expected to reach a steady state during the cyclic operation and the ratio of sodium to magnesium ions in the aqueous phase would eventually become constant. The above inference is supported by the results observed in Figure 1.43 that shows that the relative concentration of sodium and magnesium ions into the aqueous phase reaches a steady state after the fifth cycle. Thus, the osmotic advantage of the RIX-MEM process targeted via replacement of monovalent sodium by divalent magnesium ions is likely to remain low when the system is operated with high water recovery. However, the situation can be improved by replacing the strong acid cation exchanger with weak acid cation exchange resin which has higher capacity and higher affinity for polyvalent alkali metal cations.

1.2.4.3 Sustaining the Scaling-Free Desalination Process over Multiple Cycle Runs

According to Figure 1.39, calcium breakthrough occurred after 10 bed volumes leading to a logical inference that the breakthrough of barium and strontium would occur much later, whereas the forward run was only for four bed volumes. Therefore, the effluent of the column (feed to RO) in RIX-MEM process should contain a little or no calcium, barium, or strontium ions. The concentrate from the RO, used as regenerant in the reverse run, should have higher total electrolyte concentration than the influent seawater in the forward run. The separation factor values change with electrolyte concentration for the heterovalent exchange, but not for the homovalent exchange. Based on the above argument, the following points may be inferred:

1. For heterovalent exchange, selectivity decreases with increase in the total electrolyte concentration. Therefore,

$$\alpha_{Ca/Na,forward} > \alpha_{Ca/Na,reverse} \tag{1.12}$$

and

2. $\left(\dfrac{x_{Mg}}{x_{Ca}}\right)_{Reverse} > \left(\dfrac{x_{Mg}}{x_{Ca}}\right)_{Forward}$ and, $\left(\dfrac{x_{Na}}{x_{Ca}}\right)_{Reverse} > \left(\dfrac{x_{Na}}{x_{Ca}}\right)_{Forward}$. $\tag{1.13}$

For a multicomponent exchange involving Mg, Ca, and Na, the fractional calcium ion concentration in the resin phase is given by the following relationship:

$$y_{Ca} = \frac{1}{1 + \dfrac{1}{\alpha_{Ca/Mg}} \dfrac{x_{Mg}}{x_{Ca}} + \dfrac{1}{\alpha_{Ca/Na}} \dfrac{x_{Na}}{x_{Ca}}}. \tag{1.14}$$

Using relationships 1.12 through 1.14 one can conclude that

$$y_{Ca,forward} > y_{Ca,reverse}. \tag{1.15}$$

The above analysis shows that the fractional concentration of calcium in the resin phase after the forward run is greater than that at the end of reverse run. So, in an overall process consisting of a forward and reverse run incorporating a RO step in between, the cation exchange resin does not accumulate calcium ion to a significant extent and eventually, the resin phase calcium concentration should reach a steady state after some number of cycles at which there will be negligible leakage of calcium in the feed solution to the RO process. Figure 1.42 provides a proof of the concept. A similar analysis is true for barium and strontium ions.

1.2.4.4 Conclusion

It is validated in the laboratory scale that RIX-MEM process offers an energy-efficient and scaling-free desalination of sea and brackish water. It is also validated that the process does not need any extraneous regenerant or chemical for its sustenance over many number of cycles. However, for better validation of the process concept and better quantification of process benefits, a pilot scale testing involving cation exchange columns and spirally wound RO module is required. It will allow for a continuous run of the process over many number of cycles in a coupled manner. Use of weak acid cation exchanger with higher capacity and optimization of the degree of water recovery should be targeted. In fact, degree of water recovery is the single most important variable that dictates the sustainability and energy efficiency of the whole process.

References

1. USBR and SNL (United States Bureau of Reclamation and Sandia National Laboratories), Desalination and water purification technology roadmap: A report of the executive committee. Desalination and Water Purification Research and Development Report #95, United States Department of the Interior, Bureau of Reclamation, Water Treatment and Engineering Group, Denver, CO (2003).
2. Water Science and Technology Board, Division on Earth and Life Studies and National Research Council of National Academies, Review of the desalination and water purification technology roadmap, National Academies Press, Washington, DC (2004).

3. The Quality of Our Nation's Waters, A summary of the national water quality inventory: 1998, Report to Congress. EPA841-S-00-001. https://www.epa.gov/sites/production/files/2015-09/documents/2000_07_07_305b_98report_98brochure.pdf

4. R. Semiat, Desalination: Present and future, *Water Int.*, 25(1), 54–65 (2000).

5. H.M. Ettouney, H.T. El-Dessouky, R.S. Faibish, and P.J. Gowin, Evaluating the economics of desalination, *Chem. Eng. Prog.*, 98, 32–39 (2002).

6. R.F. Service, Desalination freshens up, *Science*, 313, 1088–1090 (2006).

7. K.S. Spiegler and Y.M. Al-Sayed, The energetics of desalination process, *Desalination*, 134 (1–3), 109–128 (2001).

8. W.S. Winston Ho and K.K. Sirkar (Eds), *Membrane Handbook*, Van Nostrand Reinhold, New York (1992).

9. J.R. Taylor and M.R. Wiesner, Membrane processes. In *Water Quality and Treatment*, 5th Ed., R. D. Letterman, Ed., McGraw Hill, New York, 11.1–11.71 (1999).

10. R.W. Baker, *Membrane Technology and Applications*, 2nd Ed., John Wiley and Sons, New York (2004).

11. W.J. Koros, G.K. Fleming, S.M. Jordan, T.H. Kim, and H.H. Hoehn, Polymeric membrane materials for solution-diffusion based permeation separation, *Prog. Polym. Sci.*, 13, 339–401 (1988).

12. M.R. Wiesner, The promise of membrane technology, *Environ. Sci. Technol.*, 4(9), 360A–366A (1999).

13. B.H. Jeong, E.M.V. Hoek, Y. Yan, X. Huang, A. Subramani, G. Hurwitz, A.K. Ghosh, and A. Jawor, Interfacial polymerization of thin film nanocomposites: A new concept for reverse osmosis membranes, *J. Membr. Sci.*, 294, 1–7 (2007).

14. A.K. Ghosh, B.H. Jeong, X. Huang, and E.M.V. Hoek, Impacts of reaction and curing conditions on polyamide composite reverse osmosis membrane properties, *J. Membr. Sci.*, 311, 34–45 (2008).

15. M. Kumar, M. Grzelakowski, J. Zilles, M. Clark, and W. Meier, Highly permeable polymeric membranes based on the incorporation of the functional water channel protein Aquaporin Z, *Proc. Natl. Acad. Sci.*, 104(52), 20719–20724 (2007).

16. V. Smuleac, D.A. Butterfield, and D. Bhattacharyya, Permeability and separation characteristics of polypeptide-functionalized polycarbonate track-etched membranes, *Chem. Mater.*, 16, 2762–2771 (2004).

17. T.Y. Cath, A.E. Childress, and M. Elimelech, Forward osmosis: Principles, applications, and recent developments, *J. Membr. Sci.*, 281(1–2), 70–87 (2006).

18. B. Li and K.K. Sirkar, Novel membrane and device for vacuum membrane distillation-based desalination process, *J. Membr. Sci.*, 257(1–2), 60–75 (2005).

19. J.R. McCutcheon, R.L. McGinnis, and M. Elimelech, Desalination by a novel ammonia–carbon dioxide forward osmosis process: Influence of draw and feed solution concentrations on process performances, *J. Membr. Sci.*, 278, 114–123 (2006).

20. T.Y. Cath and A.E. Childress, Membrane contactor processes for wastewater reclamation in space. II. Combined direct osmosis, osmotic distillation, and membrane distillation for treatment of metabolic wastewater, *J. Membr. Sci.*, 257, 111–119 (2005).

21. L. Song, B. Li, K.K. Sirkar, and J.L. Gilron, Direct contact membrane distillation-based desalination: Novel membranes, devices, large scale studies and a model, *Ind. Eng. Chem. Res.*, 46(8), 2307–2323 (2l007).

1.3 Forward Osmosis for Irrigation Water Supply Using Hybrid Membrane System for Draw Solution Regeneration

Ali Altaee, Saiyed R. Wahadj, Adel O. Sharif, Guillermo Zaragoza, Malak Hamdan, and Maryam Aryafar

1.3.1 Introduction

Water shortage problems have negatively affected farming and agriculture activities worldwide. Globally, agriculture water constitutes 70% of the total freshwater use (Water Use 2015). Groundwater contamination resulted in a further reduction of water resources which were available for irrigation and exacerbated the problem of water scarcity. Tapping into seawater was considered as a viable solution for freshwater supply. Reverse osmosis (RO) and thermal processes were the pioneering technologies for freshwater supply for many countries (Altaee and Sharif 2011). The current desalination technologies require high energy and are still unaffordable to many countries.

Recently, there was a recurring interest in forward osmosis (FO) process as a viable technique for seawater treatment and freshwater generation (Blandin et al., 2015; Luo et al., 2014). FO utilizes osmotic energy for freshwater extraction from seawater using high permeability and rejection rate membrane. Thermal and membrane processes were suggested for the regeneration of draw solution (Altaee and Sharif 2015; Luo et al., 2014). The current study investigates the feasibility of using an FO-membrane hybrid system for irrigation water supply. The product water generated from a series of membrane treatment should contain a sufficient concentration of nutrients. Thus, it can be directly applied to field without further treatment and dilution. A number of draw solutions consisting of single or a couple of compounds were used for freshwater extraction from seawater. The diluted draw solution from the FO process was further treated by a special membrane hybrid system for irrigation water supply (Phuntsho et al. 2016). Depending on the type of draw solution, high rejection, and water permeability, NF–BWRO membrane system was used for the draw solution treatment (Wilf and Bartels 2015). Dual-stage NF–BWRO system was proposed for seawater desalination because of its lower power consumption than RO membrane system (Altaee and Sharif 2011). For ionic species that have low rejection rate by nanofiltration (NF) and brackish water reverse osmosis (BWRO) membrane such as NO_3^-, a mixture of two chemical agents was suggested as the draw solution. The mixture contains (1) a primary draw solution, such as $MgCl_2$, of high osmotic pressure to provide the driving force for the FO process, and (2) low concentration secondary additive, such as KNO_3, to provide nutrients in the irrigation water (Figure 1.44a). The primary draw solution represents the majority of draw solution concentration, while the secondary additive forms less than 3% of the draw solution concentration.

A combination of NF and RO membranes are used for the regeneration of the draw solution mixture. Previous work showed that the dual-stage NF–BWRO membrane system is more energy efficient than conventional RO system (Altaee and Sharif 2011). Figure 1.44b shows the proposed FO-membrane hybrid system for irrigation water supply. The impact of draw solution concentration, draw solution

TABLE 1.9   Concentration and Osmotic Pressure of Draw Agents

$MgCl_2 + KNO_3$					
Concentration (M)	0.75	0.80	0.85	0.90	0.95
Osmotic pressure (bar)	54.21	57.76	61.30	64.85	68.40
Na_2SO_4					
Concentration (M)	0.75	0.80	0.85	0.90	0.95
Osmotic pressure (bar)	53.23	56.79	60.34	63.88	67.44

type, and types of NF/RO membranes in the regeneration system on the quality of irrigation water was investigated. NF and BWRO membranes of high water permeability were used for the regeneration of draw solution. Seawater, total dissolved solids (TDS) 35 g/L, was used as the feed of FO process. The performance of FO process was evaluated under (1) different concentrations of draw solutions and (2) using pressure-assisted FO (PAFO process). PAFO process was proposed previously for enhancing the process performance (Sahebi et al. 2015; Yun et al. 2012). A comparison between PAFO and FO process was carried out to underline the advantages and disadvantages of each method. Reverse osmosis system analysis (ROSA) and predeveloped FO software were used to estimate the performance of NF/BWRO and FO membranes, respectively (Altaee et al. 2014). KNO_3, Na_2SO_4, $CaNO_3$, and $MgCl_2$ were evaluated as potential draw agents of the FO process. However, not all of these chemical agents would be suitable for FO seawater treatment and subsequent regeneration membrane processes.

1.3.2 Methodology

1.3.2.1 Process Modeling

Inorganic metal salts such as KNO_3, Na_2SO_4, $CaNO_3$, and $MgCl_2$ were used as draw solutions in the FO process. These solutions have high solubility in water and contain ionic species necessary for plant growth (Iqbal et al. 2013; Phuntsho et al. 2011; Plant Nutrition 2015; Sahebi et al. 2015). As show in Figure 1.44b, draw solution enters the FO membrane for freshwater extraction from seawater feed

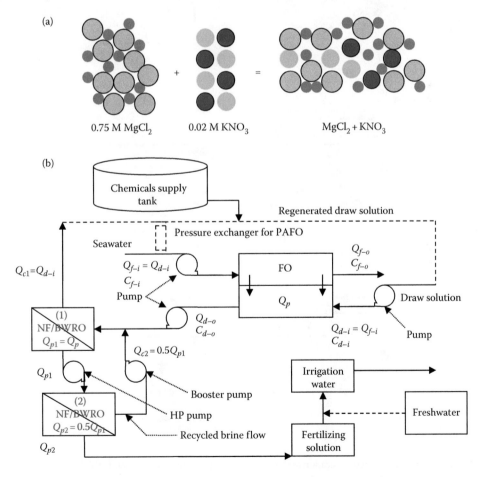

FIGURE 1.44 (a) Draw solution mixture and (b) FO system for fertilizing water supply.

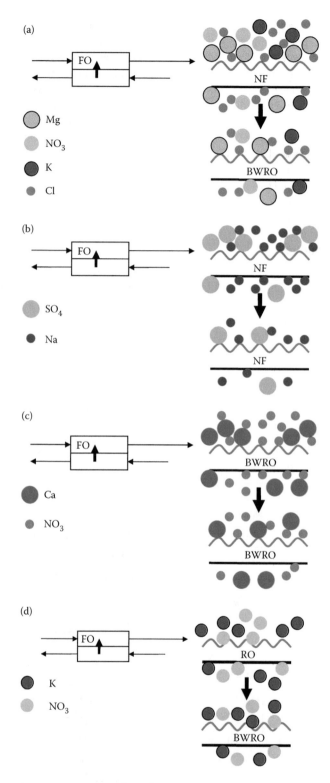

FIGURE 1.45 Draw solution regeneration and recycling process using two-stage NF/BWRO treatment: (a) $MgCl_2$ and KNO_3 draw solution, (b) Na_2SO_4, (c) $Ca(NO_3)_2$ and (d) KNO_3.

solution. Depending on the type of draw solution, two-stage membrane treatment is applied for draw solution regeneration using high permeability and water flux membranes (Figure 1.44). The concentrated brine from the first NF/BWRO membrane treatment is recycled back to the FO as the draw solution, whereas the permeate flow goes to a second membrane system for further treatment. Permeate from the second NF/BWRO membrane system forms the irrigation water, while concentrated brine is recycled to the first-stage membrane treatment. This approach will reduce the power consumption required for feed solution treatment in the regeneration stage. In the current study, two different methods were adopted to promote water flux in the FO process (1) using different draw solution concentrations (FO process) and (2) applying hydraulic pressure on the feed side to promote water permeation across the FO membrane (PAFO process). In the first method, draw solution concentrations between 0.75 and 0.95 M (osmotic pressure between 53 and 68 bar) were used to increase freshwater permeation across the FO membrane (Table 1.9). The osmotic pressure of draw solution concentrations were equivalent to the RO brine at 50% recovery rate at feed salinity 35,000 and 42,000 mg/L. In the second regeneration method, a positive hydraulic pressure, PAFO, was applied on the feed solution to enhance water permeation across the FO membrane. The range of hydraulic pressure applied on the feed side of the PAFO process was between 0 and 16 bar. Water flux, J_w (L/m²h), in the FO process was estimated from the following expression (Yip et al. 2011):

$$J_w = A_w \left(\frac{\pi_{Db} e^{\left(\frac{-J_w}{k}\right)} - \pi_{Fb} e^{(J_w K)}}{1 + \frac{B}{J_w}\left(e^{(J_w K)} - e^{\left(\frac{-J_w}{k}\right)}\right)} \right), \tag{1.16}$$

where π_{Db} and π_{Fb} are the osmotic pressures of the bulk draw and feed solution, respectively (bar), k is the mass transfer coefficient (m/s), B is the solute permeability coefficient (kg/m²h), K is the solute resistivity for diffusion within the porous support layer (s/m), and it is the ratio of membrane structure parameter, S (µm), to the solute diffusion coefficient, D (m²/s):

$$K = \frac{S}{D}. \tag{1.17}$$

For PAFO process, Equation 1.16 should include the hydraulic pressure, P (bar), on the feed side of the membrane as the following:

$$J_w = A_w \left(\frac{\pi_{Db} e^{\left(\frac{-J_w}{k}\right)} - \pi_{Fb} e^{(J_w K)}}{1 + \frac{B}{J_w}\left(e^{(J_w K)} - e^{\left(\frac{-J_w}{k}\right)}\right)} + P \right). \tag{1.18}$$

In Equation 1.18, P is the hydraulic on the feed side (bar). Reverse salt diffusion, J_{s-r} (kg/m²h), from the draw to the feed solution side is estimated from the following equation:

$$J_{s-r} = B \left(\frac{C_{Db} e^{\left(\frac{-J_w}{k}\right)} - C_{Fb} e^{(J_w K)}}{1 + \frac{B}{J_w}\left(e^{(J_w K)} - e^{\left(\frac{-J_w}{k}\right)}\right)} \right), \tag{1.19}$$

where C_{Db} and C_{Fb} are the bulk concentration of draw and feed solution, respectively (mg/L). It should be mentioned that FO and RO systems in Figure 1.44 have the same recovery rates, i.e., $Q_{pl} = Q_p$. The recovery rate of first regeneration stage is equal to that of FO system. Permeate from the first regeneration stage goes to a second NF/regeneration stage which operates at 50% recovery rate, while the concentrated brine returns to the first regeneration stage for feed flow dilution. Finally, specific power consumption, Es (kWh/m³), was estimated for the NF and RO membrane of the regeneration system according to the following equation:

$$Es = \frac{P_f}{\eta \times Re}.$$ (1.20)

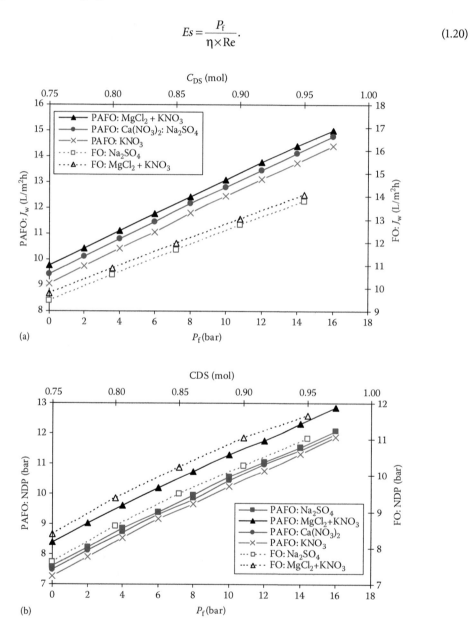

FIGURE 1.46 Impact of feed pressure and draw solution concentration on the performance of FO process: (a) water flux, (b) NDP, (c) $C_{DS\text{-out}}$ and (d) NO_3 concentration.

(Continued)

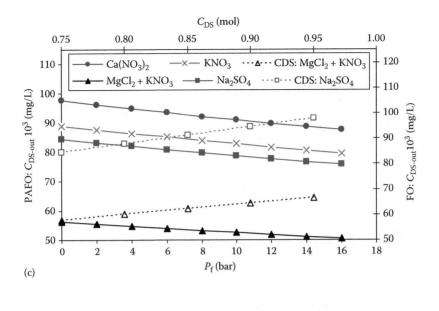

(c)

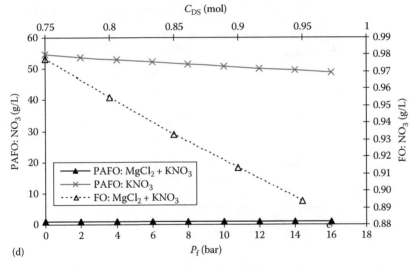

(d)

FIGURE 1.46 (Continued) Impact of feed pressure and draw solution concentration on the performance of FO process: (a) water flux, (b) NDP, (c) $C_{DS\text{-}out}$ and (d) NO_3 concentration.

A number of chemical compounds that are essential for the plants growth were considered:

- Figure 1.45a shows FO process for seawater treatment and draw solution regeneration using a mixture draw solution. NO_3^- has a moderately low rejection rate by NF and BWRO membranes. Therefore, it is introduced through a mixture of high osmotic pressure draw solution, also called a primary or carrier draw agent, and a low concentration KNO_3^- compound is added as a secondary draw agent for NO_3^- supply (Figure 1.45a). The carrier draw agent is highly rejected by NF/BWRO membrane and it constitutes the main solute in the draw solution for freshwater extraction from seawater. As shown in Figure 1.46, $MgCl_2$ and KNO_3 were introduced to the FO membrane as the primary and secondary draw agents, respectively. High permeability NF membrane is applied in the first regeneration stage for the rejection of $MgCl_2$, while most of KNO_3 will cross

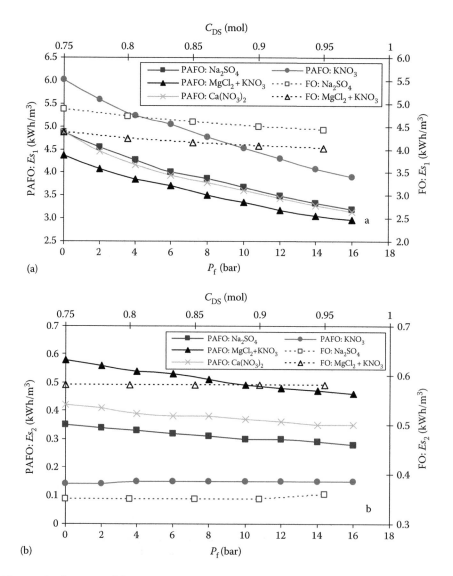

FIGURE 1.47 Performance of the regeneration system of the PAFO and FO processes: (a) specific power consumption in stage 1 and (b) specific power consumption in stage 2.

TABLE 1.10 Type of Membrane Used in the Regeneration of Draw Solutions

Draw Solution	$MgCl_2 + KNO_3$	Na_2SO_4	$Ca(NO_3)_2$	KNO_3
Stage 1 membrane	NF90-400	NF90-400	BW30LE-440	SW30HRLE-400i
Stage 2 membrane	BW30LE-440	NF90-400	BW30LE-440	BW30LE-440

the membrane. The concentrate from stage 1 is the draw solution of FO membrane, whereas the permeate goes to a second regeneration stage for further treatment. In the second regeneration stage, BWRO membrane was introduced for KNO_3 rejection and adjusting the concentration of irrigation water to an acceptable level.

- Na_2SO_4 was the main draw solution for SO_4^{2-} ions supply for irrigation water because of its high osmotic pressure and solubility in water (Figure 1.45b). Furthermore, SO_4^{2-} compound is highly rejected by NF membranes ; hence, a two-stage NF separation process was designed for Na^- and SO_4^{2-} ions separation and recycling. Using two-stage NF membranes is expected to reduce the power consumption and the regeneration cost of Na_2SO_4.

- Two-stage BWRO membrane treatment was applied for the regeneration of $Ca(NO_3)_2$ draw solutions (Figure 1.45c). $Ca(NO_3)_2$ was proposed for NO_3^- ion supply as an alternative to $MgCl_2$ + KNO_3 mixture. High-rejection BWRO membrane was used in stages 1 and 2 for draw solution regeneration.
- In Figure 1.45d, KNO_3 was used as a draw solution for NO_3^- and K^+ ions supply to irrigation water. A dual-stage RO membrane treatment was proposed for the regeneration and reuse of KNO_3 agent. This will reduce the concentrations of NO_3^- and K^+ ions in BWRO permeate.

A cellulose triacetate FO membrane, Hydration Technology Innovation (Albany, NY), was simulated for water flux in the FO unit. The membrane water permeability coefficient and rejection rate is 0.792 (L/m²h·bar) and >90%, respectively. Seawater, normal TDS 35 g/L, was the feed of FO unit (Table 1.9). Filmtech NF90-400 and BW30LE-440 membrane were used in the regeneration process. NF90-400 was selected because of its high water permeability and rejection rate to nitrate (Leist and Al-Dhaher 2000). Finally, salt diffusion coefficients of KNO_3, Na_2SO_4, $Ca(NO_3)_3$, and $MgCl_2$ were taken between 0.657×10^{-9} and 1.98×10^{-9} m²/s (Akuzhaeva et al. 2013; Annunziata et al. 2000; Daniel and Albright 1991; Leist and Al-Dhaher 2000).

1.3.3 Results and Discussions

1.3.3.1 Performance of FO Membrane System

$MgCl_2$, KNO_3, Na_2SO_4, and $Ca(NO_3)_2$ were the draw solutions of FO and PAFO processes for seawater desalination and provision of irrigation water. The impact of feed pressure, P_f (bar), concentration of feed solution, and C_{DS} (mol) on the performance of FO process was investigated (Figure 1.46). The results show that J_w in FO and PAFO processes increased with increasing the concentration of draw solution

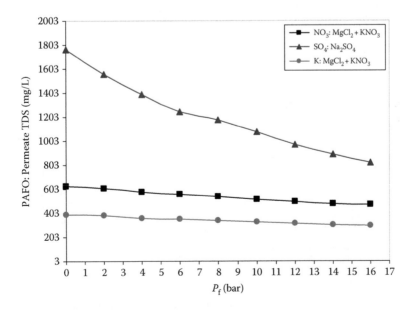

FIGURE 1.48 Concentration of K, NO_3, and SO_4 in permeate water after PAFO process; dual-stage NF90-400–BW30LE440 process was applied for NO_3 and K and NF90-400–NF90-400 for SO_4.

and the feed pressure, respectively (Figure 1.46a). In general, for MgCl$_2$ + KNO$_3$ draw solution, J_w was 15 L/m^{2}h at 16 bar feed pressure in the PAFO process and 14.1 L/m^{2}h at 0.95 mol draw solution concentration in the FO process. Evidently, J_w was higher in the PAFO than in the FO process for all draw solutions because of the higher net driving pressure (NDP) across the membrane (Figure 1.46b). It should be mentioned that feed flow in the PAFO process is pressurized through exchanging pressure with the concentrated brine from the first stage NF/RO membrane system (Figure 1.44).

As shown in Figure 1.46a, increasing the concentration of draw solution increased the water flux of the FO process, but the concentration of diluted draw solution was still high after the FO treatment

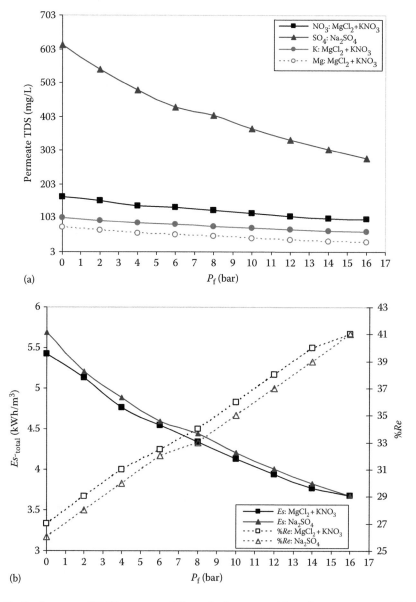

FIGURE 1.49 Performance of PAFO process at different feed pressures: (a) product water TDS and (b) power consumption and recovery rate, BW30-440i and BW30LE440 system was used for the regeneration of MgCl$_2$ + KNO$_3$ and BW30-440i–NF90-400 for the regeneration of Na$_2$SO$_4$.

TABLE 1.11 Concentration of K, NO₃, Mg, and SO₄ in Permeate Water after PAFO Process

PAFO	MgCl₂ + KNO₃			Na₂SO₄
	BW30-4401–BW30LE440			BW30-440i–NF90-400
Feed Pressure (bar)	K (mg/L)	NO₃ (mg/L)	Mg (mg/L)	SO₄
8	78	125	48	406
10	73	117	43	367
12	68	109	38	334
14	64	102	34	305
16	62	99	32	280
Recommended level (mg/L)	15–250	50–200	48–65	~321

(Figure 1.46c). The concentration of the diluted draw solution, $C_{DS\text{-}out}$, increased with increasing the concentration of draw solution, C_{DS}, from 0.75 mol to 0.95 mol (Figure 1.46c), while in the PAFO process $C_{DS\text{-}out}$ decreased with increasing the pressure of feed solution.

In the case of KNO₃, Figure 1.46d shows the concentration of NO₃ in the diluted draw solution using different draw solutions and feed pressures. Based on the concentration of NO₃ in the diluted draw solution, using MgCl₂ + KNO₃ draw solution was more relevant than KNO₃ for delivering enough NO₃ concentration to the NF/RO feed solution. Depending on the NDP across the FO membrane, using MgCl₂ + KNO₃ as a draw solution resulted in 1.4–1.59 g/L of KNO₃ concentration in the diluted draw solution (Figure 1.47d). This was almost 56 times lower than the concentration of NO₃ in the diluted draw solution when KNO₃ was used as a draw solution. Increasing the concentration of MgCl₂ or feed pressure in FO and PAFO processes resulted in 0.88 and 0.89 g/L of NO₃ in the diluted draw solution, respectively. It should be noted that high NO₃ concentration in the diluted draw solution may result in an undesirable NO₃ concentration in the irrigation water.

1.3.3.2 Performance of Regeneration System

Table 1.10 shows types of membranes used in stages 1 and 2 of the regeneration process. The regeneration system of MgCl₂ + KNO₃ consisted of NF90-400 and BW30LE-440 in stages 1 and 2, respectively. NF membrane rejection rate to MgCl₂, the primary draw solution, is more than 90% but to KNO₃ is from 20% to 67% (depending on the concentration in solution) (Kim et al. 2007; Su et al. 2006). In the second stage, BWRO membrane with 90% rejection rate to MgCl₂ and KNO₃ was used to reduce the concentration of permeate TDS close to the required level (Pure Water Products LLC 2015; Tepuš et al. 2009). Ca(NO₃)₂ rejection rate by NF membrane is about 60%, hence BWRO membranes were used in stages 1 and 2 of the regeneration process (Su et al. 2006). Finally, KNO₃ rejection rate by NF membrane is moderately low (Kim et al. 2007; Tepuš et al. 2009); therefore, RO and BWRO membranes were used in stages 1 and 2, respectively, of the regeneration process.

Figure 1.47a shows the specific power consumption, Es_1 (kWh/m³), in the first stage of the regeneration system. Specific power consumption of the PAFO process decreased with increasing the feed pressure. To a less extent, Es_1 of the FO process decreased with increasing the concentration of draw solution. On the other hand, specific power consumption of the second stage, Es_2 (kWh/m³), slightly decreased with increasing the feed pressure of the PAFO process, whereas Es_2 was almost constant in the FO process (Figure 1.47b). Regardless the draw solution type, results show that the total specific power consumption, which is equal to $Es_1 + Es_2$, in the PAFO process was lower than that in the FO process. The results show that increase in the total specific power consumption of the regeneration process in the PAFO process was in the following order: KNO₃ > Ca(NO₃)₂ > Na₂SO₄ > MgCl₂ + KNO₃. Furthermore, the results show that MgCl₂ + KNO₃ is the most cost-effective draw solution for NO₃ provision into the irrigation water. This was due to the novel application of MgCl₂ and KNO₃, a primary draw solution and a secondary additive for making the FO draw solution.

The results show that the concentrations of K^+, NO_3^-, and SO_4^{2-} in the permeate solution was higher than required in irrigation water (Fertigation Systems and Nutrients Solutions 2015; IAEA 2002). The concentrations of K^+ and NO_3^- at 16 bar feed pressure was 295 and 469 mg/L, respectively. At same feed pressure, SO_4^{2-} concentration was 821 mg/L (Figure 1.48). Depending on the crop type, the required nutrient concentration varies from 50 to 200 mg/L for NO_3^-, 15–250 mg/L for K^+, and about 321 mg/L for SO_4^{2-}. These concentrations are lower than what could be achieved by the FO and PAFO methods mentioned earlier. Therefore, further dilution using freshwater source is required to reduce the concentration of the permeate product. Freshwater source is not always available in arid areas and using desalinated water is also expensive. Alternatively, a higher rejection membrane (such as Filmtec BW30-440i) is suggested to replace NF90-400 in the first stage of the regeneration stage (Dow Water and Process Solutions 2015). BW30-440i rejection rate to monovalent ions is more than 99.5%. PAFO results show that concentration of K^+, NO_3^-, Mg^{2-}, and SO_4^{2-} dropped to a desirable level after the regeneration process (Figure 1.49a). The concentrations of K^+, NO_3^-, Mg^{2-}, and SO_4^{2-} at 16 bar feed pressure were 62, 99, 32, and 280 mg/L, respectively. These concentrations are within the level recommended for irrigation water (Table 1.11). In effect, the concentrations of K^+, NO_3^- and Mg^{2-} were within the recommended level at feed pressures between 8 and 16 bar as shown in Table 1.11. However, at 12 bar feed pressure, the concentration was about 321 mg/L, which is close to the recommended level for irrigation water. Using high-rejection rate, BW30-440i membrane affected the power consumption of the regeneration process. Total power consumption, Es_{total}, of the $MgCl_2 + KNO_3$ and Na_2SO_4 was 3.68 kWh/m³ at 16 bar feed pressure (Figure 1.49b); this was about 7% and 6% higher than Es_{total} for $MgCl_2 + KNO_3$ and Na_2SO_4 when NF90-400 membrane was used in stage 1 of the regeneration process. Interestingly, Es_{total} decreased with increasing the feed pressure as shown in Figure 1.49b. This was probably due to the higher membrane recovery rate at higher feed pressures. It should be mentioned that Es of RO seawater desalination is about 3 kWh/m³; however, the majority of RO desalination plants operates at Es close to 4 kWh/m³ (Dashtpour and Al-Zubaidy 2012). This indicates that power consumption of RO system is close to the theoretical Es of PAFO process (Figure 1.49b).

Finally, results show that PAFO process is more economical than FO process because of its lower power consumption as shown in Figures 1.49 Part a and 1.49 Part b. PAFO coupled with dual-stage membrane treatment for draw solution regeneration and reuse was able of producing fertilizer irrigation water, which can be applied directly to field. Actually, using a high-rejection rate, BW30-440i membrane enhanced the quality of irrigation water but at the expense of slight increase of Es. However, the Es of PAFO process was close to that of RO desalination process. Furthermore, it is more convenient to deliver fertilizer irrigation water ready to use without any additional treatment. These some of the advantages of PAFO system mentioned which can be offered over the conventional seawater desalination RO process.

1.3.4 Conclusion

FO was suggested for irrigation water supply in arid and semi-arid regions. The proposed system utilized a novel concept of a draw solution mixture consisting of a primary solution of high concentration and a secondary solution of low concentration for seawater desalination. The diluted draw solution was regenerated by a dual-stage NF–RO system to produce irrigation water and concentrated draw solution. The advantage of this design was to provide fertilizing water, which can be applied directly to field without further treatment. Results revealed that the TDS of fertilizing water was between 1200 and 5000 mg/L when NF–BWRO system was applied for the regeneration of draw solution. The quality of irrigation was further improved when a dual-stage BWRO membrane system was applied for the regeneration of draw solution. The specific power consumption of the proposed FO-dual-stage BWRO was 3.68 kWh/m³. This suggests that FO process has the potential for application in irrigation water supply. However, pilot plant study is required to accurately estimate the system energy demands and product water quality.

References

Akuzhaeva, G.S., S.V. Chaika, Y.Y. Gavronskaya, and V.N. Pak. 2013. Comparative characterization of the diffusion mobility of aqueous calcium salt solutions in porous-glass membranes. *Russian Journal of Applied Chemistry*. 86: 658–661.

Altaee, A. and N. Hilal. 2015. High recovery rate NF–FO–RO hybrid system for inland brackish water treatment. *Desalination*. 363: 19–25.

Altaee, A. and A.O. Sharif. 2011. Alternative design to dual stage NF seawater desalination using high rejection brackish water membranes. *Desalination*. 273: 391–397.

Altaee, A. and A.O. Sharif. 2015. Pressure retarded osmosis: Advancement in the process applications for power generation and desalination. *Desalination*. 356: 31–46.

Altaee, A., G. Zaragoza, and H. Rost van Tonningen. 2014. Comparison between forward osmosis-reverse osmosis and reverse osmosis processes for seawater desalination. *Desalination*. 336: 50–57.

Annunziata, O., J.A. Rard, J.G. Albright, L. Paduano, and D.G. Miller. 2000. Mutual diffusion coefficients and densities at 298.15 K of aqueous mixtures of NaCl and Na$_2$SO$_4$ for six different solute fractions at a total molarity of 1.500 mol dm^{-3} and of Aqueous Na$_2$SO$_4$. *Journal of Chemical and Engineering Data*. 45: 936–945.

Blandin, G., A.R.D. Verliefde, C.Y. Tang, and P. Le-Clech. 2015. Opportunities to reach economic sustainability in forward osmosis–reverse osmosis hybrids for seawater desalination. *Desalination*. 363: 26–36.

Daniel, V. and J.G. Albright. 1991. Measurement of mutual-diffusion coefficients for the system KNO$_3$–H$_2$O at 25°C. *Journal of Solution Chemistry*. 20: 633–642.

Dashtpour, R. and S.N. Al-Zubaidy. 2012. Energy efficient reverse osmosis desalination process. *International Journal of Environmental Science and Development*. 3: 4.

DOW FILMTEC™ Brackish Water Reverse Osmosis 8" Elements, Dow Water and Process Solution. http://www.dow.com/en-us/markets-and-solutions/products/DOWFILMTECBrackishWater ReverseOsmosis8Elements/DOWFILMTECBW30HR440i (Accessed May 09, 2017).

Fertigation Systems and Nutrients Solutions. 2015. http://ag.arizona.edu/ceac/sites/ag.arizona.edu.ceac/files/pls217nbCH10_0.pdf (Accessed May 01, 2015).

IAEA. 2002. Water balance and fertigation for crop improvement in West Asia. IAEA-TECDOC-1266.

Iqbal, N., A. Trivellini, A. Masood, A. Ferrante, and N.A. Khan. 2013. Current understanding on ethylene signaling in plants: The influence of nutrient availability. *Plant Physiology and Biochemistry*. 73: 128–138.

Kim, Y.-H., E.-D. Hwang, W.S. Shin, J.H. Choi, T.W. Ha, and S.J. Choi. 2007. Treatments of stainless steel wastewater containing a high concentration of nitrate using reverse osmosis and nanomembranes. *Desalination*. 202: 286–292.

Leaist, D.G. and F.F. Al-Dhaher. 2000. Predicting the diffusion coefficients of concentrated mixed electrolyte solutions from binary solution data. NaCl + MgCl$_2$ + H$_2$O and NaCl + SrCl$_2$ + H$_2$O at 25°C. *Journal of Chemical and Engineering Data*. 45: 308–314.

Luo, H., Q. Wang, T.C. Zhang, T. Tao, A. Zhou, L. Chen, and X. Bie. 2014. A review on the recovery methods of draw solutes in forward osmosis. *Journal of Water Process Engineering*. 4: 212–223.

Phuntsho, S., J.E. Kim, M.A.H. Johir, S. Hong, Z. Li, N. Ghaffour, T. Leiknes, and H.K. Shon. 2016. Fertiliser drawn forward osmosis process: Pilot-scale desalination of mine impaired water for fertigation. *Journal of Membrane Science*. 508: 22–31.

Phuntsho, S., H.K. Shon, S. Hong, S. Lee, and S. Vigneswaran. 2011. A novel low energy fertilizer driven forward osmosis desalination for direct fertigation: Evaluating the performance of fertilizer draw solutions. *Journal of Membrane Science*. 375: 172–181.

Plant Nutrition. 2015. Colorado State University. www.ext.colostate.edu/mg/gardennotes/231.html (Accessed April 07, 2015).

Pure Water Products LLC. 2015. Typical rejection percentages of thin film composite (TFC) reverse osmosis membranes. http://www.purewaterproducts.com/articles/ro-rejection-rates (Accessed April 04, 2015).

Sahebi, S., S. Phuntsho, J.E. Kim, S. Hong, and H.K. Shon. 2015. Pressure assisted fertiliser drawn osmosis process to enhance final dilution of the fertiliser draw solution beyond osmotic equilibrium. *Journal of Membrane Science.* 481: 63–72.

Su, M., D.X. Wang, X.L. Wang, M. Ando, and T. Shintani. 2006. Rejection of ions by NF membranes for binary electrolyte solutions of NaCl, NaNO$_3$, CaCl$_2$ and Ca(NO$_3$)$_2$. *Desalination.* 191: 303–308.

Tepuš, B., M. Simonič, and I. Petrinić. 2009. Comparison between nitrate and pesticide removal from ground water using adsorbents and NF and RO membranes. *Journal of Hazardous Materials.* 170: 1210–1217.

Water Use. 2015. Food and Agriculture Organization of the United Nations. http://www.fao.org/nr/water/aquastat/water_use/index.stm (Accessed April 02, 2015).

Wilf, M. and C. Bartels. 2005. Optimization of seawater RO system design. *Desalination.* 173: 1–12.

Yip, N.Y., A. Tiraferri, W.A. Phillip, J.D. Schiffman, L.A. Hoover, Y.C. Kim, and M. Elimelech. 2011. Thin-film composite pressure retarded osmosis membranes for sustainable power generation from salinity gradients. *Environmental Science and Technology.* 45: 4360–4369.

Yun, T.G., Y.J. Kim, S. Lee, and S.K. Hong. 2012. Pressure assisted forward osmosis: Effect of membrane materials and operating conditions. *Procedia Engineering.* 44: 1906.

2

Water Management

Sabla Alnouri,
Patrick Linke,
Mahmoud
El-Halwagi and
Munawar A. Shaik
*Indian Institute of
Technology (IIT) Delhi*

Subhankar
Karmakar and
Vikas Varekar
*Indian Institute of
Technology Bombay*

2.1 Toward Sustainable Water Management in Industrial Cities

Sabla Alnouri, Patrick Linke, and Mahmoud El-Halwagi

2.1.1 Introduction

The escalation of water scarcity concerns has become one of the major industrial challenges due to the rapidly expanding gap between the global water demand and limited freshwater supply (Jury and Vaux 2007). Even though the natural water cycle contributes to the replenishment of freshwater reserves, the constantly increasing freshwater demand often compels water use from underground aquifers or surface water sources, such as rivers and lakes, at rates much higher than their ability to recharge. As a result, many industries have been required to find alternative water supply strategies, especially in areas where they are unable to bring their overall water supply and demand into balance. Many recent developments have made water desalination an increasingly attractive and cost-effective solution that delivers high-quality freshwater, which could meet industrial requirements, as an alternative. However, the highly saline and dense by-product brine discharges often pose an environmental problem that should be considered when installing new desalination plants. Therefore, even though there are many challenges associated with limited freshwater sources, the environmental apprehension toward by-product brine streams that are produced as a result of desalinated freshwater alternatives is also a problem (Ahmad and Baddour 2014).

Many industries often generate different wastewater qualities, as a result of various chemical and operational activities. If treated to acceptable standards, wastewater may potentially be reused in many applications, as an alternative to freshwater supply, which would ultimately reduce the tension associated with freshwater use. Moreover, it is often necessary to ensure that industrial wastewater effluent adheres to direct discharge standards. Hence, wastewater reuse provides way for industries to comply with the increasingly stringent environmental regulations that often pertain to wastewater discharge. Additionally, the implementation of effective wastewater reuse strategies could potentially allow for increased economic benefits that are usually correlated to reduced wastewater disposal costs (Garcia and Pargament 2015).

Wastewater reuse practices may either involve direct reuse or entail a number of treatment stages before reuse. Hence, the design of integrated wastewater networks may involve a combination of wastewater reuse and regeneration opportunities. Graphical techniques (Dhole et al. 1996; Kuo and Smith 1997; Dunn and Wenzel 2001; Hallale 2002; El-Halwagi et al. 2003; Shenoy and Bandyopadhyay 2007; Wan Alwi et al. 2008; Foo, 2009), algebraic approaches (Sorin and Bedard 1999; Manan et al. 2004; AlMutlaq et al. 2005), and computer-aided methods (Takama et al. 1980; Doyle and Smith 1997; Alva-Argáez et al. 1998; Huang et al. 1999; Feng et al. 2008; De Faria et al. 2009; Poplewski et al. 2010) have been developed to assist in the design of effective wastewater reuse and regeneration networks. Methods that allow direct wastewater reuse mainly involve water targeting techniques, in which freshwater consumption and wastewater discharge can be minimized by maximizing wastewater reuse (Wang and Smith 1994a,b; Sorin and Bedard 1999; Dunn and Wenzel 2001; Manan et al. 2004). Some methods consider possible interactions between water reuse and wastewater treatment, by accounting for wastewater regeneration schemes (Wang and Smith 1995; Huang et al. 1999; El-Halwagi and Manousiouthakis 1989a,b; El-Halwagi and Manousiouthakis 1990a,b; El-Halwagi et al. 2003; Bandyopadhyay and Cormos 2008; Feng et al. 2008). While some methods are limited to systems involving a single contaminant (Olesen and Polley 1997; Savelski and Bagajewicz 2000), other methods are capable of handling systems in which multiple contaminants are involved (Doyle and Smith 1997; Alva-Argáez et al. 1998). Hence, water availability for upstream and downstream operations must always be considered, in terms of quantity and quality, when developing effective water integration schemes. Many existing methods have been successfully implemented in numerous processes, with a focus on achieving improved water integration strategies within a single stand-alone industrial facility. Later on, based on very similar principles that were applied for local water integration, many of the methods were further extended to incorporate opportunities for interplant water integration. When effectively applied, such techniques can enable efficient planning and utilization of water resources within eco-industrial parks (EIPs).

2.1.2 Water Integration in EIPs

Many industries are often confronted with significant challenges when seeking sustainable stand-alone water supplies, in addition to handling any wastewater disposal obligations. The daily operation of multiple water utility stand-alone systems often produces a heavy economic and environmental burden in many industrial areas. In case a number of water-using and water-consuming systems are within geographic proximity, this burden can possibly be reduced by allowing water-sharing schemes among a coexisting cluster of industries/plants, which are not necessarily part of the same company or organization. Resource sharing is often facilitated in EIPs, which foster plant-to-plant interactions. Effective water resource management in EIPs greatly depends on the implementation of water network design schemes, involving wastewater reuse and regeneration. Much of the economic and environmental benefits that EIPs strive for can be realized by maximizing wastewater reuse in the system, while adhering to the wastewater discharge standards dictated. The application of water integration techniques in EIPs may bring attractive economic and environmental advantages over stand-alone systems that do not allow any water exchange across different plant boundaries. As a result, the design of water exchange networks in EIPs has garnered increased attention, and interplant water integration studies have become the focus of many successive research contributions. The ways in which water is managed in EIPs have been found to have both direct and indirect consequences on the overall economic, environmental, and social development of an EIP in the long run. Various mathematical programming approaches, pinch analysis techniques, as well as a number of game theory principles have been utilized for the design of water exchange networks in EIPs. Table 2.1 outlines a number of research contributions in the area of interplant water network design.

The design of a cost-effective water exchange network is greatly influenced by the presence of fundamental infrastructure facilities that are essential for enabling and sustaining industrial symbiosis in EIPs. Hence, interplant water integration methods that could provide ways to prevent inefficient water

TABLE 2.1 Summary of Previous Work Contributions in the Area of Interplant Water Network Design

Year	Authors	Contribution
2004	Spriggs et al.	Developed a source–sink integration approach for designing EIPs in which the design challenges were classified as technical/economic or organizational/commercial/political
2007	Yoo et al.	Developed a pinch analysis technology for water and wastewater minimization. They also explored some research challenges such as simultaneous water and energy minimization, energy-pinch design, and EIPs
2007	Kim and Lee	Developed Pareto optimal networks based on the context of the benefit sharing principle among the participating entities
2007	Liao et al.	Proposed a mixed integer linear programming (MILP) model for the design of flexible water networks of individual plants, which can be applied to fixed contaminant and fixed flow operations, while being limited to a single contaminant
2008	Foo	Addressed the targeting of plant-wide integration using numerical tools for water cascade analysis. The overall freshwater and wastewater flow rates were reduced by sending water sources across different geographical zones
2008	Thillairvarrna et al.	Introduced a game theory approach for IPWI problems, in which various payoff benefits for different water network designs were assessed
2008	Chew et al.	Proposed the concept of a centralized hub topology for IPWI that can be used for collecting and redistributing water in the design of EIP water networks
2009	Chew and Foo	Utilized automated targeting for IPWI based on the concept of pinch analysis, and was formulated as a linear programming model
2009	Chew et al.	Developed a game theory scheme for designing IPWI networks, by assessing various interactions between participating companies
2009	Lovelady and El-Halwagi	Introduced a mass-integration framework and mathematical formulation for the design of EIP water networks based on a source–interception–sink-representation
2010	Lim and Park	Developed interfactory and intrafactory water network systems for an industrial park setting by utilizing opportunities for water reuse within an industrial park. Environmental and economic feasibility studies were used to demonstrate benefits from the industrial symbiosis between the companies
2010	Kim et al.	Introduced a systematic approach for optimizing utility networks in an industrial complex. Their problem was formulated as a multi-period MILP model
2010	Rubio-Castro et al.	Modeled the reuse of wastewater among different industries as a mixed integer non-linear programming (MINLP) problem. Moreover, a discretization approach was proposed as a reformulation to handle the bilinear terms in the MINLP problem as part of their global optimization strategy
2010a,b	Aviso et al.	Presented models for optimizing water and wastewater reuse among independent processing facilities in an EIP through fuzzy mathematical programming
2010a,b	Chew et al.	Introduced a new algorithm for the design of interplant resource conservation networks, by targeting minimum freshwater use and wastewater discharge
2011	Taskhiri et al.	Developed an MILP model for the design of IPWI networks, by assessing the environmental impacts of water use, energy consumption, and capital goods within an EIP in the form of a total energy indicator
2011	Rubio-Castro et al.	Proposed a global optimal formulation to design water integration networks for EIPs, by accounting for wastewater reuse both within the plant, as well as among different plants, using a superstructure approach

(Continued)

TABLE 2.1 (*Continued*)   Summary of Previous Work Contributions in the Area of Interplant Water Network Design

Year	Authors	Contribution
2012	Rubio-Castro et al.	Assessed the incorporation of different retrofitting schemes for several single-plant water networks, into an EIP using an MINLP model, by accounting for both intraplant and interplant decisions
2012	Boix et al.	Developed a methodology for designing industrial water networks using a multiobjective optimization strategy. An MILP that accounts for linearizations based on the necessary conditions of optimality defined by Savelski and Bagajewicz (2002) was proposed
2013	Montastruc et al.	Formulated a triobjective MILP, in which the freshwater flows, regenerated water flows, and the number of connections were minimized. Moreover, the flexibility of the water supply system for an EIP of any size was also investigated
2013	Lee et al.	Developed a mathematical optimization model for interplant water network synthesis, using a two-stage approach in which the individual processing units operate in a mix of both continuous and batch modes
2014	Bishnu et al.	Introduced a multiperiod approach for the design of interplant water networks. It is good to note that many of the methods developed aim to improve the overall performance of EIPs
2014	Tian et al.	Assessed the economic and environmental performance of EIPs, based on energy use, freshwater consumption, wastewater, and solid waste generation. The study highlights the importance of effectively developing interplant water network methodologies that could be applied to real-case applications

use, while accounting for infrastructure availability in industrial areas, are much needed. Pipelines are often used for the transportation of water in EIPs, since they allow for very smooth and long-lasting operations to be conducted on a regular basis. Even though major pipeline disruptions are dramatic and costly, these instances are usually very rare. Hence, featuring a well-defined assessment technique which could help identify and avoid unsound pipeline infrastructure investments for water transportation in EIPs is in fact very useful, since pipeline costs often represent one of the major expenses that need to be considered in water exchange networks. Some of the methods outlined in Table 2.1 do incorporate pipeline infrastructure costs when assessing the overall performance of an interplant water network design. However, very simple models have been utilized so far, often lacking the ability to consider interplant water network synthesis in context of a given industrial city layout, which in turn could simultaneously assess several plant-to-plant water-sharing schemes based on plant arrangement and proximity, and then identify the most cost-effective scheme. Currently, all methods either assume different pipes to be of equal length, require the length of every water pipeline to be provided upfront, or utilize a number of parameters that reflect the intrapresence of a pipeline (within a single plant), and the interpresence of a pipeline (among several neighboring plants). Additionally, none of the methods so far incorporate pipeline diameter computations, even though this often represents an essential design requirement. Water exchange networks are often designed based on a set of standard pipeline diameter sizes, depending on whichever diameter ranges are available. Ideally, each water allocation must be associated with a suitable pipeline diameter, based on the water flow rate that is assigned to be transported from source to destination. Pipe length and diameter size are also needed to establish accurate computations of network pressure drops, which could assist in predicting any necessary pumping requirements in water exchange networks. All those factors greatly affect the ability to obtain accurate predictions of interplant water network performance. Hence, infrastructure assessment considerations were found necessary to incorporate with existing interplant water integration methods. This ultimately converts interplant water network design problems in EIPs into an all-inclusive design scheme with two dimensions that need to be explored simultaneously: (1) identifying cost-effective and feasible water reuse and treatment allocations that maximize wastewater reuse and regeneration among plants, and (2) identifying cost-effective infrastructure for water piping based on the evaluation of different plant-to-plant water-sharing schemes, according to a given industrial layout. All those aspects can be addressed by developing an enhanced optimization-based framework, which accounts for a number of additional key elements to be used for conducting an improved water network performance evaluation.

2.1.3 Additional Aspects to Improve Interplant Water Network Design

Many of the methods which have been developed for water network synthesis and design entail two fundamental elements: (1) water targeting and (2) network design. The water targeting stage mainly determines the minimum freshwater and wastewater flows for a network, based on specified concentration and flow rate limits usually associated with process water sources and sinks. Network design would then involve the development of a detailed allocation strategy among the individual water-consuming and water-producing processes. Water-consuming operations (sinks) and water producing operations (sources) are matched in a way that can achieve the same flow rates obtained via the targeting stage. When effectively applied, such techniques can only result in the development of efficient strategies for water allocation and utilization in industrial cities. However, since many industrial cities rely on well-defined infrastructure boundaries for water transport, more commonly known as service corridors, enhanced methods involving network design can be developed.

An industrial city (Figure 2.1) often involves a number of coexisting plant facilities that share common infrastructure, primarily designed to facilitate any resource exchange among plants. Figure 2.2 illustrates an example of an industrial city layout scheme, with all corresponding service corridor water transportation boundaries identified. A freshwater mains delivers freshwater to plant sinks, and a wastewater mains collects any unused wastewater from plant sources, subject to meeting direct discharge standards. It is evident that water source and sink locations are often dictated by plant arrangements within the city. Assessing source–sink water allocations (including any freshwater main and wastewater main allocations) via service corridor channels certainly enriches any source–sink matching process, as well as provides a way for achieving an enhanced water infrastructure assessment, in the network design stage. Hence, capturing industrial city layouts for the network design stage is essential, such as plant arrangements, any in-between barriers, as well as water source and sink sites that can be used for the identification of cost-effective decisions of water allocation through available service corridors. Additionally, despite all research efforts that have been made so far, an interplant water integration

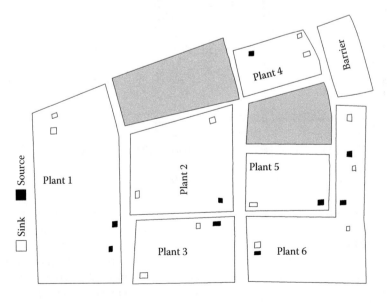

FIGURE 2.1 Industrial city layout example.

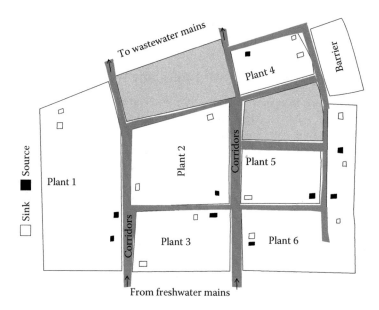

FIGURE 2.2   Industrial city layout example, with service corridors availability illustrated.

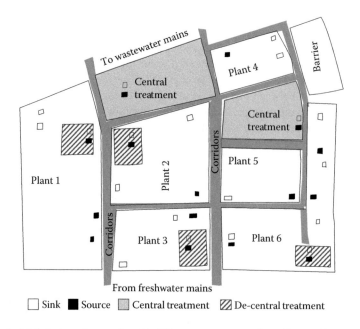

FIGURE 2.3   Industrial city layout example, with different treatment options illustrated.

methodology that explicitly addresses the different options available for the placement of intermediate water treatment interception options is also necessary in the same context. Even though most interplant water network studies that have been previously carried out do consider treatment, many of the cases that have been investigated involved introducing shared water treatment among an existing cluster of plants, with no location considerations being accounted for. Figure 2.3 illustrates an example of the same industrial city layout scheme in Figures 2.1 and 2.2, with different treatment options demonstrated (company-owned decentralized treatment units and shared centralized treatment units). Investigating an effective strategy for the integration of company-owned and shared treatment units within a water network design can be carried out more effectively once a given industrial city layout is captured. As a result, in addition to determining optimal reuse, regeneration, and allocation strategies, a water transportation problem must be formulated so as to effectively plan a cost-effective infrastructure scheme for water piping, as part of the network design stage. Hence, information on water source and sink locations within individual processing facilities must be identified, for which the shortest and most effective routing to and from can be determined via available service corridors. The development of a flexible representation that allows spatial aspects in industrial zones to be accounted for, such that information regarding the respective plant locations, corridors, and any in-between barriers, must be utilized to systematically screen all options, and identify the best pipeline routing scenarios. Additionally, the methodology must consider site information associated with any available wastewater treatment options, whether company-owned decentralized units, or centrally shared treatment units.

It has also been observed that the alternative design options that exist for interconnecting transmission and distribution networks have not been considered in water network synthesis. All existing approaches that do incorporate piping expenses in the design of interplant water networks always assign a separate pipeline for every water allocation. However, merging together common pipeline regions for the transmission of water from, or to nearby but different processing facility destinations may improve the overall water network performance not only in terms of cost efficiency but also in terms of complexity. This aspect cannot be explored without identifying the optimal routing strategies for source–sink matching in a water network design, from which plausible merging options could then be developed.

The development of several optimization-based frameworks that account for many of the aspects identified above have been addressed by some of the recent research efforts (Alnouri et al. 2014, 2015). The methods introduced mainly capitalize on previous work in the area of interplant water integration, with a focus on enhancing the techniques utilized for infrastructure assessment, in interplant water network design. Moreover, the methods aim to develop cost-efficient design strategies for managing water use and discharge in EIPs. A number of illustrative studies have been carried out to highlight the benefits of accounting for enhanced infrastructure planning features in interplant water network problems (Alnouri et al. 2014, 2015). The sections below outline the methods which have been proposed by recent research contributions (Alnouri et al. 2014, 2015).

2.1.4 Recently Proposed Methods for Interplant Water Network Synthesis

Due to the problem dependence on the layout of the industrial city (or EIP) being investigated, accounting for the spatial aspects of any industrial zone provides the necessary information that can allow an effective planning and structuring scheme for water piping among a number of coexisting plants arranged in a certain way, while being allowed to share a number of common infrastructure facilities. As it has been mentioned earlier, even though many studies describe interplant water integration problems as wastewater minimization problems, piping costs are considered an important aspect which needs to be appropriately addressed, in the course of designing cost-effective interplant water networks. Hence, water integration in EIPs must account for optimal routing and allocation of flows among the different water-using and water-consuming participants, within available city infrastructure. Such information would be utilized for the assessment of cost-effective water exchange network designs. Some of the relatively recent methods mainly focus on developing strategies that tackle a constrained water transportation problem, utilizing common water infrastructure boundaries, in addition to utilizing water exchange network design strategies which account for direct wastewater reuse and wastewater treatment. The respective water allocations in between the different plants can be planned out more effectively when a spatial representation for industrial cities is utilized, as this facilitates the integration of available water streams based on their locations (both treated and untreated). Furthermore, optimal placement strategies of wastewater treatment units onto a given industrial zone setting can also be attempted more effectively, by identifying and assessing a number of potential locations for a number of wastewater treatment units, according to a provided industrial city layout.

2.1.4.1 Spatial Information Processing from an Industrial City Map

Alnouri et al. (2014) proposed a systematic approach which enables the location of water source and sink sites within an industrial city layout scheme. This information can then be processed and converted to optimized source–sink distances, which would minimize water infrastructure costing in an interplant water network design problem. The proposed methodology was found to help generate cost-optimal water network designs, for any provided industrial city layout. First, it is imperative to accurately locate the sites for all water sources and sinks, service corridors, as well as the barriers in-between, within a given industrial city layout. The city layout is then converted to an equally spaced grid. Figure 2.4 illustrates the grid-conversion process of the city layout which was used as a demonstration in Figures 2.1 through 2.3. The number of grid squares used for a given city may be selected based on a desired grid size, which in turn can either be coarse or refined, depending on the level of distance accuracy required. Even though refined grids are computationally extensive, they are often associated with increased levels of accuracy. A city grid is then converted to an array of nodes (either active or inactive) based on the respective entities each square grid identifies. This allows for an intelligible interface, in case any additional input locations may need to be added, or any other input locations may need to be altered. All source, sink, and service corridor grid locations are classified as active grid nodes, while any other node grid locations are grouped into the inactive category. It should be noted that all treatment units are associated with a single sink site and a single source site, which are used to signify wastewater received by treatment, and treated water sent off from treatment, respectively. To illustrate these aspects, Figure 2.4 denotes the respective active and inactive search regions, for the example in Figure 2.3. All active nodes are then converted to a weighted graph problem, after identifying appropriate constraints for node-to-node connectivity. This process is explained in more detail in Alnouri et al. (2014). Having defined the appropriate graph edges by incorporating suitable node-to-node connectivity constraints, a weighted graph search can then be carried out, using an appropriate shortest-distance algorithm. The extraction of shortest source–sink information can be performed by screening through all available path options, then identifying the shortest path routing.

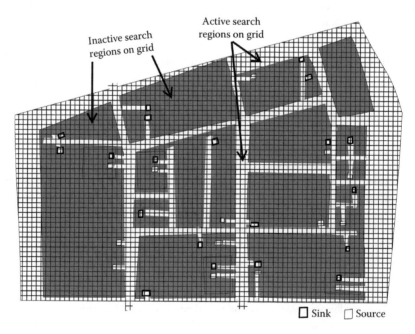

FIGURE 2.4   Grid associated with industrial city example.

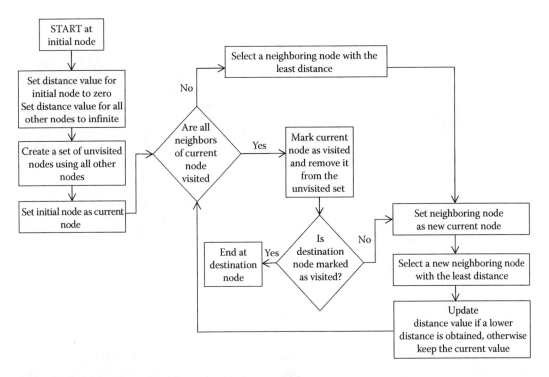

FIGURE 2.5 Dijkstra's algorithm execution flowchart (Dijkstra, 1959).

This process can be achieved using advanced tools that incorporate the use of shortest path algorithms, by means of a weighted graph. Figure 2.5 provides an outline of the shortest path search procedure which has been employed by Alnouri et al., based on Dijkstra's algorithm (Dijkstra 1959). Accordingly, the different and plausible routing options via industrial city service corridors may be obtained, from all water sources, to all water sinks. Optimized shortest distances, as well as the spatial information which constitutes the path routing for each allocation opportunity, can be used to conduct a cost-effective water network infrastructure assessment strategy. This would require incorporating the respective distance matrices obtained for all shortest routes obtained into the water exchange network design stage, which would then involve conducting a source–sink matching procedure, based on water flow and quality.

2.1.4.2 Source–Sink and Source–Interceptor–Sink Mapping

Water flow rate and quality information, associated with each source and sink, are considered essential pieces of information that need to be provided in order to achieve a water exchange network design. Data acquired in the form of water flow rate and quality, from various industrial city processing facilities, often serve as a basis for establishing feasible connectivity in water exchange networks. Having a well-classified system associated with the types of different water qualities involved is also necessary, since multiple sources and sinks of the same water quality can exist within different processing facilities, at several locations within the industrial city. As it has been described before, wastewater reuse practices may either involve direct wastewater reuse, or entail a number of regeneration or treatment stages, often referred to as interception points, before reuse. Moreover, the placement of different scenarios for the placement of intermediate water treatment interceptors may be considered as follows:

- On-site "decentralized" water treatment within each plant.
- Off-site "centralized" water treatment that can be shared among a cluster of existing industrial plants.

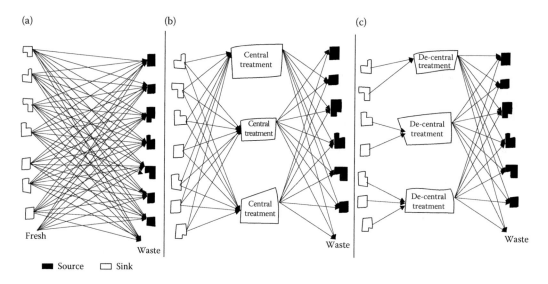

(a)   (b)   (c)

Fresh

Waste   Waste   Waste

■ Source   □ Sink

FIGURE 2.6 Water allocation category options illustrated. (a) Direct reuse allocations, (b) treatment allocations (centralized), and (c) treatment allocations (decentralized).

Identifying any high impact interprocess and intraprocess streams, which have significant effects on industrial city water use, treatment, and discharge requirements, is necessary. Moreover, since the overall objective is to develop cost-efficient water networks that can attain effective interplant water integration scenarios, while considerably reducing freshwater consumption and wastewater discharge where appropriate, the different water matching options may be one of the following scenarios:

- Source–sink allocations: a wastewater source being directly routed into a sink, as illustrated in Figure 2.6a.
- Centralized source–interceptor–sink allocations: a wastewater source being routed through a centralized treatment option before being sent to a waster sink, as illustrated in Figure 2.6b.
- Decentralized source–interceptor–sink allocations: a wastewater source being routed through a decentralized treatment option before being sent to a waster sink, as illustrated in Figure 2.6c.

The first matching option usually involves the purchase of external freshwater resources that could supplement the use of process wastewater sources, by reducing the composition of the targeted species in a way that could meet sink flow and quality requirements. The difference between centralized source–interceptor–sink allocations and decentralized source–interceptor–sink allocations is also very straightforward. The former option can receive water from any wastewater source, regardless of its respective plant boundaries, while the latter option can only receive wastewater from sources that lie within the same plant boundaries. Alnouri et al. (2014) proposed a systematic approach to assess cost-optimal water network schemes with considerations that account for both centralized and decentralized interception units, which also considers the respective locations of water source sites, water sink sites, as well as treatment interception sites, within any industrial city layout scheme.

2.1.4.3 Addressing Water Network Interconnectivity Options through Pipeline Merging

Alternative design options that exist for interconnecting transmission and distribution networks were also considered in water network synthesis, after having integrating methodologies which can account for spatial aspects within a city. Currently, all existing interplant water integration approaches assign a single pipeline for every water allocation, even though merging together common pipeline regions for the transmission of water from, or to nearby but different processing facility destinations may be useful. The complexity of any water exchange network design may be reduced significantly if pipeline interconnectivity is considered, in addition to attaining improved cost performance. Hence, Alnouri et al. (2015) also develop a representation for interplant water network synthesis, in which pipeline merging options may be explored. This work was developed as an extension to Alnouri et al. (2014) after incorporating options for synthesis and design of merged pipeline networks. In order to avoid unwanted water mixing in the merging procedure, the proposed methodology has been carried out on pipelines that may carry either treated or untreated water qualities in a separate manner. Cost-effective water exchange networks that allow the screening of less complex designs were attained by assembling together commonly existing pipe segments, using the following two schemes:

- *Forward branching scheme*: Consider the unmerged pipeline scenario in Figure 2.7a. Pipelines that apply a forward branching scheme are assembled by starting with one large pipe segment that combines all water in a given location to be distributed. Hence, forward branching corresponds to the transmission of water from a common source location, to multiple sink destinations, as demonstrated in Figure 2.7c. The segments narrow down to smaller ones as the multiple sink destinations are connected.
- *Backward branching scheme*: Consider the unmerged pipeline scenario in Figure 2.7b. Pipelines that apply a backward branching scheme are assembled by starting with multiple source pipe segments that all connect up to form a larger pipe leading to a single sink location. Hence, backward

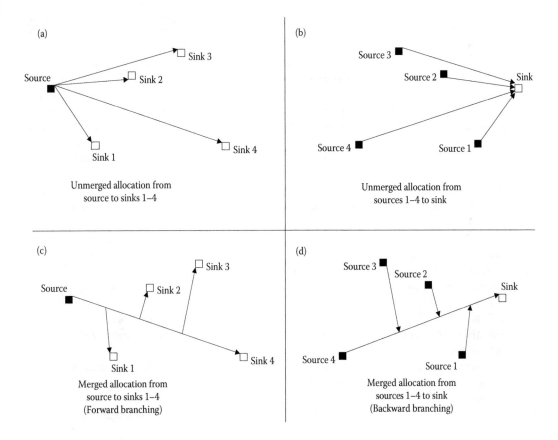

FIGURE 2.7 Pipeline merging options illustrated.

branching corresponds to the transmission of water from multiple source locations, to a single sink destination, as demonstrated in Figure 2.7d. The segments widen up and combine to larger ones as the sink destination is being approached.

Merging pipelines enables the screening of less complex pipeline networks in the course of determining optimal water integration strategies, as well as help explore decisions on how streams can be manipulated for maximum benefit in a city.

2.1.5 Conclusions

Making the best use of innovative technical methods so as to respond to the growing challenges of different aspects that influence interplant water network design has been discussed in this chapter. Cost-effective solutions are always key to ensuring adequate levels of water security in any industrial setting. Water infrastructure also creates many challenges, especially when it comes to attaining cost-effective water exchange network designs for an industrial community, so as to meet current and emerging water demand through wastewater reuse and treatment options. Hence, putting together water integration methodologies with new techniques that could assist in interplant water infrastructure planning was found imperative, due to the heavy problem dependence on the layout of industrial settings.

References

Ahmad, N. and Baddour, R. E. (2014). A review of sources, effects, disposal methods, and regulations of brine into marine environments. *Ocean and Coastal Management*, 87, 1–7.

Almutlaq, A. M., Kazantzi, V., and El-Halwagi, M. M. (2005). An algebraic approach to targeting waste discharge and impure fresh usage via material recycle/reuse networks. *Clean Technologies Environmental Policy*, 7(4), 294–305.

Alnouri, S., Linke, P., and El-Halwagi, M. M. (2014). Water integration in industrial zones—A spatial representation with direct recycle applications. *Clean Technologies and Environmental Policy*, DOI: 10.1007/s10098-014-0739-2.

Alnouri, S.Y., Linke, P. and El-Halwagi, M. (2015). A synthesis approach for industrial city water reuse networks considering central and distributed treatment systems. *Journal of Cleaner Production*, 89, 231–250.

Alva-Argáez, A., Kokossis, A. C., and Smith, R. (1998). Wastewater minimization of industrial systems using an integrated approach. *Computers and Chemical Engineering*, 22(Suppl.), S741–S744.

Aviso, K. B., Tan, R. R., and Culaba, A. B. (2010a). Designing eco-industrial water exchange networks using fuzzy mathematical programming. *Clean Technologies and Environmental Policy*, 12, 353–363.

Aviso, K. B., Tan, R. R., Culaba, A. B., and Cruz Jr., J. B. (2010b). Bi-level fuzzy optimization approach for water exchange in eco-industrial parks. *Process Safety and Environmental Protection*, 88, 31–40.

Bandyopadhyay, S. and Cormos, C.-C. (2008). Water management in process industries incorporating regeneration and recycle through a single treatment unit. *Industrial and Engineering Chemistry Research* 47, 1111–1119.

Bishnu, S. K., Linke, P., Alnouri, S. Y., and El-Halwagi, M. (2014). Multiperiod planning of optimal industrial city direct water reuse networks. *Industrial and Engineering Chemistry Research*, 53, 8844–8865.

Boix, M., Montastruc, L., Pibouleau, L., Azzaro-Pantel, C., and Domenech, S. (2012). Industrial water management by multi-objective optimization: From individual to collective solution through eco-industrial parks. *Journal of Cleaner Production*, 22(1), 85–97.

Chew, I. M. L. and Foo, D. C. Y. (2009). Automated targeting for inter-plant water integration. *Chemical Engineering Journal*, 153, 23–36.

Chew, I. M. L., Foo, D. C. Y., Ng, D. K. S., and Tan, R. R. (2010a). Flowrate targeting algorithm for interplant resource conservation network. Part 1: Unassisted integration scheme. *Industrial and Engineering Chemistry Research*, 49, 6439–6455.

Chew, I. M. L., Foo, D. C. Y., and Tan, R. R. (2010b). Flowrate targeting algorithm for interplant resource conservation network. Part 2: Assisted integration scheme. *Industrial and Engineering Chemistry Research*, 49, 6456–6468.

Chew, I. M. L., Tan, R. R., Foo, D. C. Y., and Chiu, A. S. F. (2009). Game theory approach to the analysis of inter-plant water integration in an eco industrial park. *Journal of Cleaner Production*, 17, 1611–1619.

Chew, I. M. L., Tan, R., Ng, D. K. S., Foo, D. C. Y., Majozi, T., and Gouws, J. (2008). Synthesis of direct and indirect interplant water network. *Industrial and Engineering Chemistry Research*, 47, 9485–9496.

De Faria, D. C., de Souza, A. A. U., and Ulson de Souza, S. M. A. G. (2009). Optimization of water networks in industrial processes. *Journal of Cleaner Production*, 17, 857–862.

Deng, C., Feng, X. and Bai, J., (2008). Graphically based analysis of water system with zero liquid discharge. *Chemical Engineering Research and Design*, 86, 165–171.

Dhole, V. R., Ramchandani, N., Tainsh, R. A., and Wasilewski, M. (1996). Make your process water pay for itself. *Chemical Engineering*, 103(1), 100–103.

Dijkstra, E.W. (1959). A note on two problems in connexion with graphs. *Numerische Mathematik*, 1, 269–271.

Doyle, S. J. and Smith, R. (1997). Targeting water reuse with multiple contaminants. *Transactions of International Chemical Engineering, Part B*, 75(3), 181–189.

Dunn, R. and Wenzel, H. (2001). Process integration design methods for water conservation and wastewater reduction in industry. *Clean Products and Processes*, 3, 307–318.

El-Halwagi, M. M. and Manousiouthakis, V. (1989a). Synthesis of mass exchange networks. *American Institute of Chemical Engineering Journal*, 35(8), 1233–1244.

El-Halwagi, M. M. and Manousiouthakis, V. (1989b). Design and analysis of mass-exchange networks with multicomponent targets. The American Institute of Chemical Engineering Annual Meeting, San Francisco, November 5–10.

El-Halwagi, M. M. and Manousiouthakis, V. (1990a). Automatic synthesis of mass exchanger networks with single component targets. *Chemical Engineering Science*, 45(9), 2813–2831.

El-Halwagi, M. M. and Manousiouthakis, V. (1990b). Simultaneous synthesis of mass exchange and regeneration networks. *American Institute of Chemical Engineering Journal*, 36(8), 1209–1219.

El-Halwagi, M. M., Gabriel, F., and Harell, D. (2003). Rigorous graphical targeting for resource conservation via material recycle/reuse networks. *Industrial and Engineering Chemical Research* 42(19), 4319–4328.

Feng, X., Bai, J., Wang, H. and Zheng, X. (2008). Grass-roots design of regeneration recycling water networks. *Computers and Chemical Engineering*, 32, 1892–1907.

Foo, D. C. Y. (2008). Flowrate targeting for threshold problems and plantwide integration for water network synthesis. *Journal of Environmental Management*, 88, 253–274.

Foo, D. C. Y. (2009). State-of-the-art review of pinch analysis techniques for water network synthesis. *Industrial and Engineering Chemistry Research*, 48(11), 5125–5159.

Garcia, X. and Pargament, D. (2015). Reusing wastewater to cope with water scarcity: Economic, social and environmental considerations for decision-making. *Resources, Conservation and Recycling*, 101, 154–166.

Hallale, N. (2002). A new graphical targeting method for water minimization. *Advances in Environmental Research*, 6, 377–390.

Huang, C.-H., Chang, C.-T., Ling, H.-C. and Chang (1999). A mathematical programming model for water usage and treatment network design. *Industrial and Engineering Chemistry Research*, 38, 2666–2679.

Jury, W. A. and Vaux, H. J. (2007). The emerging global water crisis: Managing scarcity and conflict between water users. *Advances in Agronomy*, 95, 1–76.

Kim, H. and Lee, T. (2007). Pareto optimality of industrial symbiosis network: Benefit sharing of wastewater neutralization network in Yeosu EIP. *Proceedings of PSE Asia August*, Xian, China.

Kim, S. H., Yoon, S.-G., Chae, S. H., and Park, S. (2010). Economic and environmental optimization of a multi-site utility network for an industrial complex. *Journal of Environmental Management*, 91, 690–705.

Kuo, W. C. J. and Smith, R. (1997). Effluent treatment system design. *Chemical Engineering Science*, 52(23), 4273.

Kuo, W.-C. J. and Smith, R. (1998). Designing for the interactions between water-use and effluent treatment. *Transactions of International Chemical Engineering Part A*, 76, 287–301.

Liao, Z. W., Wu, J. T., Jiang, B. B., Wang, J. D., and Yang, Y. R. (2007). Design methodology for flexible multiple plant water networks. *Industrial and Engineering Chemistry Research*, 46, 4954–4963.

Lim, S. R. and Park, J. M. (2010). Interfactory and intrafactory water network system to remodel a conventional industrial park to a green eco-industrial park. *Industrial and Engineering Chemistry Research*, 49, 1351–1358.

Lovelady, E. M. and El-Halwagi, M. M. (2009). Design and integration of eco-industrial parks for managing water resources. *Environmental Progress and Sustainable Energy*, 28, 265–272.

Manan, Z. A., Tan, Y. L., and Foo, C. Y. (2004). Targeting the minimum water flowrate using water cascade analysis technique. *AIChE Journal* 50(12), 3169–3183.

Montastruc, L., Boix, M., Pibouleau, L., Azzaro-Pantel, C., and Domenech, S. (2013). On the flexibility of an eco-industrial park (EIP) for managing industrial water. *Journal of Cleaner Production*, 43, 1–11.

Olesen, S. G. and Polley, G. T. (1997). A simple methodology for the design of water networks handling single contaminants. *Transactions of the Institute of Chemical Engineers*, 75, 420–426.

Poplewski, G., Walczyk, K., and Jezowski, J. (2010). Optimization-based method for calculating water networks with user specified characteristics. *Chemical Engineering Research and Design*, 88, 109–120.

Rubio-Castro, E., Ponce-Ortega, J. M., Napoles-Rivera, F., El-Halwagi, M. M., Serna-Gonzalez, M., and Jimenez-Gutierrez, A. (2010). Water integration of eco-industrial parks using a global optimization approach. *Industrial and Engineering Chemical Research*, 49, 9945–9960.

Rubio-Castro, E., Ponce-Ortega, J. M., Serna-González, M., Jiménez-Gutiérrez, A., and El-Halwagi, M. M. (2011). A global optimal formulation for the water integration in eco-industrial parks considering multiple pollutants. *Computers and Chemical Engineering*, 35, 1558–1574.

Rubio-Castro, E., Ponce-Ortega, J.M., Serna-González, M. and El-Halwagi, M.M. (2012). Optimal reconfiguration of multi-plant water networks into an eco-industrial park. *Computers and Chemical Engineering*, 44, 58–83.

Savelski, M. J. and Bagajewicz, M. J. (2000). On the optimality conditions of water utilization systems in process plants with single contaminants. *Chemical Engineering Science*, 55, 5035–5048.

Shenoy, U.V. and Bandyopadhyay, S. (2007). Targeting for Multiple Resources. *Industrial and Engineering Chemistry Research*, 46, 3698–3708.

Sorin, M. and Bedard, S. (1999). The global pinch point in water reuse networks. *Transactions of the Institution of Chemical Engineers*, 77, 305–308.

Spriggs, H. D., Lowe, E. A., Watz, J., Lovelady, E. M., and El-Halwagi, M. M. (2004). Design and development of eco-industrial parks. AIChE Spring Meeting, New Orleans, LA.

Taskhiri, M., Tan, R. and Chiu, A.F. (2011). MILP model for emergy optimization in EIP water networks. *Clean Technologies and Environmental Policy*, 13, 703–712.

Takama, N., Kuriyama, T., Shiroko, K. *and* Umeda, T. (1980). Optimal water allocation in a petroleum refinery. *Computers and Chemical Engineering*, 4, 251–258.

Thillairvarrna, S. L., Chew, I. M. L., Foo, D. C. Y., and Tan, R. R. (2008). Game theory approach to the analysis of inter-plant water integration. *Proceedings for the 2008 Regional Symposium on Chemical Engineering*, Kuala Lumpur, Malaysia.

Tian, J., Liu, W., Lai, B., Li, X., and Chen, L. (2014). Study of the performance of eco-industrial park development in China. *Journal of Cleaner Production*, 64, 486–494.

Wan Alwi, S.R., Manan, Z.A., Samingin, M.H. and Misran, N. (2008). A holistic framework for design of cost-effective minimum water utilization network. *Journal of Environmental Management*, 88, 219–252.

Wang, Y. P. and Smith, R. (1994a). Wastewater minimization. *Chemical Engineering Science*, 49(7), 981.

Wang, Y. P. and Smith, R. (1994b). Design of distributed effluent treatment systems. *Chemical Engineering Science*, 49(18), 3127.

Wang, Y. P. and Smith, R. (1995). Wastewater minimization with flow rate constraints. *Transactions of the Institution of Chemical Engineers, Part A*, 73, 889–904.

Yoo, C., Lee, T. Y., Kim, J., Moon, I., Jung, J. H., Han, C., Oh, J. M., and Lee, I. B. (2007). Integrated water resource management through water reuse network design for clean production technology: State of the art. *Korean Journal of Chemical Engineering*, 24, 567–576.

2.2 Optimal Water Network Synthesis

Munawar A. Shaik

2.2.1 Introduction

Water is precious and a vital commodity for survival of society and industries in general. Water is extensively used in several industries such as refineries, petrochemicals, metallurgy, food, pharmaceuticals, agriculture, textile, paper, electronics, and other service-oriented organizations such as hospitals, railways, and so on. Washing, steam and cooling water utility operations are some of the typical industrial activities that consume water. During these operations there are some water losses and/or several contaminants (NH_3, H_2S, hydrocarbons, metals, oils, etc.) get picked up by the water, which require either partial treatment for selective removal of some pollutants (also known as *regeneration*) or comprehensive treatment before reuse or discharge.

The freshwater resources are continuously depleting due to rapid industrialization and there is increasing water demand due to rising population growth. Thus, there is increased concern for water conservation either through rainwater harvesting or by minimizing freshwater requirement by strongly encouraging reuse and recycle of wastewater for other useful purposes at domestic, urban, and industrial scales. Additionally, the environmental regulations are increasingly becoming stringent and the industries have realized the importance of proper and effective treatment of the generated wastewater for safety, sustainability, and to minimize the environmental footprint. Decentralized wastewater treatment facilities are gaining importance due to effective application of suitable technologies depending on the type and concentration of its contaminants, relative to a centralized wastewater treatment facility where different wastewater streams are randomly mixed before common treatment of the mixed effluent.

The different water-using operations may be classified as either mass transfer-based operations (also known as fixed contaminant-load problem) or nonmass transfer-based operations (also known as fixed flow rate problem), and together these two problems are termed as global water-using operations. In fixed-load problems such as washing operations, the water stream gathers different contaminants as it flows through the process, and if water losses are negligible, then inlet and outlet stream flow rates remain the same. In fixed flow rate operations such as boilers, water is either consumed or produced. For single-contaminant processes, fixed-load problem can be converted into a fixed flow rate problem (and vice versa) by choosing the water inlet to a process as a demand point (or water sink) and the water outlet from a process as a supply point (or water source) based on *limiting water profile* data (Foo 2009), where

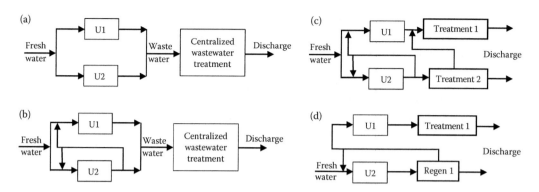

FIGURE 2.8 Different water utilization schemes. (a) Only freshwater use with centralized wastewater treatment, (b) water reuse–recycle with centralized wastewater treatment, (c) water reuse–recycle with decentralized wastewater treatment, and (d) regeneration–reuse–recycle with decentralized wastewater treatment.

maximum inlet concentration of a fixed-load problem is taken as the highest concentration limit of the water sink in the fixed flow rate problem in order to achieve maximum water recovery.

The first step for water conservation is to identify the set of water sources (producers of waste water) and water demands or sinks (consumers of fresh water) along with permissible contaminant levels in a process (or across multiple processes in a plant). Then a superstructure can be constructed depicting all possible flows across different sources to sinks. For a given set of water sources and sinks, the problem of reducing the minimum external freshwater requirement through reuse and recycle of used water is known as the Maximum Water Recovery (MWR) network or Water Allocation Network (WAN), or Water Using Network (WUN) synthesis problem. Another holistic approach is to consider the entire water management hierarchy (Wan Alwi and Manan 2006), where water conservation can be achieved through system modification through (1) source elimination, (2) source reduction, (3) direct reuse or external outsourcing, (4) regeneration, reuse, or recycling, and finally using (5) freshwater resource, rather than directly solving the MWR problem. Water network synthesis approach can be applied for water conservation either for grassroot synthesis and design of new processes or for retrofit of existing processes.

The different options for water reuse, recycle, and regeneration along with centralized versus decentralized water treatment facilities (Bagajewicz 2000; Foo 2009) are shown in Figure 2.8 for different water-using units (U1, U2, etc.).

In Figure 2.8a, there is no water conservation and only freshwater is used in all water-using units and the generated wastewater is treated in a centralized treatment facility before discharging. In Figure 2.8b, water recycle and reuse options are used to reduce the overall freshwater requirement and reduction in wastewater treatment load in the centralized treatment facility. In water reuse, the outlet streams from all other processes can be directly used or mixed with freshwater to meet the water demand, whereas in water recycle the outlet water from the same process can also be fed as input and/or mixed with freshwater. In Figure 2.8c, water recycle and reuse options are shown along with decentralized water treatment facilities. Similarly, water regeneration–reuse–recycle options are shown in Figure 2.8d. If the freshwater source is not pure then a pretreatment unit may be considered. If liquid discharge to the environment can be completely eliminated through additional capital investments for treatment leading to minimum or no intake of freshwater, then it is termed as achieving zero liquid discharge.

2.2.2 Literature Review

The different approaches for water network synthesis problems can be broadly classified as *pinch analysis*-based (or insight-based methods) and *mathematical programming*-based techniques. There are several review papers (Bagajewicz 2000; Jezowski 2008, 2010; Foo 2009; Khor et al. 2014) and recent books in the literature (Klemes et al. 2008; Majozi 2009; El-Halwagi 2012; Klemes 2013; Klemes et al. 2014; Majozi et al. 2015) that describe these approaches to handle water and/or heat integration, and in general, to handle the resource conservation network problems for process integration that can be uniformly applied across water, heat, hydrogen, and CO_2 networks based on pinch analysis and mathematical programming. Analogous to the heat exchanger network synthesis, graphical methods based on composite table algorithm and water cascade analysis are used for targeting and design of water networks. In this chapter, we focus on mathematical programming-based approaches for continuous processes.

Initially, Takama et al. (1980) presented a superstructure of water-using units and wastewater-treating units (a.k.a. total water systems or integrated water networks) for a fixed-load problem using the water recycle, regeneration, and reuse concept in a refinery. The resulting nonlinear programming (NLP) model was solved to local optimality using an iterative procedure based on Complex method for minimizing the total annual cost which includes capital and operating costs for wastewater treatment along with freshwater cost. Doyle and Smith (1997) proposed nonlinear and linear programming models for fixed mass load and fixed outlet concentration problems for multicontaminant systems involving water-using operations. Alva-Argáez et al. (1998) presented a nonconvex mixed-integer nonlinear programming problem (MINLP) for multicontaminant systems along with piping and treatment costs, and proposed a two-phase procedure to solve the same, though global optimality was not guaranteed. Alva-Argáez et al. (1999) presented some simplifying assumptions where the problem can be converted into a trans-shipment model, but the procedure is limited to small-scale problems.

Many researchers considered the water-using networks and wastewater treatment networks problems sequentially and solved them separately. Galan and Grossmann (1998) solved the distributed effluent treatment problem using mathematical programming. Savelski and Bagajewicz (2000, 2003) presented optimality conditions for single- and multicontaminant systems, where it was shown that at least one component reaches the maximum concentration limit at the outlet of a freshwater user process. Based on these conditions they could eliminate the nonlinearities arising in the model for water-using networks and converted the NLP into LP/MILP for single-contaminant systems (Bagajewicz and Savelski 2001). Gunaratnam et al. (2005) proposed a two-stage approach to solve the MINLP model for automated design of integrated water systems. Saeedi and Hosseinzadeh (2006) presented a graphical user interface (GUI) for solving the linear programming (LP) model for single contaminates and NLP for multicontaminant systems. Feng et al. (2008) presented a three-step sequential optimization for grassroots design based on minimization of freshwater consumption, regenerated water flow rate, and contaminant regeneration load.

Karuppiah and Grossmann (2006) considered simultaneous water-using and wastewater-treatment networks for fixed-load problems, and solved the resulting NLP and MINLP models using spatial branch and contract algorithm by approximating the nonconvex terms using piecewise linear under- and over-estimators. This superstructure and model were later extended by Ahmetovic and Grossmann (2011) to include multiple freshwater sources, both fixed-load and fixed flow rate operations, and feed water pretreatment operations along with piping and pumping costs. Li and Chang (2007) proposed effective initialization strategies for solving NLP and MINLP models for fixed-load water-using networks. Putra and Amminudin (2008) proposed a two-step procedure for solving the MINLP model for total water system, which can produce multiple optimal solutions.

Chew et al. (2008) considered the interplant water network problem and presented MILP/MINLP models for direct and indirect integration. Lim et al. (2009) presented an optimization model to synthesize total wastewater treatment network, which includes distributed and terminal wastewater treatment plants based on life cycle cost. Ng et al. (2010) presented an automated targeting technique for resource conservation networks based on reuse, recycle, and regeneration that effectively combines cascade analysis with optimization approach. Handani et al. (2010) proposed MILP model based on water management hierarchy for minimization of freshwater with additional options for source elimination, reduction, and outsourcing. Handani et al. (2011) proposed LP model for multicontaminate global water-using operations by assuming that all contaminate concentrations for each source and sink are

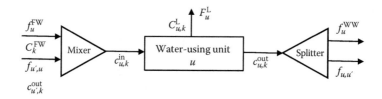

FIGURE 2.9 Superstructure representation for fixed-load water-using network.

fixed to their maximum values. Li and Chang (2011) presented sequential procedure for multiobjective optimization with NLP and MINLP models for multicontaminant global water-using network design.

2.2.3 Selected Mathematical Models

In this section, selected models from the literature are presented for optimal water network synthesis.

2.2.3.1 NLP/MINLP Models for Multicontaminate Fixed-Load Problems

For multicontaminate fixed-load problems, part of the superstructure depicting all possible flows and interconnections for each water-using operation ($u \in U$) is shown in Figure 2.9. A mixer and a splitter are shown at the inlet and outlet of each water-using unit.

The model presented here is adapted from Li and Chang (2007). Freshwater is fed to the mixer placed before each unit u with flow rate f_u^{FW} and concentration C_k^{FW} for contaminant $k \in K$. Although single freshwater source is considered, the model can be readily extended for pure or impure multiple freshwater sources. The inlet and outlet concentrations of each contaminant k in unit u are $c_{u,k}^{in}$ and $c_{u,k}^{out}$, and their maximum limits are specified ($C_{u,k}^{in,max}$ and $C_{u,k}^{out,max}$). The water loss from each unit u is F_u^L with $C_{u,k}^L$ concentration and $L_{u,k}$ is the mass load of contaminant k removed. $f_{u',u}$ is the flow rate of reuse water entering unit u from other unit u'. The flow rate of wastewater to discharge is f_u^{WW}. The NLP model is presented below for the objective of minimization of the total freshwater flow rate (F^{FW}) given in Equation (2.1).

$$\min F^{FW} = \sum_u f_u^{FW} \tag{2.1}$$

$$f_u^{FW} + \sum_{u'} f_{u',u} = \sum_{u'} f_{u,u'} + f_u^{WW} + F_u^L \quad \forall u \in U \tag{2.2}$$

$$\left(f_u^{FW} + \sum_{u'} f_{u',u} \right) c_{u,k}^{in} + L_{u,k} = \left(\sum_{u'} f_{u,u'} + f_u^{WW} \right) c_{u,k}^{out} + F_u^L C_{u,k}^L \quad \forall u \in U, k \in K \tag{2.3}$$

$$f_u^{FW} C_k^{FW} + \sum_{u'} f_{u',u} \, c_{u',k}^{out} = \left(f_u^{FW} + \sum_{u'} f_{u',u} \right) c_{u,k}^{in} \quad \forall u \in U, k \in K \tag{2.4}$$

$$c_{u,k}^{in} \le C_{u,k}^{in,max}, c_{u,k}^{out} \le C_{u,k}^{out,max} \quad \forall u \in U, k \in K \tag{2.5}$$

Equation (2.2) is the total flow balance around each water using unit u, and the corresponding component balance is in Equation (2.3). The component balance around the mixer is given in Equation (2.4), followed by bounds on inlet and outlet concentrations specified in Equation (2.5). Nonnegativity restriction applies to all continuous variables. There are bilinear terms in Equations (2.3) and (2.4) which makes it a nonconvex NLP model for which global optimum solution is not guaranteed using standard NLP solvers.

Structural or topological constraints may be added to avoid small flow rates of reuse streams from an operational point of view, as given in Equation (2.6), where $Y_{u,u'}$ is a binary variable for activation of reuse flow or connection between unit u and u', here the value of M may be chosen as the limiting water flow rate. Similar binary variables and constraints may be defined for other connections as well.

$$f_{u,u'} \leq MY_{u,u'} \quad \forall u, u' \in U \tag{2.6}$$

$$\min \sum_u \sum_{u'} Y_{u,u'} \tag{2.7}$$

Now the model becomes MINLP with the objective of minimizing the total number of interconnections given in Equation (2.7) subject to constraints given from (2.1) to (2.6), where the total freshwater flow rate in Equation (2.1) will be fixed based on the target value obtained from solving the earlier NLP model.

2.2.3.2 LP/MILP Models for Single-Contaminate Fixed-Load Problems

For single contaminates, Savelski and Bagajewicz (2000) presented optimality conditions for minimum freshwater flow rate, which states that the outlet concentration at the end of the water-using operation reaches its maximum limit, hence $c_{u,k}^{out} = C_{u,k}^{out,max}$, $K = 1$ in Equation (2.5). Therefore, the nonlinearities (bilinear terms) in Equations (2.3) and (2.4) can be eliminated leading to LP model for single contaminates. The resulting linear equations, replacing Equations (2.3) to (2.5), are given in Equations (2.8) and (2.9).

$$f_u^{FW} C_k^{FW} + \sum_{u'} f_{u',u} C_{u',k}^{out,max} + L_{u,k} = \left(\sum_{u'} f_{u,u'} + f_u^{WW} \right) C_{u,k}^{out,max} + F_u^L C_{u,k}^L \quad \forall u \in U, K = 1 \tag{2.8}$$

$$f_u^{FW} (C_k^{FW} - C_{u,k}^{in,max}) + \sum_{u'} f_{u',u} (C_{u',k}^{out,max} - C_{u,k}^{in,max}) \leq 0, \quad \forall u \in U, K = 1 \tag{2.9}$$

The resulting LP model comprises of Equations (2.1), (2.2), (2.8), and (2.9) for $K = 1$ for the objective of minimization of total freshwater flow rate. Similarly, for the objective of minimization of the total number of interconnections the resulting MILP model consists of Equations (2.2), (2.6), (2.7), (2.8), and (2.9) with $K = 1$. Nonnegativity restriction applies to all continuous variables.

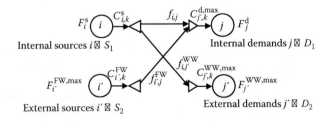

FIGURE 2.10 Superstructure representation for fixed-flow rate water-using network.

2.2.3.3 LP/MILP Models for Single or Multicontaminate Fixed-Flow Rate Problems

For fixed flow rate problems, LP model can be formulated whether it is single- or multicontaminate systems, if we have fixed contaminant concentrations specified for all internal and external sources and maximum contaminant concentrations specified for all internal and external demands/sinks. Normally, water flow rates for all internal sources and internal demands are fixed, and for external sources and external demands maximum supply rates are specified.

For multicontaminate fixed-flow rate problems, part of the superstructure depicting all possible flows and interconnections between different sources and demands is shown in Figure 2.10. Consider a set of internal water sources ($i \in S_1$) with fixed supply rates (F_i^s) and fixed contaminant concentrations ($C_{i,k}^s$); and a set of internal demands ($j \in D_1$) with fixed demand rates (F_j^d) and maximum contaminant concentrations ($C_{j,k}^{d,max}$) specified.

Similarly, we have a set of external water sources ($i' \in S_2$) with maximum supply rate ($F_{i'}^{FW,max}$), and fixed contaminant concentrations ($C_{i',k}^{FW}$) specified; and a set of external water demands ($j' \in D_2$) with maximum demand rate ($F_{j'}^{WW,max}$), and maximum contaminant concentrations ($C_{j',k}^{WW,max}$) specified. The different nonnegative continuous variables are as follows: flow rate of water from internal sources to internal demands is $f_{i,j}$, flow rate of freshwater from external sources to internal demands is $f_{i',j}^{FW}$, and the flow rate of water from internal sources to external demands is $f_{i,j'}^{WW}$. The following LP model is adapted from Li and Chang (2011) for the objective of minimization of the total freshwater flow rate.

$$\min F^{FW} = \sum_{i' \in S_2} \sum_{j \in D_1} f_{i',j}^{FW} \tag{2.10}$$

$$\sum_{j \in D_1} f_{i',j}^{FW} \leq F_{i'}^{FW,max} \quad \forall i' \in S_2 \tag{2.11}$$

$$F_i^s = \sum_{j \in D_1} f_{i,j} + \sum_{j' \in D_2} f_{i,j'}^{WW} \quad \forall i \in S_1 \tag{2.12}$$

$$F_j^d = \sum_{i \in S_1} f_{i,j} + \sum_{i' \in S_2} f_{i',j}^{FW} \quad \forall j \in D_1 \tag{2.13}$$

$$\sum_{i \in S_1} f_{i,j} C_{i,k}^s + \sum_{i' \in S_2} f_{i',j}^{FW} C_{i',k}^{FW} \leq F_j^d C_{j,k}^{d,max} \quad \forall j \in D_1, k \in K \tag{2.14}$$

$$\sum_{i \in S_1} f_{i,j'}^{WW} \leq F_{j'}^{WW,max} \quad \forall j' \in D_2 \tag{2.15}$$

$$\sum_{i \in S_1} f_{i,j'}^{WW} C_{i,k}^s \leq \left(\sum_{i \in S_1} f_{i,j'}^{WW} \right) C_{j',k}^{WW,max} \quad \forall j' \in D_2, k \in K \tag{2.16}$$

Equation (2.11) enforces the maximum supply rate for all external freshwater sources. Equation (2.12) is the total mass balance around the splitter placed after internal sources. Equations (2.13) and (2.14) are the total and contaminant mass balances around the mixer placed before internal demands. Similarly, Equations (2.15) and (2.16) are the total and contaminant mass balances at external demands. The corresponding MILP problem can also be formulated by including topological constraints similar to Equations (2.6) and (2.7). When a multicontaminate fixed-load problem is converted into a fixed flow

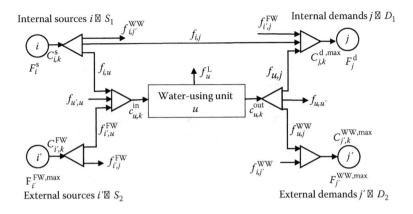

FIGURE 2.11 Superstructure representation for global water-using network.

rate problem based on limiting water flow rates, then the concentrations of sources and sinks are not fixed, which leads to an NLP model.

2.2.3.4 NLP/MINLP Models for Multicontaminate Global Water-Using Networks

Consider a system which has a combination of both fixed-load and fixed flow rate operations as shown in Figure 2.11. The nomenclature is similar to the cases presented in earlier sections.

For single contaminates, due to the optimality conditions discussed earlier, the resulting model would be LP or MILP depending on the objective function considered. For multicontaminates the combined model would be NLP or MINLP, which is presented below (Li and Chang 2011). Each internal source ($i \in S_1$) can supply water to three sinks: inlet of water-using units ($f_{i,u}$), internal demands ($f_{i,j}$), and external demands ($f_{i,j'}^{\text{WW}}$). Similarly, each internal demand ($j \in D_1$) can be supplied water from three sources: internal sources ($f_{i,j}$), outlet from water-using units ($f_{u,j}$), and external freshwater sources ($f_{i',j}^{\text{FW}}$). Accordingly, the total and contaminant mass balances are modified for both fixed-load and fixed-flow rate operations. For the objective of minimization of the total freshwater consumption rate, the NLP model is presented below (Li and Chang 2011).

$$\min F^{\text{FW}} = \sum_{i' \in S_2} \left(\sum_{j \in D_1} f_{i',j}^{\text{FW}} + \sum_{u} f_{i',u}^{\text{FW}} \right) \tag{2.17}$$

$$\sum_{i \in S_1} f_{i,u} + \sum_{i' \in S_2} f_{i',u}^{\text{FW}} + \sum_{u'} f_{u',u} = \sum_{u'} f_{u,u'} + \sum_{j \in D_1} f_{u,j} + \sum_{j' \in D_2} f_{u,j'}^{\text{WW}} + F_u^{\text{L}} \quad \forall u \in U \tag{2.18}$$

$$\left(\sum_{i \in S_1} f_{i,u} + \sum_{i' \in S_2} f_{i',u}^{\text{FW}} + \sum_{u'} f_{u',u} \right) c_{u,k}^{\text{in}} + L_{u,k} = \left(\sum_{u'} f_{u,u'} + \sum_{j \in D_1} f_{u,j} + \sum_{j' \in D_2} f_{u,j'}^{\text{WW}} \right) c_{u,k}^{\text{out}} + F_u^{\text{L}} C_{u,k}^{\text{L}} \quad \forall u \in U, k \in K$$

$$\tag{2.19}$$

$$\left(\sum_{i \in S_1} f_{i,u} \, C_{i,k}^{\text{s}} + \sum_{i' \in S_2} f_{i',u}^{\text{FW}} \, C_{i',k}^{\text{FW}} + \sum_{u'} f_{u',u} \, c_{u',k}^{\text{out}} \right) = \left(\sum_{i \in S_1} f_{i,u} + \sum_{i' \in S_2} f_{i',u}^{\text{FW}} + \sum_{u'} f_{u',u} \right) c_{u,k}^{\text{in}} \quad \forall u \in U, k \in K \tag{2.20}$$

$$c_{u,k}^{\text{in}} \le C_{u,k}^{\text{in,max}}, c_{u,k}^{\text{out}} \le C_{u,k}^{\text{out,max}} \quad \forall u \in U, k \in K \tag{2.5}$$

$$\sum_{j \in D_1} f_{i',j}^{FW} + \sum_u f_{i',u}^{FW} \le F_{i'}^{FW,max} \quad \forall i' \in S_2 \tag{2.21}$$

$$F_i^s = \sum_{j \in D_1} f_{i,j} + \sum_u f_{i,u} + \sum_{j' \in D_2} f_{i,j'}^{WW} \quad \forall i \in S_1 \tag{2.22}$$

$$F_j^d = \sum_{i \in S_1} f_{i,j} + \sum_{i' \in S_2} f_{i',j}^{FW} + \sum_u f_{u,j} \quad \forall j \in D_1 \tag{2.23}$$

$$\sum_{i \in S_1} f_{i,j} C_{i,k}^s + \sum_{i' \in S_2} f_{i',j}^{FW} C_{i',k}^{FW} + \sum_u f_{u,j} c_{u,k}^{out} \le F_j^d C_{j,k}^{d,max} \quad \forall j \in D_1, k \in K \tag{2.24}$$

$$\sum_u f_{u,j'}^{WW} + \sum_{i \in S_1} f_{i,j'}^{WW} \le F_{j'}^{WW,max} \quad \forall j' \in D_2 \tag{2.25}$$

$$\sum_u f_{u,j'}^{WW} c_{u,k}^{out} + \sum_{i \in S_1} f_{i,j'}^{WW} C_{i,k}^s \le \left(\sum_u f_{u,j'}^{WW} + \sum_{i \in S_1} f_{i,j'}^{WW} \right) C_{j',k}^{WW,max} \quad \forall j' \in D_2, k \in K \tag{2.26}$$

Structural or topological constraints may be added as given in Equations (2.6) and (2.27), where $Y_{i,j}$ is a binary variable for activation of water flow or connection between source i and demand j, where the value of big-M may be chosen as the limiting water flow rate. Similar binary variables and constraints may be defined for other connections as well. Now the model becomes MINLP with the objective of minimizing the total number of interconnections given in Equation (2.28)

$$f_{i,j} \le MY_{i,j} \quad \forall i \in S_1, j \in D_1 \tag{2.27}$$

$$\min \sum_u \sum_{u'} Y_{u,u'} + \sum_{i \in S_1} \sum_{j \in D_1} Y_{i,j} \tag{2.28}$$

2.2.4 Results and Discussion

2.2.4.1 Example 1

A case study from a refinery is presented from literature (Doyle and Smith 1997; Li and Chang 2007, 2011) to demonstrate the NLP/MINLP models for multicontaminate fixed-load problem described in previous section. The data for this case study is given in Table 2.2.

TABLE 2.2　Process Limiting Data for Example 1

Unit (u)	Limiting Flow Rate, F_u^{Lim} (t/h)	Contaminant (k)	$C_{u,k}^{in,max}$ (ppm)	$C_{u,k}^{out,max}$ (ppm)	$L_{u,k}$ (kg/h)
Distillation (U1)	45	Hydrocarbon	0	15	675
		H$_2$S	0	400	18,000
		Salt	0	35	1575
Hydrodesulphurization (U2)	34	Hydrocarbon	20	120	3400
		H$_2$S	300	12,500	414,800
		Salt	45	180	4590
Desalted (U3)	56	Hydrocarbon	120	220	5600
		H$_2$S	20	45	1400
		Salt	200	9500	520,800

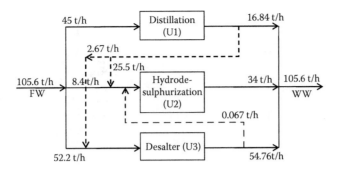

FIGURE 2.12 Optimal water network for Example 1.

TABLE 2.3 Process Limiting Data for Example 2

Unit (u)	Contaminant (k)	$C_{u,k}^{in,max}$ (ppm)	$C_{u,k}^{out,max}$ (ppm)	$L_{u,k}$ (kg/h)
U1	A	20	120	3.40
	B	300	12,500	414.80
	C	45	180	4.59
U2	A	120	220	5.60
	B	20	1000	1.40
	C	200	9500	520.80
U3	A	0	20	0.16
	B	0	60	0.48
	C	0	20	0.16
U4	A	50	150	0.80
	B	400	8000	60.80
	C	60	120	0.48
U5	A	0	15	0.75
	B	0	400	20.00
	C	0	35	1.75
U6	A	10	70	2.00
	B	200	600	100.70
	C	20	90	2.50
U7	A	25	150	1.80
	B	230	1000	6.80
	C	20	220	0.60
U8	A	5	100	3.00
	B	45	4000	102.30
	C	50	300	8.14
U9	A	13	100	4.60
	B	200	3000	200.00
	C	5	200	1.90
U10	A	10	100	4.00
	B	90	500	10.30
	C	70	800	9.00

There are three water-using units with specified limiting water flow rates, and maximum inlet and outlet concentrations given for three contaminants. Water losses in units are neglected. The mass load of contaminants removed ($L_{u,k}$) can be calculated as $L_{u,k} = F_u^{Lim}\left(C_{u,k}^{out,max} - C_{u,k}^{in,max}\right)$, as shown in Table 2.2, which is used in Equation (2.3). The resulting NLP model given by Equations (2.1) through (2.5) is solved

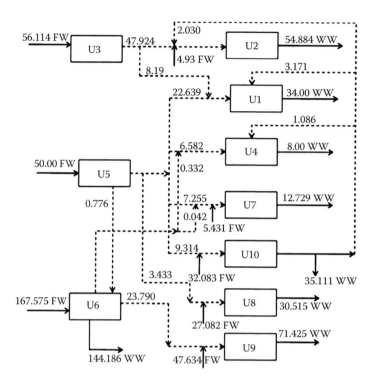

FIGURE 2.13 Optimal water network for Example 2.

to global optimality, for minimizing the total freshwater consumption rate, using OQNLP or MSNLP solvers in GAMS software using default options and without providing any initial guess. Here, it is to be noted that the other NLP solvers (such as SNOPT and CONOPT3) failed to find a feasible solution using default options (without any initial guess). The NLP model has 37 continuous variables and 28 constraints. The optimal water network design is shown in Figure 2.12. The minimum total freshwater consumption rate is 105.604 t/h, compared to 135 t/h, which is the total freshwater required without water reuse. There is no water reuse possible for distillation unit, since it requires pure water; therefore, the freshwater requirement for this unit is equal to the limiting water flow rate (45 t/h). The optimal reuse streams are $f_{u1,u2} = 25.49$ t/h, $f_{u1,u3} = 2.668$ t/h, and $f_{u3,u2} = 0.067$ t/h, with the total wastewater generation of 105.604 t/h. The small flow rate streams such as $f_{u3,u2} = 0.067$ t/h can be eliminated by introducing structural constraints given in Equation (2.6), leading to MINLP model in order to minimize objective in Equation (2.7). Then the total number of interconnections would be reduced from 9 to 8 in this case, with slight increase in the total freshwater requirement to 105.671 t/h.

2.2.4.2 Example 2

A slightly more complex example is taken from literature (Li and Chang 2007, 2011), where there are ten water-using units with maximum inlet and outlet concentrations and mass load specified for three contaminants as given in Table 2.3. Water losses in units are neglected. The resulting NLP model is solved to global optimality, for minimizing the total freshwater consumption rate, using MSNLP solver in GAMS software using default options and without providing any initial guess. Here, it is to be noted that the other NLP solvers (such as SNOPT, CONOPT3, and OQNLP) failed to find a feasible solution using default options (without any initial guess). The NLP model has 191 continuous variables and 91 constraints. The minimum total freshwater consumption rate is 390.849 t/h. The optimal water network design is shown in Figure 2.13. Here, there is no freshwater entering in units 1 and 4 and there is no wastewater generated from units 3 and 5 in the optimal solution.

2.2.5 Conclusion

In this chapter, different options such as water reuse, regeneration, and recycle are discussed for exploring water conservation opportunities in industrial water-using operations. A superstructure, depicting all possible flows from different water sources to water sinks, is first constructed followed by development of optimization model by writing appropriate total and contaminant mass balances. The optimal water network synthesis problems are discussed for reducing the total freshwater consumption rate of single- contaminate and multicontaminate systems in fixed-load and fixed-flow rate operations. Selected LP/NLP/MILP/MINLP models from literature are reviewed followed by illustration of these models through a couple of examples drawn from literature.

Acknowledgments

The author gratefully acknowledges financial support received from the Council of Scientific and Industrial Research (CSIR, India) under grant no: 22(0671)/14/EMR-II. The author acknowledges Mr. Ashutosh Gupta (M.Tech., IIT Delhi) for help in solving the case studies.

References

Ahmetovic, E., Grossmann, I.E. 2011. Global superstructure optimization for the design of integrated water networks. *AIChE J.*, 57, 434–457.

Alva-Argáez, A., Kokossis, A.C., Smith, R. 1998. Waste-water minimization of industrial systems using an integrated approach. *Comp. Chem. Eng.*, 22, S741–S744.

Alva-Argáez, A., Vallianatos, A., Kokossis, A. 1999. A multi-contaminant transshipment model for mass exchange networks and wastewater minimization problems. *Comp. Chem. Eng.*, 23, 1439–1453.

Bagajewicz, M. 2000. A review of recent design procedures for water networks in refineries and process plants. *Comp. Chem. Eng.*, 24, 2093–2113.

Bagajewicz, M., Savelski, M.J. 2001. On the use of linear models for the design of water utilization systems in process plants with a single contaminant. *Trans. IChemE*, 79 (Part A), 600–610.

Chew, I.M.L., Tan, R., Ng, D.K.S., Foo, D.C.Y., Majozi, T., Gouws, J. 2008. Synthesis of direct and indirect interplant water network. *Ind. Eng. Chem. Res.*, 47, 9485–9496.

Doyle, S.J., Smith, R. 1997. Targeting water reuse with multiple constraints. *Trans. IChemE*, 75 (Part B), 181–189.

El-Halwagi., M.M. 2012. *Sustainable Design through Process Integration*, Butterworth-Heinemann, Elsevier, Waltham, MA.

Feng, X., Bai, J., Wang, H., Zheng, X. 2008. Grass-roots design of regeneration recycling of water networks. *Comp. Chem. Eng.*, 32, 1892–1907.

Foo, D.C.Y. 2009. State-of-the-art review of pinch analysis techniques for water network synthesis. *Ind. Eng. Chem. Res.*, 48, 5125–5159.

Galan, B., Grossmann, I.E. 1998. Optimal design of distributed wastewater treatment networks. *Ind. Eng. Chem. Res.*, 37, 4036–4048.

Gunaratnam, M., Alva-Argáez, A., Kokossis, A., Kim, J.K., Smith, R. 2005. Automated design of total water systems. *Ind. Eng. Chem. Res.*, 44, 588–599.

Handani, Z.B., Wan Alwi, S.R., Hashim, H., Manan, Z.A. 2010. Holistic approach for design of minimum water networks using mixed integer linear programming technique. *Ind. Eng. Chem. Res.*, 49, 5742–5751.

Handani, Z.B., Wan Alwi, S.R., Hashim, H., Manan, Z.A., Abdullah, S.H.Y.S. 2011. Optimal design of water networks involving multiple contaminants for global water operations. *Asia-Pac. J. Chem. Eng.*, 6, 771–777.

Jezowski, J. 2008. Review and analysis of approaches for designing optimum industrial water networks. *Chem. Process. Eng.*, 29, 663–681.

Jezowski, J. 2010. Review of water network design methods with literature annotations. *Ind. Eng. Chem. Res.*, 49(10), 4475–4516.

Karuppiah, R., Grossmann, I.E. 2006. Global optimization for the synthesis of integrated water systems in chemical processes. *Comp. Chem. Eng.*, 3, 650–673.

Khor, C.S., Chachuat, B., Shah, N. 2014. Optimization of water network synthesis for single-site and continuous processes: Milestones, challenges, and future directions. *Ind. Eng. Chem. Res.*, 53, 10257–10275.

Klemes, J.J. 2013. *Handbook of Process Integration: Minimization of Energy and Water Use, Waste and Emissions*, Woodhead Publishing, Cambridge, UK.

Klemes, J.J., Smith, R., Kim, J.-K. 2008. *Handbook of Water and Energy Management in Food Processing*, Woodhead Publishing, Cambridge, UK.

Klemes, J.J., Varbanov, P.S., Wan Alwi, S.R., Manan, Z.A. 2014. *Process Integration and Intensification: Saving Energy, Water, and Resources*, Walter de Gruyter GmbH & Co KG, Berlin.

Li, B.H., Chang, C.T. 2007. A simple and efficient initialization strategy for optimizing water-using network design. *Ind. Eng. Chem. Res.*, 46, 8781–8786.

Li, B.H., Chang, C.T. 2011. Multiobjective optimization of water using networks with multiple contaminants. *Ind. Eng. Chem. Res.*, 50, 5651–5660.

Lim, S.R., Lee, H., Park, J.M. 2009. Life cycle cost minimization of a total wastewater treatment network system. *Ind. Eng. Chem. Res.*, 48, 2965–2971.

Majozi, T. 2009. *Batch Chemical Process Integration: Analysis, Synthesis, and Optimization*, Springer, Dordrecht.

Majozi, T., Seid, E.R., Lee, J.-Y. 2015. *Synthesis, Design, and Resource Optimization in Batch Chemical Plants*, CRC Press, Taylor & Francis, Boca Raton, FL.

Ng, D.K.S., Foo, D.C.Y., Tan, R.R., El-Halwagi, M. 2010. Automated targeting technique for concentration- and property-based total resource conservation network. *Comp. Chem. Eng.*, 34(5), 825–845.

Putra, Z.A., Amminudin, K.A. 2008. Two-step optimization approach for design of a total water system. *Ind. Eng. Chem. Res.*, 47, 6045–6057.

Saeedi, M., Hosseinzadeh, M. 2006. Optimization of water consumption in industrial systems using linear and nonlinear programming. *J. App. Sci.*, 6, 2386–2393.

Savelski, M.J., Bagajewicz, M.J. 2000. On the optimality conditions of water utilization systems in process plants with single contaminants. *Chem. Eng. Sci.*, 55, 5035–5048.

Savelski, M.J., Bagajewicz, M.J. 2003. On the necessary conditions of optimality of water utilization systems in process plants with multiple contaminants. *Chem. Eng. Sci.*, 58, 5349–5362.

Takama, N., Kuriyama, T., Shiroko, K., Umeda, T. 1980. Optimal water allocation in a petroleum refinery. *Comp. Chem. Eng.*, 4, 251–258.

Wan Alwi, S.R., Manan, Z.A. 2006. SHARPS: A new cost-screening technique to attain cost-effective minimum water network. *AIChE J.*, 52, 3981–3988.

2.3 Rationalization of Water Quality Monitoring Network

Subhankar Karmakar and Vikas Varekar

2.3.1 Introduction

On a global scale, aquatic ecosystems are particularly threatened due to various contaminants, improper land uses, and poor water management practices. Good water quality is considered as a vital component for sustainable development of a nation (Bartram and Balance, 1996; Strobl and Robillard, 2008). The water quality is being affected by anthropogenic activities, in the form of point and diffuse sources of pollution. It may also change due to river engineering and water use projects such as irrigation and damming (Chapman, 1996). At the same time, various natural forces involved in hydrological, physical, chemical, and biological processes may also be responsible for alteration of water quality (Bartram and Balance, 1996). Thus, the continuous degradation of water resources requires determination of ambient water quality status, which can be used to check the impact induced by anthropogenic activities. Water quality monitoring is the collection of quantitative and representative spatiotemporal information of physical, chemical, and biological characteristics of a water body. It is also the first step of effective water quality management, which acts as a connecting link between environment and policy-makers. Along with the estimation of quality degradation, the other objectives of the water quality monitoring programs are [Park et al., 2006; Central Pollution Control Board, India (CPCB), 2007]: (1) rational planning of pollution control strategies; (2) identification of nature and magnitude of pollution control measures; (3) effectiveness evaluation of existing pollution control efforts; (4) identification of the mass flow of contaminants in surface water and effluents; (5) formulation of standards and permit requirements; (6) testing of compliance with standards and classifications for waters; (7) early warning and detection of pollution; (8) understanding the trends of temporal variations in water quality parameters; (9) supporting utilization of water resources; (10) estimating pollution loads from each watershed unit in order to perform total maximum daily load analyses , and (11) establishing information systems for water resources management.

The water quality monitoring systems can be categorized in several ways including (Ward et al., 1989): (1) length of projected life of the monitoring station, which consists of long-term or fixed station monitoring and short-term or special study monitoring; (2) types of measurements to be made,

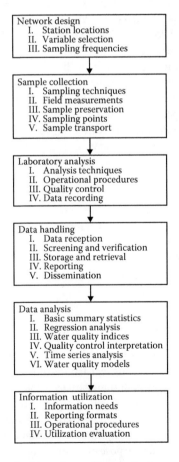

FIGURE 2.14 Principal activities involved in water quality monitoring.

i.e., physical, chemical, and biological monitoring, depending upon parameters to be sampled; (3) the monitoring systems can be classified based on the location of source, e.g., groundwater monitoring, lake monitoring, acid rain monitoring, effluent monitoring, estuary monitoring, and marine monitoring.

The initial water quality monitoring program was started in the early 1960s, with the purpose of describing the state of water quality. Most of the early monitoring efforts consist of arbitrary and inconsistent design strategies. The sampling locations and frequencies were determined by subjective criteria or convenience. Further, once a water quality monitoring system was established, there was no effectiveness and efficiency assessment of the system (Tirsch and Male, 1984; Harmancioglu et al., 1999; Strobl and Robillard, 2008). In short, a consistent design approach for water quality monitoring was missing. It is rather common to have insufficient analysis for collected water quality data due to inadequacy of proper network design methodology (Strobl and Robillard, 2008). The principal activities involved in water quality monitoring are summarized in Figure 2.14.

The design of water quality monitoring network (WQMN) is a crucial task of the monitoring program, which consists of: (1) determination of the number and spatial distribution of monitoring stations, (2) selection of a sampling frequency, and (3) selection of water quality parameters to be monitored. Defining monitoring objectives and budgetary constraints has also been found to be essential while designing WQMN (Varekar et al., 2015).

To date, various countries such as Australia (Department of Water of the Government of Western Australia, 2009), Canada (Canadian Council of Ministers of the Environment, 2006), European Union member states, Norway, and the European Commission (Water Framework Directives of the European Community, 2003), India (Ministry of Environment, Forest and Climate Change (MoEFCC), Government of India, 2005; Central Pollution Control Board, India, 2007), and, United States of America (United States Environmental Protection Agency (USEPA), 2003; United States Geological Survey, 2011) have constituted their own guidelines/frameworks for the design of WQMNs.

These guidelines are subjective, mostly based on experts' judgments and do not consider the analysis of existing water quality inventory data, watershed, and river characteristics. They may also lack a mathematical basis for design. Most of the water quality monitoring programs have been failed to define water quality due to lack of information (Ward et al., 1989). The existing monitoring systems may seem to be rich in data, but poor in information (Ward et al., 1989; Harmancioglu and Alphasan, 1992, 1994). Hence, there is a great need to develop logical and consistent rational approaches for effective data collection and efficient information extraction from the monitoring systems (Strobl and Robillard, 2008). The rationalized monitoring systems permit effective water pollution control recommendations and better financial resources allocation. Therefore, the principal objective of this chapter is to conduct a review of all past and current attempts for rationalization of WQMNs. The important components of WQMN are sampling sites, frequency of sampling, and sampling parameters. According to recent reviews (Dixon and Chiswell, 1996; Strobl and Robillard, 2008; Mishra and Coulibaly, 2009) and all available literature, various methodologies for rationalization of sampling locations, frequency, and parameters are discussed in the following sections.

2.3.2 Rationalization of Sampling Locations

The selection of sampling locations/sites is the most crucial task in designing WQMN. The common practice in almost all countries as per existing legislations (guidelines/protocols) is to locate the sampling sites at the downstream of major pollution sources for accounting the effect of anthropogenic activities (mostly in terms of point sources of pollution). According to "Uniform Protocol for Water Quality Monitoring (June, 2005)" by Ministry of Environment, Forest and Climate Change (MoEFCC), Government of India, the water quality sampling sites are classified into baseline stations, flux or impact stations, and trend stations, based on the purpose of monitoring.

2.3.2.1 Methods Based on Topology

Sharp (1971) has proposed a uniform sampling plan for rivers and streams. In this method, hierarchical topology of the river basin plays an important role. The centroids are identified by dividing the stream network into successive halves. These centroids are the optimum locations of sampling. Further, Sanders et al. (1983) have proposed a modified form of Sharp's approach, in which tributaries were replaced by the number of outfalls and pollution loads. The Sanders approach is also based on the topology of the river basin. To date, various researchers (Sharp, 1971; Sanders et al., 1983; National Institute of Hydrology India, 1996, 1997; Do et al., 2011, 2012; Cetinkaya and Harmancioglu, 2012; Varekar et al., 2012) have used these approaches for rationalization of sampling sites.

2.3.2.2 Modified Sanders Approach

Recently, Varekar et al. (2015, 2016) proposed an approach based on Sanders, (Sanders et al., 1983) and Sharp, (1971) methods, for rationalization of sampling sites. The Sanders approach (Sanders et al., 1983) is based on the uniform sampling method proposed by Sharp, in which a river network is divided into interior and exterior links. An exterior link is a tributary, has a minimum mean discharge, and is not fed by other defined streams. It is also called a first-order tributary. An interior link is not a tributary; it is formed by the intersection of two exterior tributaries and is also called a second-order tributary (Sanders et al., 1983). Each tributary or exterior link that contributes to the main stretch of a river is assigned a magnitude or weight of one, while the interior link is assigned a weight that equals the sum of the magnitudes of intersecting exterior links. The magnitude at the mouth of a river watershed equals the number of contributing exterior tributaries. After numbering the entire river network, the optimal sampling sites are selected based on the centroids of the river network. The tributaries are replaced by outfalls and pollution loads in the original Sanders approach (Sanders et al., 1983), while in the modified Sanders approach (Varekar et al., 2015, 2016), the absolute values of pollution loads at sampling sites are accounted in the selection of the optimal number and locations of sampling sites.

2.3.2.2.1 The Centroid of River Network

According to Sharp's approach, a river network is analogous to a binary tree in graph theory. The tree is a vital nonlinear structure used in computer programming (Knuth, 1968). According to Knuth (1968), the tree structure refers to the branching relations between nodes (data point) similar to those found in natural trees. The start (origin) and end points of links (i.e., the points of intersection of links and their

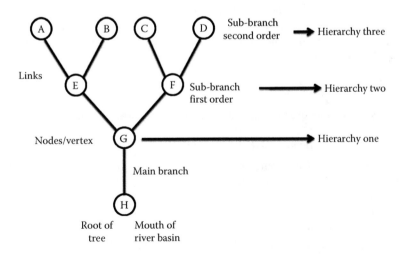

FIGURE 2.15   Modified Sanders approach for a hypothetical river tree.

terminating points) are termed vertices, and links joining vertices are the edges of a tree. The centroid is the vertex with the minimum weight (i.e., the vertex for which leading sub-trees have the minimum number of vertices). In other words, the centroid of a tree is the vertex that has approximately equal number of vertices on its upstream and downstream ends. Sharp (1971) used the concept of a centroid (Knuth, 1968) to divide a river network into approximately equal halves. They proposed a mathematical expression for determining the centroidal link, i.e., the link that divides the network in half. In nature, each river tributary generally has two sub-branches. Hence, the bifurcation ratio for a river system is considered to be two, and a river system can be considered analogous to a binary tree structure.

2.3.2.2.2 Determination of Optimum Location of Sampling Sites

A modified Sanders approach derived by Varekar et al. (2015, 2016) was used to determine the optimum locations of the sampling sites. The potential sampling locations are the nodes (data points)/vertices of the river shown in Figure 2.15.

The magnitude of each node is equal to the pollution load at the respective sampling point. The mouth of the watershed is the root of the river tree, and its magnitude is equal to the sum of the pollution loads at each node/vertex within the watershed. The first centroid, i.e., the first-order station (first hierarchy), is the node/sampling point with a magnitude closest to M_h (Equation 2.29):

$$M_h = \left[\frac{N_o + 1}{2} \right],$$
(2.29)

where M_h is the magnitude of the node/sampling location at the hth hierarchy and N_o is the total pollution load at the mouth of the river basin. The first sampling location is placed at the first centroid, identified using Equation 2.29, which indicates the main branch of the tree [the link joining the centroidal node (data point) to the root]. The first centroid is the unique node with a magnitude closest to half of the entire river network's magnitude. Hence, the first centroid is at the first hierarchy level or is the principal branch of the tree and deserves the highest priority in sampling. The successive centroids at different hierarchy levels are estimated using Equation 2.30.

$$M_{h+1} = \left[\frac{M_h + 1}{2} \right],$$
(2.30)

where M_{h+1} is the magnitude of the node/sampling location at the $(h+1)$th hierarchy level. For example, the two centroids at the second hierarchy level or the first-order sub-branch of the tree have the second highest priority in sampling. Since the river system has a binary tree structure, the maximum number of centroids used to determine sampling locations at various hierarchy levels is given as

$$n = 2^h - 1,$$
(2.31)

where n indicates the number of sampling stations at the hth hierarchy level.

2.3.2.2.3 Determination of Optimum Number of Sampling Sites

After the sampling locations have been delineated, their optimum number must be determined. Cost is the main factor that decides the optimum number of sampling locations. If the budget is known, the optimum number is determined as the total available funding divided by the operating cost for a single monitoring site. If the budget is unknown, then M_h values are first calculated as a function of the pollution load (Equations 2.29 and 2.30). Do et al. (2011) proposed that sampling should be stopped at M_h (the hth hierarchy), which is much smaller than M_1 (the first centroid), and that the optimum number of sampling locations correspond to the hierarchy level M_h (Equation 2.31). Another approach to select the optimum number of sampling sites is to consider the effluent disposal standards of wastewater. According to this approach, the number of sites corresponding to the sampling locations violating effluent disposal standards is the optimum number of sampling sites. Varekar et al. (2015, 2016) have effectively used this approach for seasonal rationalization of sampling sites based on both point and diffuse pollution loads by accounting for their seasonal variations.

2.3.2.3 Statistical Approaches

2.3.2.3.1 Statistical Entropy

The concept of statistical entropy has been applied for the rationalization of surface WQMNs. "Entropy" is a measure of the degree of uncertainty, which is derived from information theory, as described by Shannon and Weaver (1949). In this approach, a WQMN is treated as an information system. The entropy is calculated and considered in the design and assessment of sampling: sites, frequency, and parameters, in terms of cost-effectiveness and efficiency of the network. Shannon and Weaver (1949) have defined three basic information measures based on entropy, i.e., marginal and joint entropies, and transinformation (Ozkul et al., 2000). The marginal entropy, $H(X)$, of a discrete random variable X is as follows:

$$H(X) = -K \sum_{i=1}^{N} p(X_i) \log p(X_i), \tag{2.32}$$

where constant $K = 1$, if $H(X)$ is expressed in "napiers" for logarithms to the base e, N represents the number of elementary events with probabilities $p(X_i)$ $(i = 1, ..., N)$. The joint entropy of two independent random variables X and Y is equal to the sum of their marginal entropies

$$H(X, Y) = H(X) + H(Y), \tag{2.33}$$

where X and Y are stochastically dependent.

Transinformation is another entropy measure that quantifies the redundant or mutual information between X and Y. It is described as the difference between the total entropy and the joint entropy of dependent variables X and Y.

$$T(X, Y) = H(X) + H(Y) - H(X, Y). \tag{2.34}$$

The concept of entropy has been used by various past researchers (Harmancioglu and Alphasan, 1992, 1994; Harmancioglu et al., 1999; Ozkul et al., 2000; Karamouz et al., 2009a) for rationalization of water quality sampling sites.

2.3.2.3.2 Maximum Entropy-Based Hierarchical Spatiotemporal Bayesian Model

The maximum entropy-based hierarchical spatiotemporal Bayesian model was proposed by Alameddine et al. (2013). It is used for optimization of existing WQMN by identifying and retaining the sampling locations delivering maximum information. Simultaneously, new sampling locations are also identified for modification and extension of a network by using this technique. The concept of entropy is used to account the system uncertainties. In this model, dissolved oxygen standard violation entropy, total system entropy, and chlorophyll-a standard violation entropy are considered for the design of an estuarine monitoring network. A multiple attribute decision-making (MADM) framework has been proposed to consider these three competing design criteria, which results in an optimal monitoring network. The key findings and contributions of the application of maximum entropy method for optimization of WQMN are as follows:

1. The newly formulated optimization framework allows multicriteria optimization by incorporating a MADM component, giving different water quality experts the opportunity to weigh-in and give their perspective on the relative importance of different optimization criteria.
2. The proposed model is effective in modeling dynamic water bodies, viz. current driven waters, dammed rivers, estuaries, coastal waters, etc. However, the model assumes that the environmental variables can be properly modeled by a joint matrix normal distribution, which could be achieved by the variable transformation.
3. This methodology is flexible and can be easily implemented for identifying redundant stations or replacing one (or more) of the existing stations with another location based on the system's informational content application of maximum entropy-based hierarchical spatiotemporal Bayesian model.

2.3.2.3.3 *Bayesian Maximum Entropy (BME)*

The BME method of geostatistics proposes integration of water quality monitoring data with model predictions to provide improved estimates of water quality in a cost-effective manner. This information includes estimates of uncertainty and can be used to aid probabilistic-based decisions concerning the status of water (i.e., impaired or not impaired) and the level of monitoring needed to characterize the water for regulatory purposes (Lobuglio et al., 2007).

2.3.2.3.4 *Statistical Kriging*

Kriging is a geospatial method (Cressie and Christopher, 2011) used for regionalizing spatiotemporal variations, which has been widely used in various fields, such as hydrology, surveying, mining, and geology. In this approach, the spatiotemporal variation of water quality is determined by estimating variance. The number of sampling points is rationalized by delineating the sampling locations based on the estimated variance (Lo et al., 1996; Karamouz et al., 2009b).

2.3.2.3.5 *Multivariate Statistical Techniques*

Multivariate statistical techniques such as cluster analysis (CA), discriminant analysis (DA), principal component analysis (PCA), and principal factor analysis (PFA) have been used effectively for assessment of WQMNs by various researchers. These techniques are used to identify important components or factors that explain most of the variance of the system. These factors are designed to reduce the number of variables to a small number of indices while attempting to preserve the relationships present in the data (Singh et al., 2004; Ouyang, 2005; Musthafa et al., 2014; Varekar et al., 2016). The task of selecting important monitoring stations, detecting changes in the characteristics of water quality parameters, and data reduction and interpretation are accomplished using multivariate statistical techniques (Singh et al., 2004; Ouyang, 2005; Noori et al., 2010; Musthafa et al., 2012; Varol et al., 2012; Wang et al., 2012; Musthafa et al., 2014; Toochukwu, 2015; Varekar et al., 2016).

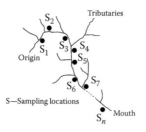

FIGURE 2.16   Hypothetical river network.

TABLE 2.4 Application of Statistical Techniques for Rationalization of Water Quality Monitoring Locations of a Hypothetical River Network

Sr. No.	Statistical Technique	Application of Statistical Techniques for Rationalization of Water Quality Monitoring Locations
1	Statistical entropy (Harmancioglu and Alphasan, 1992, 1994; Ozkul et al., 2000)	The marginal entropy $H(X_m)$ $(m = 1, ..., m)$ of the water quality variable observed at each station $(S_1, S_2, S_3, ..., S_n)$ (Figure 2.16) is computed by Equation 2.32, where m is replaced by 1 which defines the priority of sampling. The station with the highest $H(X_m)$ is denoted as the first priority station X_i; this is the location where the highest uncertainty occurs about the variable so that the highest information may be gained by making observations at this site. Later, this station is coupled with every other station in the network to select the pair that leads to the least transinformation (Equation 2.34). The station that fulfills this condition is marked as the second priority. The same procedure is continued by successively considering combinations of 3, 4, 5, ..., n stations and selecting the combination that produces the least transinformation.
2	Maximum entropy-based hierarchical spatiotemporal Bayesian model (Alameddine et al, 2013)	(1) In this approach, for a given hypothetical river system (Figure 2.16), the total entropy for water quality variables at all sampling locations $(S_1, S_2, S_3, ..., S_n)$ are estimated first. The rationalization of the monitoring design based on total system entropy focuses on minimizing residual entropy following data collection. This entails choosing to monitor locations that have maximum values of entropy given the existing data. (2) To locate additional monitoring sites that are having mostly uncertain compliance status, these are identified by calculating their violation entropies. The standard violation entropies are calculated through Bayesian hierarchal model simulation using observed water quality data. (3). Each of these criteria (total system entropy and standard violation entropies) resulted in different solutions. Therefore, a MADM process that weighs the design attributes or optimization criteria is used to generate a single optimal solution.
3	BME (Lobuglio et al., 2007)	(1) In this approach, the user has to define the water quality variables for which analysis is to be performed. (2) Hard data is in the form of observed water quality variables at existing $(S_1, S_2, S_3, ..., S_n)$ water quality monitoring locations (Figure 2.16). (3) Water quality simulation models are used to generate soft data. (4) A space/time covariance model is constructed using the existing monitoring data. (5) Preprocessing and postprocessing of both hard and soft data are accomplished with numerical tools developed to extend the BMElib implementation of BME methods within the MATLAB® programming platform. (6) The optimum monitoring plans are then delineated for three scenarios of data availability, i.e., hard data, soft data, and no data with respect to existing monitoring locations.

(Continued)

TABLE 2.4 (*Continued*) Application of Statistical Techniques for Rationalization of Water Quality Monitoring Locations of a Hypothetical River Network

Sr. No.	Statistical Technique	Application of Statistical Techniques for Rationalization of Water Quality Monitoring Locations
4	Statistical Kriging (Karamouz et al., 2009b)	The water quality monitoring data at existing sampling locations ($S_1, S_2, \ldots, S_n$) (Figure 2.16) is used for rationalization of monitoring sites by following the procedure described below: (1) First, the water quality indicators are selected considering existing spatial and temporal variations of the water quality variables and the characteristics of pollution loads for different water uses. Then, the spatial distribution of the existing water quality monitoring stations is investigated. (2) Then, the selected water quality variables are standardized, and the time series is plotted in order to determine a pattern of their temporal variations. (3) A model which best fits the time series of water quality data is selected to represent the temporal trend of the concentration of the water quality variables. Then, the parameters of this model will be regionalized using Kriging method and by fitting experimental variograms of the regional parameters to the theoretical variograms. (4) Finally, a set of potential points for the monitoring stations is proposed considering the spatial variations of Kriging estimates variances for the regionalized parameters. The locations of the main water withdrawals, the main point loads, the development projects, the length of their water quality and quantity data, the existing stations, and the main tributaries are determined. The potential monitoring stations are ranked using analytic hierarchy process.
5	Multivariate Statistical Techniques (Ouyang, 2005; Musthafa et al., 2014; Varekar et al., 2016)	The FA/PCA can be used to select the significant (principal) sampling sites out of the existing sampling location (S_n) of a hypothetical river network shown in Figure 2.16. Mathematically, FA/PCA involves five steps. (1) First, all variables $X_1, X_2, X_3, \ldots, X_n$, are coded with zero means and unit variance, i.e., measurements are normalized so that they all have equal weights during analysis. (2) The covariance matrix C is then calculated. (3) The eigenvalues $\lambda_1, \lambda_2, \lambda_3, \ldots, \lambda_n$ and corresponding eigenvectors $V_1, V_2, V_3, \ldots, V_n$ are found. (4). Components that account for only a small proportion of the variation in the data set are discarded. (5) Lastly, the factor loading matrix is developed and varimax rotation is performed on it to identify the principal stations. The factor correlation coefficients exceeding 0.9 is considered significant. Such conservative criteria are selected because the river system is highly nonlinear and dynamic. Sampling locations that did not have any factors with correlation coefficients >0.9 are considered as nonprincipal sampling sites.

Memarzadeh et al. (2013) made an innovative attempt by proposing the dynamic factor analysis followed by entropy-based approach for evaluation of sampling locations in an existing river WQMN of Karoon River, in Iran. In this approach, the dynamic factor analysis is used to extract the independent dynamic factors from time series of water quality variables. The theory of statistical entropy is applied to these dynamic factors to build up the transinformation–distance (*T–D*) curves. Then, the entropy theory is applied to the independent dynamic factors to construct *T–D* curves. The dynamic factor analysis requires significantly less computation time when the raw data is used, as the number of independent dynamic factors is usually less than that of monitored water quality variables.

Tanos et al. (2015) proposed combined cluster and discriminant analysis (CCDA) for rationalization of the WQMN of the River Tisza in Central Europe. In this approach, water quality monitoring data at 14 sampling sites with 15 water quality parameters for the time period of 1975–2005 was used for analysis. The four hydrochemical seasons, i.e., seasons distinguished by seasonal hydrochemical characteristics of the water body, were determined and characterized with unequal lengths of 2, 4, 2, and 4 months, starting with spring. Homogeneous groups of sampling sites were determined in space for every hydrochemical season, with the main separating factors being the man-made obstacles and tributaries. Similarly, an overall homogeneity pattern was determined. The final outcome resulted in the formulation of 11 homogeneous groups by 14 sampling sites based on the possibility of reducing the number of sampling locations, which makes the system cost-effective.

2.3.2.3.6 Application of Statistical Techniques to the Hypothetical River Network

Let us consider the hypothetical river network as shown in Figure 2.16. The sampling locations S_1, S_2, S_3, S_4, S_5, S_6, S_7, ..., S_n have been monitored for physical, chemical, and biological water quality variables. Table 2.4 summarizes application of various statistical techniques for rationalization of water quality monitoring locations of a hypothetical river network (Figure 2.16).

2.3.2.4 Optimization Techniques

2.3.2.4.1 Genetic Algorithm

A methodology based on a genetic algorithm was proposed for the design of WQMN by Park et al. (2006). The planning objectives were identified, and subsequently, the corresponding fitness functions were defined. The genetic algorithm was then connected with a geographic information system (GIS) database to identify the optimal sampling locations. The planning objectives, namely representativeness of a river basin, compliance with water quality standards, surveillance of pollution sources, supervision of water use, examination of water quality changes, estimation of pollution loads, and physical accessibility of sampling points, were the decision variables for selecting the optimal locations of sampling. The fitness functions were defined based on these objectives. The genetic algorithm is a nonlinear constrained optimization technique, which is heuristic in nature.

2.3.2.4.2 Linear Programming

Telci et al. (2009) have proposed the linear programming model entitled "optimal water quality monitoring network for river systems." This optimization model considered two decision variables, the first was to minimize the average detection time of contamination event and the second was to maximize reliability of the monitoring system. Maximum performance of the monitoring system was achieved when the number of detections for potential scenarios increased and the time between the start of a contamination event and its detection was decreased. The model is demonstrated for the Altamaha river system in the State of Georgia, which shows that it can be effectively used for rationalization of water quality monitoring locations. Further, Kao et al. (2012) have proposed two new linear optimization models to minimize the deviation of the cost values expected to identify the possible pollution sources based on uniform cost (UC) and coverage elimination UC (CEUC) schemes. The proposed models are applied to the Derchi reservoir catchment in Taiwan. The author reported that the global optimal WQMN can be effectively determined by using the UC or CEUC scheme, for which both results are better than those from simulated annealing method, especially when the number of stations becomes high.

2.3.2.4.3 Dynamic Programming Approach (DPA)

DPA is an optimization technique, which is used to transform complex problems into a series of simple problems. The simple problems are then solved and their solutions are combined to reach the overall solution of the main problem. In this approach, the decision variables are the total number of stations retained in a basin, number of primary basins, number of stations retained in a primary basin, and uniformized total attribute value of the station combination in primary basin. The objective function is to maximize the total number of stations retained in a basin (Harmancioglu et al., 1999). Cetinkaya and Harmancioglu (2012) have used the DPA and Sanders et al. (1983) modification of Sharp's approach for rationalization of WQMN in the Gediz River basin along the Aegean coast of Turkey. The Sanders et al. (1983) modification of Sharp's approach was used for selection of sub-basins, and dynamic programming was used to select the optimum number of sampling stations. Asadollahfardi et al. (2014) have used the DPA technique for assessment of the existing water quality sampling stations of the Sefid-Rud River, located in the north of Iran. The results obtained show that, out of 21, eight sampling stations on upstream and seven sampling stations on downstream of the basins should be retained.

2.3.2.5 Multiple Criteria Analysis

Chang and Lin (2014a) have proposed the multiple criteria analysis, to check the suitability of the water quality monitoring design in Taipei Water Resource Domain in northern Taiwan. Seven criteria viz. percentage of farmland area, amount of diffuse source pollution, green cover ratio, percentage of built-up area, landslide area ratio, density of water quality monitoring stations, and overutilization on hillsides were selected for the multiple criteria analysis. These criteria were weighted and normalized. The weighted method was applied to score the sub-basins. The density of water quality monitoring stations depends on the score of sub-basins. The prioritization of water quality monitoring stations of higher density was carried out by the fuzzy sets theory.

2.3.2.5.1 VIKOR Method

Chang and Lin (2014b) have applied the "VIKOR" method, a multiple criteria analysis approach, to evaluate the design of the WQMN in the Taipei Water Resource Domain in Northern Taiwan. For multiple criteria analysis, five criteria were selected: over-utilization area ratio of hillsides, green cover ratio, landslide area ratio, density of water quality monitoring stations, and diffuse source pollution.

2.3.2.6 Fuzzy-Based Approach with Hydrological Simulation Model

Strobl et al. (2006a,b) introduced a methodology for identifying the critical sampling points within a watershed using an approach that integrated a hydrological simulation model and concepts of fuzzy logic in a GIS platform. In this approach, hydrological factors and factors pertaining to soils, vegetation, topography, and land use are taken into consideration in the selection of critical sampling locations.

2.3.2.7 Matter Element Analysis

Wang et al. (2015) proposed a matter element analysis for the rationalization of river WQMN. In this approach, the matter element analysis and gravity distance were applied to optimize the monitoring sections. This methodology was applied for the rationalization of monitoring network in the Taizihe River, northeast China. The rationalized monitoring sections were 13 out of 17; hence, this optimization model could be effectively used for the optimal design of monitoring networks in the river systems.

2.3.2.8 Optimization Approach Based on Shannon Entropy

Lee et al. (2015) proposed a new optimization approach, which is significantly simplified via an analogy with the formulation of Shannon entropy. The algorithm was formulated for efficient determination of optimal location of sampling sites. It is also possible to use the proposed algorithm in conjunction with a heuristic optimization algorithm such as a genetic algorithm. The proposed algorithm filters only competitive candidates and reduces the problem size significantly. The performance of the proposed algorithm was demonstrated by its application for the Logan and Albert River networks, which were rationalized in earlier studies by Dixon et al. (1999) and Ouyang et al. (2008), respectively. The computation time required for the present algorithm is short, as compared to that of the earlier rationalization techniques applied.

2.3.2.9 Hybrid Techniques

2.3.2.9.1 Reference Point Approach (RPA)

An easily applicable and flexible rationalization methodology, aiming to the "performance" assessment of existing network was proposed by Cetinkaya and Harmancioglu (2014). This approach is termed as the RPA and used for stream flow monitoring network assessment. The reference point approach is derived from reference point and goal programming interactive multicriteria decision-making and optimization procedures. Both of these approaches search optimal decision alternatives in a vector space of objectives and criteria, which should satisfy decision makers' goals. Hybrid technique stands for a rationalization approach combining more than one rationalization technique. RPA method is incorporating both reference point, i.e., multicriteria decision-making and optimization techniques, hence it is a hybrid technique. It seems to be an effective technique possible to use for rationalization of WQMN. This methodology was demonstrated for the Gediz Basin, along the Aegean coast of Turkey. This approach permits ranking of the hydrometric stations with respect to the desired attributes of the basin network and the multiple objectives of monitoring. It facilitates the involvement of the decision-makers in methodology application for a more interactive assessment procedure between the network designer and monitoring agency. Also, it assists decision-making in cases with limited data and meta-data. Hence, the proposed method is simple and easy to apply in contrast to more complicated computational techniques.

2.3.2.9.2 Mixed-Effect State Space Model

The mixed-effect state space model was proposed by Costa and Monteiro (2016) for rationalization of the water quality monitoring sites of the Vouga River, in Portugal. The dissolved oxygen concentration dataset was used for analysis. In this approach, trend and seasonal components are extracted in a linear mixed-effects state space model. Both maximum likelihood method and distribution-free estimators are used for parameter estimation. The application of Kalman smoother algorithm allows obtaining the predictions of the structural components as seasonality and trend. The rationalization of water quality monitoring sites was done through the structural components by a hierarchical agglomerative clustering procedure. This procedure identified different homogenous groups relative to the seasonality and trend components. The characteristics of the hydrological basin are presented in order to support the results.

2.3.3 Rationalization of Sampling Frequency

The selection of sampling frequency is also one of the governing factors in rationalization of WQMNs. It affects the cost of monitoring program and depends on: the objectives of monitoring, relative importance of sampling locations, and variability of water quality at each sampling site (Canter, 1985). Once the representativeness of water samples has been established in space through the placement of optimum sampling sites, then the sampling frequency must be defined to achieve the sample representativeness in time. The frequency should not be too less such that important information is missed out, nor should it be too high such that redundant information is collated. Sampling frequency is such that the samples are representative of time. The frequency depends upon the objectives of sampling and the available resources (Strobl and Robillard, 2008).

The common practice is to conduct a monthly (infrequent) sampling in fixed station monitoring networks. Additionally, in most of the cases, the sampling frequency is equally divided among the existing sampling sites and decided on the basis of the available capacity of an analytical laboratory (World Meteorological Organization, 1994; Strobl and Robillard, 2008). Sampling frequency should be rationalized in such a way that the water quality information can be gathered with a minimum of sampling efforts. Although automatic monitoring has been applied in some network designs, it should be noted that a large portion of the costs for operating a monitoring network as well as for data management and QA/QC (including metadata) is directly related to the frequency of sampling. Limited financial resources, most likely, will not allow the same sampling frequency of all parameters at every sampling site.

Increasing the number of samples collected can reduce the standard error of the mean value. As the standard error of the mean is inversely proportional to the square root of the number of samples collected, an increase in the number of observations brings a small change in results. "The mean value is the most reported statistical parameter. The selected sampling frequency gives an estimate of the mean within a given confidence limit. The confidence limit of the mean quantify the choice of sampling frequency by relating sampling frequency to the water quality variation" (Strobl and Robillard, 2008).

In India, the CPCB guidelines are followed by government agencies and stack holders to design sampling frequency. The sampling frequency is governed by the level of variation in water quality for a given source. If variations are large in a short duration of time, a larger frequency is required to cover such variations. On the other hand, if there is no significant variation in water quality, frequent collection of sample is not required (CPCB, 2007). Researchers have proposed various approaches for rationalization of sampling frequency (Sanders and Adrian, 1978; Loftis and Ward, 1980; Sanders et al., 1983; Harmancioglu et al., 1999; Ozkul et al., 2000).

Sanders and Adrian (1978) proposed a sampling frequency criterion for river WQMN. The criterion was based on the assumption that the primary objectives of the future river quality monitoring networks are the determination of ambient water quality conditions and an assessment of yearly trends rather than detection of violations of stream or effluent standards. Sampling frequency is derived as a function of the random variability of river flow. The criterion was satisfactorily applied for the Massachusetts portion of the Connecticut River in the United States.

Loftis and Ward (1980) formulated a dynamic programming code for the purpose of assigning sampling frequencies throughout a regulatory WQMN, in order to rationalize the statistical performance of the network while operating within a fixed budgetary constraint. The proposed dynamic code was used to assign sampling frequencies to nine stations in Illinois from which historical data have been obtained and analyzed. The optimal design was produced, showing the effective use of dynamic programming code.

The concept of statistical entropy has been successfully demonstrated, for rationalization of sampling sites by Harmancioglu et al. (1999) and Ozkul et al. (2000). In this analysis, the assessment of reduction in marginal entropy (as discussed in Section 2.3.2.3) due to the presence of dependence within the data series is used for rationalization of sampling frequencies. This reduction, if any, is equivalent to the redundant information of measurements. The analysis has been carried out for a given site and parameter. Hence, the result obtained indicates the decrease or increase in sampling frequency of a particular variable. Further, it aids decision-making of either termination or decrease in sampling frequency for a given water quality parameter at a particular sampling site if the information produced is redundant in a time scale.

2.3.4 Rationalization of Water Quality Parameters

The selection of water quality monitoring locations and frequency are influenced by water quality parameters being monitored. Hence, the selection of specific water quality parameters of concern is a vital decision-making task for rationalization/design and operation of WQMN. The selection of water quality variables mainly depends upon the monitoring objectives. It is required that the sampling parameters are selected very carefully and effectively because some parameters are extremely expensive to monitor, either in terms of specialized sampling/preservation techniques or analytical laboratory cost. Hence, during rationalization, attempts have been made to reduce the sampling parameters without substantial information loss. It may be possible to save time and effort by analyzing fewer parameters, and based on observed values, one may analyze and establish correlations among the rest of the water quality parameters (Sanders et al., 1983; Canter, 1985; Steele, 1987; Strobl and Robillard, 2008).

The water quality parameters are mainly classified based on physical, chemical, and biological properties of water. The physical parameters include discharge, temperature, specific conductance, turbidity, suspended sediment concentration, particle size distribution, odor, and taste. Chemical water quality parameters are classified in terms of organic and inorganic characteristics, while the biological parameters are macroinvertebrates, fish, and plants. In some cases, chemical, physical, and biological characteristics must be observed, which are determined on the basis of various water usages like drinking, domestic, industrial, and agricultural (World Health Organization—United Nations Educational, Scientific and Cultural Organization, 1978).

In an Indian scenario, a combination of general parameters, nutrients, oxygen consuming substances, and major ions should be analyzed at all stations on a routine basis, i.e., once in a month throughout the year. The parameters like micropollutants, pesticides, or other site-specific variables may be monitored at lower frequency (seasonally, bi-annually, quarterly) based on the industrial and anticipated activities, i.e., bathing, washing of clothes, and possible ungauged discharges (CPCB, 2007).

Even though the classification of water quality parameters is useful and convenient, the interplay between the variables needs to be considered. This is so, because it has been noted that in order to increase the economical efficiency of water quality monitoring program, assessment and detection of these interactions facilitate identification of variables that might potentially serve as indicators of a wider category of variables (Strobl and Robillard, 2008). The monitoring for all water quality variables is practically not viable. Hence, rationalization of sampling parameters is strongly recommended and various rationalization techniques have been proposed by researchers (Sanders et al., 1983; Harmancioglu and Alpaslan, 1994; Ouyang, 2005; Khalil et al., 2010; Noori et al., 2010; Zhang et al., 2011; Musthafa et al., 2012; Varol et al., 2012; Wang et al., 2012; Musthafa et al., 2014; Toochukwu, 2015).

Sanders et al. (1983) have proposed several techniques, viz. statistical regression, frequency distribution analysis, multivariate statistical techniques [Factor analysis (FA), PCA, CA, Canonical discriminant analysis(CDA)], time-trend analysis, regional pattern and distribution, and water quality indices for rationalization of water quality variables. These statistical techniques have been used for transforming data into information. Sanders et al. (1983) reported that there is a great need to understand the interactions between water quality variables for effectively solving the water quality-related problems.

Harmancioglu and Alpaslan (1994) reported the use of an objective-based criterion in the selection of water quality variables. Two basic objectives are defined for a monitoring network, i.e., water use and impact assessment for evaluating point and nonpoint sources. Consideration of water use purposes results in a list of gross variables to be monitored at every station in the network. Further, selection criteria such as cost and ease of measurements, statistical properties, and common occurrence in more than one water use objectives are proposed to reduce the long list of gross variables. The reduction in the number of variables to be sampled may be attained by analyzing information transfer between the already selected variables. With respect to the second objective, i.e., impact assessment, a group of specific variables may be defined to assess point and/or nonpoint pollution by domestic, industrial, and

TABLE 2.5 Important Water Quality Variables Used for Rationalization of WQMN

Sr. No.	Rationalization Approach	Researcher(s)	Rationalization of Water Quality Sampling Location/Frequency/Parameter	Water Quality Variables Used for Rationalization of WQMN
1	Methods based on Topology	Sanders et al. (1983) and Do et al. (2011)	Sampling location	Biochemical oxygen demand (BOD), nitrate and phosphate
2	Modified Sanders approach	Varekar et al. (2015, 2016)	Sampling location	BOD, nitrate and phosphate
3	Statistical techniques	Harmancioglu and Alphasan (1992, 1994), Harmancioglu et al. (1999), Ozkul et al. (2000), Ouyang (2005), Karamouz et al. (2009a,b), Alameddine et al. (2013), Lobuglio et al. (2007), Musthafa et al. (2014), and Varekar et al. (2016)	Sampling location, sampling frequency, and sampling parameter	Total suspended solids (TSS), phosphate, NO_3-N, temperature, turbidity, dissolved oxygen (DO), pH, total dissolved solids (TDS), electrical conductivity (EC), chloride, sulfate, nitrate, chemical oxygen demand (COD), total petroleum hydrocarbons, heavy metals: Fe, As, Hg, Zn, Pb, Cd, Cr, Cu, Mn, Fecal coliform, total coliform, chlorophyll a, transparency, color, alkalinity, salinity, total nonfilterable residue, total organic carbon, calcium, magnesium, pheophytin a, Ca and Mg hardness, sodium
4	Optimization techniques	Park et al. (2006), Telci et al. (2009), Cetinkaya and Harmancioglu (2012), and Asadollahfardi et al. (2014)	Sampling location and sampling frequency	Total phosphorus (TP), NH_3-N, BOD_5, COD, NO_3, NH_3, PO_4, EC, TDS, TSS, pH, temperature, DO, Fe, As, Pb, Cd, Cr, Cu, Mn, Zn
5	Other approaches	Strobl et al. (2006a,b), Chang and Lin (2014a,b), Cetinkaya and Harmancioglu (2014), Lee et al. (2015), and Wang et al. (2015)	Sampling location, sampling frequency, and sampling parameter	Suspended solids, TP, pH, DO, BOD, COD, NH_3-N, Cu, Zn, Cr, Hg, Pb, Cd

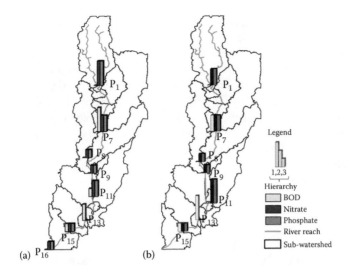

FIGURE 2.17 Effect of seasonal variation on design of sampling locations: (a) Monsoon season and (b) nonmonsoon season.

agricultural effluents. As such, the list of specific variables is often shorter than that of gross variables. The most significant feature of this group is that they include station-based variables, which do not need to be observed at every station in the network.

Khalil et al. (2010) developed a statistical approach for assessment and selection of the optimal combination of water quality variables. The purposed approach overcame the deficiencies in the conventional correlation-regression method. In this approach, record-augmentation procedures were integrated with correlation and CA to identify highly associated water quality variables. This approach was applied for rationalization of the water quality variables in the Nile delta surface WQMN in Egypt. Results indicated that the proposed approach represents a useful decision support tool for the optimized selection of water quality variables.

The multivariate statistical techniques (FA/PCA), as discussed in Section 2.3.2.1.3, were also used for rationalization of water quality variables. In this analysis, water quality sampling stations were the cases (observations), while water quality parameters were the variables. The PCA was followed by FA and the water quality parameters with an absolute factor correlation coefficient value greater than designed thresholds are considered to be the most decisive and important parameters in contribution to variations of water quality (Sanders et al., 1983; Ouyang, 2005; Noori et al., 2010; Musthafa et al., 2012; Varol et al., 2012; Wang et al., 2012; Musthafa et al., 2014; Varekar et al., 2016; Toochukwu, 2015). Table 2.5 summarizes all the important water quality variables used in the literature.

2.3.5 Implementation of Rationalized WQMN and Seasonal Rationalization

Mishra and Coulibaly (2009) have presented an extensive review on rationalization techniques for various design components of hydrometric network. The reviewed approaches are also possible to apply for rationalization of WQMNs. In 2010, Mishra and Coulibaly carried out and implemented the rationalization of hydrometric network for Canadian watersheds using the theory of statistical entropy. Most of the time researchers have proposed many rationalization approaches, but very few were implemented. Hence, along with the investigations for various rationalization techniques there is a need to implement these techniques for solving real-world problems.

Water quality varies seasonally, especially in tropical countries like India where precipitation is in the form of rain. At these places, surface runoff is mainly active in the monsoon season and it affects surface water quality. In the monsoon season, surface runoff brings two major changes in pollution load entering into the surface water bodies: (1) dilution of point pollution loads and (2) generation of diffuse pollution loads. Hence, there is a need to study the effect seasonal variations of pollution load has on rationalization of WQMNs. Varekar et al. (2015, 2016) made the first successful attempt for seasonal rationalization of river water quality monitoring sites. In this analysis, a generalized framework was proposed for seasonal rationalization of sampling sites considering seasonal variation of point and diffused pollution loads. Modified Sanders approach was used for the rationalization of the sampling sites. The application of the framework was demonstrated for the Kali River basin in India. The results obtained are shown in Figure 2.17 (figure is self-explanatory). Further, there is a need to explore the effect of seasonal variations on rationalization of sampling frequency and variables.

2.3.6 Summary

In this chapter, various aspects of design and rationalization of WQMN were discussed to pave new research priorities. Applications of various rationalization methods based on topology, modified Sanders approach, multivariate statistics, and optimization and hybrid techniques, with their chronological development have been reported. Importance of selecting a suitable approach based on the quantity, quality, and availability of datasets has been pointed out. For example, if less data of water quality parameters, but sufficient information pertaining to watershed characteristics are available then the modified Sanders/Topology-based approaches might be more suitable. However, the statistical techniques might be the best choice for rationalization when sufficiently large and consistent water quality data are available. Also, very few efforts have been reported for rationalization of sampling frequency and parameters as compared to that of sampling sites. Hence, there is a need to explore performance of existing rationalization techniques for optimal selection of sampling frequency and parameters.

Most of the past attempts focused on pollution loads from point sources while designing/rationalizing monitoring network, rather giving importance to diffuse pollution loads. However, in tropical countries pollution loads from nonpoint/diffuse sources are more due to high precipitation and agricultural activities. Hence, there is a need to consider both point and nonpoint sources of pollution, for effective rationalization of WQMNs.

Seasonal variation of water quality, particularly monsoonal effect in the tropics may cause a significant change in pollution loading, which will consequentially affect the design of WQMN. Therefore, the seasonal variations of water quality need to be accounted for premonsoon and postmonsoon evaluation and rationalization of WQMN. Impacts of seasonal variation and anthropogenic activities may continuously affect the water quality particularly for surface water bodies, which necessitate a periodic assessment and redesign of WQMN.

Both qualitative and quantitative characteristics of water are equally responsible for water pollution and must be considered in any hydrometric analysis. Most of the monitoring programs are primarily focused on water quality characteristics, while the water flow/quantity is ignored, particularly in statistical rationalization approaches. The flow of water governs the fate of pollutants and controls water quality characteristics. Therefore, both water quality and quantity should be considered for accurate estimation of pollution loads for effective and efficient rationalization of monitoring networks.

The advanced geospatial techniques viz. GIS and remote sensing can be implemented for effective rationalization of monitoring network. The GIS tools have been used for various aspects of rationalization including delineation of watershed, identification of potential locations for point and nonpoint sources, and precise representation of sampling sites. The remote sensing products can be used for delineation of land use maps, which quantifies the anthropogenic activities in terms of change in land use applications. The advancement in remote sensing techniques, i.e., use of thermal and optical sensors on satellites can provide spatiotemporal information of water quality parameters. The recent launches of satellites with advanced spatial and spectral resolution sensors are facilitating the use of remote sensing techniques for water quality assessment. The decision makers can use a geo-referenced database for strategic planning and management of water bodies. Therefore, exploration of GIS and remote sensing application will definitely facilitate rationalization of monitoring the network.

Research studies on water quality monitoring involve multidisciplinary science, which needs concepts of hydrology, geology, morphology, topology, statistics, optimization, hydrometeorology, and geospatial technology. The existing techniques of rationalization are rigorous and mathematically complicated. In realistic situations, mostly the watershed managers and other stakeholders may not have knowledge of all these interdisciplinary branches and will be unable to implement rationalization algorithms in the field. Thus, there is a need for development of software packages, which will facilitate efficient rationalization and operations of water quality monitoring program.

There is no universally accepted unique approach for rationalization of monitoring network; however, it is a prime responsibility of central and state-level pollution control agencies to conduct rationalization of network for a sustainable water quality management; considering (1) data availability issues of both quantity and quality, (2) seasonal variability and periodical assessment with redesign, (3) impacts of both point and diffuse pollution loadings, and (4) application of GIS and remote sensing tools.

References

Alameddine, I., Karmakar, S., Qian, S. S., Paerl, H. W. and Reckhow, K. H. 2013. Optimizing an estuarine water quality monitoring program through an entropy-based hierarchical spatiotemporal Bayesian framework. *Water Resour Res* 49: 1–13.

Asadollahfardi, G., Asadi, M., Nasrinasrabadi, M. and Faraji, A. 2014. Dynamic programming approach (DPA) for assessment of water quality network: A case study of the Sefīd-Rūd River. *Water Pract Technol* 9(2): 135–149.

Bartram, J. and Balance, R. 1996. *Water Quality Monitoring: A Practical Guide to the Design and Implementation of Freshwater Quality Studies and Monitoring Programmes*. Chapman & Hall, London.

Canadian Council of Ministers of the Environment, Water Quality Task Group, Canada (CCME). 2006. A Canada-wide framework for water quality monitoring. http://www.ccme.ca/assets/pdf/wqm_framework_1.0_e_web.pdf (Accessed May 27, 2015).

Canter, L. W. 1985. *River Water Quality Monitoring*. Lewis Publishers, Chelsea, MI.

Central Pollution Control Board, India (CPCB). 2007. Guidelines for water quality monitoring. http://mpcb.gov.in/envtdata/GuidelinesforWQMonitoring%5B1%5D.pdf (Assessed December 17, 2014).

Cetinkaya, C. P. and Harmancioglu, N. B. 2012. Assessment of water quality sampling sites by a dynamic programming approach. *J Hydrol Eng* 17(2): 305–317.

Cetinkaya, C. P. and Harmancioglu, N. B. 2014. Reduction of streamflow monitoring networks by a reference point approach. *J Hydrol* 512: 263–273.

Chang, C. and Lin, Y. 2014a. A water quality monitoring network design using fuzzy theory and multiple criteria analysis. *Environ Monit Assess* 186: 6459–6469.

Chang, C. and Lin, Y. 2014b. Using the VIKOR method to evaluate the design of a water quality monitoring network in a watershed. *Int J Environ Sci Technol* 11: 303–310.

Chapman, D. (Ed.) 1996. *Water Quality Assessments. A Guide to the Use of Biota, Sediments and Water in Environmental Monitoring*. Chapman & Hall, London.

Costa, M. and Monteiro, M. 2016. Discrimination of water quality monitoring sites in River Vouga using a mixed-effect state space model. *Stoch Environ Res Risk Assess* 30: 607–619.

Cressie, N. and Christopher, K. W. 2011. *Statistics for Spatio-Temporal Data*. John Wiley & Sons, Hoboken, NJ.

Department of Water, Government of Western Australia. 2009. Water quality monitoring program design: A guideline for field sampling for surface water quality monitoring programs. http://www.water.wa.gov.au/PublicationStore/first/87153.pdf (Accessed May 27, 2015).

Dixon, W. and Chiswell, B. 1996. Review of aquatic monitoring program design. *Water Res* 30(9): 1935–1948.

Dixon, W., Smith, G. K. and Chiswell, B. 1999. Optimized selection of river sampling sites. *Water Res* 33(4): 971–978.

Do, H. T., Lo, S., Chiueh, P. and Thi, L. 2012. Design of sampling locations for mountainous river monitoring. *Environ Model Softw*, 27(28): 62–70.

Do, H. T., Lo, S. L., Chiueh, P. T., Thi, L. A. and Shang, W. T. 2011. Optimal design of river nutrient monitoring points based on an export coefficient model. *J Hydrol* 406: 129–135.

European Union Member States, Norway and the European Commission. 2003. Guidance on monitoring for the water framework directive. http://www.nve.no/PageFiles/4661/Monitoring_Guidance.pdf?epslanguage=no (Accessed May 27, 2015).

Harmancioglu, N. B. and Alphasan, N. 1992. Water quality monitoring network design: A problem of multi-objective decision making. *Water Resour Bull* 28(1): 179–192.

Harmancioglu, N. B. and Alphasan, N. 1994. Basic approach in design of water quality monitoring networks. *Water Sci Technol* 30(10): 49–56.

Harmancioglu, N. B., Fistikoglu, O., Ozkul, S. D., Singh, V. P. and Alpasan, M. N. 1999. *Water Quality Monitoring Network Design*. Kluwer Academic Publishers, London.

Kao, J., Li, P. and Hu, W. 2012. Optimization models for siting water quality monitoring stations in a catchment. *Environ Monit Assess* 184: 43–52.

Karamouz, M., Nokhandan, A. K., Kerachian, R. and Maksimovic, C. 2009a. Design of on-line water quality monitoring system using the entropy theory: A case study. *Environ Monit Assess* 155: 63–85.

Karamouz, M., Nokhandan, A. K., Kerachian, R. and Maksimovic, C. 2009b. Design of river water quality monitoring networks: A case study. *Environ Monit Assess* 14: 705–714.

Khalil, B., Ouarda, T. B. M. J., St-Hilaire, A. and Chebana, F. 2010. A statistical approach for the rationalization of water quality indicators in surface water quality monitoring networks. *J Hydrol* 386: 173–185.

Knuth, D. E. 1968. *The Art of Computer Programming, Vol. 1, Fundamental Algorithms*, 1st edition, Addison-Wesley, San Francisco, CA.

Lee, C., Paik, K., Yoo, D. G. and Kim, J. H. 2015. Efficient method for optimal placing of water quality monitoring stations for an ungauged basin. *J Environ Manag* 132: 24–31.

Lo, S. L., Kuo, J. T. and Wang, S. M. 1996. Water quality monitoring network design of Keelung River, Northern Taiwan. *Water Sci Technol* 34 (12): 49–57.

Lobuglio, J. N., Characklis, G. W. and Serre, M. L. 2007. Cost-effective water quality assessment through the integration of monitoring data and modeling results. *Water Resour Res* 43: 1–16.

Loftis, J. C. and Ward, R. C. (1980) Cost-effective selection of sampling frequencies for regulatory water quality monitoring. *Environ Int* 3: 297–302.

Musthafa, M. O., Karmakar, S. and Harikumar, P. S. 2014. Assessment and rationalization of water quality monitoring network: A multivariate statistical approach to the Kabbini River (India). *Environ Sci Pollut Res Int* 21: 10045–10066.

Memarzadeh, M., Mahjouri, N. and Kerachian, R. 2013. Evaluating sampling locations in river water quality monitoring networks: Application of dynamic factor analysis and discrete entropy theory. *Environ Earth Sci* 70: 2577–2585.

Ministry of Environment, Forest and Climate Change, Government of India (MoEFCC). 2005. Uniform protocol on water quality monitoring order, 2005. http://envfor.nic.in/legis/2151e.pdf (Accessed May 27, 2015).

Mishra, K. and Coulibaly, P. 2009. Developments in hydrometric network design: A review. *Rev Geophys* 47: 1–24.

Mishra, K. and Coulibaly, P. 2010. Hydrometric network evaluation for Canadian watersheds. *J Hydrol* 380: 420–437.

Musthafa, O. M., Karmakar, S. and Harikumar, P. S. 2012. Evaluation of river water quality monitoring network: A multivariate statistical approach to Kabbini River catchment. In: *Proceedings of IWA World Water Congress and Exhibition 2012*, International Water Association, Busan, Korea, September 16–21.

National Institute of Hydrology (NIH), Jal Vigyan Bhawan, Roorkee, India. 1996. Identification of water quality monitoring sites on the Narmada river basin. http://www.indiawaterportal.org/sites/indiawaterportal.org/files/Identification%20of%20sampling%20sites%20for%20water%20quality%20monitoring%97.pdf (Accessed May 27, 2015).

National Institute of Hydrology (NIH), Jal Vigyan Bhawan, Roorkee, India. 1997. Identification of water quality monitoring sites on the Kshipra River. http://www.indiawaterportal.org/sites.indiawaterportal.org/files/Salinity_modelling_of_groundwater_in_Saharanpur_and_Hardwar_district_NIH_98.pdf (Accessed May 27, 2015).

Noori, R., Sabahi, M. S., Karbassi, A. R., Baghvand, A. and Zadeh, H. T. 2010. Multivariate statistical analysis of surface water quality based on correlations and variations in the data set. *Desalination* 260: 129–136.

Ouyang, Y. 2005. Evaluation of river water quality monitoring stations by principal component analysis. *Water Res* 39: 2621–2635.

Ouyang, H. T., Yu, H., Lu, C. H. and Lou, Y. H. 2008. Design optimization of river sampling network using genetic algorithms. *J Water Resour Plann Manag* 134(1): 83–87.

Ozkul, S., Harmancioglu, N. B. and Singh, V. P. 2000. Entropy-based assessment of water quality monitoring networks. *J Hydrol Eng* 5(1): 90–100.

Park, S. Y., Choi, J. H., Wang, S. and Park, S. S. 2006. Design of a water quality monitoring network in a large river system using the genetic algorithm. *Ecol Model* 99: 289–297.

Sanders, T. G. and Adrian, D. D. 1978. Sampling frequency for river quality monitoring. *Water Resour Res* 14(4): 569–576.

Sanders, T. G., Ward, R. C., Loftis, J. C., Steele, T. D., Adrian, D. D. and Yevjevich, V. 1983. *Design of Networks for Monitoring Water Quality*. Water Resources Publications, Highlands Ranch, CO.

Shannon, C. E. and Weaver, W. 1949. *The Mathematical Theory of Communication*. University of Illinois Press, Urbana, IL.

Sharp, W. E. 1971. A topologically optimum water-sampling plan for rivers and streams. *Water Resour Res* 7(6): 1641–1646.

Singh, K. P., Malik, A., Mohana, D. and Sinha, S. 2004. Multivariate statistical techniques for the evaluation of spatial and temporal variations in water quality of Gomti river (India): A case study. *Water Res* 38: 3980–3992.

Steele, T. D. 1987. Water quality monitoring strategies. *Hydrol Sci J* 32: 207–213.

Strobl, R. O. and Robillard, P. D. 2008. Network design for water quality monitoring of surface freshwaters: A review. *J Environ Manag* 87: 639–648.

Strobl, R. O., Robillard, P. D., Day, R. L., Shanon, R. D. and McDonnell, A. J. 2006a. A water quality monitoring network design methodology for the selection of critical sampling points: Part II. *Environ Monit Assess* 122: 319–334.

Strobl, R. O., Robillard, P. D., Day, R. L., Shanon, R. D. and McDonnell, A. J. 2006b. A water quality monitoring network design methodology for the selection of critical sampling points: Part I. *Environ Monit Assess* 122: 137–158.

Tanos, P., Kovács, J., Kovács, S., Anda, A. and Hatvani, I. G. 2015. Optimization of the monitoring network on the River Tisza (Central Europe, Hungary) using combined cluster and discriminant analysis, taking seasonality into account. *Environ Monit Assess* 187: 575.

Telci, I. T., Nam, K., Guan, J. and Aral, M. M. 2009. Optimal water quality monitoring network design for river systems. *J Environ Manag* 90: 2987–2998.

Tirsch, F. S. and Male, J. W. 1984. River basin water quality monitoring network design: Options for reaching water quality goals. In: Schad, T. M. (Ed.), *Proceedings of the Twentieth Annual Conference of American Water Resources Associations*. AWRA Publications, Middleburg, VA.

Toochukwu, C. O. 2015. Use of multivariate statistical techniques for the evaluation of temporal and spatial variations in water quality of the Kaduna River, Nigeria. *Environ Monit Assess* 187:137, 1–17.

United States Environmental Protection Agency (USEPA). 2003. Elements of a state water monitoring and assessment program. http://www.epa.gov/owow/monitoring/elements/elements03_14_03.pdf (Accessed May 27, 2015).

United States Geological Survey (USGS). 2011. Water-quality sampling by the U.S. Geological Survey: Standard protocols and procedures. http://pubs.usgs.gov/fs/2010/3121/fs2010–3121.pdf (Accessed May 27, 2015).

Varekar, V., Karmakar, S. and Jha, R. 2012. Seasonal evaluation and redesign of surface water quality monitoring network: An application to Kali River basin, India, AOGS-AGU joint assembly 2012, Resorts World Convention Centre, Singapore, August 13–17, 2012.

Varekar, V., Karmakar, S. and Jha, R. 2016. Seasonal rationalization of river water quality sampling locations: A comparative study of the modified Sanders and multivariate statistical approaches. *Environ Sci Pollut Res* 23: 2308–2328.

Varekar, V., Karmakar, S., Jha, R. and Ghosh, N. C. 2015. Design of sampling locations for river water quality monitoring considering seasonal variation of point and diffuse pollution loads. *Environ Monit Assess* 187: 376.

Varol, M., Gokot, B., Bekleyen, A. and Sen, B. 2012. Spatial and temporal variations in surface water quality of the dam reservoirs in the Tigris River basin, Turkey. *Catena* 92: 11–21.

Wang, X., Cai, Q., Lin, Y. and Qu, X. 2012. Evaluation of spatial and temporal variation in stream water quality by multivariate statistical techniques: A case study of the Xiangxi River basin, China. *Quaternary Int* 282: 137–144.

Wang, H., Liu, Z., Sun, L. and Luo, Q. 2015. Optimal design of river monitoring network in Taizihe river by matter element analysis. *PLoS One* 10(5): e0127535.

Ward, R. C., Loftis, J. C. and McBride, G. B. 1989. *Design of Water Quality Monitoring Systems*. Van Nostrand Reinhold, New York.

World Health Organization (WHO)—United Nations Educational, Scientific and Cultural Organization (UNESCO). (1978). Water quality surveys: A guide for the collection and interpretation of water quality data. https://hydrologie.org/BIB/Publ_UNESCO/SR_023_1978.pdf (Accessed May 17, 2017).

World Meteorological Organization (WMO). 1994. *Guide to Hydrological Practice*. WMO-Number 168, WMO, Geneva.

Zhang, X., Wang, Q., Liu, Y., Wu, J. and Yu, M. 2011. Application of multivariate statistical techniques in the assessment of water quality in the Southwest New Territories and Kowloon, Hong Kong. *Environ Monit Assess* 173: 17–27.

3

Wastewater Treatment

Ligy Philip and
Bhallamudi
S. Murty
*Indian Institute of
Technology Madras*

Channarong
Puchongkawarin,
Miao Guo, Nilay
Shah, David C.
Stuckey, and
Benoit Chachuat
Imperial College London

Yannic Vaupel
RWTH Aachen University

Sarojini Tiwari,
Chitta Ranjan
Behera, and Babji
Srinivasan
*Indian Institute of
Technology, Gandhinagar*

Chedly Tizaoui,
Olajumoke Ololade
Odejimi, and
Ayman Abdelaziz
Swansea University

3.1 Appropriate Interventions and Technologies for Providing Safe Drinking Water to Rural and Underprivileged Communities

Ligy Philip and Bhallamudi S. Murty

3.1.1 Introduction

Water is an essential commodity for human well-being. The World Health Organization (Howard and Bartram, 2003) has estimated that a person needs at least 7.5 L of water per day for drinking, food, and personal hygiene. A person requires 50 L of water per day to meet other needs. A poor water supply can affect health either directly or indirectly. Incidents of many water-connected diseases can be reduced noticeably by providing sufficient quantity of potable water (Fewtrell et al., 2005). Pathogens from human and animal excreta are transmitted through soil, surface and groundwater, and by hands, flies, and other vectors (Figure 3.1). Finally, humans get exposed to these pathogens either through consumption of contaminated water, food, or through unsanitary contact.

Figure 3.1 illustrates how various barriers such as (1) sanitation and source water protection, (2) hygiene and food safety, and (3) water treatment and safe distribution can be introduced to disrupt the transmission of pathogens, and thus prevent infection. Supply of an adequate quantity of good quality water is a necessary precondition for the maintenance of personal and domestic hygiene and food safety. Therefore, water quality has been receiving significant attention in the water, sanitation, and hygiene (WASH) field (Esrey et al., 1991). It has been well established that developing nations can significantly reduce expenditure in the health sector by providing a sufficient quantity of safe water. It has been estimated by Haller et al. (2007) that every US$1 spent on water supply and sanitation services can lead to an economic return of between $5 and $46, depending on the region. It has also been found that, among all interventions, access to piped water supply and sewage connections on plot had the

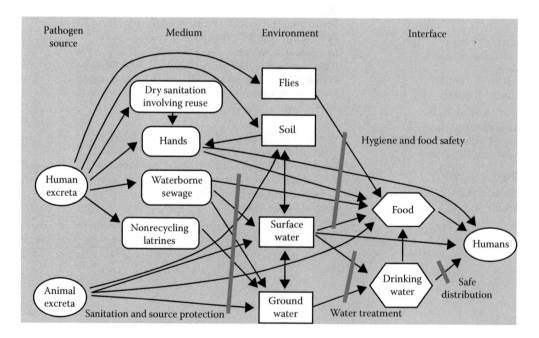

FIGURE 3.1 Pathways and barriers in the transmission of fecal-oral pathogens. (Modified from Prüss, A. et al., *Environ. Health Perspect.*, 110(5), 537–542, 2002.)

largest health impact in the study area. On the other hand, household water treatment was found to be the most cost-effective intervention. In the above context, it cannot be overemphasized that appropriate sanitation is absolutely essential for protection of the quality of ever dwindling drinking water sources. In India, rampant open defection and absence of appropriate measures for treating sullage and septage from burgeoning septic tanks have contaminated many of the surface water bodies. For example, leachate from poorly constructed and maintained septic tanks has resulted in contamination of more than 80% of water from wells in the state of Kerala, India. Therefore, an integrated approach of water quality, hygiene, and sanitation is essential to have a sustainable solution for providing safe drinking water to all.

As per the Millennium Development Goal (MDG) 7 (United Nations, 2007), the proportion of population without sustainable access to drinking water and basic sanitation should be halved by 2015. This MDG also affects the meeting of other MDGs related to poverty, education, gender equality, and child mortality. However, recent status reports indicate that although the world is mostly on track to meet the drinking water target, accelerated and targeted efforts are needed to meet the target in rural areas. There has been significant progress in many sectors in India in the last couple of decades. However, this is not so in case of infrastructure, particularly in relation to water supply, sewerage systems, and sanitation, due to the tremendous stress created by rapid population growth, industrialization, and urbanization. Although India has already achieved the MDG water target (United Nations, 2015), insufficient availability of water, inequity in its access, and sustainability of water sources are cause for concern. Of the 89% of urban population who have access to improved water supply, only 48% has piped on to premises water supply. Of more than 80% of rural population who have access to improved water supply, only 18% have access to piped water supply. Women and children spend a significant amount of their time and energy daily fetching potable water in many rural areas. There will also be a significant demand–supply gap in the near future. As per the estimates made by the Ministry of Water Resources, Government of India, water availability in India will reduce to $1341\,m^3$ per capita per year in the year 2025, while the anticipated demand will be 1093 billion cubic meters (MoWR, 2011). According to the 12th Five Year Plan report of the Government of India, groundwater accounts for nearly 80% of domestic water needs, but almost 60% of districts in India suffer from issues relating to groundwater quality or availability. Many of the drinking water sources are vulnerable because only a small percentage of domestic wastewater (20% in Class I cities and <5% in Class II cities) is treated, and wastewater management is virtually absent in rural areas.

The magnitude of the challenges are particularly disconcerting given that provision of safe drinking water and sanitation appears to be articulated and prioritized in the country at the national and subnational levels.

As India makes efforts to provide safe drinking water to more than a billion people, it should be recognized that the problems encountered in underprivileged communities in urban areas are quite unique and different from those in privileged communities. Recent studies have shown that percentage of income spent by poor people in urban areas on water is substantially more than the percentage of income spent by relatively well-to-do people (Amit and Subhash, 2015). On a similar note, the problems encountered in providing safe drinking water in rural areas are very different from those faced in urban areas. These differences need to be understood while making technology choices and interventions, so that our efforts to provide safe water to all the people do not go wasted.

3.1.2 Providing Safe Drinking Water in an Underprivileged Urban Community

Slums are very common in the urban landscape of the developing world. Economic deprivation and social marginalization significantly increase human health risks arising from unsafe water and environmental degradation in slums. The most serious health risk faced by people living in slums is infectious waterborne diseases (Graf et al., 2008). South Asia has the largest share of slum-dwellers in the world, with almost 56% of the urban population living in slums (UNFPA, 2007). This region accounts for an estimated 24% of total child mortality due to the high degree of urban poverty (Black et al., 2003).

Although development of alternative approaches for sanitation in low-income settings has received much attention in the recent past, it is not so in the case of provision of safe drinking water. Conventional centralized water treatment and distribution systems have failed to meet the expectations in many rapidly expanding cities and towns. Ineffective governance within urban local bodies is coming in the way of building and maintaining effective urban water infrastructure (Biswas and Tortajada, 2009). In the absence of piped water supply, people in slums usually get the water from four major sources: (1) groundwater, (2) local surface water, (3) vendors, and (4) illegal tapping from nearby municipality water main. Water from all these sources is usually unfit for consumption, and this in turn is resulting in persistence of diarrheal diseases.

The vulnerability in centralized water systems arises mostly from the distribution component. Therefore, alternative safe water systems for slums in urban areas could be based on distributing the treatment capacity to points where drinking water is consumed. Decentralized systems are increasingly being advocated for slum areas because (1) they are amenable to treating water drawn from different sources, (2) the capital costs could be substantially low, (3) they fill the safe water gap in the absence of piped water supply, and (4) they may be able to achieve immediate public health goals among underprivileged populations (Montgomery and Elimelech, 2007). Decentralized water systems can be implemented either at the household or at the community level. Both household and community systems are differently suited to different circumstances. Suitability also depends upon local water quality and its variability. Main advantages of a household system are as follows: (1) it does not require physical infrastructure and capital investment for the creation of a small-scale treatment facility, (2) operation and maintenance costs are nil, (3) risk of recontamination is lower, and (4) it does not require a significant level of community mobilization for system's sustainability. They are also better suited in slums where many individuals do not have title to the land they inhabit because such individuals can be displaced at any time and are less likely to invest in its improvement (Davis, 2006). Advantages of community-based systems are (1) economies of scale can be realized, (2) it is easier to ensure a universal standard of water quality (Luby et al., 1999), (3) they are better suited to integrate the government in development and operation, and (4) they have associated benefits relating to community organization and development. Participatory process involving all stakeholders is the best way to take the decision as to which system should be adopted.

Sobsey et al. (2008) critically assessed several point of use (PoU) technologies for their suitability based on multiple criteria, which included (1) microbial efficacy, (2) health impacts, (3) quantity of water produced, (4) robustness of treatment, (5) ease of use, (6) time required to treat water, (7) cost of treatment, and (8) supply chain requirements. They concluded that ceramic and bios and household filters are the most effective. Hunter (2009) found that ceramic filters performed much better than other systems in the long term. However, chlorination is the most cost-effective, followed closely by solar disinfection (Clasen and Haller, 2008). It should be mentioned here that there is quite a lot of difference among various studies, and adoption of a particular technology should be context-specific.

3.1.2.1 Sustainability of Safe Water Systems in Slums

Development and sustainable implementation of appropriate technologies for safe water provision in slums requires consideration of different factors such as (1) effectiveness in producing potable water, (2) affordability and willingness to pay, (3) simplicity and ease of use, (4) ready availability of materials and parts, (5) reliability, (6) robustness of the process, (7) safety against recontamination, and (8) capacity to produce adequate quantities. More importantly, the system should be appropriate to the economic and sociocultural context in which it is situated. System's management, community mobilization, and the approach taken to system's development are affected by this criterion. It is also obvious that this criterion necessitates a participatory approach to the system's development, which should be appropriate to the realities of the community in which it is situated (Kelly and Farahbakhsh, 2008). Reller et al. (2003) conducted an epidemiological field study of a system, utilizing the proprietary coagulation–disinfection technology of Procter & Gamble in rural Guatemala. Their study showed significant reductions in diarrheal diseases among the intervention population in the initial years. However, a follow-up study by Luby et al. (2008) revealed that despite initial successes, the relatively high cost of the commercial product resulted in the rapid decline of its use once institutional supports were removed. This raises important issues on sustainable operation and maintenance of decentralized safe water systems in slums (Clasen, 2009). Purely market-based approaches give rise to complexities in low-income settings, especially if one seeks to address sustainability (Harris, 2005). Approaches that integrate community involvement can contribute to better service delivery and to its "democratization" (Ahmed and Ali, 2006).

Socio-cultural factors also affect the sustainability of the decentralized systems. For example, *Moringa oleifera* (drumstick tree) is widely grown in India. Mature seeds of the fruit are good adsorbents and used in PoU applications throughout India. However, it is widely used as a food material in South India, resulting in the nonavailability of mature seeds. On the other hand, in Northern India, the moringa fruit is not consumed as food and the seeds are widely available and inexpensive. Thus, the use of *M. Oleifera* for safe water applications in Northern India would be more sustainable. A participatory process, which integrates a range of stakeholders, can identify such sociocultural matters and help the design. Gender is also a critical issue for making decentralized safe water systems sustainable because women are the primary collectors and managers of water in the household, and they should be involved in the development and implementation of safe water systems, including the structuring of O&M frameworks. Finally, attention must also be given to marginalized groups within a given community because they may be unable to effectively self-advocate.

3.1.3 Case Study of an Alternative Water Supply Project

A case study, conducted in a typical slum in South Asia, is briefly discussed in this section to illustrate how participatory approach can be used for effective design of treatment systems for the provision of safe water system in peri-urban low-income settlements. The overall goal of the Alternative Water Systems Project was the participatory development and assessment of an alternative model of safe water provision in urban slums in order to improve drinking water quality. Three of the specific objectives of the project were: (1) selection of the appropriate level of approach (i.e., household, community, or both), (2) development of the appropriate water treatment processes through stakeholder participation, within the social, economic, cultural, and political context of the community, and (3) assessment of the in-field performance of the PoU treatment system with regard to pathogen removal efficiency and appropriateness to context.

3.1.3.1 The Social, Economic, and Cultural Context of the Community

The study has been conducted in a low-income settlement, known as Mylai Balaji Nagar, which is located on the southern periphery of Chennai city in India. Informal exploratory discussions were conducted with community members, government agencies, and other stakeholders on the water, environment, and public health situation to develop a basic understanding of the issue and the history of the study location. The interests, needs, and constraints of the various stakeholder groups were identified. WASH education was delivered to the community to enhance community awareness through schools and community meetings. These sessions were used for mobilizing people around community development initiatives and develop project activities with the community. Stakeholder forums were used to bring all stakeholders together for discussions on water, environment, and public health challenges facing the community and to explore the key issues. The forums encouraged discussion between the stakeholder groups, explored group perspectives and conflicts and encouraged community and governmental support, developed community organization around self-identified development priorities, and refined the project. Baseline survey was made using an extended sanitary risk questionnaire and a health status survey in order to assess the relative importance of the various fecal–oral pathogen transmission pathways in the study community. Additionally, a rapid survey was performed to get a preliminary understanding of community demographics. Key informant interviews were undertaken in order to develop a fuller contextual understanding.

The estimated population of the slum is around 10,000. Housing materials vary from thatch to permanent brick or concrete construction. Community members are involved in a variety of employment and livelihood activities such as labor, transport, retailing, education, etc. There are some 70 women's self-help groups, comprising of approximately 20 members each. Typical of many other slums in South Asia, although this slum originated as resettled area for people displaced from somewhere else, at present it is a community of original settlers and recent arrivals who bought plots/houses in informal market. The land tenure status is highly complex and ambiguous in the study community. Therefore, various levels of government are reticent to extend material or bureaucratic support to the community.

3.1.3.2 Water, Environment, and Public Health Conditions

Public water supply comes from a highly contaminated lake, with excessive levels of fecal and total coliforms. The water is piped from the lake and distributed through a network of public standpipes, each serving between five and 20 homes. The water is mostly untreated except for ad hoc and intermittent addition of bleaching powder by Panchayat agents. Supply is intermittent and controlled by a representative from the Panchayat. The supply of water via standpipes and water tankers is sporadic—sometimes once in 2 weeks. Women from each household collect water in several open-mouthed plastic containers or other large containers, which are then stored in the home. Water conflicts often break out during summer because the lake dries up and the water is trucked very infrequently. Some 70% of the households purchase canned water and another 10% obtain their drinking water from water tankers. Some households, which are made of concrete, are also equipped with roof-top rainwater collection tanks. With regard to drinking water treatment, water is consumed as it is although some households (38%) boil their water using domestic cooking gas. Risk of recontamination is high because water is mostly stored in open-mouthed containers or drums into which glasses are dipped. Water for all other purposes (bathing, washing, laundry, etc.) is sourced mainly from the public water supply (96%).

Water samples were analyzed for two biological and nine chemical parameters. Data collected between August 2009 and December 2010 was compared against permissible limits defined by the Bureau of Indian Standards (BIS 10500: 1991). Of the 10 parameters tested, chemical oxygen demand (COD), biological oxygen demand (BOD) (Figure 3.2), turbidity, and microbiological quality of the public water supply were found to be problematic. COD, BOD, and turbidity levels in the lake samples were almost always in excess of the permissible limits of 20, 5 mg/L, and 5 NTU, while the levels in household (lane) samples were largely within the permissible limits. However, the microbiological quality of the water was found to be severely degraded in both lake and stored household water samples. Total coliform levels were consistently higher in lake water samples than in household water samples, but in both cases, always far greater than the permissible limit of <1 MPN (Figure 3.3).

Health services in the area are abysmal. Most illnesses, from records of a small health outpost operated by a community-based organization, include diarrheal diseases, infectious hepatitis, typhoid fever, parasitic infections, and other gastrointestinal disorders (Chellappan 2007). Diarrheal disease

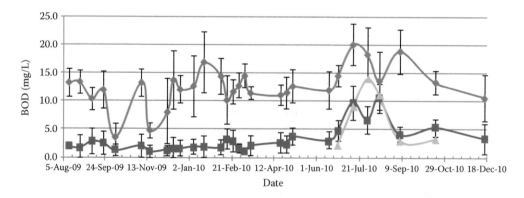

FIGURE 3.2   Temporal variation in BOD from the baseline survey: Lake (blue); household (red); infiltration well (green). (Adapted from MacDonald et al., 2012.)

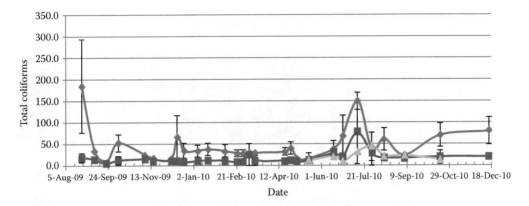

FIGURE 3.3 Temporal variation in total coliforms from the baseline survey: Lake (blue); household (red); infiltration well (green). (Adapted from MacDonald et al., 2012.)

prevalence across all age groups was found to be 2.9%. Only few roads are paved. Numerous potholes and stagnant pools provide favorable breeding grounds for mosquitoes and other insects. The waste management services are not adequate. Solid waste is simply dumped in various places, and often gets collected in the canals, turning the canals into swamps. Thirty-one percent of households had a toilet in the home, 60% used public latrines, 6% used improvised personal outhouses, and 4% practiced open defecation. A few public toilet blocks existing in the slum are underutilized due to inadequate water supply, mismanagement, and sporadic cleaning.

3.1.3.3 Development of Safe Water System with Stakeholder Involvement

Design of safe water system should be contextual. Baseline monitoring indicated that main water quality concerns are (1) organic contamination, (2) turbidity, and (3) microbiological contamination. A water treatment system was developed in a participatory manner, for the provision of safe water in the low-income community in order to reduce the burden of water-related diseases. Application of the decision-making support tool developed by Ali (2010) indicated that a household level safe water system would be more appropriate, given the community's specific circumstances. Six different filter designs were modeled and evaluated for the purification of household drinking water in order to meet the needs expressed by the residents. These included (1) the four steel pot system, (2) the four Plastic pot system, (3) earthen pot filter, (4) the PVC pipe filter, (5) the simple plastic bucket filter, and (6) the alum method. Uniform grade sand (0.3/0.5 mm in diameter) and crushed charcoal (3–5 mm in diameter) were used as filter media in all the filters. Extensive laboratory pilot studies were conducted using influent water from the lake in the study area to assess the efficacy of treatment. Water quality analysis was performed on both influent and effluent samples to assess removal efficiencies of turbidity, organic content (COD), and total coliforms.

One of the earliest filter designs, in the form of a tower, consisted of a series of four vertically aligned pots (Figure 3.4a and b). Each pot was responsible for a different stage of treatment. Raw untreated water was added to first pot in the top shelf, and it was subsequently drained by gravity through second, third, and fourth pots through holes in the bottoms of first three pots. These pots played their respective roles in sand filtration, charcoal filtration, and finally the collection of filtered water. Flow rate was maintained at 2.5 L/h.

An earthen pot design (Figure 3.4c) of approximately 5 L capacity was also considered. A sand and charcoal layer measuring almost 3 cm in height was used as the filter material, and flow rate was maintained at 2.5 L/h. A subsequent filter was fabricated using a 40 cm long section of PVC pipe with a diameter of 16 cm (Figure 3.4d). A tap was fixed to the base of the filter for the retrieval of purified water into the pot shown in the figure. The filter media consisted of (1) a bottom gravel bed, (2) middle layer of crushed charcoal with a bed depth of 13 cm, and (3) a top layer of sand with a bed depth of 13 cm. The

(a)   (b)   (c)   (d)   (e)   (f)

FIGURE 3.4 Four different filters: (a) Four metal pot system, (b) four plastic pot system, (c) earthen pot system, (d) PVC pipe filter system, (e) plastic bucket filter, and (f) alum treatment method. (Adapted from MacDonald et al., 2012.)

three layers were separated by nylon cloth to avoid mixing of the filter media. Five liters of water samples were added to the filter and allowed to stand, and filtered water samples were collected from the tap after 30 min for analyses.

In the fifth type of filter (Figure 3.4e), a normal plastic bucket of approximately 20 L capacity was used. The filter consisted of a dual media (sand and charcoal) filter bed on the top of a gravel base with nylon cloth spacers. Laboratory assessment of this design indicated that bacterial removal efficiency was not satisfactory. The baseline survey indicated that a family of five requires an average of 20 L of drinking water per day. In the sixth alternative, i.e., the alum method (Figure 3.4f), an attempt was made to meet this need by using a bucket of 50 L capacity to coagulate suspended solids and distil out harmful waterborne pathogens. A tap was fixed at 7 cm height from the bottom. Raw water was pretreated with 2 g of alum, stirred well, and allowed to settle for 1 h, after which the water was screened through a cotton sari cloth and then treated with two chlorine tablets. Thirty minutes of chlorination time was allowed before the water was sampled and analyzed.

Community meetings were held to determine their level of satisfaction with each of the aforementioned prototypes. Several key criteria for the design of an appropriate water treatment system, as presented in Table 3.1, evolved after several rounds of meetings and interviews. It was found that a bucket

TABLE 3.1 Design Criteria for Treatment Technologies, Evolved through Participatory Approach

	Description	Designs Meeting Criteria
Size	Suitable for storage in households with limited space	Earthen pot/PVC pipe/bucket filter
Safety	Stable and minimize falling hazards	Bucket filter
Treatment effectiveness	Effective in removing biological contaminants and improving overall aesthetics of water	All designs
Capacity	More than 20 L of water in a single application	Bucket filter
Affordability	Use of locally available materials, a low cost system in relation to the household income	Earthen pot/PVC pipe/bucket filter
Ease of use	Uncomplicated design	All designs
Durability	Use of least breakable materials	PVC pipe/vertical tower/final adapted bucket filter

type of filter, housed in a single container, with the capacity to treat 20 L of water in a single application would meet the needs of the average family. The need for placing the filter on a raised platform, which may pose a falling hazard, was avoided by placing a pipe manifold in the base of the filter, which directs water outflow to an elevated tap on the side of the bucket (Figure 3.5a). It was also found that the pipe manifold increased the efficiency of filtration by increasing the bed contact time.

The filter consists of (1) gravel base of depth 7 cm, (2) crushed charcoal (3 mm diameter) bed of 15 cm depth, and (3) a sand bed (0.3 mm diameter) of 15 cm. The pipe manifold is placed in the bottom most gravel bed, and each layer is separated by a nylon cloth. The filter head is large enough to accommodate 20–25 L of influent water. Moreover, final disinfection by chlorination performs extremely well by either killing or rendering inactive the majority of waterborne bacterial pathogens.

The bacterial removal efficiency in the first 2 months of field assessment of the bucket filter was 83.1% (95% CI 75.4–90.9). However, the users were not satisfied with the filter and tended to abandon the system entirely because of two reasons: (1) the plastic buckets housing the filter media were fragile and developed small fractures, and (2) the infrequent supply of water to the community resulted in stagnation of water in the filter for days at a time, leading to odor development. The situation called for technological adaptation. The modification involved the use of durable high-density polyethylene (HDPE) bins instead of the plastic bucket for housing the filter (Figure 3.5b). Also, a drainage spout (1.3 cm diameter) was added to the base of the bin through which residual water from the filter can be drained during times of water scarcity. This prevents the growth of odor forming algae and bacteria. Tests were conducted to ensure that the bins do not leach toxic plasticizers and are suitable for water storage, meeting the guidelines for drinking water quality set by the World Health Organization. The new bin was INR 25 cheaper and could treat 55% more water in a single application. Also, users could move the filter freely within their homes without fear of damaging it. These simple modifications increased user satisfaction and improved behavioral uptake. This in turn improved the likelihood of system sustainability and potential long-term health benefits. Field assessment of the safe water system indicated the systems acceptability and showed the need to properly educate all recipients on filter use and application, especially on filter flushing.

Field assessment of the safe water system was carried out also to answer the following key questions: (1) is there a significant difference in the microbiological quality of drinking water between treated and untreated water samples retrieved from intervention households; (2) is there a significant difference in the microbiological quality of drinking water between control and intervention households? Filters were distributed to 150 households who were using tap water for drinking purposes. Another criterion for

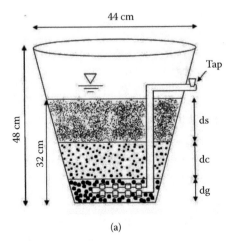

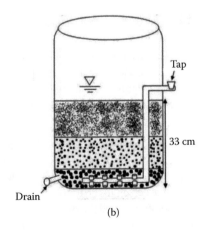

FIGURE 3.5 Evolution of bucket type filter: (a) Original plastic bucket filter and (b) modified HDPE bin filter.

selecting the houses was the presence of a child in the house. Also, equal number of households were selected as control sample.

There was a significant improvement in all biological and chemical parameters of interest in the samples from intervention homes. Mean values of COD of samples of treated, untreated, and control were 8.83 (SEM=1.16), 17.46 (SEM=2.42), and 15.17 mg/L (SEM=2.45), respectively. Similarly, mean values of total coliform were 936.5 cfu (SEM=249.7) and 15698.9 cfu (SEM=1806.4) for treated and untreated samples, respectively. Control samples demonstrated total coliform concentrations similar to those reported in intervention untreated samples (M=19229.6 cfu, SEM=2427.8). Although the bacterial load was high in the raw water, the safe water system resulted in an average log reduction value of 1.53 (95% CI 0.88–2.19) for total coliforms and 1.00 (95% CI 0.19–1.81) for *Escherichia coli*. It was found that treated water was stored for extended periods of time in some households, which increased the possibility of recontamination. It was found from a heath survey that disease prevalence rates (M=0.02%, SEM=0.02) among children from households using the safe water system were slightly lower than that among children from control households (M=0.31%, SEM=0.16). Similar trends in disease prevalence were identified among adults and children over 5 years.

3.1.4 Effective Intervention to Improve Rural Water Supply Quality

In this section, a case study is discussed to illustrate how simple and novel interventions can be made in rural areas for significant improvement in the water quality. Philip (2013) carried out a project for scaling up of community-based water quality monitoring and sanitary surveillance in several blocks of Krishnagiri and Hosur Districts of Tamil Nadu, India. Drinking water systems already exist in the intervening villages of these blocks, but the water quality at household level was poor. Also, the water-related disease prevalence was high. The overarching goal of the project was to make an intervention in these villages to improve the water quality situation, and thereby reduce the disease prevalence. The specific objective was to enhance the capacity of the Panchayat Level Convergence Committee (PLCC) of the block to undertake a community-based water quality monitoring and sanitary surveillance system. It was hypothesized that if the PLCC members are (1) aware of the importance of water quality and sanitation for good health, (2) capable of testing for basic water quality parameters without dependence on external actors, and (3) know how good or bad their water is, then they would act as agents to put pressure on local government officials to undertake remedial actions at the earliest, and also spread the awareness on water supply, sanitation, and hygiene among the local population. This, in turn would have a long-term positive effect on the public health in the intervening blocks. The specific scope of the project included the following action items:

- Facilitate baseline water quality and sanitary surveys.
- Supply of water quality kits.
- Impart training on water quality monitoring system for PLCC.
- Undertake bimonthly visit to monitor and review the progress and results.
- Cross-check the quality of water samples to conform whether the test findings provided by the village volunteers were in line with the procedures.
- Analyze WQ data forwarded by the village volunteers and provide technical support to carry out appropriate action.

The intervention was designed to include the participation and contribution of the local water supply service provider, i.e., the Tamil Nadu Water Supply and Drainage (TWAD) Board so that the intervention is effective even after the exit of the external actors. The responsibilities of the TWAD Board included: (1) deputation of one resource person to the training to share about the functions of the TWAD board, importance of water quality monitoring, and the support that can be extended by the board for scaling the project; (2) facilitating monitoring visits and conducting sample water quality tests; (3) extending support for the establishment of a system for the replenishment of reagents for field test kits; and (4) helping in organizing refresher training.

Intervention was made in six blocks of Krishnagiri district (Tamil Nadu), namely (1) Veppanapalli, (2) Bargur, (3) Thally, (4) Krishnagiri, (5) Kaveripatnam, and (6) Mathur blocks. Prior to this intervention water quality in the region was assessed in the year 2008 through two campaigns: premonsoon and postmonsoon. A total of 87 samples were collected from bore wells, hand pumps, and water taps, and were analyzed for water quality parameters, which included (1) bacteriological quality, (2) fluoride, (3) total dissolved solids (TDS), (4) pH, (5) iron, (6) chloride, and (7) nitrate. Results from bacteriological tests are shown in Figure 3.6. It was observed that almost 100% of the samples from bore wells had bacteriological contamination, while 29% of samples from hand pumps were contaminated. Similarly it was observed that 27% of samples from hand pumps had a fluoride concentration more than 3.0 mg/L, while this was 18% for bore wells. The TDS was more than 500 mg/L in 85% of hand pump samples and 95% of bore well samples.

Several 3-day training programs were conducted for PLCC members. Three volunteers from each Panchayat participated in these programs. Similarly, several 2-day training programs (about 30 participants each) were conducted for Panchayat Presidents, block development officers, ADWs, CFVP coordinators, and Panchayat Secretaries. Finally, 1-day refresher programs were conducted for selected volunteers and Panchayat Presidents.

A major objective of these training programs was to provide easy-to-use water quality test kits to each village, and block coordinators and Water Analyst of the TWAD Board. Training was imparted in the use of these easy-to-use field kits to monitor water quality. Training was also given on the development of such kits. These test kits could be used for testing for nine parameters: H_2S strip, fluoride, chloride, residual chlorine, nitrate, alkalinity, hardness, total dissolved solids, and iron. Each kit had sufficient chemicals to perform 100 water sample analyses. Ten H_2S strip bottles were provided with each kit. Each month, H_2S strip bottles and chemicals were supplied to volunteers as per their requirement. Kits were also provided to all participants in the refresher program. Another major focus of this training program was to make people aware about the importance of using safe and palatable water for daily use and the adverse effects of consuming contaminated water and good sanitary practices.

Most of the trained panchayat volunteers carried out the water quality analyses of almost all water sources in the selected villages. They also carried out the survey pertaining to sanitary practices in the respective panchayats along with the inventory of all drinking water sources. Two samples from each volunteer were analyzed in the District TWAD Board Lab. It was noticed that almost 80% of the results matched well. Wherever there was a discrepancy in the results, possible reasons for the discrepancy were discussed and volunteers were instructed by the project coordinator on the remedial measures to be taken. A coordinating team visited the villages once in 2 months to conduct review meetings. The coordinating team discussed with the volunteers about the problems they have faced in the field and offered advice on how to surmount them. They also examined the correctness of the analyses by comparing with the results from the TWAD Board. There was a good correlation between the two analyses results

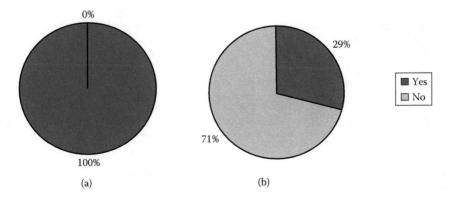

FIGURE 3.6 Bacteriological test results: (a) Bore wells and (b) hand pumps.

in most of the cases. New bottles of all the chemicals, along with H$_2$S bottles sufficient for 2 months of water sample analyses were also provided during these meetings.

The water quality parameters including bacteriological quality of water were monitored for 6 months. A decreasing trend in bacteriological contamination was clearly visible as shown in Figure 3.7. To assess the improvement in health of people in intervened blocks, secondary data from primary health centers were collected and analyzed. It was found that there was a visible decrease in the number of cases of dysentery and cholera (Figure 3.8). Finally, a stakeholder workshop was conducted to evolve a plan for scaling up and sustainability of the intervention.

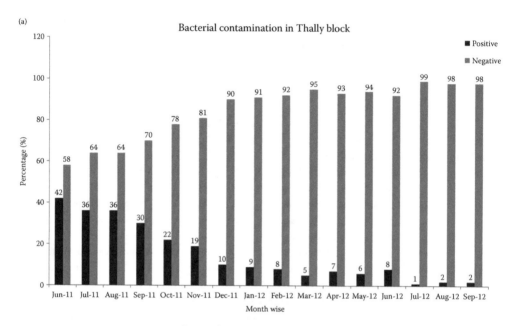

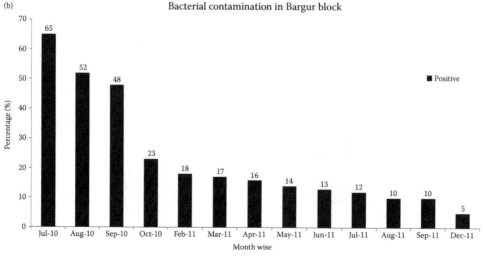

FIGURE 3.7 (a) Bacterial contamination in water samples analyzed from Thally block. (b) Bacterial contamination in water samples analyzed from Bargur block.

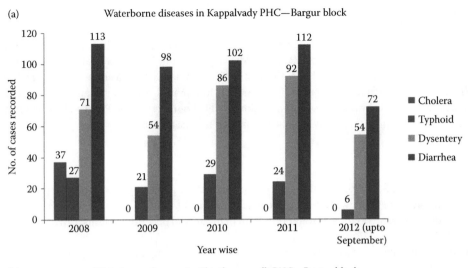

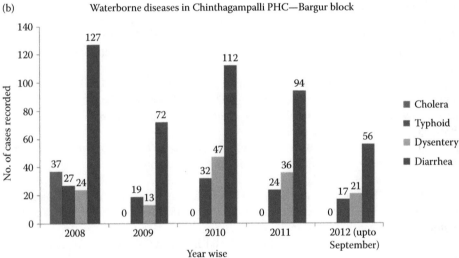

FIGURE 3.8 (a) Waterborne diseases reported in Kappalvady PHC-Bargur block. (b) Waterborne diseases reported in Chinthagampalli PHC-Bargur block.

3.1.5 Summary

In this chapter, issues related to provision of safe water in urban slums and rural areas have been discussed. At present, marginalized communities in urban and peri-urban areas are not provided with many urban services including safe water due to various political, technical, economical, and social factors, resulting in an overburden of water-related diseases in these communities. It is illustrated, through a case study in an urban slum, how technical and economic barriers can be overcome and a cost-effective decentralized safe water system can be provided to an underprivileged peri-urban community. The importance of participatory approach and the protocol for such an approach for developing a safe water system have been brought out clearly. It is also illustrated through a different case study the importance and effectiveness of awareness creation and periodical monitoring of water quality for improving safe drinking water supply in rural areas in India.

Acknowledgments

The authors acknowledge IDRC Canada and UNICEF for funding the study. They also acknowledge the entire project team consisting of Ms. Jincy, Ms. G. Vinothini, Mr. A. Sambath, Mr. S. Srinivasan, Ms. F. Begum, and Mr. Pradeep, Mr. D. Kumaran, Mr. Soundar, Mr. Balasubramanian, Mr. Morgan, Mr. Imran Ali, Dr. Prema Rajagopalan, D. Kumaran, Ramprasad, Balasubramaniam, and others for the contributions made by them. The authors acknowledge Engineering and Physical Sciences Research Council (EPSRC; grant number EP/M017141/1) for their financial support.

References

Ahmed, S. and Ali, S. (2006), People as partners: Facilitating people's participation in public–private partnerships for solid waste management, *Habitat Int.* 30, 781–796.

Ali, S.I. (2010), Alternatives for safe water provision in urban and peri-urban slums, *J. Water Health* 8(4), 720–734.

Amit, R.K. and Subhash, S. (2015), Coping strategies and coping costs for accessing safe water in Chennai, India, Report prepared for SANDEE.

Biswas, A. and Tortajada, C. (2009), *Water Supply of Phnom Penh: A Most Remarkable Transformation.* Singapore: Institute of Water Policy, Lee Kuan Yew School of Public Policy.

Black, R., Morris, S. and Bryce, J. (2003), Where and why are 10 million children dying every year? *Lancet* 361, 2226–2234.

Chellappan, F. (2007), Personal communication: A walking tour of Mylai Balaji Nagar Slum with Community Research and Development (CRAD) Trust, Chennai, India.

Clasen, T. (2009), *Scaling Up Household Water Treatment among Low-Income Populations.* Geneva, Switzerland: World Health Organization.

Clasen, T. and Haller, L. (2008), *Water Quality Interventions to Prevent Diarrhoea: Cost and Cost-Effectiveness.* Geneva, Switzerland: World Health Organization.

Davis, M. (2006), *Planet of Slums.* London: Verso.

Esrey, S.A., Potash, J.B., Roberts, L. and Shiff, C. (1991), Effects of improved water supply and sanitation on ascariasis, diarrhoea, dracunculiasis, hookworm infection, schistosomiasis, and trachoma, *World Health Organ.* 69(5), 609–621.

Fewtrell, L., Kaufmann, R.B., Kay, D., Enanoria, W., Haller, L. and Colford Jr., J.M. (2005), Water, sanitation, and hygiene interventions to reduce diarrhoea in less developed countries: A systematic review and meta-analysis, *Lancet Infect. Dis.* 5, 42–52.

Graf, J., Meierhofer, R., Wegelin, M. and Mosler, H. (2008), Water disinfection and hygiene behaviour in an urban slum in Kenya: Impact on childhood diarrhoea and influence of beliefs, *Int. J. Environ. Health Res.* 18(5), 335–355.

Haller, L., Hutton, G. and Bartram, J. (2007), Estimating the costs and benefits of water and sanitation improvements at global level, *J. Water Health* 5, 467–480.

Harris, J. (2005), Challenges to the commercial viability of point-of-use (POU) water treatment systems in low-income settings. School of Geography and Environment, Oxford University, MSc Thesis, 78.

Howard, G. and Bartram, J. (2003), Domestic water quantity, service level and health. Geneva, Switzerland: World Health Organisation. http://whqlibdoc.who.int/hq/2003/WHO_SDE_WSH_03.02.pdf. Accessed May 2009.

Hunter, P. (2009), Household water treatment in developing countries: Comparing different intervention types using meta-regression, *Environ. Sci. Technol.*, 43(23), 8991–8997.

Kelly, B. and Farahbakhsh, K. (2008), Innovative knowledge translation in urban water management: An attempt at democratizing science, *Int. J. Technol. Knowl. Soc.* 4, 73–84.

Luby, S., Mendoza, C., Keswick, B., Chiller, T. and Hoekstra, R. (2008), Difficulties in bringing point-of-use water treatment to scale in rural Guatemala, *Am. J. Trop. Med. Hyg.* 78(3), 382–387.

Luby, S.P., Syed, A.H., Atiullah, N., Faizan, M.K. and Fisher-Hoch, S. (1999), Limited effectiveness of home drinking water purification efforts in Karachi, Pakistan, *Int. J. Infect. Dis.* 4, 3–7.

Montgomery, M. and Elimelech, M. (2007), Water and sanitation in developing countries: Including health in the equation, *Environ. Sci. Technol.* 41(1), 17–24.

MoWR, Ministry of Water Resources, Government of India. (2011), Strategic plan for Ministry of Water Resources.

Philip, L. (2013), Final report on scaling up of the community based water quality monitoring and sanitary surveillance in Veppanapalli Block, Krishnagiri District, Tamil Nadu, UNICEF Chennai Office for Tamil Nadu and Kerala.

Prüss, A., Kay, D., Fewtrell, L. and Bartram, J. (2002), Estimating the burden of disease from water, sanitation, and hygiene at a global level. *Environ. Health Perspect.* 110(5), 537–542.

Reller, M.E., Mendoza, C.E., Beatriz Lopez, M., Alvarez, M., Hoekstra, R.M., Olson, C.A., Baier, K., Keswick, B.H. and Luby, S. (2003), A randomized controlled trial of household-based flocculant-disinfectant drinking water treatment for diarrhea prevention in rural Guatemala, *Am. J. Trop. Med. Hyg.* 69(4), 411–419.

Sobsey, M., Stauber, C., Casanova, L., Brown, J. and Elliott, M. (2008), Point of use household drinking water filtration: A practical, effective solution for providing sustained access to safe drinking water in the developing world, *Environ. Sci. Technol.* 42, 4261–4267.

UNFPA. (2007), *State of the World Population 2007: Unleashing the Potential for Urban Growth.* New York: United Nations Population Fund.

United Nations. (2007), The millennium development goals report. New York: United Nations. http://www.un.org/millenniumgoals/pdf/mdg2007.pdf. Accessed May 2009.

United Nations. (2015), India and the MDGs: Towards a sustainable future for all. Report prepared by United Nations ESCAP.

3.2 Toward the Synthesis of Wastewater Recovery Facilities Using Enviroeconomic Optimization

Channarong Puchongkawarin, Yannic Vaupel, Miao Guo, Nilay Shah, David C. Stuckey, and Benoît Chachuat

The wastewater treatment industry is undergoing a major shift toward a proactive interest in recovering materials and energy from wastewater streams, driven by both economic incentives and environmental sustainability. With the array of available treatment technologies and recovery options growing steadily, systematic approaches to determining the inherent trade-off between multiple economic and environmental objectives become necessary, namely enviroeconomic optimization. The main objective of this chapter is to present one such methodology based on superstructure modeling and multiobjective optimization, where the main environmental impacts are quantified using life cycle assessment (LCA). This methodology is illustrated with the case study of a municipal wastewater treatment facility. The results show that accounting for LCA considerations early on in the synthesis problem may lead to dramatic changes in the optimal process configuration, thereby supporting LCA integration into decision-making tools for wastewater treatment alongside economical selection criteria.

3.2.1 Introduction

Untreated sewage presents a threat to human health and the environment. For the most part, wastewater treatment design retains its foundations in engineering traditions established in the early twentieth century (Daigger, 2009). The aerobic treatment processes used to produce an effluent that complies with the discharge standards are often energy intensive and may be significant contributors to greenhouse gas (GHG) emissions including carbon dioxide (CO_2), methane (CH_4), and nitrous oxide (N_2O) (Bufe, 2008). Moreover, large quantities of sludge may be produced as a by-product of aerobic treatment, and wastewater treatment facilities can also be land demanding. However, a paradigm shift is underway toward making wastewater treatment facilities more sustainable, driven by a range of sustainability issues including increase in electricity demand and price (International Energy Agency, 2012), and long-term nutrient scarcity and high extraction costs (Batstone et al., 2015; Mehta et al., 2015). In this new paradigm, wastewater is regarded as a renewable resource from which water, materials, and energy can be recovered, thereby transitioning to resource recovery facilities (Guest et al., 2009).

Because wastewater treatment is often regarded as being an end-of-pipe technology, the design of wastewater treatment facilities tends to focus on minimizing the environmental impacts of given con-taminants present in the effluent, such as organic matter (chemical oxygen demand, COD), nitrogen (N), and phosphorus (P). Indeed, many environmental regulations still focus on the removal of these tar-geted contaminants without consideration of the broader environmental issues. Nowadays, there is a greater awareness that sustainability objectives, beyond receiving water quality, should be accounted for in the design and operation of wastewater treatment facilities, in response to which there has been much research emphasis on life cycle assessment (LCA) in the wastewater treatment industry in recent years.

LCA is a holistic, cradle-to-grave standardized approach to evaluating the environmental impacts of products and services (ISO 14040, 2006). As far as wastewater treatment is concerned, LCA was first applied in the 1990s for identifying the environmental impacts of various small-scale wastewater treat-ment technologies (Emmerson et al., 1995). Since then, it has been increasingly used in this industry as a means of comparing different wastewater treatment technologies (Gallego et al., 2008; Foley et al., 2010; Rodriguez-Garcia et al., 2011) and sludge management technologies (Suh and Rousseaux, 2002), and for evaluating the main environmental impacts associated with specific wastewater treatment processes (Tangsubkul et al., 2006; Pasqualino et al., 2009). The sensitivity of the LCA results to various impact assessment methods has also been investigated (Renou et al., 2008). As pointed out in a recent review paper (Corominas et al., 2013), however, there is a need for better linking LCA with economic and societal assessments in order to provide a more complete and accurate sustainability picture for decision makers.

An approach to incorporating LCA evaluation into a knowledge-based decision-support system to design wastewater treatment plants (WWTPs) has recently been presented in Garrido-Baserba et al. (2014). The results demonstrate the potential of LCA for decision making, although the approach is largely dependent on the data quality and their specifications (Khiewwijit et al., 2015). Moreover, this approach may not provide further information with regards to the optimal (or near-optimal) solutions. In essence, superstructure optimization (Biegler et al., 1997) provides an ideal framework to iden-tify optimal solutions of those design problems having a large number of alternative processes, and it may be combined with multiobjective optimization in the presence of multiple conflicting objec-tives, e.g., economic and environmental performance indicators. This approach has been successful in various application areas, including the synthesis of water networks (Ahmetović and Grossmann, 2011; Khor et al., 2014; Quaglia et al., 2014), as well as wastewater treatment and resource recovery systems (Rigopoulos and Linke 2002; Bozkurt et al., 2015; Puchongkawarin et al., 2015). Because it is computa-tionally demanding, the key to its success is the development and selection of mathematical models for the units that are simple enough for the optimization problems to remain tractable, yet provide reliable estimates of their performances and associated costs.

The main objective of this chapter is to present a modeling methodology for decision in wastewater treatment and resource recovery systems in order to arrive at WWTP designs that are both environ-mentally sustainable and economically viable, namely enviroeconomic optimization. This methodology is based on superstructure modeling and multiobjective optimization, where the main environmen-tal impacts are quantified using LCA. Moreover, these developments are illustrated with a simple case study in municipal wastewater treatment. The rest of this chapter is organized as follows: a review of resource recovery and WWTP design is first presented (Section 3.2.2), followed by the methodology (Section 3.2.3) and illustrative case study (Section 3.2.4), before drawing conclusions.

3.2.2 Backgrounds

3.2.2.1 Resource Recovery from Wastewater

This subsection reviews various technologies for recovery of energy and materials from wastewater with an emphasis on proven technologies. For further details, we refer the reader to the recent survey/perspective articles (Batstone et al., 2015; Mehta et al., 2015).

The organic compounds present in municipal (and many industrial) wastewaters can be converted into a CH_4-rich biogas via anaerobic digestion. It is estimated that about 30–60 L/day of CH_4 per capita can be generated from a typical municipal wastewater by transforming all of the biodegradable organic matter into biogas (Owen, 1982). Moreover, anaerobic digestion can be adopted in conjunction with downstream resource recovery units due to its minimal effect on ammonia or phosphate removal. In contrast to biogas generation from high-strength wastewater and wastewater sludge that has been employed for many years, direct anaerobic treatment of low-strength wastewater has not been widely practiced so far, especially in temperate climates where wastewater temperature is in the range of 5°C–15°C. Innovations in reactor design to maintain elevated biomass inventories, such as the upflow anaerobic sludge blanket and anaerobic membrane bioreactor (AnMBR), have mitigated some of the limitations and have extended the range of applications of anaerobic treatment (Liao et al., 2006; Lew et al., 2009; McCarty et al., 2011). Particularly, promising configurations include the submerged anaerobic membrane bioreactor (SAnMBR; (Hu and Stuckey, 2006; Lin et al., 2011)) and, more recently, the anaerobic fluidized membrane bioreactor (Kim et al., 2011). Research is underway to develop improved membranes and reactor designs that reduce membrane fouling and enhance dissolved CH_4 recovery (Smith et al., 2012; Stuckey, 2012). In urban water systems, thermal energy can be recovered through the use of heat pumps or heat exchangers. Although in the form of low-grade energy due to small temperature differences, this energy may be suitable for heating buildings (EPA, 2007). Besides biogas and thermal energy, promising technologies for electricity or hydrogen generation from wastewater are also emerging (Logan et al., 2006; Kim and Logan, 2011).

Nitrogen (N), phosphorus (P), and potassium (K) are critical nutrients for intensive agriculture, whereas there has been increasing concern about the long-term scarcity or high extraction cost associated with these nutrients; Phosphorus is a nonrenewable mineral resource, that is fast depleting. It is estimated that the worldwide phosphorus demand will outstrip supply within a few decades as a consequence of an expanding population, which has major implications in terms of global food security as 90% of the phosphate (PO_4^{3-}) rock reserves are located in just five countries (Mehta et al., 2015). There has been little discussion regarding K as a macronutrient target for recovery so far, although its price is projected to rise substantially in the next decade due to a limited geological distribution (Batstone et al., 2015). Fertilizer production is also highly reliant on the energy-intensive and natural gas-dependent ammonia production process. The rising natural gas market price in past decades is projected to continue (doubling by 2025), which will directly affect the price and supply of N fertilizers (Batstone et al., 2015). In this context, much research has been dedicated to N and P recovery from nutrient-rich wastewater in recent years (Doyle and Parsons, 2002; Le Corre et al., 2009; Stolzenburg et al., 2015). In the presence of a precipitating or fixing agent, a majority of the phosphates, which accounts for 50%–80% of the total P compounds in municipal wastewater (Tchobanoglous et al., 2003), can be recovered. Moreover, technologies for recovering soluble N compounds from municipal wastewater, around 50%–80% of the total N content (Tchobanoglous et al., 2003), are becoming economically viable. Ion exchange, for instance, can recover ammonia, nitrate, or phosphates by passing the secondary effluent through adsorbent columns and chemical regeneration (Liberti et al., 2001; Johir et al., 2011), and the resulting nutrient-enriched solutions can be further processed into a saleable product (e.g., fertilizer). Nonetheless, key challenges for the wider deployment of ion exchange units include their potential for fouling with suspended solids, the limited exchange capacity of certain adsorbents, the limited ion selectivity, and their high capital costs (Miladinovic and Weatherley, 2008). A range of absorbents are available, which include zeolites (e.g., clinoptilolite) for ammonium ions (NH_4^+) fixation (Aiyuk et al., 2004; Wang and Peng, 2010) and polymeric resins with higher exchange capacities. Research is also underway to develop anion adsorbents with high phosphate selectively and easy regeneration characteristics, including hydrotalcite (Kuzawa et al., 2006) and polymeric resins with hydrated ferric oxide nanoparticles (Martin et al., 2009). An alternative promising technology for P recovery is reactive filtration, which combines physical filtration of particulate P compounds with coprecipitation and adsorption of soluble P compounds onto coated sand in a moving bed filter—up to 95% P recovery by using hydrous ferric oxide coated sand (Newcombe et al., 2008). In addition, adoption of crystallization for P recovery from concentrated wastewater streams has attracted substantial interest to produce reusable compounds such as calcium phosphate ($Ca_3(PO_4)_2$) and struvite ($MgNH_4PO_4$) (Le Corre et al., 2009). The recovery technology involves precipitation in either stirred tanks or fluidized bed reactors (Bhuiyan et al., 2008), with the latter being the most commonly applied in struvite crystallization. Note, however, that most nutrient recovery technologies to date have been applied to sludge liquors with PO_4^{3-} concentrations of 50–100 mg/L, for which 80% removal efficiencies or higher have been reported (Ueno and Fujii, 2001). For dilute streams, such as secondary effluents with PO_4^{3-} concentrations of 4–12 mg/L, struvite crystallization may be combined with adsorbent columns and fed with the enriched solutions from the adsorbent regeneration, e.g., **REM-NUT** process reported in Liberti et al. (2001). Besides N and P recovery, organic carbon too can be recovered as polyhydroalkanoates (Coats et al., 2007), whereas heavy metals can be recovered via adsorption, membrane filtration, or chemical precipitation (Fu and Wang, 2011).

3.2.2.2 Design of Wastewater Treatment Facilities

The traditional rules and guidelines for design of wastewater treatment facilities—see, e.g., Tchobanoglous et al. (2003)—are being challenged nowadays by tighter economic and environmental constraints. Moreover, with the extra degrees of freedom offered by advanced treatment and separation technologies, the plant synthesis/design problem becomes much more complex, especially when it comes to selecting the most sustainable wastewater treatment facilities in a given regional context (Rivas et al., 2008).

Mathematical modeling has been a powerful tool in assisting decision making in WWTP design since the 1990s. Not only have the computational capabilities and numerical solution technology improved dramatically, but the mathematical models too have become more predictive, now enabling plant-wide simulation routinely using a range of commercial simulators (GPS-X, BioWin, WEST, etc.). Conventional model-based WWTP design starts with the selection of a plant layout and then focuses on the detailed design and analysis of this particular layout. The selection of an appropriate design that meets the specified objectives and constraints involves comparing the capital and operating costs of multiple plant configurations, thus requiring repetitive model simulations alongside comprehensive process knowledge. As the number of processes and configurations increases, this design approach becomes more tedious and ultimately unmanageable.

In order to deal with large numbers of treatment or separation units and possible interconnections, a system engineering approach is the most useful. Systematic methods for the synthesis of complex chemical plants and biorefineries based on superstructure modeling and optimization are well developed (Biegler et al., 1997; Kokossis and Yang, 2010; Liu et al., 2011). These approaches are also increasingly applied to water network synthesis in process plants in order to minimize fresh water consumption and wastewater generation through regeneration, recycle, and reuse (Faria and Bagajewicz, 2009; Khor et al., 2014). Regarding municipal wastewater facilities, the need for systematic approaches has been emphasized (Balkema et al., 2001; Hamouda et al., 2009), but relatively few studies have been published to date (Rigopoulos and Linke, 2002; Alasino et al., 2007, 2010; Bozkurt et al., 2015; Puchongkawarin et al., 2015). These studies provide insight into the potential of the systematic optimization-based approaches for wastewater treatment design, but they are nonetheless limited to optimizing a given process or selecting the most appropriate process among a small number of alternatives mainly based on economical considerations.

Another challenging task for WWTP design is satisfying multiple conflicting objectives simultaneously, while meeting the discharge regulations. One example is the trade-off between nutrient discharge targets and plant-wide energy consumption. Traditionally, these problems have been addressed using single objective optimization through weighting of the various contributions in an overall cost index. In contrast, multiobjective optimization provides a means of describing multiple objectives separately by determining the so-called Pareto solution set of nondominated solutions (Hakanen et al., 2013). This approach has not been applied widely in WWTP design to date, perhaps due to the complexity of wastewater treatment processes. A decision-making tool to support the design of WWTPs based on multicriteria evaluation was proposed in Flores et al. (2007). The selection of different process alternatives in this tool is based on an overall degree-of-satisfaction index, as obtained through the weighting of selected criteria and objectives, and it relies on a mix of mathematical modeling and qualitative knowledge. Another interactive multiobjective optimization platform coupled with model-based simulation, called NIMBUS, was presented in Hakanen et al. (2013). This platform allows a decision maker to simultaneously consider the design of WWTPs from different standpoints and to balance between the different objectives. More recently, an integrated framework combining LCA with dynamic simulation to compare different treatment processes was developed in Bisinella de Faria et al. (2015), with a focus on source separation and energy/nutrient recovery. On the whole, these existing approaches are certainly heading in the right direction, but the exploration remains limited to a small number of process configurations nonetheless. In contrast, the following sections present and illustrate a superstructure modeling and optimization methodology, which addresses some of these limitations.

3.2.3 Methodology

The combination of traditional wastewater treatment technology with advanced energy and materials recovery solutions offers considerable promise to improve the sustainability and reduce the cost of wastewater facilities. Clearly, the ultimate goal here is a closed cycle, energy-sufficient process, where all waste streams are recycled and the only output is saleable/valuable products. This section investigates the question as to how to select and interconnect, from a wide variety of unit operations, those units that will lead to the most sustainable wastewater treatment systems—a problem known as synthesis or flow sheeting in process engineering.

The synthesis problem statement starts with the specification of the following data:

- A set of wastewater streams of given flow rates and compositions.
- A set of water sinks with known maximum concentration limits or financial penalties as defined by local/federal authorities.
- A set of treatment and separation units with given performance for targeted compounds and associated costs and environmental burdens.

These specifications can be represented by a generic superstructure, which considers every possible interconnection in a fixed network topology. One such superstructure is illustrated in Figure 3.11 for a simple network topology that consists of a single wastewater stream, two sinks (treated effluent and biosolids), and a range of treatment/separation units. The objective of the synthesis problem is to determine an optimal resource recovery facility in terms of (1) its units, (2) the piping interconnections between the units, and (3) the flow rates and compositions in the interconnections. Here, *optimal* is understood in terms of maximizing the net present value (NPV) and minimizing the given environmental impacts.

3.2.3.1 Surrogate-Based Optimization

Mathematically, superstructure optimization problems can be formulated as optimization models with two types of decisions:

- The discrete, usually binary, decisions on the units that should be included in the system along with their interconnections, here denoted by y.
- The continuous decisions that define the flows and compositions as well as certain design and operating parameters, here denoted by x.

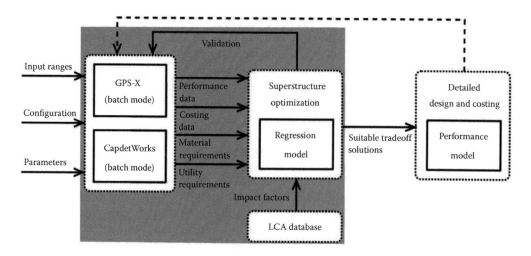

FIGURE 3.9 Illustration of the proposed methodology based on multiobjective superstructure optimization and surrogates models. From Vaupel (2015.)

This leads to multiobjective mixed-integer nonlinear programs (MO-MINLP) in the form:

$$\min_{x,y} \ [\text{KPI}_1(x, y), \text{KPI}_2(x, y), \ldots]$$

$$\text{s.t.} \ h(x) = 0$$

$$g(x,y) \le 0$$

$$x \le 0, y \in \{0,1\}. \tag{3.1}$$

The objective of the MO-MINLP consists of minimizing two or more key performance indicators (KPIs), which are functions of both types of variables. These variables must also satisfy restrictions of the form $g(x, y) \le 0$, either design specifications in terms of discharge allowance and physical operating limits or logical constraints for the existence of piping interconnections with nonzero flows or the sequencing of certain units. Last, but not least, the continuous variables x must obey material balance equations of the form $h(x) = 0$, where usually $\dim(h) < \dim(x)$, describing models of the physical units.

A key element of the superstructure optimization approach is that these latter models should remain as simple as possible in order to comply with state-of-the-art algorithms and computational capabilities, thereby calling for a surrogate-based approach for the synthesis problem. Previous work by Puchongkawarin et al. (2015) advocates for the use of surrogate models constructed from state-of-the-art WWTP simulators as depicted in Figure 3.9. In order to determine solutions that are accurate enough, the most promising process alternatives determined from the superstructure optimization problem (P) are then validated against the simulator-projected performances and costs. Typically, this would create an iteration between the superstructure optimization and the simulator for refining the surrogates. In particular, recent developments in surrogate-based optimization can provide guarantees that the iterations will converge to a (local) optimum with minimum recourse to high-fidelity models (Agarwal and Biegler, 2013; Biegler et al., 2014). Finally, the selected process candidates can be considered for detailed performance and cost analyses, including integration options and operability issues. To account for additional design and operational constraints, further iterations with the superstructure optimization block may then be necessary.

The following subsection presents an approach to deriving surrogate models for predicting plant-wide performance. Then, the computation of economic and environmental performance indicators is discussed in Sections 3.2.3.3 and 3.2.3.4. Finally, numerical solution considerations are discussed in Section 3.2.3.5.

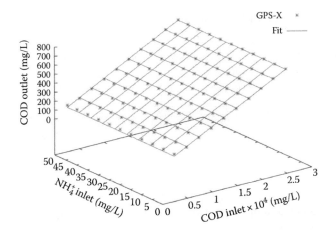

FIGURE 3.10   Illustration of a linear surrogate model obtained from GPS-X simulated data.

3.2.3.2 Plant-Wide Performance Surrogates

Performance models in the superstructure optimization problem (P) are based on the material balances on flows (F) and concentrations (X) around the sources, the units, and the sinks. For instance, the material balances for a given species c in the treatment unit k can be formulated as Puchongkawarin et al. (2015):

$$F_k^{\text{in}} = \sum_{i=1}^{N_{\text{source}}} F_{i \to k} + \sum_{k'=1}^{N_{\text{unit}}} F_{k' \to k}$$

$$X_{k,c}^{\text{in}} F_k^{\text{in}} = \sum_{i=1}^{N_{\text{source}}} F_{i \to k} X_{i,c} + \sum_{k'=1}^{N_{\text{unit}}} F_{k' \to k} X_{k',c}^{\text{out}}$$

$$X_{k,c}^{\text{in}} F_k^{\text{in}} (1 - \rho_{k,c}) = \sum_{k'=1}^{N_{\text{unit}}} F_{k \to k'} X_{k',c}^{\text{out}} + \sum_{j=1}^{N_{\text{sink}}} F_{k \to j} X_{k,c}^{\text{out}}$$

where the superscripts "in" and "out" refer to flows/concentrations entering or leaving the unit, respectively, and $\rho_{k,c}$ stands for the removal efficiency of species c in unit k.

This latter removal efficiency is key to the accuracy of the superstructure optimization model, yet the direct use of complex biodegradation models (such as ADM1 (Batstone et al., 2002) and ASM1-3 (Henze et al., 2000)) or complex crystallization/adsorption/filtration models in the MO-MINLP (P) is currently computationally intractable. As already mentioned, our approach considers surrogate models constructed from input–output data predicted by state-of-the-art process simulators instead. Specifically, the performance—either at steady state or averaged over a cyclic steady state—of a given unit can be computed for various influent compositions (COD, NH_4^+, etc.) and given operation parameters (hydraulic rentention time (HRT); sludge retention time (SRT) etc.). Then, simple regression models can be fitted to the simulated data points, e.g., in the form of linear, piecewise-linear, or polynomial input–output relationships, as appropriate. For instance, Figure 3.10 shows the outlet COD concentration of an anaerobic digester predicted by the ManTIS3 model in GPS-X for various inlet COD and NH_4^+, along with the corresponding linear regression model used in the superstructure optimization problem (P). In order to limit the number of variables in the surrogate models, multiple instances of the same unit can be considered as part of the superstructure, which correspond to different sets of operating parameters; for example, two instances of an anaerobic digester with SRTs of 15 and 20 days.

3.2.3.3 Economic Performance Indicators

The superstructure optimization problem (P) involves minimizing a number of KPIs representing a particular plant configuration's performance. As far as economical performance is concerned, one can consider the NPV over the project's lifetime, given by

$$KPI_{NPV} = -CAPEX + \sum_{yr=1}^{LT} \frac{(SALES-OPEX)}{(1+DISC)^{yr}}, \tag{3.1}$$

where LT denotes the project lifetime, typically 20 years; SALES represents revenues from energy/nutrient sales; CAPEX and OPEX denote the costs caused by WWTP capital investment and operation, respectively; and DISC is the discount factor over the whole project lifetime, namely the rate at which future payoffs are discounted to the present value. Alternatively, one may consider a life-cycle costing indicator (Rebitzer et al., 2003).

Reliable costing information needed to compute such indicators, including capital and operating costs, can be obtained from preliminary costing software, such as CapdetWorks. The use of a common source and methodology for costing various technologies presents the advantage of consistency, although such data may not be widely available or reliable for the newer technologies. Similar to the approach outlined previously in Section 3.2.3.2 for plant-wide performance surrogates, one approach involves deriving costing surrogates based on data generated from these computer programs for each

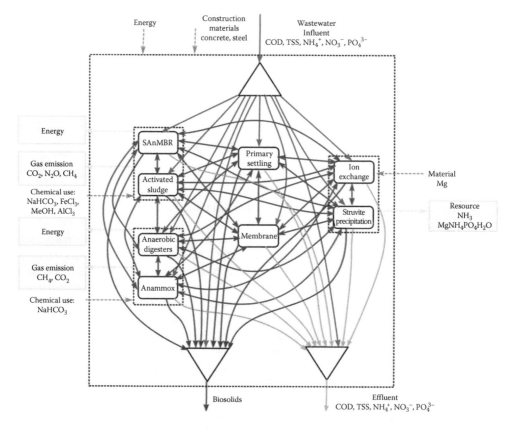

FIGURE 3.11 Schematic representation of the system boundary and superstructure in the case study. (Puchongkawarin, C., Decision-making for sustainable wastewater treatment: Model-based methodology. PhD thesis, Imperial College London, 2015.).

unit by varying their size and regressing these data as appropriate. Since the residence time in a given unit is fixed here, their volume is indeed proportional to the inlet flow rate.

3.2.3.4 Environmental Performance Indicators

The main environmental impacts associated with each process configuration in the superstructure can be estimated using LCA—an illustration of possible subprocesses modeled within the boundary of a WWTP system are shown in Figure 3.11, which includes the WWTP infrastructure, operational inputs, and emissions over the plant's lifetime. To carry out the inventory analysis, a rather natural choice for the functional unit is "a unit volume of wastewater influent over a given time period" (Suh and Rousseaux, 2001). Moreover, where a given product is attributed multiple functions, allocation can be made preferentially by substitution; for instance, treated effluents plus nutrients recovered from the WWTP would be accounted for as fertilizer replacement and green electrical power generated from a combined heat and power (CHP) system would be exported to the grid. This way, a wastewater treatment facility is credited with the avoided environmental burdens, such as GHG emissions or resource depletion that would otherwise be incurred by generating the corresponding amount of fertilizer or electricity through conventional routes.

Impact categories that are most relevant for wastewater treatment facilities include the global warming potential (GWP100) and the eutrophication potential. The overall load for a given category j can be quantified as follows:

$$\text{KPI}_j = \text{INFRA}_j + \text{LT} \cdot (\text{OPER}_j + \text{WWEFF}_j + \text{WWSLU}_j - \text{CRED}_j), \qquad (3.2)$$

where INFRA_j, OPER_j, WWEFF_j, WWSLU_j, and CRED_j represent the individual loads associated with the required infrastructure, annual operation of the plant, discharged effluent, discharged sludge, and obtained credit, respectively. All these loads may themselves be computed as combinations of a list of "elementary" environmental burdens corresponding, but not limited, to the use of steel, concrete, or electrical power; the emissions of CO_2 or CH_4 from the treatment units; the release of COD, ammonia, phosphates, or suspended solids with the treated effluent; and the utilization of N, P, or magnesium resources. In particular, these elementary impacts can be obtained from the EcoInvent database, which is available through LCA software such as SimaPro. Aggregation of these various burdens into the loads in Equation 3.2 relies on inventory data predicted by the plant-wide performance surrogates (Section 3.2.3.2). As a first approximation, the loads may be assumed to scale linearly with the inventory flows. Further details regarding the impact assessment can be found in Puchongkawarin (2015) and Vaupel (2015).

3.2.3.5 Numerical Solution Strategies

The superstructure optimization problem (P) yields a nonconvex optimization model due to the presence of bilinear terms that arise in the material balances of the units as a result of contaminant mixing, in addition to other nonlinearities in certain performance and costing expressions. This nonconvexity can lead to multiple local optimal solutions, thereby calling for the implementation of global optimization techniques to guarantee a reliable solution. Recent work in water network synthesis (Ahmetović and Grossmann, 2011; Khor et al., 2012) demonstrates that deterministic global optimization solvers such as BARON (Tawarmalani and Sahinidis, 2004) or ANTIGONE (Misener and Floudas, 2014) are now able to provide global optimality certificates within reasonable computational times for such problems.

With regards to the multiobjective nature of the problem, popular (deterministic) approaches to computing Pareto sets include the weighted, ε-constraints, and goal programming methods (Rangaiah and Bonilla-Petriciolet, 2013). For instance, the weighted-sum approach combines a set of objectives into a unique objective by using weights that are varied in order to describe the solution set, whereas the ε-constraint method converts all but one objective into inequality constraints whose right-hand side values are varied to describe the solution set.

TABLE 3.2   Composition of the Municipal Wastewater in the Case Study

Total COD	Soluble COD	TSS	volatile suspended solids (VSS)	volatile fatty acids (VFA)
569 mg L^{-1}	129 mg L^{-1}	259 mg L^{-1}	231 mg L^{-1}	10 mg L^{-1}
Total N	Ammonia	Total P	Phosphate	Alkalinity
51.6 mg L^{-1}	38 mg L^{-1}	7.6 mg L^{-1}	4.1 mg L^{-1}	253 mg CaCO$_3$ L^{-1}

3.2.4   Case Study

The main objective of the present case study is to illustrate the methodology outlined in Section 3.2.3, along with typical results. Due to space restrictions, the reader is referred to Puchongkawarin (2015) and Vaupel (2015) for a more detailed account of the results and further discussion.

The case study considers the synthesis of a wastewater treatment/recovery facility for a 10,000 m^3/day municipal wastewater stream with the average compositions reported in Table 3.2. In agreement with the EU Directive 91/271/EEC on Urban Wastewater Treatment, the targeted maximum concentrations for the treated effluent stream are 142.3 mg/L total COD, 7.6 mg/L NH$_4^+$, 10.3 mg/L NO$_3^-$, 0.82 mg/L PO$_4^{3-}$, and 25.9 mg/L total suspended solids (TSS), which correspond to minimum abatements of 75% total COD, 80% total N and total P, and 90% TSS. Besides the inlet and outlet wastewater streams, the case study also considers an outlet biosolids stream in connection to sludge production. The superstructure shown in Figure 3.11 is comprised of the following treatment/separation units—see Puchongkawarin et al. (2015) and Vaupel (2015) for further details:

- *Biological treatment*, including three activated sludge processes [with (1) nitrification, (2) nitrification and denitrification, and (3) enhanced P removal], one SAnMBR unit, two digesters (with solids retention times of 15 and 20 days), and one Sharon–Anammox process;
- *Resource recovery*, including one struvite precipitation unit for P recovery and one ion exchange unit for N recovery;
- *Physical separation*, with one membrane unit and one primary clarifier (besides the secondary clarifiers of the activated sludge processes).

The focus is on optimizing two KPIs, namely NPV and GWP100 over a period of 20 years.

In order to carry out the computations, the performance of each treatment unit is predicted by the ManTIS3 simulator within GPS-X, which includes the most common biological, physical, and chemical processes in WWTPs. Surrogate models are developed for the biological treatment units based on performance and GHG emission projections by the simulator. For the membrane units and the primary clarifier, simple models based on split fractions for the solids are used, assuming no biological reactions and a perfect split for the soluble species. Finally, performance predictions for the ion exchange units are based on literature data, regressed by the Langmuir isotherm model (Puchongkawarin et al., 2015; Vaupel, 2015). In addition, the CAPEX and OPEX of all the units, but membranes, are estimated using CapdetWorks, and regressed with simple linear models as a function of the unit volume and/or processed flow rate (within the operational range). The CAPEX and OPEX of the membrane units, on the other hand, are set based on expert's recommendations (Puchongkawarin et al., 2015). The LCA model is implemented in Simapro 7.3, with a midpoint approach CML 2 baseline 2000 (V2.05) applied as characterization method at the life-cycle impact assessment (LCIA) stage. Statistics on the UK country-level N and P fertilizer composition (International Fertilizer Industry Association 2014) and UK average electricity grid mix (Department of Energy and Climate Change, 2015) are adopted in this study to credit the WWTP system with the avoided production of fertilizers and electricity. The inventory for generic fertilizer production and fuel combustion is derived from the EcoInvent database (v2.2).

The optimal configuration shown in Figure 3.12 is obtained by maximizing the NPV of the treatment facility only, without consideration to environmental impacts; that is, it corresponds to one of the endpoints of the Pareto set (single-objective optimization). About 91% of the wastewater stream is processed in the

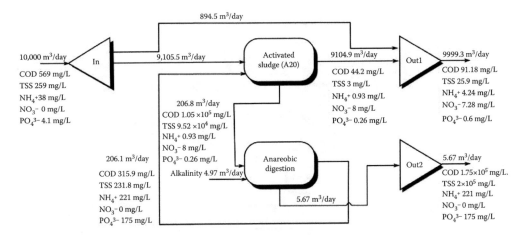

FIGURE 3.12   Optimal plant configuration for the NPV maximization problem.

TABLE 3.3   Economic and Environmental Performance in the Case Study Problems

Flow Sheet	Maximal NPV	Maximal Enviroeconomic Value
CAPEX, M£	−5.42	−7.70
OPEX, M£/year	−0.45	−0.64
SALES, M£/year	0.24	0.42
NPV, M£	−7.69	−10.07
GWP100, × 10^4 tCO_2 e	23.2	1.72
Net profit, M£	−14.3	−10.6

activated sludge process with enhanced P removal (A2O), before mixing with the remaining part of the wastewater stream and discharging into the environment. The sludge produced by the activated sludge treatment is processed into the anaerobic digester (SRT 20 days), whose supernatant stream is returned to the activated sludge unit and the biosolids stream is disposed into a landfill. The treated wastewater meets all the discharge requirements, and it is the minimum TSS abatement of 90% that happens to be the most restrictive here. The NPV for this optimal configuration is estimated as M£-7.69, with the breakdown costing analysis including CAPEX, OPEX, and SALES shown in Table 3.3. Because CAPEX is indeed the largest contributor to NPV, the WWTP configuration for maximizing NPV is comprised of a minimal number of treatment units in order to comply with the discharge constraints. Moreover, a longer SRT is selected for the anaerobic digester in order to increase the amount of biogas produced and mitigate the sludge disposal cost.

In the enviroeconomic optimization problem, the conflicting objectives of NPV and GWP100 are considered simultaneously. The plant configuration shown in Figure 3.13 is one particular point on the optimal Pareto solution set (not shown); if monetization was used for GWP100, this solution would correspond to a carbon trading price of £28.5/tCO_2, currently DECCs central scenario (Department of Energy and Climate Change, 2014). Observe that this plant configuration is markedly different from the economically optimum one in Figure 3.12. The SAnMBR unit treats about 96% of the incoming wastewater stream: the sludge produced by this unit is processed in the anaerobic digester (SRT 20 days), whose digester cake is sent for disposal, incurring a significant cost due to the landfill tax. Furthermore, the digester liquor is mixed with both the SAnMBR outlet stream and the bypass stream, and passed through the ion exchange and struvite precipitation units for N and P recovery. Here again, the treated wastewater meets all the discharge requirements, but it is now the minimum phosphate abatement of 80% that becomes the most restrictive. As shown in Table 3.3, the NPV is greater by about M£2.4 due to a high CAPEX, but the GHG emissions are reduced by over 13 fold, thus defining an enviroeconomic

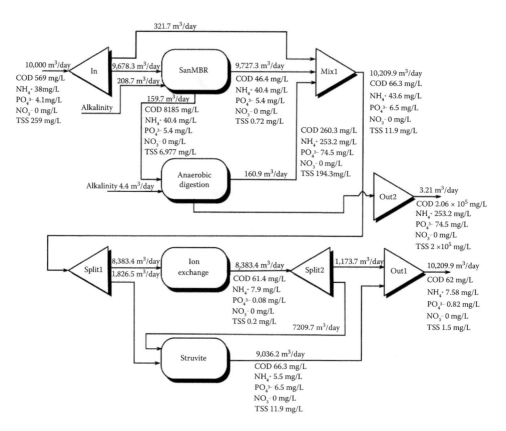

FIGURE 3.13 Optimal plant configuration for the enviroeconomic optimization problem.

trade-off. The SAnMBR unit becomes part of the optimal configuration in combination with both N and P recovery units due to their lower GHG emissions compared with conventional activated sludge treatment, both CO_2 emissions from electrical power consumption and N_2O evolved from biological nitrification/denitrification.

3.2.5 Conclusions

This chapter has presented and illustrated a systematic optimization-based methodology for incorporating LCA alongside economic criteria into a multiobjective optimization methodology for the synthesis of sustainable WWTPs. This methodology relies on surrogate models as a means for overcoming the limitation of current global optimization technology, which does not allow for optimizing complex plant configurations (complex biological processes, multiple scales, time dependence, etc.) all in a single step. A key requirement in applying this methodology nonetheless is the availability of reliable performance models for the treatment and separation units, on one hand, and reliable costing and environmental impact data, on the other hand. Our work advocates the use of state-of-the-art wastewater treatment simulators for deriving simple response-surface models, which are general enough to be independent of detailed design choices and keep the superstructure optimization model computationally tractable, and the use of LCA state-of-the-art databases to assess the main environmental impacts likewise.

Overall, this methodology should be regarded as a decision-support system for identifying, among hundreds or even thousands of alternatives, a number of promising wastewater treatment and resource recovery systems for a given wastewater stream and regional context. The preselected plant configurations can be considered for detailed design analysis and optimization in a subsequent step. The case study results demonstrate that the proposed framework can provide valuable insights for decision making in WWTP design (Puchongkawarin, 2015; Vaupel, 2015) and that LCA integration into decision-making tools for wastewater treatment alongside economical considerations may lead to radical changes in the design of tomorrow's wastewater treatment facilities.

Acknowledgments

Channarong Puchongkawarin is grateful to the Royal Thai Government Scholarship program and to the Centre of Process Systems Engineering of Imperial College for financial support. Benoît Chachuat gratefully acknowledges financial support by ERC career integration grant PCIG09-GA-2011-293953.

References

Agarwal, A. and Biegler, L. T. (2013). A trust-region framework for constrained optimization using reduced order modeling. *Optimization and Engineering*, 14(1):3–35.

Ahmetović, E. and Grossmann, I. E. (2011). Global superstructure optimization for the design of integrated process water networks. *AIChE Journal*, 57(2):434–457.

Aiyuk, S., Amoako, J., Raskin, L., van Haandel, A., and Verstraete, W. (2004). Removal of carbon and nutrients from domestic wastewater using a low investment, integrated treatment concept. *Water Research*, 38(13):3031–3042.

Alasino, N., Mussati, M. C., and Scenna, N. J. (2007). Wastewater treatment plant synthesis and design. *Industrial and Engineering Chemistry Research*, 46(23):7497–7512.

Alasino, N., Mussati, M. C., Scenna, N. J., and Aguirre, P. (2010). Wastewater treatment plant synthesis and design: Combined biological nitrogen and phosphorus removal. *Industrial and Engineering Chemistry Research*, 49(18):8601–8612.

Balkema, A. J., Preisig, H. A., Otterpohl, R., Lambert, A. J., and Weijers, S. R. (2001). Developing a model based decision support tool for the identification of sustainable treatment options for domestic wastewater. *Water Science and Technology*, 43(7):265–270.

Batstone, D. J., Hülsen, T., Mehta, C. M., and Keller, J. (2015). Platforms for energy and nutrient recovery from domestic wastewater: A review. *Chemosphere*, 140:2–11.

Batstone, D. J., Keller, J., Angelidaki, I., Kalyuzhny, S. V., Pavlostathis, S. G., Rozzi, A., Sanders, W. T. M., Siegrist, H., and Vavilin, V. A. (2002). *Anaerobic Digestion Model No. 1 (ADM1)*. IWA Publishing, London.

Bhuiyan, M. I., Mavinic, D. S., and Koch, F. A. (2008). Phosphorus recovery from wastewater through struvite formation in fluidized bed reactors: A sustainable approach. *Water Science and Technology*, 57(2):175–181.

Biegler, L. T., Grossmann, I. E., and Westerberg, A. W. (1997). *Systematic Methods of Chemical Process Design*. Prentice Hall, Upper Saddle River, N.J.

Biegler, L. T., Lang, Y. D., and Lin, W. (2014). Multi-scale optimization for process systems engineering. *Computers and Chemical Engineering*, 60:17–30.

Bisinella de Faria, A. B., Spérandio, M., Ahmadi, A., and Tiruta-Barna, L. (2015). Evaluation of new alternatives in wastewater treatment plants based on dynamic modelling and life cycle assessment (DM-LCA). *Water Research*, 84:99–111.

Bozkurt, H., Quaglia, A., Gernaey, K. V., and Sin, G. (2015). A mathematical programming framework for early stage design of wastewater treatment plants. *Environmental Modelling and Software*, 64:164–176.

Bufe, M. (2008). Getting warm? Climate change concerns prompt utilities to rethink water resources, energy use. *Water Environment and Technology*, 20(1):29–32.

Coats, E. R., Loge, F. J., Wolcott, M. P., Englund, K., and McDonald, A. G. (2007). Synthesis of polyhydroxyalkanoates in municipal wastewater treatment. *Water Environment Research*, 79(12):2396–2403.

Corominas, L., Foley, J., Guest, J. S., Hospido, A., Larsen, H. F., Morera, S., and Shaw, A. (2013). Life cycle assessment applied to wastewater treatment: State of the art. *Water Research*, 47(15):5480–5492.

Daigger, G. T. (2009). Evolving urban water and residuals management paradigms: Water reclamation and reuse, decentralization, and resource recovery. *Water Environment Research*, 81(8):809–823.

Department of Energy and Climate Change (2014). Updated short-term traded carbon values used for UK public policy appraisal. https://www.gov.uk/government/uploads/system/uploads/attachment_data/file/360277/Updated_short-term_traded_carbon_values_used_for_UK_policy_appraisal__2014_.pdf (Accessed September 30, 2015).

Department of Energy and Climate Change (2015). Digest of UK energy statistics (DUKES). https://www.gov.uk/government/collections/digest-of-uk-energy-statistics-dukes (Accessed July 7, 2017).

Doyle, J. D. and Parsons, S. A. (2002). Struvite formation, control and recovery. *Water Research*, 36(16):3925–3940.

Emmerson, R. H. C., Morse, G. K., Lester, J. N., and Edge, D. R. (1995). The life-cycle analysis of small-scale sewage-treatment processes. *Water and Environment Journal*, 9(3):317–325.

EPA (2007). Opportunities for and benefits of combined heat and power at wastewater treatment facilities. Report #EPA-430-R-07–003, U.S. Environmental Protection Agency.

Faria, D. C. and Bagajewicz, M. J. (2009). Profit-based grassroots design and retrofit of water networks in process plants. *Computers and Chemical Engineering*, 33(2):436–453.

Flores, X., Bonmati, A., Poch, M., Roda, I. R., Jimenez, L., and Banares-Alcantara, R. (2007). Multicriteria evaluation tools to support the conceptual design of activated sludge systems. *Water Science and Technology*, 56(6):85–94.

Foley, J., de Haas, D., Hartley, K., and Lant, P. (2010). Comprehensive life cycle inventories of alternative wastewater treatment systems. *Water Research*, 44(5):1654–1666.

Fu, F. and Wang, Q. (2011). Removal of heavy metal ions from wastewaters: A review. *Journal of Environmental Management*, 92(3):407–418.

Gallego, A., Hospido, A., Teresa, M., and Feijoo, G. (2008). Environmental performance of wastewater treatment plants for small populations. *Resources, Conservation and Recycling*, 52(6):931–940.

Garrido-Baserba, M., Hospido, A., Reif, R., Molinos-Senante, M., Comas, J., and Poch, M. (2014). Including the environmental criteria when selecting a wastewater treatment plant. *Environmental Modelling and Software*, 56:74–82.

Guest, J. S., Skerlos, S. J., Barnard, J. L., Beck, M. B., Daigger, G. T., Hilger, H., Jackson, S. J., Karvazy, K., Kelly, L., Macpherson, L., Mihelcic, J. R., Pramanik, A., Raskin, L., Van Loosdrecht, M. C., Yeh, D., and Love, N. G. (2009). A new planning and design paradigm to achieve sustainable resource recovery from wastewater. *Environmental Science and Technology*, 43(16):6126–6130.

Hakanen, J., Sahlstedt, K., and Miettinen, K. (2013). Wastewater treatment plant design and operation under multiple conflicting objective functions. *Environmental Modelling and Software*, 46:240–249.

Hamouda, M. A., Anderson, W. B., and Huck, P. M. (2009). Decision support systems in water and wastewater treatment process selection and design: A review. *Water Science and Technology*, 60(7):1757–1770.

Henze, M., Gujer, W., and Mino, T. (2000). *Activated Sludge Models ASM1, ASM2, ASM2d and ASM3*. IWA Publishing, London.

Hu, A. and Stuckey, D. C. (2006). Treatment of dilute wastewaters using a novel submerged anaerobic membrane bioreactor. *Journal of Environmental Engineering*, 132(2):190–198.

International Energy Agency (2012). Technology roadmap: Bioenergy for heat and power. http://www.iea.org/publications/freepublications/publication/technology-roadmap-bioenergy-for-heat-and-power-.html (Accessed September 30, 2015).

ISO 14040 (2006). Environmental management—Life cycle assessment—Principles and framework.

Johir, M. A. H., George, J., Vigneswaran, S., Kandasamy, J., and Grasmick, A. (2011). Removal and recovery of nutrients by ion exchange from high rate membrane bio-reactor (MBR) effluent. *Desalination*, 275(1–3):197–202.

Khiewwijit, R., Temmink, H., Rijnaarts, H., and Keesman, K. J. (2015). Energy and nutrient recovery for municipal wastewater treatment: How to design a feasible plant layout? *Environmental Modelling and Software*, 68:156–165.

Khor, C. S., Chachuat, B., and Shah, N. (2012). A superstructure optimization approach for water network synthesis with membrane separation-based regenerators. *Computers and Chemical Engineering*, 42:48–63.

Khor, C. S., Chachuat, B., and Shah, N. (2014). Optimization of water network synthesis for single-site continuous problems: Milestones, challenges, and future directions. *Industrial and Engineering Chemistry Research*, 53(25):10257–10275.

Kim, J., Kim, K., Ye, H., Lee, E., Shin, C., McCarty, P. L., and Bae, J. (2011). Anaerobic fluidized bed membrane bioreactor for wastewater treatment. *Environmental Science and Technology*, 45(2):576–581.

Kim, Y. and Logan, B. E. (2011). Hydrogen production from inexhaustible supplies of fresh and salt water using microbial reverse-electrodialysis electrolysis cells. *Proceedings of the National Academy of Sciences of the United States of America*, 108(39):16176–16181.

Kokossis, A. C. and Yang, A. (2010). On the use of systems technologies and a systematic approach for the synthesis and the design of future biorefineries. *Computers and Chemical Engineering*, 34(9):1397–1405.

Kuzawa, K., Jung, Y., Kiso, Y., Yamada, T., Nagai, M., and Lee, T. (2006). Phosphate removal and recovery with a synthetic hydrotalcite as an adsorbent. *Chemosphere*, 62(1):45–52.

Le Corre, K. S., Valsami-Jones, E., Hobbs, P., and Parsons, S. A. (2009). Phosphorus recovery from wastewater by struvite crystallization: A review. *Critical Reviews in Environmental Science and Technology*, 39(6):433–477.

Lew, B., Tarre, S., Beliavski, M., Dosoretz, C., and Green, M. (2009). Anaerobic membrane bioreactor (AnMBR) for domestic wastewater treatment. *Desalination*, 243(1–3):251–257.

Liao, B. Q., Kraemer, J. T., and Bagley, D. M. (2006). Anaerobic membrane bioreactors: Applications and research directions. *Critical Reviews in Environmental Science and Technology*, 36(6):489–530.

Liberti, L., Petruzzelli, D., and De Florio, L. (2001). REM NUT ion exchange plus struvite precipitation process. *Environmental Technology*, 22:1313–1324.

Lin, H., Chen, J., Wang, F., Ding, L., and Hong, H. (2011). Feasibility evaluation of submerged anaerobic membrane bioreactor for municipal secondary wastewater treatment. *Desalination*, 280(1–3):120–126.

Liu, P., Georgiadis, M., and Pistikopoulos, E. (2011). Advances in energy systems engineering. *Industrial and Engineering Chemistry Research*, 50(9):4915–4926.

Logan, B. E., Hamelers, B., Rozendal, R., Schröder, U., Keller, J., Freguia, S., Aelterman, P., Verstraete, W., and Rabaey, K. (2006). Microbial fuel cells: Methodology and technology. *Environmental Science and Technology*, 40(17):5181–5192.

Martin, B. D., Parsons, S. A., and Jefferson, B. (2009). Removal and recovery of phosphate from municipal wastewaters using a polymeric anion exchanger bound with hydrated ferric oxide nanoparticles. *Water Science and Technology*, 60(10):2637–2645.

McCarty, P. L., Bae, J., and Kim, J. (2011). Domestic wastewater treatment as a net energy producer—Can this be achieved? *Environmental Science and Technology*, 45(17):7100–7106.

Mehta, C. M., Khunjar, W. O., Nguyen, V., Tait, W., and Batstone, D. J. (2015). Technologies to recover nutrients from waste streams: A critical review. *Critical Reviews in Environmental Science and Technology*, 45(4):385–427.

Miladinovic, N. and Weatherley, L. R. (2008). Intensification of ammonia removal in a combined ion-exchange and nitrification column. *Chemical Engineering Journal*, 135(1–2):15–24.

Misener, R. and Floudas, C. A. (2014). ANTIGONE: Algorithms for continuous/integer global optimization of nonlinear equations. *Journal of Global Optimization*, 59(2–3):503–526.

Newcombe, R. L., Strawn, D. G., Grant, T. M., Childers, S. E., and Möller, G. (2008). Phosphorus removal from municipal wastewater by hydrous ferric oxide reactive filtration and coupled chemically enhanced secondary treatment: Part I—Performance. *Water Environment Research*, 80(3):238–247.

Owen, W. F. (1982). *Energy in Wastewater Treatment*. Prentice-Hall, Englewood Cliffs, NJ.

Pasqualino, J. C., Meneses, M., Abella, M., and Castells, F. (2009). LCA as a decision support tool for the environmental improvement of the operation of a municipal wastewater treatment plant. *Environmental Science and Technology*, 43(9):3300–3307.

Puchongkawarin, C. (2015). Decision-making for sustainable wastewater treatment: Model-based methodology. PhD thesis, Imperial College.

Puchongkawarin, C., Gomez-Mont, C., Stuckey, D. C., and Chachuat, B. (2015). Optimization-based methodology for the development of wastewater facilities for energy and nutrient recovery. *Chemosphere*, 140:150–158.

Quaglia, A., Pennati, A., Bogataj, M., Kravanja, Z., Sin, G., and Gani, R. (2014). Industrial process water treatment and reuse:A framework for synthesis and design. *Industrial and Engineering Chemistry Research*, 53(13):5160–5171.

Rangaiah, G. P. and Bonilla-Petriciolet, A. (2013). *Multi-Objective Optimization in Chemical Engineering: Developments and Applications*. John Wiley & Sons, New York.

Rebitzer, G., Hunkeler, D., and Jolliet, O. (2003). LCC—The economic pillar of sustainability: Methodology and application to wastewater treatment. *Environmental Progress*, 22(4):241–249.

Renou, S., Thomas, J. S., Aoustin, E., and Pons, M. N. (2008). Influence of impact assessment methods in wastewater treatment LCA. *Journal of Cleaner Production*, 16(10):1098–1105.

Rigopoulos, S. and Linke, P. (2002). Systematic development of optimal activated sludge process designs. *Computers and Chemical Engineering*, 26(4–5):585–597.

Rivas, A., Irizar, I., and Ayesa, E. (2008). Model-based optimisation of wastewater treatment plants design. *Environmental Modelling and Software*, 23(4):435–450.

Rodriguez-Garcia, G., Molinos-Senante, M., Hospido, A., Hernández-Sancho, F., Moreira, M., and Feijoo, G. (2011). Environmental and economic profile of six typologies of wastewater treatment plants. *Water Research*, 45(18):5997–6010.

Smith, A. L., Stadler, L. B., Love, N. G., Skerlos, S. J., and Raskin, L. (2012). Perspectives on anaerobic membrane bioreactor treatment of domestic wastewater: A critical review. *Bioresource Technology*, 122:149–159.

Stolzenburg, P., Capdevielle, A., Teychené, S., and Biscans, B. (2015). Struvite precipitation with MgO as a precursor: Application to wastewater treatment. *Chemical Engineering Science*, 133:9–15.

Stuckey, D. C. (2012). Recent developments in anaerobic membrane reactors. *Bioresource Technology*, 122:137–148.

Suh, Y. J. and Rousseaux, P. (2001). Considerations in life cycle inventory analysis of municipal wastewater treatment systems. In *Oral Presentation at COST 624 WG Meeting*, Bologna, Italy.

Suh, Y. J. and Rousseaux, P. (2002). An LCA of alternative wastewater sludge treatment scenarios. *Resources, Conservation and Recycling*, 35(3):191–200.

Tangsubkul, N., Parameshwaran, K., Lundie, S., Fane, A., and Waite, T. (2006). Environmental life cycle assessment of the microfiltration process. *Journal of Membrane Science*, 284(1–2):214–226.

Tawarmalani, M. and Sahinidis, N. V. (2004). Global optimization of mixed-integer non-linear programs: A theoretical and computational study. *Mathematical Programming*, 99(3):563–591.

Tchobanoglous, G., Burton, F. L., and Stensel, H. D. (2003). *Wastewater Engineering: Treatment and Reuse*. McGraw-Hill, New York.

Ueno, Y. and Fujii, M. (2001). Three years experience of operating and selling recovered struvite from full-scale plant. *Environmental Technology*, 22(11):1373–1381.

Vaupel, Y. (2015). Superstructure-optimisation-based methodology for the development of processes for resource recovery from wastewater. Master's thesis, RWTH Aachen University.

Wang, S. and Peng, Y. (2010). Natural zeolites as effective adsorbents in water and wastewater treatment. *Chemical Engineering Journal*, 156(1):11–24.

3.3 Water Conservation, Reuse, and Challenges: A Case Study Performed at Amul Dairy

Sarojini Tiwari, Chitta Ranjan Behera, and Babji Srinivasan

3.3.1 Introduction

India is now the largest producer of milk (Deparment of Dairying Fisheries and Animal Husbandary 2013) owing to "Operation Flood" [A. Banerjee (Exeutive Director of National Dairy Development Board) 2001], a project launched by National Dairy Development Board (NDDB) in 1970 and led by late Dr. Verghese Kurien. The common dairy products, such as butter, cheese, ghee, and curd, are the only acceptable sources of animal protein for the vegetarians. The enormous vegetarian population makes India the largest consumer of its own dairy products. Dairying is a persistent source of income for the rural areas in India. As per NDDB, the Indian dairy industry is all set to experience high growth rates in the next 8 years with demand likely to reach 200 million tons by 2022 (National Dairy Development Board 2014).

Dairy industries, similar to other food processing industries, consume substantial amounts of water predominantly for cleaning in place (CIP), tankers, crates, silos, and floor wash. The present annual consumption of water in the dairy sector is roughly 62 billion m^3 and is all set to rise well above 400 billion m^3 by 2025 (Sustainability Outlook 2014). The major sources of water in India for industrial and agricultural consumptions are river and groundwater. The Central Pollution Control Board (CPCB) report on the status of water quality shows alarming decrease in the quality of surface water in India (CPCB India 2011). According to a report by the World Bank, about 60% of aquifers in India will be in a critical condition in another 20 years (World Bank Reports 2010). A report by Columbia Water Centre states that the overexploitation of groundwater in northern Gujarat is unsustainable and may exhaust this valuable water resource if left unchecked (Narula et al. 2011). The proliferating water demand by the booming dairy industry would create a considerable impact on the water sustainability in India (Amarasinghe et al. 2012). The depleting water resources and quality necessitates the initiation and application of sound water and wastewater management systems. This management system should not only be able to address water scarcity but also be able to minimize waste water production by endorsing reclamation and reuse.

The unorganized and mishandled practices in water usage in dairy industries are leading to adverse impacts on our environment. Feasible water and waste water management practices are the need of the hour (Brião and Tavares 2007; Carawan et al. 1972). Although many water and waste water management protocols in dairy industries have been advised in literature, there are few case studies to observe the outcomes of their application (Patil et al. 2014). Alongside better water management practices, new technologies on treatment of dairy waste effluent are coming up with the sole purpose of reclamation and reuse (Sarkar et al. 2006; Vourch et al. 2008). There are still many uncertainties on the application of such technologies due to the lack in their cost-effectiveness. Hence, a proficient water and waste water management is best practice to minimize water usage and maximize water reclamation. A detailed review of the state-of-the-art water reclamation and energy saving practices in dairy industries is provided by S.J. Rad and M.J. Lewis (Rad and Lewis 2014). Although techniques like pinch analysis and water reclamation are widely used in dairy industries at developed countries (Rad and Lewis 2014), there is hardly any followed in India. Also, the challenges faced in Indian dairy industries are different with a huge emphasis on cost resulting in fewer measurements (minimum sensors installed) and less emphasis on waste water treatment. This work aims to study Indian dairy industries and develop a framework for their water and energy sustainability while understanding the constraints in their process operations.

A comprehensive method to analyze the efficiency in all the sectors viz. water, energy, and waste of a food industry is to determine its performance indices in those fields. Following are the key performance indices (KPI) that are commonly used:

- Water used per kg or liter of product formed
- Waste water generated per kg or liter of product formed

- Solid waste production
- Energy consumption
- Chemicals usage

The excessive consumption of water in dairy industries leads to generation of waste water to a great extent. Thus, water consumed or waste water produced per liter of milk processed is the best KPI to assess the water usage at a dairy industry. The volume of waste water effluent is 10 L for each liter of milk processed in Indian dairy industries (Kolhe et al. 2002). The same ratio for Australian dairy industries is 1.9 (Dairy Australia 2011) and varies from 0.5 to 2.5 for Denmark (Food Efficiency 2011). A comprehensive list of KPI in terms of volume ratio of water and waste water with the amount of milk processed of various countries is listed in Table 3.4. The comparison clearly concludes that Indian dairy industries are relatively producing a large quantity of waste water.

Indian dairy industries are operated by a large number of independent cooperative bodies. The effective usage of water to sustain this valuable resource is still not a serious concern in various sectors including dairy. An inclusive and accurate report on the overall water usage exclusively for the dairy industries is not yet available. However, it can be concluded that the water consumption for every liter of milk processed is more in some developed countries since waste water production is highest.

The subject of this case study, Kaira District Co-operative Milk Producers Union Limited, widely known as Amul Dairy, is located in Anand, Gujarat. Amul is the largest food brand in India majoring in production of milk and milk products. In 1946, this dairy was the significant center of "white revolution." It is run by collection of milk from around thousand villagers.

The collected milk is first filtered to remove unwanted solid particles. The filtrate is chilled to 3°C–4°C and then stored in a buffer tank before it is sent to a centrifugal clarifier. This clarification using centrifugal force ensures the removal of microscopic spores from the raw milk. The clarified milk is stored in raw milk silos. This purified milk is then separated and stored in cream and skimmed milk

TABLE 3.4 Performance Indices of Dairy Industries around the World

Country	Key Performance Indices		
	Average Values and Range	Units	References
Australia	1.0–4.5	$\dfrac{L\ Water}{L\ milk\ processed}$	Dairy Australia (2011)
	1.9	$\dfrac{L\ Waste\ water}{L\ milk\ processed}$	
Canada	1.0–5.0	$\dfrac{L\ Water}{L\ milk\ processed}$	Wardrop Engineering Inc. (1997)
	0.6–2.7	$\dfrac{L\ Water}{kg\ milk\ processed}$	
Denmark	0.5–2.5	$\dfrac{L\ Wastewater}{L\ milk\ processed}$	Food Efficiency (2011)
USA	3.57	$\dfrac{kg\ Water}{kg\ of\ total\ product}$	Carawan et al. (1979)
France	0.2–10.0	$\dfrac{L\ Wastewater}{L\ milk\ processed}$	Vourch et al. (2008)

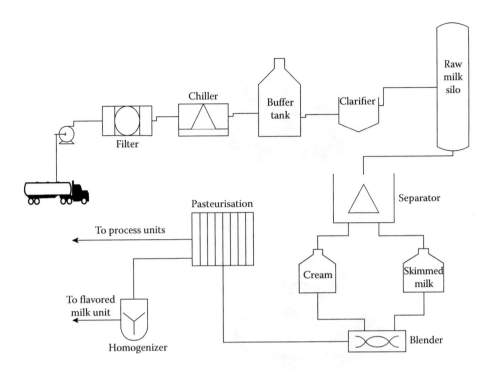

FIGURE 3.14 Raw milk receiving and processing at Amul Dairy.

tanks. These two are blended as per the fat content requirement of the products. The blended product is then pasteurized and transported to the respective process unit to manufacture the required product. A small amount of the same pasteurized milk is homogenized and transported to the flavored milk manufacturing section. Figure 3.14 summarizes the entire process. With a daily milk handling capacity of 4.5 million L, Amul Dairy manufactures ghee, butter, processed milk, milk powder, and flavored milk.

This chapter lists down a framework for water reclamation for this dairy industry. The proposed approach minimizes water consumption by reclamation of the generated waste water. At the same time, some modifications on its effluent treatment plant (ETP) have been suggested to improve its performance in terms of energy consumption and treated effluent quality. Finally, possibilities of nitrous oxide emission from the ETP, an important greenhouse gas, are investigated.

3.3.2 Water Usage Assessment at Amul Dairy

Amul Dairy, apart from processing the collected raw milk, manufactures various milk products such as butter, ghee, milk powder, skimmed milk and powder, and flavored milk. It has a separate milk packaging unit where the processed milk is packed in pouches before being sold in markets. The chart in Figure 3.15 gives a complete description of distribution of water usage in various process units within the dairy. Operational processes involve a daily usage of hot and chilled water mostly in heat exchanger units used in ghee and butter manufacture and milk pasteurization. Crate wash and railway tanker wash together take up 2.4% of the total water used per day. The make-up water requirement for the cooling tower and boiler feed adds to 62.5% of the total. CIP and floor wash account for maximum consumption of water which is 75% of the total. A small fraction (1.55%) of total water used per day in research and development labs and other similar areas are clubbed in the "other section" of the chart.

CIP is essentially a three-stage process. The first stage is rinsing of equipment with soft water. The second stage is to wash the equipment with hot lye (sodium hydroxide solution) to wash out the unwanted microorganisms. The third stage involves washing out leftover lye in the equipment with soft water.

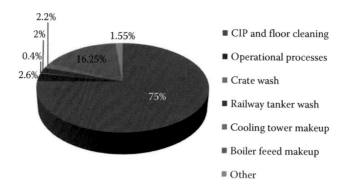

FIGURE 3.15   Breakdown of water usage at Amul Dairy.

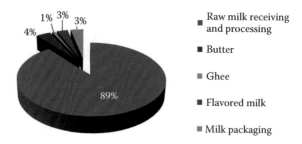

FIGURE 3.16   Breakdown of CIP and floor cleaning water in process units.

Automatic CIP systems usually reutilizes the washed out water used in its third stage for rinsing. A fourth and a final cycle of washing out the equipment with nitric acid followed by soft water is also carried out once in a week. On an average, about 75% of the overall water intake is being used by the dairy for CIP and floor wash.

Figure 3.16 shows the distribution of the same in various process units of Amul Dairy. Raw milk receiving and processing department (RMRD) of Amul Dairy utilizes 89% of the total water used for CIP and floor cleaning, possibly because of the large amount of milk handled every day. The milk packaging section packs the pasteurized milk into pouches utilizing 3% of the total water used per day for CIP and floor wash. The flavored milk section, a product made by adding edible flavors to homogenized milk, takes up the same amount as milk packaging section. The butter section is the largest consumer of water next to RMRD which utilizes 4% of water used daily for CIP. Ghee, a form of clarified butter, comparatively takes up the least amount of water for CIP and floor cleaning owing to the small sized equipment required for its manufacture.

Figure 3.17 shows the overall water flow and effluent sources within the dairy. This water is fed to cooling tower and boiler section as a make-up feed and to the processing units for CIP and manual floor cleaning. The processing units are further divided as shown in the figure. The dotted lines indicate the flow of waste water effluent generated after the cleaning. All the waste water is diverted to the dairy's own ETP. A small quantity of treated water is used for toilet flushing and the rest is disposed without any significant usage. Amul Dairy has two powder plants, one manufacturing powder from normal milk, while the other from skimmed milk. Both the powder plants generate a large amount of condensed water which is of good quality. The condensates are used up as boiler make-up feed and as make-up water in the cooling tower. The circulation of fresh water along with condensates is shown in Figure 3.17.

Apparently, the waste water effluents from various sections/process units are being mixed to be treated at ETP. It was obvious that not all the effluents from various process units have the same pollution level.

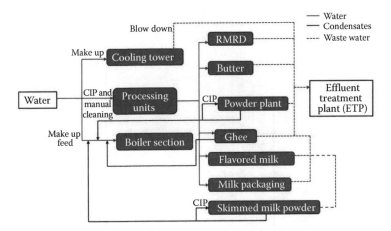

FIGURE 3.17 Waste water effluent sources at Amul Dairy.

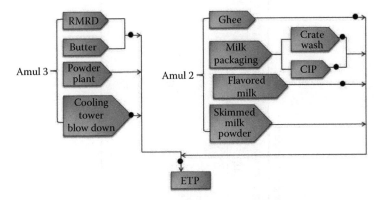

FIGURE 3.18 Sampling points indicated by dots to identify pollution level at various process units.

Hence, to monitor and compare the pollution level of the effluents, waste water samples were collected from different process units of Amul Dairy. The dots shown in Figure 3.18 indicate the sampling points. There is a common waste water effluent discharge line for RMRD and butter section. Thus, only mixed effluent samples could be collected for analysis. The cooling tower blow-down water was the next sampling point. The milk packaging section has two waste water discharge lines, one for crate wash effluent and the other for effluent generated from CIP. This section of Amul Dairy does not have an automatic CIP system. The water used in the third stage is not reused for rinsing. This water sample was collected along with the effluent from crate wash and analyzed separately.

The rest of the effluent samples were collected from the ghee and flavored milk manufacturing units. The final sample was the mixed waste water effluent entering the ETP. The waste water parameters chosen to analyze the above effluent samples were total COD, total suspended solids (TSS), conductivity, pH, total kjeldahl nitrogen (TKN), total alkalinity, and hardness. The waste water analysis techniques were referred from American Public Health Association (Walter 1961).

3.3.3 Waste Water Segregation, Treatment, and Reuse Scheme

The analysis result of the waste water samples collected from various process units at the dairy are shown in Table 3.5. The results clearly show that the waste water effluents have different pollution levels. A scheme, shown in Figure 3.19, is created to categorize the effluent based on its pollution level. The effluent

TABLE 3.5 Average (Avg) and Standard Deviation (SD) Values of Tabulated Parameters

Process Units	Total COD (mg/L)		TKN (mg N/L)		Total Alkalinity (mg CaCO₃/L)		Hardness (mg CaCO₃/L)		pH		Conductivity (mS)		TSS (mg/L)	
	Avg	SD	Avg	SD	Avg	SD	Avg	SD	Avg	SD	Avg	SD	Avg	SD
Influent to ETP	4747.2	1006	106	32	1700	400	504	134	4.74	0.4	2.25	0.3	1408	134
Amul-3	3807	856	98	39	2115	963	597	187	8.94	0.5	3.95	1.6	1414.8	109
Flavored milk	1573.8	497	23.8	9	1394	124	557	80.6	6.83	0.3	1.26	0.02	791.8	142
Ghee section	1099.4	400	16.8	3	1967	414	475	50	6.98	0.2	1.25	0.02	691	132
Packaging CIP	633.6	352	12.6	6	717	160	295	41	7.31	1.6	0.65	0.15	300.6	52
Crate wash	549	287	16	3	1031	331	465	147	6.67	0.6	1.29	0.02	749.8	142
Cooling tower (blow-down)	626	202	4	1.5	2133	642	508	173	9.20	0.2	4.77	1.40	987.6	119

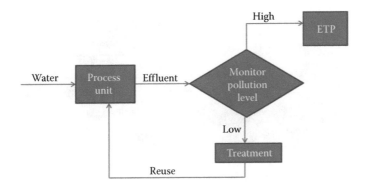

FIGURE 3.19 Waste water segregation, treatment, and reuse scheme.

is monitored and then divided into high and low pollution level categories. This categorization of waste water can then be used to decide the level of its treatment. If the waste water is less polluted, it should be treated at the level where it is produced and reused within that process unit. Mixing of all the waste effluents often increase its pollution intensity. Following are the important inferences made from the analysis results:

RMRD and butter section—The effluent from these sections do not have separate pipeline reaching to the collection tank at ETP. Thus, the waste water is mixed and gets highly polluted.

- The effluent from ghee section has high oily and fatty matter content. This contributes the most to the total COD of waste water.
- CIP water from milk packaging—The analysis result clearly shows that this water is least polluted and mixing it with high and medium polluted effluent is an open wastage of a valuable resource.
- Cooling tower blow-down water—Although this effluent is comparatively less polluted, the conductivity and TSS values are high and greatly variable.

3.3.4 Suggestions for Water Reclamation

Experiments at lab scale have shown that ghee CIP effluents when treated by coagulation followed by carbon adsorption, the total COD can reduce by 96% (Ahmad et al. 2014). In addition, TDS, TSS, and total inorganic carbon get reduced considerably. This methodology removes the oil and fatty matter using optimum dosage of trivalent metallic salts as coagulants. The sludge thus formed is removed by adsorption on powdered carbon. Recent case studies (Atman et al. 2012) have established that blow-down water from cooling tower can be treated through membrane filters such as reverse osmosis. This can drastically reduce TSS and conductivity of the water and make it suitable for reuse. Based on the above suggestions on water reclamation, a treatment and reuse scheme was formulated as shown in Figure 3.20. The waste water effluent from the ghee section and blow-down water from cooling tower can be treated as suggested at the process unit instead of mixing it with highly polluted effluent from other process units. The CIP water in the milk packaging section is least polluted. It can be reused for rinsing of the equipment.

Assuming that the water used by the above process units is equal to the total waste water produced, more than 7% of the total water used for CIP and floor wash can be reclaimed every day by implementation of the aforementioned techniques.

3.3.5 Effluent Treatment Plant and Suggested Modifications

The ETP at Amul Dairy has a combination of anaerobic digesters and activated sludge systems to treat the dairy waste water effluent. The layout of the dairy's ETP is shown in Figure 3.21.

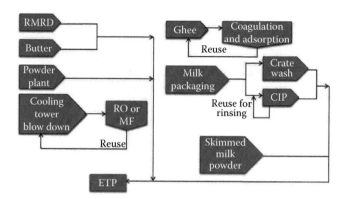

FIGURE 3.20 Waste water segregation, treatment, and reuse scheme for Amul Dairy.

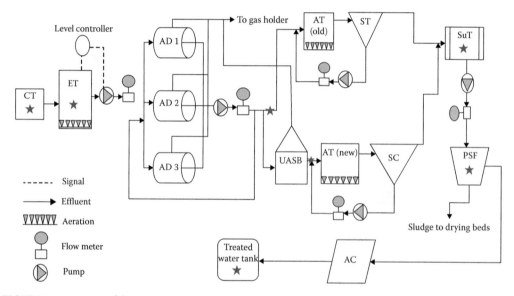

FIGURE 3.21 Layout of the ETP.

Table 3.6 gives the description of abbreviations used in the figure. The waste water is accumulated in a collection tank and then fed to an equalization tank. The purpose of this tank is (1) to improve the pH by addition of sodium bicarbonate, (2) homogenization through continuous aeration, and (3) to maintain a constant effluent flow rate of 50 m³/h. The effluent then enters three parallel tank digesters at the outlet of which it splits into three parts. The effluent is recycled to the tank digesters at the rate of 5 m³/h. The effluent enter the upflow anaerobic sludge blanket (UASB) at the rate of 35 m³/h, and the rest is fed to the old aeration tank (AT) followed by settling tank. The partially treated effluent from UASB enters the new AT followed by secondary clarifier. The supernatant from both the clarifiers is stored in supernatant tank. The effluent is then passed through a pressure sand filter to remove the suspended solids in it. It is then fed to activated carbon unit which removes the color and odor from the treated water. The treated water is finally stored in a tank for reuse in gardening and flushing. The biogas produced is blazed in the gas holder.

The real ETP at Amul Dairy is nonoptimized and operated manually. The efficiency of the reactors was determined by analyzing influent and effluent samples from every reactor of the ETP. The parameters, total COD, filtered COD, TKN, and TSS were analyzed using standard methods. Total COD gives a good estimate of pollution reduction capacity of a biological reactor, while filtered COD can be used to estimate the

TABLE 3.6 Description of Abbreviations and Residence Time of the Reactors in Figure 8.8

Abbreviations	Units	Effective Volume (m³)	Residence (h)
ET	Equalization tank	1000	12 (minimum)
AD 1, 2, 3	Anaerobic digesters 1,2,3	750 (each)	40
UASB	Uplift anaerobic sludge blanket	1205	24
Anoxic	Anoxic tank	840	9
AT(new)	Aeration tank (new)	840	9
SC	Secondary clarifier	452.16	7.5
SuT	Supernatant tank	210	4.2
PSF	Pressure sand filter	19.625	NA

TABLE 3.7 Efficiency of the Biological Reactors at ETP in Terms of Percentage Reduction of Tabulated Parameters

Reactor	Reduction in Percentage			
	COD (%)	Filtered COD (%)	TKN (%)	TSS (%)
Equalization tank	52	54	−41	15
Anaerobic digester	46	26	58	17
UASB	42	55	50	37
Aeration tank	61	63	40	25
Pressure sand filter	63	50	30	67
Activated carbon	−28	−47	−14	33

biodegradable matter present in the waste water effluent. TKN value reports the sum of total organic and ammoniacal nitrogen accounting for the approximate nutrient concentration in the effluent. Composite waste water samples were collected from the collection tank, equalization tank, mixed effluent from the three anaerobic digesters, effluent from UASB, supernatant tank, effluent from the pressure sand filter and final treated tank. The sampling points are indicated in Figure 3.21. The analyzed data were compared and percentage reduction in parameters values was calculated using the following equation:

$$P = \frac{P_i - P_O}{P_i} \times 100,$$

where P is the percent reduction in the value of the parameter such as total COD, filtered COD, TKN, and TSS. P_i is the parameter value of the waste water entering the reactor and P_O is the parameter value of the treated effluent from the reactor. The results are shown in Table 3.7. The negative percentage indicates the increase in the parameter values of the treated effluent from the reactors. Following are the important inferences made from the analyzed data:

- There is a 52% decrease in the total COD in effluent from the equalization tank. This occurs due to the continuous aeration meant for homogenization of the waste water in the tank. The increase in TKN might be due to the formation of organic compounds containing nitrogen by hydrolysis of complex matter.
- The COD reduction efficiency in all the reactors is low. The highest reduction of 63% occurs in the pressure sand filter due to higher removal of solid particles.
- There is an increase in total COD, filtered COD, and TKN in the effluent from the activated carbon filter by 28%, 47%, and 14%, respectively. This inefficiency indicates low maintenance of the activated carbon unit or usage of low grade carbon for adsorption.

Based on the above inferences, the ETP can be modified to enhance the characteristics of treated water, biogas production, and energy saving in the following suggested ways:

- The aeration in the equalization tank is taking up a lot of carbon sources that can later contribute to enhanced biogas production from the anaerobic digesters. The aeration can be replaced by slow stirrers to bring about homogenization.
- Removal of old AT—The old AT comparatively treats a smaller quantity of waste water. The removal of this tank saves energy by shutting down the blower running for 24 h. The waste water stream, instead of splitting into two, enters UASB. Thus, the biodegradable matter entering this reactor increases which in turn improves biogas production.
- A clarifier behind UASB—Removal of the old AT can reduce the hydraulic retention time (HRT) of UASB. A clarifier that allows the sludge to settle and recycled to the reactor can ensure its better performance with lower HRT.
- A separate anoxic tank—The new AT may be replaced by two tanks, one anoxic and the other oxic. The anoxic conditions bring about denitrification of the nitrate and nitrite ions and hence ensure better removal of total nitrogen from waste water.
- Maintenance of the reactors—A timely removal of dead sludge from the biological reactors will ensure their longevity and better performance. The carbon filters in the activated carbon unit should be replaced frequently to improve the quality of treated water. A timely cleaning of the pipelines is necessary to prevent algae or bacterial formation.

3.3.6 Green House Gas (GHG) Emissions

Methane (CH_4), carbon dioxide (CO_2), and nitrous oxide (N_2O) are the GHGs known to be released during waste water treatment processes. According to intergovernmental panel on climate change (IPCC) fourth assessment report, GHG emissions from postconsumer waste water in the year 2004–2005 was 3% of global anthropogenic GHG emissions. CH_4 from waste water and landfills accounted for 18% of the total anthropogenic methane emissions (Bogner et al. 2008). Countries such as Canada and China have evaluated the total GHG emissions from their municipal waste water treatment systems. Canada reports 1600 Mg/year CH_4 and 669,100 Mg/year of CO_2 on-site emissions (Sahely et al. 2006). China has estimated an average of 5.68×10^5 kg emissions of CO_2 per day from their treatment plants (Tao and Wang 2012). Nitrous oxide (N_2O) is another greenhouse gas with a global warming potential 310 times more than carbon dioxide. It contributes significantly to the depletion of the ozone layer (IPCC 2007). According to an estimate, biological nutrient removal processes from waste water treatment plants contribute about 3.2% of the total nitrous oxide emissions (Kampschreur et al. 2009).

A comparative study on anaerobic and aerobic reactors has shown that efficient methane recovery from anaerobic digesters reduces their potential to release GHGs. Aerobic reactors, on the other hand, release a considerable amount of GHG when the influent BOD becomes greater than 700 mg/L (Cakir and Stenstrom 2005). The ETP at Amul Diary releases CO_2 and CH_4 mixture directly to the atmosphere. The lack of awareness and infrastructure to utilize the gas mixture is causing wastage of a valuable energy resource.

There is a possibility of N_2O emissions from the aerobic reactors operating the ETP. Ammonia oxidizing bacteria (AOB) are known to produce nitrous oxide when exposed suddenly to aerobic environment from anoxic conditions (Rassamee et al. 2011). Although the dairy's ETP do not have an anoxic tank, the secondary clarifier of the activated sludge system may provide the anoxic conditions to the AOBs. An activated sludge system consists of an AT followed by a secondary clarifier. The sludge is then allowed to settle after which it is recycled to the AT. The AT at the dairy's ETP runs for an average of 31 h providing the oxic conditions to the nitrifying bacteria. Next, the effluent enters the clarifier that is in anoxic condition and has a retention time of 7.5 h.

The recycled sludge is, therefore, exposed to an oxic environment for 31 h followed by anoxic condition for the next 7.5 h. Since the recycled activated sludge usually has rich microorganism population, AOBs are exposed to recurrent oxic and anoxic conditions. The feed to the AT also has high ammoniacal nitrogen concentration. The above conditions may lead to the production of nitrous oxide gas in considerable quantity. Minimization of nitric oxide emission during activated sludge processes is an important concern for the environmental engineers.

3.3.7 Conclusion

The current work provides a framework for water and energy sustainability in diary industries in India. A detailed analysis of the sources of waste water at Amul Dairy is conducted to identify the quality of these effluents and a unique scheme has been suggested for water reclamation. The modifications suggested to the ETP can considerably improve the treated effluent quality and reduce its expenses by saving energy. However, detailed simulation and experimental studies need to be performed to understand the potential of the suggested modifications. Finally, the possibilities of nitrous oxide emissions call for further investigations of the ETP to analyze the gas production and incorporate mitigation techniques.

The proposed unit level approach for water reuse and reclamation can be extended to develop a holistic framework that could foster water and energy sustainability. This framework can be realized by an optimization approach that minimizes the energy and water usage using the reclamation techniques while accounting for cost constraints and plant dynamics through appropriate models. The reclamation techniques suggested here are best applicable to dairy effluents and can be further generalized for various industries to promote water and energy sustainability.

Acknowledgment

The authors wish to gratefully acknowledge Mr S. S. Sundaran (OSD, Public Relations) and Mr V. B. Dholakia (Engineer-in-charge of ETP) of Amul Dairy, Anand for granting us necessary permissions.

References

Ahmad, W., Hussain, R., Nafees, M., and Hussain, A. (2014). Optimization of Wastewater Treatment Process in Industry "A Case Study of Hattar Industrial Estate Haripur." *Pakistan Journal of Analytical and Environmental Chemistry*, 15(1), 7.

Altman, S. J., Jensen, R. P., Cappelle, M. A., Sanchez, A. L., Everett, R. L., Anderson, H. L., and McGrath, L. K. (2012). Membrane treatment of side-stream cooling tower water for reduction of water usage. *Desalination*, 285, 177–183.

Amarasinghe, U. A., T. Shah, and V. Smakhtin. 2012. Water–milk nexus in India: A path to a sustainable water future? *International Journal of Agricultural Sustainability* 10 (1): 93–108. doi:10.1080/14735 903.2012.620336.

Banerjee, A. (NDDB Exeutive Director). 2001. Dairying systems in India. FAO Corporate Document Repository. http://www.fao.org/docrep/T3080T/t3080T07.htm

Bogner, J., R. Pipatti, S. Hashimoto, C. Diaz, K. Mareckova, L. Diaz, P. Kjeldsen S. Monni, A. Faaij, Q. Gao and T. Zhang 2008. Mitigation of global greenhouse gas emmision from waste: Conclusions and strategies from IPCC fourth assessment report (working group III). *Waste Management and Research*. doi:10.1177/0734242X07088433.

Brião, V. B. and C. R. G. Tavares. 2007. Effluent generation by the dairy industry: Preventive attitudes and opportunities. *Brazilian Journal of Chemical Engineering* 24 (4): 487–97. doi:10.1590/S0104-66322007000400003.

Cakir, F. Y. and M. K. Stenstrom. 2005. Greenhouse gas production: A comparison between aerobic and anaerobic wastewater treatment technology. *Water Research* 39 (17): 4197–4203. doi:10.1016/j.watres.2005.07.042.

Carawan, R. E., V. A. Jones, and A. P. Hansen. 1972. Water and wastewater management in dairy processing, North Carolina Cooperative Extension Service 1.

Carawan, R. E., V. A. Jones, and A. P. Hansen. 1979. Water use in a multiproduct dairy. *Journal of Dairy Science* 62 (8): 1238–42. doi:10.3168/jds.S0022-0302(79)83406-2.

Central Pollution Control Board (CPCB) India. 2011. Status of water quality in India—2010. http://cpcb.nic.in/WQSTATUS_REPORT2010.pdf

Dairy Australia. 2011. Australian Dairy Manufacturing Environmental sustainability report—Australian dairy, 2010–11. http://www.dmsc.com.au/dmsc/reports/DMSC%20Sustainability%20Report%202010-11%20_%20FINAL%20111013.pdf

Deparment of Dairying Fisheries and Animal Husbandary. 2013. Annual Report 2012–13. *Ministry of Agriculture, Government of India* 15 (1): 1. doi:10.1016/j.parkreldis.2008.12.001.

Food Efficiency. 2011. Water usage, waste water and biogas potential within the dairy manufacturing sector 1–4. http://www.foodefficiency.eu/

IPCC. 2007. Climate change 2007: Impacts, adaptation and vulnerability. Contribution of Working Group II to the Fourth Assessment Report of the Intergovernmental Panel. Genebra, Suíça.

Kampschreur, M. J., H. Temmink, R. Kleerebezem, M. S. M. Jetten, and M. C. M. van Loosdrecht. 2009. Nitrous oxide emission during wastewater treatment. *Water Research.* doi:10.1016/j.watres.2009.03.001.

Kolhe, A. S., S. R. Ingale, and R. V. Bhole. 2002. Effluent of dairy technology. *International Research Journal* II (5): 459–61.

Narula, K., R. Fishman, V. Modi, and L. Polycarpou. 2011. Addressing the water crisis in Gujarat, India Columbia Water Center, New York.

National Dairy Development Board. 2014. Annual Report, 2013–14. India. http://www.nddb.org/about/report.

Patil, S. A., V. V. Ahire, and M. H. Hussain. 2014. Dairy wastewater—A case study. *International Journal of Research in Engineering and Technology* 3 (9): 30–34.

Rad, S. J. and M. J. Lewis. 2014. Water utilisation, energy utilisation and waste water management in the dairy industry: A review. *International Journal of Dairy Technology* 67 (1): 1–20. doi:10.1111/1471-0307.12096.

Rassamee, V., C. Sattayatewa, K. Pagilla, and K. Chandran. 2011. Effect of oxic and anoxic conditions on nitrous oxide emissions from nitrification and denitrification processes. *Biotechnology and Bioengineering* 108 (9): 2036–45. doi:10.1002/bit.23147.

Sahely, H. R., H. L. Maclean, H. D. Monteith, and D. M. Bagley. 2006. Comparison of on-site and upstream greenhouse gas emissions from Canadian municipal wastewater treatment facilities. *Journal of Environmental Engineering Science* 5: 405–15. doi:10.1139/S06-009.

Sarkar, B., P. P. Chakrabarti, A. Vijaykumar, and V. Kale. 2006. Wastewater treatment in dairy industries—Possibility of reuse. *Desalination* 195 (1–3): 141–52. doi:10.1016/j.desal.2005.11.015.

Sustainability Outlook, India. 2014. Water : A business risk to the food and water: A business risk to the food and beverages industry ? Quantifying lifecycle water risk and embodied value for water in food and beverage products in India.

Tao, X. and C. Wang. 2012. Greenhouse gas emissions from wastewater treatment plants. *Journal of Tsinghua University (Science and Technology)* 52: 473–477. http://en.cnki.com.cn/Article_en/CJFDTOTAL-QHXB201204010.htm.

Vourch, M., B. Balannec, B. Chaufer, and G. Dorange. 2008. Treatment of dairy industry wastewater by reverse osmosis for water reuse. *Desalination* 219 (1–3): 190–202. doi:10.1016/j.desal.2007.05.013.

Walter, W. G. 1961. Standard methods for the examination of water and wastewater (11th Ed.). *American Journal of Public Health and the Nations Health* 51 (6): 940. doi:10.2105/AJPH.51.6.940-a.

Wardrop Engineering Inc. 1997. Guide to energy efficiency opportunities in the dairy processing industry—Canada. National Dairy Council of Canada, Mississauga, ON. http://www.fpeac.org/dairy/EnergyEfficiencyOpportunities-DairyProcessing.pdf.

World Bank Reports. 2010. Deep wells and prudence: Towards pragmatic action for addressing groundwater overexploitation in India 120. http://documents.worldbank.org/curated/en/272661468267911138/Deep-wells-and-prudence-towards-pragmatic-action-for-addressing-groundwater-overexploitation-in-India

3.4 Occurrence, Effects, and Treatment of Endocrine-Disrupting Chemicals in Water

Chedly Tizaoui, Olajumoke Ololade Odejimi, and Ayman Abdelaziz

3.4.1 Introduction

The effects of environmental contaminants on humans as well as animals have long been a great concern. In the past decades, some of these contaminants have been found to interfere with the endocrine system, leading to adverse effects on humans and aquatic life (The Royal Society, 2000; Ning et al., 2007; Diamanti-Kandarakis et al., 2009).

Endocrine-disrupting chemicals (EDCs) can be defined as substances that interfere with the endocrine system, causing undesirable health effects in an intact creature. They can act in several ways and affect different components of the body. They can act by (1) mimicking the structure of natural hormones in the body, (2) reducing or preventing the production of hormones in the endocrine glands, and (3) increasing or decreasing the metabolic rate of hormones.

Most of the organic compounds present as EDCs are produced as a result of human activities, while others occur naturally. There is a broad list of chemicals listed as having endocrine-disrupting effects, most of which are only present in minute quantity, making their quantification and analysis very difficult. Some of the known EDCs include agricultural chemicals such as pesticides, fungicides, and dioxins; phthalates such as butyl benzyl phthalate and diethyl phthalate; biphenyls such as bisphenol A (BPA; Gore, 2007; Jackson and Sutton, 2008); pharmaceutical drugs such as tamoxifen (Leffers et al., 2001); and steroid estrogens such as the natural sex hormones estrone (E_1) and 17β-estradiol (E_2) and the synthetic hormone 17α-ethinyl estradiol (EE_2) (Nghlem et al., 2003; Kanda and Churchley, 2008; Diamanti-Kandarakis et al., 2009). Figure 3.22 shows the chemical structures of some of the known EDCs.

EDCs have been detected in wastewater, surface water, groundwater, and even drinking water (Ying et al., 2002; Nakada et al., 2006; Benotti et al., 2009). Recent studies and surveys have shown that estrogenic EDCs are present in quantifiable amounts in rivers and wastewater treatment plants (WWTPs) across the globe (Tables 3.8 and 3.9). These studies have also reported that steroid EDCs are the most dominant EDCs (Desbrow et al., 1998; Auriol et al., 2006). For instance, an estimated concentration of 0.2–1.0 ng/L of E_2, which have a relative estrogen receptor binding affinity of 100 (Sullivan and Krieger, 2001), and 1 ng/L of EE_2 can cause a male rainbow trout to produce vitellogenin, which is normally associated only with mature females (Purdom et al., 1994). Taking into consideration that estrogenic EDCs have adverse effects on the endocrine system even at low concentrations, several water industries and environmental agencies around the globe have shown great interest in the issue, thereby investing time and money in conducting research on the occurrence, fate, effects, and treatment of EDCs. The National Demonstration Programme (NDP) in the UK has recently evaluated the performance and costs of several treatment technologies to reduce the concentration of EDCs in the rivers in the UK. On completion of its first phase, the NDP has provided robust information related to the occurrence, treatment, and release of various EDCs from conventional WWTPs (Huo and Hickey, 2007; Kanda and Churchley, 2008; UKWIR, 2009). In addition, the EU has recently established a dynamic watch list (Priority Substances Directive 2013/39/EU) as a mechanism to monitor the emission of potential priority substances, and it has so far included three substances (pharmaceutical and steroid estrogens) [E_2, EE_2, and diclofenac (a pain killer)] in the first watch list; the monitoring data for these three substances will support further regulatory developments under the Water Framework Directive.

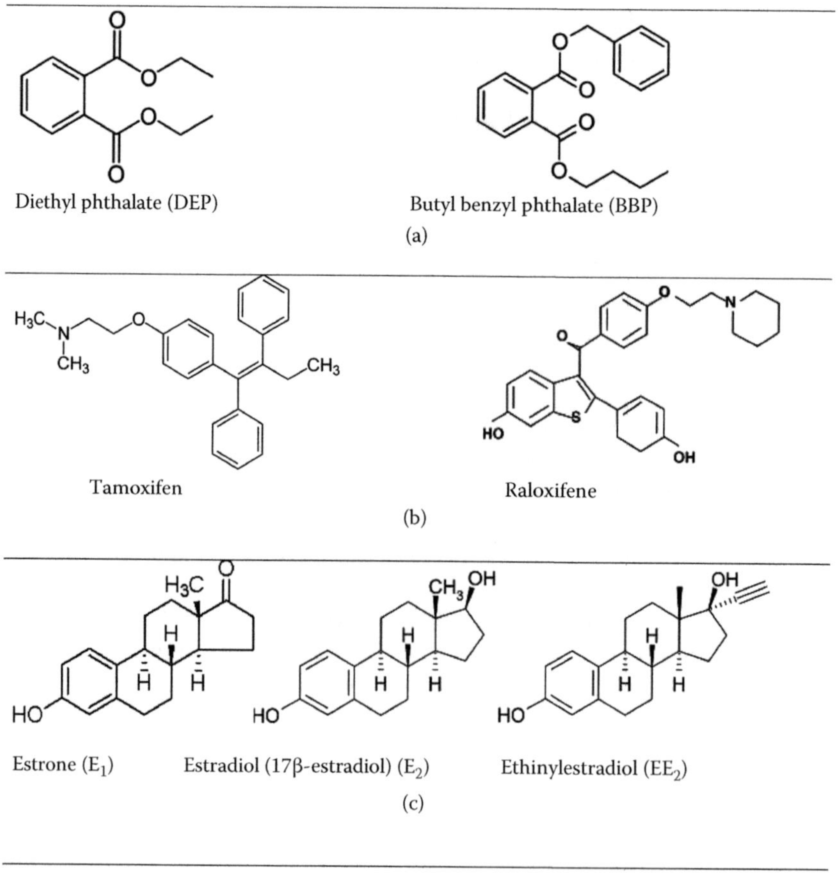

Diethyl phthalate (DEP)   Butyl benzyl phthalate (BBP)

(a)

Tamoxifen   Raloxifene

(b)

Estrone (E_1)   Estradiol (17β-estradiol) (E_2)   Ethinylestradiol (EE_2)

(c)

FIGURE 3.22 Chemical structure of some EDCs: (a) phthalates, (b) pharmaceutical drugs; and (c) natural and synthetic estrogens.

TABLE 3.8 Concentrations of Estrogenic EDCs in Water Media

Estrogen	Sampling Sites	Concentration (ng/L)	References
E_1	Shijing River (China)	79	Zhao et al. (2009)
	Ekbatan dam (Hamadan, Iran)	0.9–6	Jafari et al. (2009)
	Southern Taiwan	7.4–1267	Chen et al. (2010)
	French rivers	0.8–3.9	Cargouet et al. (2004)
E_2	Shijing River (China)	7.7	Zhao et al. (2009)
	Ekbatan dam (Hamadan, Iran)	1–2	Jafari et al. (2009)
	Southern Taiwan	N/D–313.6	Chen et al. (2010)
	French rivers	0.8–3.6	Cargouet et al. (2004)
EE_2	Ekbatan dam (Hamadan, Iran)	0.004–2	Jafari et al. (2009)
	French rivers	0.6–3.5	Cargouet et al. (2004)

The main objective of this chapter is to discuss the recent advances in the issue of EDCs in water, with the focus on the occurrence and fate of EDCs and the treatment techniques used for EDCs removal from water and wastewater. Although the list of chemicals that cause endocrine disruption is vast and getting larger and larger by the day, particular focus will be given to three steroid estrogens, namely the hormones E_1, E_2, and EE_2, of greatest concern due to their severe effects.

TABLE 3.9 Concentrations of Estrogens Found in WWTPs

Estrogen	Concentration (ng/L)	Influent or Effluent	WWTP Location	References
E_1	0.1–47	Effluent	The Netherlands	Belfroid et al. (1999)
	1.6–62.5	Effluent	UK	Thorpe et al. (2009)
	4.3–40.7	Influent	Beijing	Chang et al. (2011)
	0.1–9.6	Effluent		
	19–78	Influent	Across Canada	Servos et al. (2005)
	1–96	Effluent		
	44.7	Influent	UK	Kanda and Churchley (2008)
	1.22	Effluent		
E_2	1–12	Effluent	The Netherlands	Belfroid et al. (1999)
	7.22	Influent	New York	McAvoy (2008)
	N/D–7.2	Influent	Beijing	Chang et al. (2011)
	N/D–1.5	Effluent		
	2.4–26	Influent	Across Canada	Servos et al. (2005)
	0.2–14.7	Effluent		
	29.0	Influent	UK	Kanda and Churchley (2008)
	0.23	Influent		
EE_2	<0.2–7.5	Effluent	The Netherlands	Belfroid et al. (1999)
	0.65	Influent	UK	Kanda and Churchley (2008)
	0.63	Effluent		

3.4.2 Steroid Estrogens: E_1, E_2, and EE_2

The natural estrogens (E_1 and E_2) are female sex hormones secreted in both humans and animals and men secrete these hormones in small amounts. They are secreted by the ovaries and placenta in females and by the testes in males. On the other hand, EE_2 is a synthetic estrogen, which is mainly used as an active ingredient in oral contraceptives. It is also used in treating menopausal symptoms. Figure 3.22c shows the molecular structures of E_1, E_2, and EE_2. These molecules are biosynthesized from the precursor cholesterol molecule, and some of their properties are given in Table 3.10. E_1 has high persistency, high biological potency, and moderate concentration in wastewater and is a metabolic product of E_2. It is also secreted in a minute amount by the male reproductive system.

Natural and synthetic estrogens cause the greatest endocrine-disrupting activity in domestic wastewater effluents (Desbrow et al., 1998), with E_1, E_2, and EE_2 being the most active EDCs, because they are able to exert their physiological effects at much lower concentrations than other estrogens (Arnon et al., 2008). EE_2 has the highest estrogenic activity followed by E_2. Although E_1 has only half the potency of E_2, it is often found at more than double the concentration of E_2 in wastewater effluents (Johnson and Sumpter, 2001). Due to its high concentration in the aquatic environment, E_1 is potentially the most important endocrine disruptor among all the estrogens (D'Ascenzo et al., 2003). The total "potential no effect concentration" (PNEC) for natural and synthetic estrogens in surface waters was 1.0 ng/L, which was calculated as a weighted sum of E_1, E_2, and EE_2 [PNEC = (10EE_2 + E_2 + E_1)/3]. The PNEC values for

TABLE 3.10 Properties of the Estrogenic EDCs (Essandoh et al., 2012)

Compound	Molecular Formula	Molecular Mass[a] (g/mol)	Log K_{ow}	Aqueous Solubility (mg/L)
E_1	$C_{18}H_{22}O_2$	270.37	3.43	12.42
E_2	$C_{18}H_{24}O_2$	272.39	3.94	12.96
EE_2	$C_{20}H_{24}O_2$	296.40	4.15	4.83

E_1, E_2, and EE_2 are 3.0, 1.0, and 0.1 ng/L, respectively (Young et al., 2002). Tables 3.8 and 3.9 give typical concentrations of E_1, E_2, and EE_2 in various water matrices.

3.4.3 Sources, Exposure Routes, and Effects of EDCs

Humans are exposed to EDCs usually through eating contaminated food or drinking contaminated water. Exposure to wastes from industrial processes and consumer products, for example, via dermal absorption and eating foods that contain natural estrogens, such as soya and yams, are also other routes that lead to exposure to EDCs. Animals are exposed to EDCs through water, air, and food as a result of discharges of industrial and domestic effluents into the environment.

The principal source of estrogenic EDCs in the environment, especially water bodies, is human excretion (Braga et al., 2005). As the conventional WWTPs are not specifically designed to remove EDCs, estrogens are incompletely removed during treatment, and it is not surprising that WWTP effluents have become the main source of estrogens in the environment. Other sources of estrogens in water bodies include runoff from farmlands used for rearing livestock (Matthiessen et al., 2006). Low concentrations of estrogens have also been detected in aquifers recharged with wastewater effluents or in areas where animal manure or WWTP sludge is applied on the overlying agricultural land in the catchment area. They have been detected in spring water from some aquifers indirectly recharged with wastewater through irrigation practices (Gibson et al., 2007) or animal husbandry (Peterson et al., 2000). Hohenblum et al. (2004) also reported the presence of estrogens in groundwater due to the agricultural practices and contamination from domestic sewage. Septic tank systems have also been identified as a contributor to estrogens found in groundwater (Swartz et al., 2006). Avbersek et al. (2011) investigated the dynamics of steroid estrogen daily concentrations in hospital effluent and reported that in the hospital of their study, interday concentrations in the hospital effluent were between 8.6 and 31.3 ng/L for E_1 and 4.2 ng/L for E_2. It has also been established that women on a daily basis excrete on average ~32 µg E_1 and 14 µg E_2, while pregnant women excrete 100 times more (D'Ascenzo et al., 2003; also see Table 3.11). Other sources of estrogenic EDCs in the environment include agricultural runoff of chemicals such as pesticides and leakage from landfill areas. Johnson and Williams (2004) investigated the various ways in which estrogens enter the environment through humans, and the result of the investigation reinforces the fact that most of the estrogenic EDCs are female hormones, with 34% of E_1 from pregnant women, 26% from menstrual females, 10% from males, 4% from females on contraceptives, and 24% formed by conversion of E_2. Some of the sources of E_1, E_2, and EE_2 are summarized in Table 3.12.

The effects of EDCs can be immediate or delayed and reversible or irreversible, depending on the type of chemical, kind of tissue exposed, the dose, timing and duration of exposure, as well as metabolism and elimination rate from the body. The major categories of adverse biological effects linked to exposure to endocrine disruptors are reproductive problems (such as problem in the sperm quality and sperm count decline; Toppari et al., 1996; Swan et al., 1997), developmental disruptions (such as delayed sexual development; Gore, 2007), cancer (such as testicular cancer in males and breast cancer

TABLE 3.11 Excretion of Estrogens by Human Population per Day

Group	Amount Excreted in Microgram per Day (µg/day)		
	E_1	E_2	EE_2
Menopausal females	1.8	56.1	—
Menstrual females	11.7	3.2	—
Pregnant females	550	393	—
Females on contraceptives	—	—	10.5
Males	2.6	1.8	—

Source: Johnson, A.C., Williams, R.J., *Environ. Sci. Technol.*, 38, 3649–3658, 2004.

TABLE 3.12   Some Sources of E_1, E_2, and EE_2

Estrogenic EDC	Sources
Estrone (E_1)	Protein foods (e.g., eggs, fish, meat), agricultural waste, industrial discharge, agricultural runoffs, sewage sludge, and Sewage treatment plant (STP) effluents
17β-estradiol (E_2)	Protein foods (e.g., eggs, fish, meat), agricultural waste, industrial discharge, agricultural runoffs, sewage sludge, and STP effluents
17α-ethinylestradiol (EE_2)	Oral contraceptives, sewage sludge, and STP effluents

Source: Langford, K. and Lester, J.N., *Endocrine Disrupters in Wastewater and Sludge Treatment Processes*, CRC Press, Boca Raton, FL, 2003.

TABLE 3.13   Some Known EDCs and Their Effects on the Endocrine System

Chemical	Use	Mechanism	Health Effect	References
Bisphenol A (BPA)	Plasticizer (e.g., compact discs, safety helmets)	Mimics estrone; estrogen agonist	Altered mammary gland development	Rubin (2011) and Acevedo et al. (2013)
Dioxins	Flame retardants, by-product of industrial waste	Increases estrogen metabolism	Delayed puberty, change of sexual behavior in males to that of females	Bjerke et al. (1994), Kulkarni et al. (2008), and Korrick et al. (2011)
Phthalates (e.g., butyl benzyl phthalate)	Plasticizer, artificial leather	Mimic estrogen	Feminization of males, low sperm count	Heudorf et al. (2007) and Wittassek et al. (2009)
Dichlorodiphenyltrichloroethane (DDT)	Pesticide	Estrogen agonist	Delayed puberty, abnormal ovary cycle	Roy et al. (2009)
Vinclozolin	Fungicide	Androgen receptor antagonist	Feminization and nipple development in males	Uzumcu et al. (2004)
Diethylstilbestrol (DES)	Synthetic estrogen taken by pregnant women	Estrogen receptor antagonist	Vaginal and cervical cancers	Wang and Baskin (2008)

in females), immunological effects, and neurological effects (The Royal Society, 2000; Hotchkiss et al., 2008). Table 3.13 shows some known EDCs and their effects on the endocrine system.

3.4.4 Removal of EDCs in Water

Although EDCs have a range of potential sources, WWTP effluents are the major contributor of EDCs in the aquatic environment (Kolpin et al., 2002; Snyder et al., 2003; Tan et al., 2007). This is due to the fact that conventional wastewater treatment facilities are not specifically designed to remove EDCs. Even though the conventional primary and secondary treatment steps in a WWTP may remove some EDCs, they are not capable of reducing significantly the EDCs to concentrations at which no effect is observed. On the other hand, conventional drinking water treatment processes, which generally involve a sequence of flocculation, sedimentation, filtration, and disinfection, have also been found ineffective for substantially reducing the concentrations of EDCs. However, a significant amount of research using bench scale or pilot plant experiments has been done to assess the performance of a range of potential treatment techniques for removing EDCs. The various methods that have potential can be classified into physical methods, biological treatment, and advanced oxidation processes (AOPs). Cost as well as carbon footprint remain as very important criteria for the selection of a given treatment method regardless how effective it is.

3.4.4.1 Physical Methods

3.4.4.1.1 Adsorption by Activated Carbon (AC)

AC is a high surface area carbon produced by activating a carbon source like coal, wood, and petroleum. It is generally used in powdered or granular form (Temmink and Grolle, 2005; Liu et al., 2009). Adsorption of wastewater contaminants on AC has been widely used over the years as a means of treatment, and it is a versatile technology (Suzuki et al., 1996; Huang and Su, 2010; Oller et al., 2010). It has been successfully used by different researchers, but one disadvantage is that the AC is very expensive, making the process not economically feasible.

Zhang and Zhou (2005) examined the removal of E_1 and E_2 using adsorption through the use of various adsorbents such as chitin, ion exchange resin, granular activated carbon (GAC), and chitosan. Results obtained from their experiments showed that chitosan has the lowest adsorption capacity, with AC having the highest. A drawback to this method is that the adsorption was affected by environmental conditions. Rowsell et al. (2009) deduced, from their study on the removal of estrogens using granulated carbon, that reactivated carbon has more removal efficiency (81% estrogen removal achieved) than virgin carbon (which achieved only 65% estrogen removal).

Recently, a study was carried out by Grover et al. (2011) for the removal of estrogenic EDCs, E_1, E_2, and EE_2, from wastewater using AC, and reported that after adsorption, the concentrations of E_1, E_2, and EE_2 decreased from 0.6–3.1, <1.2–5.4, and <0.4–1.7 to <0.6–2.0, 1.2, and 0.4 ng/L, respectively. Churchley et al. (2011) have studied the removal of estrogens from a nitrifying activated sludge plant with rapid gravity sand filters using a GAC pilot plant containing eight filters, each having a capacity of 2.5 m^3 carbon. Their results showed that breakthrough for the steroid estrogens (E_1, E_2, and EE_2) occurred at 12–18 months operation. However, their operating cost analysis showed that GAC was expensive to operate.

Despite the fact that there is a significant reduction in concentrations of the EDCs after treatment using AC, the effluent concentrations of these estrogenic EDCs are still sufficient to cause dangerous endocrine-disrupting effects in humans and wildlife (Sullivan and Krieger, 2001; Cui et al., 2006); hence, there is a need for other more efficient removal methods.

3.4.4.1.2 Use of Nanofiltration/Reverse Osmosis (NF/RO) Membranes

A successful method for EDCs removal using membranes depends on a variety of factors such as the level of EDCs rejection by the membrane, adsorption of the EDCs on the membrane, water flux, retention, and resistance to fouling. NF and RO membranes were used by Schäfer et al. (2003) to remove E_1 from aqueous solutions. This particular work focused on the effect of the operating conditions on the retention of E_1 such as estrogen's concentration, pH, and ionic strength. It was concluded that size segregation and adsorptive effects are helpful in maintaining high retention on a variety of NF and RO membranes. However, a drawback to this method is that deprotonation of estrogens leads to a significant decrease in retention as a result of strong electrostatic repulsion with the membrane surface and decreasing the potential for adsorptive retention. This method only removed about 90% of the estrogens. In another study, Comerton et al. (2008) investigated the membrane's rejection of 22 EDCs, some of which were E_1, E_2, and EE_2, using NF and RO. The study indicated that the rejection values were high for RO membranes and low for loose NF membranes. The method only rejected about 90% of the EDCs, but the remaining 10% EDCs still have the capacity to cause endocrine-disrupting effects (Baronti et al., 2000; Pauwels and Verstraete, 2006).

3.4.4.2 Biological Treatment

Biological treatment of wastewater is commonly carried out using a microbial population capable of degrading the organic pollutants under defined conditions. Biodegradation rapidly converts aqueous organic substances into biomass, reducing their biological oxygen demand (BOD) in the process. Biological treatments are usually carried out aerobically by biological oxidation in trickling filters or activated sludge systems during secondary treatment in conventional WWTPs. Other systems, such as soil aquifer treatment, have also been used to remove estrogens using a biological population (Essandoh et al., 2012). In order to decide whether biodegradation is suitable for removal of estrogens, it is essential to comprehend the partitioning of the estrogens between water and the sediment; the higher the partitioning coefficients K_{ow}, the easier it is to remove the estrogens (Lai et al., 2000).

Various researchers have performed different kinds of experiments on the biodegradation of EDCs (Cajthaml et al., 2009; Robinson and Hellou, 2009; Shi et al., 2010). These investigations have anonymously reported that of all the EDCs, E_2 is the most easily and rapidly degraded EDC, while E_1 is the least removed EDC using this method (Miya et al., 2007). One of such studies was carried out by Robinson and Hellou (2009), in which the aerobic biodegradation of BPA, E_2, and EE_2 was measured. It was found that although biodegradation removed a significant amount of EDCs from wastewater, detectable levels of EDCs still remained after a period of 2 weeks. The drawbacks of biodegradation of estrogenic EDCs are (1) the long residence time taken to remove the estrogens, especially E_1 (E_1 is also a biodegradation product of E_2; hence, E_1 takes a longer time to be removed); (2) EE_2 exhibits slow microbial degradation during the treatment process; (3) long retention times of 10–14 h (Cargouet et al., 2004) or sometimes weeks (Robinson and Hellou, 2009); and (4) potential distribution of EDCs to biosolids used on agricultural soils. Miya et al. (2007) have studied the biological removal of estrogenic compounds over a period of 1 year and reported an average removal rate of 69.2% and 94.7% for E_1 and E_2, respectively. This reinforces the point that E_1 is present in double fold during biodegradation (i.e., it is present in influent and as a biodegradation product of E_2), and consequently, it is least removed by biodegradation. A much lower average total estrogen removal rate of 50% was reported by Cargouet et al. (2004).

Although biological treatment achieves a significant removal of estrogenic EDCs from wastewater, activated sludge is still not strong enough to completely break down or eliminate estrogens (Ternes et al., 1999; Johnson and Sumpter, 2001; Auriol et al., 2006; Pauwels and Verstraete, 2006; Koh et al., 2008). EDCs still have the capacity to cause endocrine-disrupting effects at concentrations as low as ng/L (Baronti et al., 2000; Pauwels and Verstraete, 2006).

3.4.4.3 Advanced Oxidation Processes

The basic feature of AOPs is that they generate highly reactive species by dissipating high energy (chemical, electrical, or radiating) into the water body. The most important species produced by AOPs are hydroxyl radicals, oxygen radicals, ozone, and hydrogen peroxide. These oxidizing agents are used to destroy resistant organic compounds into water, carbon dioxide, and inorganic compounds (Zhou and Smith, 2002; Comninellis et al., 2008; Poyatos et al., 2010). The most potent radicals are hydroxyl radicals ($\cdot$OH), which are extremely unstable species characterized by a one-electron deficiency. Hydroxyl radicals tend to react unselectively with the first chemical they come into contact leading to complete oxidation of dissolved organic contaminants. The different oxidation reagents and systems used in AOPs include photochemical degradation processes (e.g., O_3/UV), photocatalysis (e.g., TiO_2/UV), and chemical oxidation processes (e.g., chlorination and ozonation) (Rosenfeldt and Linden, 2004; Poyatos et al., 2010; Frontistis et al., 2015). Typical oxidation processes are briefly discussed in the following sections. The advantages and disadvantages of each process are outlined in Table 3.14.

3.4.4.3.1 Chlorination

For decades, chlorination has proved an effective way of treating water and has been widely used by most countries in the world for water disinfection. It works by destroying microorganisms such as bacteria. To treat water by chlorination, chlorine gas is dissolved in water and it hydrolyzes according to the following reaction (Deborde and Von Gunten, 2008):

TABLE 3.14 Advantages and Disadvantages of the Various Treatment Processes

	Advantages	Disadvantages
Physical Treatment		
Membranes		Expensive and need high maintenance (Auriol et al., 2006).
Adsorption by AC	Offers a reasonably high EDC removal (Auriol et al., 2006). It can also be used to control tastes and odors (Snyder et al., 2007).	ACs are expensive and only transfers EDCs from one medium to another (Zhang and Zhou, 2008). Adsorption capacity declined with increasing operation year (Choi et al., 2005).
Biological Treatment		
Biodegradation	It is a well-established method	E_1 is also a biodegradation product of E_2; E_1 takes a longer time to be removed, EE_2 exhibits slow microbial degradation and long retention times of 10–14 h (Cargouet et al., 2004) or sometimes weeks (Robinson and Hellou, 2009).
Oxidation Processes		
Chlorination	EE_2 reacts rapidly with HOCl and is completely removed (Auriol et al., 2006) Cost-effective (Yang, 2004)	By-products formed can be carcinogenic (Yang, 2004, Auriol et al., 2006). It cannot be generated on site, and because of this, it requires storage as a concentrated liquid.
Ozonation	Ozonation leaves no harmful by-products in water as it readily converts back to oxygen after use, does not require storage as it can be generated in situ from oxygen or air and oxidation breaks down a wide range of contaminants in solution (Ward et al., 2004).	Since it is produced from oxygen, it requires a considerable amount of electricity for its generation, thus increasing capital cost (Ward et al., 2003; Gogate and Pandit, 2004). Ozone does not dissolve extensively in water and due to its strong oxidizing power, it has to be handled with care (Ward et al., 2003).
Photolysis	It offers a high possibility of treating a wide range of other organic pollutants present in wastewater as it is independent of sample matrix and initial chemical concentration (Zhang and Zhou, 2008). Easy to handle and no chemical additives (Freese and Nozaic, 2004).	
Non-thermal plasma	Easy to set up. Does not leave products requiring further treatment (e.g., removal of catalysts used in other AOPs). It can be used for a wide range of organic pollutants, due to the generation of a variety of reactive species. The energy efficiency is higher than that used in other AOPs (e.g., UV) (Banaschik et al., 2015).	The treated volume is small. Scaling up and conversion to pilot process could present some technical challenges. It is still in the research stage. NTP can affect the physical and chemical properties of water. Some by-products could be generated.

$$Cl_2 + H_2O \leftrightarrow HOCl + Cl^- + H^+ \tag{3.3}$$

$$HOCl \leftrightarrow ClO^- + H^+ \tag{3.4}$$

Hypochlorous acid (HOCl), the product of chlorine hydrolysis in water, is the most reactive form of chlorine, which then reacts with organic or inorganic compounds in water by electrophilic substitution reaction. Chlorine attacks the phenol ring of the EDCs.

An important factor in favor of the use of chlorine in water treatment is the fact that it is cost-effective and above all a known technology in the water industry. Chen et al. (2007) reported that only 20%–40% of E_1, E_2, and EE_2 were removed by chlorination. However, oxidation of EE_2 has been revealed to be poor during chlorination of wastewater containing high concentrations of ammonia due to the rapid conversion of chlorine to ammonium chloride (Lee et al., 2008). Chlorination of estrogens is time- and dose-dependent, with higher removal efficiencies having been achieved with longer contact times and higher doses.

In recent years, chlorination of water has been a global concern to governments, governing bodies of the water industry, and water utilities; this is because there are a number of known health risks associated with chlorine usage in water as it produces trace amounts of carcinogenic organic molecules and the matter is further complicated by the fact that potential toxic estrogenic by-products are produced when chlorine reacts with EDCs. Another risk associated with chlorination is storage; since it is not generated on site, it has to be stored as a concentrated liquid. Another disadvantage of chlorination is that it requires long retention times to remove estrogenic activities (minimum of 2 h for E_2 and 24 h for EE_2; Shappell et al., 2008). Therefore, due to these concerns regarding the use of chlorine in wastewater treatment, other methods are now used to treat wastewater.

3.4.4.3.2 Ozone-Based Processes

Ozone wastewater treatment is an oxidation treatment, which is becoming increasingly popular in the world due to the increasing stringent environmental regulations. Ozone is a strong oxidizing agent, an allotrope of oxygen, and a very reactive gas (Zhou and Smith, 2002; Poyatos et al., 2010). It has some advantages, one of which is that it is produced on site from oxygen, thereby eliminating the need for storage and transportation. Ozone is very effective in removing most organic compounds as well as color-, odor-, and taste-causing compounds. Besides, compared to chlorine, ozone leaves fewer harmful by-products in water [e.g., bromate and *N*-nitrosodimethylamine (Marti et al., 2015)] and easily converts back to oxygen after use. Furthermore, ozone is preferred to chlorine because it has a higher redox potential, making it a stronger oxidizing agent.

The conventional water ozonation is by dissolving ozone directly into water. Once dissolved in water, ozone reacts via two possible mechanisms: (1) molecular ozone or (2) indirect reaction involving radicals such as hydroxyl radicals produced from the decomposition of ozone in water. The extent of each mechanism depends on various parameters but particularly pH and the chemical composition of water. The direct molecular ozone reaction may proceed via cycloaddition, electrophilic, or even nucleophilic reactions. The molecular structure of the organic molecule defines which type of reaction takes place. In contrast, the indirect reaction proceeds via hydroxyl radicals. It takes place according to the following steps: (1) initiation, (2) radical chain reaction or propagation, and (3) termination. Reactions that possibly take place during the initiation, propagation, and termination steps are shown below (Staehelin and Hoigne, 1982; Tomiyasu et al., 1985; Chelkowska et al., 1992).

Initiation Step

Compounds that lead to the formation of a superoxide ion $O_2^{\cdot-}$ from ozone are termed as initiators. Examples of initiators include hydroxide ions, hydrogen peroxide, metal ions, and UV.

$$O_3 + OH^- \rightarrow O_2^{\cdot-} + HO_2^{\cdot} \quad k = 70\ M^{-1}s^{-1} \tag{3.5}$$

$$HO_2^{\bullet} \leftrightarrow O_2^{\bullet -} + H^+ \quad pK_a = 4.8 \tag{3.6}$$

Propagation Step

$$O_3 + O_2^{\bullet -} \rightarrow O_3^{\bullet -} + O_2 \quad k = 1.6 \times 10^9 \ M^{-1}s^{-1} \tag{3.7}$$

$$HO_3^{\bullet} \leftrightarrow O_3^{\bullet -} + H^+ \quad pK_a = 6.2 \tag{3.8}$$

$$HO_3^{\bullet} \rightarrow {}^{\bullet}OH + O_2 \quad k = 1.1 \times 10^5 \ s^{-1} \tag{3.9}$$

$${}^{\bullet}OH + O_3 \rightarrow HO_4^{\bullet} \quad k = 2.0 \times 10^9 \ M^{-1}s^{-1} \tag{3.10}$$

$$HO_4^{\bullet} \rightarrow HO_2^{\bullet} + O_2 \quad k = 2.8 \times 10^4 \ s^{-1} \tag{3.11}$$

With the formation of new hydroperoxyl radical ($HO_2^{\bullet}$) (Equation 3.11), the chain reaction can proceed again (Equations 3.6 and 3.7). Any substance in solution that can convert the hydroxyl radicals into superoxide radicals (i.e., $HO_2^{\bullet}/O_2^{\bullet -}$) acts as a promoter of the radical chain reaction. Examples of promoters include organic molecules with aryl groups, primary and secondary alcohols, and humic acids. In the above reactions, the promoter was ozone (Equation 3.10).

Termination Step

Certain organic or inorganic substances may react with ${}^{\bullet}OH$ to produce secondary radicals that do not produce $HO_2^{\bullet}/O_2^{\bullet -}$, hence inhibiting (i.e., terminating) the chain reaction. These substances are termed as inhibitors or scavengers. Examples include carbonate and bicarbonate ions, phosphates, tertiary alcohols, and humic substances. Termination reactions involving carbonate and bicarbonate ions and $HO_2^{\bullet}$ are represented by the following equations:

$${}^{\bullet}OH + CO_3^{2-} \rightarrow OH^- + CO_3^{\bullet -} \quad k = 4.2 \times 10^8 \ M^{-1}s^{-1} \tag{3.12}$$

$${}^{\bullet}OH + HCO_3^- \rightarrow OH^- + HCO_3^{\bullet} \quad k = 1.5 \times 10^7 \ M^{-1}s^{-1} \tag{3.13}$$

$${}^{\bullet}OH + HO_2^{\bullet} \rightarrow O_2 + H_2O \quad k = 3.7 \times 10^{10} \ M^{-1}s^{-1} \tag{3.14}$$

Overall Reaction

The overall ozone reaction that leads to the formation of hydroxyl radicals is shown in Equation 3.15.

$$3O_3 + OH^- + H^+ \rightarrow 2{}^{\bullet}OH + 4O_2 \quad k = 3.7 \times 10^{10} \ M^{-1}s^{-1} \tag{3.15}$$

Lin et al. (2009) carried out an investigation on the ozonation of E_1, E_2, and EE_2 in waters and reported that at a pH of 9.0, E_1 reached a degradation efficiency of 94% after 8 min, EE_2 was 100% degraded after only 6 min, and E_2 was also 100% degraded after 4 min; this proves that ozone reactions with the EDCs proceed at fast rates. Ozone is an effective oxidant for EDCs treatment. In their large pilot plant studies on real wastewater effluent, Churchley et al. (2011) have found that ozone was the most effective treatment in comparison to GAC and chlorine dioxide in removing estrogens. They have also made an economic appraisal of these three treatment techniques, and the results showed that ozone, applied at 1 mg/L, had the lowest operating cost. The authors suggested that ozone may be an economically viable option when considering advanced treatment for EDCs removal in sewage effluent. Moreover, ozonation in conjunction with AC have been adopted in Switzerland as the preferred processes to control the discharge into the natural environment of trace organic compounds including EDCs (Audenaert et al., 2014).

Nevertheless, there are also drawbacks coupled with the use of ozone such as generation of bromate, which is a potential carcinogenic compound, from bromide-containing waters and some dissolved toxic compounds are resistant to removal by ozone. Besides, good mass transfer and mixing of ozone with the wastewater must be established in order to achieve effective ozonation. Since the ozone plant is made of a complex process involving several unit operations including gas treatment, ozone generation, ozone contactors, and ozone destruction units, its operation requires highly skilled personnel.

3.4.4.3.2.1 Ozone Combined Solvents and Adsorbents Due to the low solubility of ozone in water, researchers have been interested to develop new ways by which the ozonation of water can be enhanced. For this, ozone dissolved in an inert, but nontoxic, solvent, which has high ozone solubility, has been suggested for wastewater treatment (Ward et al., 2003, 2004, 2005). This ozonation technique has been termed as the liquid/liquid–ozone (LLO) system by Tizaoui et al. (2008). The basic principle of the LLO system is shown in Figure 3.23. From their studies, improvement in wastewater ozonation was achieved through, for example, increased percentage of organic pollutants removed compared with the conventional ozonation process (i.e., direct ozone bubbling in water). Figure 3.24 shows a comparison between the change in E_1 degradation using a conventional gas/liquid system and LLO (work carried out by the author but was not published before). Figure 3.24 shows that E_1 reacted very fast with gas ozone, but when the LLO system was used, E_1 was removed almost instantaneously (half-life time $< 3\,s$). The enhanced removal rate of E_1 with LLO was due to a reactive extraction process resulting from a synergistic effect of both liquid–liquid extraction and chemical reaction. Moreover, given that the solvent has higher ozone absorption capacity than water and at the same time has high partition coefficient for the estrogens, the reaction rates are enhanced. Other studies have used suitable adsorbents (i.e., a solid phase) to concentrate ozone before it is used for the removal of organic compounds (Tizaoui and Slater, 2003a,b).

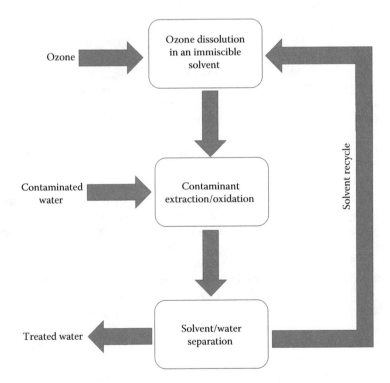

FIGURE 3.23 Principle of the LLO system.

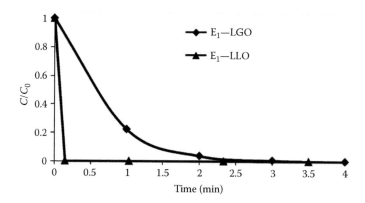

FIGURE 3.24 Removal of estrone with liquid/gas-ozone and LLO systems.

3.4.4.3.2.2 Ozone–Hydrogen Peroxide (O_3/H_2O_2) In aqueous solutions, hydrogen peroxide partially dissociates to its conjugate base; hydroperoxide anion HO_2^- reacts quickly with ozone to initiate the chain reaction and enhance the formation of hydroxyl radicals according to the following reactions (Staehelin and Hoigne, 1982; Glaze, 1987; Merenyi et al., 2010):

$$H_2O_2 + H_2O \rightarrow HO_2^- + H_3O^+ \quad pK_a = 11.8 \tag{3.16}$$

$$O_3 + HO_2^- \rightarrow HO_2^{\bullet} + O_3^{\bullet-} \quad k = 2.2 \times 10^6 \text{ M}^{-1}\text{s}^{-1} \tag{3.17}$$

$$H_2O_2 + O_3 \rightarrow H_2O + 2O_2 \quad K < 10^{-2} \text{ M}^{-1}\text{s}^{-1}, \text{ hence negligible} \tag{3.18}$$

$$2O_3 + H_2O_2 \rightarrow 2\,{}^{\bullet}OH + 3O_2 \quad \text{Overall reaction} \tag{3.19}$$

An increase in the pH leads to reduction in the half-life of H_2O_2, giving an outcome of increased ozone decomposition rate (Staehelin and Hoigne, 1982). The process is usually carried out at peroxide to ozone ratios ranging from 0.3:1 to 0.6:1 depending on the disinfection or oxidation requirements (Zhou and Smith, 2002). An advantage of this process is that it does not depend on ultraviolet radiation to activate the peroxide or ozone molecules, enabling it to work well even in cloudy water matrices. Apart from disinfection, O_3/H_2O_2 also has an added advantage of removing color (Zhou and Smith, 2002).

Zwiener and Frimmel (2000) enhanced their experiment on oxidative treatment of pharmaceuticals in water using the O_3/H_2O_2 process. They reported that at a very low concentration of 1 mg/L of ozone, clofibric acid and ibuprofen in distilled water can be degraded to about half of their initial concentration; hence, degradation efficiency is improved. Nakonechny et al. (2008) have shown that E_1 reacts at very high rates in an O_3/H_2O_2 system and suggested that O_3/H_2O_2 can be used as an effective treatment technique for E_1-contaminated waters. In a comparative study using various ozone- and UV-based AOPs, Sarkar et al. (2014) have shown that O_3/H_2O_2 was second to O_3/UV in terms of degradation rates of E_1. The O_3/H_2O_2 system was also found effective to remove E_2 and EE_2 as well as their estrogenic activity in water (Maniero et al., 2008).

3.4.4.3.2.3 Ozone–Ultraviolet (O_3/UV) The main mechanism of O_3/UV process is the use of photons in the UV spectrum to convert ozone to hydrogen peroxide and oxygen in the presence of water (Equation 3.20) (Peyton and Glaze, 1988; Poyatos et al., 2010). This is then followed by initiation reactions, leading to hydroxyl radicals via photolysis of the formed H_2O_2 and/or its reaction with ozone. The O_3/UV system is the most complete ozone-based AOP since it is possible to have three possible initiation reactions for the generation of hydroxyl radicals [i.e., O_3/H_2O_2, UV/H_2O_2, and other initiating species present in water e.g., OH^-)]. This is in addition to other possible three direct oxidation/photolysis reactions (i.e., direct ozone reaction, direct oxidation with H_2O_2, and direct photolysis). Therefore, it is important to consider all possible removal mechanisms when evaluating the O_3/UV AOP. Ozone absorbs maximum UV light at a wavelength of 253.7 nm with a high extinction coefficient of 3300 L/mol/cm; therefore, it is very important to use low- to medium-pressure UV lamps, which generate UV light in the wavelength range of 200–280 nm (Zhou and Smith, 2002; Poyatos et al., 2010).

$$O_3 + H_2O \xrightarrow{UV} H_2O_2 + O_2 \quad (\lambda < 385nm) \tag{3.20}$$

Irmak et al. (2005) have shown that coupling UV with ozone enhanced the degradation of E_2 and decreased ozone consumption by 22.5%. This resulted in considerable reduction of the degradation time from that when ozone alone was used. The O_3/UV system was also found effective in removing E_1 and was the fastest in doing so compared with other ozone- and UV-based systems (Sarkar et al., 2014).

3.4.4.3.3 Hydrogen Peroxide–Ultraviolet (H_2O_2/UV)

The H_2O_2/UV process is a chain reaction involving the photolysis of H_2O_2 to produce hydroxyl radicals as shown below, which then reacts with the organic compounds in water. Because H_2O_2 absorbs UV radiation, a major disadvantage of this process is that if organic compounds are present in water, part of the UV radiation will be absorbed by the organic compounds resulting in reduced production yield of hydroxyl radicals.

Initiation

$$H_2O_2 \xrightarrow{UV} 2\,^{\bullet}OH \quad \varepsilon_{254nm} = 18.6\ M^{-1}cm^{-1} \tag{3.21}$$

$$HO_2^- \xrightarrow{UV} {}^{\bullet}OH + O^{\bullet -} \quad \varepsilon_{254nm} = 240\ M^{-1}cm^{-1} \tag{3.22}$$

$$O^{\bullet -} + H_2O \rightarrow {}^{\bullet}OH + OH^- \tag{3.23}$$

Propagation

$$H_2O_2 + {}^{\bullet}OH \rightarrow HO_2^{\bullet} + H_2O \tag{3.24}$$

$$HO_2^{\bullet} + H_2O_2 \rightarrow {}^{\bullet}OH + O_2 + H_2O_2 \tag{3.25}$$

Termination

$$2\,^{\bullet}OH \rightarrow H_2O_2 \tag{3.26}$$

$$HO_2^{\bullet} + {}^{\bullet}OH \rightarrow H_2O + O_2 \tag{3.27}$$

In terms of stoichiometry and as seen in Equation 3.21, a higher initial concentration of H_2O_2 produces increased hydroxyl radicals, leading to decomposition of more organic contaminants. However, an optimal H_2O_2 concentration is essential as a very high H_2O_2 concentration will result in H_2O_2 competing for hydroxyl radicals, thereby reducing the overall effectiveness of the process.

Rosenfeldt and Linden (2004) applied the H_2O_2/UV process in their experiment in removing BPA, E_2, and EE_2 from three different water sources. Their results signify that addition of H_2O_2 gives a greater removal of the EDCs (90% of BPA, 99% of E_2, and ~95% of EE_2 were removed). Another research study using H_2O_2/UV process was carried out by Chen et al. (2006) in the degradation of BPA, noting that H_2O_2/UV was effective in reducing BPA concentration to below detectable limit (97% removal percentage was achieved). Sarkar et al. (2014) have shown that the addition of 20 mg/L of H_2O_2 to the UV reactor has increased the degradation rate of E_1 by almost 30% compared with UV alone.

3.4.4.3.4 *Photocatalysis (TiO₂/UV)*

Photocatalytic oxidation works on the basis of using UV radiation for the photoexcitation of a semiconductor catalyst in the presence of an electron acceptor (e.g., oxygen), thereby generating hydroxyl radicals. Titanium dioxide (TiO_2) in its anatase form is one of the commonly used semiconductor catalysts in photocatalytic processes for water treatment. TiO_2 absorbs UV light at wavelengths below ~400 nm.

Nakashima et al. (2002) studied the decomposition of E_2, BPA, and 2,4-dichlorophenol (2,4-DCP) in water using TiO_2 photocatalyst and reported that each EDC was decomposed quickly from the same initial concentrations of 90 µg/L to ~3.5 µg/L for E_2, ~9 µg/L for DCP, and ~2 µg/L for BPA. Coleman et al. (2000) studied the photocatalytic degradation of E_2 using immobilized titanium dioxide and found that E_2 was readily destroyed, but the degradation mechanism has intermediate breakdown products, which might exhibit endocrine-disrupting effects. The photocatalytic degradation of a multicomponent system containing E_1, E_2, EE_2, and estriol (E_3) was studied using UVA and UVC in the presence of TiO_2 (Degussa P25) suspensions (Li Puma et al., 2010). A model that predicted appropriately the degradation profiles of the estrogens in multicomponent mixtures was developed in this study, taking into account of the explicit effect of photon absorption. A study by Mboula et al. (2015) highlights that the photocatalytic degradation of E_2 leads to the formation of intermediates still containing the phenol group and thereby still exhibiting estrogenic activity.

3.4.4.3.5 *Nonthermal Plasma (NTP)*

NTP has also been studied as an alternative AOP method for wastewater treatment. It is an individual group of AOPs and it is considered as a promising technology for water and wastewater treatment. Ins laboratory, NTP is generated using a high electric field (about 30 kV/cm in air and more than 1 MV/cm in water) applied between two electrodes (Ruma et al., 2013). The interest in NTP is that it combines the contribution of reactive chemical species, energetic electrons, intense UV radiation, high electric fields, and shock waves; therefore, it is considered as a very effective approach in water treatment. This combination of a multitude of effects is difficult or impossible to obtain in the traditional water treatment technologies. The most important chemical reactive species generated in NTP are O_3, ·OH, H_2O_2, and UV, which all play a vital role in the water treatment process due to their high oxidation potential. However, in some reactors, the high electric field and the shock wave processes compete with the role of these chemical reactive species, especially for killing microorganisms in water (Šunka, 2001; Chen et al., 2008).

For the water and wastewater treatment, plasma discharge could be generated directly in water, in gas above the aqueous solution, or in bubbling inside the water. Various configurations of plasma discharge reactors have been studied for the treatment of water. Pulsed corona discharge, dielectric barrier discharge, and gliding discharge are the most popular methods applied for water treatment. A detailed review of plasma types and plasma reactors for water remediation can be found in Jiang et al. (2014).

In the case of plasma generating above the aqueous solution, the surrounding gas above the aqueous solution determines the reactive species generated. These reactive species diffuse into the aqueous solution and react with the pollutant molecules or generate another reactive species, which in turn react with the pollutants. Therefore, the distance between the plasma source and the surface of the aqueous solution is one of the important parameters, since it affects the amount and the rate of diffusion of the reactive species into water. It also affects the performance of the electric field and the shock wave on the treatment of water.

In the case of bubbling discharge, the most important feature is that the reactive species are generated in the gas phase and introduced directly via the bubbles into the water. On the other hand, the most reactive species generated in the direct plasma discharge in water are H_2O_2 and ·OH, which are generated from the decomposition of H_2O molecules. In addition to the role of the ·OH radicals and H_2O_2 in the degradation process, the strong electric field and the shock wave play an important role in the treatment of water by the direct plasma discharge in water.

In the last two decades, NTP has been used to degrade a range of organic pollutants and to inactivate microorganisms in water, which affect human health and aquatic life. NTP has been particularly used to treat waters contaminated with EDCs in aqueous solutions, including pharmaceuticals, personal care products, surfactants, and estrogens. Conventional wastewater treatment methods using biological methods cannot effectively eliminate these compounds due to their tolerance to biological degradation. Recently, various NTP techniques and reactors have been developed for the degradation of different pharmaceutical compounds in aqueous solution. The degradation of sulfadiazine was investigated using wetted-wall corona discharge (Rong and Sun, 2014) and water falling film dielectric barrier discharge (Rong et al., 2014). Pentoxifylline degradation has also been investigated using a dielectric barrier discharge in coaxial configuration and operated in a pulsed regime (Magureanu et al., 2010). Nonsteroidal anti-inflammatory drugs, such as diclofenac [2-(2,6-dichloranilino)phenylacetic acid] and ibuprofen, have been degraded in water using a pulsed corona discharge above the water surface (Dobrin et al., 2013) and wetted-wall corona discharge (Zeng et al., 2015), respectively. Carbamazepine degradation was investigated using dielectric barrier discharge (Liu et al., 2012).

Kim et al. (2015) investigated the degradation of three of sulfonamides antibiotics [sulfathiazole (STZ), sulfamethazine (SMT), and sulfamethoxazole (SMZ)] by dielectric barrier discharge reactor with air or oxygen. The antibiotics exhibited different responses to the plasma treatment. SMT was the most easily degradable compound, while SMZ was the most recalcitrant to degradation. The degradation of the compounds in the presence of oxygen was much faster than that in the presence of air. This is due to higher ozone generation in oxygen than in air, which plays a vital role in the decomposition process, either via the direct pathway or via the indirect pathway involving ·OH radical reactions. Krause et al. (2009) obtained different degradation efficiencies for different pollutants using a similar reactor. Moreover, Hijosa-Valsero et al. (2013) used two different dielectric barrier discharge reactors (a batch reactor and a coaxial thin-falling-water-film reactor) to assess the degradation of four organic pollutants in aqueous solution. The difference in the energy efficiencies between the two reactors was about one order of magnitude. These studies concludeed that the degradation efficiency and the energy yield of the degradation process using NTP reactors depend significantly on the type of the plasma reactor as well as the type of pollutants studied.

In NTP technique, the most important parameters that influence the degradation of pollutants in water are the input power, the frequency of the applied voltage, initial concentration of the compound, pH of the aqueous solution, and the conductivity of the aqueous solution. The decomposition efficiency decreases with the increase in the initial concentration of most pollutants, although the energy yield may be improved for some compounds (Magureanu et al., 2010; Zeng et al., 2015). The decrease in degradation efficiency of the pollutants at a higher concentration could be due to the competition between the degradation of the initial reactant and degradation of the reaction intermediates by the active species, which becomes more important as the initial concentration of the pollutant is increased. However, the degradation rate increases with the increase in the initial concentration of the pollutants, whether the decomposition efficiency decreased (Zeng et al., 2015) or increased. The frequency of the applied voltage has also an important effect. Indeed, as the frequency of the applied voltage increases, the degradation rate also increases primarily due to increased input power resulting from increased frequency, which leads to increase in the amount of the reactive species in the reactor. However, the behavior of the energy efficiency for the degradation process does not follow the same trend as the degradation efficiency. It also depends on the type of the pollutant molecule and the reactor configuration as has been mentioned earlier. Zeng et al. (2015) found that among three frequency values (50, 75, and 100 Hz) applied to the wetted-wall corona discharge reactor, the energy efficiency of the degradation of ibuprofen is higher at 100 Hz compared with the other lower frequencies.

Studies carried out so far on the treatment of water using NTP concluded that plasma might be suitable for pretreatment of wastewater containing recalcitrant pharmaceutical compounds. Although the degradation of many chemical compounds has been studied using NTP, only very few studies have investigated the degradation of E_1, E_2, and EE_2 using this technique. Gao et al. (2013) investigated the decomposition of E_2 by generating plasma over the aqueous solution using dielectric barrier discharge. This study concluded that the O_3 and $\cdot OH$ generated during the process were the species responsible for the degradation of E_2. Moreover, the decomposition efficiency of E_2 was significantly influenced by the applied voltage and the pH of the solution. They found that ozone played a vital role in the degradation process when E_2 solution is acidic, whereas the degradation process was dominated by $\cdot OH$ radicals in the neutral and the alkaline solution. Panorel et al. (2013) investigated the oxidation of β-estradiol in pulsed corona discharge reactor and evlauted the influence of adding urea and sucrose to the solution. The presence of urea and sucrose slowed down the degradation process and increased the energy demand of the system. In the case of β-estradiol mixed with urea and sucrose, the energy demand was 1240 Wh m^{-3} at 75% degradation while the energy demand was only 25 Wh m^{-3} for β-estradiol alone.

3.4.5 Conclusions

Chemicals that have potential to cause endocrine disruption such as pharmaceuticals and hormones are ubiquitous in wastewaters. Their occurrence in the aquatic environment is becoming a serious problem due to the negative effects they have on humans and the aquatic environment. The list of substances that are found to disrupt the endocrine system is increasing steadily, but the natural hormones E_1 and E_2 and the synthetic hormone EE_2 appear to have received significant attention due to their severe effects. The principal source of EDCs in the environment, especially water bodies, is through human excretion. Unlike common wastewater contaminants (e.g., particles, BOD, and nutrients), EDCs are not easy to treat in conventional WWTPs without the need for new treatment technologies. Hence, as WWTPs were not designed to remove EDCs, WWTP effluents are the main sources of EDCs in the aquatic environment. As the effects of EDCs are becoming more understandable and as the public is becoming more aware of the issues linked to exposure to EDCs, regulatory directives aiming to control their spread into the aquatic environment are increasing. For example, the EU has introduced a watch list as part of the Priority Substances Directive 2013/39/EU, aiming to control emissions of the substances included in this watch list, which already contains a pharmaceutical substance diclofenac (a generic painkiller) and the hormones E_2 and EE_2. To remove these substances before they reach the aquatic environment, several treatment techniques based on physical separation, chemical oxidation, and biological means have been researched using bench scale and pilot plant experiments. Of the many techniques studied, ozone-based treatment appears to be the most effective solution to control EDCs. As research continues to advance in this area, focus should be on new innovative methods to control EDCs at source before they reach the wastewater systems instead of end-of-pipe treatment. Eventually, a multidisciplinary approach involving engineers of various disciplines, chemists, biologists, environmentalists, social scientists, and economists is required to tackle this growing problem.

Acknowledgement

Financial support received from the Engineering and Physical Sciences Research Council (EPSRC; grant number EP/M017141/1) is acknowledged.

References

Acevedo, N., Davis, B., Schaeberle, C. M., Sonnenschein, C. and Soto, A. M. 2013. Perinatally administered bisphenol A as a potential mammary gland carcinogen in rats. *Environmental Health Perspectives*, 121, 1040–1046.

Arnon, S., Dahan, O., Elhanany, S., Cohen, K., Pankratov, I., Gross, A., Ronen, Z., Baram, S. and Shore, L. S. 2008. Transport of testosterone and estrogen from dairy-farm waste lagoons to groundwater. *Environmental Science and Technology*, 42, 5521–5526.

Audenaert, W. T. M., Chys, M., Auvinen, H., Dumoulin, A., Rousseau, D. and Hulle, S. W. H. V. 2014. (Future) Regulation of trace organic compounds in WWTP effluents as a driver of advanced wastewater treatment. Ozone News. The Newsletter of the International Ozone Association.

Auriol, M., Filali-Meknassi, Y., Tyagi, R. D., Adams, C. D. and Surampalli, R. Y. 2006. Endocrine disrupting compounds removal from wastewater, a new challenge. *Process Biochemistry*, 41, 525–539.

Avbersek, M., Soemen, J. and Heath, E. 2011. Dynamics of steroid estrogen daily concentrations in hospital effluent and connected waste water treatment plant. *Journal of Environmental Monitoring*, 13, 2221–2226.

Banaschik, R., Lukes, P., Jablonowski, H., Hammer, M. U., Weltmann, K. D. and Kolb, J. F. 2015. Potential of pulsed corona discharges generated in water for the degradation of persistent pharmaceutical residues. *Water Research*, 84, 127–135.

Baronti, C., Curini, R., D'Ascenzo, G., Di Corcia, A., Gentili, A. and Samperi, R. 2000. Monitoring natural and synthetic estrogens at activated sludge sewage treatment plants and in a receiving river water. *Environmental Science and Technology*, 34, 5059–5066.

Belfroid, A. C., Van Der Horst, A., Vethaak, A. D., Schäfer, A. J., Rijs, G. B. J., Wegener, J. and Cofino, W. P. 1999. Analysis and occurrence of estrogenic hormones and their glucuronides in surface water and waste water in The Netherlands. *Science of the Total Environment*, 225, 101–108.

Benotti, M. J., Trenholm, R. A., Vanderford, B. J., Holady, J. C., Stanford, B. D. and Snyder, S. A. 2009. Pharmaceuticals and endocrine disrupting compounds in US drinking water. *Environmental Science and Technology*, 43, 597–603.

Bjerke, D. L., Brown, T. J., Maclusky, N. J., Hochberg, R. B. and Peterson, R. E. 1994. Partial demasculinization and feminization of sex behavior in male rats by in utero and lactational exposure to 2,3,7,8-tetrachlorodibenzo-*p*-dioxin is not associated with alterations in estrogen receptor binding or volumes of sexually differentiated brain nuclei. *Toxicology and Applied Pharmacology*, 127, 258–267.

Braga, O., Smythe, G. A., Schäfer, A. I. and Feltz, A. J. 2005. Steroid estrogens in primary and tertiary wastewater treatment plants. *Water Science and Technology*, 52, 273–278.

Cajthaml, T., Kresinova, Z., Svobodova, K. and Moeder, M. 2009. Biodegradation of endocrine-disrupting compounds and suppression of estrogenic activity by ligninolytic fungi. *Chemosphere*, 75, 745–750.

Cargouet, M., Perdiz, D., Mouatassim-Souali, A., Tamisier-Karolak, S. and Levi, Y. 2004. Assessment of river contamination by estrogenic compounds in Paris area (France). *Science of the Total Environment*, 324, 55–66.

Chang, H., Wan, Y., Wu, S., Fan, Z. and Hu, J. 2011. Occurrence of androgens and progestogens in wastewater treatment plants and receiving river waters: Comparison to estrogens. *Water Research*, 45, 732–740.

Chelkowska, K., Grasso, D., Fábián, I. and Gordon, G. 1992. Numerical simulations of aqueous ozone decomposition. *Ozone: Science and Engineering*, 14, 33–49.

Chen, T. S., Chen, T. C., Yeh, K. J., Chao, H. R., Liaw, E. T., Hsieh, C. Y., Chen, K. C., Hsieh, L. T. and Yeh, Y. L. 2010. High estrogen concentrations in receiving river discharge from a concentrated livestock feedlot. *Science of the Total Environment*, 408, 3223–3230.

Chen, C. W., Lee, H.-M. and Chang, M. B. 2008. Inactivation of aquatic microorganisms by low-frequency AC discharges. *IEEE Transactions on Plasma Science*, 36, 215–219.

Chen, P.-J., Linden, K. G., Hinton, D. E., Kashiwada, S., Rosenfeldt, E. J. and Kullman, S. W. 2006. Biological assessment of bisphenol A degradation in water following direct photolysis and UV advanced oxidation. *Chemosphere*, 65, 1094–1102.

Chen, C.-Y., Wen, T.-Y., Wang, G.-S., Cheng, H.-W., Lin, Y.-H. and Lien, G.-W. 2007. Determining estrogenic steroids in Taipei waters and removal in drinking water treatment using high-flow solid-phase extraction and liquid chromatography/tandem mass spectrometry. *Science of the Total Environment*, 378, 352–365.

Choi, K. J., Kim, S. G., Kim, C. W. and Kim, S. H. 2005. Effects of activated carbon types and service life on removal of endocrine disrupting chemicals: amitrol, nonylphenol, and bisphenol-A. *Chemosphere*, 58, 1535–1545.

Churchley, J., Drage, B., Cope, E., Narroway, Y., Ried, A., Swierk, T., Alexander, K. and Kanda, R. 2011. Performance of ozone for EDC removal from sewage effluent. *20th IOA World Congress—6th IUVA World Congress. Ozone and UV Leading-Edge Science and Technologies*. Paris, France: International Ozone Association.

Coleman, H. M., Eggins, B. R., Byrne, J. A., Palmer, F. L. and King, E. 2000. Photocatalytic degradation of 17-β-oestradiol on immobilised TiO_2. *Applied Catalysis B: Environmental*, 24, L1–L5.

Comerton, A. M., Andrews, R. C., Bagley, D. M. and Hao, C. 2008. The rejection of endocrine disrupting and pharmaceutically active compounds by NF and RO membranes as a function of compound and water matrix properties. *Journal of Membrane Science*, 313, 323–335.

Comninellis, C., Kapalka, A., Malato, S., Parsons, S. A., Poulios, L. and Mantzavinos, D. 2008. Advanced oxidation processes for water treatment: advances and trends for R&D. *Journal of Chemical Technology and Biotechnology*, 83, 769–776.

Cui, C. W., Ji, S. L. and Ren, H. Y. 2006. Determination of steroid estrogens in wastewater treatment plant of a controceptives producing factory. *Environmental Monitoring and Assessment*, 121, 409–419.

D'Ascenzo, G., Di Corcia, A., Gentili, A., Mancini, R., Mastropasqua, R., Nazzari, M. and Samperi, R. 2003. Fate of natural estrogen conjugates in municipal sewage transport and treatment facilities. *Science of the Total Environment*, 302, 199–209.

Deborde, M. and Von Gunten, U. 2008. Reactions of chlorine with inorganic and organic compounds during water treatment—Kinetics and mechanisms: A critical review. *Water Research*, 42, 13–51.

Desbrow, C., Routledge, E. J., Brighty, G. C., Sumpter, J. P. and Waldock, M. 1998. Identification of estrogenic chemicals in STW effluent. 1. Chemical fractionation and in vitro biological screening. *Environmental Science and Technology*, 32, 1549–1558.

Diamanti-Kandarakis, E., Bourguignon, J. P., Giudice, L. C., Hauser, R., Prins, G. S., Soto, A. M., Zoeller, R. T. and Gore, A. C. 2009. Endocrine-disrupting chemicals: An endocrine society scientific statement. *Endocrine Reviews*, 30, 293–342.

Dobrin, D., Bradu, C., Magureanu, M., Mandache, N. B. and Parvulescu, V. I. 2013. Degradation of diclofenac in water using a pulsed corona discharge. *Chemical Engineering Journal*, 234, 389–396.

Essandoh, H. M. K., Tizaoui, C. and Mohamed, M. H. A. 2012. Removal of estrone (E1), 17β-estradiol (E2) and 17α-ethinylestradiol (EE2) during soil aquifer treatment of a model wastewater. *Separation Science and Technology*, 47, 777–787.

Freese, S. D. and Nozaic, D. J. 2004. Chlorine: Is it really so bad and what are the alternatives? *Water SA*, 30, 566–572.

Frontistis, Z., Kouramanos, M., Moraitis, S., Chatzisymeon, E., Hapeshi, E., Fatta-Kassinos, D., Xekoukoulotakis, N. P. and Mantzavinos, D. 2015. UV and simulated solar photodegradation of 17α-ethynylestradiol in secondary-treated wastewater by hydrogen peroxide or iron addition. *Catalysis Today*, 252, 84–92.

Gao, L., Sun, L., Wan, S., Yu, Z. and Li, M. 2013. Degradation kinetics and mechanism of emerging contaminants in water by dielectric barrier discharge non-thermal plasma: The case of 17β-estradiol. *Chemical Engineering Journal*, 228, 790–798.

Gibson, R., Becerril-Bravo, E., Silva-Castro, V. and Jiménez, B. 2007. Determination of acidic pharmaceuticals and potential endocrine disrupting compounds in wastewaters and spring waters by selective elution and analysis by gas chromatography-mass spectrometry. *Journal of Chromatography A*, 1169, 31–39.

Glaze, W. H. 1987. Drinking-water treatment with ozone. *Environmental Science* and *Technology*, 21, 224–230.

Gogate, P. R. and Pandit, A. B. 2004. A review of imperative technologies for wastewater treatment I: Oxidation technologies at ambient conditions. *Advances in Environmental Research*, 8, 501–551.

Gore, A. C.2007. *Endocrine-Disrupting Chemicals from Basic Principle to Clinical Practice.* Totowa, NJ: Humana Press.

Grover, D. P., Zhou, J. L., Frickers, P. E. and Readman, J. W. 2011. Improved removal of estrogenic and pharmaceutical compounds in sewage effluent by full scale granular activated carbon: Impact on receiving river water. *Journal of Hazardous Materials,* 185, 1005–1011.

Heudorf, U., Mersch-Sundermann, V. and Angerer, J. 2007. Phthalates: Toxicology and exposure. *International Journal of Hygiene and Environmental Health,* 210, 623–634.

Hijosa-Valsero, M., Molina, R., Schikora, H., Müller, M. and Bayona, J. M. 2013. Removal of priority pollutants from water by means of dielectric barrier discharge atmospheric plasma. *Journal of Hazardous Materials,* 262, 664–673.

Hohenblum, P., Gans, O., Moche, W., Scharf, S. and Lorbeer, G. 2004. Monitoring of selected estrogenic hormones and industrial chemicals in groundwaters and surface waters in Austria. *Science of the Total Environment,* 333, 185–193.

Hotchkiss, A. K., Rider, C. V., Blystone, C. R., Wilson, V. S., Hartig, P. C., Ankley, G. T., Foster, P. M., Gray, C. L. and Gray, L. E. 2008. Fifteen years after "Wingspread"—Environmental endocrine disrupters and human and wildlife health: Where we are today and where we need to go. *Toxicological Sciences,* 105, 235–259.

Huang, C.-C. and Su, Y.-J. 2010. Removal of copper ions from wastewater by adsorption/electrosorption on modified activated carbon cloths. *Journal of Hazardous Materials,* 175, 477–483.

Huo, C. X. and Hickey, P. 2007. EDC demonstration programme in the UK Anglian Water's approach. *Environmental Technology,* 28, 731–741.

Irmak, S., Erbatur, O. and Akgerman, A. 2005. Degradation of 17β-estradiol and bisphenol A in aqueous medium by using ozone and ozone/UV techniques. *Journal of Hazardous Materials,* 126, 54–62.

Jackson, J. and Sutton, R. 2008. Sources of endocrine-disrupting chemicals in urban wastewater, Oakland, CA. *Science of the Total Environment,* 405, 153–160.

Jafari, A. J., R. P. Abasabad and Salehzadeh, A. 2009. Endocrine disrupting contaminants in water resources and sewage in Hamadan city of Iran. *Journal of Environmental Health Science* and *Engineering,* 6, 89–96.

Jiang, B., Zheng, J., Qiu, S., Wu, M., Zhang, Q., Yan, Z. and Xue, Q. 2014. Review on electrical discharge plasma technology for wastewater remediation. *Chemical Engineering Journal,* 236, 348–368.

Johnson, A. C. and Sumpter, J. P. 2001. Removal of endocrine-disrupting chemicals in activated sludge treatment works. *Environmental Science* and *Technology,* 35, 4697–4703.

Johnson, A. C. and Williams, R. J. 2004. A model to estimate influent and effluent concentrations of estradiol, estrone, and ethinylestradiol at sewage treatment works. *Environmental Science* and *Technology,* 38, 3649–3658.

Kanda, R. and Churchley, J. 2008. Removal of endocrine disrupting compounds during conventional wastewater treatment. *Environmental Technology,* 29, 315–323.

Kim, K.-S., Kam, S. K. and Mok, Y. S. 2015. Elucidation of the degradation pathways of sulfonamide antibiotics in a dielectric barrier discharge plasma system. *Chemical Engineering Journal,* 271, 31–42.

Koh, Y. K. K., Chiu, T. Y., Boobis, A., Cartmell, E., Scrimshaw, M. D. and Lester, J. N. 2008. Treatment and removal strategies for estrogens from wastewater. *Environmental Technology,* 29, 245–267.

Kolpin, D. W., Furlong, E. T., Meyer, M. T., Thurman, E. M., Zaugg, S. D., Barber, L. B. and Buxton, H. T. 2002. Pharmaceuticals, hormones, and other organic wastewater contaminants in US streams, 1999–2000: A national reconnaissance. *Environmental Science and Technology,* 36, 1202–1211.

Korrick, S. A., Lee, M. M., Williams, P. L., Sergeyev, O., Burns, J. S., Patterson, D. G., Turner, W. E., Needham, L. L., Altshul, L., Revich, B. and Hauser, R. 2011. Dioxin exposure and age of pubertal onset among Russian boys. *Environmental Health Perspectives,* 119, 1339–1344.

Krause, H., Schweiger, B., Schuhmacher, J., Scholl, S. and Steinfeld, U. 2009. Degradation of the endocrine disrupting chemicals (EDCs) carbamazepine, clofibric acid, and iopromide by corona discharge over water. *Chemosphere,* 75, 163–168.

Kulkarni, P. S., Crespo, J. G. and Afonso, C. A. M. 2008. Dioxins sources and current remediation technologies—A review. *Environment International*, 34, 139–153.

Lai, K. M., Johnson, K. L., Scrimshaw, M. D. and Lester, J. N. 2000. Binding of waterborne steroid estrogens to solid phases in river and estuarine systems. *Environmental Science and Technology*, 34, 3890–3894.

Langford, K. and Lester, J. N. 2003. Fate and behavior of endocrine disrupters in wastewater treatment processes. In: Birkett, J. W. and Lester, J. N. (eds) *Endocrine Disrupters in Wastewater and Sludge Treatment Processes*. Boca Raton, FL: CRC Press.

Lee, Y., Escher, B. I. and Von Gunten, U. 2008. Efficient removal of estrogenic activity during oxidative treatment of waters containing steroid estrogens. *Environmental Science and Technology*, 42, 6333–6339.

Leffers, H., Naesby, M., Vendelbo, B., Skakkebae, N. E. and Jorgensen, A. 2001. Oestrogenic potencies of zeranol, oestradiol, diethylstilboestrol, bisphenol-A and genistein: Implications for exposure assessment of potential endocrine disrupters. *Human Reproduction*, 16, 1037–1045.

Lin, Y., Peng, Z. and Zhang, X. 2009. Ozonation of estrone, estradiol, diethylstilbestrol in waters. *Desalination*, 249, 235–240.

Li Puma, G., Puddu, V., Tsang, H. K., Gora, A. and Toepfer, B. 2010. Photocatalytic oxidation of multicomponent mixtures of estrogens (estrone (E1), 17β-estradiol (E2), 17α-ethynylestradiol (EE2) and estriol (E3)) under UVA and UVC radiation: Photon absorption, quantum yields and rate constants independent of photon absorption. *Applied Catalysis B: Environmental*, 99, 388–397.

Liu, Z.-H., Kanjo, Y. and Mizutani, S. 2009. Removal mechanisms for endocrine disrupting compounds (EDCs) in wastewater treatment—Physical means, biodegradation, and chemical advanced oxidation: A review. *Science of the Total Environment*, 407, 731–748.

Liu, Y., Mei, S., Iya-Sou, D., Cavadias, S. and Ognier, S. 2012. Carbamazepine removal from water by dielectric barrier discharge: Comparison of ex situ and in situ discharge on water. *Chemical Engineering and Processing*, 56, 10–18.

Magureanu, M., Piroi, D., Mandache, N. B., David, V., Medvedovici, A. and Parvulescu, V. I. 2010. Degradation of pharmaceutical compound pentoxifylline in water by non-thermal plasma treatment. *Water Research*, 44, 3445–3453.

Maniero, M. G., Bila, D. M. and Dezotti, M. 2008. Degradation and estrogenic activity removal of 17β-estradiol and 17α-ethinylestradiol by ozonation and O(3)/H(2)O(2). *Science of the Total Environment*, 407, 105–115.

Marti, E. J., Pisarenko, A. N., Peller, J. R. and Dickenson, E. R. V. 2015. N-nitrosodimethylamine (NDMA) formation from the ozonation of model compounds. *Water Research*, 72, 262–270.

Matthiessen, P., Arnold, D., Johnson, A. C., Pepper, T. J., Pottinger, T. G. and Pulman, K. G. T. 2006. Contamination of headwater streams in the United Kingdom by oestrogenic hormones from livestock farms. *Science of the Total Environment*, 367, 616–630.

Mboula, V. M., Hequet, V., Andres, Y., Gru, Y., Colin, R., Dona-Rodriguez, J. M., Pastrana-Martinez, L. M., Silva, A. M. T., Leleu, M., Tindall, A. J., Mateos, S. and Falaras, P. 2015. Photocatalytic degradation of estradiol under simulated solar light and assessment of estrogenic activity. *Applied Catalysis B: Environmental*, 162, 437–444.

McAvoy, K. 2008. Occurrence of estrogen in wastewater treatment plant and waste disposal site water samples. *Clearwaters: Contaminants of Emerging Concern*, 38, 28–34.

Merenyi, G., Lind, J., Naumov, S. and Von Sonntag, C. 2010. Reaction of ozone with hydrogen peroxide (peroxone process): A revision of current mechanistic concepts based on thermokinetic and quantum-chemical considerations. *Environmental Science and Technology*, 44, 3505–3507.

Miya, A., Onda, K., Nakamura, Y., Takatoh, C., Katsu, Y. and Tanaka, T. 2007. Biological treatment of estrogenic substances. *Environmental Sciences*, 14, 89–94.

Nakada, N., Tanishima, T., Shinohara, H., Kiri, K. and Takada, H. 2006. Pharmaceutical chemicals and endocrine disrupters in municipal wastewater in Tokyo and their removal during activated sludge treatment. *Water Research*, 40, 3297–3303.

Nakashima, T., Ohko, Y., Tryk, D. A. and Fujishima, A. 2002. Decomposition of endocrine-disrupting chemicals in water by use of TiO$_2$ photocatalysts immobilized on polytetrafluoroethylene mesh sheets. *Journal of Photochemistry and Photobiology A: Chemistry*, 151, 207–212.

Nakonechny, M., Ikehata, K. and El-Din, M. G. 2008. Kinetics of estrone ozone/hydrogen peroxide advanced oxidation treatment. *Ozone: Science and Engineering*, 30, 249–255.

Nghlem, L. D., Schäfer, A. I., Waite, T. D. and Iwa Programme, C. 2003. Membrane filtration in water recycling: Removal of natural hormones. *3rd World Water Congress: Efficient Water Supply and Water Reuse*. Melbourne, Australia.

Ning, B., Graham, N., Zhang, Y., Nakonechny, M. and El-Din, M. G. 2007. Degradation of endocrine disrupting chemicals by ozone/AOPs. *Ozone: Science and Engineering*, 29, 153–176.

Oller, I., Malato, S. and Sanchez-Perez, J. A. 2010. Combination of advanced oxidation processes and biological treatments for wastewater decontamination: A review. *Science of the Total Environment*, 409, 4141–4166.

Panorel, I., Preis, S., Kornev, I., Hatakka, H. and Louhi-Kultanen, M. 2013. Oxidation of aqueous pharmaceuticals by pulsed corona discharge. *Environmental Technology*, 34, 923–930.

Pauwels, B. and Verstraete, W. 2006. The treatment of hospital wastewater: An appraisal. *Journal of Water and Health*, 4, 405–416.

Peterson, E. W., Davis, R. K. and Orndorff, H. A. 2000. 17β-Estradiol as an indicator of animal waste contamination in mantled karst aquifers. *Journal of Environmental Quality*, 29, 826–834.

Peyton, G. R. and Glaze, W. H. 1988. Destruction of pollutants in water with ozone in combination with ultraviolet-radiation. 3. Photolysis of aqueous ozone. *Environmental Science and Technology*, 22, 761–767.

Poyatos, J. M., Munio, M. M., Almecija, M. C., Torres, J. C., Hontoria, E. and Osorio, F. 2010. Advanced oxidation processes for wastewater treatment: State of the art. *Water Air and Soil Pollution*, 205, 187–204.

Purdom, C. E., Hardiman, P. A., Bye, V. V. J., Eno, N. C., Tyler, C. R. and Sumpter, J. P. 1994. Estrogenic effects of effluents from sewage treatment works. *Chemistry and Ecology*, 8(4), 275–285. doi:10.1080/02757549408038554.

Robinson, B. J. and Hellou, J. 2009. Biodegradation of endocrine disrupting compounds in harbour seawater and sediments. *Science of the Total Environment*, 407, 5713–5718.

Rong, S. and Sun, Y. 2014. Wetted-wall corona discharge induced degradation of sulfadiazine antibiotics in aqueous solution. *Journal of Chemical Technology and Biotechnology*, 89, 1351–1359.

Rong, S.-P., Sun, Y.-B. and Zhao, Z.-H. 2014. Degradation of sulfadiazine antibiotics by water falling film dielectric barrier discharge. *Chinese Chemical Letters*, 25, 187–192.

Rosenfeldt, E. J. and Linden, K. G. 2004. Degradation of endocrine disrupting chemicals bisphenol A, ethinyl estradiol, and estradiol during UV photolysis and advanced oxidation processes. *Environmental Science and Technology*, 38, 5476–5483.

Rowsell, V. F., Pang, D. S. C., Tsafou, F. and Voulvoulis, N. 2009. Removal of steroid estrogens from wastewater using granular activated carbon: Comparison between virgin and reactivated carbon. *Water Environment Research*, 81, 394–400.

Roy, J. R., Chakraborty, S. and Chakraborty, T. R. 2009. Estrogen-like endocrine disrupting chemicals affecting puberty in humans—A review. *Medical Science Monitor*, 15, RA137–RA145.

The Royal Society. 2000. Endocrine Disrupting Chemicals (document 06/00). https://royalsociety.org/media/Royal_Society_Content/policy/publications/2000/10070.pdf.

Rubin, B. S. 2011. Bisphenol A: An endocrine disruptor with widespread exposure and multiple effects. *Journal of Steroid Biochemistry and Molecular Biology*, 127, 27–34.

Ruma, P. L., Aoki, N., Spetlikova, E., Hosseini, S. H. R., Sakugawa, T. and Akiyama, H. 2013. Effects of pulse frequency of input power on the physical and chemical properties of pulsed streamer discharge plasmas in water. *Journal of Physics D: Applied Physics*, 46, 125202.

Sarkar, S., Ali, S., Rehmann, L., Nakhla, G. and Ray, M. B. 2014. Degradation of estrone in water and wastewater by various advanced oxidation processes. *Journal of Hazardous Materials*, 278, 16–24.

Schäfer, A. I., Nghiem, L. D. and Waite, T. D. 2003. Removal of the natural hormone estrone from aqueous solutions using nanofiltration and reverse osmosis. *Environmental Science and Technology*, 37, 182–188.

Servos, M. R., Bennie, D. T., Burnison, B. K., Jurkovic, A., Mcinnis, R., Neheli, T., Schnell, A., Seto, P., Smyth, S. A. and Ternes, T. A. 2005. Distribution of estrogens, 17β-estradiol and estrone, in Canadian municipal wastewater treatment plants. *Science of the Total Environment*, 336, 155–170.

Shappell, N. W., Vrabel, M. A., Madsen, P. J., Harrington, G., Billey, L. O., Hakk, H., Larsen, G. L., Beach, E. S., Horwitz, C. P., Ro, K., Hunt, P. G. and Collins, T. J. 2008. Destruction of estrogens using Fe-TAML/peroxide catalysis. *Environmental Science and Technology*, 42, 1296–1300.

Shi, W., Wang, L., Rousseau, D. P. L. and Lens, P. N. L. 2010. Removal of estrone, 17 α-ethinylestradiol, and 17-estradiol in algae and duckweed-based wastewater treatment systems. *Environmental Science and Pollution Research*, 17, 824–833.

Snyder, S. A., Adham, S., Redding, A. M., Cannon, F. S., Decarolis, J., Oppenheimer, J., Wert, E. C. and Yoon, Y. 2007. Role of membranes and activated carbon in the removal of endocrine disruptors and pharmaceuticals. *Desalination*, 202, 156–181.

Snyder, S. A., Westerhoff, P., Yoon, Y. and Sedlak, D. L. 2003. Pharmaceuticals, personal care products, and endocrine disruptors in water: Implications for the water industry. *Environmental Engineering Science*, 20, 449–469.

Staehelin, J. and Hoigne, J. 1982. Decomposition of ozone in water—Rate of initiation by hydroxide ions and hydrogen-peroxide. *Environmental Science and Technology*, 16, 676–681.

Sullivan, J. B. and Krieger, G. R. 2001. *Clinical Environmental Health and Toxic Exposures*, Philadelphia, USA: Lippincott Williams and Wilkins.

Šunka, P. 2001. Pulse electrical discharges in water and their applications. *Physics of Plasmas*, 8, 2587–2594.

Suzuki, Y., Mochidzuki, K., Takeuchi, Y., Yagishita, Y., Fukuda, T., Amakusa, H. and Abe, H. 1996. Biological activated carbon treatment of effluent water from wastewater treatment processes of plating industries. *Separations Technology*, 6, 147–153.

Swan, S. H., Elkin, E. P. and Fenster, L. 1997. Have sperm densities declined? A reanalysis of global trend data. *Environmental Health Perspectives*, 105, 1228–1232.

Swartz, C. H., Reddy, S., Benotti, M. J., Yin, H., Barber, L. B., Brownawell, B. J. and Rudel, R. A. 2006. Steroid estrogens, nonylphenol ethoxylate metabolites, and other wastewater contaminants in groundwater affected by a residential septic system on Cape Cod, MA. *Environmental Science and Technology*, 40, 4894–4902.

Tan, B. L., Hawker, D. W., Muller, J. F., Leusch, F. D., Tremblay, L. A. and Chapman, H. F. 2007. Modelling of the fate of selected endocrine disruptors in a municipal wastewater treatment plant in South East Queensland, Australia. *Chemosphere*, 69, 644–654.

Temmink, H. and Grolle, K. 2005. Tertiary activated carbon treatment of paper and board industry wastewater. *Bioresource Technology*, 96, 1683–1689.

Ternes, T. A., Stumpf, M., Mueller, J., Haberer, K., Wilken, R. D. and Servos, M. 1999. Behavior and occurrence of estrogens in municipal sewage treatment plants—I. Investigations in Germany, Canada and Brazil. *Science of the Total Environment*, 225, 81–90.

Thorpe, K. L., Maack, G., Benstead, R. and Tyler, C. R. 2009. Estrogenic wastewater treatment works effluents reduce egg production in fish. *Environmental Science and Technology*, 43, 2976–2982.

Tizaoui, C., Bickley, R. I., Slater, M. J., Wang, W. J., Ward, D. B. and Al-Jaberi, A. 2008. A comparison of novel ozone-based systems and photocatalysis for the removal of water pollutants. *Desalination*, 227, 57–71.

Tizaoui, C. and Slater, M. J. 2003a. The design of an industrial waste-water treatment process using adsorbed ozone on silica gel. *Process Safety and Environmental Protection*, 81, 107–113.

Tizaoui, C. and Slater, M. J. 2003b. Uses of ozone in a three-phase system for water treatment: Ozone adsorption. *Ozone-Science and Engineering*, 25, 315–322.

Tomiyasu, H., Fukutomi, H. and Gordon, G. 1985. Kinetics and mechanism of ozone decomposition in basic aqueous-solution. *Inorganic Chemistry*, 24, 2962–2966.

Toppari, J., Larsen, J. C., Christiansen, P., Giwercman, A., Grandjean, P., Guillette, L. J., Jegou, B., Jensen, T. K., Jouannet, P., Keiding, N., Leffers, H., Mclachlan, J. A., Meyer, O., Muller, J., Rajpertdemeyts, E., Scheike, T., Sharpe, R., Sumpter, J. and Skakkebaek, N. E. 1996. Male reproductive health and environmental xenoestrogens. *Environmental Health Perspectives*, 104, 741–803.

UKWIR. 2009. Endocrine disrupting chemicals national demonstration programme: Assessment of the performance of WwTW in removing oestrogenic substances (09/TX/04/16).

Uzumcu, M., Suzuki, H. and Skinner, M. K. 2004. Effect of the anti-androgenic endocrine disruptor vinclozolin on embryonic testis cord formation and postnatal testis development and function. *Reproductive Toxicology*, 18, 765–774.

Wang, M.-H. and Baskin, L. S. 2008. Endocrine disruptors, genital development, and hypospadias. *Journal of Andrology*, 29, 499–505.

Ward, D. B., Tizaoui, C. and Slater, M. J. 2003. Ozone-loaded solvents for use in water treatment. *Ozone: Science and Engineering*, 25, 485–495.

Ward, D. B., Tizaoui, C. and Slater, M. J. 2004. Extraction and destruction of organics in wastewater using ozone-loaded solvent. *Ozone: Science and Engineering*, 26, 475–486.

Ward, D. B., Tizaoui, C. and Slater, M. J. 2005. Continuous extraction and destruction of chloro-organics in wastewater using ozone-loaded Volasil (TM) 245 solvent. *Journal of Hazardous Materials*, 125, 65–79.

Wittassek, M., Angerer, J., Kolossa-Gehring, M., Schaefer, S. D., Klockenbusch, W., Dobler, L., Guensel, A. K., Mueller, A. and Wiesmueller, G. A. 2009. Fetal exposure to phthalates—A pilot study. *International Journal of Hygiene and Environmental Health*, 212, 492–498.

Yang, C. Y. 2004. Drinking water chlorination and adverse birth outcomes in Taiwan. *Toxicology*, 198, 249–254.

Ying, G. G., Kookana, R. S. and Ru, Y. J. 2002. Occurrence and fate of hormone steroids in the environment. *Environment International*, 28, 545–551.

Young, W. F. 2002. Proposed predicted-no-effect-concentrations (PNECs) for natural and synthetic steroid oestrogens in Surface waters, RandD Technical Report P2-T04/1. Environment Agency, UK.

Zeng, J., Yang, B., Wang, X., Li, Z., Zhang, X. and Lei, L. 2015. Degradation of pharmaceutical contaminant ibuprofen in aqueous solution by cylindrical wetted-wall corona discharge. *Chemical Engineering Journal*, 267, 282–288.

Zhang, Y. P. and Zhou, J. L. 2005. Removal of estrone and 17β-estradiol from water by adsorption. *Water Research*, 39, 3991–4003.

Zhang, Y. and Zhou, J. L. 2008. Occurrence and removal of endocrine disrupting chemicals in wastewater. *Chemosphere*, 73, 848–853.

Zhao, J.-L., Ying, G.-G., Wang, L., Yang, J.-F., Yang, X.-B., Yang, L.-H. and Li, X. 2009. Determination of phenolic endocrine disrupting chemicals and acidic pharmaceuticals in surface water of the Pearl Rivers in South China by gas chromatography–negative chemical ionization–mass spectrometry. *Science of the Total Environment*, 407, 962–974.

Zhou, H. and Smith, D. W. 2002. Advanced technologies in water and wastewater treatment. *Journal of Environmental Engineering and Science*, 1, 247–264.

Zwiener, C. and Frimmel, F. H. 2000. Oxidative treatment of pharmaceuticals in water. *Water Research*, 34, 1881–1885.

II

Food

4

Advances in Cereal Processing: An Approach for Energy and Water Conservation

P. Srinivasa
Rao, Soumya
Ranjan Purohit,
and Lakshmi E.
Jayachandran
*Indian Institute of
Technology, Kharagpur*

4.1 Introduction

Worldwide, cereals are the main source of energy and nutrition according to the food habits of consumers in different regions. Cereal grains undergo many processing and unit operations for better utilization, value addition, and manifestation of its inherent functionality. At par with advancement in science and technology, processing methods are also endowed with several advances in the existing and new processing methods. Significant improvement has been made in the domain of technologies for postharvest processing and product development. Some of them are elaborated in this context, namely pneumatic polishing, freeze milling, high-pressure processing (HPP), arabinoxylan (AX) extraction, rice analogues, reduced retrogradation of starch, modified starch, and gluten-free (GF) bread.

Pneumatic polishing is one of the best examples of technology advancement till now in rice processing sector, which reduces broken percentage of rice and consumes less energy compared with traditional polishing systems. However, further research is needed for commercialization of this technology. Similarly, freeze milling is another emerging improved milling technology, which yields fine-sized particles with better flour characteristics, modified starch properties, and enhanced extractability of bioactive compounds from valuable cereal sources. The by-product of milling and polishing (bran) can be utilized to extract AXs to achieve zero discharge. The AXs of different molecular weight fractions have a diverse range of bioactivity, which can be obtained by different new extraction methods.

Recently, HPP has gained much attention in rice processing because of its potential to solubilize allergenic proteins in the pressure range of 300–400 MPa. Besides, pressure-treated rice shows improved brightness, color, flavor, and texture. A similar improvement has been observed in wheat gluten with HPP domain of 200–800 MPa, which results in novel texture, increased elasticity, and hardness.

Apart from this, HPP also showed potential to alter the enzymatic activity in wheat and barley. The enzyme activity increases in the lower pressure range, whereas beyond 600 MPa it reduces. In addition to pressure-induced starch gelatinization, HPP preserves the starch integrity, which in turn reduces swelling and amylose leaching out of the granule.

In the functionality domain of cereal starch from different food products, the starch modification has received much attention on low viscous products. Extrusion, annealing, heat–moisture treatment, and application of ionizing radiation are some of the potential technologies to modify the starch in different ways. Extrusion and ionizing radiation involve defragmentation of starch, which reduces the viscosity of resulting dispersion. In contrast, annealing and heat–moisture treatment increases the shear and thermal stability by physical reorganization of starch. In extrudate products, further development can be marked well with rice analogues and starch products having reduced retrogradation rate. An improved extrusion cooking technology (IECT), characterized by low screw speed, low temperature, and high residence time, has been used for the production of starch products with a low rate of retrogradation, which avoids staling phenomena of starch products. On the other hand, rice analogues have opened up the way for reformation of rice or imitation rice using flours from broken rice and other cereals along with added nutrients like proteins, vitamins, and antioxidants. Also, it will pave the way for engineered rice and dal analogues. Further, to overcome gluten intolerance, GF baked products have been developed. But GF dough has low elasticity and exhibits poor baking quality due to the absence of extensive gluten network. However, this issue can be resolved by using hydrocolloids [carboxymethyl cellulose (CMC), xanthan gum], enzymes (transglutaminase, protease), fermentation, and HPP.

4.2 Advances in Cereal Processing Technologies

4.2.1 Pneumatic Polishing

The rice polishing process transforms crude brown rice into a consumer-acceptable form of milled rice. Polishing causes removal of oil-rich bran layer, which creates a smooth surface of the rice and imparts better digestibility. The bran separated is considered as a good source of edible oil, but the nutritive value of the polished rice kernel reduces considerably in the form of insoluble fiber and essential fatty acids (Delcour and Hoseney, 2010). Usually, polishing operation performed by abrasive rubbing of brown rice kernel by a mechanical body leads to harsh and excessive work done on the rice kernel, resulting in reduced head rice yield (whole kernel; Bond, 2004). This may lead to huge economical loss, although broken rice is useful in some alternative processes.

Rice polishing is carried out by exposing the exterior part of rice to abrasive and high friction surfaces. Rice polishing systems consist of heavy abrasive rollers/cones, revolving at a high speed to achieve polishing. Thus, specific energy (SE) consumption of the polishing system is very high to overcome frictional counterpart, whereas minimal energy is consumed for actual polishing operation. Therefore, the polishing operation is considered to be the most energy-consuming unit operation in the rice milling process (Mohapatra and Bal, 2007). The equipment used for this operation imparts high shearing action on the rice (5–10 kPa), involving intergranular attrition and high friction rate. Such a series of physicomechanical changes result in the generation of high temperature due to friction and develop thermal stress in kernels, leading to fissure or fracture in the kernel (Bhattacharya, 1969; Rao and Juliano, 1970). In view of this, the ability of rice kernel to resist mechanically induced thermal stresses depends on the severity of the polishing operation and its intrinsic properties like moisture and kernel density (Rao et al., 2007).

To overcome the aforementioned issues, pneumatic polishing of rice has been developed recently. According to Prakash et al. (2014), in this new technique of rice polishing, "the rice grains are allowed to move very fast over a surface coated with the abrasive material, using two-phase flow (rice + air) system." In other words, the process includes conveying of rice with air at a velocity higher than its terminal velocity ($u > u_t$). The air–grain mixture moves through a straight pipe having internal wall coated with hard and rough (emery) material. The high-speed movement of rice over abrasive surface results in removal of the bran without development of severe thermal and mechanical stresses. Additionally, exclusion of heavy rotating units helps reduce breakage of rice, and the system consumes lesser energy. Mohapatra and Bal (2007) reported the SE consumption during polishing of three rice varieties such as Pusa Basmati, Swarna, and ADT37. Significant increase in SE can be positively correlated with higher degree of polishing (DOP). However, beyond a DOP limit, SE did not show significant change. The cereal bran is composed of mainly fibrous materials like cellulose and hemicellulose and a fraction of fat and minerals. Thus, it is easily removed by any mechanical action, whereas the endosperm part is more rigid and resistant to mechanical actions. For ADT37 variety at 5% and 12% DOP, SE was reported to increase by 0.51 kJ per percentage of DOP. Bran composition in rice is about 8%, thus extended polishing leads to removal of endosperm material, which consumes higher SE. Again, depending on the variety, hardness of brown rice varied over a significant range, which results in lower bran removal and higher SE consumption. Lesser bran removal was reported for Pusa Basmati which was more harder (73 N) than Swarna variety (46 N) As far as pneumatic polishing is concerned, detailed energy consumption in pneumatic polishing of rice has not yet reported in the literature. However, this technology is supposed to render a significant reduction in energy consumption, as far as its design and operations are concerned. The factors affecting pneumatic polishing are moisture content, shape and size of the kernel, residence time, length of the pipe, and the number of recycling loops (Prakash et al., 2014).

4.2.1.1 Effect of Recycling Loops on DOP at Different Pipe Lengths

The DOP is the governing parameter to achieve uniform polishing, concerning whiteness, and transparency of kernel. Prakash et al. (2014) reported pneumatic polishing of rice in a 4 m long pipe, results in DOP, whiteness, and transparency of 9.97, 14.1, and 0.72%, respectively. Figure 4.1 depicts the variation in DOP with the number of recycling loops and the pipe length. The DOP increases with an increase in the pipe length and the number of recycling loops. However, the difference in DOP is less initially but increased as the number of recycling loops increased.

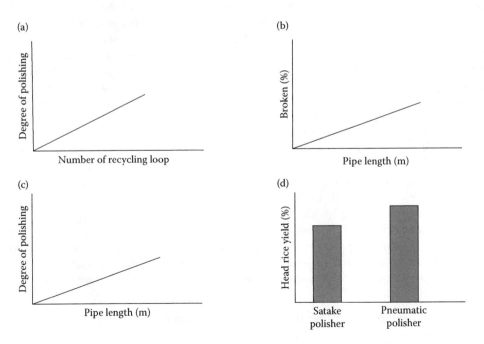

FIGURE 4.1 (a) Effect of the number of recycling loop on DOP, (b) effect of pipe length on broken percentage, (c) effect of pipe length on DOP, and (d) effect of polishing module on head rice yield.

4.2.1.2 Effect of Pipe Length on Breakage of Kernels

The head rice yield is greatly affected by the pipe length of the pneumatic polishing system. Broken rice content varied between 2.18% (1 m) and 8.52% (4 m). A study comparing the laboratory polisher (Satake Corporation, Japan; Model: TM 05) and pneumatic polishing system at same DOP (10%) showed that the content of broken rice produced by both was around 20% and 8.5%, respectively. Lower breakage in pneumatic polishing system might be due to lower level of thermal and mechanical stresses in the kernel. Higher dissipation of heat with large air flow rate and gentle handling of the grain inside the polishing zone are also responsible for these low stresses (Prakash et al., 2014).

4.2.2 Freeze Milling

Conventional milling techniques include either dry or wet grinding, which has certain limitations. The wet milling consists of a five-step process including soaking, addition of excess water, filtering, drying, and sieving, thus involving many machines and manpower. Although the wet grinding yields fine-sized particles, it is associated with high water and energy consumption, hassles of wastewater treatment, and loss of volatile components and minerals. In contrast, the dry milling produces larger sized particles, compromising on the quality of the final product, and as a result of this, the extractability of comminuted particles is also hindered. Alternatively, low-temperature freezing of grain/kernel makes the material brittle and easy to break, thus aiding trouble-free grinding process. In comparison to dry grinding, wet grinding process demonstrates significantly higher SE consumption (13,868 KJ/kg) due to the large consumption of electrical energy by several machines in the process. However, the energy consumption during freeze grinding was similar to that of dry grinding (Ngamnikom and Songsermpong, 2011). The freeze grinding technique has also been adopted as an alternative to the particle size reduction and homogenization of the breakfast cereal samples. Freeze grinding or cryomilling is also recognized as a nonchemical way of modifying the starch structure, which finds wide applications as filling material in confectionaries and baby foods (Devi et al., 2009). It modifies the starch by causing a loss of crystallinity, which reduces its pasting temperature and increases water absorption, followed by reduction in the viscosity and increased solubility. This evolved in a time of strict regulations on the use of chemically modified starches in food products and its associated environmental concerns on wastewater generation. Freeze milling can extract low–molecular weight bioactive compounds from the product without addition of enzymes, acid, etc., which can be applied as food and pharmaceutical ingredients, especially with much higher concentration due to its higher solubility arising from low molecular weight. The oats β-glucan prepared by freeze milling has lower viscosity, which facilitates its inclusion into beverages (Harasym et al., 2015). Samples with high moisture content can also be easily pulverized at very low-temperature conditions. This is due to the increased brittleness of the material, which enables easy comminution, supporting the extensive size reduction followed by enhanced mass transfer and extractability of desirable elements. Freeze grinding of cereals with the hammer mill significantly reduced both the average particle size and damaged starch content and resulted in higher yield after sieving, compared with dry grinding using an identical grinder. The damaged starch content in rice flour was reported to be 3%, 6%, and 11% for wet, freeze, and dry ground samples, respectively. The yield of the rice flour was observed to be maximum for dry grinding (96%), followed by freeze (86%) and wet (79%) grinding processes (Ngamnikom and Songsermpong, 2011). The loss of yield during freeze grinding is because of the spurting of the product while adding the liquid nitrogen in the intermediate grinding stages. Various applications and advantages of liquid nitrogen are summarized below (Wang et al., 2015a,b):

1. Liquid nitrogen is cheaper than the cost incurred in wastewater treatment during conventional wet grinding.
2. It reduces the average particle size and damaged starch content due to the extremely low temperature of the sample before grinding.

3. Minimal usage of water in freeze milling.
4. Enhances the solubility and extractability of protein from different sources.
5. Keeps plant surroundings clean.
6. Increases the speed and effectiveness of procedure for sample comminution and homogenization in preparation of breakfast cereals.

Limitations of liquid nitrogen are as follows:

1. Contamination of the product with the grinder material content. However, this issue is associated with any grinder with metallic parts of contact.
2. To be tested with the high-performance grinding machine.

4.2.3 HPP of Cereals

HPP in the discipline of cereals and cereal-based products is an interesting field. Several scientific reports are available about the effect of HPP on the cereal-specific components such as starch and gluten.

HPP treatment has been investigated for its macromolecular solubilizing ability. The ingestion of rice protein has been considered to be one of the reasons for allergic disorder like asthma and dermatitis. Particularly, 16 kDa albumin, 26 kDa α-globulin, and 33 kDa globulin (Baldo and Wrigley, 1984) have been previously identified as major rice allergens. In the past decade, HPP has been explored for its ability to solubilize the allergenic proteins in rice grains, without any alteration in physical and sensory attributes of the treated material (Limas et al., 1990). For instance, rice grains (Akitakomachi cv.) soaked in distilled water and subjected to HPP treatment of 100–400 MPa resulted in solubilization and release of allergenic proteins (Kato et al., 2000). In this case, the pressure acted as a catalyst to enable the release of a considerable amount of allergenic proteins, which can be positively correlated with ultra high pressure up to a certain limit (Estrada-Girón et al., 2005). Further use of protease enzyme to enhance the effect of pressure treatment on solubilization and release of allergenic protein has been investigated. A remarkable increase in the penetration ability of protease into the rice kernel under pressure has been reported by Watanabe et al. (1990). In addition, pressure treatment not only reduces the allergens in rice grains but also improves the quality of cooked rice, imparting better color, flavor, and texture to the rice grains.

The majority of wheat is used for human consumption as flour, which is later processed to bread, sweet doughs, cakes, biscuits, doughnuts, crackers, etc. The inherent enzymes like amylase, protease, and lipase cause undesirable changes in wheat flour. HPP has been tested for its efficient inactivation of these cereal enzymes (Gomes et al., 1998). Amylase is the prime enzyme, which is well distributed throughout all cereal commodities. Commercially, amylase has got immense applications in many industrial processes for starch hydrolysis (Purohit and Mishra, 2012), as it cleaves starch to its simpler form. This results in the formation of thinner starch slurry compared with its native counterpart. In this respect, HPP has been investigated for its effect on either activation or inactivation of amylases. Significant increase in amylolytic activity on the starch occurs during HPP treatment at 400–600 MPa due to starch gelatinization, which makes starch accessible to amylase for hydrolysis. Such behavior is supposed to be due to inability of pressure (up to 400 MPa) to completely inactivate enzymes; rather it supports starch gelatinization, thereby increasing enzyme activity in terms of its easy accessibility to starch. However, the pressures above 600 MPa have been demonstrated for their ability to either modify the enzyme active site or unfolding, leading to inactivation of the enzyme (Seyderhelm et al., 1996).

Wheat gluten was often used to improve wheat flours low in gluten and to develop novel textured products due to its special cohesive and viscoelastic properties. The gels formed during pressure/heat treatment of wheat gluten in the range 200–800 MPa at temperatures 20°C–60°C, with holding times from 20 to 60 min, were incredibly different from that from heat-processed gluten since they exhibit a more marked elasticity with high values of moduli of elasticity. The hardness of the wheat gluten treated at ambient temperature under pressure shows an increasing trend with time. Time has a profound effect as far as HPP is concerned for novel texture and product development (Gomes et al., 1998).

In addition to the aforementioned applications, HPP is pertinent in starch processing, which is well known for inducing gelatinization below the gelatinization temperature (T_{Gel}) of the products. Studies reported that ambient temperatures (20°C–25°C) and pressure treatment of 400 and 900 MPa are required for gelatinization of wheat and potato starch. It has been reported that combination of heat and pressure in the treatments enhance gelatinization of food products (Ahromrit et al., 2007). The characteristic features of thermally gelatinized and pressure-gelatinized starches are easily distinguishable due to their distinct behavior under heat and pressure treatment. The compression action of high pressure leads to reduced swelling of starch granules compared with heat-gelatinized starch. In contrast to this, higher swelling has been observed in case of waxy maize starch and tapioca starch. During HPP processing, the aforementioned phenomenon modifies starch. Investigations have been made on HPP-treated sorghum batter for evaluating the potential of pressure-treated sorghum as a gluten replacement in production of sorghum bread. The effect of HPP in delaying the staling effect of GF products has been reported by Vallons et al. (2016).

The major high pressure–induced modifications in cereals are listed below.

1. Polymorphic transition of A-type to B-type X-ray diffraction pattern, whereas B-type starch remains unaffected.
2. Retains the granular integrity and limits amylose leaching into medium during gelatinization.
3. Preserve the molecular weight distribution of starch components (amylose:amylopectin).
4. Modification of the paste viscosity and gel formation due to restricted swelling power of the starch granules and low solubility of amylose.
5. The retrogradation of pressure-induced gels is supposed to occur within starch granules, which make the starch gels less sensitive to aging.

Nevertheless, substantial research is still essential for commercialization of HPP process and to achieve suitable processing conditions for product development with improved nutritional and sensory characteristics. However, HPP process is not a cost-effective process, but proper design of the process and product may be helpful in commercializing the technology under the current industrial scenario.

4.2.4 Extraction of AX

AXs are pentosan complex of arabinose and xylan (hemicellulose), commonly found in the outer layer (bran) of cereal grain. AX and its extraction from resources has received much attention in recent research scenarios due to its functional qualities. The extraction method has a significant effect on characteristics, functionality, and yield of AXs, as shown in Table 4.1 (Xu et al., 2006).

Saeed et al. (2011) reported higher pentosan or AX content in bran layer of major cereal grains. AX has wide industrial applications due to its diverse physical attributes like higher viscosity, water-holding capacity, and thickening behavior. Subsequently, AXs have been reported to bear diverse biological functionalities, such as lowering of serum cholesterol, reducing the risk of coronary heart disease and the blood sugar level, has antioxidant activity, reducing glycemic response, and immunity enhancement.

Cereal bran is considered to be a rich source of AXs, and being the by-product of cereal milling operation, it is a cost-effective raw material for AXs extraction. Previous studies have demonstrated that the bioactivities of AXs may be associated with their specific molecular characteristics like degree of branching, molecular weight, and chain alignment. Wheat bran AXs with low Mw (6.6×10^4 Da) have potential prebiotic properties in vitro, and the modified rice bran with Mw 30–50 kDa has shown immunomodulating activities in both in vitro and in vivo studies. In contrast to this, AXs with higher Mw have the ability to lower the postprandial glycemic response (Zheng et al., 2011).

Different methods have been explored for the extraction and purification of AXs from cereal by-products, which are potential enough to affect the extraction yields and macromolecular characteristics of AXs. The improved extraction methods include alkaline and acidic conditioning, enzyme hydrolysis, microwave treatment, ultrasonication, steam explosion, and twin-screw extrusion (Zhang et al., 2014a).

TABLE 4.1 Different Sources of AXs and Their Extraction Routes

Raw Material	Extraction Type	Extraction Aid	Advantages	Disadvantages
Barley flour, rye flour Wheat flour	Water extractions	Water	Environmentally friendly No changes to molecules	Low extraction yields of AXs
Wheat bran, corn bran, barley husks, wheat straws	Chemical treatments	NaOH/HCl	Highly efficient treatment	Break down functional groups Decrease AX, low branched degree
Wheat bran, barley husk Corn husk, rye flour	Enzymatic treatments	Endoxylanase	Eco-friendly Controlled degradation of AX molecules	Process cost
Wheat/corn bran and straw	Mechanical treatments (extrusion, microwave, steam explosion)	NA	Efficient extraction	Uncontrolled debranching

Source: Zhang et al. (2014b).

As far as the economical use of water and energy is concerned, the feasibility aspect of abovementioned major process can be evaluated based on their extraction conditions. For water extraction of AXs, a mild heating condition (below 100°C) has been considered to have the inability to break complex intermolecular bonding within AXs (Izydorczyk and Biliaderis, 1992). Thus, hydromechanical treatments are advisable in order to increase the extraction yield. Mok and Antal (1992) reported that major amounts of hemicellulose could be solubilized from different plant sources using hot compressed liquid water in 0–15 min at 200°C–230°C. On the other hand, chemical and enzymatic methods are more efficient as far as yield is concerned. Most of the chemical and enzymatic treatments are operated at ambient temperature, in which mild mixing is the only requisite to facilitate the changes desirable for extraction. From the above viewpoint, enzymatic and chemical methods are more energy efficient compared with water extraction method. However, use of water as reaction medium is unavoidable in these circumstances. Interestingly, extrusion processing is one of the old technologies, but emerging as most efficient and eco-friendly tool, used in food and bioprocessing. A major advantage of this technology is the ability to disrupt the tissue-based structure, which enhances the extractability of useful components. Besides, the system has the potential to process the materials at low moisture content, which is a big relief as far as water consumption is concerned. Therefore, extrusion-assisted enzymatic or chemical extraction could be the most efficient method in terms of energy and water consumption.

4.2.5 Gelatinization of Paddy Using Far Infrared Radiation (FIR)

Parboiling involves soaking, steaming, and drying of paddy, which results in starch gelatinization and disintegration of the protein bodies within the grain. This imparts hardness, translucency, and higher milling quality of the paddy. Soaking supplies the grains with water required for gelatinization; high heat content generated during steaming accelerates and completes the gelatinization of paddy; and drying removes the excess moisture from the paddy. High-temperature steam is known to destroy the natural antioxidants in paddy and thus increasing the incidence of grain rancidity. The potential of FIR in gelatinization of rice in parboiling process has been recently established (Dissanayake et al., 2015). The FIR system has high rate of heat transfer, requires less space than that of a steaming setup, saves energy, is easily automated, preserves the nutrients in the food, and prevents solute migration from inner to outer layers of the grain. Das et al. (2004) evaluated the quality of infrared dried parboiled rice in terms of head rice yield, color, percent-gelatinized kernel, and SE consumption using five levels of radiation intensity and four levels of grain bed depths in a vibration-aided infrared dryer. However, the gelatinization of rice using FIR is an innovative approach. From previous studies, it has been established that the radiation intensity and the grain depth are the key factors affecting the quality of the FIR-treated products. In this technology, the cleaned paddy is soaked in cold water

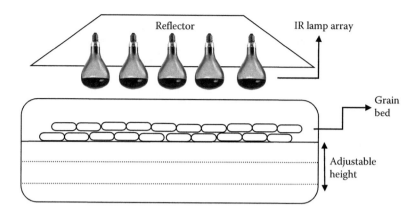

FIGURE 4.2 Schematic diagram of the IR processing system for gelatinization of rough rice.

for 36 h, changing the water every 12 h. When the moisture content of the paddy was 30% (wet basis), the water was drained and the paddy was exposed to FIR (Figure 4.2), for different time intervals from 5 to 30 s. This ensured quick gelatinization of the starch and was followed by the final convective oven drying at 40°C to a moisture content of 14% (w.b.). The quality of the FIR-treated rice was analyzed in terms of moisture content, degree of gelatinization, and yellowness index. The moisture content decreased with increase in exposure time, which involved an increase in temperature as well. The rate of drying decreased and showed a constant value with increase in exposure time. The degree of gelatinization of rough rice was 40% within 20 s of FIR heating. Marshall et al. (1993) reported that degree of starch gelatinization of ≈40% was sufficient to obtain a maximal head rice yield for microwave parboiled rough rice. L^{*} (the tristimulus values namely L^{*}, a^{*} and b^{*} which represents the lightness, red-green and blue-yellow were significantly affected during FIR treatment) values were decreased with FIR radiation exposure time and hence lightness of rice reduced. b^{*} had positive values, and it implied that color of FIR-exposed rough rice had turned to a yellowish color.

Yellowness index was in the range of 22–35. Therefore, it can be concluded that the simultaneous gelatinization and drying of rice could be effectively carried out in a very short period using FIR.

4.3 Advances in Cereal Products

4.3.1 Rice Analogues

The milling process of rice leads to the production of broken rice kernels, which are not preferred by consumers. Further, broken kernels are utilized for preparation of alternative value-added products. In an innovative product development context, these broken kernels could be ground to flour and gelatinized/mashed into dough form, followed by addition of desired additives and reforming into rice shape by extrusion. This could be spelled out as reconstituted rice or rice analogues. During the preparation of rice and pulse analogues, fortification can be done with additives like proteins, vitamins, minerals, and fibers to achieve superior functionality. Otherwise, being a staple food, such product can be tailored as per the requirement of specific consumer community, which could satisfy hunger and food functionality (Mishra et al., 2012).

Extrusion technology has been demonstrated for its ability to develop the aforementioned products with desirable attributes (Bett-Garber et al., 2004). The process comprises basic steps of grinding of the broken rice kernel, dough formation, and moisture conditioning followed by extrusion. High temperature and pressure in extrusion cause the rice starch to gelatinize and melt, which facilitates the molten mass to reform into native rice shape (Figure 4.3). Further treatment of the extrudates with binder or cross-linking agents helps stabilize the product. Finally, drying of these extrudates results in rice analogues at desired moisture content for safe storage of the product. Various studies have suggested that

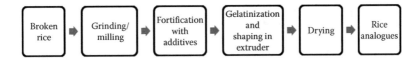

FIGURE 4.3 Process flow for preparation of rice analogues.

rice analogues are potentially very useful means of improving nutrition in rice-consuming societies (Mishra et al., 2012). Even engineered rice and pulse analogues can be prepared with fortification and subsequently added to regular products in desired composition (Alavi, 2008).

4.3.2 Reduced Retrogradation of Starch

The IECT, a new gelatinization technology, is based on traditional single-screw extruders. The transformed single-screw extruder contains a longer screw (1950 cm) and has longer residence time (40–68 s), higher die pressure (13.4–19.1 MPa), lower temperature (70°C–120°C), and lower screw speed (20.1–32.6 rpm) than the traditional extrusion cooking machines. The percentage of retrogradation in these systems turned out to be low. However, during storage, the rate of retrogradation of rice starch is low, with the pattern of rice starch changed from A-type to amorphous and B-type after the IECT and the retrogradation processes, respectively. With such potential, IECT is an applicable and promising technique for preparing rice starch products with low percentage as well as low rate of retrogradation in the food industry (Zhang et al., 2014a).

4.3.3 Modified Starches

Major starch modification aspects target to modify the rheological behavior of starch in order to prepare starch dispersion with reduced viscosity (Zhang et al., 2014a). Extrusion is considered as physical means of starch modification. Extrusion increases the solubility of a dispersion as the material experiences high shearing action between the screws rotating at high speed. Such harsh treatment leads to complete melting of starch at high pressure (shear)/temperature domain along with molecular fragmentation of starch polymer, which results in shorter polymeric chain length and reduced viscosity in the dispersion.

Annealing and heat moisture treatment are the two physical processes executed one after another for starch modification purposes. These processes are carried out at a temperature higher than the glass transition temperature (T_g) but below the gelatinization temperature (T_{gel}) (Kiatponglarp et al., 2015). The properties of resulting starch can be characterized by improved thermal stability with an increase in gelatinization temperature, which further retards gelatinization and amylose leaching. Such changes result in the production of low viscous starch dispersion. The abovementioned phenomenon is due to the reorganization of the starch macromolecular structure during heat and moisture treatment. Further, cooling below annealing temperature results in increased crystallinity (Dundar and Gocmen, 2013).

Ionizing radiation, such as gamma ray, electron beam, and ultraviolet ray, has the ability to induce photon-based energy dissipation on the target material, which results in breaking of chemical bonds (Figure 4.4). Cleavage of glycosidic bond in starch results in molecular changes and fragmentation of starch. This principle is often used for the preparation of starch dispersion with reduced viscosity (Bhat and Karim, 2009).

In extrusion processing, feed rate and speed of the screw always contribute toward specific mechanical energy (SME) consumption. Fayose and Huan (2015) have recently reported the SME consumption in the processing of major starchy crops like cassava, maize, wheat, etc. For the feed rate of 10 kg/h, SME was reported to be the function of the type of source, moisture content, and screw speed. The

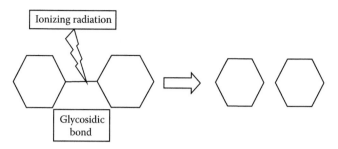

FIGURE 4.4   Fragmentation of starch by ionizing radiation.

highest SME was reported to be 42.7 KJ/kg for cassava, and the lowest SME was 3.95 KJ/kg for maize. Contrastingly, screw speeds were reported to be 84 and 100,rpm respectively, for cassava and maize. These data highlight the importance of source material and its composition.

4.3.4   GF Bread

A GF diet is the only effective treatment for a fraction of the global population with celiac disease and gluten intolerance. The formulation of high-quality GF cereal-based products represents a challenging task for both the cereal technologist and bakers, especially due to the low baking quality of GF flours. Gluten is the major structure forming protein in the wheat flour and is responsible for the elastic nature of the dough. It also contributes to the appearance and crumb structure of many baked products. Several alternatives, such as starches, dairy products, gums and hydrocolloids, other nongluten proteins, prebiotics, and combinations thereof, are being used in place of gluten to improve the structure, mouthfeel, acceptability, and shelf life of GF bakery products. To overcome this issue, use of starches and gum/hydrocolloids is important in the development of GF cereal-based products to achieve desirable characteristics, such as high loaf volume, crumb softness, and appearance properties along with sensory acceptability (McCarthy et al., 2005). However, right selection and combination of hydrocolloids are crucial to obtain breads with a desirable quality. Replacement of 30% of wheat flour with rice flour results in bread with acceptable quality, whereas more than 30% of gluten substitutes are necessary. An optimized combination of hydroxypropyl methyl cellulose (HPMC) and CMC renders high-quality baked products similar to wheat bread. Different hydrocolloids like HPMC improve gas retention and water absorption, mimicking gluten characteristic. Similarly, xanthan gum imparts a good crumb structure in the absence of gluten, and CMC, agarose, or β-glucan promotes loaf volume (Dwivedi et al., 2014). Wheat and nonwheat starches (rice, potato, corn, cassava, buckwheat) can also be incorporated in the GF products for better quality baked products.

Dairy powders (whey powder, milk solids, protein isolates, etc.) can be added for the fortification of the GF products, ensuring balanced nutrient supply to the consumers. These powders enhance the flavor and texture of the GF bread, besides improving the water absorption characteristics and prolonging the storage life (Kenny et al., 2001). Incorporation of dietary sources is yet another area on which extensive research is being conducted. People of GF diet have been found to be deficient in their dietary intake compared with their counterparts on normal diet. The inclusion of inulin, amaranth flour, and quinoa powder has been established as an effective way of increasing the dietary fiber content of the bread (Taylor and Parker, 2002). This leads to increase in loaf volume, sliceability, dough stability, crumb softness, and the nutritional profile. GF pasta has been successfully formulated using pea flour and fat sources, resulting in better flavor and texture after cooking (Wang et al., 1999). GF biscuits based on nonwheat starches and fat sources are now available in the market, and the quality of these biscuits is similar to that of wheat biscuits (Arendt et al., 2002). GF pizza bases have also been formulated using corn starches, guar gum, and fat sources.

4.4 Conclusion

The novel technologies and products in the field of cereal processing have indeed improved the yield and quality of the final value-added products. Several technologies are still under research, and every year, the food technologists and engineers are coming up with innovative concepts, adding to the pool of advanced technologies for cereal processing. Pneumatic polishing of grains has already gained the upper hand over abrasive polishing due to less broken percentage and low level of thermomechanical stresses. Freeze grinding/cryomilling has found applications in the size reduction of grains, such as rice and oats, with less consumption of energy and water compared with the wet milling process. It reduces the average particle size and the damaged starch content in the grains, resulting in starch modification. With respect to the processing of cereal grains and legumes, HHP technology offers many advantages over traditional techniques involving thermal treatment for reducing the microbial population of spoilage microorganisms, inactivating unwanted food enzymes and compounds, and consequently increasing the shelf life of products in addition to the development of novel cereal-based products. Various methods have been developed for the extraction and purification of AXs from the cereal by-products, which include techniques like alkaline and acid conditioning, enzymatic hydrolysis, microwave treatment, ultrasonication, steam explosion, and twin-screw extrusion. FIR offers the possibility of quick gelatinization of rice compared with the existing methods, with simultaneous gelatinization and drying of rice using FIR waves. Rice analogues are also the new trends, improving the nutritional status of the rice-consuming societies. Modified starches with their altered structure and improved properties find wider applications in the food industry. IECT is a promising technique for preparing rice starch products with a low rate of retrogradation in the food industry. The use of starches, gums, and hydrocolloids represent the most widespread approach used to mimic gluten in the manufacture of GF bakery products, due to their structure-building and water-binding properties. Novel approaches to the application of dietary fibers and alternative protein sources are also new emerging areas, and the process can be optimized with response surface methodology.

References

Ahromrit, A., Ledward, D. A., and Niranjan, K. (2007). Kinetics of high pressure facilitated starch gelatinisation in Thai glutinous rice. *Journal of Food Engineering, 79*(3), 834–841.

Alavi, S. (2008). Rice fortification in developing countries: A critical review of the technical and economic feasibility. Edited report. Academy for Educational Development, Washington, DC.

Arendt, E. K., O'Brien, C. M., Schober, T., Gormley, T. R., and Gallagher, E. (2002). Development of gluten-free cereal products. *Farm and Food, 12,* 21–27.

Baldo, B. A. and Wrigley, C. M. (1984). Allergies to cereals. *Advances in Cereal Science and Technology, 6,* 289–356.

Bett-Garber, K. L., Champagne, E. T., Ingram, D. A., and Grimm, C. C. (2004). Impact of iron source and concentration on rice flavour using a simulated rice kernel micronutrient delivery system. *Cereal Chemistry, 81*(3), 384–388.

Bhat, R. and Karim, A. A. (2009). Impact of radiation processing on starch. *Comprehensive Reviews in Food Science and Food Safety, 8*(2), 44–58.

Bhattacharya, K. R. (1969). Breakage of rice during milling, and effect of parboiling. *Cereal Chemistry, 46,* 478–485.

Bond, N. (2004). Rice milling. In *Rice Chemistry and Technology,* Ed. E. T. Champagne. St. Paul, MN: AACC.

Das, I., Das, S. K., and Bal, S. (2004). Specific energy and quality aspects of infrared (IR) dried parboiled rice. *Journal of Food Engineering, 62*(1), 9–14.

Delcour, J. and Hoseney, R. C. (eds) (2010). Rice and oats processing. In *Principles of Cereal Science and Technology.* St. Paul, MN: AACC.

Devi, A. F., Fibrianto, K., Torley, P. J., and Bhandari, B. (2009). Physical properties of cryomilled rice starch. *Journal of Cereal Science, 49*(2), 278–284.

Dissanayake, T. M. R., Amarathunga, K. S. P., Thilakaratne, B. M. K. S., Bandara, D. M. S. P., and Fernando, A. J. (2015). Gelatinization of rough rice using far-infrared (FIR) radiation. *Tropical Agricultural Research, 26*(4), 707–713.

Dundar, A. N. and Gocmen, D. (2013). Effects of autoclaving temperature and storing time on resistant starch formation and its functional and physicochemical properties. *Carbohydrate Polymers, 97*(2), 764–771.

Dwivedi, M., Chakraborty, S., Deora, N. S., and Mishra, H. N. (2014). Use of response surface methodology to optimize the formulation of rice based gluten free bread and its characterization. *Research and Reviews: Journal of Food Science and Technology, 3*(3), 1–12.

Estrada-Girón, Y., Swanson, B. G., and Barbosa-Cánovas, G. V. (2005). Advances in the use of high hydrostatic pressure for processing cereal grains and legumes. *Trends in Food Science and Technology, 16*(5), 194–203.

Fayose, F. and Huan, Z. (2015). Energy consumption and efficiency in single screw extrusion processing of selected starchy crops. *African Journal of Agricultural Research, 10*(7), 710–719.

Gomes, M. R. A., Clark, R., and Ledward, D. A. (1998). Effects of high pressure on amylases and starch in wheat and barley flours. *Food Chemistry, 63*, 363–372.

Harasym, J., Suchecka, D., and Gromadzka-Ostrowska, J. (2015). Effect of size reduction by freeze-milling on processing properties of β-glucan oat bran. *Journal of Cereal Science, 61*, 119–125.

Izydorczyk, M. S. and Biliaderis, C. G. (1992). Effect of molecular size on physical properties of wheat arabinoxylan. *Journal of Agricultural and Food Chemistry, 40*(4), 561–568.

Kato, T., Katayama, E., Matsubara, S., Omi, Y., and Matsuda, T. (2000). Release of allergic proteins from rice grains induced by high hydrostatic pressure. *Journal of Agricultural and Food Chemistry, 48*, 3124–3126.

Kenny, S., Wehrle, K., Auty, M., and Arendt, E. K. (2001). Influence of sodium caseinate and whey protein on baking properties and rheology of frozen dough. *Cereal Chemistry, 78*, 458–463.

Kiatponglarp, W., Tongta, S., Rolland-Sabaté, A., and Buléon, A. (2015). Crystallization and chain reorganization of debranched rice starches in relation to resistant starch formation. *Carbohydrate Polymers, 122*, 108–114.

Limas, G., Salinas, M., Moneo, I., Fischer, S., Wittmann-Liebold, B., and Mendez, E. (1990). Purification and characterization of ten new rice NaCl-soluble proteins. *Planta, 181*(1), 1–9.

Marshall, W. E., Wadsworth, J. I., Verma, L. R., and Velupillai, L. (1993). Determining the degree of gelatinization in parboiled rice: Comparison of a subjective and an objective method. *Journal of Cereal Chemistry, 70*, 226–230.

McCarthy, D. F., Gallagher, E., Gormley, T. R., Schober, T. J., and Arendt, E. K. (2005). Application of response surface methodology in the development of gluten-free bread. *Cereal Chemistry, 82*(5), 609–615.

Mishra, A., Mishra, H. N., and Rao, P. S. (2012). Preparation of rice analogues using extrusion technology. *International Journal of Food Science and Technology, 47*(9), 1789–1797.

Mohapatra, D. and Bal, S. (2007). Effect of degree of milling on specific energy consumption, optical measurements and cooking quality of rice. *Journal of Food Engineering, 80*(1), 119–125.

Mok, W. S. L. and Antal, Jr., M. J. (1992). Uncatalyzed solvolysis of whole biomass hemicellulose by hot compressed liquid water. *Industrial and Engineering Chemistry Research, 31*(4), 1157–1161.

Ngamnikom, P., and Songsermpong, S. (2011). The effects of freeze, dry, and wet grinding processes on rice flour properties and their energy consumption. *Journal of Food Engineering, 104*(4), 632–638.

Prakash, K. S., Someswararao, C., and Das, S. K. (2014). Pneumatic polishing of rice in a horizontal abrasive pipe: A new approach in rice polishing. *Innovative Food Science and Emerging Technologies, 22*, 175–179.

Purohit, S. R. and Mishra, B. K. (2012). Simultaneous saccharification and fermentation of overnight soaked sweet potato for ethyl alcohol fermentation. *Advanced Journal of Food Science and Technology*, 4(2), 56–59.

Rao, P. S., Bal, S., and Goswami, T. K. (2007). Modelling and optimization of drying variables in thin layer drying of parboiled paddy. *Journal of Food Engineering*, 78, 480–487.

Rao, S. N. R. and Juliano, B. O. (1970). Effect of parboiling on some physicochemical properties of rice. *Journal of Agricultural and Food Chemistry*, 18(2), 289–294.

Saeed, F., Pasha, I., Anjum, F. M., and Sultan, M. T. (2011). Arabinoxylans and arabinogalactans: A comprehensive treatise. *Critical Reviews in Food Science and Nutrition*, 51(5), 467–476.

Seyderhelm, I., Boguslawski, S., Michaelis, G., and Knorr, D. (1996). Pressure induced inactivation of selected food enzymes. *Journal of Food Science*, 61(2), 308–310.

Taylor, J. R. N. and Parker, M. L. (2002). Quinoa. In *Pseudocereals and Less Common Cereals, Grain Properties and Utilization Potential*, Eds P. S. Belton and J. R. N. Taylor (pp. 93–122). Berlin: Springer Verlag.

Vallons, K. J., Ryan, L. A., Koehler, P., and Arendt, E. K. (2010). High pressure–treated sorghum flour as a functional ingredient in the production of sorghum bread. *European Food Research and Technology*, 231(5), 711–717.

Wang, N., Bhirud, P. R., Sosulski, F. W., and Tyler, R. T. (1999). Pasta like product from pea flour by twin-screw extrusion. *Journal of Food Science*, 64, 671–678.

Wang, T., Liu, F., Wang, R., Wang, L., Zhang, H., and Chen, Z. (2015a). Solubilization by freeze-milling of water-insoluble subunits in rice proteins. *Food and Function*, 6(2), 423–430.

Wang, T., Zhang, H., Wang, L., Wang, R., and Chen, Z. (2015b). Mechanistic insights into solubilization of rice protein isolates by freeze–milling combined with alkali pretreatment. *Food Chemistry*, 178, 82–88.

Watanabe, M., Miyakawa, J., Ikezawa, Z., Suzuki, Y., Hirano, T., and Yoshizawa, T. (1990). Production of hypoallergenic rice by enzymatic decomposition of constituent proteins. *Journal of Food Science*, 55, 781–783.

Xu, F., Liu, C. F., Geng, Z. C., Sun, J. X., Sun, R. C., and Hei, B. H. (2006). Characterisation of degraded organosolv hemicelluloses from wheat straw. *Polymer Degradation and Stability*, 91(8), 1880–1886.

Zhang, Y., Liu, W., Liu, C., Luo, S., Li, T., Liu, Y., and Zuo, Y. (2014a). Retrogradation behaviour of high-amylose rice starch prepared by improved extrusion cooking technology. *Food Chemistry*, 158, 255–261.

Zhang, Z., Smith, C., and Li, W. (2014b). Extraction and modification technology of arabinoxylans from cereal by-products: A critical review. *Food Research International*, 65, 423–436.

Zheng, X., Li, L., and Wang, X. (2011). Molecular characterization of arabinoxylans from hull-less barley milling fractions. *Molecules*, 16(4), 2743–2753.

5

Clean Energy Technologies for Sustainable Food Security

Aprajeeta Jha and
P. P. Tripathy
Indian Institute of
Technology Kharagpur

5.1 Introduction

Sufficient, safe, and nutritious food is not only the basic necessity of life but also the right of every individual living on the planet. According to the projections of FAO, WFP, and IFAD (2015), nearly 842.3 million people in the world were estimated to be chronically undernourished and do not have access to sufficient food for maintaining an active and healthy life. The substantial growth in agricultural sector is an effective strategy for eradicating hunger and achieving food security. Being food secure is ensuring adequate food for everyone at all times, which can be achieved only by proper accessibility to food. This, in turn, depends on the distribution network starting from harvesting stage, postharvest processing stage, transportation stage, and distribution stage until the food reaches the consumer. However, energy is an inevitable requirement in all stages of the agri-food supply chain as can be seen from Table 5.1.

The food production volume is directly linked to irrigation management practices. It was observed that surface water is not sufficiently available for optimum use in agriculture due to seasonal fluctuations and quality degradations. Hence, the groundwater is normally adopted as the principal source of irrigation. In addition, the global primary energy demand is also expected to rise by 56% from 2010 to 2040, thus increasing the cost of energy and agricultural production (IEO, 2013). So, securing energy

TABLE 5.1   Energy Use at Various Stages of Food Production

Utility Stages	Energy Use (quads)
Farm level (fuel, chemical, equipment, irrigation)	3.6
Processing and packaging	3.5
Distribution and retailing	3.9
Residential utility	7.1

Source: Finley, J. W. and J. N. Seiber, *Journal of Agricultural and Food Chemistry* 62, no. 27: 6255–6262, 2014.

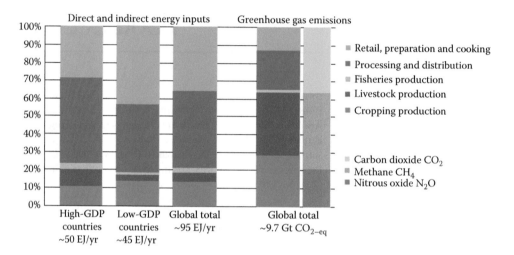

FIGURE 5.1   Energy consumption along with GHG emissions at different stages of agri-food processing. (From FAO, Energy-smart food for people and climate. www.fao.org/docrep/014/i2454e/i2454e00.pdf.)

and food supply is a key challenge for achieving future food security. Globally, the agri-food chain consumes around 30% of the world's available energy, met largely with fossil fuels and produces about 20% of the world's greenhouse gas (GHG) emissions (FAO, 2011). The energy consumption and GHG emissions of different parts of the food system for developing [low gross domestic product (GDP)] countries and developed (high GDP) countries is shown in Figure 5.1.

The future food security is intrinsically linked to energy, water, and climate issues, and therefore must be managed in an integrated way. Securing energy and water supply is a fundamental cue to feed the global population, whereas mitigating climate change is the decisive challenge toward a sustainable development. Therefore, government and industries all around the world must look into cutting-edge energy-efficient, low-emission technologies and management practices for the agricultural sector. The present chapter makes an attempt to explain the role of clean energy sources in agri-food processing as a means of enhancing sustainable food security. In this context, specific applications of solar energy, wind energy, geothermal energy, and biomass energy in agriculture are discussed, and some emerging opportunities of clean energy applications in food processing have been highlighted.

5.2   Role of Clean Energy for Sustainable Food Security

The requirement of heat, electricity, and transportation services within the agri-food sector is accomplished by conventional sources of energy utilizing the fossil fuels. However, the fast depletion of fossil

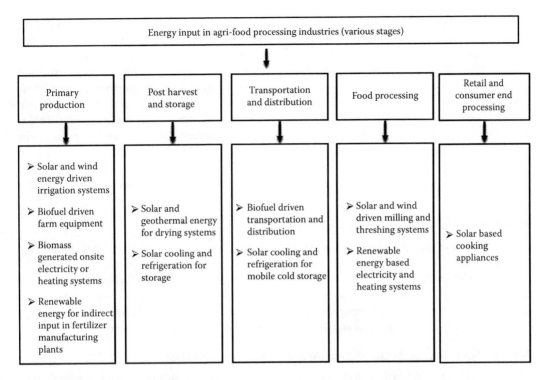

FIGURE 5.2 Process flow chart of clean energy utilization in food processing chain.

fuel energy along with their rising cost of production and adverse environmental impacts has emphasized the need on the utilization of clean energy sources. FAO (2014) proposes "energy smart" program in the agri-food chain by improving access to clean energy services and enhancing energy efficiency. Clean energy can be used, either directly or indirectly, to provide on-site or centralized energy supply. For example, on-site anaerobic digesters are used in the farms that utilize a wide variety of agricultural crop residues and animal and food wastes to generate usable energy in the form of electricity or boiler fuel for space or water heating. It was estimated that nearly 840 gigawatt-hours (GWh) of energy was generated by anaerobic digesters placed on farms of the United States (IRENA, 2015). This type of practice will positively affect the economic, social, and environmental security of the farmers, landowners, and communities across all major segments of the agri-food chain.

The widely used clean energy resources, such as solar, wind, geothermal, and biomass, can be exploited to meet the energy requirements. Figure 5.2 represents a schematic diagram of the potential of clean energies at various stages of the agri-food processing chain.

5.3 Application of Clean Energy Technologies in Agri-Food Processing

5.3.1 Solar Energy

In recent years, solar energy has gained tremendous growth in agri-food processing industries. Several attempts have been made to harness solar energy with the help of solar collectors, sun trackers, and giant mirrors in order to gather solar radiation, store it, and use it for air or water heating in domestic, commercial, or industrial purposes. Solar collectors are basically used to heat air or water that acts as the heat transfer medium for different processes. They are classified into flat-plate collectors for low- to medium-temperature applications and concentrating and sun-tracking parabolic trough collectors (PTCs) for very high temperature applications (>250°C). The two axes tracking collectors are used for power generation and stationary (nontracking) and one-axis PTCs are mainly used in industrial heat processes. According to REN21 (2014), solar thermal systems alone have contributed 326 GW thermal energy, which is approximately equivalent to 24.5 million tons of petroleum oil. Similarly, electricity generation using solar radiation is either achieved by employing photovoltaic (PV) panels or by concentrating solar power (CSP) systems. In the power sector, the PV systems and CSP has been reported to achieve largest growth rate of 55% and 48%, respectively in 2013 (IRENA, 2015). Offering tremendous opportunity for growth and easy accessibility, solar energy proves to be one of the most important renewable sources to combat the energy crisis in the agriculture and food processing sector as compared to wind, geothermal, and biomass-derived energy by several orders of magnitude (IPCC, 2011). The major applications of solar energy in the agri-food processing sector have been highlighted as follows.

5.3.1.1 Solar Energy-Based Pumping Solutions

Irrigation is one of the most energy-intensive operations, and it consumes almost 15–20% of total electricity production in the developing countries (Bazilian et al., 2011). Solar PV-powered water pumps are one of the most promising options, as compared to grid- or diesel-based irrigation pump sets, to deal with the energy and water crisis for poor farmers. According to the Ministry of New and Renewable Energy, India (MNRE, 2014), PV pump sets could lead to a saving of nearly 18.7 gigawatts energy, which is equivalent to 10 billion L of diesel and 26 million tons of carbon dioxide (CO_2) emissions.

5.3.1.2 Solar Energy Technologies for Water and Space Heating

Water heating and cooling are the basic requirement in any agri-food processing industry where different unit operations, namely, parboiling, cleaning, pasteurization, blanching, etc., are carried out on a daily basis. All these processes account for a relatively high share of the energy budget and solar thermal or solar PV systems can potentially solve the energy demand issues related to water and space heating. The solar energy captured by thermal collector can be used to heat the cold water, store it in hot water storage tank, and then send to the industrial units for suitable applications as shown in Figure 5.3.

5.3.1.3 Solar Desalination

According to a review report of FAO, IFAD and WFP (2015), around one-sixth of the world population has limited access to sufficient water for satisfying their daily needs. The Middle East and

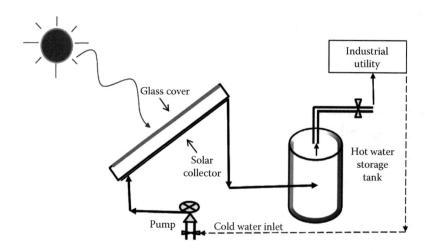

FIGURE 5.3 Layout of solar thermal system for water and space heating.

North Africa are the most water-scarce regions in the world and the freshwater level is depleting day by day in these areas. Currently, around 60% of the seawater desalination process is done by reverse osmosis and 27% is contributed by multistage flash desalination process. According to IRENA (2015), desalination consumes 75.2 terawatt hours (TWh) of electricity per year, equivalent to 0.4% of global electricity consumption. Hence, the solar PV-generated electricity integrated with the concentrated solar power (CSP) can prove to be a booming technique for high-temperature thermal energy (>200°C) generation in order to compensate the huge energy demand of electricity for desalination.

5.3.1.4 Solar Refrigeration and Cooling

Refrigeration is one of the effective unit operations employed in food processing industries and cold storage chains to prolong the shelf life of agricultural produce. Solar refrigeration technology offers a wide range of cooling techniques powered by solar collector-based thermally driven cycles and PV-based electrical cooling systems. In few available design of cold stores used for storage of fruits and vegetables, the mechanism of evaporative cooling is employed to reduce the temperature of the system between 10°C and 25°C, and hence minimizes food wastage (Dienst et al., 2011). Solar refrigeration is a very expensive technology, and its detail implementation needs further research.

5.3.1.5 Solar Energy for Food Drying

Drying is one of the most essential unit operations in agri-food processing industries, and generally, it is performed by conventional dryers. Nowadays, much emphasis has been given on the utilization of solar energy for drying of agricultural food products which can provide the low-temperature heating required for drying. A variety of agricultural commodities such as fresh fruits and vegetables, tea, spices, herbs, fish, candy, pickles, and also green leafy vegetables such as curry leaves, gogu (roselle), drumstick, and mint are dehydrated using solar energy drying systems (Bayrakcı et al., 2012; Eswara and Ramakrishnarao, 2013).

5.3.2 Wind Energy

Wind is simply the movement of air due to heat gradient caused by uneven heating during day and night time. When this wind is used to generate any form of mechanical or electrical energy, it is referred to as "wind energy." It is eco-friendly and freely available and therefore stands as a potential energy-generating technology to meet the present world energy demand. Average annual growth of wind energy systems has shown an increase from 12.4% to 21% during 2008–2013 (IRENA, 2015). Wind turbines operate on simple principles, i.e., when wind blows, the rotor blades of the turbine move; as a result, it drives the electric generator and converts the kinetic energy of the wind into electrical energy, which can be used for various applications. Wind turbines are classified into two different types according to energy-generation capacity and design of the system components, as shown in Figure 5.4.

The onshore wind turbine systems are one of the ancient and most mature renewable energy technologies being used since the 1970s. In contrast, offshore wind power plants are constructed inside the water bodies to generate maximum electricity as stronger wind speeds are available offshore than on land. Additionally, they mitigate the noise factor associated with onshore wind mills but their installation and maintenance cost is very large.

In order to harness electricity from wind turbines for different process applications, the wind turbine systems are integrated to grid, as depicted in Figure 5.5. The wind turbines are a source of power in agriculture for pumping water, grinding grains and legumes, saw milling, watering livestock, drainage, irrigation, and satisfying household needs.

It was estimated that each mega-watt hour power generated by wind energy helps to reduce 0.8 to 0.9 tons of greenhouse gas emissions per year that are produced by fossil fuel energy (Chel and Kaushik, 2010).

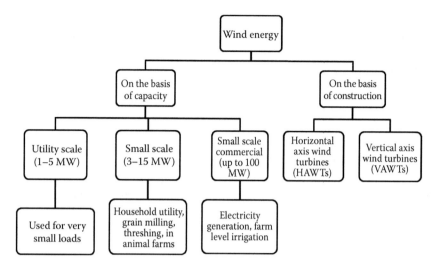

FIGURE 5.4   Classification of wind energy systems.

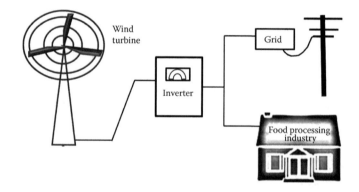

FIGURE 5.5   Grid connected wind turbine system for food processing industries.

5.3.3 Geothermal Energy

Geothermal energy is high-temperature thermal energy, stored in both rocks and trapped steam or hot water below the earth's surface. This geothermal energy is used to produce thermal and electrical energy and has direct application in heating, aquaculture, and industrial processes. On the basis of temperature generation, geothermal power plants can be classified into three categories as follows (Nguyen et al., 2015):

Low-temperature (20°C–70°C) geothermal plants
Intermediate temperature (70°C–149°C) geothermal plants
High-temperature (150°C–300°C) geothermal plants.

The low- and medium-temperature geothermal plants with temperatures less than 150°C are employed in the agri-food processing sector (Ogola, 2013). More than 73 countries around the world make direct use of a total geothermal energy output of 75.9 terawatt hours (TWh) annually (Mburu, 2009). Some important uses of geothermal energy are greenhouse heating, soil heating, aquaculture (fish farming and algae production), etc. Greenhouse heating is one of the most widely used processes in agriculture which consumes huge amounts of low enthalpy energy. To maintain this low-temperature heating in greenhouses, the geothermal energy is employed to heat water from 40°C to 100°C (Nguyen et al., 2015).

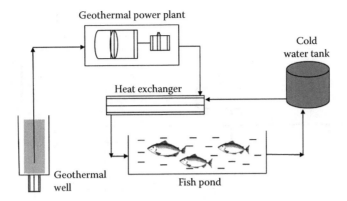

FIGURE 5.6   Schematic diagram of geothermal energy integrated to aquaculture.

In industrial food processing operations like drying, pasteurization, evaporation, distillation, cleaning, blanching, parboiling, etc., the geothermal energy can be employed with ease. Geothermal heating systems are relatively simple to install and maintain and can be used for successful drying of onions, tomatoes, and garlic (Lund, 2005). The steam generated from geothermal energy can be used for milk pasteurization, evaporation, and the ultrahigh temperature processing.

Geothermal energy is used to maintain a stable soil temperature of around 25°C–30°C in the agricultural fields resulting in enhanced yield (Kumoro and Kristanto, 2003). In addition, this can be successfully applied to aquaculture (fish farming and algae production) for cheap and profitable breeding throughout the year. The water heated with geothermal energy is mixed with the cold water until it reaches a suitable temperature around 20°C–30°C and then the water is pumped into the fish pond, as shown in the Figure 5.6.

Even though geothermal energy seems promising and opens a vast opportunity to meet energy demands in the agri-food processing industry, still its implementation faces tremendous challenges. Technical expertise is crucial for developing the system and proper infrastructure in terms of transportation pathway and communication networks to support geothermal systems.

5.3.4   Bioenergy

This is produced by anaerobic fermentation of biodegradable materials such as biomass, municipal and agricultural waste, and plant material into useful forms of energy such as heat, electricity, and liquid fuels (biofuels). Biomass can be converted into heat and electrical energy by thermochemical conversion processes like gasification and pyrolysis. Dry solid biomass can be directly burned to obtain energy; it can also serve as a feedstock for conversion into liquid or gas fuels (biofuels), which can substitute conventional oils for transportation. Bioenergy can be classified into two categories such as biomass-based energy and biofuel, and they are again subclassified into different sections depending on the way of utilization of energy as shown in Figure 5.7.

Biomass indeed is renewable, easily available, and has many positive solutions to organic waste utilization. The global capacity of installed biomass plant increased from 66 GW in 2010 to 72 GW by the end of 2011 with annual average growth rate of about 5% in 2012 (IRENA, 2015). There are significant obstacles to bioenergy amenities, the major demerit in using excessive bioenergy leads to air pollution. Moreover, electricity generation technology using biomass energy is well established, but the price paid for electricity seldom offsets the full cost of the biomass fuel (Ellabban et al., 2014).

The prime source of income in developing countries is agriculture, and the sustainable development of any country is solely dependent on the agricultural production. Although the agricultural production has gained considerable attention from researchers and policymakers, still food wastage and storage has

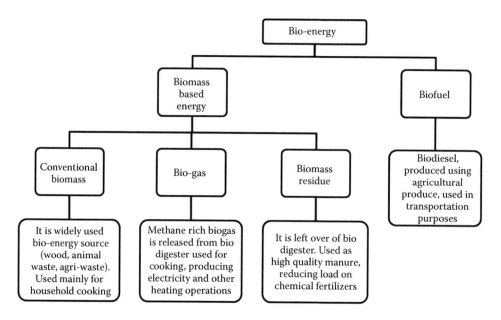

FIGURE 5.7 Classification of bioenergy resources.

received very less attention. In spite of huge production of fresh fruits and vegetables, there is an estimated food loss of around 33% in the world due to improper handling and poor storage facilities at the production catchment area (IRENA, 2015). These postharvest and storage losses are a major quandary, and this needs to be addressed in due diligence. Sustainable methods for food preservation are the need of the hour and solar drying is one of the best choices in this context. Proper solar drying technologies combined with improved food storage facilities involving solar refrigeration systems will lead to significant improvements in achieving food security in a sustainable manner. Hence, a detailed description of various solar drying and refrigeration technologies as a sustainable way of food processing is outlined here.

5.4 Prospects of Solar Energy Drying Systems for Achieving Food Security

The oldest and most widely employed application of solar energy in agri-food processing is drying. The drying process utilizing solar energy ranges from open sun drying to advanced solar dryers. Figure 5.8 illustrates a systematic classification of solar energy drying systems according to their heating modes and the manner in which the solar heat is utilized. It is mainly classified as open sun drying and solar dryers.

Open sun drying has been used since early ages for drying of various vegetables, fruits, and grains to prolong their shelf life. The food products are spread on a mat under open sun in the day time and due to the natural circulation of air around the food products drying is achieved. This traditional method offers uncontrolled drying due to uneven heat and mass transfer process leading to longer drying times. Additionally, the food products are highly prone to dirt, pests, and contamination leading to food, energy, and water wastage. In order to avoid these disadvantages, advanced solar dryers are gaining importance in recent years for achieving controlled drying rate.

The solar dryers have the advantage of providing a reduction of drying time and improvement of product quality in terms of color, taste, and texture as compared to open sun drying (Mahapatra and

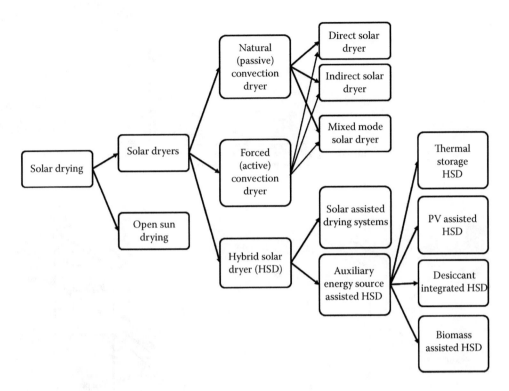

FIGURE 5.8   Classification of solar energy drying systems.

Imre, 1990). The solar dryers are further classified into two major groups, namely, passive solar dryers and active solar dryers.

Passive solar dryer: This is also known as natural convection solar energy dryer. In such a system, solar-heated air is circulated through the food product by buoyancy forces. Since this dryer does not require any electricity/fossil fuels to run the motorized fan necessary for air flow, it appears to be the most attractive option for use in remote rural locations where such facilities are not available.

Active solar dryer: This dryer employs an external motorized fan and/or pump for circulation of air inside the dryer (Prakash and Kumar, 2014). The solar-heated air flows through stacked food trays made of wire mesh placed inside the drying chamber. The dryer is recognized to be suitable for drying higher moisture foodstuffs such as papaya, kiwi fruits, brinjal, cabbage, and cauliflower slices (Chua and Chou, 2003).

There are three distinct subclasses of either passive or active solar drying systems, and they can be identified as direct type, indirect type, and mixed-mode type solar dryer.

5.4.1 Direct Solar Dryer

In this dryer, the food product is kept on a wire mesh tray placed inside the drying chamber with a transparent cover on the top, as shown in Figure 5.9. The incident solar radiation falls on the glass cover and a part of this is reflected back to the atmosphere. The remaining part is transmitted into the dryer, increasing the temperature inside the drying cabinet due to "greenhouse effect." The rise in temperature along with the hot air circulating inside the cabinet helps in removal of moisture from the food products. Direct dryers are generally used for drying fresh fruits and vegetables like banana, pineapple, potato, carrots, and french beans (Ezekoye and Enebe, 2006).

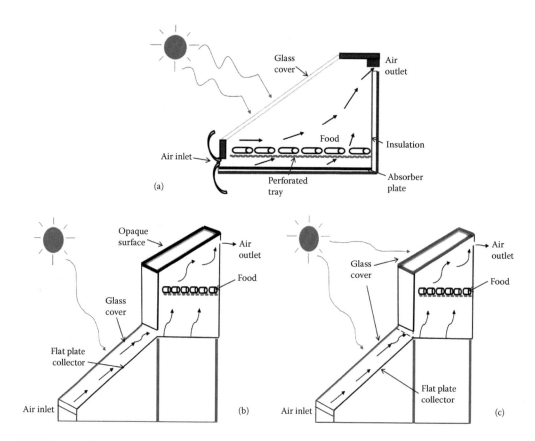

FIGURE 5.9   Schematic diagram of (a) direct, (b) indirect, and (c) mixed-mode type solar dryer.

5.4.2 Indirect Solar Dryer

In this dryer, a separate unit termed as solar air heater is used for heating of entering air. The air heater is connected in series to a drying chamber with an insulation cover at the top. The heated air from the collector is allowed to flow through moist food placed on wire mesh trays kept inside the drying chamber as shown in Figure 5.9. Here the food products are not directly exposed to sun; hence the dried products are rich in vitamins, minerals, and also retain the original color. Indirect type solar dryers are widely used for drying seedless grapes, apples, figs, green peas, onions, and tomatoes (El-Sebaii and Shalaby, 2012).

5.4.3   Mixed-Mode Solar Dryer

A typical mixed-mode type solar energy dryer has the same structural features as the indirect type dryer (i.e., solar air heater, separate drying chamber, etc.). In addition, the top of the drying chamber is glazed so that solar radiation can impinge directly on the food product. It works on the principle of combined action of solar radiation incident directly on the product to be dried and heated air from solar collector (Figure 5.9). Simate (2003) compared the performance of different designs of solar dryer and concluded that the mixed-mode solar dryer is found to be the most effective in terms of product drying rate and drying cost. This system is employed to dry various agricultural crops like cluster beans, thymus, grapes, and mint (Pardhi and Bhagoria, 2013; Saravanan et al., 2015).

There are several designs of the solar dryers investigated by different researchers for drying of agricultural commodities (Rathore and Panwar, 2010; Belessiotis and Delyannis, 2011; Sulaiman et al., 2013;

Ringeisen et al., 2014). Although these dryers are most effective for drying of several fruits and vegetables, yet there are two major limitations: (1) temperature control inside the drying chamber and (2) dependence on the availability of solar radiation intensity. In order to combat these issues, solar dryers are supplemented by other energy sources for continuous drying of food products, these are types of dryers termed hybrid solar dryers.

5.4.4 Hybrid Solar Dryers

In hybrid solar drying systems, either solar energy is coupled to conventional drying systems in order to reduce energy budget or other auxiliary sources of energy are integrated to the solar dryer to enhance drying potential of the system. Different types of robust hybrid solar drying technologies that can extract energy from other sources like heat storage media, biomass, PV, and desiccant materials are gaining pace with new generation drying facilities. Some of the upcoming hybrid solar drying technologies are discussed in this section.

5.4.4.1 PV-Assisted Hybrid Solar Dryer

The hybrid PV–thermal (PV/T) solar drying system is a combination of PV modules and solar air heater, which produces both electricity and heat from the integrated component, as shown in Figure 5.10. PV module converts 4%–17% of the incoming solar radiation into electricity, which is used as energy source to heat the inlet air that is flowing into the dryer. This electricity is stored in chemical form in the batteries and is used at night or off sunshine hours. A PV-integrated HSD generally comprises of solar PV panels connected to inverter, blower, and to the heat exchanger/heater for air heating.

This resilient system has gained much importance and has a remarkable market growth among all renewable systems i.e., 39% to 54% since 2008–2013 (REN21, 2014). The major challenges faced in employing PV plants at larger scale are the area required for installation and initial cost of the hybrid system.

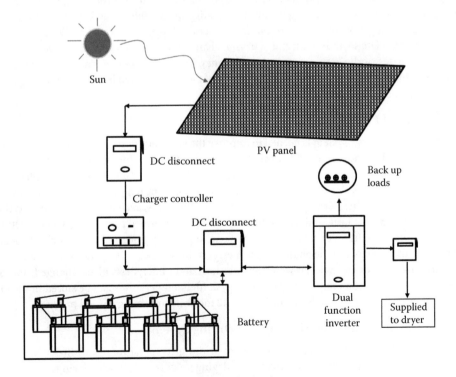

FIGURE 5.10 Schematic sketch of a PV-integrated hybrid solar dryer.

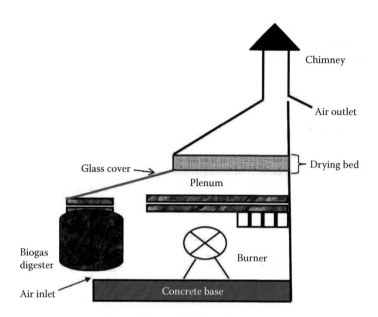

FIGURE 5.11 Schematic diagram of a biogas-integrated hybrid solar dryer.

5.4.4.2   Biomass Integrated Hybrid Solar Dryer

The use of processed fruit and vegetable wastes to generate biogas is used to support the drying process. The methane gas released from biogas is utilized to heat air during off sunshine or cloudy days. The advantages of this type of drying system are that it provides uniform control or regulation of air temperature and thus helps in maintaining the product quality. Biogas-integrated HSD is also a renewable-renewable type conjugation of energy resources. Generally biogas-integrated HSD consists of solar dryer unit with flat plate solar air collector, a drying chamber, fans, and a heat exchanger. The biomass plant unit comprises of digester, connected to chimney in which a stove is installed through which biogas is burnt for heating air. The schematic diagram of biomass-assisted hybrid solar drying system is depicted in Figure 5.11.

5.4.4.3   Thermal Energy Storage Hybrid Solar Dryer

The thermal energy storage system is needed to improve the efficiency of solar dryer and simultaneously provides higher quality end product. There are two known systems for thermal energy storage, i.e., sensible heat storage system and latent heat storage system. Latent heat storage system uses phase change materials (PCMs) and has the ability to store energy at a constant temperature and provides a high-energy storage density. The thermophysical properties of the PCM, such as high density and thermal conductivity, high specific heat, and heat of fusion, should be identified for its use (Bal et al., 2011). The thermal heat storage medium for solar dryers often uses liquid water or solid sands/rocks through sensible heating without changing the phase. Sometimes, certain chemicals like paraffin wax and inorganic salts are also used as storage material which absorbs heat and undergoes a phase change from solid to liquid state at a desired temperature. Thermal energy at high temperatures can be stored in the form of sensible heat in a mass of sand as shown in Figure 5.12 for uniform drying of food products.

5.4.4.4   Desiccant-Integrated Hybrid Solar Drying System

Desiccant materials are capable of producing low- to medium-temperature hot air inside the dryer and thus are integrated with solar dryers to enhance the drying process. This type of drying system is mostly used for heat-sensitive food material. The desiccant materials can either be liquid or solid type and some

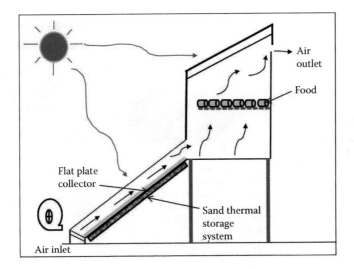

FIGURE 5.12 Schematic representation of a thermal storage hybrid solar dyer.

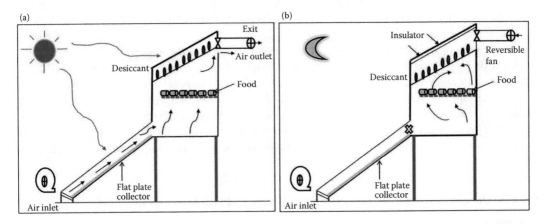

FIGURE 5.13 Schematic sketch of a desiccant-integrated hybrid solar dryer (a) sunshine period and (b) off sunshine period.

of the solid desiccant materials used for drying purposes are clay $CaCl_2$, bentonite-$CaCl_2$, and silica gel. Similarly, lithium chloride (LiCl), $CaCl_2$ solution, and solutions of polyethylene glycol are used as liquid desiccants. Construction of desiccant-integrated solar dryers is similar to indirect or mixed mode dryers with the exception of an additional desiccant bed over the drying chamber as given in Figure 5.13.

Proper insulation is provided on the bottom side of desiccant chamber during day time and at the top during the night hours in order to avoid the heat loss. During day time, the solar-heated air helps in removing the moisture from the food products, and simultaneously, the desiccant materials also absorb the radiation. At night hours, the reversible fan circulates air, heated by desiccant materials inside the drying chamber for drying.

5.4.5 Factors Influencing Solar Drying Process

The prediction of drying kinetics of food product under variable drying conditions is an important tool for the proper design of solar drying system. The solar drying process depends on a number of external

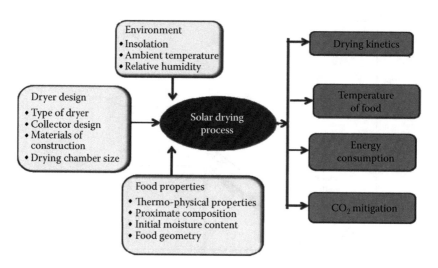

FIGURE 5.14   Factors influencing solar drying process.

and internal parameters which ultimately affects the drying kinetics and temperature of the food product. The detail flow chart showing the effect of process parameters on the solar drying kinetics of food products is depicted in Figure 5.14.

The environmental factors (solar insolation, ambient air temperature, and relative humidity of air) along with physical features of the dryer (type, size, shape, collector area, dryer capacity, etc.) and internal properties of the food material (thermophysical properties and proximate composition, geometry, and initial moisture content) largely affect the solar drying process. In the past, extensive experimental studies on influence of various drying variables like air temperature, relative humidity, and air flow rate on the drying kinetics of food product were reported in the literature. The studies indicated that these variables significantly affect the moisture distribution in food during drying (Vega-Gálvez et al., 2012; Udomkun et al., 2015).

5.4.6 Potential of Solar Dryers in Mitigating Carbon Dioxide Emissions

The Clean Development Mechanism (CDM) under the Kyoto Protocol program is the main target to implement carbon dioxide emission mitigation technologies in developing countries. Such technologies can earn saleable certified emission reduction (CER) credits for meeting Kyoto targets. Nowadays, solar energy is promoted as a promising climate-resilient technology under the Kyoto Protocol program for most of the developing countries. The solar dryers are more relevant in view of CDM projects as they can be directly linked in the reduction of CO_2 emissions while contributing to sustainable development (Singh and Kumar, 2013). Several researchers have established the link between CO_2 emission mitigation potential of solar dryers while drying of agricultural commodities. A theoretical approach has been proposed by Kumar and Kandpal (2005) by assuming some basic input data to estimate CO_2 emissions mitigation during drying of various foods using an indirect type solar dryer. Piacentini and Mujumdar (2009) estimated annual CO_2 emissions of 14.77 tonnes during drying of agricultural products by considering the electricity consumption of 100 kWh/day for the UK conditions. Tripathy (2015) investigated the CO_2 mitigation potential of mixed-mode solar dryers for drying of potato and suggested that the replacement of coal with solar energy for drying resulted in mitigating maximum carbon dioxide from the atmosphere. It was also predicted in the study that by the year 2020, 23% of CO_2 emissions can be mitigated by the use of mixed-mode solar dryers for drying of agricultural products.

5.5 Prospects of Solar Refrigeration Systems for Achieving Food Security

Lack of proper postharvest storage and transit facilities accounts for huge loss of agricultural produce in the production catchment area. Hence, highly perishable commodities require a cold chain arrangement to maintain quality and to extend the shelf life of the products. Reducing the temperature of produce is one of the most effective ways of preserving primary produce as well as processed food products. It is the most widely used technology at storage, distribution, and consumer level in the food supply chain. However, similar to drying, refrigeration is also an energy-intensive operation and it was estimated that about 8% of the worldwide electric energy utilization is reported to be consumed in refrigeration operations performed in food industries, which is approximately equal to 1300 TWh (Guilpart, 2008). Therefore, researchers are currently focusing on mature technologies that utilize clean energy to power refrigeration and at the same time respond to the cooling needs of agricultural produce. Based on the utilization of available solar energy, the refrigeration system intended to cool the agricultural produce can be broadly classified into two types as depicted in Figure 5.15.

Cooling can be achieved through two basic technologies: First, solar PV technology that includes vapor compression cooling and thermoelectric cooling. Second, solar thermal technology that includes vapor absorption and vapor jet ejection methods. These systems can also be recognized as open/closed refrigeration systems and thermomechanical technology-based systems (Sarbu and Sebarchievici, 2013).

5.5.1 Solar PV Refrigeration System

The solar PV vapor compression refrigeration system comprises of solar panels, inverter, battery system, and air conditioning unit. The vapor compression type system is a conventional type refrigerator device, which utilizes electricity harnessed by PV panels (Figure 5.16). The second type of PV refrigeration is also known as Thermoelectrical cooling that employs "Peltier effect" for cooling (Sarbu and Sebarchievici, 2013). The thermoelectric refrigerator comprises of a small number of thermocouples producing low thermoelectric power, but can produce high electric current. This system has the advantage of operating with a low-level heat source, and is therefore helpful in converting solar energy into electricity. The thermoelectric refrigerator is a unique cooling system, in which the electron gas serves as the working fluid, and therefore, it can successfully replace conventional domestic refrigeration systems which ulitizes CFCs. However, updated information on mature designs that appropriately respond to the cooling needs of agricultural produce are very scarce. Few designs of solar PV refrigerators are

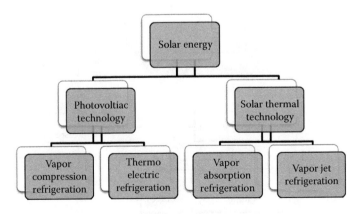

FIGURE 5.15 Classification of solar energy refrigeration system.

commercially available (Fan et al., 2007; Pero et al., 2015). These refrigeration systems are used in dairy as well as in food processing industries for preservation of milk, vegetables, fruits, fish, and meat for a longer period.

5.5.2 Solar Thermal Refrigeration System

Solar thermal referigeration systems are more popular as they utilize solar thermal collectors for harnessing the solar energy and convert it into heat energy. Sorption technology is also integrated in thermal refrigeration systems to enhance the overall performance. Different types of solar sorption refrigeration systems are classified on the basis of the dessicant type used, i.e., solid, liquid, or gas. In the first type, a thin layer of solid dessicant is used to absorb water from the incoming air stream causing dehumidification. Further, this dessicant is regenerated by using solar energy for removal of absorbed moisture. Water is sprayed into the dehumidified air stream, thus lowering its temperature and providing a cooling effect. The second type of thermally driven system is absorption cooling, in which the refrigerant vapor is absorbed into a liquid, thus allowing its pressure to be economically increased by a pump, rather than by a vapor compressor that requires much more mechanical input. Generally, NH_3-H_2O and H_2O-aqueous LiBr type dessicants or referigirants are utilized in absorption referigeration cycles (Fan et al., 2007; Otanicar et al., 2012). The third and final type of thermally driven system is the adsorption cycle, where the refrigerant vapor is adsorbed onto the surface of a solid adsorbent, which when heated by solar energy, desorbs the vapor, and thus pressurizes the vessel in which the vapor is contained. This, in effect, creates a "thermal compressor" that replaces a conventional electrically driven compressor. Activated carbon, silica gel, and zeolite are most widely used adsorbents, whereas water, methanol (ethanol), or ammonia are most widely used adsorbates in solar-powered adsorption refrigeration. Layout of solar dessicant cooling system implemented in various agri-processing operation is represented in Figure 5.17. Solar thermal refrigeration system is mostly employed in air-conditioning (8°C–15°C) for spaces and cold storages, refrigeration (0°C–8°C) for food products and vaccine storage, and to some extent in freezing (<0°C) for ice making (Sarbu and Sebarchievici, 2013; Claudio Zilio, 2014).

5.6 Scaling Up Clean Energy Technologies

Climate change and the fast depletion of fossil fuel energy are two major problems the world is facing in the twenty-first century. Clean energy technologies have the potential to solve these problems at a greater extent and on proper platform; these technologies could prove to be genuinely competitive and a self-sustaining investment proposition. Regardless of acknowledging magnificent growth over the past

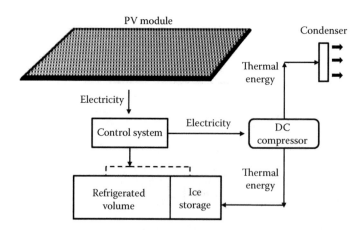

FIGURE 5.16   Schematic diagram of PV refrigeration system (Pero et al., 2015).

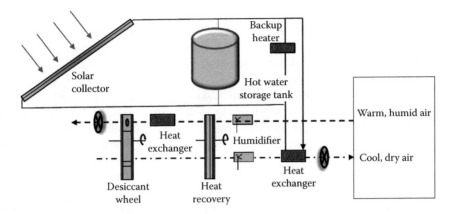

FIGURE 5.17 Layout of solar-assisted dessicant cooling system implemented in various agri-processing operations.

few years, the clean energy technologies persist to confront hurdles and will need more than just public and private finance to scale up. The major challenges faced for scaling up of clean energy technologies are enlisted as follows:

1. Absence of long-term planning due to ineffective communication between government and regulatory bodies, leading to delay in clean energy project approval and formulation.
2. Lack of stakeholder involvement in decision making.
3. Limited grid infrastructure in areas where clean energy resources are most abundantly available, and this creates a present and future obstacle to increased power generation.
4. Shortfall of experienced technical staff and certification.
5. Lack of operation and maintenance facilities.

It is very imperative to understand the impact of regulatory and infrastructural challenges for possible scaling up of clean energy technologies. However, the major challenges faced by different clean energy technologies along with possible recommendations for scaling up at a high level are discussed in length.

Solar energy, being a clean and free source of energy, has gained much focus of researchers as well as industries to combat energy demands. Solar thermal and solar PV power are two booming technologies to harness solar energy in the form of direct heat or as electricity. In recent times, different CSP technologies have attained higher degrees of availability in commercial markets. Similarly, the widely available PV panels in the world market are of two types: (1) crystalline silicon-based and (2) thin film technologies (amorphous silicon, cadmium telluride, and copper–indium–gallium–diselenide). Polycrystalline-silicon PV panels typically have efficiencies in the range of 14%–16%, which increases the cost of this technology, and therefore, scaling up becomes more tedious (Palit, 2013). One of the major drawbacks of this technology is the need of efficient storage systems, such as batteries, inverters, and other power-conditioning equipment, which is still in research and development. Still many mature solar PV and solar thermal projects have been realized in the past few years, and therefore, PV-installed capacity is predicted to rise up to 1845 GW by 2030 (Palit, 2013). A projection of Bloomberg New Energy Finance (2015) highlighted that the world will invest around USD 12.2 trillion in clean power-generating capacity over a period of 2015–2040. The projections also highlighted that nearly one third of the cumulative global investment over this period will be spent on solar power, followed by onshore wind power. Recently, Kazusatsurumai Solar Sharing Project, Japan has been realized in which 348 PV panels are installed in a 750 m² farm. Different food products like peanuts, yams, eggplants, cucumber, tomatoes, and cabbages are grown under the panels. In India, large-scale solar energy cooking operations have been installed and currently operated at different pilgrimage centers, such as at Tirumala-Tirupati Devasthanams, Tirupati, Andhra Pradesh, and Shirdi Sai Baba Temple in Shirdi town, Maharashtra, where large-scale free meal distribution is a common practice on a daily basis (Eswara and Ramkrishnarao, 2013). IEA (2013) projected that the sun could be the world's largest source of electricity by 2050, ahead of fossil fuels, wind, hydropower, and nuclear. In addition, *IEA roadmaps also predict that solar PV systems could generate up to 16% of the world's electricity by 2050, while electricity harnessed from CSP plants can provide an additional 11% electrical energy.*

Like solar energy, wind energy offers a great potential as clean energy resource. Around 67% of total renewable energy exploited till date is contributed by *wind energy* at a cost of 2–3 Rs per kWh, which is comparable to conventional energy resources. However, the gap between real time usage and the existing potential of wind energy to serve as a clean and cheap energy resource is huge; about 81 GW of available wind energy resources remains unutilized (Tripathi et al., 2016). The major limitations to full volume utilization of wind energy are technical barriers, know-how to install, scale up the technology, and high operation and maintenance costs, especially in offshore wind turbines. Technically, the wind harnessing technology depends on various factors like number of blades and their configuration (horizontal or vertical), pitch of the blades (fixed or variable), and speed of turbine (fixed speed or variable speed). Generally, horizontal axis is used at utility scale farms due to its easy installation, handling, and higher efficiency, and variable pitch offers better adjustment, hence is preferred over fixed type in larger wind turbines (Njiri and Soffker, 2016). For mega utility-scale wind turbines that are manufactured nowadays are variable-speed, variable-pitch, and horizontal-axis turbines (Njiri and Soffker, 2016). But, manufacturing turbines with massive sizes can lead to problems related to structural loads and poor quality of generated power and hence, scaling up of the technology up to required capacity is under research and development. The other challenge in scaling up wind energy is variability in wind speed which causes the fluctuations in power generation. The complete dependence on climatic conditions is a major obstacle in this case which arises due to ecological and geographical limitations and this is encountered by high-end generator and variable speed turbines. Noise pollution along with the requirement of larger land area hinders the establishment of wind farms near populated areas. Mayurappriyan et al. (2014) have designed a 250 kW wind mill with horizontal axis turbine in which technical modification was done in terms of yaw and hub hydraulic systems for better power production, to reduce machine stoppage, and reduced operation and maintenance costs. *IEA technology roadmap (2013) detailed the possible actions government will set for scaling up the available wind power for generating electricity from current 2.6% to 18% by 2050.*

Another promising alternative source of energy, *bioenergy* in the form of traditional biomass is largely used in agriculture-based countries due to its easy availability in rural areas. This has a great disadvantage of polluting the environment with release of hazardous gases like CO_2, CO, etc., and hence, new technologies of biogas and biofuel have emerged in the past few decades. Biofuels mainly come from first-generation edible crops like soybean, rapeseed, palm, and sunflower oils. However, the limited production of these edible crops presents a challenge to food security, resulting in imbalance in the competitive global food market, thereby making biodiesel production even more expensive. On the other hand, the second-generation biofuels, such as jatropha, mahua, jojoba oil, tobacco seed, salmon oil, have led to large-scale deforestation in many countries resulting in the destruction of natural environment. Therefore, the major limitation associated with the use of second-generation biofuels is the issue of sustainability. Alternatively, microalgae production is being employed to cultivate biodiesel in open pond systems at larger scale but the recovery of biomass is too costly, which offsets the advantages by cost of production. Biogas production is another clean technology to meet the energy demands. The major challenge faced by this technology is the decentralized waste collection and treatment system along with higher installation, operational, and maintenance cost (Kothari et al., 2010). In spite of these limitations, large-scale biodigester has been set up in developing countries for continuous supply of biogas for several purposes such as heating, cooking, and lighting. A layout of biodigester system for sustainable waste to energy management is represented in Figure 5.18a and photograph of a large-scale biodigester of 1100 m³ capacity, installed in Beijing, China, for household biogas supply for cooking purposes is shown in Figure 5.18b (Hojnacki et al., 2011). In India, a large-scale biodigester

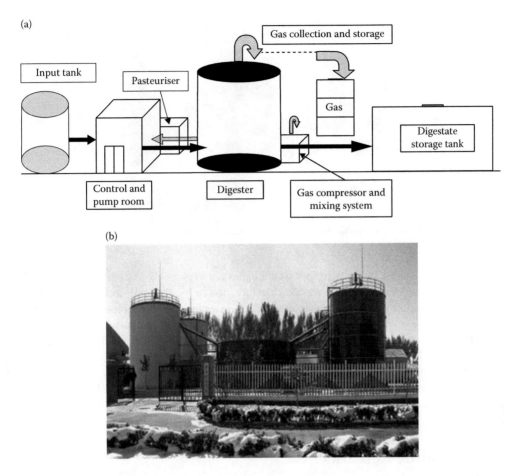

FIGURE 5.18 (a) Layout of biodigester system and (b) photograph of large-scale bio-digester Beijing, China.

plant of "Deenbandhu design" is run by a NGO, Sulabh International Academy of Environmental Sanitation (SIAES), in which the toilet complex is attached to a biogas digester to supply biogas to locals in Shirdi, Nasik District of Maharashtra.

Similarly, the *geothermal technologies* are one of the most underused clean energy resources due to its significantly high power plant costs, drilling costs, resource quality, and financing costs. Another major issue in scaling up this technology is lack of storage technology, which could enable power to be "banked" during periods of low demand for later dispatch during periods of high demand. Research and development of appropriate storage system is under process to enhance the ability of geothermal plants to operate in a flexible manner. Currently, three types of geothermal power plants are in operation: dry steam plants, flash steam plants, and binary-cycle plants. *According to the annual report of geothermal energy association (2015), the establishment of global geothermal industry is expected to reach between 14.5 and 17.6 GW by 2020 and 27–30 GW by the early 2030s.* Iceland is one of the leading countries in utilization of geothermal energy in the agricultural sector. The total surface area of geothermal greenhouses in Iceland is estimated to be around 175,000 m², out of which 55% is used for growing vegetables such as tomatoes, carrots, cucumbers, and paprika (Nguyen et al., 2015). A large-scale set up of greenhouse powered by geothermal energy for growing cucumber and tomato in Iceland is shown in Figure 5.19.

FIGURE 5.19   Geothermal greenhouse for growing cucumber in Iceland.

Large-scale installations of clean energy technologies are very costly as compared to incumbent technologies, such as conventional resources. Therefore, nowadays, the clean technology fund (CTF) helps in empowering transformation in developing countries by providing resources to scale up the demonstration, deployment, and transfer of clean energy technologies with a significant potential for saving long-term greenhouse gas emissions. It is evident from the foregoing study that the clean energy technologies have the potential to provide long term and secure energy for the agricultural sector and food processing industries. The major constraints impeding the use of clean energy technologies in the agri-food processing sector are policy, regulatory, technical, and financial barriers. Hence, the government should consider these constraints and challenges before implementation of real technological applications.

5.7 Concluding Remarks

In this chapter, we have discussed the present status and scope of clean energy (solar, wind, geothermal and bioenergy) technologies and its inevitable linkage with food security. Through this chapter updated information related to applications, benefits, and investments of these technologies in agri-food processing is presented. The positive impacts of green energy on environmental issues, like climate change, CO_2 mitigation, global warming, are also highlighted. In the current scenario, solar energy applications have proven to be the most promising technology for food processing due to its easy access and availability worldwide. A systematic arrangement of different solar drying and refrigeration systems is outlined. In spite of being a sustainable, clean, and free source of energy, implementation of all these resources is still lagging behind its optimum capacity due to the technological gap which is still under research and development. The challenges faced by different clean energy technologies along with the possible recommendations for scaling up the technologies are also addressed for optimum use of these technologies. Clean energy technologies should be given wider publicity so as to create awareness amongst the people which will help the society to generate income groups in rural livelihoods and in achieving global food security in a sustainable manner.

References

Bal, L. M., S. Satya, S. N. Naik, and V. Meda. 2011. Review of solar dryers with latent heat storage systems for agricultural products. *Renewable and Sustainable Energy Reviews* 15, no. 1: 876–880.

Bayrakcı, A. G. and G. Kocar. 2012. Utilization of renewable energies in Turkey's agriculture. *Renewable and Sustainable Energy Reviews* 16, no. 1: 618–633.

Bazilian, M., H. Rogner, M. Howells, S. Hermann, D. Arent, D. Gielen, and K. K. Yumkella. 2011. Considering the energy, water and food nexus: Towards an integrated modelling approach. *Energy Policy* 39, no. 12: 7896–7906.

Belessiotis, V. and E. Delyannis. 2011. Solar drying. *Solar Energy* 85, no. 8: 1665–1691.

Bloomberg New Energy Finance. 2015. New energy outlook 2015. www.about.bnef.com/content/uploads/sites/4/2015/06/BNEF-NEO2015_Executive-summary.pdf.

Chel, A. and G. Kaushik. 2011. Renewable energy for sustainable agriculture. *Agronomy for Sustainable Development* 31: 91–118.

Chua, K. J. and S. K. Chou. 2003. Low-cost drying methods for developing countries. *Trends in Food Science and Technology* 14, no. 12: 519–528.

Dienst, C., W. Ortiz, J. C. Pfaff, and D. Vallentin. 2011. Food issues: Renewable energy for food preparation and processing. http://epub.wupperinst.org/frontdoor/index/index/docId/3909.

Ellabban, O., H. Abu-Rub, and F. Blaabjerg. 2014. Renewable energy resources: Current status, future prospects and their enabling technology. *Renewable and Sustainable Energy Reviews* 39: 748–764.

El-Sebaii, A. A. and S. M. Shalaby. 2012. Solar drying of agricultural products: A review. *Renewable and Sustainable Energy Reviews* 16, no. 1: 37–43.

Eswara, A. R. and M. Ramakrishnarao. 2013. Solar energy in food processing—A critical appraisal. *Journal of Food Science and Technology* 50, no. 2: 209–227.

Ezekoye, B. A. and O. M. Enebe. 2006. Development and performance evaluation of modified integrated passive solar grain dryer. *The Pacific Journal of Science and Technology* 7, no. 2: 185–190.

Fan, Y., L. Luo, and B. Souyri. 2007. Review of solar sorption refrigeration technologies: Development and applications. *Renewable and Sustainable Energy Reviews* 11, no. 8: 1758–1775.

FAO. 2011. Energy-smart food for people and climate. FAO, Rome. www.fao.org/docrep/014/i2454e/i2454e00.pdf.

FAO. 2014. FAO and the post-2015 development agenda issue papers: Theme: Energy. www.fao.org/fileadmin/user_upload/post2015/14_themes_Issue_Papers/EN/14_themes__december_2014_/Energy-8.pdf.

FAO, IFAD, and WFP. 2015. The state of food insecurity in the world 2015. http://www.fao.org/3/a-i4646e.pdf.

Finley, J. W. and J. N. Seiber. 2014. The nexus of food, energy, and water. *Journal of Agricultural and Food Chemistry* 62, no. 27: 6255–6262.

Guilpart, J. 2008. Froidet alimentation: Sécurité, sûretéouprocédé. In *Conférence Centenaire du froid*, Paris.

Hojnacki, A., L. Li, N. Kim, C. Markgraf, and D. Pierson. 2011. Bio digester global case studies. D-Lab waste. https://colab.mit.edu/sites/default/files/D_Lab_Waste_Biodigester_Case_Studies_Report.pdf.

IEA. 2013. World energy outlook 2013 factsheet, www.iea.org/media/files/WEO2013_factsheets.pdf.

IEO. 2013. International energy outlook 2013 with projections to 2040. U.S. Energy Information Administration, Washington, DC. http://www.eia.gov/forecasts/ieo/pdf/0484(2013).pdf.

IPCC (Intergovernmental Panel on Climate Change). 2011. Renewable energy sources and climate change mitigation: Summary for policy makers and technical summary. www.ipcc.ch/pdf/special-reports/srren/SRREN_FD_SPM_final.pdf.

IRENA. 2015. Renewable energy in water, energy and food nexus. http://www.irena.org/DocumentDownloads/Publications/IRENA_Water_Energy_Food_Nexus_2015.pdf.

Kothari, R., V. V. Tyagi, and A. Pathak. 2010. Waste-to-energy: A way from renewable energy sources to sustainable development. *Renewable and Sustainable Energy Reviews* 14, no. 9: 3164–3170.

Kumar, A. and T. C. Kandpal. 2005. Solar drying and CO_2 emissions mitigation: Potential for selected cash crops in India. *Solar Energy* 78, no. 2: 321–329.

Kumoro, A. C. and D. Kristanto. 2003. Preliminary study on the utilization of geothermal energy for drying of agricultural product. In *International Geothermal Conference*, September 2003, Reykjavík, Iceland.

Lund, J. W., R. G. Bloomquist., T. L. Boyd., and J. Renner. 2005. Geothermal and mineral resources of modern volcanism areas. In *Proceedings of the International Kuril-Kamchatka Field Workshop*, 25–50. The United States of America Country Update.

Mahapatra, A. K. and L. Imre. 1990. Role of solar agricultural-drying in developing countries. *International Journal of Ambient Energy* 11, no. 4: 205–210.

Mayurappriyan, P. S., J. Jerome., and T. C. Raj. 2014. Performance improvement in an Indian wind farm by implementing design modifications in yaw and hub hydraulic systems—A case study. *Renewable and Sustainable Energy Reviews* 32: 67–75.

Mburu, M. 2009. Geothermal energy utilisation. Short course IV on exploration for geothermal resources, UNU-GTP, KenGen and GDC, at Lake Naivasha, Kenya, pp. 1–22.

Ministry of New and Renewable Energy (MNRE). 2014. Solar energy based dual pump piped water supply scheme. http://mnre.gov.in/file-manager/UserFiles/solar-energy-based-dual-pump-pipedwater-supply-scheme.pdf.

Nguyen, M. V., S. Arason, M. Gissurarson, and P. G. Pálsson. 2015. Uses of geothermal energy in food and agriculture: Opportunities for developing countries. FAO, Rome. http://www.fao.org/3/a-i4233e.pdf.

Njiri, J. G. and D. Söffker. 2016. State-of-the-art in wind turbine control: Trends and challenges. *Renewable and Sustainable Energy Reviews* 60: 377–393.

Ogola, P. F. A. 2013. The power to change: Creating lifeline and mitigation-adaptation opportunities through geothermal energy utilisation. PhD thesis, University of Iceland.

Otanicar, T., R. A., Taylor, and P. E., Phelan. 2012. Prospects for solar cooling: An economic and environmental assessment. *Solar Energy* 86, no. 5: 1287–1299.

Palit, D. 2013. Solar energy programs for rural electrification: Experiences and lessons from South Asia. *Energy for Sustainable Development* 17, no. 3: 270–279.

Pardhi, C. B. and J. L. Bhagoria. 2013. Development and performance evaluation of mixed-mode solar dryer with forced convection. *International Journal of Energy and Environmental Engineering* 4, no. 1: 1–8.

Pero, C. D., F. M. Butera, M. Buffoli, L. Piegari, L. Capolongo, and M. Fattore. 2015. Feasibility study of a solar photovoltaic adaptable refrigeration kit for remote areas in developing countries. In *International Conference on Clean Electrical Power (ICCEP), IEEE*, pp. 701–708.

Piacentini, R. D. and A. S. Mujumdar. 2009. Climate change and drying of agricultural products. *Drying Technology* 27, no. 5: 629–635.

Prakash, O. and A. Kumar. 2014. Solar greenhouse drying: A review. *Renewable and Sustainable Energy Reviews* 29: 905–910.

Rathore, N. S. and N. L. Panwar. 2010. Experimental studies on hemi cylindrical walk-in type solar tunnel dryer for grape drying. *Applied Energy* 87, no. 8: 2764–2767.

REN21 (Renewable Energy Policy Network for 21st Century). 2014. Renewables 2014: Global status report. www.ren21.net/Portals/0/documents/Resources/GSR/2014/GSR2014_full%20report_low%20res.pdf.

Ringeisen, B., D. M. Barrett, and P. Stroeve. 2014. Concentrated solar drying of tomatoes. *Energy for Sustainable Development* 19: 47–55.

Saravanan, P., T. Balusamy, and R. Srinivasan. 2015. Design, fabrication and testing of mixed mode solar dryer for vegetables. *International Journal of Research and Innovation in Engineering Technology* 1, no. 11: 27–32

Sarbu, I. and C. Sebarchievici. 2013. Review of solar refrigeration and cooling systems. *Energy and Buildings* 67: 286–297.

Simate, I. N. 2003. Optimization of mixed-mode and indirect-mode natural convection solar dryers. *Renewable Energy* 28, no. 3: 435–453.

Singh, S. and S. Kumar. 2013. Solar drying for different test conditions: Proposed framework for estimation of specific energy consumption and CO_2 emissions mitigation. *Energy* 51: 27–36.

Sulaiman, F., N. Abdullah, and Z. Aliasak. 2013. Solar drying system for drying empty fruit bunches. *Journal of Physical Science* 24, no. 1: 75–93.

Tripathi, L., A. K. Mishra, A. K. Dubey, C. B. Tripathi, and P. Baredar. 2016. Renewable energy: An overview on its contribution in current energy scenario of India. *Renewable and Sustainable Energy Reviews* 60: 226–233.

Tripathy, P. P. 2015. Investigation into solar drying of potato: Effect of sample geometry on drying kinetics and CO_2 emissions mitigation. *Journal of Food Science and Technology* 52, no. 3: 1383–1393.

Udomkun, P., D. Argyropoulos, M. Nagle, B. Mahayothee, S. Janjai, and J. Müller. 2015. Single layer drying kinetics of papaya amidst vertical and horizontal airflow. *LWT-Food Science and Technology* 64, no. 1:67–73.

Vega-Gálvez, A., K. Ah-Hen., M. Chacana., J. Vergara., J. Martínez-Monzó, P. García-Segovia, R. Lemus-Mondaca, and K. Di Scala. 2012. Effect of temperature and air velocity on drying kinetics, antioxidant capacity, total phenolic content, colour, texture and microstructure of apple (var. Granny Smith) slices. *Food Chemistry* 132, no. 1: 51–59.

Zilio, C. 2014. Moving toward sustainability in refrigeration applications for refrigerated warehouses. *HVAC&R Research* 20, no. 1: 1–2.

6

Bioenergy and Food Production: Appropriate Allocation for Future Development

Nishith B. Desai
and Santanu
Bandyopadhyay
*Indian Institute of
Technology Bombay*

6.1 Introduction

Renewable energy technologies can contribute to the reduction of greenhouse gases emission and the growing energy need. In recent years, the renewable energy sources such as solar, biomass, geothermal, and wind have been used widely for electrical power generation. The developing countries still use far less energy compared to the developed countries, and therefore finding alternative energy strategies is vital to ensure climate change targets (GEA 2012). Biomass, material derived from living/recently living organisms, is one of the oldest and the most important renewable energy resource. The traditional use of the biomass is for cooking. It is also used for heating, production of fuels and chemicals, structural materials, and generation of electricity. Sources of biomass include agricultural and forest residues, livestock wastes, dedicated plantation for energy, and the organic part of waste streams (biogas, landfill gas, municipal solid waste, etc.) (GEA 2012). Biomass has a significant potential to be used for sustainable energy needs of the future.

Due to population growth, change in consumption patterns and dietary preferences as well as postharvest losses, agriculture faces challenges in delivering food security. The additional increase in agricultural demand for bioenergy is considered as a threat to food security. The other view is that bioenergy can be considered as a driver for increase in agricultural production and productivity, which may result in rural development and poverty reduction. The impact of increase in utilization of land for bioenergy production needs to be analyzed before its allocation, and therefore energy policy decisions of developing countries play an important role.

6.2 Bioenergy Potential

Historically, the dominance of biomass was overtaken by coal in the first half of the twentieth century, giving way to oil around 1970 (GEA 2012). Today, oil is the largest contributor of global primary energy need, and biomass is the world's fourth largest energy source (after oil, coal, and natural gas), contributing to about 10% of the world's primary energy demand. In developing countries, its contribution to the national primary energy demand is higher and usually used in unsustainable ways. Figure 6.1 shows the evolution of contributions of the different energy sources for world's total primary energy demand. In Figure 6.1, biomass refers to traditional biomass (agricultural and forest residues, livestock wastes, etc.) until the most recent decades, and fossil fuels refers to coal, gas, and oil. It may be observed that the share of fossil fuels, for primary energy demand, decreased in recent years. This is mainly due to increase in the share of hydro, nuclear, and other renewable energy sources. In recent years, modern biomass has become more popular and accounts for almost one-fourth of the total biomass energy (GEA 2012). It may be noted that the new renewable energy sources (such as wind and solar) are also visible in the past few decades. However, their contribution is low compared to fossil fuels.

The conversion of biomass to energy can be achieved in many ways (see Figure 6.2). There are mainly two types of conversions: one thermochemical route and the other biochemical route. In thermochemical conversion, charcoal, oil, gas, or syngas (H_2, CO) is produced, and ethanol, butanol, or biogas is generated in biochemical conversion. Bioenergy, all energy derived from biofuels, could play a vital role in developing countries, especially those with very good agricultural sector. Today, bioenergy growth is higher for many countries to improve energy access, energy security, and reduce greenhouse gas emissions. Moreover, the consumption of biomass for combined heat and power (Desai and Bandyopadhyay 2016) and production of biofuel (Winchester and Reilly 2015; Popp et al. 2014) has increased significantly in developed as well as developing countries.

Distribution of global biomass consumption in various sectors (as on year 2008) is shown in Figure 6.3 (GEA 2012). It may be noted that biomass provides about 50 EJ/year of primary energy. In recent years, the consumption of biomass has increased for applications other than traditional cooking. Distribution

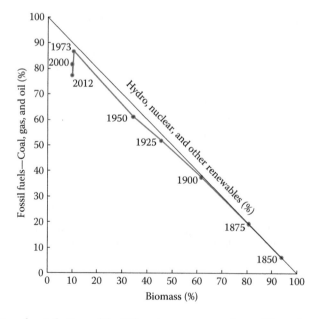

FIGURE 6.1   Evolution of contributions of the different energy sources for world's total primary energy demand.

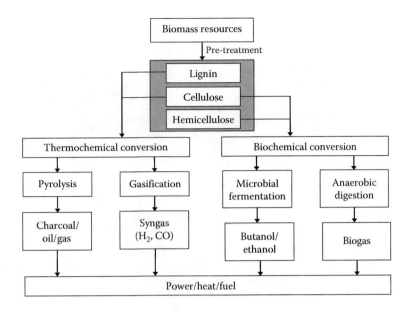

FIGURE 6.2 Conversion of biomass to energy. (From Chung, J.N., *Front. Energy Res.*, 1, 1–4, 2013.)

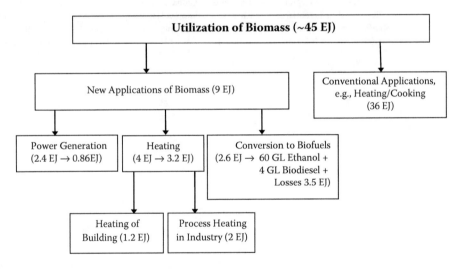

FIGURE 6.3 Distribution of global biomass consumption in various sectors as on year 2008. (From GEA, Global energy assessment—Toward a sustainable future, Cambridge University Press, Cambridge, UK and New York and the International Institute for Applied Systems Analysis, Laxenburg, Austria, 2012.)

of biomass sources as a primary energy source is given in Figure 6.4. Biomass resources are very diverse. The consumption of fuel wood is largest (67%) among all the categories. The agriculture consumption includes energy crops, agriculture by-products, and animal by-products.

Worldwide theoretical biomass potential for bioenergy production is given in Table 6.1 (GEA 2012). The theoretical bioenergy potential is about 1126 EJ/year. This would lead to the clearing of all forests and there will be no biomass left in the ecosystems to build up and maintain long-lasting carbon stocks. As a result, the greenhouse gas balance will be destroyed. Practical bioenergy potential is about 793 EJ/year (about 70% of the total), and this does not include the production of algae in the coastal seawaters, open

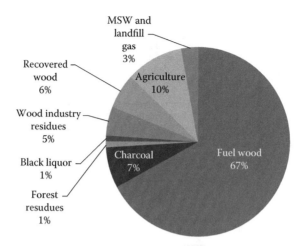

FIGURE 6.4 Distribution of biomass sources as a primary energy source. (From GEA, Global energy assessment—Toward a sustainable future, Cambridge University Press, Cambridge, UK and New York and the International Institute for Applied Systems Analysis, Laxenburg, Austria, 2012.)

TABLE 6.1 Worldwide Theoretical Potential of Biomass for Bioenergy Production

	Total Terrestrial Surface Area (1000 km²)	Aboveground Potential of Vegetation (EJ/year)	Aboveground Current Vegetation (EJ/year)	Global Human Biomass Harvest for Food, Feed, Fiber (EJ/year)	Theoretical Total Bioenergy Potential (EJ/year)	Theoretical (practical) Bioenergy Potential (EJ/year)
Asia (excluding Former Soviet Union)	26,697	239	236	67	172	126
Africa	22,537	273	247	23	250	168
Australia, New Zealand, and other Oceania	7,913	60	58	5	55	40
Europe (excluding Former Soviet Union, including Turkey)	10,337	64	65	22	43	32
Former Soviet Union	21,614	179	165	9	170	117
North America	20,698	162	158	26	137	100
South America and the Caribbean	20,295	331	312	31	299	210
Total	130,091	1,308	1,241	183	1,126	793

Source: GEA, Global energy assessment—Toward a sustainable future, Cambridge University Press, Cambridge, UK and New York and the International Institute for Applied Systems Analysis, Laxenburg, Austria, 2012.

sea, and industrial installations (GEA 2012). This huge potential of biomass along with other renewable energy sources can meet the future energy demand in sustainable ways.

Global technical potentials of bioenergy from various biomass resources in 2050 are given in Table 6.2 (GEA 2012). The total global bioenergy potential is expected to be in the range of 162–267 EJ/year in 2050 (GEA 2012). It may be noted that among all the biomass resources the energy crop and forestry are highly uncertain and depends on the policy of the country and other usages. The other sources of biomass (e.g., crop residues, manures, municipal solid waste, etc.) have low uncertainty. This biomass can be used for different heat requirements (e.g., cooking, process, etc.), electricity generation, or biofuel production.

TABLE 6.2 Global Technical Potentials of Bioenergy from Various Biomass Resources in 2050 (GEA 2012)

Biomass Resource	Estimates (EJ)	Remarks
Forestry	19 (minimum); 35 (maximum)	Depends on other uses
Bioenergy crop	44 (minimum); 133 (maximum)	Highly uncertain: depends on production and technology, food need of the society, government policies
Crop residues	49	Conservation of the soil balance needs to be addressed
Manures	39	Low uncertainty
Municipal solid waste (MSW)	11	Low uncertainty
Total	162 (minimum); 267 (maximum)	

Source: GEA, Global energy assessment—Toward a sustainable future, Cambridge University Press, Cambridge, UK and New York and the International Institute for Applied Systems Analysis, Laxenburg, Austria, 2012.

TABLE 6.3 Estimates of Technical Potential of Biomass Supply for Electricity and Liquid Transportation Fuels in 2050

Region	Potential of Biomass Supply in 2050 (EJ/year)						Electricity (TWh/year)		Biofuels (EJ/year)
	Crop Residues	Animal Waste	MSW	Forest Residues	Energy Crops	Total	Small Scale	Large Scale	
Asia (excluding Former Soviet Union)	23	15	5	4	17	63	4,329	5,773	21
Africa	6	6	1	1	18	31	2,201	2,934	10
Australia, New Zealand, and other Oceania	0.7	1.7	0.1	0.5	5.1	8	561	748	2.6
Europe (excluding Former Soviet Union, including Turkey)	3	4	1	7	7	22	1,610	2,146	7
Former Soviet Union	2	2	0	3	6	13	966	1,288	4
North America	4	4	1	9	14	31	2,396	3,193	11
South America and the Caribbean	11.3	7.9	1.8	3	22.7	46.6	3,390	4,519	16.1
Total	49	39	11	27	88	215	15,453	20,601	72

Source: GEA, Global energy assessment—Toward a sustainable future, Cambridge University Press, Cambridge, UK and New York and the International Institute for Applied Systems Analysis, Laxenburg, Austria, 2012.

Worldwide estimates of technical potential of biomass supply for electricity and liquid transportation fuels in 2050 is given in Table 6.3 (GEA 2012). The total electricity generation potential is calculated in such a way that all sustainably producible biomass is used for electricity generation. If all the biomass-based electricity generation is from small-scale plants, then the potential is 15,453 TWh/year. However, for large-scale plants, the potential is 20,601 TWh/year. Alternatively, if all of the biomass were to be converted to transport fuel, then its potential is about 72 EJ/year. It may be noted that the worldwide generation of electricity from all the sources in 2008 is about 20,183 TWh/year, and the transport fuel production is about 95 EJ/year (GEA 2012).

6.3 Bioenergy and Constraints

Several nongovernmental organizations (NGOs) have criticized the conversion of food to fuel because a lot of people worldwide are still suffering from malnutrition. This is also considered as a main reason for increase in price of vegetable oils and fats worldwide. The Bioenergy and Food Security Project is also initiated by the Food and Agriculture Organization of the United Nations, with funding from the Government of Germany, to develop technical understanding of impacts of bioenergy development on food security (FAO 2015). This will facilitate the policy makers to arrive at a decision about bioenergy. This project consists of mainly five modules as follows:

Module 1: Biomass Potential
Module 2: Biofuel Chain Production Costs
Module 3: Agriculture Markets Outlook
Module 4: Economy-Wide Effects
Module 5: Household-Level Food Security

Use of biomass for generation of electricity or transportation fuels is considered as a carbon-neutral process because CO_2 produced during the combustion process is actually fixed while growing the feedstock. However, in some cases, biomass fuels are more carbon positive than fossil fuels as discussed by Johnson (2009). Moreover, large scale biomass usage may lead to problems related to food security, water scarcity, deforestation, loss of biodiversity, etc. Different techniques are used to analyze the effects of excessive usage of biomass on the different footprints. One such technique used for analyzing bioenergy production system with different footprint constraints is Pinch Analysis.

6.4 Pinch Analysis and Other Methods

Pinch Analysis is a system-oriented approach to industrial process design with an objective of sustainable development. Pinch Analysis has established itself as a tool for analyzing and developing efficient processes through appropriate integration between various processes/systems. Techniques of Pinch Analysis have been applied for analyzing heat exchanger networks (Linnhoff 1993; Shenoy 1995), utility systems (Shenoy et al. 1998), mass exchanger networks (El-Halwagi and Manousiouthakis 1989; Hallale and Fraser 1998), water networks (Wang and Smith 1994; Prakash and Shenoy 2005a and 2005b), distillation columns (Bandyopadhyay et al. 1999, 2003, and 2004), fired heaters (Varghese and Bandyopadhyay 2007), production planning (Singhvi and Shenoy 2002; Singhvi et al. 2004), renewable energy systems (Kulkarni et al. 2007; Arun et al. 2007; Roy et al. 2007), carbon-constrained energy sector planning (Tan and Foo 2007), isolated power systems (Arun et al., 2008), etc. Recently, applications of Pinch Analysis have been extended to optimal planning of energy systems, subject to different footprint constraints such as carbon (Foo et al. 2008), water (Tan et al. 2009a), and land (Tan and Foo 2013).

Apart from Pinch Analysis, there are many methodologies and software tools available for evaluation of different footprints. Čuček et al. (2012) presented an overview of the definitions and units of measurement associated with different footprints (environmental, social, and economic). Currently, there are about 80 carbon footprint calculators available (Čuček et al. 2012). The tools for other different indicators (e.g., ecological footprint, water footprint, nitrogen footprint, health footprint, etc.) have been reviewed by Čuček et al. (2012). Sandholzer and Narodoslawsky (2007) developed a software tool (SPIonExcel) for calculation of ecological footprint based on processes and products. Kettl et al. (2011) developed a software tool (RegiOpt) by combining the process network synthesis (Friedler et al. 1995) and sustainable process index methodologies, which creates the economically optimal regional energy technology networks subject to the environmental impacts. Dipolar Pty Limited and Centre for Integrated Sustainability Analysis developed a software package (BottomLine), which uses more than 100 indicators (e.g., ecological footprint, carbon footprint, greenhouse gases, energy and resource usages, air pollutants, material flows, etc.). Process network synthesis solution, a graphical tool, performs the optimization of footprints and economics of product or process or complete supply chain (Lam et al. 2011).

Mathematical programming tools that use mathematical methods, such as linear programming, nonlinear programming, mixed-integer nonlinear programming, or other approaches, can also be used for life cycle and sustainability analyses. Using mathematical optimization, the system can be optimized based on different footprints, impacts, or performances separately or simultaneously (Čuček et al. 2012). In case of a linear programming problem, with special mathematical structures, the Pinch Analysis is computationally more efficient and provides physical insight to the problem.

Ludwig et al. (2009) used the supply chains composite curves (Singhvi and Shenoy 2002; Singhvi et al. 2004) to determine various strategies for biomass productions with seasonal variations in demands. A Pinch Analysis technique to production planning gives a simple and effective tool for analyzing production strategies with seasonal demands. Improvement in the efficiency of biomass supply chains is an important aspect, and one of the earliest works on systematic regional bioenergy planning, based on Pinch Analysis, was done by Lam et al. (2010). Recently, Krishnan and McCalley (2016) presented a planning model for energy derived from biomass (bioenergy and biofuel), and many planning scenarios are studied to assess the impact of increasing penetration of biomass for power and fuel on the national long-term planning cost and emissions.

Carbon footprint, which is the total amount of CO_2 and other greenhouse gases emitted over the full life cycle of a process or product, is commonly used in assessing the environmental impacts of biomass process and supply chain. Different life cycle analyses demonstrated that the carbon footprint for biofuels may be more than that of traditional fuels, under some unfavorable conditions (Nonhebel 2005). Recently, several studies on carbon-constrained planning with focus on optimization within single facility (Hashim 2005; Mirzaesmaeeli 2010) as well as focus on regional-level targeting (Tan and Foo 2007; Foo et al. 2008; Tan et al. 2009b; Atkins 2010) have been reported. Pękala et al. (2010) developed a mathematical model to handle optimum biofuel production with consideration of multiple footprints and deployment of CO_2 capture and storage (CCS) retrofit with concern for cost-effectiveness. Foo et al. (2013) proposed a linear programming model to optimally allocate biomass, for combined heat and power (CHP) plants, between sources and sinks. The model also gives the capacity of biomass-powered CHP plants. A mixed integer linear programming model, to minimize carbon footprint of the transportation of biomass by accounting different practical considerations (uncertainty) in biomass logistics, was also developed by Foo et al. (2013). Ooi et al. (2013) applied the mathematical programming approach called the automated targeting model, based on the concept of Pinch Analysis, for bioenergy with CCS systems, which offer the possibility of negative emissions.

Pearman (2013) evaluated the biophysical limits of biofuels and biosequestration of carbon based on the availability of the solar radiation and the efficiencies of conversion of energy to biomass for the natural ecosystems and agricultural systems. Lecksiwilai et al. (2016) compared the net energy ratio and life cycle greenhouse gases for the production of dimethylether from five agricultural residues (rice straw, palm empty fruit bunches, sugar cane tops and leaves, cassava rhizome, and maize stem) with dimethylether derived from lignite coal and natural gas. The bio-dimethylether produced from sugar cane tops and leaves, rice straw, and maize stem have lower life cycle greenhouse emissions as well as high net energy ratio compared to fossil fuel-based production (Lecksiwilai et al. 2016).

In recent years, it has been documented that water availability may be a limiting constraint to the production of biomass for energy (Berndes 2002; Gerbens-Leenes 2009; Tan et al. 2009a). Gerbens-Leenes (2009) reported that the usage of water for the production of primary energy from biomass is twice or thrice the demand of water for the production of the same amount of primary energy from fossil fuels. Therefore, the water usages of the nations (water footprint) have been studied for sustainable production of the bioenergy crop (Hoekstra and Hung 2005; Hoekstra and Chapagain, 2007; Tan et al. 2009a). The reported results strongly suggest that water requirements for the bioenergy crop are a major constraint for the future policies on bioenergy. A novel graphical targeting tool based on the Pinch Analysis technique, for the analysis of water footprints, has been proposed by Tan et al. (2009a). The approach uses the water footprint pinch diagram and identifies the water footprint pinch point. Based on the energy demands and water resources availability, the allocations of energy crop are done.

The uncontrolled growth of bioenergy may increase the problems related to food security. Lam et al. (2010) performed the systematic regional bioenergy planning and proposed a hybrid optimization-graphical algorithm, where geographic limitations on land availability were addressed using the Pinch Analysis. Tan and Foo (2013) presented the generalized Pinch principles for energy sector planning, which involves the optimal matching of sources and sinks, using energy qualities as carbon footprint, agricultural land footprint, water footprint, energy, and inoperability. The generalized problem statement of quality-constrained energy sector planning and case study based on land footprint is discussed in the following sections. Brien et al. (2015) presented an overview of footprint accounting methods, a review of actual activities to calculate land footprints, and global land use accounting approach for European policies regarding bioeconomy, land, and resource targets. Khoo (2015) reviewed the bioconversion pathways of lignocelluloses to ethanol based on the land footprint projections.

6.4.1 Quality-Constrained Energy Sector Planning

The generalized problem statement of the quality-constrained energy sector planning is given as follows:

- Energy streams for a given system are measured in power or energy units.
- Energy streams are also quantified by a quality index that conforms to a linear mixing rule and follows an inverse scale. The highest quality refers to zero and the highest positive numerical values refer to lower quality levels. The given system contains m internal sources. Each source i is able to supply an energy stream of quantity S_i at a quality Q_i. Unutilized potential of internal source is termed as waste (W).
- System contains n demands. Each demand j requires energy at a quantity D_j and within a quality index limit of $Q_{j,\max}$.
- The external energy resource (F) is available with unlimited supply and quality level Q^{ext}. The quality of the external energy resource is considered to be better compared to the internal sources. It is desirable to minimize the use of total external energy resource (R).

The generalized quality-constrained energy sector planning problem can be represented as a linear programming problem. The conservation equations in the optimization model are given as follows:

$$W_i + \sum_{j=1}^{n} E_{ij} = S_i \text{ for every internal source } i. \tag{6.1}$$

The energy demand balance is given by

$$F_j + \sum_{i=1}^{m} E_{ij} = D_j \text{ for every internal demand } j. \tag{6.2}$$

The quality load balance at the demands is given by

$$Q^{\text{ext}} F_j + \sum_{i=1}^{m} Q_i E_{ij} \leq Q_j^{\max} D_j \text{ for every internal demand } j. \tag{6.3}$$

The objective is to minimize the external energy resource requirement.

$$\text{minimize } R = \sum_{j=1}^{n} F_j. \tag{6.4}$$

It may be noted that all system variables are nonnegative. The optimization problem is to minimize the external energy resource requirement, given by Equation 6.4, subject to Equations 6.1 through 6.3.

6.4.2 Case Study: Energy Sector Planning Based on Agricultural Land Footprint

This case study is taken from Tan and Foo (2013), and it involves the production planning of ethanol from sugarcane and corn to meet the requirements of three different regions. It may be noted that this case study was originally solved algebraically by Foo et al. (2008). Due to geographic limitations, the trade of ethanol among these different regions is not allowed. The production of ethanol is limited by the availability of the agricultural land, which is considered as an energy quality index in the present case. The production of each feedstock is represented as millions of liters (ML) of ethanol equivalent. The extra requirement of the ethanol is met through external resource in such a way that there is no land footprint in any of the regions, and therefore, $Q^{ext}=0$. The data for the case study are given in Table 6.4 (Tan and Foo 2013).

The given problem can be solved by various graphical or algebraic techniques. The graphical approach based on the material recovery pinch diagram, which was developed independently by El-Halwagi et al. (2003) and Prakash and Shenoy (2005a) and Shenoy (2005a) is used for targeting in the present case study. The steps involved in the methodology are as follows:

- First, all the sources are arranged in decreasing quality.
- For each source, the quality load is calculated by multiplying the quantity with its quality.
- To obtain the source composite curve, cumulative energy quantity is plotted on horizontal axis and the cumulative quality load on vertical axis. Each source is plotted to form a segment of the composite curve.
- Similarly, the demand (or sink) composite curve is plotted, as shown in Figure 6.5. It may be noted that this is an initial infeasible solution.

TABLE 6.4 Source and Demand Data for Case Study

Source	Sugarcane ethanol	Corn ethanol
S_i (ML)	499.8	27.5
Q_i (hectares/ML)	238	1429
Demand	Region I	Region II
D_j (ML)	422	105
$Q_{j,max}$ (hectares/ML)	177.7	214.3

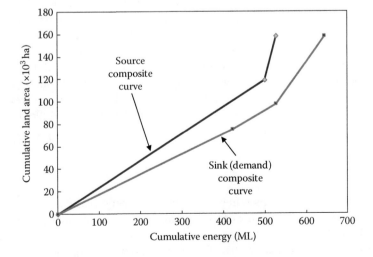

FIGURE 6.5 Energy planning composite curve with initial infeasible solution.

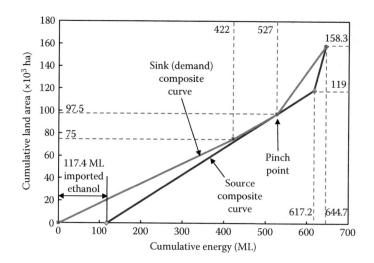

FIGURE 6.6 Energy planning composite curve with optimum solution.

TABLE 6.5 Optimal Allocation of Energy for the Case Study (ML)

Source	Region I	Region II	Region III	Unutilized Energy Source	Total
Sugarcane ethanol	315.1	94.5	90.1	0	499.7
Corn ethanol	0	0	27.5	0	27.5
Imported ethanol	106.9	10.5	0	0	117.4
Total	422	105	117.6	0	

- A feasible solution is achieved when the source composite curve is exactly below and to the right side of the demand composite curve. Source composite curve is shifted along the locus, with slope Q^{ext} in such a way that it lies just below and tangent to the demand composite curve. This is demonstrated in Figure 6.6, and the point where the two composite curves touches is called the pinch point.
- The horizontal distance of the shifted source composite curve from its initial position gives the external resource target, which is 117.4 ML of ethanol in the present case study.
- Similar to the conventional Pinch Analysis-based methodologies, there exists a golden rule based on the location of the pinch point. In the present case study, the external energy resource requirement should be below pinch only. The above and below pinch regions should be analyzed separately and there should not be any cross pinch energy transfer. However, the energy source that lies at the pinch point should be divided to supply the requirements of both the regions.
- The allocation of different sources to demands may be calculated using nearest neighbor algorithm, which was developed by Prakash and Shenoy (2005a). The optimal allocation of the present case study is given in Table 6.5.

6.5 Conclusions

Biomass is one of the oldest and the most important renewable energy resources. Biofuels are considered as carbon-neutral energy sources because CO_2 produced during the combustion process is actually fixed while growing the feedstock. Today, biomass is the world's fourth largest energy source (after oil, coal, and natural gas), contributing to about 10% of the world's primary energy demand, and the practical theoretical bioenergy potential is about 793 EJ/year. The additional increase in agricultural demand for bioenergy is considered as a threat to food security. The impact of increase in utilization of land for bioenergy production needs to be analyzed before its allocation, and therefore energy policy decisions of developing countries play an important role. Large scale of biomass usage may also lead to problems related to water scarcity, deforestation, and loss of biodiversity. Pinch Analysis techniques, a simple and effective tool, have been applied to bioenergy systems subject to carbon, water, and land footprints constraints. Improvement in the efficiency of biomass supply chains is an important aspect, and Pinch Analysis-based techniques are also applied for such problems. Apart from Pinch Analysis, there are many methodologies and software tools available for evaluation of different footprints for sustainable development.

References

Arun, P., Banerjee, R., and Bandyopadhyay, S. 2007. Sizing curve for design of isolated power systems. *Energy for Sustainable Development* 11(4): 21–28.

Arun, P., Banerjee, R., and Bandyopadhyay, S. 2008. Optimum sizing of battery integrated diesel generator for remote electrification through design-space approach. *Energy* 33(7): 1155–1168.

Atkins, M. J., Morrison, A. S., and Walmsley, M. R. 2010. Carbon emissions pinch analysis (CEPA) for emissions reduction in the New Zealand electricity sector. *Applied Energy* 87(3): 982–987.

Bandyopadhyay, S., Malik, R. K., and Shenoy, U. V. 1999. Invariant rectifying–stripping curves for targeting minimum energy and feed location in distillation. *Computers and Chemical Engineering* 23(8): 1109–1124.

Bandyopadhyay, S., Malik, R. K., and Shenoy, U. V. 2003. Feed preconditioning targets for distillation through invariant rectifying–stripping curves. *Industrial and Engineering Chemistry Research* 42(26): 6851–6861.

Bandyopadhyay, S., Mishra, M., and Shenoy, U. V. 2004. Energy-based targets for multiple-feed distillation columns. *AIChE Journal* 50(8): 1837–1853.

Berndes, G. 2002. Bioenergy and water—The implications of large-scale bioenergy production for water use and supply. *Global Environmental Change* 12(4): 253–271.

Brien, M. O., Schütz, H., and Bringezu, S. 2015. The land footprint of the EU bioeconomy: Monitoring tools, gaps and needs. *Land Use Policy* 47: 235–246.

Chung, J. N. 2013. Grand challenges in bioenergy and biofuel research: Engineering and technology development, environmental impact, and sustainability. *Frontiers in Energy Research* 1: 1–4.

Čuček, L., Klemeš, J. J., and Kravanja, Z. 2012. A review of footprint analysis tools for monitoring impacts on sustainability. *Journal of Clean Production* 34: 9–20.

Desai, N. B. and Bandyopadhyay, S. 2016. Biomass-fueled organic rankine cycle-based cogeneration system. In *Process Design Strategies for Biomass Conversion Systems*, eds Ng, D. S. K., Tan, R. R., El-Halwagi, M. M., and Foo, D. C. Y. John Wiley & Sons, Chichester, UK. doi: 10.1002/9781118699140. ch10.

El-Halwagi, M. M. and Manousiouthakis, V. 1989. Synthesis of mass exchange networks. *AIChE Journal* 35(8): 1233–1244.

El-Halwagi, M. M., Gabriel, F., and Harell, D. 2003. Rigorous graphical targeting for resource conservation via material reuse/recycle networks. *Industrial and Engineering Chemistry Research* 42: 4319–4328.

FAO. 2015. Food and Agriculture Organization of the United Nations. http://www.fao.org/energy/befs/en/ (Accessed September 14, 2015).

Foo, D. C. Y., Tan, R. R., and Ng, D. K. S. 2008. Carbon and footprint-constrained energy planning using cascade analysis technique. *Energy* 33: 1480–1488.

Foo, D. C. Y., Tan, R. R., Lam, H. L., Abdul Aziz, M. K., and Klemeš, J. J. 2013. Robust models for the synthesis of flexible palm oil-based regional bioenergy supply chain. *Energy* 55: 68–73.

Friedler, F., Varga, J. B., and Fan, L. T. 1995. Decision-mapping: A tool for consistent and complete decisions in process synthesis. *Chemical Engineering Science* 50: 1755–1768.

GEA, 2012. *Global energy assessment—Toward* a sustainable future, Cambridge University Press, Cambridge, UK and New York and the International Institute for Applied Systems Analysis, Laxenburg, Austria.

Gerbens-Leenes, P. W., Hoekstra. A. Y., and Van der Meer, T. 2009. The water footprint of energy from biomass: A quantitative assessment and consequences of an increasing share of bio-energy in energy supply. *Ecological Economics* 68(4): 1052–1060.

Hallale, N., and Fraser, D. M. 1998. Capital cost targets for mass exchange networks. A special case: Water minimization *Chemical Engineering Science* 53(2): 293–313.

Hashim, H., Douglas, P., Elkamel, A., and Croiset, E. 2005. Optimization model for energy planning with CO_2 emission considerations. *Industrial and Engineering Chemistry Research* 44 (4): 879–890.

Hoekstra, A. Y., and Hung, P. Q. 2005. Globalisation of water resources: International virtual water flows in relation to crop trade. *Global Environmental Change* 15(1): 45–56.

Hoekstra, A. Y., and Chapagain, A. K. 2007. Water footprints of nations: Water use by people as a function of their consumption pattern. *Water Resources Management* 21(1): 35–48.

Johnson, E. 2009. Goodbye to carbon neutral: Getting biomass footprints right. *Environmental Impact Assessment Review* 29: 165–168.

Kettl, K. H., Niemetz, N., Sandor, N., Eder, M., Heckl, I., and Narodoslawsky, M. 2011. Regional Optimizer (RegiOpt)—Sustainable energy technology network solutions for regions. *Computer Aided Chemical Engineering* 29: 1959–1963.

Khoo, H. H. 2015. Review of bio-conversion pathways of lignocellulose-to-ethanol: Sustainability assessment based on land footprint projections. *Renewable and Sustainable Energy Reviews* 46: 100–119.

Krishnan, V. and McCalley, J. D. 2016. The role of bio-renewables in national energy and transportation systems portfolio planning for low carbon economy. *Renewable Energy* 91: 207–223.

Kulkarni, G. N., Kedare, S. B., and Bandyopadhyay, S. 2007. Determination of design space and optimization of solar water heating systems. *Solar Energy* 81(8): 958–968.

Lam, H. L., Klemeš, J. J., Kravanja, Z., and Varbanov, P. S. 2011. Software tools overview: Process integration, modelling and optimisation for energy saving and pollution reduction. *Asia-Pacific Journal of Chemical Engineering* 6(5): 696–712.

Lam, H. L., Varbanov, P. and Klemeš, J. 2010. Minimising carbon footprint of regional biomass supply chains. *Resources, Conservation and Recycling* 54: 303–309.

Lecksiwilai, N., Gheewala, S. H., and Sagisaka, M. 2016. Net energy ratio and life cycle greenhouse gases (GHG) assessment of bio-dimethyl ether (DME) produced from various agricultural residues in Thailand. *Journal of Cleaner Production* 134: 523–531. doi: 10.1016/j.jclepro.2015.10.085.

Linnhoff, B. 1993. Pinch analysis: A state-of-the-art overview. *Chemical Engineering Research and Design* 71(A5): 503–522.

Ludwig, J., Treitz, M., Rentz, O., and Geldermann, J. 2009. Production planning by pinch analysis for biomass use in dynamic and seasonal markets. *International Journal of Production Research* 4: 2079–2090.

Mirzaesmaeeli, H., Elkamel, A., Douglas, P. L., Croiset, E., and Gupta, M. 2010. A multi-period optimization model for energy planning with CO_2 emission consideration. *Journal of Environmental Management* 91(5): 1063–1070.

Nonhebel, S. 2005. Renewable energy and food supply: Will there be enough land? *Renewable and Sustainable Energy Reviews* 9(2): 191–201.

Ooi, R. E. H., Foo, D. C. Y., Tan, R. R., Ng, D. K. S., and Smith, R. (2013). Carbon constrained energy planning (CCEP) for sustainable power generation sector with automated targeting model. *Industrial and Engineering Chemistry Research.* 52: 9889–9896.

Pearman, G. I. 2013. Limits to the potential of bio-fuels and bio-sequestration of carbon. *Energy Policy* 59: 523–535.

Pękala, L. M., Tan, R. R., Foo, D. Y. C., and Jeżowski, J. M. 2010. Optimal energy planning models with carbon footprint constraints. *Applied Energy* 87(6): 1903–1910.

Popp, J., Lakner, Z., Harangi-Rákos, M., and Fári, M. 2014. The effect of bioenergy expansion: Food, energy, and environment. *Renewable and Sustainable Energy Reviews* 32: 559–578.

Prakash, R. and Shenoy, U. V. 2005a. Targeting and design of water networks for fixed flowrate and fixed contaminant load operations. *Chemical Engineering Science* 60(1): 255–268.

Prakash, R. and Shenoy, U. V. 2005b. Design and evolution of water networks by source shifts. *Chemical Engineering Science* 60(7): 2089–2093.

Roy, A., Arun, P., and Bandyopadhyay, S. 2007. Design and optimization of renewable energy based isolated power systems. *SESI Journal* 17(1–2): 54–69.

Sandholzer, D. and Narodoslawsky, M. 2007. SPIonExcel—Fast and easy calculation of the sustainable process index via computer. *Resources, Conservation and Recycling* 50: 130–142.

Shenoy, U. V. 1995. *Heat Exchanger Network Synthesis: Processes Optimization by Energy and Resource Analysis*. Gulf Publishing Company, Houston, TX.

Shenoy, U. V., Sinha, A., and Bandyopadhyay, S. 1998. Multiple utilities targeting for heat exchanger networks. *Chemical Engineering Research and Design* 76(3): 259–272.

Singhvi, A. and Shenoy, U. V. 2002. Aggregate planning in supply chains by pinch analysis. *Chemical Engineering Research and Design* 80(6): 597–605.

Singhvi, A., Madhavan, K. P., and Shenoy, U. V. 2004. Pinch analysis for aggregate production planning in supply chains. *Computers and Chemical Engineering* 28(6–7): 993–999.

Tan, R. R. and Foo, D. Y. C. 2007. Pinch analysis approach to carbon-constrained energy sector planning. *Energy* 32: 1422–1429.

Tan, R. R., Foo, D. Y. C., Aviso, K. B., and Ng, D. S. K. 2009a. The use of graphical pinch analysis for visualizing water footprint constraints in biofuel production. *Applied Energy* 86(5): 605–609.

Tan, R. R., Ng, D. K. S., and Foo, D. C. Y. 2009b. Pinch analysis approach to carbon-constrained planning for sustainable power generation. *Journal of Cleaner Production*, 17(10), 940–944.

Tan, R. R. and Foo, D. C. Y. 2013. Pinch analysis for sustainable energy planning using diverse quality measures. In *Handbook of Process Integration*, ed. Klemeš, J. J., Woodhead Publishing, Cambridge, UK.

Varghese, J. and Bandyopadhyay, S. 2007. Targeting for energy integration of multiple fired heaters. *Industrial and Engineering Chemistry Research* 46(17): 5631–5644.

Wang, Y. P. and Smith, R. 1994. Wastewater minimization. *Chemical Engineering Science* 49(7): 981–1006.

Winchester, N. and Reilly, J. M. 2015. The feasibility, costs, and environmental implications of large-scale biomass energy. *Energy Economics* 51: 188–203.

7

Uses of Water and Energy in Food Processing

Mohammad Shafiur
Rahman and
Nasser Al-Habsi
Sultan Qaboos University

7.1 Water in Food Processing

Water and energy interact in several ways in the production of food, and this is one of the most energy-intensive industries. Food production will need to be sufficient to support a global population of 9–10 billion by 2050 (FAO report, 2015). Starling (2015) points out that a UK group identified that there are 2000–5000 L of water embedded in food every day per person, which is equivalent to 730,000–1,825,000 million L annually. Total annual global water requirements are expected to jump from 4500 to 6900 billion m³ by 2030 (Hoekstra, 2015). The FAO (2015) reported that agriculture will continue to be the largest user of water resources in most countries. It was mentioned that 90% of all freshwater is used by agriculture (70%) and industry (20%), leaving 10% for domestic use (Starling, 2015). The report "Feeding Ourselves Thirsty" ranked 37 major food companies on their corporate water management and risk response. This report also identified that the food industry is unprepared for global water shortages (Michail, 2015).

Governmental and institutional actions are needed urgently to reduce the estimated 1.8 million tons of water per person used in food production around the globe. The Institution of Chemical Engineers (IChemE) said regulations and incentives are needed to meet a targeted reduction of 20% by 2020 or face major water shortages by 2050. It was forecasted that €677m investment is needed for development of the infrastructure, technology, and other measures for the "water footprint." Agriculture will need around 19% more water to produce 60% extra food by 2050. The current production methods are unsustainable and more efficient water technology, political will, major investment, and lifestyle changes are needed (FAO, 2015; Starling, 2015). Governments need to reward for alternative and sustainable water users, such as water in food, rainfall, and saltwater (Starling, 2015). Moreover, policy interventions are needed to reduce water withdrawals and pollution. The FAO (2015) report point out that in most countries 70% or more of water withdrawals will be from rivers, lakes, and aquifers. In many regions, farmers will be under pressure to increase production with less water available. This will need adaptation, education, and training on water management. Moreover, policy interventions are needed to reduce water withdrawals and pollution. Seventeen Sustainable Development Goals (SDGs) have been proposed and two of them are aligned closely with issues regarding water and food security (Maurice, 2013). There would be more favorable national food security impact if cash transfer is combined with an increase in agricultural research and development (Anderson and Strutt, 2014). Water balance (i.e., rates of consumptive use and water losses with the amount of water available within a season and over time) needs to be performed and this analysis is necessary to understand interactions involving users in upstream and downstream settings management plans. Arjen Hoekstra, founder of the Water Footprint Network, pointed out that food and beverage companies could play a major role in helping reduce global water consumption through their supply chains (Hoekstra, 2015). He coined the term "waterfootprint" and founded the international Water Footprint Network in 2008. He emphasized that taking action is more necessary rather talking positively on the sustainable use of water by the food industry. Governments also need to play a part in addressing food industry water uses by setting benchmarks and set targets for water footprint and applying indirect taxes. A meat tax would address the significant energy and water resources used in cattle farming and meat production. This, in turn, could help change consumption patterns, encouraging consumers to eat more vegetables and fruits and less consumption of meat and dairy. The change in the consumption pattern could make a much larger impact on overall water consumption than taking shorter showers. The consumption of crop products instead of meat could save 800 L of water per person in a day. By comparison, shorter shower times or reducing the number of showers taken saves relatively low water (Hoekstra, 2015). However, this comparison should not be misunderstood by considering that we should not take efforts to save water from showers. Food industry uses water for mainly three purposes: as a raw material to be incorporated in foods, as a thermal fluid for heating/cooling purposes, and as cleaning agent to eliminate undesirable compounds. For each usage, it is important to apply different strategies to reduce the use of water.

7.2 Water as Food Ingredient

The largest water footprint of a beverage company, for example, is not bottled water. Instead, it can be traced to the main agricultural ingredients used in flavored drinks, such as sugar, oranges, or barley (Hoekstra, 2015). Milk production shows a water footprint three times greater than that for growing vegetables while beef production shows a footprint 48 times higher (Hoekstra, 2015). The required mass of water to harvest or produce a kilogram of potatoes, rice, chicken, and beef is 500–1500, 1900–5000, 3500–5700, and 15,000–70,000 kg, respectively (Gleick, 2000). Jalava et al. (2014) suggest that diet change is a possible solution to combat water scarcity. It is commonly suggested that limiting animal product consumption could alleviate water scarcity. Alternative protein sources with lower water footprint (i.e., crop based) could considerably decrease the agricultural green water footprint. However, savings in the blue water would be somewhat smaller. The environmental cost of beef production is nearly ten times more than any other form of meat or livestock production. The elimination of beef, or replacing it with relatively efficient animal-based alternatives such as eggs, can achieve an environmental improvement. This could be comparable to switching to plant food source (Gray, 2014). White and Brady (2014) suggest that an incentive-based adoption by the consumers could play a major role in saving water and could reduce environmental impact. They identified that a 10% premium could be the ideal willingness to pay, and this could result in the reduction of water usage up to 41.4 L. Brabeck-Letmathe (2013) pointed to the role of free trade on regional water security. He stated that "The concept of 'virtual water' was introduced by Tony Allan in the early 1990s and refers to the water that is required for the production of agricultural commodities, or in other words, the water 'embedded' in agricultural products. Thus, international food trade can be seen as 'virtual water trade,' which implies that the corresponding amount of 'virtual water' is transferred from the export country to the import country through international food trade." It was expressed that more than four-fifths of food and consumer goods companies will face a fundamental risk to their businesses if there is shortage of water. It was found that 81% of those in the food and consumer goods sector were concerned about water availability and quality, and one in five (20%) companies involved in the report predicted water shortage would inhibit business growth (Scott-Thomas, 2014). The detailed quality parameters of water for food industry are given by International Life Sciences Institute (ILSI, 2008).

7.3 Water as Processing Aid

Washing, cleaning, and sanitizing are common water-intensive operations in food processing, and thus adequate measures need to be taken to restrict their overuse and waste. Sanitation includes cleaning which removes undesirable material (e.g., soils, fats, starch, protein, mineral salt, chemicals, and biological contaminants) from the surfaces, product residues, foreign bodies, and cleaning chemicals; and disinfection includes the killing of microorganisms. The efficacy of washing depends on the surface characteristics and the mechanism of bacterial attachment to plant tissue surfaces. Multistage washing can reduce the use of water and it includes the following stages: wetting of the soil and surface by the cleaning chemical, reaction of the chemical to facilitate removal from the surface, prevention of redeposition, and disinfection of residual microbes (Koopal, 1985; Holah, 1992; Gibson et al., 1999). Factors that affect sanitation are: chemical energy, mechanical/kinetic energy, temperature/thermal energy, surface characteristics (i.e., energy and topography related), and time (Gibson et al., 1999; Mauermann et al., 2009). In the cleaning phase, the chemicals break down soils and reduce their attachment strength to facilitate removal from the surface. In the disinfection phase, the chemicals reduce the viability of the microbes remaining after cleaning. Mechanical or kinetic energy, such as manual brushing, scraping, automated scrubbing, pressure-jet washing, and circulation of fluid in clean-in-place systems, is applied to remove soils from the surface. Temperature enhances the chemical effects and facilitates the removal of fats and oils by lowering their viscosities. However, high temperatures could denature protein-based soils and may cause strong attachments. The time of contact is important for reaction or interaction with chemical and soils; it could be achieved by soaking and using foam or gels (Gibson et al., 1999). The use of chemical compounds for sanitization presents some concerns related to their disposal and workers' safety. In the case of equipment, the cleanability depends on the types of contaminants and their surface energies. In addition, surface modifications showed different abrasion resistance and different levels of resistance to repeated stress with detergents. The surface modification could be a potential option for the reduction of cleaning costs in the food industry (Mauermann et al., 2009).

The amount of water used could be reduced by applying different options, such as use of new technologies for efficient cleaning and processing and use of management or control tools such as the adherence to good manufacturing practices and the application of Hazard Analysis and Critical Control Point (HACCP) plans. It is also important to develop safe, efficient, and environmentally compatible detergent formulae for the food industry in order to ensure adequate levels of quality and safety, while saving energy, water, and time (Jurado-Alameda et al., 2012).

In the case of spray cleaning, the kinetic energy of the spray droplets produces the cleaning action. Increased spray pressure increases energy. If the energy is too great, produce may be physically damaged. If the energy is low, the surface may not be cleaned. The advantages of spray energy washing over washing by dipping, soaking, or gravity rinse are (1) reduced volume of water used and wastewater generation and (2) reduced water uptake by produce (Pordesimo et al., 2002). Appropriate pressure and temperature of the water and resident time of the produce in the sprayer need to be used for best possible cleaning and disinfection action. It is also important to maintain the qualities of spray such as droplet spectrum, droplet velocity, angle of droplet impingement, number and orientation of nozzles, and spray rate (Pordesimo et al., 2002). In many instances, detergent plays an important role. For example, the use of an alkaline, acidic, or neutral detergent prior to spray with high-pressure water increases the removal of attached bacteria on the surface and reduces cell viability (Gibson et al., 1999).

Sanitizers are used for processing aids and these exclude equipment and utensils. These are not consumed as food ingredients and must be removed from the food or inactivated after their intended technological uses. The presence of trace quantities of these substances or their derivatives may be admitted in the final product. The efficient use of sanitizers in washing and cleaning could reduce excessive use of water. The commonly used sanitizers are chlorine, bromine, iodine, trisodium phosphate, quaternary ammonium compounds, acids, hydrogen peroxide, and ozone (Beuchat, 2000). The three major groups of antimicrobial chlorine compounds are liquid chlorine, hypochlorites, and chlorine dioxide. The effectiveness of the sanitizers depends on their concentration, pH, temperature, and contact time. In fruits and vegetables, chlorine is commonly used at 200 ppm at pH < 8.0 with contact time of about 1–2 min. A pH of 6.0–7.5 is more appropriate in order to avoid corrosive damage of the processing equipment at low pH. Water at about 4°C is more effective in processing equipment due to the maximum solubility of chlorine in water. However, in the case of fruits and vegetables, chlorinated water above 10°C is more appropriate since low temperature creates a vacuum inside the produce and causes suction of wash water through the stems, thus contaminating it. Addition of 100 ppm of a surfactant (Tween 80) to hypochloride washing solution enhances lethality, but adversely affects sensory quality. Sanitizers containing solvent could remove the waxy cuticle layer causing greater effectiveness in microbial decontamination without adversely affecting sensory characteristics (Beuchat, 2000).

Washing or dipping treatments can significantly reduce the pesticide residue in fresh fruits and vegetables. In the case of cucumber, 94% reduction of pesticide (i.e., imidacloprid) was observed when it was washed with citric acid solution with concentration of 9/100 g solution, while the reduction was 73% in the case of capsicum (Randhawa et al., 2014). It was also observed that combination of citric and acetic acids was more effective in removing the pesticides. A combination of citric and acetic solutions acids each at concentration of 1.5 g/100 g solution reduced 91% of pesticide residue in the case of cucumber. In the case of fish fillet, Hajeb and Jinap (2009) observed that the overall optimal condition resulting in maximum mercury reduction (81% reduction) was at pH 2.79, sodium chloride concentration of 0.5%, and exposure time of 13.5 min. It was also observed that the reduction of mercury in fish flesh significantly depended on the pH of the solution used. The washing of lettuce followed by dipping treatment with 20% vinegar gave comparable microbial decontamination when compared with the treatment of sodium hypochlorite (200 ppm of free chlorine) (de Oliveira et al., 2012).

Electrolyzed water and ozone are two alternative sanitizing technologies that generate the active oxidizing component on site of food processing and do not use toxic chemical substances. They are safe for handling, distribution, and more environmental friendly as compared to conventional chlorine sanitizers (Deza et al., 2003, 2005; Kim et al., 2003; Yang et al., 2013). Electrochemically activated water is an electrolyzed water sanitizer used for food and food equipment. Electrolysis of dilute sodium chloride solutions generates two distinct fractions, catholyte and anolyte (i.e., sanitizing fraction, which contains different forms of chlorine including hypochlorous acid) (Hricova et al., 2008). The effectiveness of sanitation depends on free available chlorine, oxidation–reduction potential, and pH (Yang et al., 2013). Different types of electrolyzed water are reported to actively kill various foodborne pathogens (Guentzel et al., 2008).

Ozone (O_3), a strong oxidant and potent disinfectant, is formed from oxygen (O_2) using different types of energy, such as photochemical (i.e., ultraviolet radiation), electric discharge (i.e., corona discharge), chemical, thermal, chemonuclear, and electrolytic methods (Guzel-Seydim et al., 2004; Karaca and Velioglu, 2007; Emer et al., 2008; Novak et al., 2008). The advantages of the ozone-based cleaning are (1) it can be applied in gaseous (continuous or intermittently) or ozonated water (washing or dipping) for sanitizing, (2) it can be spontaneously decomposed into a nontoxic product (i.e., oxygen) without disinfectant or toxic residues, (3) it can reduce the chemical oxygen demand (COD) of the wastewater from cleaning process, since ozone can partially oxidize the organic matter and surfactants molecules, (4) it can reduce membrane fouling on the ultrafilter for the wastewater treatment, (5) it is more soluble at lower temperature and it enables operation with a reduction of washing temperature, (6) it can be generated on site, thus reducing the need for storage unlike other cleaning agents, (7) it is considered as Generally recognized as safe (GRAS) as the FDA declared in 1997, (8) it can effectively remove protein, fats, and oils from the surface (Park, 2002; Guzel-Seydim et al., 2004; Karaca and Velioglu, 2007; Pascual et al., 2007; Vurma et al., 2009; Perry and Yousef, 2011; Jurado-Alameda, 2012). Spraying of disinfectants with gaseous and ozonated water, acetic acid, acidic calcium sulfate, and combinations on buckwheat grain in a fluidized bed was also effective in lowering the microbial load (Dhillon et al., 2012).

Ozone could be used to reduce aflatoxin in peanuts and cottonseed meals (Dwarakanath et al., 1968). Karaca and Velioglu (2007) reviewed the reduction of mycotoxins, especially aflatoxins, ochratoxin A, and patulin in many other food products. Ozonated water can also reduce the pesticide and fungicide residue in foods (Karaca and Velioglu, 2007). In a model system, maximum degradation rate of azinophos-methyl was observed at 83%, while captan was completely removed (Ong et al., 1996). The use of 0.25 ppm ozonated water dipping treatment to apples resulted in reducing levels of azinophos-methyl, captan, and formetanate hydrochloride on the surface of apples in the ratio of 75%, 72%, and 46%, respectively. Karaca and Velioglu (2007) reviewed the benefits of ozonated water as well as their concern; thus, treatment conditions should be specifically determined for all kinds of products for effective and safe use of ozone.

Dry separation and cleaning, for example of seeds and grain, can avoid the use of water. In this case, sieve cleaning is the most common. Electrostatic separation is also used by creating electrostatically charged particles first and then passing the seeds or grains through the electrostatic separator. It takes advantage of differences in the electrical characteristics of seeds to accomplish separation that cannot be made with conventional seed-cleaning equipment (Basiry and Esehaghbeygi, 2012). UV light and the use of ultrasound could also reduce the consumption of water in cleaning and disinfecting.

The environment could be affected if wastewater from the food industry is not properly treated. The reuse of water after treatment is another option to reduce net water consumption. Water reuse during food production and processing will likely be increased in the future (Kirby et al., 2003) and conservation by its reuse contributes to sustainable development and market competitiveness of the food industry (Wu et al., 2013). However, the practice of reusing water in the food industry faces a great challenge due to production costs, required technical expertise, and issues of maintaining hygienic quality (Katherine et al., 2001; Casani et al., 2005). In 2001, The Codex Committee on food hygiene has provided guidelines for the hygienic processing of reclaimed water in food processing plants (CODEX, 2001). In Thailand and South Africa, water reuse rate is greater than 40%, which reduces costs, avoids environment pollution, and conserves natural water (Frankel and Phongsphetraratana, 1986; Nozaic, 2000). However, quality of the treated wastewater needs to possess high standard in terms of their toxic contaminants and microbial level; and this could ensure the safety of treated water from waste (Casani et al., 2005). Applications of Hazard Analysis and Critical Control Point (HACCP) in the water recycling could ensure water safety (Kirby et al., 2003). Many countries law and regulations need to be updated to enforce the industry to reuse wastewater after purification treatments. Wu et al. (2013) designed a water reclamation system where the discharged water from mandarin transportation using a conveyor belt was collected in a pool, chlorinated, filtered by active carbon, and then UV sterilized. This water was then reused after alkaline treatment. Ozone is also used in wastewater treatment for its reuse, ability to lower biological oxygen demand (BOD) and chemical oxygen demand (COD) of food plant waste (Guzel-Seydim et al., 2004).

7.4 Energy in Food Processing

Energy is consumed in the industrial sector by a diverse group of industries including manufacturing, agriculture, mining, and construction. It is used in processing, assembly, space conditioning, and lighting. An industrial sector uses more energy than any other end-use sector, and currently this sector consumes about 37% of the world's total delivered energy. The use of energy management, energy-saving technologies, and energy-saving policies are the keys to achieve the saving (Abdelaziz et al., 2011). Energy management is the strategy of meeting energy demand when and where it is needed. This can be achieved by adjusting and optimizing energy using systems and effective procedures, thus resulting in the reduction of total production costs (Petrecca, 1992). Considering the rising price of energy and depletion of world energy resources, energy management began to be considered as one of the main priorities for the industry (Petrecca, 1992). The energy management programs include four main parts (Petrecca, 1992; Kannan and Boie, 2003; Abdelaziz et al., 2011: (1) analysis of historical data, (2) energy audit and accounting, (3) engineering analysis and investments proposals based on feasibility studies, and (4) personnel training and information. Energy audit is an inspection, survey, and analysis of energy flows in the system to enable the reduction of energy input without negatively affecting the output. The energy audit is a reliable, rational, and systematic approach in quantifying energy usage and energy efficiency and is the key for decision-making in the area of energy management (Saidur, 2010). The food manufacturing sector is one of the high energy-intensive manufacturing industries (Hazarika, 2014) and the most common methods of processing are drying, canning, and freezing. There are two ways to save energy: (1) develop the existing process by using improved energy saving through updated design and improving process control and operations and (2) the use of alternative technologies to save the energy, such as high-pressure processing, ohmic heating, and pulsed electric field processing.

7.5 Energy-Efficient Design

There are different forms of energy used in more than 150 unit operations related to food processing. Mechanical operations require electrical energy for driving motors such as liquid pumping, solids conveying, peeling, cutting, size reduction, pressing, centrifugation, mixing, extrusion, and high-pressure processing. Heating process includes direct or indirect heating such as cooking, frying, blanching, baking, roasting, pasteurization, sterilization, thawing, evaporation, drying, and steam generation. The cooling, freezing, and crystallization require extraction of heat by refrigeration, which consumes electric energy for the compressor (Hazarika, 2014).

Steam is required in many unit operations such as heating and blanching. A steam system comprises primarily a steam generation system (i.e., boiler) and a distribution system (i.e., process heat). For improving boiler energy efficiency, the focus is primarily on reducing the heat loss, enhancing heat recovery, and improving process control. It is recommended to use custom-designed boilers instead of "off-the-shelf" boilers in order to have energy-efficient practices. Process heat distribution is potentially a major contributor to energy losses, and this can be improved by reducing heat losses through the system and recovering useful heat (Hazarika, 2014). Hazarika (2014) has listed a number of the possible options to save energy. The use of integrated system, such as recovering heat rejected in a cooling process and using it in process heating, could significantly improve energy efficiency (Das, 2000). Motors are used throughout food processing facilities and 5–15% electricity saving could be achieved by using efficient motor-driven systems (Hazarika, 2014). He has also compiled a selection procedure for motor systems and pumps with adequate monitoring and control systems. Abdelaziz et al. (2011) have explained the use of variable speed drive (VSD), energy economizer, waste recovery, high-efficiency motors, leak prevention in air compressor, reducing pressure drop and regulations, and how these standards could save industrial energy. In the cases of motors, different energy savings strategies could be used, such as use of high-efficient motor, VSD, and capacitor bank to improve the power factor. In addition, different policy measures (i.e., regulatory, voluntary, and incentives based) could be applied to save motor energy. Computer tools (such as MotorMaster+, Canadian motor selection tool (CanMOST), EuroDEEM international) to analyze the energy used by electric motors, energy audit, and cost parameters for carrying out economic analysis (i.e., payback period for different energy saving strategies) need to be applied (Saidur, 2014).

The energy efficiency in refrigeration units could be improved by refrigeration system management considering possible cooling load reduction, heat recovery from compressors, condensers, and evaporators; and efficient piping design of the refrigeration system. The cooling load (i.e. heat removal capacity) of the refrigeration system could be reduced by avoiding infiltration of heat in the cold storage area, proper insulation, using physical or air curtain at the doors, switching off the heat-generating equipment when not needed (e.g., lighting controls can be motion activated to dim the lights when not needed and energy-saving bulbs and lighting strips, heat generating fans), adequate ventilated compressors, removal of surface water before freezing, free cooling, and hydro-cooling. Proper management of the refrigeration system, such as control, adjustable-speed drive, raising suction pressure, indirect lubricant cooling and heat recovery from compressor and maintenance of the compressor often leads to energy efficiency. The reduction of freezing time could also save energy from central process and heat recovery from the thawing process could also lead to good energy management (Hall and Howe, 2012; Hazarika, 2014).

Efficient steam blanchers are designed to retain heat properly, provide efficient distribution of heat in the product, and reduce steam losses. The energy efficiency could be achieved by (1) minimizing steam leakage by using steam seals at the entrance and exit, (2) reducing heat losses through insulation of the blanching chamber, (3) reducing the blanching time through forced convection steam, (4) optimizing the flow of steam by properly designed process controls, and (5) recovering the condensate (Hazarika, 2014). The exiting heat in hot condensate could be recovered by a heat exchanger and used for boiling water or cleaning water (Lund, 1986). The uncondensed steam could also be recirculated by multiple pass of steam in the chamber, thus reducing steam requirement (Masanet et al., 2012). Mysorewala et al. (2015) proposed a novel energy-aware approach for locating leaks in water pipelines using a wireless sensor network and noise pressure sensor data.

Drying is one of the most energy-intensive processes in the food industry. The legislation on pollution, sustainable and environmental friendly technologies created greater demand for energy-efficient drying processes in the food industry. Improving energy efficiency by only 1% could result in as much as 10% increase in profits (Beedie, 1995). The heat energy consumption of dryers can be reduced by (1) avoiding heat loss with exhaust air and dried products, (2) avoiding over drying of products (i.e., optimum drying time with proper control), (3) avoiding leakage of air through doors and seals, (4) increasing heat transfer efficiency by checking fouling, and (5) avoiding improper and damaged insulation around dryer, air ducts, heat exchanger, and burner (Rahman and Perera, 2007; Hazarika, 2014). Table 7.1 shows the possible energy savings for drying of walnuts. Applying different improved options of drier design (i.e., sectioning of drier and operating temperature, fuel switching, hot air recirculation, thermal insulation, and alternate flow control to the drying system) can increase the energy efficiency (Kudra, 2004; Tippayawong et al., 2008).

The prior mechanical or other dewatering, such as filtration by mechanical compression, centrifugal force and gravity, and osmotic process, could significantly reduce the energy required by a drying process. Kanda et al. (2008) developed energy-efficient dewatering techniques by physical contact with liquefied dimethyl ether (DME) at room temperature, and DME is separated by flash distillation. In the separation process, DME vapor is compressed and cooled in a heat exchanger, and the latent heat of condensation is reused to vaporize the DME in the heat exchanger.

The primary fuel requirement can be reduced by 35%–45% as compared to indirect heating by air. However, it is important to check the quality of the products when combustion gas is used. Heat recovery from drying system is another aspect of higher energy efficiency, and heat from the exit air could be recovered by preheating the inlet air by direct mixing or by heat exchanger. In addition, energy recovery could be achieved by heat recovery from heat exchangers, thermal wheels, and heat pipe insulation. These methods recover

TABLE 7.1 Energy Savings for Walnut Dehydration

Method	Possible Savings (%)
Preventing over drying	25–33
Recirculation of drying air	25
Reducing airflow rate	≤25
Improved burner design and operation	≤10
Insulation of drying	3–4

Source: Strumillo, C., Lopez-Cacicedo, C. 1987. *Handbook of Industrial Drying.* Marcel Dekker, New York.

TABLE 7.2   General Comparison of Heat Pump Dryer with Vacuum and Hot Air Drying

Parameter	Hot Air Drying	Vacuum Drying	Heat Pump Drying
Specific moisture evaporation rate (kg water/kWh)	0.12–1.28	0.72–1.2	1.0–4.0
Drying efficiency (%)	35–40	≤70	95
Operating temperature range (°C)	40–90	30–60	10–65
Operating relative humidity range (%)	Variable	Low	10–65
Capital cost	Low	High	Moderate
Running cost	High	Very high	Low

Source: Rahman, M. S., Perera, C. O. 2007. *Handbook of Food Preservation*. CRC Press, Boca Raton, FL.

mainly the sensible heat from the exhaust, while most of the heat is lost with the latent heat of water vapor in the exhaust air. The exit air could be desiccated or dehumidified by condensing out water vapor before its reuse in the inlet (Strumillo and Lopez-Cacicedo, 1987; Hazarika, 2014). The latent heat of water vapor can be recovered by condensing out water using a refrigeration system although extra power needs to be used in the refrigeration unit. The heat pump drying method has been developed based on this principle (Strumillo and Lopez-Cacicedo, 1987; Perera and Rahman, 1997; Rahman and Perera, 2007). In this system, dry air is supplied continuously to collect moisture, thus humidifying it. This humid air passes over the evaporator of the heat pump (i.e., refrigerator's evaporator) where it condenses, thus giving up the latent heat of water vapor for vaporization of the refrigerant. This heat is used to reheat the cool dry air passing over the hot condenser of the heat pump (i.e., refrigerator's condenser). The use of the heat pump dryer offers several advantages: (1) higher energy efficiency, (2) better product quality, (3) ability to operate independent of outside ambient weather conditions, and (4) ability to cause minimal environmental impact (Rahman and Perera, 2007). Comparison of the heat pump with vacuum and hot air drying are presented in Table 7.2. Similarly, heat from the dried product could be recovered by cooling it with air and could be used to preheat the inlet air. In addition, proper process control and tempering step could also reduce the energy requirement (Hazarika, 2014). Similarly, the energy savings in evaporation process are reviewed by Hazarika (2014). Mujumdar and Law (2010) have summarized the potential renewable energy sources for drying such as solar and wind energy, biomass as fuel, and geothermal energy. In addition, there is a need to advance thermal and electrical energy storage as an intermittent to solar and wind energy.

7.6 Nonthermal Processing

A number of nonthermal technologies are emerging to produce quality foods by avoiding damage caused by high-temperature heat treatment, which is an energy-intensive process. These are high-pressure processing, ultrasound, irradiation, plasma treatment, pulsed and UV light. High-pressure processing (HPP) is a novel nonthermal food processing technique. It can inactivate pathogenic microorganisms in food at room temperature and extend the shelf life of foods. The main advantage is that it reduces the damage of the heat-sensitive food components caused by high temperatures (Huang et al., 2014). The food materials are sealed in packages and placed in an enclosed and insulated container. An ultrahigh pressure of 100–600 MPa is generated using a liquid as a transfer medium (Bermudez-Aguirre and Barbosa-Canovas, 2011), and the pressure in the container is rapidly and evenly transferred to the food product undergoing high-pressure treatment. The temperature of the product is raised with the increase of pressure (i.e., adiabatic process), thus a combination of heat and pressure could also be achieved. This shows synergistic effect in pasteurization and sterilization. However, a cooling system needs to be used if the process requires isothermal processing. This process has been developed primarily to produce high-quality fresh products. However, detailed comparisons of the capital and operating costs as compared to intensive thermal process needs to be explored. The effect of low-frequency (20–1000 KHz) high-power ultrasound on physical, biochemical, and microbial properties of foods have also been studied (Mason et al., 1996; Torley and Bhandari, 2007). The primary bases for the application of ultrasound power in food processing are the chemical and mechanical effects of cavitation (Terefe et al., 2015). Hall and Howe (2012) point out that utilization of hurdle technology by reducing the severity of one hurdle (i.e., intense thermal energy process or low-temperature freezing process) could be also an option to save energy. Many food products stored at room temperature could be developed without severe heat treatment and do not need to be stored at freezing temperature when preservation hurdles are used intelligently (Leistner, 1978). Since most of the novel or innovative processes are developed to address the quality improvement as alternatives, there is minimal data available on the comparisons of energy usage between conventional and new technologies.

7.7 New Heating Technologies

New heating technologies such as ohmic heating, pulsed electric field, microwave heating, superheated-steam heating, magnetic field heating, and infrared heating are also emerging in order to be more energy efficient and maintain product quality. Microwave heating of foods is attractive due to its volumetric origin, rapid increase in temperature, controllable heat deposition, and easy cleanup opportunities. The volumetric heat generation can significantly reduce the total heating time and energy, thus reducing severity of elevated temperature (Decareau, 1985). In addition, high heating efficiencies (i.e., 80% or higher) can be achieved. This technique is currently being used for a variety of domestic and industrial food preparations and processing applications. The very low time required to reach desired process temperature makes it preferable over conventional heating in pasteurization, sterilization, drying, and baking. Similarly, infrared heating in drying could heat the product directly instead of heating air as in conventional drying, which causes heat loss with the exit air (Hazarika, 2014). In the drying of apple slices, the energy cost could be saved by reducing drying time by up to 50% (Nowak and Lewicki, 2004).

In ohmic heating, alternating electric current is passed through food material and heat is internally generated within the material owing to its resistance to applied electric current. In conventional heating, heat transfer occurs from a heated surface or medium to the product by convection and conduction; thus, it is slow and can cause surface burning. Ohmic heating is volumetric in nature and has potential to reduce overprocessing. The energy conversion efficiencies are high (Lima, 2007). Ohmic heating shows potential in thermal processing operations, such as pasteurization, dehydration, blanching, evaporation, fermentation, and extraction (Knirsch et al., 2010). It could increase freeze-drying rate up to 25% and so saving energy could be possible by reducing processing time (Lima et al., 2002). Pulsed electric field is another nonthermal processing which requires lower cooling requirement (Hazarika, 2014). Rodriguez-Gonzalez et al. (2015) have compared four alternative new technologies i.e., HPP, membrane filtration (MF), pulsed electric fields (PEF), and ultraviolet radiation (UV) with conventional high-temperature short-time (HTST) processing for their energy evaluation. They reviewed, from published papers, three levels of energy evaluation for each technology including internal energy, applied energy, and consumed energy. The inactivation of *Escherichia coli* in apple juice was considered and it was concluded that MF and UV showed the potential to consume less specific energy than HTST, PEF, and HPP. In addition, the differences in energy consumption within each group of technologies illustrate that there is potential for improvement in the most new technologies.

Earlier food packaging materials used to provide only barrier and protective functions whereas now various types of active substances can be incorporated into the packaging material to improve its functionality and to give new or extra function. Recently, the concerns of the ecological dimension of packaging are emerging. This means that packaging has to satisfy the required physical, chemical, and biological criteria during their life cycle (active protection function) and once the original function has been fulfilled, the packaging should decay without polluting the environment (passive protection function). Food companies are exploring green manufacturing methods and sustainable or renewable packaging applications in order to reduce energy costs with environment protection (Rahman, 2014). Nachay (2008) reviewed different solutions to address the issues and points out that any single solution would not solve the problem immediately, but the proposed changes need to take place to compel more companies to save precious resources and to protect the environment. It is not easy and simple to be green, and innovation is the key to be successful in sustainability (Sauers and Mitra, 2009). It is important to initiate science-based standards within the food industry to communicate clearly with the consumers looking for green and sustainable products (Bruhn, 2009).

7.8 Energy from Food Wastes

Food production, processing, transportation, storage, distribution, and marketing have an impact on the environment and have potential to cause environmental pollution. Waste from the food industry carries waste disposal issues and can cause problems if not handled properly. In addition, unutilized waste is a loss of valuable biomaterials, and waste could be utilized to produce edible food, feed, or industrial chemicals or biochemical compounds (Kroyer, 1995). However, innovations in processing are required to develop new value-added products from waste and to reduce environmental impact. Kroyer (1995) reviewed the general characteristics and treatment operations of by-products, wastes, and effluents from different categories of the food processing industry and their utilization into different value-added products could reduce the detrimental environmental impact of waste. Hossain et al. (2014) reviewed possibilities of developing wide varieties of value-added products from date pits, which are waste from the date fruit processing industry. Anaerobic digestion can be used to produce energy from waste, and this could assist in achieving energy security in a sustainable manner (Hall and Howe, 2012).

Energy generation from the food industry waste could have potential to reuse the waste as a form of renewable energy. Currently, in many instances, there is uncertainty in global petroleum oil production, supply, and price. In addition, environmental concerns are also a major issue for fossil fuels. Over the next 25 years, conventional supply of oil and gas is unlikely to meet the growth of energy demand. In this regard, considerable attention has been given to the production of biodiesel (Chhetri et al., 2008). The main advantages are environmental benefits for their renewable sources, low pollutants emission, and mostly closed carbon dioxide cycles (Chhetri et al., 2008). Chhetri et al. (2008) developed a process of biodiesel (ethyl ester) from waste cooking oil collected from restaurants. They used ethyl alcohol with sodium hydroxide as a catalyst for the transesterification process. Hecht (2009) indicated that it is important to advance science and technology, implement government regulations and policies as well as green business, if we want to advance in sustainability.

7.9 Energy Saving by Selecting Efficient Meals

Carlsson-Kanyama et al. (2003) explained that life cycle changes could save energy when energy-efficient meals are selected. Energy inputs in food life cycles vary from 2 to 220 MJ per kg due to multiple factors related to animal or vegetable origin, degree of processing, choice of processing, and preparation technology and transportation distance. The total life cycle energy inputs for food per person vary from 13 to 51 MJ/day. It was identified that up to a third of the total energy input is related to the items with little nutritional value (i.e., snacks, sweets, and drinks). It is important to give attention to these items considering the environmental consequences and to compose a diet compatible with energy efficiency and equal global partition of energy resources (Carlsson-Kanyama et al., 2003).

References

Abdelaziz, E. A., Saidur, R., Mekhilef, S. 2011. A review on energy saving strategies in industrial sector. *Renewable and Sustainable Energy Reviews*. 15: 150–168.

Anderson, K., Strutt, A. 2014. Food security policy options for China: Lessons from other countries. *Food Policy*. 49, 50–58.

Basiry, M., Esehaghbeygi, A. 2012. Cleaning and charging of seeds with an electrostatic separator. *Applied Engineering in Agriculture*. 28: 143–147.

Beedie, M. 1995. Energy savings—A question of quality. *South African Journal of Food Science and Technology*. 48(3), 14 and 16.

Bermudez-Aguirre, D., Barbosa-Canovas, G. 2011. An update on high hydrostatic pressure, from the laboratory to industrial applications. *Food Engineering Reviews*. 3: 44–61.

Beuchat, L. R. 2000. Use of sanitizers in raw fruit and vegetable processing. In: *Minimally Processed Fruits and Vegetables*. Alzamora, S. M., Tapia, S. T., López-Malo, A. Aspen Publishers, Gaithersburg, MD. pp. 63–78.

Brabeck-Letmathe, P. 2013. Water scarcity and food security: A role for free trade? https://www.linkedin. com/pulse/20130814122257-230883806-water-scarcity-and-food-security-a-role-for-free-trade (accessed November 7, 2015).

Bruhn, C. M. 2009. Understanding 'green' consumers. *Food Technology*. 63(7): 41–46.

Carlsson-Kanyama, A., Ekstromb, M. P., Shanahan, H. 2003. Food and life cycle energy inputs: Consequences of diet and ways to increase efficiency. *Ecological Economics*. 44: 293–307.

Casani, S., Rouhanyb, M., Knochel, S. A. 2005. A discussion paper on challenges and limitations to water reuse and hygiene in the food industry. *Water Research*. 39: 1134–1146.

Chhetri, A. B., Watts, K. C., Islam, M. R. 2008. Waste cooking oil as an alternate feedstock for biodiesel production. *Energies*. 1: 3–18.

CODEX. 2001. Codex Alimentarius Commission, Codex committee on food hygiene. Proposed draft guidelines for the hygienic reuse of processing water in food plants. Joint FAO/WHO Food Standards Programme, 34th Session, Bangkok, Thailand.

Das, F. 2000. Integrated heating and cooling in the food and beverage industry. Centre for Analysis and Dissemination of Demonstrated Energy Technologies (CADDET) Newsletter, Number 2.

de Oliveira, A. B. A., Ritter, A. C., Tondo, E. C., Cardoso, M. I. 2012. Comparison of different washing and disinfection protocols used by food services in Southern Brazil for lettuce (*Lactuca sativa*). *Food and Nutrition Sciences*. 28: 28–33.

Decareau, R. V. 1985. *Microwaves in the Food Processing Industry*. Academic Press, New York.

Deza, M. A., Araujo, M., Garrido, M. J. 2003. Inactivation of *Escherichia coli* O157:H7, *Salmonella enteritidis* and *Listeria monocytogenes* on the surface of tomatoes by neutral electrolyzed water. *Letters in Applied Microbiology*. 37: 482–487.

Deza, M. A., Araujo, M., Garrido, M. J. 2005. Inactivation of *Escherichia coli*, *Listeria monocytogenes*, *Pseudomonas aeruginosa* and *Staphylococcus aureus* on stainless steel and glass surfaces by neutral electrolysed water. *Letters in Applied Microbiology*. 40: 341–346.

Dhillon, B., Wiesenborn, D., Sidhu, H., Wolf-Hall, C. 2012. Improved microbial quality of buckwheat using antimicrobial solutions in a fluidized bed. *Journal of Food Science*. 77: E98–E103.

Dwarakanath, C. T., Rayner, E. T., Mann, G. E., Dollear, F. G. 1968. Reduction of aflatoxin levels in cottonseed and peanut meals by ozonation. *Journal of American Oil and Chemical Society*. 45: 93–95.

Emer, Z., Akbas, M. Y., Ozdemir, M. 2008. Bactericidal activity of ozone against *Escherichia coli* in whole and ground black-peppers. *Journal of Food Protection*. 71: 914–917.

FAO Report. 2015. Towards a water and food secure future: Critical perspectives for policy-makers. Food and Agricultural Organization of the United Nations, FAO and World Water Council (WTO), Rome and Marseille.

Frankel, R. J., Phongsphetraratana, A. 1986. Effects of water reuse, recycling and resource recovery on food processing waste treatment in Thailand. *Water Science and Technology*. 18: 23–33.

Gibson, H., Taylor, J. H., Hall, K. E., Holah, J. T. 1999. Effectiveness of cleaning techniques used in the food industry in terms of the removal of bacterial biofilms. *Journal of Applied Microbiology*. 87: 41–48.

Gleick, P. H. 2000. *The World's Water 2000–2001: The Biennial Report on Freshwater Resources*. Island Press, Washington, DC.

Gray, N. 2014. The true cost of beef: Report warns of high environmental cost of beef production. http://www.foodnavigator.com/content/view/print/949132 (accessed August 1, 2014).

Guentzel, J. L., Lam, K. L., Callan, M. A., Emmons, S. A., Dunham, V. L. 2008. Reduction of bacteria on spinach, lettuce, and surfaces in food service areas using neutral electrolyzed oxidizing water. *Food Microbiology*. 25: 36–41.

Guzel-Seydim, Z. B., Greene, A. K., Seydim, A. C. 2004. Use of ozone in the food industry. *Food Science and Technology*. 37: 453–460.

Hajeb, P., Jinap, S. 2009. Effects of washing pre-treatment on mercury concentration in fish tissue. *Food Additives and Contaminants*. 26: 1354–1361.

Hall, G. M., Howe, J. 2012. Energy from waste and the food processing industry. *Process Safety and Environmental Protection*. 90: 203–212.

Hazarika, M. K. 2014. Energy-efficient food processing: Principles and practices. In: *Introduction to Advanced Food Process Engineering*. Sahu, J. K. ed. CRC Press, Boca Raton, FL. pp. 631–673.

Hecht, A. 2009. Government perspectives on sustainablity. *Chemical Engineering Progress*. 105, 41–46.

Hoekstra, A. 2015. Shrinking agriculture's water footprint. http://futurefood2050.com (accessed June 10, 2015).

Holah, J. T. 1992. Industrial monitoring: Hygiene in food processing. In *Biofilms: Science and Technology*. Melo, L.F., Bott, T.R., Fletcher, M., Capdeville, B. Kluwer Academic Publishers, Dordrecht. pp. 645–660.

Hossain, M. Z., Waly, M. I., Singh, V., Sequeira, V., Rahman, M. S. 2014. Chemical composition of date-pits and its potential for developing value-added product—A review. *Polish Journal of Food and Nutrition Sciences*. 64(4): 215–226.

Hricova, D., Stephan, R., Zweifel, C. 2008. Electrolyzed water and its application in the food industry. *Journal of Food Protection*. 71: 1934–1947.

Huang, H., Yang, B. B., Wang, C. Y. 2014. Effects of high pressure processing on immunoreactivity and microbiological safety of crushed peanuts. *Food Control*. 42: 290–295.

ILSI-Report. 2008. Considering water quality for use in the food industry, ILSI Europe Report Series, Belgium.

Jalava, M., Kummu, M., Porkka, M., Siebert, S., Varis, O. 2014. Diet change—A solution to reduce water use? *Environmental Research Letters* 9: 1–14 (article 074016).

Jurado-Alameda, E., Garcia-Roman, M., Altmajer-Vaz, D., Jimenez-Perez, J. L. 2012. Assessment of the use of ozone for cleaning fatty soils in the food industry. *Journal of Food Engineering*. 110: 44–52.

Kanda, H., Makino, H., Miyahara, M. 2008. Energy-saving drying technology for porous media using liquefied DME gas. *Adsorption* 14: 467–473.

Kannan, R., Boie, W. 2003. Energy management practices in SME-case study of a bakery in Germany. *Energy Conversion and Management*. 44: 945–959.

Karaca, H., Velioglu, Y. S. 2007. Ozone applications in fruit and vegetable processing. *Food Reviews International*. 23: 91–106.

Katherine, H., Ann, S., Miranda, S., Stefan, H. 2001. The challenge of waste minimisation in the food and drink industry: A demonstration project in East Anglia, UK. *Journal of Cleaner Production*. 9: 57–64.

Kim, C., Hung, Y. C., Brackett, R. E., Lin, C. S. 2003. Efficacy of electrolyzed oxidizing water in inactivating *Salmonella* on alfalfa seeds and sprouts. *Journal of Food Protection*. 66: 208–214.

Kirby, R. M., J. Bartram, R. Carr. 2003. Water in food production and processing: Quantity and quality concerns. *Food Control* 14: 283–299.

Knirsch, M. C., dos Santos, C. A., Martins, A. A., Vicente, O. S., Vessoni Pena, T. C. 2010. Ohmic heating—A review. *Trends in Food Science and Technology*. 21: 436–441.

Koopal, L. K. 1985. Physicochemical aspects of hard surface cleaning 1. Soil removal mechanisms. *Netherlands Milk and Dairy Journal*. 39: 127–154.

Kroyer, G. T. 1995. Impact of food processing on the environment an overview. *Food Science and Technology*. 28: 547–552.

Kudra, T. 2004. Energy aspects in drying. *Drying Technology*. 22(5): 917–932.

Leistner, L. 1978. Hurdle effect and energy saving. In: *Food Quality and Nutrition*. Downey, W. K. ed. Applied Science Publishers, London, UK. pp. 553–557.

Lima, M. 2007. Food preservation aspects of Ohmic heating. In: *Handbook of Food Preservation*. Rahman, M. S. ed. CRC Press, Boca Raton, FL. pp. 741–750.

Lima, M., Zhong, T., Rao, L. N. 2002. Ohmic heating: A value-added food processing tool. *Louisiana Agricultural Magazine*, Fall Semester.

Lund, D. B. 1986. Low-temperature waste-heat recovery in the food industry. In: *Energy in Food Processing*. Singh, R. P. ed. Elsevier, Amsterdam.

Masanet, E., Therkelsen, P., Worrell, E. 2012. Energy efficiency improvement and cost saving opportunities for the baking industry: An energy star guide for energy and plant managers. Lawrence Berkeley National Laboratory, Berkeley, CA. Report LBNL-6112E.

Mason, T. J., Paniwnyk, L., Lorimer, J. P. 1996. The uses of ultrasound in food technology. *Ultrasound and Sonochemistry*. 3: S253–S260.

Mauermann, M., Eschenhagen, U., Bley, T., Majschak, J. P. 2009. Surface modifications—Application potential for the reduction of cleaning costs in the food processing industry. *Trends in Food Science and Technology*. 20: S8–S15.

Maurice, J. 2013. New goals in sight to reduce poverty and hunger. *The Lancet*. 382, 383–384.

Michail, N. 2015. Food industry unprepared for water shortage. http://www.foodnavigator.com (accessed May 13, 2015).

Mujumdar, A. S., Law, C. L. 2010. Drying technology: Trends and applications in postharvest processing. *Drying Technology*. 3: 843–852.

Mysorewala, M., M. Sabih, L. Cheded, M. T. Nasir, M. Ismail. 2015. A novel energy-aware approach for locating leaks in water pipeline using a wireless sensor network and noisy pressure sensor data. *International Journal of Distributed Sensor Networks* (ID675454): 1–10.

Nachay, K. 2008. In search of sustainable. *Food Technology*. 62, 38–49.

Novak, J., Demirci, A., Han, Y. 2008. Novel chemical processes: Ozone, supercritical CO_2, electrolyzed oxidizing water, and chlorine dioxide gas. *Food Science and Technology International*. 14: 437–441.

Nowak, D., Lewicki, P. P. 2004. Infrared drying of apple slices. *Innovative Food Science and Emerging Technologies*. 5(3): 353–360.

Nozaic, D. J. 2000. Risk and safety associated with the reuse of food processing water. Water Reuse Symposium, Pretoria, South Africa.

Ong, K. C., Cash, J. N., Zabik, M. J. Siddiq, M., Jones, A. L. 1996. Chlorine and ozone washes for pesticide removal from apples and processed apple sauce. *Food Chemistry*. 55 (2): 153–160.

Park, Y. G. 2002. Effect of ozonation for reducing membrane-fouling in the UF membrane. *Desalination*. 147: 43–48.

Pascual, A., Llorca, I., Canut, A. 2007. Use of ozone in food industries for reducing the environmental impact of cleaning and disinfection activities. *Trends in Food Science and Technology*. 18: S29–S35.

Perera, C. O., Rahman, M. S. 1997. Heat pump drying. *Trends in Food Science and Technology*. 8(3): 75.

Perry, J. J., Yousef, A. E. 2011. Decontamination of raw foods using ozone-based sanitization techniques. *Annual Review of Food Science and Technology*. 2: 281–298.

Petrecca, G. 1992. *Industrial Energy Management: Principles and Applications*. Kluwer Academic Publisher, Norwell, MA.

Pordesimo, L. O., Wilkerson, E. G., Womac, A., Cutter, C. N. 2002. Process engineering variables in the spray washing of meat and produce. *Journal of Food Protection*. 65(1): 222–237.

Rahman, M. S. 2014. Innovations in food packaging. In: *Introduction to Advanced Food Process Engineering*. Sahu, J. K. ed. CRC Press, Boca Raton, FL. pp. 293–314.

Rahman, M. S., Perera, C. O. 2007. Drying and food preservation. In: *Handbook of Food Preservation*. Rahman, M. S. ed. CRC Press, Boca Raton, FL. pp. 403–432.

Randhawa, M. A., Anjum, M. N., Butt, M. S., Yasin, M., Imran, M. 2014. Minimization of imidacloprid residues in cucumber and bell pepper through washing with citric acid and acetic acid solutions and their dietary intake assessment. *International Journal of Food Properties*. 2, 93–99.

Rodriguez-Gonzalez, O., Buckow, R., Koutchma, T., Balasubramaniam, V. M. 2015. Energy requirements for alternative food processing technologies—Principles, assumptions, and evaluation of efficiency. *Comprehensive Reviews in Food Science and Food Safety*. 14: 536–554.

Saidur, R. 2010. A review on electrical motors energy use and energy savings. *Renewable and Sustainable Energy Reviews*. 14: 877–898.

Sauers, L., Mitra, S. 2009. Sustainability innovation in the consumer products industry. *Chemical Engineering and Processing*. 105, 36–40.

Scott-Thomas, C. 2014. Water challenges could restrict business growth, major companies predict. http://www.foodnavigator.com/content/view/print/987375 (accessed November 6, 2014).

Starling, S. 2015. 'Regulations and incentives' needed to reduce trillions of tonnes of hidden water use globally. http://www.foodnavigator.com (accessed January 9, 2015).

Strumillo, C., Lopez-Cacicedo, C. 1987. Energy aspects in drying. In: *Handbook of Industrial Drying*. Mujumdar, A. S. ed. Marcel Dekker, New York. pp. 823–862.

Terefe, N. S., Buckow, R., Versteeg, C. 2015. Quality-related enzymes in plant-based products: Effects of novel food-processing technologies part 3: Ultrasonic processing. *Critical Reviews in Food Science and Nutrition*. 55: 147–158.

Tippayawong, N., C. Tantakitti, S. Thavornun. 2008. Energy efficiency improvements in longan drying practice. *Energy* 33: 1137–1143.

Torley, P. J., Bhandari, B. R. 2007. Ultrasound in food processing and preservation. In: *Handbook of Food Preservation*. Rahman, M. S. ed. CRC Press, Boca Raton, FL. pp. 713–739.

Vurma, M., Pandit, R. B., Sastry, S. K., Yousef, A. E. 2009. Inactivation of *Escherichia coli* O157:H7 and natural microbiota on spinach leaves using gaseous ozone during vacuum cooling and simulated transportation. *Journal of Food Protection*. 72: 1538–1546.

White, R. R., Brady, M. 2014. Can consumers' willingness to pay incentivize adoption of environmental impact reducing technologies in meat animal production? *Food Policy*. 49: 41–49.

Wu, D., Chu, Y., Chen, J., Hong, J., Gao, H., Fang, Z., Shen, L., Lin, M., Liu, D., Ye, X. 2013. Quality monitoring for a water reclamation system in a mandarin orange canning factory. *Desalination and Water Treatment*. 51: 3138–3144.

Yang, H., Feirtag, J., Diez-Gonzalez, F. 2013. Sanitizing effectiveness of commercial "active water" technologies on *Escherichia coli* O157:H7, *Salmonella enterica* and *Listeria monocytogenes*. *Food Control*. 33: 232–238.

8

A General Model for Food Cooking Undergoing Phase Changes

Davide Papasidero,
Sauro Pierucci, and
Flavio Manenti
*Polytechnic University
of Milan*

Laura Piazza
University of Milan

8.1 Introduction and Scope

Cooking has been addressed to influence the human species, evolution, firstly due to the impact on the energy balance related to food processing and associated with higher digestibility of many foods (including starchy food and protein-rich food like meat). In their brilliant research paper, Carmody and Wrangham (2009) try to quantify the impact of cooking from the discovery of fire to the energetic aspects related to cooking, suggesting many factors promoting its benefits on human development from an anthropological perspective and through the analysis of very different scientific fields. Coming back to food engineering and science, food thermal treatments are involved in many transformations affecting

- Microbial and enzymatic inactivation, favoring food preservation
- Modifications in the structure of proteins and carbohydrates, favoring digestibility and texture evolution, with impact on tenderness (e.g., with meat) or crispiness (e.g., bakery products)
- Chemical reactions of food components, with impact on aroma, taste, flavor, and visual aspect

Within this context, many researchers are focusing their attention to model a tool for optimizing food processes. It is not only for the energy saving and economic aspects, but also to achieve the final food products' texture, flavor, and nutritional value. Modeling is also a valid instrument for better understanding the different scales of the cooking process. Quantitative models can be used for many activities and applications. A computational model is a computer tool that can make predictions of growth and inactivation kinetics, or predict dynamics over a food chain. One can also argue that a computer tool is not the model, but the mathematical equations that are implemented in the program. Others will state that the model is not the mathematical equations, but the set of assumptions that are made, that result in mathematical equations. A further conceptualization is that a model is a simplified representation of reality. This is defined as a set of assumptions that will result in mathematical equations, which can be programmed in a computer tool. A concise overview of models used in food science and engineering is given by Van Boekel (2008). Within this context, a brilliant overview from Zwietering and den Besten (2011) mentions the points a model can be helpful to

- Structurally store data
- Quantitatively describe specific phenomena
- Test significant kinetic differences
- Investigate mechanisms and correlations (to understand phenomena and/or detect markers)
- Design experiments and sampling plans
- Determine the order of magnitude of the effect of interventions
- Combine models to develop decision support systems for monitoring, controlling, and optimizing processes
- Combine various models within quantitative risk assessments
- Impress or even confuse

From this perspective, it is extremely important to define the appropriate degree of detail one needs to achieve. This has to be compared to the experimental possibilities, since specific experiments need to be designed to get a specific degree of detail. One good example of this kind is the use of X-ray microtomography as a tool for the prediction of the food transport and microstructural properties (Warning et al. 2014), as well as the use of mechanistic models to deduce them (Gulati and Datta 2015).

It is of utmost relevance to underline that food is usually very far from an ideal system: Both experimental techniques and models used for homogeneous systems as well as those for heterogeneous ones with two or three phases for many chemical systems could be useless for most of the practical applications (Ho et al. 2013). This leads to model's simplification and multiscale simulations with a hierarchy of interconnected submodels, where homogenization is almost mandatory for reasonable process viewpoints (Quang et al. 2011).

For this reason, some major assumptions have been considered in the development of the approach presented in this work, consisting in heat and mass balance equations for a homogenized multicomponent porous media undergoing heat treatment with phase transition. Some case studies are presented to support the validity of the approach.

8.2 Physical Model

8.2.1 Governing Equations

The purpose of this paragraph is that of introducing the physical model for the description of the temperature trend and the changes affecting the water component, considered relevant variables in many food thermal processing applications. In the general case of a frozen food, the process variables during the heating process are ice, liquid water, and vapor mass fraction and temperature. In case of nonfrozen food, the ice component can be considered null. Water vapor can be often assumed to be an ideal gas. As a major simplification to achieve a simple model that can cover the basic cases of food with a relevant content of water with respect to other components affected by phase changes (e.g., fats). Other process phenomena, such as fat migration and volume change, are considered to be negligible for many applications. Fat melting and migration can be included with an analogous approach by explicitly taking into account a fat concentration and the relative phase change temperature, according to the available data and references sources.

The resulting partial differential equation (PDE) system is built based on the conservation principles and takes into account an explicit formulation for the melting and evaporation terms, assuming a direct dependence on the heat flux and on the food matrix temperature. By doing so, one can identify different situations, occurring locally, based on the temperature in the matrix:

- $T < T_{m,\text{ice}}$. In this case, the matrix is assumed to rise in temperature without any phase change from solid to liquid water (melting), which is assumed not to occur, since the temperature is below the ice fusion one. Phase change dealing with fat melting can be taken into account by implementing a similar condition with respect to the melting temperature. Liquid water is not produced and temperature rises due to the heat flux. $T > T_{m,\text{ice}}$ and $< T_{v,\text{water}}$, $C_w^i > 0$. When the ice melting temperature is reached, ice starts to melt and liquid water replaces the ice. The system temperature remains constant locally. Liquid water starts to form, replacing the ice, until ice has completely melted. Evaporation has not started yet.
- $T > 0°C$ and $< T_{v,\text{water}}$, $C_w^i = 0$. Ice has completely melted; then, the temperature can rise due to the heat flux. Evaporation is negligible, and liquid water state change does not occur.
- $T > T_{v,\text{water}}$, $C_w^l > 0$. Assuming an atmospheric pressure, it is possible to consider a water boiling temperature of 100°C. When the boiling temperature is reached, it remains constant and the matter enters the evaporation regime. Liquid water starts to evaporate into water vapor, until it finishes.
- $T > T_{v,\text{water}}$, $C_w^l = 0$. The last case occurs when liquid water has completely evaporated. In this case, the temperature starts to rise again, approaching the oven one. The evaporative term is not present anymore in the energy balance.

The melting/freezing point, $T_{m,\text{ice}}$, which represents the chemical potential in equilibrium between liquid and solid phases, can be different depending on the food kind, due to the solutes present in the food matrix, causing a freezing point depression compared to the ideal case of pure water. The reason of this depression is the fact that solutes decrease the partial vapor pressure of the water in food, affecting the equilibrium condition that can be achieved only by a reduction in temperature (Sahagian and Goff 1996). The techniques to estimate the freezing point and tables with the values for different foods are reported in the book chapter by Rahman et al. (2009).

It is possible to take these phases into account with the introduction of two step functions for melting, κ_M and κ_I, and two step functions for evaporation, κ_T and κ_C. These functions are defined as follows:

$$\kappa_M = \begin{cases} 0 & \text{if } T < T_m \\ 1 & \text{if } T \geq T_m \end{cases} \tag{8.1}$$

$$\kappa_I = \begin{cases} 0 & \text{if } C_w^i = 0 \\ 1 & \text{if } C_w^i > 0 \end{cases} \tag{8.2}$$

$$\kappa_T = \begin{cases} 0 & \text{if } T < T_v \\ 1 & \text{if } T \geq T_v \end{cases} \tag{8.3}$$

$$\kappa_C = \begin{cases} 0 & \text{if } C_w^l = 0 \\ 1 & \text{if } C_w^l > 0 \end{cases} \tag{8.4}$$

All of the balances will include these functions to manage the three cases, as shown below.

The ice melting term I_m is then defined as the liquid water produced by the local heat flux melting the ice:

$$I_m = \frac{Q}{H_{\text{mel}}}, \tag{8.5}$$

where H_{mel} represents the enthalpy of ice melting. As a first approximation, it could assume a constant value of 335 kJ/kg, despite being a function of temperature (it decreases when temperature decreases). This is also a mathematic assumption to handle the melting time, that could also be addressed to a lower heat capacity or partial melting due to the nonhomogeneity. Several models have been proposed to describe the latent heat of ice melting in food as a function of temperature. Riedel (1978) provided a parabolic equation for its calculation:

$$H_{mel} = 334.1 + 2.05T - 4.19 \times 10^{-3} T^2 \tag{8.6}$$

while Schwartzberg (1976) introduced a formula accounting also for the water and ice content:

$$H_{mel} = L_w^0 - (C_w - C_i)(T_f - T). \tag{8.7}$$

The evaporation term I_v is then defined as the vapor produced by the local heat flux:

$$I_v = \frac{Q}{H_{ev}}, \tag{8.8}$$

where H_{ev} represents the enthalpy of evaporation of water, and is assumed to have a constant value of 2272 kJ/kg, due to the constant pressure hypothesis. In this way, the mass and energy balances, functions of the considered variables, can be written as follows.

8.2.1.1 Ice Balance

$$\frac{\partial C_w^i}{\partial t} = \kappa_M \kappa_I (-I_m) \tag{8.9}$$

8.2.1.2 Liquid Water Balance

$$\frac{\partial C_w^l}{\partial t} + \nabla \cdot \mathbf{n}_w^l = \kappa_M \kappa_I I_m + \kappa_T \kappa_C (-I_v) \tag{8.10}$$

8.2.1.3 Water Vapor Balance

$$\frac{\partial C_w^v}{\partial t} + \nabla \cdot \mathbf{n}_w^v = \kappa_T \kappa_C I_v \tag{8.11}$$

8.2.1.4 Energy Balance

$$\rho c_p \frac{\partial T}{\partial t} + \nabla \cdot (\lambda \nabla T) = \kappa_T \kappa_C I_v H_{ev} + \kappa_M \kappa_I I_m H_{mel} + R_{react} \tag{8.12}$$

For temperatures higher than 0°C ($\kappa_M = 1$) if ice is still present ($\kappa_I = 1$), and for temperatures higher than 100°C ($\kappa_T = 1$) and in case the liquid water is still present ($\kappa_C = 1$), the heat flux Q, equal to the conduction flux ($\lambda \nabla T$) (plus the convection term, here neglected for simplification due to the different orders of magnitude), brings the energy balance to approximately become:

$$\rho c_p \frac{\partial T}{\partial t} = 0 \tag{8.13}$$

thus representing the condition of temperature plateau. An effective diffusivity is introduced for the water and vapor fluxes, as a function of concentration, while the ice phase is assumed to be fixed.

$$\mathbf{n}_w^i = 0 \tag{8.14}$$

$$\mathbf{n}_w^l = D_w^l \nabla C_w^l \tag{8.15}$$

$$\mathbf{n}_w^v = D_w^v \nabla C_w^v \tag{8.16}$$

A reaction term, R_{react}, is provided to take into account the impact of some reactions to the energy balance. Those reactions could be endothermic (e.g., starch gelatinization and proteins denaturation) or exothermic (such as oxidations) (Rahman 2009a). To appropriately describe the kinetics of those reactions, it could be necessary to know (and model) kinetic laws functions of both temperature and water content (or water activity). Both variables are then relevant to be characterized in time and locality. Using few data points or global measurements could deeply affect the reliability of the model and its predictive capabilities. A first approximation for the model development could be the description of those variables from the energy viewpoint (then, represented by heat of reaction), without the use of concentration variables.

8.2.2 Initial and Boundary Conditions

8.2.2.1 Initial Conditions

The initial temperature and water concentration have to be measured accurately to provide enough information for the model solution. The initial water vapor concentration could be set as zero as a first assumption in case of ambient (or low) temperature. Otherwise, it could be calculated from a water vapor–liquid equilibrium relationship. In case of frozen food, the initial liquid water is set to a minimum value, since the ideal case of 100% ice is not realistic, with a maximum of 20–30% water still remaining in the liquid phase even at very low temperature [i.e., −40°C or below (Rahman 2009b)]. The ice fraction w_{ice}, takes this phenomenon into account, as well as the possibility to describe a partially frozen status due to a local temperature in the range of the melting one. Frozen food being initially in temperature range close to melting temperature could locally present a small quantity of liquid water. That liquid water fraction in icy phase could sometimes be used as a fitting parameter in order to better describe the melting time and temperature profiles. The prediction of ice content can be obtained using different methods. Sakai and Hosokawa (1984) grouped those methods according to the strategy they use: freezing point method, freezing curve method, methods using the Clausius–Clapeyron equation, and methods using enthalpy value. A detailed description of them can be found in the chapter by Rahman (2009b), which also reports the typical values of unfreezable water content in foods. A measurement of the total water content can be provided from a drying experiment with the standard procedures. One example of initial conditions for a (partially) frozen food is reported here:

$$\begin{cases} C_w^i = C_{w,0}^i = C_{w,0}^{\text{total}} W_{\text{ice}} \\ C_w^l = C_{w,0}^l = C_{w,0}^{\text{total}} (1 - W_{\text{ice}}) \\ C_w^v = 0 \\ T = T_0 \end{cases} \tag{8.17}$$

8.2.2.2 Boundary Conditions

The boundary condition for the energy balance should take into account the heat transfer by convection (i.e., by forced or natural convection in the oven) and radiation. For this reason, an overall heat transfer coefficient can be introduced due to the complexities that the rigorous dissertation requires. This permits to avoid the emissivity and view factor estimations, needed when dealing with heat irradiation (effectively included in the heat transfer coefficient). This is still not difficult to insert in the model with a boundary condition dependent to the fourth power of the irradiating surface temperature (Equation 8.18). This boundary condition permits to consider a homogeneous environment, whereas a coupled CFD approach with a detailed description of the oven geometries and characteristics can lead to more accurate boundary conditions, going to the detriment of the computational effort.

$$Q \cdot n = h \cdot (T_{ext} - T) + \varphi \sigma \left(T_{wall}^4 - T^4 \right). \tag{8.18}$$

The free liquid water is often assumed not to leave the surface of the food matrix, as well as the ice, especially in case a cooking pan is used, insulating from dripping. In this case:

$$\mathbf{n}_w^i \cdot \mathbf{n} = 0 \tag{8.19}$$

$$\mathbf{n}_w^l \cdot \mathbf{n} = 0. \tag{8.20}$$

Nonetheless, in some cases the local (internal, volumetric) or surface water evaporation (or ice sublimation) have to be taken into account until the boiling condition is reached. In the second case, one can assume that most of the evaporation occurring for nonequilibrium takes place in proximity of the surface, due to intense air convection at the surface. In both contexts, evaporation can have a specific volumetric term or a dedicated boundary condition. In the latter situation, the difference of water partial pressure with the equilibrium pressure can be the evaporation driving force. Otherwise, avoiding the description of a pressure field, often difficult to be included due to the lack of knowledge on the pore scale information and water activity, an effective evaporation boundary condition proportional to the water concentration can be included (Equation 8.21).

$$\mathbf{n}_w^l \cdot \mathbf{n} = K_{vap} C_w^l. \tag{8.21}$$

The water vapor leaves the surface due to concentration difference between the bulk and the food piece surface. Since the bulk concentration is considerably lower than the internal one, it can be neglected, bringing to the following formulation of the last boundary condition:

$$\mathbf{n}_w^v \cdot \mathbf{n} = K_m (C_w^v - C_w^{v,ambient}) \tag{8.22}$$

In this expression, K_m is an overall mass transfer coefficient in analogy with the heat transfer boundary condition, Equation (8.18). This condition is not applicable where the presence of a blocking surface (e.g., an oven pan) imposes a wall condition blocking the outflow due to direct contact. When the ambient vapor concentration is much less than that released from the food load (or than the equilibrium concentration between liquid and vapor water at the surface), it could be negligible for model purposes.

8.2.3 Material Properties

Due to the high nonhomogeneity of food systems, it is impossible, or at least very difficult (Datta et al. 2012), to get the exact food material properties for a specific sample, especially with a very complex food matrix like a meat pie. For this reason, it is common to find modeling and simulation works where the material properties are tailored for the specific case.

For the development of the current work, the homogenization approach has been considered. According to this approach, a "porous media formulation homogenizes the real porous material and treats it as a continuum where the pore scale information is no longer available." This approach defines the food matrix as a multiphase continuum, with material balances regarding some components in three phases: a "vapor phase," usually bringing to mass balances on water vapor and sometimes CO_2, a "liquid water phase," and a "solid" component is chosen as an averaged bulk material (Jury et al. 2007), sometimes including ice. The present work considered the use of well-known empirical correlations (Choi and Okos 1986a) for calculating effective properties for the food matrix (density, heat capacity, and thermal conductivity), based both on composition and on the local temperature. It is worth to noticing a big limitation of the approach from Choi and Okos, whose model is only for pure food components. For this reason, appropriate equations are provided after those equations to describe the case of multicomponent food with a porous structure. This is an enormous advantage when considering reacting mixtures that are the main reasons for this choice; indeed, one of the aims of this work is to provide a model ready to be applied to cases where chemical reactions bring substantial modifications to the sensorial profile, e.g., browning, starch gelatinization, and CO_2 formation. Macro-components like proteins, carbohydrates, ashes, fibers, fats are considered for the solid phase, but this approach can be extended to more specific species. Together with this, already introduced in the previous works (i.e., bread and meat modeling activities), the ice component is added, to specifically address the differences in considering water in the solid and liquid states.

8.2.3.1 Density

Density is one of the most important properties for food modeling and it is involved both in the mass and in the energy balances. In this work, there is a difference between the intrinsic compound density and concentration, as introduced before in the mass balances. In fact, the former is the mass per unit volume of the pure compound (Equation 8.22), while the latter is the compound mass per unit volume of the mixture (Equation 8.23) and considers the fact that the medium is made of three phases, with a certain porosity and pore saturation. The total mass for each component can be therefore deduced by integrating the mass concentration in the total volume domain.

$$\rho_j = \frac{m_j}{V_j} \tag{8.23}$$

$$C_j = \frac{m_j}{V} \tag{8.24}$$

The intrinsic densities of the macro-components are described as polynomial functions of temperature. The related mixture density is calculated with a parallel model weighted on the mass fractions:

$$\rho = \frac{1}{\sum \frac{\omega_j}{\rho_j}}. \tag{8.25}$$

These equations are valid for nonporous foods. Total (or apparent) porosity expresses the volume of fluid or void per material volume:

$$\varepsilon = 1 - \frac{\rho_{fluid}}{\rho_{tot}}. \tag{8.26}$$

In case of porous food, porosity needs to be included as follows (Michailidis et al. 2009a):

$$\rho = \frac{1-\varepsilon}{\sum \dfrac{\omega_j}{\rho_j}}. \tag{8.27}$$

8.2.3.2 Heat Capacity

Using a similar approach, heat capacity is first defined for the single components and then calculated using a mass fractions averaged mixing rule:

$$C_p = \sum C_{p,j}\omega_j. \tag{8.28}$$

It is important to underline that the mass fraction is considered as a local property and not a global one, in order to take into account the nonhomogeneity of the final product (e.g., it is affected by the local temperatures). In this sense, the definition can be deduced from the component mass of an infinitesimal, exploiting the aforementioned concentration definition, Equation (8.24):

$$\omega_j = \frac{\delta m_j}{\delta m} = \frac{\delta m_j}{\sum \delta m_j} = \frac{C_j \delta V}{\sum C_j \delta V} = \frac{C_j}{\sum C_j}. \tag{8.29}$$

8.2.3.3 Thermal Conductivity

Thermal conductivity follows the aforementioned approach. The pounded weighted average on volume fractions (parallel model), as defined by Choi and Okos, could be assumed to be adequate for the conductivity estimation.

$$\lambda = \sum \lambda_j \varphi_j. \tag{8.30}$$

For a more general purpose of describing porous media thermal conductivity, Carson et al. (2006) introduced a complex model for unfrozen porous foods. This represents a modified Maxwell model with a weighing parameter (f) to weigh the differences between different porous media, quantitatively representing the heat conduction pathways in the material. Due to the complexity of the approach, the authors prefer the reader to refer to the original paper or to the interesting review by Gulati and Datta (2013).

The chosen formula for the volume fraction of the compounds can be deduced starting from the infinitesimal volumes and substituting the concentration and density definitions (respectively Equations 8.24 and 8.23):

$$\varphi_j = \frac{\delta V_j}{\delta V} = \frac{\delta m_j}{\rho_j} \frac{1}{\delta V} = \frac{C_j \delta V}{\rho_j} \frac{1}{\delta V} = \frac{C_j}{\rho_j}. \tag{8.31}$$

The component's thermal conductivity is then calculated by using other polynomial expressions, which account for the temperature dependency (Gulati and Datta 2013).

Semiempirical and empirical models have been used to describe the thermal conductivity of frozen foods (Carson 2006; Choi and Okos 1986b; Gulati and Datta 2013) The application of a porous model for thermal conductivity has been successfully obtained by Hamdami et al. (2004), who used it for partially baked French bread as a high porosity model food.

8.2.3.4 Diffusivities

A reasonable averaged constant value for liquid water diffusivity could range between 10^{-9} and $10^{-7}\,\text{m}^2/\text{s}$, and could be used as a parameter to fit the total moisture loss from experiments.

Water vapor diffuses faster than liquid water. The chosen value for vapor diffusivity can usually range between 0.02 and $0.04\,\text{cm}^2/\text{s}$.

8.2.3.5 Process Specifics

The representation of the oven environment, that is, the air flow and temperature, affecting heat and mass transfer, could be made by choosing an oven model respecting the real geometries or, at least, simplified geometries approximating the real ones. Anyway, it is really uncommon to find applications of computational fluid dynamics (CFD) to baking processes, when focusing on the building of a food model (Chhanwal et al. 2011; van der Sman 2013). Indeed, this approach has some advantages. For instance, it allows taking into account nonhomogeneities in the airflow distribution and temperature, as well as possible local radiation effects due to particularly shaped resistances. At the same time, it also brings the need of more equations for the description of the fluid flow (with momentum, mass, and energy conservation) and a need for more finite elements, resulting in a higher computational effort. Finally, some special instrumental analyses should be performed to get reasonable details for the simulations.

For these reasons, it is more common to find a representation of the oven environment with the use of effective heat and mass transfer coefficients for the boundary conditions (Purlis and Salvadori 2009b).

8.2.3.6 Heat Transfer Coefficient

Heat transfer depends on the heating regime of the oven. In this case, forced convection is the predominant phenomenon for energy transfer. As introduced in Equation (8.18), an effective heat transfer coefficient has to be inserted in the model for the energy boundary condition. Literature studies for baking have typical values in the range between 10 and $20\,\text{W/m/K}$ (Nicolas et al. 2014; Purlis and Salvadori 2009b; Zhang and Datta 2006). This coefficient could come directly from the system Nusselt number, according to its definition:

$$Nu = \frac{hd}{\lambda}, \tag{8.32}$$

where h is the heat exchange coefficient, λ is the fluid thermal conductivity, and d is the characteristic length. The Nusselt number is generally deduced from semiempirical correlations as a function of the Reynolds (or Grashof, in case of natural convection) and Prandtl numbers, according to the particle shape.

In our case, heat transfer coefficient will be deduced from the experiments and used as a fitting parameter. The influence of different oven pans and of the measured oven temperatures can be applied to the boundary conditions so that the temperature fitting could be as general as possible. Different heat transfer coefficients may be chosen according to the considered oven pan and oven sides.

8.2.3.7 Mass Transfer Coefficient

A fundamental theory for the transport phenomena introduced the analogy in heat mass and momentum transfer (Bird et al. 2007). Chilton and Colburn proposed a quantification of this analogy which is named after them, expressed by the equation:

$$j_H = j_M = \frac{f}{2}, \tag{8.33}$$

where the first term is the heat transfer factor, the second one is the mass transfer factor, and the last one on the right is the Fanning friction factor. The heat and mass transfer factors are dependent on adimensional numbers (such as Schmidt, Sc, and Prandtl, Pr, numbers) and of transfer coefficients (heat, h, and mass transfer, K_m, coefficients, respectively). In cases of air–water mixtures (e.g., with drying), when Sc and Pr are approximately equal to each other (Sc $\approx$ Pr $\approx$ 0.8), the mass transfer coefficient can be deduced from the heat transfer one based on the following equation:

$$K_m = \frac{h}{\rho_{air} C_{p,air}},$$
(8.34)

where $\rho_{air} C_{p,air}$ are the density and the specific heat of air, depending on the process conditions. In general, the analogy allows to calculate one coefficient when the other is available.

A specific equation to calculate jM for different processes and food material has been introduced by Krokida et al. (2001) as a function of Reynolds number (*Re*). The values of the constants are reported in the book of Saravacos and Maroulis (2001) and a detailed discussion on the topic can be found in the chapter from Michailidis et al. (2009b).

8.2.4 Numerical Implementation

The presented PDE system can be implemented in a finite element software using some numerical precautions in order to solve the model.

8.2.4.1 Numerical Step Functions

As described before, four step functions are introduced in the model in order to directly take into account the evaporation and melting phenomena. Nonetheless, if those functions were written as in their original formula, it would be very hard to solve the PDE system. In fact, applying the spatial discretization and then Newton method-based algorithms to solve the PDE system, many solvers have to integrate the resulting ordinary differential equations (ODE) system for every time step. This presupposes the Jacobian to be calculable. In case of discontinuities, as for the analytical step functions (1.3) and (1.4), it is not possible to calculate the Jacobian in the discontinuity points; then, the Newton method is not applicable. For this reason, a numerical formulation for the step functions has been deployed (see the following equations). In this way, an analytical sigmoid-like expression substitutes the former formulations, eliminating the discontinuities and permitting the calculation of the derivatives and then of the ODE system Jacobian.

$$\kappa_M^{num} = \frac{1}{(1+\exp(-(T-T_{mel})/(k_m)))}$$
(8.35)

$$\kappa_I^{num} = \frac{1}{(1+\exp(-(C_w^i - C_{w,shift}^i)/(k_I)))}, \text{ with } C_w^i \geq 0$$
(8.36)

$$\kappa_T^{num} = \frac{1}{(1+\exp(-(T-T_{vap})/(k_T)))}$$
(8.37)

$$\kappa_C^{num} = \frac{1}{(1+\exp(-(C_w^l - C_{w,shift}^l)/(k_C)))}, \text{ with } C_w^l \geq 0$$
(8.38)

As functions of the parameters k_T and k_C, these expressions are more or less smoothed. The choice of these parameters strongly affect the numerical solution, since, even if the Jacobian become calculable, it could have some matrix elements with very high values (due to almost vertical trend and then to derivatives that tend to infinite), being almost numerically singular and bringing either to a need for a smaller time step or not solving at all. If the maximum iteration number or the minimum time step is reached, that brings to a termination in the model solution.

In addition, the analyses of the experimental temperatures can often suggest to use smoothed step functions to take into account the presence of "nonstraight plateau" regions in correspondence with the melting temperature, probably caused by nonhomogeneities, either in the formulation or in the distribution of compounds and heat.

8.2.4.2 Oven Temperature

The temperature trend in the oven should be measured and inserted for the respective boundary condition. The easiest way to insert the oven temperature in a model is that of interpolating or fitting (by regression) the experimental data with a reasonably easy-to-calculate form (e.g., with spline interpolation). For instance, in case of a cold oven being heated from ambient temperature to the set point temperatures of 180°C and 220°C, the temperature trend is definitely not constant, driving a change in time for the energy balance boundary condition. An exponential or polynomial expression that tends to the set point temperature (e.g., Equation 8.38) can be set in the model according to parameter regression from experimental data. In the cases of food cooking in an oven pan or subject to different heating sources/intensity from the different faces, it is possible to insert as many temperatures and heat transfer coefficients as it is reasonable to do. The uncertainty dealing with the recipe and the data for the food piece can make the team decide not to use the same assumption for each specific case.

$$T_{\mathrm{env}} = T_{\mathrm{start}} + \left(T_{\mathrm{sp}} - T_{\mathrm{start}}\right)\left[1 - \exp\left(-\frac{t}{\tau}\right)\right] \tag{8.39}$$

T_{env} is the average oven temperature, T_{start} is the initial oven temperature, while T_{sp} is the set point temperature, and τ is a parameter that takes into account the temperature/time dependence. A similar trend is seen from the use of third and fourth grade polynomials, giving oven temperatures as a function of cooking time.

8.3 Case Study: Bread Baking

8.3.1 Introduction

- Due to its wide consumption, bread is one of the most studied foodstuffs, whose estimated production is about 9 billion kg/year (Heenan et al. 2008). Bread baking is one of the most clear examples of a food process that involves several chemical and physical phenomena (Dewettinck et al. 2008; Gellynck et al. 2009; Mondal and Datta 2008). For this reason, the study of this process has been for decades at the attention of several researchers, that focus on the general problem or on specific issues such as the flavor generation (Mondal and Datta 2008), the browning development (Schieberle and Grosch 1989), the water and heat transport (Purlis and Salvadori 2007), the influence on oven design (e.g., for CFD-based design (Chhanwal et al. 2011; Purlis 2012) for a process perspective), and the volume expansion (Purlis 2011). One of the most important considerations on bread baking is that the final quality and the many phenomena influencing it are direct consequences of the heat and mass transfer regime. From this perspective, the baking operating conditions modify the state variables with a hierarchical dependency. As stated in Hadiyanto et al. (2007), some relevant quality parameters like crumb formation, softness and crispness, volume, color, and aroma are related to transformations which have a direct relationship with the heat and mass transfer problem. For this reason, the present paragraph focuses on the transport phenomena involved in the bread-baking phase, as a major example of a physics problem that affects the quality development. It is worth underlining that protein thermosetting, starch gelatinization, Maillard and caramelization reactions, etc. can be deduced with appropriate models only when the initial stage and the dynamic evolution of temperature and water content are sufficiently characterized. In case of adequate characterization, the presented model can be integrated with other models of the product properties transformations. To complete the modeling framework, this model is developed under some simplifying assumptions, which could provide more details when appropriately addressed and removed. Volume expansion is considered constant: this limits the transport phenomena since the distances are varying with time during baking. The use of the approximate final volume as the model volume could be a reasonable approximation, considering that most of the volume variation occurs in the first part of baking (Pojić et al. 2016).

- Starch gelatinization and gluten denaturation are endothermic reactions. In this case, the energy variation is considered to be mostly influenced by conduction and water evaporation, due to the difficulty of properly quantifing the impact of the reactions.
- The effective diffusivity takes into account several mass transport phenomena, including that related to the pressure (Darcy's flow). The accurate description of this would require the estimation of more parameters (e.g., gases permeability and viscosity), needing more data. Nonetheless, this would allow coupling with mechanical models to describe volume expansion, as shown in some references (Nicolas et al. 2014; Zhang and Datta 2006).
- The latter consideration could also be related to the texture dynamics and the related properties' variation, which could give interesting insights on the quality side.

8.3.2 Experimental Measurements

The validation of the bread baking model needed to perform baking experiments for getting temperature vs. time data and weight loss measurements. The baking test was repeated three times, with a couple of analogue cases and a third case with different initial weight for a sensitivity analysis. Since the experimental data are consistent between the series of experiments, only one configuration is presented and discussed in detail.

8.3.2.1 Bread Samples

Samples were prepared using a standard recipe for bread: wheat flour (100%), water (58%), salt (2%), dry yeast (2%). The flour composition is (g per 100 g): carbohydrates (70.8), proteins (12.0), fats (1.5), fibers (3), and water (12.7). Dough was made by mixing the ingredients manually, and then underwent double leavening process for a total time of about 1 h at ambient temperature. The individual sample of about 50 g (approximate half oblate ellipsoid, ca. 7 cm diameter and 4 cm height) was formed and placed on a grid covered by a piece of oven paper to hold the dough avoiding any drip on the oven base and minimizing the fluid dynamics and heat distribution effects of the support.

8.3.2.2 Baking Tests

The domestic oven (KitchenAid, USA) was preheated to the set point temperature of 200°C. Then, the grid with the sample was positioned in the central zone of the oven to achieve homogeneous air distribution. The sample was baked under forced convection ($v = 2$ m/s) for about 40 min, terminating when a golden-brown crust formed on the bread. The temperature was measured all along the test in the oven and inside the bread, while weight was measured before and after the baking process.

8.3.2.3 Temperature

The temperature trend for the bread center, crust, and for the oven is reported in Figure 8.1.

One can see that the oven temperature increases until the set point temperature is reached. The oven controls this parameter with an average oscillation of ±6.8°C in the steady state phase.

Bread core temperature grows slowly in the first 5 min (lag phase due to thermal gradient not yet established), then a ramp-like growth can be seen in the next 10 min, until a plateau phase is reached at about 97°C to 99°C. This trend can be found in other literature sources (Nicolas et al. 2014; Zhang and Datta 2006).

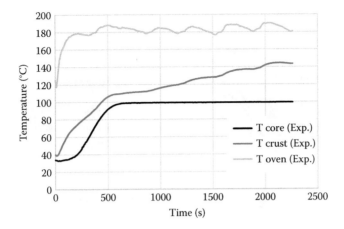

FIGURE 8.1 Temperature profiles for oven (cavity) and bread (sample).

It is worth underlining that the oven temperature starts from about 120°C due to the door opening, then quickly rises back to 185°C. In this sense, the main target of the experimentation is to analyze bread baking in a domestic oven. This can require preheating and door opening, not necessarily common in the industrial practice.

8.3.2.4 Weight and Water Loss

Baked bread weight measure (34.9 g from an average of three measurements) evidences a loss of about 15.1 g (30.2%) with respect to the initial dough mass (50 g). This is certainly due to the drying process, as expected and confirmed from literature (Purlis and Salvadori 2009a).

8.3.3 Results

The approach presented in Section 2 has been applied to model the experimental results from the presented test. Based on the operating conditions, the authors deduced the model parameters that represent them. Those are reported in Table 8.1.

8.3.3.1 Bread Temperature

Applying the model to the test case to simulate the bread core heating, the obtained results on temperature prevision (Figure 8.2) seem to follow the experimental trends, despite a noticed delay time. This is first due to the thermal properties calculation. In fact, the advantage of predicting many foods properties with simple formulas for the pure compounds and then applying a mixing rule (always to be verified

TABLE 8.1 Model Parameters

Parameter	Description	Value
H	Heat exchange coefficient	19 W/m/K
k_{mat}	Material exchange coefficient	1.e − 3 m/s
D_w^l	Liquid water diffusion coefficient	1.e − 8 m²/s
D_w^v	Water vapor diffusion coefficient	1.e − 4 m²/s
C_{shift}	Shift concentration for numerical step function parameter	0.012 kg/m³
k_T	Temperature numerical step function parameter	3°C
k_C	Concentration numerical step function parameter	1.5e − 3 kg/m³
τ	Oven temperature trend parameter	60 s

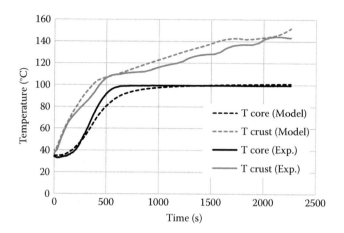

FIGURE 8.2   Temperature trend. Model and experimental data comparison.

in a wide range of mixing rules (Carson 2006)) becomes a disadvantage when compared to specifically measured properties.

Then, the assumption of constant volume can entail several consequences. One of them is the possible movement of the thermocouple, which could have modified the predicted core position. One other consequence is related to the fact that constant volume affects the prediction of density and thermal properties, influencing heat and mass transfer. Anyway, the simulated trend is reasonable (and reasonably general) enough to justify the approach. The experimental trend of the crust temperature is good as well, considering that a millimetre difference can lead to 10°C–20°C or more in the final crust temperature. For this reason, a sensitivity analysis on the probe height has been performed to have a best fit, resulting in a distance of 3 mm from the top as the best point to put the (model) probe. Since the point is reasonably close to the surface, the plateau phase is too short to be seen in this case, differently from other literature sources (Purlis and Salvadori 2009a). Anyway, in some cases and with a finer computational mesh, this can still occur.

8.3.3.2 Weight Loss

As discussed earlier, initial and final data on bread weight are available. The model results and the experimental weight data are reported in Figure 8.3. The bread water loss, taken into account from the model, is mainly responsible for weight loss. From the comparison of the two charts, an initial plateau trend can be evidenced: This respects the initial heating of the bread, when surface temperatures are lower than 100°C and surface drying is small. After that (approximately at 5 min from the baking start), the most important part of the drying process occurs, and weight loss appears to be almost linear till the end of the process.

Comparing the simulation data on weight loss to the experimental ones, the model seems to slightly overestimate the final weight loss. The impact of liquid water diffusion and some other parameters (e.g., heat transport coefficient and bound/free water ratio) influence both the data and the results, and have to be further investigated. Anyway, a possible interpretation for an overestimated weight loss is that the heat required for drying a porous material (i.e., the latent heat of vaporization to be used in the formula) can be higher than that of pure water due to the water–macromolecule interaction (e.g., between water and starch molecules), as found in literature (Wang and Liapis 2012). In addition, the aforementioned volume expansion can be, again, responsible for changes in the thermal properties, influencing the vaporization ratio prevision. Indeed, volume expansion is related to pore volume increase due to carbon dioxide production and expansion, while the solid phase density can be assumed to be almost constant. According to the change in volume, both density and conductivity of the mixture are modified.

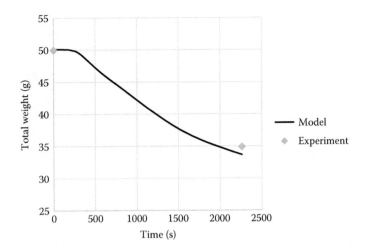

FIGURE 8.3 Weight loss. Model (solid line) and experimental data gray-shaded diamond comparison.

In particular, lower conductivity can imply less heat transfer, then less internal vaporization. At the same time, a longer distance for heat transfer would influence the water vaporization as well, affecting the heat transfer.

8.3.3.3 Mesh and Computational Effort

The physical model is made of three highly nonlinear PDEs in the variables water vapor, liquid water, and temperature. For this reason, a reasonably accurate mesh was chosen, with about 1200 elements and a number of freedom degrees of about 6200. The computational time with an Intel core7 i7-3770 processor was about 21 min, reasonable enough to perform a sensitivity study without taking a huge time amount. Some parameters have been chosen to be constant in order not to favor further nonlinearities in the equations and not to bring to singularity-like behaviors.

8.3.3.4 Conclusions

A model for the description of bread baking and its temperature and weight loss trends has been implemented, which consists a coupled PDEs system (energy conservation and water vapor and liquid conservation) and has been validated on experimental data (temperature, weight loss). An explicit formulation of water vaporization has been introduced and applied, as never been applied to bread baking. The considered material properties, depending on the bread macro-components mixture, can permit to represent structural properties variations with bread formulation and to consider the chemical and physical modifications occurring during baking, directly affecting the whole model and giving instruments to control quality issues (e.g., caramelization and Maillard reaction in the crust, starch gelatinization, and gluten coagulation in the crumb).

8.4 Case Study: Roast Meat

8.4.1 Introduction

Meat is a very common foodstuff in every part of the world. Indeed, meat is a nutritious food, rich in protein (typically ranging from 16% to 22% w/w) and essential nutrients including iron, zinc, and vitamin B12 (McAfee et al. 2010) which make it scarcely replaceable (Speedy 2003). Every day, tons of several meat kinds are globally consumed. Meat spoilage due to improper processing or conservation can then be a problem. Dealing with this topic, meat cooking is one of the processes that can extend the storage life and prevent damages to the consumer by the destruction of a considerable number of microorganisms, according to the time and temperature of the process (i.e., it is well known that bacteria are gradually destroyed at temperatures above 70°C. See, for instance, the paper from van Asselt and Zwietering (2006). Together with this, meat cooking involves some other transformations (Boles 2010):

- Water decreases, especially in the surface, leading to a low water activity with a consequent extended shelf life
- Texture results to be modified inducing a tenderness variation
- Meat proteins coagulate, altering their solubility
- Flavor gets intensified increasing the palatability
- Endogenous proteolytic enzymes are inactivated, preventing proteolysis to occur and the off-flavors to be produced
- Color changes due to chemical reactions in the colored compounds (e.g., myoglobins).

The study of the cooking process could then comprehend some of these transformations, which are dependent on the processing conditions (e.g., temperature, equipment, heat sources, and time) as well as on the raw material (e.g., meat kind, preprocessing, conservation, and preservatives).

Within this context, food modeling could be helpful to suppliers and equipment producers by giving them the possibility to assess the qualitative and quantitative description of specific markers for process control and optimization purposes (Papasidero et al. 2014). It can also enable model-based design of products with specific advantageous characteristics, to investigate the possible physical and chemical mechanisms that lay behind a process (Bessadok-Jemai 2013) and favor the engineering of highly automated, high-performance processing devices (Ryckaert et al. 1999).

Dealing with this concept, a series of computational and experimental studies related to conventional and combined oven are into development (e.g., Papasidero, Manenti, and Pierucci (2015). Since roast beef is a very common dish, it has been considered as a good base case for meat cooking.

Since almost all transformations are dependent on the temperature and on the water content of the roast beef, every model should be based on heat and mass transfer models. As a first assumption, one can think to simplify the problem by discarding the influence of the chemical reactions on the energy and mass balances (e.g., avoiding reaction terms, taking constant properties or temperature-dependent properties instead of composition-dependent properties). Then, all the further transformations can be described as hierarchically dependent on the heat and mass transfer problem. This approach, for instance, has been successfully applied to the bread baking process with interesting results (Hadiyanto et al. 2007). Heat and mass (water) transfer are then considered in the current work to develop a basic model to be further advanced in the future.

In the matter of basic models with heat and mass transfer, some literature has to be cited. First of all, the model by van der Sman (2007) applies a linearization of the Flory-Rehner theory for the free energy of elastic polymer gels to consider the moisture transport due to swelling pressure (following Darcy's law). In that paper, shrinking and deformations are not considered, evaporation is assumed to occur only at the meat surface, the permeability of meat is constant but anisotropic (fibers constitute a preferential direction for the moisture transport). This model has been complicated in a further works (van der Sman 2013) with the application of the model to industrial tunnel ovens. A very comprehensive model for meat cooking is provided in a work of Datta's group from Cornell University (Dhall, Halder, and Datta 2012): They describe beef patties as a multicomponent and multiphase system subject to evaporation and to moisture and fat unsaturated flow in a hygroscopic porous medium. One of the first works in modeling meat cooking, with particular reference to roast beef, is that of Obuz et al. (2002), where the authors consider water evaporation to be driven by the moisture difference between the meat surface and the air. They include latent heat of evaporation in the boundary condition for the energy balance by adding a term that is multiplied by the surface mass flow, coupling the mass and energy balances. Goni and Salvadori (2010) pay more attention on the difference between dripping loss and evaporative loss. The latter is dependent on equilibrium formulation, while the first is assumed to be dependent on the difference between the initial liquid water content and the so-called water holding capacity, function of the meat temperature, representing the response of the protein matrix (i.e., contraction) to heat. One of the most interesting aspects of their work is the experimental measure of these different losses, which involves the collection of the liquid loss in a roasting pan filled by water. Finally, Kondjoyan et al. (2013) further examine these phenomena, extending the investigation to different processes and meat pieces.

The model presented in this chapter is certainly less complicated than some of those mentioned earlier. Anyway, it constitutes a basis for further developments, and it is meant to be simple enough for possible control applications. Indeed, very complex models can require a considerable computational time. The compromise between accuracy and time consumption can be faced by choosing the appropriate assumptions.

8.4.2 Experimental Measurements

The experiments for the development and validation of the model have been carried out in a commercial domestic oven. A piece of bovine muscle whose dimensions were about that of a cylinder with 18 cm height and 10 cm diameter and that weighted about 1.4 kg was selected for the roasting process.

The procedure consisted of the following phases:

- The meat piece was extracted from the fridge and placed on a grid.
- It was then equipped with 3 thermocouples (T-type) to monitor the center of the roast, a point at 5 mm under the surface and a point in-between them (see Figure 8.4).
- The oven was then turned on with a set point temperature of 180°C. The initial oven temperature was 22°C.
- All the temperatures were continuously measured with thermocouples in the oven and recorded with a data logger.
- The cooking process was stopped when the core temperature reached 65°C.

The procedure has been replicated for three times to validate the experiment.

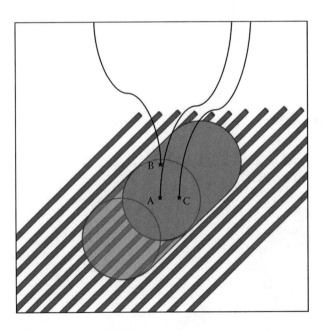

FIGURE 8.4 Schematic representation of the experimental layout. Grid and thermocouples in the three points for the central section: (A) Core, (B) below surface, and (C) intermediate position.

8.4.3 Results

The model has been implemented in a commercial software for the solution of partial derivative equation systems (PDE systems) with finite elements discretization (COMSOL-AB 2012). A cylinder that approximates the real dimensions represented the meat piece. Axial symmetry was then considered to simplify the model solution. The environment temperature was modeled with a piecewise function to approximate the average conditions in the oven.

The results related to heat transfer (see Figure 8.5) show a good trend for the description of the surface and core temperatures. Actually, the experimental difference between the intermediate point temperature and the core one seems not to be that relevant. By the way, the model is very sensitive to minimum variations, and it is possible that a variation of only 5 mm could be responsible of a temperature difference of 5°C–10°C. Except from that, the model seems to reasonably reproduce the experimental trend.

A reasonable prediction has also been reached dealing with water loss (see Figure 8.6). The model overestimates the weight loss by 25% compared to the experiment result. This can be attributed either to the mixture properties calculation or to the estimation of the mass transfer parameters. Additional attention should be paid on estimation of parameters. Furthermore, a more accurate description of the experimental results can be obtained through the addition of moisture transport with pressure gradients.

The model seems to reasonably agree with experiments, despite more attention on the thermal properties and diffusion coefficients estimation should be done. A good comparison with the existing literature models and experiments can give a great contribution to the development of the model, in order

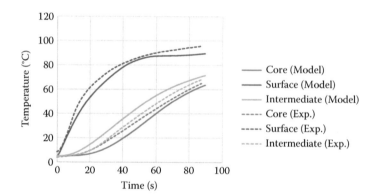

FIGURE 8.5   Roast meat temperature. Model and experiment comparison.

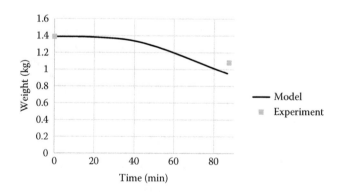

FIGURE 8.6   Meat water content. Model and experiment comparison.

to better represent several process conditions and meat kinds, with particular reference to the collagen denaturation, the diffusion of fat, and the combined oven processing. For instance, fat melting and diffusion can influence system by modifying the diffusion of the liquid phase, then constituted by a fat–water emulsion. Further developments could involve the color and flavor description through the use of kinetic models. These would require adequate experiments and techniques for the validation process.

8.5 Case Study: Frozen Meat Pie

8.5.1 Introduction

Frozen, multicomponent food is an interesting food category from the process viewpoint, since it needs for a preparation before the consumption, including a heating stage (with few exceptions of food to be consumed in the frozen stage, like popsicles and ice creams). This heating stage can be performed at ambient temperature or can require the use of specific equipment, often an oven. The present paragraph applies to the latter case through the selected test case, a pre-cooked frozen meat pie going through thermal processing. It is made of several ingredients, layers with different structure, both water and fatty components. Due to the complexity of the medium, the homogenization assumption can still be valid within a range of uncertainty.

8.5.2 Experimental Measurements

8.5.2.1 Experimental Setup

The heating experiments have been executed in a domestic oven (KitchenAid) in forced convection regime. A data logger has been connected to the thermocouples for the registration of the data into a flash memory drive. The thermocouples were located in the oven in several positions to evaluate the temperature distribution and possible relevant differences between the oven zones.

A grid has been arranged to hold the food load in the middle of the oven, then the thermocouples have been positioned in the food load. Formulation and specifics

The composition of the meat pie (macro-component average content) is reported in Table 8.2:

Based on the dosage below, the thermal properties have been estimated. The initial water content was adapted to be the same as the measured one, instead of being the one calculated from the label.

8.5.2.2 Temperature Measurements

The temperature trend for a meat pie case is reported in Figure 8.7. This chart includes the temperatures trends for three thermocouples, one in the food load center and the other two apart from it in the diagonal and perpendicular direction.

Analogous test cases have been performed showing similar trends with respect to Figure 8.7. Some of them have been discarded due to high nonhomogeneity and probable thermocouples movement during the heating phase. These two phenomena are assumed to influence the temperature expected from a melting process.

TABLE 8.2   Meat Pie Composition

Macro-Component	g per 100 g Total
Carbohydrates	12.0
Proteins	10.5
Fats	7
Fibers	1
Salt	0.7
Water	68.8

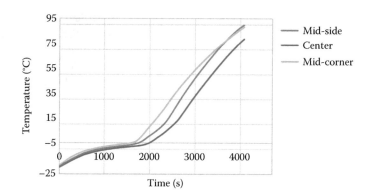

FIGURE 8.7   Temperature trend, meat pie.

8.5.2.3 Weight Loss Measurements

The meat pie weight has been measured at the end of the cooking. The initial weight for the presented case was 0.562 kg. After the heating process, the total mass measured was 0.496 kg, with a decrease in weight of 11.7% due to surface drying.

8.5.2.4 Model Implementation

Based on the operating conditions, the authors deduced the model parameters for the test case. The parameters are variable according to the operating conditions, including oven temperatures, initial ice fraction, cooking time, and oven pan. The initial temperature and the cooking time are taken from the experimental data from the beginning of the heating phase to the end of the experimental run, for a total cooking time of 1 h and 8 min. The heat transfer coefficients are coming from data fitting and experimental conditions. Within this context, the top part gets a higher heat flux of 14 W/m²/K, while the sides and the bottom of the food load get partial insulation due to the oven pan contact and/or for the limited air convection, then the chosen heat transfer coefficient is lower (8 W/m²/K). The average initial temperature is −19°C and the food load is assumed to be completely frozen (assuming that the temperature is low enough for the fat component to be solid, too). The model has been implemented in a finite-element modeling software to solve the PDEs for ice, liquid water, vapor, and temperature (energy).

8.5.3 Results and Discussion

The model temperatures are compared with the experimental ones in Figure 8.8. The temperature trend and the slope of the temperature curves in the heating part are consistent with the experimental data. The initial

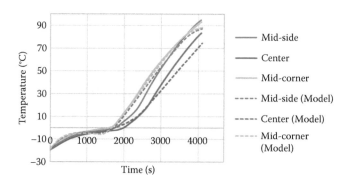

FIGURE 8.8   Temperature trend, top layer, experiment (solid lines), and simulation results (dotted lines).

temperature was a key factor and nonhomogeneities in that aspect (both giving different local temperatures and melting degree) made the data to be sometimes spread on a large time/temperature range. The possible thermocouples movement (due to the melting) and uncertainty on the location could have led to similar effects, leading to difficulties in the fitting of single points temperatures.

As described before, the heat transfer was assumed to be different on the top, bottom, and door, according to the experimental data. The authors referred to the experimental data set to have more information on the temperature distribution in the oven.

The dotted lines represent the model probes, while the solid lines represent the experimental data.

Due to the lack of measurements of the dynamic ice and water distribution, the model can give an estimation of the total (and local) ice and water content, predicting the final mass. In this case, a surface evaporation boundary condition (Equation 8.21) has been specified to get the appropriate weight loss, since the low surface temperature does not allow intense boiling, while the food load has a substantial decrease in weight.

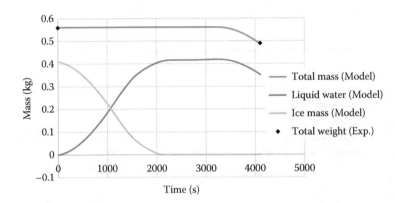

A physical model that takes into account the time/temperature evolution of a meat pie heating test in a domestic oven has been developed and validated upon experiment. A water transport and phase change model based on mass conservation equations has been coupled with the energy equation and brought to a reasonable trend from several experimental datasets in different conditions. The experimental conditions were described with effective heat and mass transfer coefficients to avoid the use of computational fluid dynamics (CFD), which would have brought to a high computational effort. The simulation parameters are deduced based on data fitting. The influence of the oven pan was taken into account by considering different heat and mass transfer boundary conditions. The nonhomogeneity of the food matrix led us to the need for simplification: Meat pie was always considered as a homogenized food load. This simplification could have led to under and overpredictions, as well as the uncertainty due to the thermocouples position: The melting of ice and fat could have influenced their position, giving fluctuations in the temperature trends. Further experiments in different experimental conditions could be useful for model integrations including other quality parameters. The simplification of the food matrices to include more than one homogeneous layer (e.g., with a homogeneous center of mashed potato and meat and an external layer of dough) can favor the representation of more processing details, despite the complication on the resulting mathematics. The latter approach could give a better description of the reality, if coupled with a good thermal and mass transfer properties estimation.

8.6 Conclusions

A general model for the thermal processing of food matrices based on heat and mass transfer equations and a PDEs approach have been presented, implemented on a commercial finite-elements software and validated on different test cases. The macro-component approach used for the research seems to be valid for the estimation of the transport properties, even with the simplification coming from the homogenization assumption. Certainly, a deeper investigation of the single phases could lead to accurate results and a deeper investigation of the thermal process. Nonetheless, since the purpose of the model is to be general enough to describe many cases, we agree that the achieved results are reasonably good for a further step of investigation toward the integration of these models with other equation systems and conditions. For instance, the model could be complicated to include microwave heating by adding a volumetric microwave source, depending either on a Lambert's law (then, on the incident radiation and a penetration depth) or on the direct solution of the Maxwell's equations for electromagnetic heating. A CFD model can also be included to better model the thermo-fluid dynamics of a real oven, going to the detriment of the ease of solution (and of the computational time). Design and control studies can benefit of these models, while the need for iteration of the process model solutions could need either further simplifications or the use of cluster computing. The model can be further integrated with color development kinetics (associated, for instance, to the Maillard reaction and tomato lycopene degradation) and, in general, with chemical kinetics and quality development models.

References

Bessadok-Jemai, A. 2013. Characterizing the drying kinetics of high water content agro-food particles exhibiting non-fickian mass transport. *Chemical Engineering Transactions* 32:1759–1764. doi: 10.3303/Cet1332294.

Bird, R. B., W. E. Stewart, and E. N. Lightfoot. 2007. *Transport Phenomena*. New York: Wiley.

Boles, J. A. 2010. Thermal processing. In *Handbook of Meat Processing*, 169–183. Ames, IA: Wiley-Blackwell.

Carmody, R. N. and R. W. Wrangham. 2009. The energetic significance of cooking. *Journal of Human Evolution* 57 (4):379–91. doi: 10.1016/j.jhevol.2009.02.011.

Carson, J. K. 2006. Review of effective thermal conductivity models for foods. *International Journal of Refrigeration* 29 (6):958–967. doi: 10.1016/j.ijrefrig.2006.03.016.

Carson, J. K., S. J. Lovatt, D. J. Tanner, and A. C. Cleland. 2006. Predicting the effective thermal conductivity of unfrozen, porous foods. *Journal of Food Engineering* 75 (3):297–307. doi: 10.1016/j.jfoodeng.2005.04.021.

Chhanwal, N., D. Indrani, K. S. M. S. Raghavarao, and C. Anandharamakrishnan. 2011. Computational fluid dynamics modeling of bread baking process. *Food Research International* 44 (4):978–983. doi: 10.1016/j.foodres.2011.02.037.

Choi, Y. and M. R. Okos. 1986a. Effects of temperature and composition on the thermal properties of foods. *Food Engineering and Process Applications* 1:93–101.

Choi, Y. and M. R. Okos. 1986b. Effects of temperature and composition on the thermal properties of foods. In *Food Engineering and Process Applications*, edited by M. LeMaguer and P. Jelen, 93–101. London: Elsevier Applied Science Publishers.

COMSOL-AB. 2012. COMSOL Multiphysics reference guide, version 4.3b, Burlington, MA, USA.

Datta, A. K., R. van der Sman, T. Gulati, and A. Warning. 2012. Soft matter approaches as enablers for food macroscale simulation. *Faraday Discussions* 158:435–459. doi: 10.1039/C2fd20042b.

Dewettinck, K., F. Van Bockstaele, B. Kuhne, D. V. de Walle, T. M. Courtens, and X. Gellynck. 2008. Nutritional value of bread: Influence of processing, food interaction and consumer perception. *Journal of Cereal Science* 48 (2):243–257. doi: 10.1016/j.jcs.2008.01.003.

Dhall, A., A. Halder, and A. K. Datta. 2012. Multiphase and multicomponent transport with phase change during meat cooking. *Journal of Food Engineering* 113 (2):299–309. doi: 10.1016/j.jfoodeng.2012.05.030.

Gellynck, X., B. Kuhne, F. Van Bockstaele, D. Van de Walle, and K. Dewettinck. 2009. Consumer perception of bread quality. *Appetite* 53 (1):16–23. doi: 10.1016/j.appet.2009.04.002.

Goni, S. M. and V. O. Salvadori. 2010. Prediction of cooking times and weight losses during meat roasting. *Journal of Food Engineering* 100 (1):1–11. doi: 10.1016/j.jfoodeng.2010.03.016.

Gulati, T. and A. K. Datta. 2013. Enabling computer-aided food process engineering: Property estimation equations for transport phenomena-based models. *Journal of Food Engineering* 116 (2):483–504. doi: 10.1016/j.jfoodeng.2012.12.016.

Gulati, T. and A. K. Datta. 2015. Mechanistic understanding of case-hardening and texture development during drying of food materials. *Journal of Food Engineering* 166:119–138. doi: 10.1016/j.jfoodeng.2015.05.031.

Hadiyanto, H., A. Asselman, G. van Straten, R. M. Boom, D. C. Esveld, and A. J. B. van Boxtel. 2007. Quality prediction of bakery products in the initial phase of process design. *Innovative Food Science and Emerging Technologies* 8 (2):285–298. doi: 10.1016/j.ifset.2007.01.006.

Hamdami, N., J.-Y. Monteau, and A. Le Bail. 2004. Heat and mass transfer in par-baked bread during freezing. *Food Research International* 37 (5):477–488. doi: 10.1016/j.foodres.2004.02.011.

Heenan, S. P., J. P. Dufour, N. Hamid, W. Harvey, and C. M. Delahunty. 2008. The sensory quality of fresh bread: Descriptive attributes and consumer perceptions. *Food Research International* 41 (10):989–997. doi: 10.1016/j.foodres.2008.08.002.

Ho, Q. T., J. Carmeliet, A. K. Datta, T. Defraeye, M. A. Delele, E. Herremans, L. Opara, H. Ramon, E. Tijskens, R. van der Sman, P. Van Liedekerke, P. Verboven, and B. M. Nicolai. 2013. Multiscale modeling in food engineering. *Journal of Food Engineering* 114 (3):279–291. doi: 10.1016/j.jfoodeng.2012.08.019.

Jury, V., J. Y. Monteau, J. Comiti, and A. Le-Bail. 2007. Determination and prediction of thermal conductivity of frozen part baked bread during thawing and baking. *Food Research International* 40 (7):874–882. doi: 10.1016/j.foodres.2007.02.006.

Kondjoyan, A., S. Oillic, S. Portanguen, and J. B. Gros. 2013. Combined heat transfer and kinetic models to predict cooking loss during heat treatment of beef meat. *Meat Science* 95 (2):336–344. doi: 10.1016/j.meatsci.2013.04.061.

Krokida, M. K., Z. B. Maroulis, and M. S. Rahman. 2001. A structural generic model to predict the effective thermal conductivity of granular materials. *Drying Technology* 19 (9):2277–2290.

McAfee, A. J., E. M. McSorley, G. J. Cuskelly, B. W. Moss, J. M. W. Wallace, M. P. Bonham, and A. M. Fearon. 2010. Red meat consumption: An overview of the risks and benefits. *Meat Science* 84 (1):1–13. doi: 10.1016/j.meatsci.2009.08.029.

Michailidis, P. A., M. K. Krokida, and M. S. Rahman. 2009a. Data and models of density, shrinkage, and porosity. In *Food Properties Handbook*, edited by M. S. Rahman, 759–810, Boca Raton, FL: CRC Press.

Michailidis, P. A., M. K. Krokida, and M. S. Rahman. 2009b. Surface heat transfer coefficient in food processing. In *Food Properties Handbook*, 2nd ed., edited by M. S. Rahman, 759–810, Boca Raton, FL: CRC Press.

Mondal, A. and A. K. Datta. 2008. Bread baking—A review. *Journal of Food Engineering* 86 (4):465–474. doi: 10.1016/j.jfoodeng.2007.11.014.

Nicolas, V., P. Salagnac, P. Glouannec, J. P. Ploteau, V. Jury, and L. Boillereaux. 2014. Modelling heat and mass transfer in deformable porous media: Application to bread baking. *Journal of Food Engineering* 130:23–35. doi: 10.1016/j.jfoodeng.2014.01.014.

Obuz, E., T. H. Powell, and M. E. Dikeman. 2002. Simulation of cooking cylindrical beef roasts. *Food Science and Technology* 35 (8):637–644. doi: 10.1006/fstl.2002.0940.

Papasidero, D., F. Manenti, and S. Pierucci. 2015. Bread baking modeling: Coupling heat transfer and weight loss by the introduction of an explicit vaporization term. *Journal of Food Engineering* 147:79–88. doi: 10.1016/j.jfoodeng.2014.09.031.

Papasidero, D., F. Manenti, M. Corbetta, and F. Rossi. 2014. Relating bread baking process operating conditions to the product quality: A Modelling approach. *Chemical Engineering Transactions* 39 (Special Issue):1729–1734.

Pojić, M., M. Musse, C. Rondeau, M. Hadnađev, D. Grenier, F. Mariette, M. Cambert, Y. Diascorn, S. Quellec, and A. Torbica. 2016. Overall and local bread expansion, mechanical properties, and molecular structure during bread baking: Effect of emulsifying starches. *Food and Bioprocess Technology* 9 (8):287–1305.

Purlis, E. 2011. Bread baking: Technological considerations based on process modelling and simulation. *Journal of Food Engineering* 103 (1):92–102. doi: 10.1016/j.jfoodeng.2010.10.003.

Purlis, E. 2012. Baking process design. In *Handbook of Food Process Design*, 743–768. Oxford, UK: Wiley-Blackwell.

Purlis, E. and V. O. Salvadori. 2007. Bread browning kinetics during baking. *Journal of Food Engineering* 80 (4):1107–1115. doi: 10.1016/j.jfoodeng.2006.09.007.

Purlis, E. and V. O. Salvadori. 2009a. Bread baking as a moving boundary problem. Part 1: Mathematical modelling. *Journal of Food Engineering* 91 (3):428–433. doi: 10.1016/j.jfoodeng.2008.09.037.

Purlis, E. and V. O. Salvadori. 2009b. Bread baking as a moving boundary problem. Part 2: Model validation and numerical simulation. *Journal of Food Engineering* 91 (3):434–442. doi: 10.1016/j.jfoodeng.2008.09.038.

Quang, T. H., P. Verboven, B. E. Verlinden, E. Herremans, M. Wevers, J. Carmeliet, and B. M. Nicolai. 2011. A three-dimensional multiscale model for gas exchange in fruit. *Plant Physiology* 155 (3):1158–1168. doi: 10.1104/pp.110.169391.

Rahman, M. S. 2009a. Food properties: An overview. In *Food Properties Handbook*, edited by M. S. Rahman, Chapter 1. Boca Raton, FL: CRC Press.

Rahman, M. S. 2009b. Prediction of ice content in frozen foods. In *Food Properties Handbook*, edited by M. S. Rahman, 193–206. Boca Raton, FL: CRC Press.

Rahman, M. S., K. M .Machado-Velasco, M.E. Sosa-Morales, and J. F. Velez-Ruiz. 2009. Freezing point: Measurement, data, and prediction. In *Food Properties Handbook*, edited by M. S. Rahman, 153–192. Boca Raton, FL: CRC Press.

Riedel, L. 1978. Eine formel zur berechnung der enthalpie fettarmer lebensmittel in abhangigkeit von wassergehalt und temperatur. *Chemie Mikrobiologie Technologie der Lebensmittel* 5 (5):129–133.

Ryckaert, V. G., J. E. Claes, and J. F. Van Impe. 1999. Model-based temperature control in ovens. *Journal of Food Engineering* 39 (1):47–58.

Sahagian, M. E. and H. D. Goff. 1996. Fundamental aspects of the freezing process. In *Freezing Effects on Food Quality*, edited by L. E. Jeremiah, 1–50. New York: Marcel Dekker, Inc.

Sakai, N. and A. Hosokawa. 1984. Comparison of several methods for calculating the ice content of foods. *Journal of Food Engineering* 3 (1):13–26. doi: 10.1016/0260-8774(84)90004-9.

Saravacos, G. D. and Z. B. Maroulis. 2001. *Transport properties of foods*. New York : Marcel Dekker.

Schieberle, P. and W. Grosch. 1989. Bread flavor. In *Thermal Generation of Aromas*, 258–267. Washington, DC: AIChE.

Schwartzberg, H. G. 1976. Effective heat capacities for the freezing and thawing of food. *Journal of Food Science* 41 (1):152–156. doi: 10.1111/j.1365-2621.1976.tb01123.x.

Speedy, A. W. 2003. Global production and consumption of animal source foods. *Journal of Nutrition* 133 (11):4048s–4053s.

van Asselt, E. D. and M. H. Zwietering. 2006. A systematic approach to determine global thermal inactivation parameters for various food pathogens. *International Journal of Food Microbiology* 107 (1):73–82. doi: 10.1016/j.ijfoodmicro.2005.08.014.

Van Boekel, M. A. J. S. 2008. Kinetic modeling of food quality: A critical review. *Comprehensive Reviews in Food Science and Food Safety* 7 (1):144–158. doi: 10.1111/j.1541-4337.2007.00036.x.

van der Sman, R. G. M. 2007. Soft condensed matter perspective on moisture transport in cooking meat. *AIChE Journal* 53 (11):2986–2995. doi: 10.1002/Aic.11323.

van der Sman, R. G. M. 2013. Modeling cooking of chicken meat in industrial tunnel ovens with the Flory-Rehner theory. *Meat Science* 95 (4):940–957. doi: 10.1016/j.meatsci.2013.03.027.

Wang, J. C. and A. I. Liapis. 2012. Water-water and water-macromolecule interactions in food dehydration and the effects of the pore structures of food on the energetics of the interactions. *Journal of Food Engineering* 110 (4):514–524. doi: 10.1016/j.jfoodeng.2012.01.008.

Warning, A., P. Verboven, B. Nicolai, G. van Dalen, and A. K. Datta. 2014. Computation of mass transport properties of apple and rice from X-ray microtomography images. *Innovative Food Science and Emerging Technologies* 24:14–27. doi: 10.1016/j.ifset.2013.12.017.

Zhang, J. and A. K. Datta. 2006. Mathematical modeling of bread baking process. *Journal of Food Engineering* 75 (1):78–89. doi: 10.1016/j.jfoodeng.2005.03.058.

Zwietering, M. H. and H. M. W. den Besten. 2011. Modelling: One word for many activities and uses. *Food Microbiology* 28 (4):818–822. doi: 10.1016/j.fm.2010.04.015.

III

Energy

9

Fossil Fuel

Emilio
Diaz-Bejarano and
Sandro Macchietto
Imperial College London

Andrey V. Porsin
*Boreskov Institute of
Catalysis and UNICAT Ltd*

Davide Manca and
Valentina Depetri
Politecnico di Milano - Italy

9.1 Energy Efficient Thermal Retrofit Options for Crude Oil Transport in Pipelines

Emilio Diaz-Bejarano, Andrey V. Porsin, and Sandro Macchietto

9.1.1 Introduction

Pipelines are used to transport large amounts of crude oil over large distances (either overland or subsea), representing the most economical alternative. Flow assurance faces two main problems: viscosity increase due to gradual cooling of the oil along the pipeline and fouling deposition. These problems are especially important in very cold environments (Russia, Alaska, North Sea, deep oceanic waters, etc.) and when dealing with nonconventional oils, usually heavy or extra-heavy oil and waxy oils. In many cases, the depletion of deposits in conventional oil reservoirs is gradually leading to more extraction of these types of feedstock from remote locations. All these situations result in pipeline transport difficulties such as increased pumping costs, reduced flow rates, and the possibility of flow inhibition or blockage, with potentially major economic impact (Correra et al., 2007; Martínez-Palou et al., 2011).

Crude oils are usually classified according to their American Petroleum Institute (API) gravity, which relates specific gravity of oil at 60 F to that of water at 60 F, as shown in this chapter, Table 9.1 (Riazi, 2005). Normally, the heavier the oil, the more viscous it is. Petroleum viscosity (μ) increases exponentially as temperature decreases. Pressure drop along a pipe of given diameter increases with viscosity.

277

TABLE 9.1 Categorization of Crude Oils according to API Gravity

	Light	Conventional	Heavy	Extra-Heavy
API	>35	35–20	20–10	<10
μ (cP) at room T	<10	10–100	100–10,000	Up to 10^6

Hence, if crude oil is cooled down to low temperatures it becomes very difficult to transport. The problem is even more significant when transporting heavy and extra-heavy oil, which present high viscosity (10^3–10^6 cP) even at mild temperature conditions, normally due to high asphaltenes content (Martínez-Palou et al., 2011). A pipeline designed for low viscosity oils may not be able to handle the transition to heavier oils without some adaptation.

Oils with high content of heavy paraffinic hydrocarbons (waxy oils) are also problematic (Aiyejina et al., 2011). At temperatures below the cloud point (or wax appearance temperature, WAT), waxes start solidifying, leading to a significant increase in viscosity. When cooling is restricted to the wall of the pipe, these solids precipitate forming a fouling layer. If cooling affects the bulk of the fluid, it may lead to flow blockage due to gelling (Arnold and Gebhart, 2000). The modeling of wax deposition has drawn the attention of many researches over the past years and many mathematical models have been presented (Kok and Saracoglu, 2000; Singh et al., 2001; Correra et al., 2007; Edmonds et al., 2008; Eskin et al., 2014). In the case of offshore subsea pipelines, the high pressure required by the deep location, combined with the cold temperature of deep waters, may lead to additional deposition problems due to the formation of hydrates that can cause pipe blockage (Mehta et al., 2006).

Current and under development techniques for drag reduction and wax prevention in pipelines are reviewed in the following section. In this work, a thermal strategy using single- and multiple point heating at intermediate locations between pumping stations is studied. A model of buried, insulated pipelines is used to estimate the pressure drop and heat losses between heating points. A novel flameless and modular catalytic heating technology is proposed as an alternative to current methods for heating at each of the heating stations. Two case studies based on real pipelines of great geopolitical and economic importance are then presented, together with results of a preliminary feasibility analysis, where the objectives are, respectively, (1) to manage the transition from light to progressively heavier oils and (2) to prevent wax deposition.

9.1.2 Drag Reduction and Prevention of Wax Deposition

The main methods for drag reduction and wax deposition prevention are listed in Table 9.2, together with respective advantages and disadvantages. The most widely used strategies for *drag reduction* consist on reducing the oil viscosity (Rana et al., 2007; Martínez-Palou et al., 2011):

- Thermal: Temperature conservation and/or increase (point heating or heated pipelines).
- Dilution or mixing with light products (e.g., lighter stock oil, condensates from natural gas production, etc.) by adding up to 20–30_{vol}% of diluent.
- Oil-in-water emulsification by dispersing oil in water (up to 30_{vol}%).
- *In situ* oil upgrading (i.e., near the reservoir) by producing synthetic oil with lower viscosity and density (API).
- Newer, promising alternatives for cold oil transportation include oil-in-water core annular flow, internal polymeric coatings, and drag reduction agents.

The main approaches used to *prevent wax deposition*, which also help reduce viscosity, are (Aiyejina et al., 2011):

- Thermal: temperature conservation and/or increase (heating).
- Addition of chemical inhibitors or pour point depressants.
- Other alternatives, still at the development stage, include the development of wax repellent surfaces and the so-called *waxy cold flow* technology (Merino-Garcia and Correra, 2008;

TABLE 9.2 Comparison between Techniques for Drag Reduction and Flow Assurance in Oil Pipelines

Technology	Main Application	Advantages	Disadvantages
		Nonthermal Techniques	
Oil-in-water emulsification	Drag reduction	Reduced viscosity No need of high temperature No impact on pipeline infrastructure Ability to handle turndown, start-up, or shutdown scenarios	Large water demand (20%–30%). Potential corrosion problems Substantial investment in pumping and large pipeline due to increase in transport volume or reduced oil throughput Costly emulsification process: energy intensive, costly additives for emulsification and stability, and need of emulsifying infrastructure Potential difficulty of operations such as dehydration and desalting Stability issues for extra-heavy oils (further development required)
Dilution or mixing with light products	Drag reduction	Reduced viscosity w/o heating or with mild heating No impact on pipeline infrastructure Potential facilitation of operations such as dehydration and desalting Advantageous if light hydrocarbon produced on site	Large demand of light crude oils or condensate (20%–30%) Substantial investment in pumping and large pipeline due to increase in transport volume or reduced oil throughput Mixing infrastructure. Potential need of auxiliary pipeline if light oil needs to be reused (sent back) or transported from somewhere else Potential incompatibility problems leading to deposition
Partial upgrading	Drag reduction	Reduced viscosity and content of pollutants	Limited by location and need for refining infrastructure
Drag reducing additives (polymers, surfactants, fibers)	Drag reduction	Reduced friction between pipe and fluid No need of high temperature	Costly chemicals Limited performance with heavy oils with high asphaltene content Stability problems
Chemical inhibitors or pour point depressants	Wax prevention	Reduction of pour point, also leading to reduction of viscosity (drag reduction) Reduced need of heating	Costly chemicals Limited efficacy System specific
Wax repellent surfaces	Wax prevention	Removes need for inhibition and removal	Under development. Further understanding required
Cold flow	Wax prevention	Waxes transported as solid dispersion within fluid	Under development. Further understanding required
		Thermal Techniques	
Passive: pipe insulation—burial	Drag reduction and wax prevention	Cost-effective, widely used	Not enough on its own; to be used in combination with other techniques
Hot water: direct (annulus circulation) and indirect (pipe-in-pipe)	Drag reduction and wax prevention	Direct heating along the entire pipeline Low risk of cold spots Able to handle turndown, start-up, or shutdown Core annular flow acting as lubricant	Requires water supply, heater, and pumping facilities Complex flow pattern formation and instability of core (addition of chemicals to improve stability) Injection and extraction units, corrosion problems

(Continued)

TABLE 9.2 (*Continued*)   Comparison between Techniques for Drag Reduction and Flow Assurance in Oil Pipelines

Technology	Main Application	Advantages	Disadvantages
Hot water: indirect	Wax prevention	Surface heating along the entire pipeline Low risk of cold spots Able to handle turndown, start-up, or shutdown	Requires water supply, heater, and pumping facilities Complex, costly configuration of pipe-in-pipe design
Electrical continuous heating (direct, induction, skin effect current trace)	Wax prevention	Uniform heat along entire length of the line Low risk of cold spots Ability to handle turndown, start-up, or shutdown Smaller pipeline (lesser cost) than water methods	Requires electricity generation facilities every few kilometers and fuel supply (e.g., auxiliary gas pipeline). Difficult to retrofit Heating in two steps: electricity production+dissipation leading to low overall performance
Point heating (heat exchanger/fired heater)	Drag reduction	Compact with high increase in temperature Well-known design Viscosity reduction	High pressure drop Location restricted to pumping/processing terminals No control of temperature along the line Poor temperature distribution leading to vapor production and fouling
Electromagnetic heater	Drag reduction	Viscosity reduction Compact point heating with high increase in temperature Location anywhere along the pipe	High energy intensive process Electricity supply Nonmetal pipe wall required (especial pipe design) Under development (initial stage)
Catalytic heating	Drag reduction and wax prevention	Minimum impact on pipe infrastructure. Low pressure drop High thermal efficiency. Low NO_x emissions and complete combustion Adaptability to different types of gas fuels Modular construction of heating stations with freedom to choose number of stations, size, and location Easy setup and dismantling. Flexibility for progressive adaptation to requirements, retrofit, seasons, maintenance, or even mobility	Heating intensity limited by maximum pipe wall temperature. Especial pipe designs may be required for heavy/extra-heavy oils Stations with large number of heaters may be required for thick oils Location only limited by gas supply as fuel for heaters Auxiliary gas pipeline may be required for multipoint heating Catalyst cost and limited life Need of displacement oil for shutdown/restart

Margarone et al., 2013), which consist of inducing precipitation in the "cold flow unit" to form a slurry stable enough to be transported along the pipeline without further deposition.

It is important to remark that there is not a unique solution to flow assurance and drag transportation issues. The best strategy depends upon the oil type, the location of the pipeline, environmental conditions, and available resources and infrastructure. Consequently, the above techniques are generally used in combination.

9.1.2.1 Thermal Techniques

The thermal approach is widely used for both drag reduction and wax prevention. The main objective of a thermal strategy for drag reduction is to maintain oil temperature high enough to avoid significant viscosity increase and pressure drop. The oil viscosity should be maintained at all times below the maximum recommended values: 500 cP in South America (Guevara et al., 1998) and 250 cP in Europe and North America (Dehkissia et al., 2004). The main objective of a thermal strategy for wax prevention is to maintain oil temperature above the oil WAT.

Pipe insulation is the simplest and most cost-effective strategy to minimize heat losses and keep the temperature as close as possible to that at which oil is produced and inserted into the line. For land pipelines, it is common practice to bury the pipeline as a means of achieving additional insulation. Insulation, however, is not usually sufficient and reheating is commonly needed to compensate for the heat losses along the pipeline (Martínez-Palou et al., 2011; Dunia and Edgar, 2012). Compared to alternative technologies, the full capacity of the pipeline is available to transport the actual oil (as opposed to water or diluents), but significant energy input may be required.

A number of techniques are available for pipeline heating (Table 9.2). These may be classified into *continuous* and *point* heating. For onshore (land) pipelines, heaters are typically installed in the same locations as the pumping stations. A schematic diagram of an onshore pipeline segment with a point heating and pumping station at either end is shown in Figure 9.1, together with typical temperature, viscosity, and pressure drop profiles along the pipeline. Pipeline heaters include (Martínez-Palou et al., 2011; Dunia and Edgar, 2012):

- Direct fired heaters: These have relatively low pressure drop, but poor temperature distribution (high temperature near the pipe wall), which leads to vapor production inside the pipes and requires later separation (i.e., a flash unit). The vapors thus separated may be used as fuel for the heater.
- Shell-and-tube heat exchangers: These have a higher pressure drop than fired heaters, but better temperature distribution. Steam is typically used as the heating fluid, normally on the tube side, to ensure a smaller pressure drop for the oil on the shell side. Fouling may be an issue (Macchietto et al., 2011).

The most common continuous heating techniques, mainly used for offshore (subsea) pipelines, involve (Easton et al., 2002; Dominguez, 2008; Martínez-Palou et al., 2011; SINTEF, 2013):

- Hot water system: This involves direct (annulus circulation) or indirect heating, both requiring a hot water supply, heater, and pumping system.
- Electrical heating: Some technologies are direct electrical heating, induction heating, and skin-effect current trace heating, all of which supply heat along the entire pipeline.

Alyeska in Alaska (Saniere et al., 2004) and Chad-Cameroon (Dehkissia et al., 2004; Dunia and Edgar, 2012) represent examples of pipelines with point heating at pumping stations for drag reduction. Mangala in India, the longest continuously heated pipeline in the world (Chakkalakal et al., 2014), is the most representative example of onshore heated pipeline for prevention of wax deposition. The *continuous* electrical heating in the Mangala pipeline is intended to maintain the oil temperature above WAT, but also to enable shutdown/start-up without *displacement oil* by heating the static oil within the pipeline above the wax dissolution temperature (WDT) before start-up. This feature represents the main advantage of a continuous heating for transport of waxy oils.

The choice of heating technology and the optimal structure of a heated pipeline will depend on the location of the pipeline, the availability of resources, the main objective (e.g., drag reduction or wax prevention), the

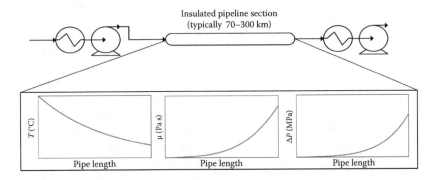

FIGURE 9.1 Schematic representation of a pipeline segment with point heating at pumping stations at either end and qualitative temperature, viscosity, and pressure drop profiles between stations.

various factors involved in the economic trade-off (value of throughput, costs of pumping, heating (installation and operation), and insulation (Dominguez, 2008)), and on the sensitivity of viscosity to temperature at the specific operating conditions, flow regime, and environmental conditions (Arnold and Gebhart, 2000).

Thermal-hydraulic pipeline models and optimization techniques are useful to analyze such systems. An example is the study by Dunia and Edgar (2012), who proposed the use of electromagnetic heaters for point heating between stations to reduce pressure drop for the Keystone pipeline. With a simple optimization formulation, the study showed that the length of pipeline segment between pumping stations could be potentially increased by 30% by optimally locating a single-point heating station.

Some disadvantages of current heating technologies (Table 9.2) are high pressure drop (e.g., with heat exchangers) and high energy requirements. In particular, methods that require indirect utilization of energy (e.g., electricity generation for electrical heating, steam generation for heat exchangers, or heating of water for hot water systems) have low overall energy efficiency. Also, the ability to control, modulate, and direct the amount of heating (heat flux) where it is most needed is important. Finally, an important further consideration (with all methods) is the ease or otherwise of modifying a design after first installation. This may arise when additional heating duty is required, for example, to transport heavier, more viscous oil. In such case, a different location to the original heating station may be better for this additional duty or even for the original station.

Difficulties in retrofitting (e.g., upgrading, dismantling, or relocating) installed heating stations in view of new utilizations typically lead to two situations: either significant overdesign has to be built-in when the pipeline is first built, at increased capital costs, or the risk is taken that less of 100% (possibly even no) throughput will be possible with the future utilizations.

The above limitations are leading to intense research for the improvement of current technologies and the development or application of other technologies, such as the previously mentioned electromagnetic heaters (Dunia and Edgar, 2012) or the flameless catalytic heaters proposed in this work.

9.1.2.1.1 Catalytic Heating as an Alternative Point Heating Method

As discussed by Diaz-Bejarano et al. (2016), an ideal point/multipoint pipeline heating technology should use a variety of fuels with high energy efficiency (hence low emissions and operating costs) and have low capital cost. It should also have a modular construction such that heating duty could be flexibly and incrementally set up, dismantled, and moved for retrofits, without requiring significant modifications to the oil pipeline infrastructure.

A design for industrial heating devices with the above features based on the flameless (catalytic) combustion of gas hydrocarbons was developed in the UNIHEAT Project (UNIHEAT, 2015), a joint project between Imperial College London (UK) and the Boreskov Institute of Catalysis (Russia) funded by the Skolkovo Foundation and BP Russia, aimed at increasing the energy efficiency in the downstream Oil and Gas sector (Macchietto and Coletti, 2015). The research generated a patent portfolio that covers the design of a flameless industrial heating device and various applications (Coletti et al., 2014; Kulikov et al., 2014; Nizovsky et al., 2014). The device, schematically shown in Figure 9.2a, is a closed, well-insulated system that completely surrounds the pipeline. Heating mainly occurs in a radiant section, in which fuel gas is burned in a catalytic bed. The combustion, which is clean and complete, occurs in catalytic regime, allowing flameless operation. The heat of combustion is directly transferred to the pipeline outer surface without compromising safety and without indirect carriers or heat transfer. A convective section may be optionally included to further recover energy from the hot combustion effluents, giving high overall energy efficiency (>70%). Lower alkanes (methane, propane, butane, natural gas) are the preferred fuel, although liquid fuels could also be used following a vaporization stage (Porsin et al., 2015). For applications in which a single heating unit is not sufficient to satisfy the duty requirements, a modular design permits the installation of multiple heaters in series, as schematically shown in Figure 9.2b. The main advantages and disadvantages of catalytic heaters for pipeline applications are indicated in Table 9.2. A particular advantage is the ability to turn up and down the heat duty in a module (via its control system) and to easily add (or remove) incremental modules (heat duty).

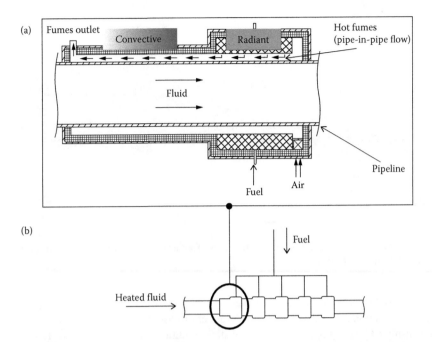

FIGURE 9.2 Schematic representation of (a) first design of catalytic heating unit; (b) multiple heating units in series installed on a pipeline. (Adapted from Kulikov, A. et al., Apparatus for heating of local sections of pipelines, Russian Patent No. 2564731; Application no. PCT/RU2014/000378, 2014.)

As part of the research, methods and simulation models for the design of such catalytic heaters for pipeline applications have been developed that take into account fuel types, operating ranges of the heaters, pipeline dimensions, operating conditions inside the pipeline, and physical properties of the pipelined fluid (not shown here due to confidentiality). A key parameter in the design is the maximum wall temperature allowable for the pipe, T_w. Pipelines are normally designed to work at temperatures up to 120°C (Arnold and Gebhart, 2000). However, higher temperatures may be achieved using special metallurgy and construction just for the length of pipe sections within the heating station. This may increase the capital cost of a heating station, although the lengths involved are typically short and it is expected this will have little impact on overall cost.

In this work, the models and design criteria for flameless catalytic heaters are used in conjunction with thermo-hydraulic pipeline models to demonstrate the applicability of the catalytic heating technology to pipeline drag reduction and wax deposition. The pipeline module predicts the heat and temperature losses and pressure drop along a (unheated) pipeline segment. The catalytic heater model calculates the temperature increase in a heating station, given pipe diameter, oil properties, and flow rate, the catalytic heater design and gas fuel used (or alternatively, it calculates the design, duty and number of modules required for a given oil temperature increase).

9.1.3 Thermo-Hydraulic Model of Pipelines with Intermediate Point Heating

The thermo-hydraulic model of a pipeline segment by Diaz-Bejarano et al. (2016) for unheated pipeline sections was used to predict heat losses, pressure drop, and temperature profiles along the pipeline sections. Four spatial domains are considered in the model (oil flow, pipe wall, insulation, and environment), as shown in the cross-sectional view at the bottom left in Figure 9.3. The physical properties of the oil are evaluated as function of key oil characteristics (e.g., API parameters (Riazi, 2005)) and the local temperature along the pipeline. The model permits choosing different configurations, including optional

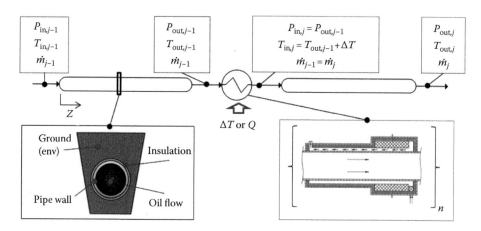

FIGURE 9.3 Schematic representation of the modeling framework for simulation of buried pipelines with intermediate point catalytic heating.

insulation layer and location (e.g., subsea, buried, or over ground), for which the corresponding boundary conditions are automatically set up. The main equations for the model configuration corresponding to a buried, insulated pipeline operating in steady-state conditions are summarized in Table 9.3. The symbols are defined in the Nomenclature section. It is noted that although here the model is used in steady state, the full model by Diaz-Bejarano et al. (2016) is dynamic and allows simulating transient scenarios.

The pipeline section model can be used as a module to simulate multisection pipelines. Each section may have a different configuration or be subject to different external cooling conditions. Models for the processing at intermediate stations (e.g., heating, pumping, and flow diversion) may also be included in a simulation of a whole pipeline, according to its structure.

The above models were used to simulate the effect of single- and multipoint heating at intermediate locations between pumping stations for two applications: drag reduction and wax prevention. For this

TABLE 9.3 Main Equations for a Buried, Insulated Pipeline Section in Steady-State Operation

Feature	Equation
Pressure drop inside the pipeline	$-\dfrac{dP(z)}{dz} = \dfrac{1}{4R_{w,i}} f(z)\rho(z)u(z)^2 = \dfrac{2\tau_w(z)}{R_{w,i}}$
Heat balance for the pipelined fluid (oil)	$\rho(z)u(z)\dfrac{\partial e(z)}{\partial z} = \dfrac{2}{R_{w,i}}\left(h(z)\big(T_w(z) - T(z)\big) + q''_{fr}(z) \right)$
Mass flow rate	$\dot{m} = \pi R_{w,i}^2 \rho(z)u(z)$
Energy balance for the insulation layer	$\dfrac{1}{r}\dfrac{\partial}{\partial r}\left(r\lambda_{ins}\dfrac{\partial T_{ins}(z, r)}{\partial r} \right) = 0$
Energy balance for the pipeline wall	$\dfrac{1}{r}\dfrac{\partial}{\partial r}\left(r\lambda_w\dfrac{\partial T_w(z, r)}{\partial r} \right) + q''' = 0$
Heat flux lost to the environment (for a buried pipeline)	$q''_{env}(z) = \dfrac{\lambda_{gr}\left(T_{ins}\big\|_{r=R_{ins,o}}(z) - T_{env} \right)}{R_{ins,o}\cosh^{-1}(\delta/R_{ins,o})}$

Source: Diaz-Bejarano, E. et al., *Appl. Therm. Eng.*, 105, 170–179, doi:10.1016/j.applthermaleng.2016.05.150, 2016.

purpose, a number of pipeline modules may be connected in series. For example, a simple system with two unheated, insulated pipeline sections in series and a single intermediate heating station is shown schematically in Figure 9.3. Continuity of mass flow rate, temperature, and pressure is imposed at the junction between two modules, with a heating station providing a certain heat duty and temperature increase input to the next section (similarly for pumping stations, not shown). For simplicity, here the pressure drop in the catalytic heating unit is neglected, as it is quite small relative to the pressure drop in the pipeline itself. In addition to characteristic oil properties, typical conditions imposed include inlet conditions, external temperatures, pipe segment diameters and lengths, and duties (or temperature increases) at heating stations. Typical constraints are the maximum total pressure drop allowable, the maximum temperature increase in each heating station, and maximum allowable bulk oil temperature.

As a result, it is possible to assess different heating strategies and establish whether they satisfy the system operation constraints. Once the location of and heat duty at each heating point has been identified, the methods and models for design of catalytic heaters are applied to establish the number of modules at each heating station and their detailed design. The overall problem may be solved either in sequence, in an iterative procedure, or simultaneously if some of the design variables are externally imposed.

In combination, the models represent a multiscale framework that goes from the detailed description of the heating elements to the calculation of the thermal-hydraulic behavior of the pipeline throughout long unheated sections. This framework, together with cost evaluation and optimization techniques, can be potentially used at different scales, from design and control of individual heating units to design of complete thermal strategies for entire pipeline systems.

9.1.4 Case Studies

Two case studies are presented based on realistic pipelines. The first considers the use of single and multiple heating points for the retrofit of an existing unheated pipeline based on the China leg of the Eastern Siberia–Pacific Ocean (ESPO) pipeline in Russia (Henderson, 2011; Kandiyoti, 2012). Here the main goal is to identify suitable retrofit options enabling the transport of substantially heavier oils. The second case study considers the flameless point heating method as an alternative to the current continuous electrical heating system used in the Mangala pipeline (Chakkalakal et al., 2014) where the main goal is prevention of wax deposition. In both cases, the above pipeline model is used to evaluate the thermal-hydraulic profile between heating points. Heating requirements at each heating station are identified and a design of modular catalytic heaters is developed. A preliminary comparison is made of the operating costs.

9.1.4.1 Drag Reduction: Retrofit of China–Russia Pipeline to Enable Transport of Heavier Oils

This case study considers the scenario of a gradual change in the quality of the transported oils from a light-conventional crude oil to increasingly heavier, more viscous ones. If a pipeline was originally designed for oil with low viscosity, it may not be able to handle such transition without some adaptation. Russia is an example where this kind of transition is likely to occur in the future. Russia initially focused on the exploitation of light oil reservoirs and constructed a huge pipeline network to export crude to Europe, China, and the Far East (Kononczuk, 2012). However, Russia has also the third highest reserves (after Canada and Venezuela) of heavy oils and bitumen, which remain almost untouched (Saniere et al., 2004). The China–Russia pipeline, the latest extension of the ESPO pipeline, was recently completed and allows export of 15 million tonnes per annum (Mtpa) of oil from Russia to China. This new infrastructure, of great economic and geopolitical importance (Henderson, 2011; Kandiyoti, 2012), was designed to transport light ESPO oil in cold conditions (Li et al., 2010).

A case study based on the China–Russia pipeline, presented by Diaz-Bejarano et al. (2016), addressed the scenario of a gradual transition from light ESPO oil to heavier feedstock. The oils and properties considered are shown in Table 9.4.

TABLE 9.4 Transition in Oil Properties

Oil	Original		Time		
	ESPO	1	2	3	4
API	34	24	22	20	18
$\nu_{38°C}$ (mm^2 s^{-1})	5.1	120	150	300	450

The 32″ China–Russia pipeline is buried and insulated to prevent extreme cooling of the oil, which could lead to wax deposition (the pour point of the oil is −36°C). This configuration allows cold transportation despite the extreme environmental conditions (−40°C and below in winter). The case study considers an unheated section of 300 km located between two pumping stations in the China section of the pipeline. The minimum oil inlet temperature to the China segment is about −6°C. The pipeline model in Section 9.1.3 was used to evaluate the temperature and pressure drop along the pipeline section. For the ESPO oil, maximum throughput (15 Mtpa) and worst-case cooling conditions, the maximum pressure drop along the 300 km section was estimated at 4.8 MPa. This value was taken as the maximum pressure drop achievable in the pipeline section with the current pumping infrastructure.

As the viscosity of the oil increases, it is no longer possible to maintain the maximum throughput whilst satisfying the pressure drop constraints. Diaz-Bejarano et al. (2016) studied the impact of increasing oil viscosity on throughput. In order to minimize throughput loss, point heating of the oil is proposed so as to decrease the viscosity of the oil and reduce drag along the line. In each heating station, the temperature increase (hence heat duty) is only limited by the maximum bulk oil temperature allowed, here set to 90°C. Several retrofit options were evaluated and the main results are reported in the following (the reader is referred to the cited reference for full details on all the retrofit cases considered).

9.1.4.1.1 Thermal Retrofit Options That Minimize Throughput Loss

The temperature, viscosity, and pressure drop profiles for some of the retrofit options discussed below are shown in Figure 9.4. The retrofit options are explained in a logical order for a gradual adaptation of the pipeline to the oil transition:

1. *Inlet heating at the pumping station*: A new heating station is located at the pumping station before the pipeline section inlet. This option is preferred since it takes advantage of infrastructure already available at the pumping station (access, utilities, etc.), thus reducing the capital cost of the new heating station. Heating the oil to the maximum allowable temperature of 90°C will result in the temperature, viscosity, and pressure drop profiles along the 300 km section shown by the dashed lines in Figure 9.4. This solution enables transport of oils with viscosities up to that of Oil 1 without reducing throughput. For heavier oils, throughput must be reduced to satisfy the maximum pressure drop requirement by 2.7%, 8.8%, and 15.8% for Oil 2, 3, and 4, respectively (equivalent to 0.4, 1.3, and 2.4 Mtpa for each case).

2. *Single intermediate point heating*: A reheating station (in addition to the inlet heating station) is installed at an intermediate location between the pumping stations. This option enables transport of Oil 2 without reducing throughput. Locations in the first half of the section were shown to require lower heat duty than locations at the pipeline midpoint and beyond. For instance, 7.4 MW is required if the station is located at 100 km from the inlet (Case B), while 10.2 MW is required if it is located at the midpoint (150 km from inlet, Case A). However, the crude outlet temperature from heating stations placed before the midpoint is close to the maximum allowable (90°C), which will pose a bottleneck for further upgrading of the heating station. Temperature and other profiles for Case A are shown in Figure 9.4 as a dash-dotted line. Retrofit options for other locations (Cases C, D, E) are reported in Diaz-Bejarano et al. (2016).

   For Oils 3 and 4, this solution is insufficient and a reduction in throughput is inevitable even when reheating the oil to 90°C. In both cases, the optimal location (minimum throughput loss) of

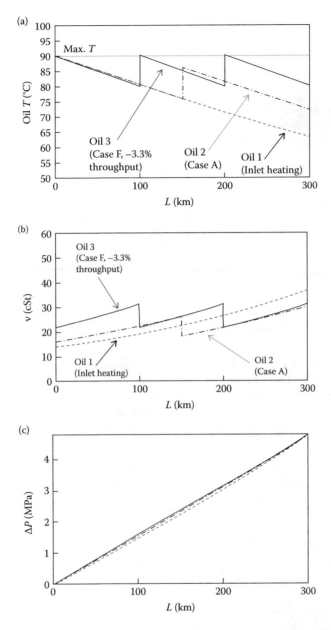

FIGURE 9.4 Temperature (a), viscosity (b), and pressure drop (c) along the pipeline section for Oil 1 (inlet heating), Oil 2 (Case A), and Oil 3 (Case F, 3.3% throughput loss).

a single station is in the vicinity of the section midpoint. The minimum throughput loss are 4.5% and 10.6% (equivalent to 0.675 and 1.59 Mtpa) for Oil 3 and 4, respectively, representing significant savings (recovery of throughput) compared to the *inlet heating* option—case 1) above.

3. *Multipoint intermediate heating*: Here, two or three heating stations are considered, placed at equal intervals. With multiple reheating points it is possible to maintain the temperature at an average higher value, reducing the overall viscosity and drag. With this solution, significant savings in throughput are obtained for Oils 3 and 4 with respect to the *single-point heating* option, albeit with greater capital and operating costs. With two intermediate heating points (at 100 and

TABLE 9.5 Thermal Power Consumption, Savings in Throughput, Economic Costs and Savings and Characteristics of Catalytic Heating Stations for Various Retrofit Options

Oil		2		3		4	
Case		A	B	F	G	F	G
No. of stations		1	1	2	3	2	3
Required duty Q/station (MW/station)		10.2	7.4	9.4	7.1	6.1	7
Savings in throughput w.r.t. inlet heating (%)		2.7	2.7	5.5	6.1	6.6	7.3
Savings in throughput w.r.t. inlet heating (Mtpa)		0.40	0.41	0.83	0.92	0.99	1.10
Total duty/station (70% eff.) (MW/station)		13.7	9.9	12.6	9.5	8.2	9.4
Cost of fuel/station (m$/year/station)		2.7	1.9	2.5	1.9	1.6	1.8
Cost of total fuel (m$/year)		2.7	1.9	4.9	5.6	3.2	5.5
Savings in throughput (m$/year)		134.7	138.0	279.4	309.7	333.3	370.3
Heaters per MW input (1/MW)							
$T_w=120°C$		128		159		159	
$T_w=150°C$		68		81		81	
Length heating station/1 MW (m/MW)							
$T_w=120°C$		103		119		119	
$T_w=150°C$		74		81		81	

200 km from inlet, Case F) the throughput loss is reduced to 3.3% and 9.2% for Oil 3 and Oil 4, requiring 9.4 and 9.2 MW per station, respectively. With three stations (at 75, 150, and 225 km from the inlet, Case G), the throughput loss is reduced to 2.7% and 8.5% for Oils 3 and 4, requiring 7.1 and 7.0 MW per station, respectively. The incremental improvement becomes smaller as the number of stations increase. The temperature and other profiles for Case F are shown in Figure 9.4 as a continuous black line.

The number of heating stations, required thermal duty, and savings in throughput achieved are summarized in Table 9.5 for the retrofits to manage the transition from Oil 1 to Oil 4 listed in Table 9.4 (the transition from ESPO oil to Oil 1 being managed by a new heating station at the pipeline inlet).

9.1.4.1.2 Catalytic Heaters as Point Heating Technology

The above results show that the best retrofit option using intermediate heating stations (heat duty, number of stations, and location) depends on the oil quality. A progressive change in oil from light ESPO oil to Oil 4 would require a gradual adaptation of the heating system as follows: first, the addition of a heating station at the inlet pumping station, then the addition of a single point heating station, followed by an upgrade of such station to higher duty, and finally the installation of new stations at different locations. In order to minimize the loss in throughput and additional duty, relocation (either complete partial) of the heat duty would be most advantageous.

Modular heating stations, such as those comprising of a number of catalytic heaters provide a flexible solution suitable for this retrofit. This solution would be particularly suitable for intermediate (remote) heating stations where auxiliary facilities (e.g., steam and electricity generation) are likely lacking. For the inlet heating, where a substantial increase in oil temperature may be required and auxiliary facilities may already exist, other options such as fired heaters or heat exchanger may be considered.

Individual catalytic heaters are of small size, and a number of units in series can be combined to provide the required duty at intermediate stations. Based on the duties required for each retrofit option, the configuration was determined of each intermediate heating station required at each time, and the duty (i.e., modules) to add, remove, and relocate to the most suitable location in each retrofit.

9.1.4.1.3 *Results and Discussion*

Taking into account the operating conditions at the heating stations (temperature, flow rate) and the oil properties, the maximum power achievable by the catalytic modules and the number of heaters required were calculated assuming in the first instance that the heaters are perfectly insulated and complete combustion is achieved. An average module size for the heater unit within the design range and a maximum wall temperature T_w of 120°C and 150°C were considered. The results are summarized in Table 9.5. With T_w=120°C, the typical constraint for standard oil pipeline construction, for Oil 2, about 128 heaters are required per MW of heat duty, with an approximate length of the heating station of 103 m/MW. This value increases to 159 heaters (119 m) per MW for Oils 3 and 4. For the heavier oils, with lower flow rate and poorer heat transfer characteristics, less power per unit is achieved in order to avoid overheating of the pipe wall. A relatively large number of modules are needed to satisfy the duty requirements, leading to a relatively long section of pipeline under heating (some hundreds of meters).

For a metallurgy in the heater units with a higher maximum allowable pipe temperature of 150°C, the number of heaters (and pipe length) is decreased to about half, resulting in a more compact design. Upgrading of the pipeline section where the heating station is to be installed should be considered against the additional capital cost.

The total duty needed and the cost of fuel, assuming a (conservative) 70% combustion efficiency, are reported in Table 9.5. The economic savings of managing to maintain throughput are also reported. Costs assume that the heating units are fuelled with natural gas, and consider Russian natural gas and average oil prices in October 2015 (IMF, 2015). The economic benefit of managing to maintain throughput is clearly dominant with respect to the operating cost of fuel. For example, for Case A, the total cost of fuel to the heaters is $2.7m/year, while saving 2.7% of the throughput leads to total (operating) savings of $134.7m/year. Subject to the capital cost of the heating stations, this indicates that the benefits should be high enough to provide a fast return on the investment.

The flexibility of this type of system permits adaptation to changing heat demand by installing additional heaters as required, and avoids unnecessary capital investments as a result of overdesign to cover all possible scenarios during the useful life of the pipeline. The unique capability to gradually adapt and retrofit the system should enable gradual investment based on shorter term, more predictable forecast of the future quality of the feedstock.

9.1.4.2 Transport of Waxy Oils: The Mangala Pipeline

The Mangala Pipeline, at 670 km is the longest continuously heated pipeline in the world (Pipelines International, 2010; Chakkalakal et al., 2014). It is a 24″ pipeline with capacity for 175,000 bpd (Hydrocarbons-Technology, 2012). It is heated with a skin-effect heat management system (SEHMS) continuous electrical heating along the entire pipeline, in sections of 18 km. Electricity supply for the heating system is produced using gas turbine engine generators at the beginning of each section, providing up to 40 W/m of heat flux. An auxiliary gas pipeline provides the fuel for the power stations. A schematic representation of the pipeline and auxiliary infrastructure is shown in Figure 9.5. The aim of this heating system is to keep the temperature of the waxy Mangala oil above its WAT (65°C) to prevent undesired deposition and eventual clog-up. The pipeline is insulated with a layer of foam and buried at least 1 m deep to minimize heat losses.

9.1.4.2.1 *Model Setup and Heating System Operation*

The specifications of the Mangala pipeline were extracted from various online sources (Cairn India, 2009, 2011; Pipelines International, 2010), including dimensions, insulation material and thickness, location of pumping stations, and crude oil properties. The continuous heating system (SEHMS) was simulated by including a heat source term in the wall model heat balance (q''') of the pipeline model in Table 9.3.

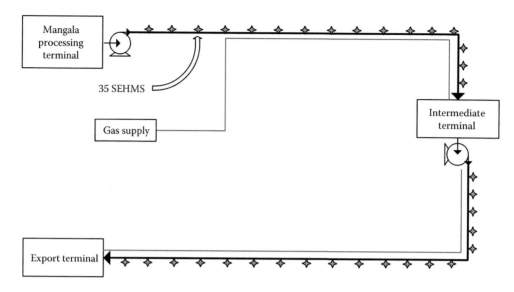

FIGURE 9.5 Mangala pipeline schematic. (Adapted from Cairn India, Cairn India: Rajasthan site visit. Cairn India Ltd, https://www.cairnindia.com, 2009.)

In this case study, it was assumed that the oil is produced and introduced into the pipeline at 65°C, and the SEHMS is used to maintain that temperature along the line (in practice, the heat input would be adjusted depending on the environmental conditions and throughput). It should be noted that, according to some references (Cairn India, 2011; IFC, 2013), inlet heating was proposed at the processing terminal and, as a result, SEHMS is only necessary for extreme operating conditions (very low flow, very cold environmental conditions) and for shutdown/start-up activities (i.e., heating of cold gelled oil above the WDT = 75°C).

9.1.4.2.2 *Alternative Multipoint Heating System*

A multipoint heating system is considered as an alternative to the SEHMS system used in the Mangala Pipeline. A total of 35 intermediate heating stations are considered (in the same locations every 18 km as the original electricity generation stations), with heating only supplied at the heating stations instead of continuously along the line. Each station comprises a number of catalytic heating units in series as described in Figure 9.2 b. The auxiliary infrastructure required to fuel these devices would be very similar to that required for SEHMS, in particular the auxiliary gas pipeline. The main difference is that SEHMS uses gas turbines in a first stage to produce electricity, which is then transmitted along the 18 km section dissipating energy to heat up the pipe wall. In the case of the heating elements, gas is directly used in a single stage to provide heat to the pipe wall. The former has the advantage of providing heat along the entire line, but the disadvantage of having lower energy efficiency due to the electricity generation step. The structure of the two heating systems is schematically presented in Figure 9.6 for comparison.

9.1.4.2.3 *Results and Discussion*

First, the pipeline model of Section 9.1.3 was used to determine the likely worst case scenario conditions (oil flow and ambient temperature) requiring full power of the continuous system (40 W/m) to maintain throughout the temperature of the oil at 65°C (assuming that this is the inlet temperature to the pipeline). The ambient temperature in winter was reported to reach values down to 5°C.

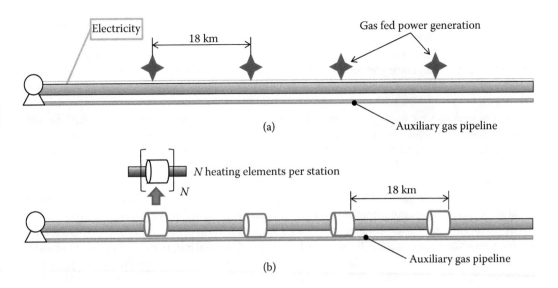

FIGURE 9.6 Continuous SEHMS system in Mangala Pipeline (a), and alternative multipoint system using catalytic heating in the same locations as original electricity generation stations (b).

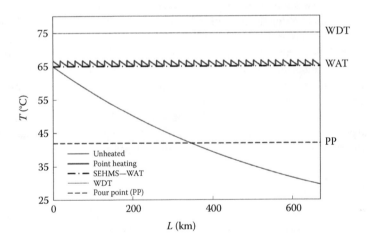

FIGURE 9.7 Temperature profile along pipeline for heated and unheated cases under worst case scenario conditions.

For this environment temperature, the flow rate for which SEHMS has to operate at full power to maintain the oil temperature was determined to be 80% of design flow rate. Under these worst-case conditions, in an unheated pipeline the temperature of the oil would drop along the line more than 30°C in spite of the insulation, as shown in Figure 9.7 (continuous black line), no doubt leading to wax deposition and, eventually, catastrophic flow inhibition (the Pour Point is about 42°C). In the case of SEHMS, the oil temperature is kept at 65°C along the continuously heated 18 km section (dot-dashed line in Figure 9.7).

With a multipoint heating system, heat is provided by a heating station at the entrance of each section to make up for the losses along the subsequent 18 km. The duty in each station is 0.744 MW/section. Using the pipeline model for an unheated 18 km section with an end point temperature of 65°C indicated that the temperature at each station must be raised from 65°C to 66.6°C. The temperature profile along the entire pipeline with multipoint heating has a characteristic saw shape (red line in Figure 9.7).

These conditions were used to determine the configuration of the multipoint heating stations with catalytic heating modules. For an average size of heating device 1 m long including radiant and convective sections and a maximum wall temperature of 120C, 72 units are required per station to match the heat provided by the SEHMS throughput per 18 km section. This would correspond to approximately 72 m of heated pipeline per station. Note that the heating station locations in this design were not optimized but simply correspond to the original gas turbine locations. Alternative, more compact designs could be considered, for example, with fewer intermediate stations in different locations.

The gas turbine engines used to supply electricity to the SEHMS system have been reported to achieve an electrical efficiency of 48.7% (GE Energy, 2008). The fuel, natural gas, is mainly composed of methane (LHV 50.05 MJ/kg). Assuming that electricity generation is the only source of inefficiency, the total power per station (at full capacity, 0.744 MW) with the current system is 1.57 MW, or a total 54.95 MW for the 35 stations.

With the point heating system, assuming perfect insulation and complete combustion in the heaters, the power consumption per station is 0.79 MW (27.65 MW for the 35 stations). This is about 50% less fuel than the equivalent SEHMS system. In practice, some sources of inefficiency are expected in the heaters. As in the first case study, assuming a conservative 70% efficiency the power consumption per station is 1.06 MW (37.10 MW for the 35 stations), which represents 32% less fuel than the equivalent SEHMS system. Total savings of 17.85 MW, or about $14.35 \times 10^6 \, m^3$ of methane per year, corresponds to a monetary saving of $3.16m/year [at the price of Russian natural gas in Germany in October 2015 of $6.5/MBtu (IMF, 2015)]. It is important to remark that these calculations correspond to normal winter operation (with 5°C ambient temperature and 80% throughput capacity) and do not account for the effect of seasons. Less heating would be required in summer.

On the other hand, start-up from cold conditions would not be possible with this technology, and low viscosity displacement oil would need to be introduced during shutdown to enable later operation restart. This is the main disadvantage with respect to the SEHMS system.

9.1.5 Conclusions

With current heating technologies, which are rather difficult to upgrade and retrofit, there are two alternatives: (1) to design and install a heating system for the scenario in hand, accepting the risk that it could become a bottleneck if the oil quality was to change significantly in the future, leading to loss in throughput and consequent economic losses; (2) to greatly overdesign and install heating stations at the outset (at larger capital cost), so as to be able to handle a wide range of scenarios which in fact might never occur or occur years later.

Two case studies show that point and multipoint thermal strategies based on flameless catalytic heaters represents a potential solution for the retrofit of oil pipelines, in applications aimed at reducing drag and preventing wax deposition. Modular stations comprising a number of catalytic heaters in series permit direct transfer of the heat of combustion from a variety of gas fuels to the pipeline surface, providing an energy efficient, NO_x free, and flexible alternative to current technologies.

In the first case study, the use of point heating demonstrated the ability to incrementally add heating capacity so as to enable the transport of increasingly viscous oils in the same pipeline. At each stage of the transition, the optimal configuration of the heating system, in terms of number of stations, duty and location, varies depending on the oil quality. The flexibility of catalytic heating modular stations has the significant advantage of permitting gradual addition, adaptation, or readjustment of the duty deployed as the oil quality changes. This militates against overdesign to cover all possible scenarios or the risk of throughput reductions. The economic benefit given by ensuring that throughput is maintained is likely to provide a fast return on the investment. The main limitation of the catalytic heating technology is the relative small size of individual units and the maximum pipe wall temperature. Upgraded pipeline metallurgy/construction in the heated sections would permit more compact designs.

In the second case study, the use of multipoint heating as alternative to continuous electrical system has been illustrated. Preliminary results show that it would be possible to save up to 32% of the fuel consumption with the proposed method, as a result of removing the electricity production stage. On the other hand, displacement oil may be necessary for shutdown and start-up activities, while a continuous electrical system is able to handle such operations with waxy oil in place.

Acknowledgment

This research was performed under the UNIHEAT project. The authors wish to acknowledge the Skolkovo Foundation and BP for financial support and Hexxcell Ltd for providing part of the modeling framework and software used in this work.

Nomenclature

API	API gravity
C_p	Specific heat capacity ($J\ kg^{-1}\ K^{-1}$)
E	Specific enthalpy ($J\ kg^{-1}$)
ESPO	Eastern Siberia–Pacific Ocean
f	Friction factor (no unit)
h	Convective heat transfer coefficient ($W\ m^{-2}\ K^{-1}$)
L	Pipeline length (m)
LHV	Lower heating value (J/kg)
MeABP	Mean average boiling point (°C)
$\dot{m}$	Mass flow rate ($kg\ s^{-1}$)
P	Pressure (Pa)
PP	Pour point
Q	Heat duty (W)
q''	Heat flux ($W\ m^{-2}$)
q'''	Heat generation ($W\ m^{-3}$)
R	Radius (m)
r	Radial coordinate (m)
SEHMS	Skin-effect heat management system
T	Temperature (K)
t	Time (s)
u	Linear velocity ($m\ s^{-1}$)
WAT	Wax appearance temperature
WDT	Wax disappearance temperature
z	Axial coordinate (m)

Subscripts

env	Environment
fr	Friction
gr	Ground surrounding the pipeline
i	Inner
in	Inlet
ins	Insulation
j	Pipeline module number
losses	Losses through pipe inner surface
max	Maximum
o	Outer
out	Outlet
vol	Volume
w	Pipeline wall

Greek letters

ΔP	Pressure drop (MPa)
ΔT	Temperature increase between pipeline modules
δ	Distance ground surface to pipe center (m)
ρ	Density (kg m^{-3})
λ	Thermal conductivity (W m^{-1} K^{-1})
μ	Dynamic viscosity (Pa s)
ν	Kinematic viscosity (mm^2 s^{-1})
$\nu_{38°C}$	Kinematic viscosity at 38°C (mm^2 s^{-1})
τ_w	Wall shear stress (N m^{-2})

References

Aiyejina, A., D.P. Chakrabarti, A. Pilgrim, and M.K.S. Sastry. 2011. Wax formation in oil pipelines: A critical review. *International Journal of Multiphase Flow* 37(7): 671–694.

Arnold, C.L., and L.A. Gebhart. 2000. Oil systems piping. In *Piping Handbook*, edited by M.L. Nayyar, 7th ed. New York: McGraw-Hill.

Cairn India. 2009. Cairn India: Rajasthan site visit. Cairn India Ltd. https://www.cairnindia.com.

Cairn India. 2011. Skin effect heat management system: Cairn India. Cairn India Ltd. http://www.cairnindia.com.

Chakkalakal, F., M. Hamill, and J. Beres. 2014. Building the world's longest heated pipeline a technology application review. In *2014 IEEE Petroleum and Chemical Industry Technical Conference (PCIC)*, San Francisco, CA: IEEE, September 8–10, 481–489. doi: 10.1109/PCICon.2014.6961915 481–489.

Coletti, F., E. Diaz-Bejarano, S. Macchietto, V.A. Kulikov, and A.V. Porsin. 2014. Heating device for high temperature fouling rig. Russian Patent No. 2564377; Application no. PCT/RU2014/000380 (May 26, 2014).

Correra, S., A. Fasano, L. Fusi, and D. Merino-Garcia. 2007. Calculating deposit formation in the pipelining of waxy crude oils. *Meccanica* 42(2): 149–165.

Dehkissia, S., F. Larachi, D. Rodrigue, and E. Chornet. 2004. Lowering the viscosity of Doba–Chad heavy crude oil for pipeline transportations—The hydrovisbreaking approach. *Energy and Fuels* 18: 1156–1168.

Diaz-Bejarano, E., A.V. Porsin, and S. Macchietto. 2016. Enhancing the flexibility of pipeline infrastructure to cope with heavy oils: Incremental thermal retrofit. *Applied Thermal Engineering* 105: 170–179. doi:10.1016/j.applthermaleng.2016.05.150.

Dominguez, J.C. 2008. Transporte de crudo pesado a través de oleoducto: evaluación de diferentes tecnologías. *Ingenieria Quimica* 465: 54–65 (in Spanish).

Dunia, R., and T.F. Edgar. 2012. Study of heavy crude oil flows in pipelines with electromagnetic heaters. *Energy and Fuels* 26(7): 4426–4437.

Easton, S., R. Sathananthan, and H. Subsea. 2002. Enhanced flow assurance by active heating within towed production systems. Offshore. http://www.offshore-mag.com.

Edmonds, B., T. Moorwood, R. Szczepanski, and X. Zhang. 2008. Simulating wax deposition in pipelines for flow assurance. *Energy and Fuels* 22: 729–741.

Eskin, D., J. Ratulowski, and K. Akbarzadeh. 2014. Modelling wax deposition in oil transport pipelines. *The Canadian Journal of Chemical Engineering* 92(6): 973–988.

GE Energy. 2008. GE to supply engines for Indian crude pipeline project. Downstreamtoday.com. http://www.downstreamtoday.com.

Guevara, E., J. Gonzalez, and G. Nuñez. 1998. Highly viscous oil transportation methods in the Venezuela oil industry. In *15th World Petroleum Congress*, Beijing, China: John Wiley & Son, October 12–17, 495–502.

Henderson, J. 2011. *The Strategic Implications of Russia's Eastern Oil Resources*. Oxford, UK: Oxford Institute for Energy Studies.

Hydrocarbons-Technology. 2012. Mangala development pipeline (MDP), Rajasthan, Gujarat, India. Hydrocarbons-Technology.com. http://www.hydrocarbons-technology.com.

IFC. 2013. Cairn India II: Environmental and social review summary. Project 26763. International Finance Corporation. https://disclosures.ifc.org/#/enterpriseSearchResultsHome/26763

IMF. 2015. Commodity Market Monthly October 2015. International Monetary Fund. http://www.imf.org.

Kandiyoti, R. 2012. *Pipelines: Flowing Oil and Crude Politics*. London: I. B. Tauris & Co Ltd.

Kok, M.V., and R.O. Saracoglu. 2000. Mathematical modelling of wax deposition in crude oil pipelines (comparative study). *Petroleum Science and Technology* 18(9–10): 1121–1145.

Kononczuk, W. 2012. Russia's best ally: The situation of the Russian oil sector and forecast for its future. *OSW Studies* 39: 1–67.

Kulikov, A., V. Rogozhnikov, A. Porsin, E. Diaz-Bejarano, and S. Macchietto. 2014. Apparatus for heating of local sections of pipelines. Russian Patent No. 2564731; Application no. PCT/RU2014/000378 (May 26, 2014).

Li, G., Y. Sheng, H. Jin, W. Ma, J. Qi, Z. Wen, B. Zhang, Y. Mu, and G. Bi. 2010. Forecasting the oil temperatures along the proposed China–Russia crude oil pipeline using quasi 3-D transient heat conduction model. *Cold Regions Science and Technology* 64(3): 235–242.

Macchietto, S., and F. Coletti. 2015. Innovation: Better together. *The Chemical Engineer* 893: 24–27.

Macchietto, S., G.F. Hewitt, F. Coletti, B.D. Crittenden, D.R. Dugwell, A. Galindo, G. Jackson, R. Kandiyoti, S.G. Kazarian, P.F. Luckham, O.K. Matar, M. Millan-Agorio, E.A. Müller, W. Paterson, S.J. Pugh, S.M. Richardson, D.I. Wilson, 2011. Fouling in crude oil preheat trains: a systematic solution to an old problem. *Heat Transfer Engineering*, 32 (3): 197-215.

Margarone, M., A. Bennardo, C. Busto, and S. Correra. 2013. Waxy oil pipeline transportation through cold flow technology: Rheological and pressure drop analyses. *Energy and Fuels* 27 (4): 1809–1816.

Martínez-Palou, R., M. de Lourdes Mosqueira, B. Zapata-Rendón, E. Mar-Juárez, C. Bernal-Huicochea, J. de la Cruz Clavel-López, and J. Aburto. 2011. Transportation of heavy and extra-heavy crude oil by pipeline: A review. *Journal of Petroleum Science and Engineering* 75(3–4): 274–282.

Mehta, A., J. Walsh, and S. Lorimer. 2006. Hydrate challenges in deep water production and operation. *Annals of the New York Academy of Sciences* 912(1): 366–373.

Merino-Garcia, D., and S. Correra. 2008. Cold flow: A review of a technology to avoid wax deposition. *Petroleum Science and Technology* 26 (4): 446–459.

Nizovsky, A.I., V.A. Kulikov, A.V. Porsin, V.N. Rogozhnikov, E. Diaz-Bejarano, and S. Macchietto. 2014. A method for flameless starting up of a catalytic device. Application no. PCT/RU2014/000377 (May 26, 2014).

Pipelines International. 2010. Heating up in India: The Mangala to Salaya oil pipeline. *Pipelines International*, March. http://pipelinesinternational.com.

Porsin, A.V., A.V. Kulikov, I.K. Dalyuk, V.N. Rogozhnikov, and V.I. Kochergin. 2015. Catalytic reactor with metal gauze catalysts for combustion of liquid fuel. *Chemical Engineering Journal* 282: 233–240.

Rana, M.S., V. Sámano, J. Ancheyta, and J.A.I. Diaz. 2007. A review of recent advances on process technologies for upgrading of heavy oils and residua. *Fuel* 86(9): 1216–1231.

Riazi, M.R. 2005. *Characterization and Properties of Petroleum Fractions*, 1st ed. Philadelphia, PA: ASTM.

Saniere, A., I. Hénaut, and J.F. Argillier. 2004. Pipeline transportation of heavy oils, a strategic, economic and technological challenge. *Oil and Gas Science and Technology* 59(5): 455–466.

Singh, P., R. Venkatesan, H.S. Fogler, and N.R. Nagarajan. 2001. Morphological evolution of thick wax deposits during aging. *AIChE Journal* 47(1): 6–18.

SINTEF. 2013. Gas and oil flows with pipeline heating. SINTEF Homepage. http://www.sintef.no.pipeline-heating.

UNIHEAT. 2015. UNIHEAT. http://www.uniheat-project.com.

9.2 Process Industry Economics of Crude Oil and Petroleum Derivatives for Scheduling, Planning, and Feasibility Studies

Davide Manca and Valentina Depetri

9.2.1 Introduction

"Contrary to what most people believe, oil supply capacity is growing worldwide at such an unprecedented level that it might outpace consumption. This could lead to a glut of overproduction and a steep dip in oil prices." This is what Leonardo Maugeri of the Harvard Kennedy School wrote in his report on "Oil—The next revolution" in June 2012.

Since then, a number of political, economic, and financial events have happened worldwide with significant consequences on the quotations of crude oil (CO) price. Maugeri (2012) observed an unprecedented upsurge of oil production and forecast a possible reduction of CO prices. Indeed, the price of CO ranged from 80 to 125 USD/bbl throughout 2012. Two years later, on January 2014, the CO price was still around 110 USD/bbl but a few months later, at the end of 2014, both Brent and West Texas Intermediate (WTI) reached their lowest prices since 2009 at 60 USD/bbl. In April 2015, Brent topped its highest price of 2015 at 65 USD/bbl. In November 2015, it dropped to 50 USD/bbl and closed the year at 35 USD/bbl. A few weeks later, on January 20, 2016, Brent price broke the psychologically threatening threshold of 30 USD/bbl and hit 27 USD/bbl, the lowest price since November 2003. This means that in less than 9 months (i.e., April 2015, January 2016) CO saw a 2.4-fold decrease in its value. An even more abrupt fall of CO quotations happened in 2008 triggered by the US subprime mortgage crisis. In the very first days of July 2008, Brent and WTI prices were both above 140 USD/bbl. At the end of 2008, 5 months later, CO quotations were below 40 USD/bbl after a steep and relentless price drop that dramatically affected the world economies for several quarters to come. The sudden price fall of last months of 2008 arrived unexpectedly after more than 12 quarters of almost continuous increases of CO price. Indeed, in March 2005, an oil barrel cost a little more than 50 USD, which increased to 79 USD in mid-2006 and to 90 USD in October 2007. In March 2008, CO reached 110 USD/bbl and topped 145 USD/bbl in July 2008, which is the Brent historical record. Worldwide, the finance experts were predicting bullish scenarios with probable breaking of the 180–200 USD/bbl threshold in a matter of few months. As reported a few lines above, less than 15 days later, on July 15, 2008, the growth trend of CO quotations suddenly inverted direction and started collapsing to touch 32 USD/bbl at the end of December 2008. These lows and highs in the CO quotations are triggered by a number of causes that range from geopolitical to natural events, from wars to discovery of new oil fields, from new extraction techniques to financial crises, and from economic indicators to targeted embargos. Most of the aforementioned causes can have either a positive or a negative fate and thus they increase even more the multifaceted panorama of possible future scenarios of CO quotations.

9.2.1.1 ID Card of CO

Google for "CO" and you will find the first 20–30 results pointing to CO prices, CO charts, and CO futures, which signifies that the focus of the Internet community is on economy and finance with chemistry, geology, exploitation, extraction, and refinement as secondary elements. However, before discussing in detail the CO markets and the economic elements of CO, it is worth presenting the main features that characterize this compound and allow understanding the major role played by CO on the industrial economy.

CO is also known as petroleum from the Latin *petra* (rock) and *oleum* (oil), i.e., "oil from the rocks." CO is a liquid that accumulates inside geological formations beneath the earth surface, with the lighter and heavier fractions that can be, respectively, gas and solid. The gas/liquid/solid phases of CO depend on the current temperature and pressure of the reservoir and the phase diagram of the mixture of compounds that constitute CO.

CO is regularly extracted through oil drilling in onshore or offshore oil wells and then distilled into various cuts:

- Refinery gases (bottled gas)
- Gasoline (fuel for cars)
- Naphtha (for petrochemical industry, to make chemicals)
- Kerosene (aircraft fuel)
- Diesel (fuel for cars, trucks, and buses)
- Fuel oil (fuel for ships, furnaces, and power stations)
- Bitumen (for roads and roofs)
- Heavy residues (for visbreaking, gasification, and other recovery processes)

Hydrocarbons account for 93%–99% of CO weight composition with minor contributions from nitrogen-, oxygen-, and sulfur-derived compounds and metal traces of copper, iron, nickel, and vanadium.

A barrel of CO (bbl) is equivalent to 42 gallons or 159 liters. Worldwide, the CO consumption is 80–90 Mbbl/d as a function of bearish/bullish industrial trends. The estimated oil reserves are 1200 Gbbl (i.e., 1.2 Tbbl) that is equivalent to 191 km^3. As the yearly consumption of CO is 4.9 km^3/y, one could extrapolate that in less than 40 years the CO reserves will be completely zeroed. However, besides the conventional oil reserves that account for only 30% of the known petroleum reserves, one has to consider also the heavy, extra heavy, and oil sands reserves that significantly contribute to the worldwide expected CO reserves. Oil sands are geological formations that trap the oil fraction into a semisolid mixture of sand and oil. Oil sands are primarily present in Canada and Venezuela, respectively, in the Athabasca and Orinoco reservoirs. Due to the high viscosity of oil sands (aka tar sands, bitumen), they must be heated and/or diluted before being treated, which causes an increase of the extraction costs. This is why only above a proper CO price threshold the processing of oil sands and extra heavy oils becomes economically feasible. Indeed, just before the August 2008 oil crisis, the investments in Canadian oil sands exploitation were huge and projected over long-term horizons.

By the summer of 2009, IEA (the Information Energy Agency) reported a total of 16 oil sands projects being suspended, delayed, or even aborted (Crooks 2011) as the CO feasibility threshold was over 100 USD/bbl, while the CO prices were as low as 30–40 USD/bbl.

Oil sand reservoirs are also called unconventional resources to differentiate respect to the traditional extraction process (i.e., by means of oil wells). Experts estimate that the oil sands reservoirs of Canada and Venezuela contain some 570 km³ of CO in the form of bitumen and extra heavy oil, which is almost twice the volume of world's conventional reserves. This would increase significantly the expected life of CO exploitation to more than 120 years (at present consumption rate). In addition, the increased capability of extracting oil from oil wells and finding new oilfields (often offshore and at higher depths) is moving forward the time when the last drop of oil will be distilled in the world. Paragraph 4.1 provides further details on the expected production curve of CO subjected to finite resources in a market economy.

Besides the oil sands, there is also another form of unconventional oil reserve, aka oil shale reserve. The hydrocarbons trapped in the oil shale reserves are in contact with sedimentary rocks, but they did not reach the temperature and pressure conditions that in a million-year natural process allowed the fossilized organic materials (e.g., zooplankton and algae) to become conventional CO. Conversely, these hydrocarbons impregnate and constitute the shales and rocks as an organic solid, aka kerogen. Once extracted from the oil shales, the kerogen is artificially converted into shale oil by pyrolysis, hydrogenation, or thermal dissolution. These processes can reproduce in a short time what happened naturally over geologic time scales. The largest reservoirs of oil shale are in the United States. The oil shale has been well known for centuries and found large application in the 19th century. Afterward, the discovery of large conventional oilfields reduced the interest in oil shale. Eventually, the significant increase of CO quotations at the beginning of 21st century led to a renewed interest in oil shale exploitation that was even more revived by the introduction of new extraction technologies. The opportunity of reducing the political and strategic dependency from OPEC (Organization of Petroleum Exporting Countries) pushed the United States to test and implement new extraction methods such as advanced hydraulic fracturing (aka fracking) and horizontal drilling. Fracking allows also to increase the yield of older wells and extract natural gas from the wellbore.

9.2.1.2 Chapter Structure and Motivation

This chapter presents and discusses the economic features and forecasting techniques of CO and petroleum derivatives for process systems engineering (PSE)/computer-aided process engineering (CAPE) applications, such as (i) scheduling and planning (over short- and medium-term horizons), (ii) conceptual design (CD), and feasibility studies of chemical plants (over long-term horizons) subject to price/cost variability originated by the supply and demand law, market uncertainties, and quotations volatility. CO plays a key role in most industrial and financial markets as it is the reference component of both chemical and oil and gas sectors. Starting from these sectors a number of derived industrial activities and interconnected economic endeavors are connected to and depend on the economics of CO.

This chapter offers an extended introduction to CO and specifically on (i) its fundamentals (i.e., production, consumption, reserves, and refinery capacities), (ii) historical trends and political events that influenced the dynamics of its quotations, and (iii) economic, strategic, and political players of international markets.

This chapter reports the most significant econometric, economic, and hybrid models of the scientific literature, which are able to forecast the evolution of CO quotations over different time horizons. These price models can be used in PSE/CAPE applications to simulate a probabilistic distribution of future scenarios for the economic assessment of industrial activities.

Several figures and tables allow the reader understanding of the main features of CO and how to model the price of CO and derivatives (i.e., commodities and utilities) over short-, medium-, and long-term horizons. These models can forecast possible economic scenarios and quantify their impact on the operative expenditures (OPEX) terms of industrial plants.

9.2.2 Introduction to CO

CO is the most basic and globally distributed raw material of the process industry. It is processed in oil refineries to separate by distillation hydrocarbon fractions, which are further treated in petrochemical plants to obtain various derivatives. The BP Statistical Review of World Energy (BP 2015) maintains that CO was the world's dominant fuel in 2014. Fluctuations in the CO prices have both direct and indirect impact on the global economy. Indeed, both investors and process designers/managers observe and follow very closely the CO prices because CO is a reference component for both the oil and gas and petrochemical supply chains, and plays a central role in a number of industrial utilities (e.g., electric energy, fuel oil, hot water, and steam).

The quotations of CO are either directly (i.e., distillates) or indirectly (i.e., derived commodities) taken into account for the economic assessment and feasibility study of PSE problems such as the scheduling and planning of supply chains and the design of chemical plants. As a function of the specific problem to be solved, the time interval chosen for the economic assessment can cover a short-, medium-, or long-term horizons (i.e., from hours/days to months/years) as discussed in Paragraph 7. Manca (2013) and Mazzetto et al. (2013) showed how CO economics influences the quotations of commodities and utilities, which on their turn play a major role in the economic assessment of OPEX terms in both chemical and biochemical processes/markets.

Both CO and distillate prices can be affected by exogenous events that have the potential to disrupt the flow of oil and products to the markets. These events may include geopolitical and weather-related incidents (Hamilton 2005; Zhang et al.2008) and either lead to actual disruptions or create uncertainty about future supply or demand, which on their turn may lead to higher volatility of prices. The aforementioned variations are driven by short-term imbalances on supply and demand terms and by uncertainties originated by political, economic, and financial contributions. This is the main problem of short-term horizon models, involved in scheduling problems, which cover time horizons spanning from days to a few weeks. On the other hand, the medium- and long-term horizon problems of CD are difficult to solve due to the need to forecast the different variables involved (e.g., levels of supply, demand, production, and capacity storage) for longer periods (i.e., from a few months to some years). The feasibility study of chemical plants depends partially on the purchase costs of raw materials and the selling prices of products. In recent years, the terms dynamic conceptual design (DCD) and predictive conceptual design (PCD) have been proposed by Grana et al. (2009), Manca and Grana (2010), Manca et al. (2011), Manca (2013), and Barzaghi et al. (2016) to account for variable prices/costs over different time horizons. It results rather conveniently that the price/cost evaluation of commodities and utilities should not rely on customized models specifically carried out for each of them. On the contrary, it is worth identifying a reference component and measuring the price/cost of commodities in respect to it. CO is the precursor of a number of commodities and utilities, and its cost is well known, largely available in several databanks [e.g., Energy Information Administration (EIA), IEA, ICIS] and periodically updated. Therefore, CO is the natural candidate to play the role of reference component.

9.2.2.1 The Hubbert Curve

According to Hubbert (1956), CO production follows the so-called "Peak Oil Theory," because it adheres to the production of finite resources in a market economy (i.e., that resource depletes faster than it can be replaced). The Peak Oil Theory predicts that the observed quantity initially grows in production, reaches a maximum peak, and eventually declines to zero (Bardi 2009). The intrinsic features of the oil resource support this dynamic. Actually, the extraction of an abundant and cheap resource leads initially to an economic growth and subsequently to further extraction investments. Gradually, the cheap resource depletes and extraction costs increase due to lower quality deposits. Consequently, investments cannot keep pace with these rising costs. The key factors that generate Hubbert's curve are the positive feedback that derives from the reinvestment of the profits generated by the resources and the negative feedback that derives from the gradual depletion of cheap resources (Bardi and Lavacchi 2009).

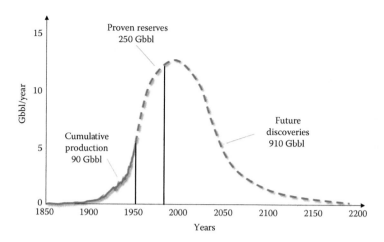

FIGURE 9.8 Ultimate world CO production based upon initial reserves of 1250 billion barrels as originally reported in Hubbert (1956).

Figure 9.8 shows the typical bell curve of an energy resource that is originated by the energy cost, aka energy return on energy investment, which quantifies the dynamics of the net extraction energy (Miller and Sorrell 2014). High exploration rates lead to extreme and premature production peaks, whereas low exploration rates produce the opposite effect.

9.2.2.2 Recent Historical Trends of CO Quotations

The interaction among oil price, oil supply, and oil demand is eccentric and responds to different exogenous events that may occur in the market at a specific historical period such as military conflicts, economic recessions, and monetary policies (Hamilton 2005). This point is particularly crucial, as prices do not explicitly respond to real events, but rather to their perception. Most of times, prices rise because there *might* be a shortage of oil, not because there *is* a real one. Prices fall only when that perception changes. This issue is extensively discussed by Hamilton (2005), Chevallier (2014), Davis and Fleming (2014), and Dowling et al. (2016).

Table 9.6 reports a qualitative list of events (e.g., global tensions, local conflicts, and rumors) that played a crucial role in determining the recent fluctuations of oil markets. A CO price model should account for and possibly forecast the contributions introduced by those conflicts, tension, and events.

TABLE 9.6   List of Events from 2008 to 2015 That Affected the CO Prices

Time	Event Description	ΔPrice (%)	ΔTime (months)	Absolute Values (USD/bbl)
July–December 2008	Financial crisis	−69.9	6	From 132.72 to 39.95
February 2011	War in Libya	10.5	1	From 103.72 to 114.64
March 2011	Tsunami in Japan/Fukushima nuclear accident	7.5	1	From 114.64 to 123.26
November 2011–March 2012	Political tensions with Iran/strikes of oil workers in Nigeria	13.3	5	From 110.77 to 125.45
May–July 2012	End of the tensions/slow growth in China	−6.9	3	From 110.34 to 102.62
June–August 2013	Threat of an American attack to Syria	8.1	3	From 102.92 to 111.28
June 2015	Greek crisis	−8	1	From 61.48 to 56.56
July 2015	Withdrawal of Iranian embargo	−17.6	1	From 56.56 to 46.58
July–August 2015	Chinese crisis of stock exchanges	−15.8	2	From 56.56 to 47.62

Source: EIA, Energy Information Administration, http://www.eia.gov, 2016.

For instance, in the first months of 2011, the conflicts in Libya and the tsunami in Japan, combined with the following Fukushima nuclear disaster, played a primary role in the increase of CO prices by 18%. Similar comments can also be made for the political situation of Libya, Iran, Iraq, and Syria, which significantly affected the quotations of CO in recent years.

Agnoli (2014) estimated that a drop of 10 USD/bbl transfers roughly half point of global gross domestic product (GDP) from producer to consumer countries. According to Goldman Sachs (2014), the recent highly variable trend of CO prices produced not only a global impact, but also made oil companies afraid of breaking their own neutrality threshold between incomes and outcomes. Indeed, oil producing countries and companies count on a certain price level to cover operative expenses and financial commitments.

Nowadays, besides this economic instability, new scenarios are emerging thanks to the production of shale oil in the United States (which has recently approached the output capacity of one of the biggest petroleum producers, i.e., Saudi Arabia) that has already overtaken Russian gas production and issued the first CO exporting licenses. For instance, the trend in the United States oil production is the key variable in the oil market and several oil and natural gas industry executives believe that the United States will achieve energy independence in the next 5–10 years (Bassett 2014). In 2014, the United States exported a record of 3.8 Mbbl/d of petroleum products, up 347,000 bbl/d from the previous year, according to EIA (2015a). The Oil and Gas Journal (2015) reported that the production increase was driven by record-high refinery runs, which averaged 16.1 Mbbl/day in 2014, higher global demand for petroleum products, and export of motor gasoline, propane, and butane especially toward Central and South America, followed by exports to Canada and Mexico. Meanwhile, exports of distillates declined for the first time since 2004. Most of that decrease can be ascribed to declines in exports to Western Europe and Africa, where distillate exports fell by 61,000 bbl/d and 8,700 bbl/d, respectively, in 2014 (EIA 2015b). Likewise, in the second half of 2014, the production increase by European refineries together with exports from recently upgraded Russian refineries, and enhanced refinery capacity in the Middle East improved the supply to European distillate markets. This reduced the need for distillate from the United States.

9.2.2.3 The 2008 CO Price Crash

The first noteworthy event of recent CO quotation history is the financial crisis of 2008 when CO prices fell all of a sudden due to the presence of excessive speculation (Chevallier 2014). Indeed, it is possible to observe that the CO quotation curve of Figure 9.9 in the second semester of 2008 saw a tremendous financial and economic calamity that was triggered by the US subprime mortgage crisis. After having trespassed the 145 USD/bbl value in July 2008, West Texas Intermediate (WTI) CO price crashed to 36 USD/bbl in December 2008 and eventually bounced back to 76 USD/bbl in November 2009. For these reasons, the analyses and models carried out in this chapter start from January 2010 to avoid the impact that the financial crisis of 2008 had on petroleum markets. This anomalous trend in the last 20 years deals with the problem that CO prices are usually traded on futures market and that financial fundamentals (e.g., the role of exchange and interest rates, or the commodity indexes) contribute to the petroleum market (Chevallier 2014).

9.2.2.4 Brent vs WTI and the Shale Oil Revolution

Kao and Wan (2012) define a CO benchmark as "the market from which the price changes first appear, and toward which the prices of other crude oils equilibrate." The most important global CO benchmarks are Brent and WTI. Liu, Schultz, and Swieringa (2015) provided some details on those benchmarks. The quotations of Brent and WTI specialize in European and American markets, respectively. Indeed, to describe consistent scenarios according to different markets, it is reasonable to choose Brent quotations for European refineries and WTI quotations for North America refineries (Rasello and Manca 2014).

Brent is the original name for the oil extracted from specific fields and collected through a pipeline that arrives to the Sullom Voe terminal in the Shetland Islands of Scotland, UK. Declining supplies

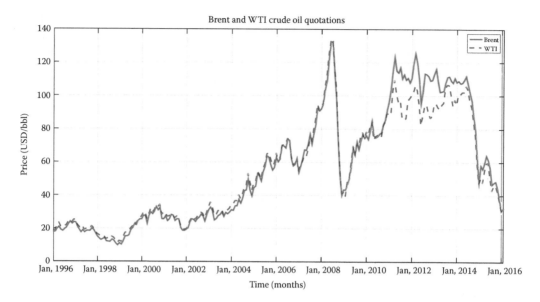

FIGURE 9.9 Brent and WTI monthly quotations from January 1996 to February 2016. (From EIA, Energy Information Administration 2016, http://www.eia.gov, 2016.)

from the original Brent fields have led to blending with oil from the Ninian fields, which widened the benchmark definition to include oil from Forties, Oseberg, and Ekofisk fields (hence the acronym BFOE).

WTI is a light and sweet CO that is collected from wells in Texas, New Mexico, Kansas, and Oklahoma and conveyed to the storage facilities in Cushing, Oklahoma (USA). The larger trading volumes of the New York Mercantile Exchange (NYMEX, i.e., the main commodity futures exchange for energy products), the American CO contracts, and the fact that WTI contracts can efficiently incorporate London's information on Brent data (thanks also to five to six time zones between London and New York) into their dynamics allowed WTI to become more influential than Brent regarding the quotations of other oils (Kao and Wan 2012).

As shown in Figures 9.9 and 9.10, Brent and WTI quotations lost their mutual consistency as of 2011 (Kao and Wan 2012; Sen 2012), and (Dowling et al. 2016)). Current pipeline constraints on the United States side (in particular at Cushing, Oklahoma) and the shale oil boom resulted in the divergence between Brent and WTI quotations (Liu et al. 2015). Indeed, Cushing has been the pricing point for WTI contracts since 1983 and nowadays the reservoir covers more than 9 square miles and has a CO storage capacity of 71 Mbbl (EIA 2016). Even if the Brent–WTI price differential should be around 8–12 USD/bbl, which is the price that makes rail freight to US Gulf Coast economic, Figure 9.10 concludes that recent differential values between Brent and WTI monthly quotations have been substantially higher. The reason for the price differential involves a number of distinct but correlated reasons.

Since 2005, production of CO from conventional extraction means has not grown concurrently with demand growth, so the oil market switched to a new and different state, which can be coined as "phase transition." Indeed, current manufacturing is "inelastic," which means that it is unable to follow the demand fluctuations and this pushes prices to oscillate significantly because the resources of other fossil fuels (e.g., tar sands, oil sands, shale oil, and unconventional natural gas) do not seem capable to fill the gap in the supply chain. The capacity to maintain and grow global supplies is attracting an increasing concern. In particular, the CO production curve of the United States shows a trend reversal: the curves of CO production and of net imports intersected in 2013 (Figure 9.11) with the American production approaching the Saudi Arabian one (Figure 9.12).

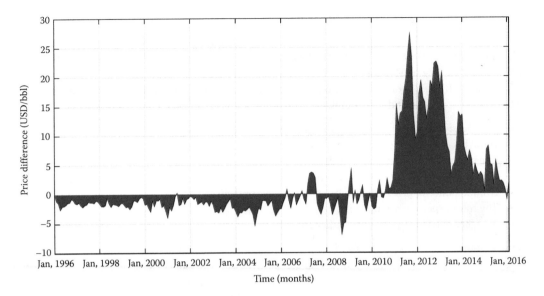

FIGURE 9.10 Brent–WTI differential values between Brent and WTI monthly quotations. (From EIA, Energy Information Administration 2016, http://www.eia.gov, 2016.)

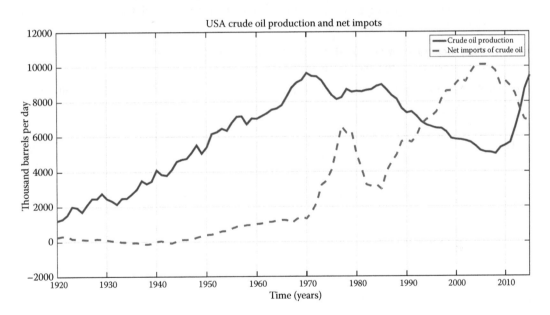

FIGURE 9.11 US CO annual production and net imports from 1920 to 2015. (From EIA, Energy Information Administration 2016. http://www.eia.gov, 2016.)

Shale oil is diffused especially in the United States, Australia, and Brazil. The extraction of shale oil in the United States, which counts for 2 trillion barrels of these unconventional reserves, started at the beginning of the last century, but since 2005 resulted in a substantial growth of total CO production, which increased from 5.1 Mbbl/d in 2008 to 9.6 Mbbl/d in 2015 (EIA 2016). Shale oil output reached a record of 5.47 Mbbl/d in March 2015 thanks to technology improvements, although the number of oilrigs was the lowest since 2013. Liu et al. (2015) observed that the sudden abundance of American and Canadian shale oil produced a substantial discount of WTI with respect to Brent quotations.

9.2.2.5 The Price Crash of December 2014

The second semester of 2014 saw a sharp drop of Brent price from 106.7 USD/bbl to 62.34 USD/bbl, and of WTI price from 103.59 USD/bbl to 59.29 USD/bbl (Figure 9.13). In the first quarter of 2015, the price plummeted from 76.4 to 54 USD/bbl for Brent and from 73.2 to 48.5 USD/bbl for WTI.

The roots of this price crash lie on different factors, which can be distinguished in supply and demand factors, macroeconomic and financial issues, and political concerns (Davis and Fleming 2014):

- The massive Chinese reduction of import growth (from 16% in 2010 to 5% in 2014), even if China could become the world's largest importer by 2017 to outpace its domestic production (Li 2014).
- The stagnation of oil demand in Western countries and Japan.
- The oversupply of CO due to the growth of Canadian oil sands and the US fracking boom.
- An imbalance in price ratios between oil and natural gas.
- The role of speculative investors.
- A higher dollar with respect to other currencies.

In front of these emerging scenarios, Middle East producers decided not to cut their quotas at the end of 2014 in order to maintain their production competitive with the global CO spot market. The highest authorities of Saudi Arabia declared acceptable that prices remained low for long periods if that would reduce investments in shale oil and rebalance global markets. The recent high volatility of CO prices made oil companies and some traditional producers (i.e., Iran, Venezuela, and Russia) afraid of breaking their neutrality threshold between incomes and outcomes. Indeed, the collapse of CO quotations can hurt the national budgets of these producer and exporter countries, which traditionally pay public administration subsidies by means of CO business revenues.

9.2.2.6 Very Recent Events: Greek Crisis, Withdrawal of Iranian Embargo, and Chinese Crisis of Stock Exchanges

A bunch of recent anthropic events further upset the CO markets in 2015 and 2016. Both geopolitical and financial issues were the main causes.

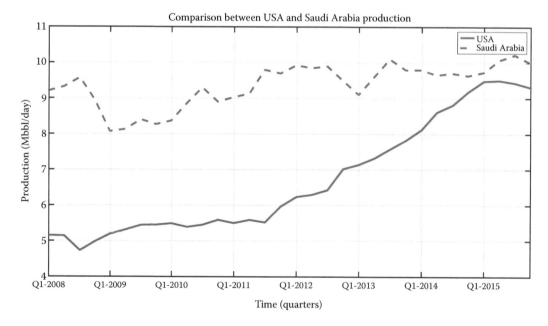

FIGURE 9.12 Comparison between the USA CO production and Saudi Arabia production from Q1-2008 to Q4-2015. (Quarterly from EIA, Energy Information Administration 2016. http://www.eia.gov, 2016.)

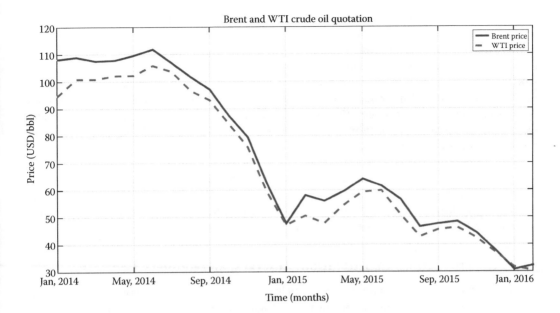

FIGURE 9.13 Brent and WTI CO monthly quotations from January 2014 to February 2016. (From EIA, Energy Information Administration 2016. http://www.eia.gov, 2016.)

The Greek depression began in late 2009 because of the turmoil of global recession, structural economic weaknesses (e.g., government spending, tax evasion, lack of budget compliance), and a sudden crisis in confidence among lenders due to a misreporting of the government debt levels. By 2012, the Greek debt-to-GDP ratio rose to 175%, nearly three times the EU's limit of 60%. After three bailout programs from (i) European Commission, (ii) European Central Bank, and (iii) International Monetary Fund, Greece incurred in 2015 a solvency problem and a national bankruptcy crisis that dragged down global equity and commodity markets. Greece is not a major oil producer or consumer, so it does not have a direct impact on CO markets, but this situation had two immediate effects on oil. First, there were currency fluctuations. The rising prospect that Greece might leave the Eurozone damaged the Euro currency to benefit the USD. Second, such a calamity scared the investors who feared a broader contagion with a consequent decrease of oil prices.

Another disruptive event, which turned out to be a bigger threat to the global economy, was the stock market rout in China, with the entailed political, social, and economic risks. Unlike the majority of stock markets, where investors are mostly institutional, more than 80% of Chinese investors are small retail investors. Although the Chinese economy lost steam in recent years (as reported by Yao (2014) its GDP growth rate halved from 14% in 2007 to 7.4% in 2014), the last semester of 2015 saw a record number of businesses listed on the Shanghai exchanges. This countercurrent event was fueled by liberalization laws that made easier for funds to invest and for companies to offer shares to the public for the first time. At the center of the dramatic stock market slide there were individual investors, and the explosion of the so-called margin lending, which is a stock system where the broker can make a demand for more cash or other collaterals if the securities price has fallen. Consequently, shares plunged 30% in 3 weeks, as of mid-June 2015, hundreds of companies suspended their dealings, and fears that the slump would spill over into other markets grew unrestrainedly. Likewise, Brent quotations collapsed below 60 USD/bbl for the first time since April 2015 and closed at their lowest level in nearly 3 months on July 1, 2015 (i.e., at 51.56 USD/bbl). This fact is particularly interesting as it allows understanding better the historical background, considering China is the second largest oil consumer in the world. Indeed, commodities were sucked into this market turmoil and the Chinese stock market plunged the world economy and fuel demand.

Another noteworthy event was the Iranian nuclear deal, which had the capability of bringing the markets to CO and petroleum products oversupply. On July 14, 2015, Iran and the six world powers (i.e., the United States, the United Kingdom, France, China, Russia, and Germany) struck a landmark agreement to curb Tehran's nuclear activities in exchange for lifting the crippling economic and financial sanctions imposed in 2012. After this decision, Iran was allowed to sell large volumes of polymers to Europe, while China, which became Iran's major market, when the sanctions were imposed, saw a 30% decrease of supply from that country (Fadhil 2015). According to a Dubai-based petrochemical trader, CO prices were set for more volatility with Iran back, since its reentry in the European market raised oversupply concerns (Fadhil and Dennis 2015).

The first month of 2016 saw a further decrease of CO quotations due to exceptionally high production in Middle Eastern countries (e.g., Saudi Arabia) and the oil shale boom in the United States. These elements contributed to a large oversupply of oil also triggered by a reduced demand by developing nations. This supply and demand imbalance caused the CO price to drop below the psychological threshold of 30 USD/bbl to hit a 12-year low as of January 20, 2016, at 27 USD/bbl.

9.2.3 CO Production, Consumption, Reserves, and Refinery Capacities

As mentioned above, CO is a mixture of unrefined hydrocarbons deposits that exist as a liquid in natural underground reservoirs, remains a liquid when brought to the surface, and can produce usable products, such as gasoline, diesel, kerosene, asphaltenes, and various forms of petrochemicals (e.g., ethylene, propylene, aromatics).

CO markets are characterized by the existence of a cartel of producers (i.e., OPEC) together with independent producers. The main players in the CO markets are clustered into OPEC (Organization of Petroleum Exporting Countries), OECD (Organization for Economic Cooperation and Development), and BRIC countries (i.e., Brazil, Russia, India, and China).

OPEC is an intergovernmental organization that was created at the Baghdad Conference in 1960 by Iraq, Kuwait, Iran, Saudi Arabia, and Venezuela. In the following years, nine governments joined OPEC: Libya, United Arab Emirates, Qatar, Indonesia, Algeria, Nigeria, Ecuador, Angola, and Gabon. Figure 9.14 shows the main importers of OPEC CO. Oil consumption comes especially from Western countries and Asia, even though Chinese and Japanese flat economic growths have seen a stagnation of global CO demand.

OECD and BRIC countries largely represent the consumers' side. OECD is an international economic organization of 34 countries that was founded in 1961 to stimulate economic progress and world trade. OECD gathers the world's most developed countries and therefore is the principal consumer of CO and derived products. The United States is obtaining a rising power on the producer side: the increasing interest in oil sands both in the United States and Canada has given a new power to the North

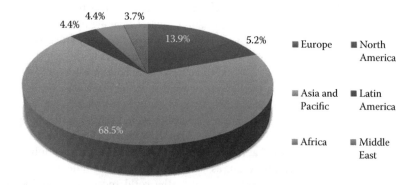

FIGURE 9.14 OPEC exports by destination in 2013. (From BP, Statistical review of world energy, 2014.)

American economy that could become independent of OPEC decisions about quotas and production in the near future. Equally, the economic background has changed and is more complex than the one of 5–10 years ago. For instance, OPEC and OECD do not include the BRIC countries, which are considered to be at a similar stage of both newly advanced economic development and emerging countries such as Argentina, South Africa, and Indonesia. On its turn, Indonesia abandoned OPEC from 2008 to June 2015 when it rejoined OPEC, because of the increased relations with the oil companies of Saudi Arabia (Suratman 2015).

The following graphs contain recent information on CO proved reserves, production, consumptions, and refining capacity. The last two decades have seen an increase of OPEC and BRIC proved reserves, especially to the detriment of OECD reserves (Figure 9.15).

The distribution of OPEC reserves has become more homogeneous, and Saudi Arabia has lost its record for the benefit of Venezuela (Figure 9.16). Nonetheless, Saudi Arabia remains the leading oil producer within OPEC and is the world's largest oil exporter.

Among OECD, Canada is the country that has more increased its proved reserves (Figure 9.17), while BRIC countries have seen a contribution improvement from Russian and Brazilian reserves (Figure 9.18).

OPEC countries are the leaders in terms of CO production (Figure 9.19), even if the oil produced by BRIC countries has seen a constant increase in last 20 years (Figure 9.20).

Among OECD countries, the United States saw an increasing production from 37% in 1998 to 49% in 2013. Meanwhile, Canadian oil sands conferred a new power to North America CO output at the expense of Mexican, Danish, and British production percentages (Figure 9.21).

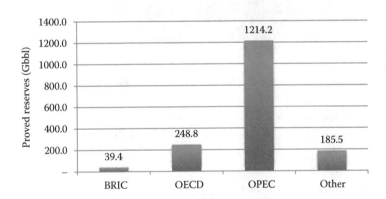

FIGURE 9.15   Global proved reserves in 2013. (From BP, Statistical review of world energy, 2014.)

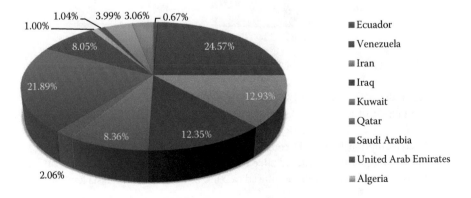

FIGURE 9.16   OPEC proved reserves in 2013. (From BP, Statistical review of world energy, 2014.)

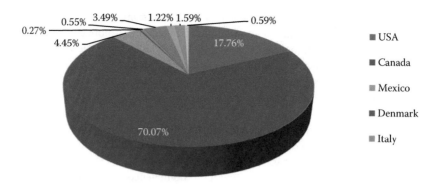

FIGURE 9.17   OECD proved reserves in 2013. (From BP, Statistical review of world energy, 2014.)

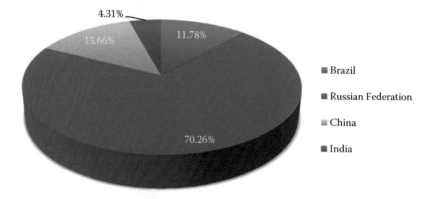

FIGURE 9.18   BRIC proved reserves in 2013. (From BP, Statistical review of world energy, 2014.)

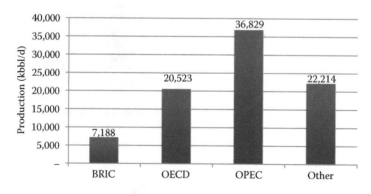

FIGURE 9.19   Global CO production in 2013. (From BP, Statistical review of world energy, 2014.)

A different situation characterizes OPEC countries, which redeployed their own CO production for the benefit of Saudi Arabia, while Venezuela, Iran, and Nigeria cut their output fraction (Figure 9.22).

The CO consumption by OECD countries has increased in last decades (Figure 9.23), and BRIC countries have enlarged their demand slice due to the Chinese growth (Figure 9.24). For the sake of correctness, the Chinese consumption slowed in 2013, while India and Russia kept on increasing.

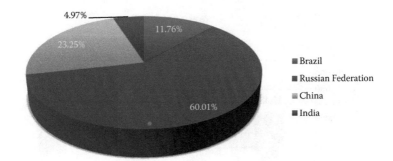

FIGURE 9.20 BRIC CO production in 2013. (From BP, Statistical review of world energy, 2014.)

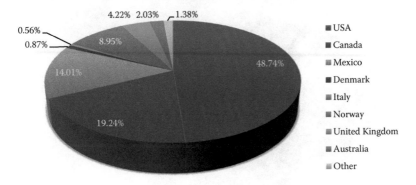

FIGURE 9.21 OECD CO production in 2013. (From BP, Statistical review of world energy, 2014.)

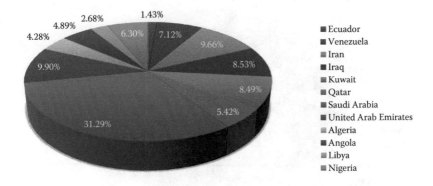

FIGURE 9.22 OPEC CO production in 2013. (From BP, Statistical review of world energy, 2014.)

According to BP (2014), the CO consumptions from 1965 to 2013 distributed better in most OECD countries than in the United States, which since 1998 accounts for 40% of OECD crude consumptions (Figure 9.25).

The refinery capacity data of Figure 9.26 signify the sum of reported atmospheric crude distillation and condensate splitting capacity. The capacity term refers to the amount of raw materials that a refinery can process under usual operating conditions with scheduled downtimes included. As for production

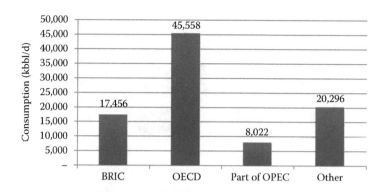

FIGURE 9.23   Global CO consumption in 2013. (From BP, Statistical review of world energy, 2014.)

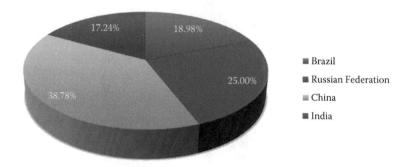

FIGURE 9.24   BRIC CO consumption in 2013. (From BP, Statistical review of world energy, 2014.)

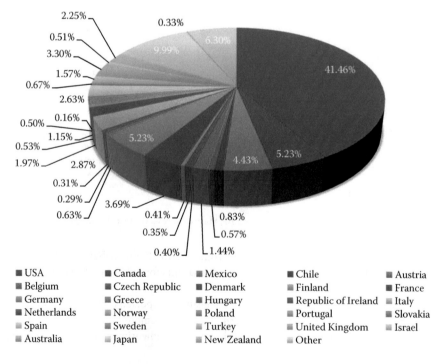

FIGURE 9.25   OECD CO consumption in 2013. (From BP, Statistical review of world energy, 2014.)

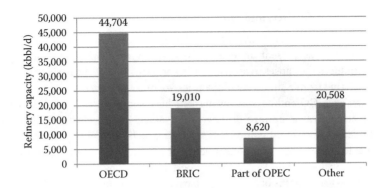

FIGURE 9.26 Global refinery capacity in 2013. (From BP, Statistical review of world energy, 2014.)

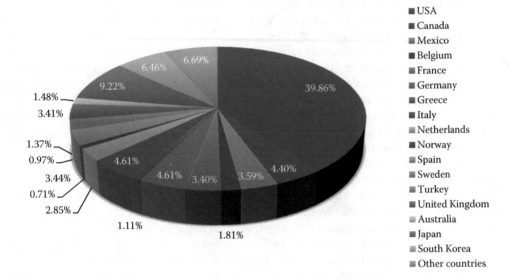

FIGURE 9.27 OECD refinery capacity in 2013. (From BP, Statistical review of world energy, 2014.)

and consumption, BRIC countries increased their refining capacity with respect to OECD nations (Figure 9.26), which have seen since 1965 a reduction of the United States capacity to benefit Mexico, South Korea, and Japan.

Conversely, the distribution of OECD refinery capacity has not seen significant variations in the last two decades (Figure 9.27).

9.2.4 The Fundamentals of CO and Derivatives Markets

As already discussed in Paragraph 4, DCD and PCD approaches have been proposed by Grana et al. (2009), Manca and Grana (2010), Manca et al. (2011), Manca (2013), and Barzaghi et al. (2016) to account for the dynamics of raw materials purchase costs and products selling prices. Indeed, it is rather convenient and it is recommended to identify a reference component (i.e., CO) and evaluate/model the price/cost of commodities with respect to such a component. Due to the high degree of volatility induced by historical, economic, and political contingencies, the real price of CO is difficult to model. The following paragraphs will discuss the main characteristics of CO and its distillates, petrochemical derivatives, and utility markets in terms of volatility and supply and demand law, in order to propose suitable models for each of them.

9.2.4.1 Fluctuations, Volatility, and Uncertainties

A number of scientific papers have investigated the dynamics of oil prices and their volatilities. Paragraphs 7 and 8 categorize, present, analyze, and discuss several econometric and economic models. This paragraph introduces the causes, reasons, and features (that are at the basis of such models) more systematically by starting from a summary of the price volatility evidences.

Today, more than ever, CO and derivatives markets have shown their dynamism with unexpected oscillations that reached drastic upward or downward values. Indeed, fluctuations and uncertainties characterize both raw material and commodity/utility markets with either simultaneous or delayed trends with respect to the reference component behavior (Manca 2013, 2016). As for the recent trend of CO prices, Paragraph 3 showed how that second semester of 2014 saw a sharp drop. A similar behavior was experienced in the first months of 2016 with a barrel of Brent valued at 27 USD on January 20, 2016 (with a 7 USD/bbl decrease from December 2015 and the lowest monthly average price since December 2003). The markets world is becoming complicated as shown in Figure 9.13. Brent technical floors of 70 USD/bbl and 54 USD/bbl for the price, dating back to 2010 and 2006, respectively, were shown to be inconsistent with the recent market trend, which points out that no real support level exists in the current oil marketplace (Davis and Fleming 2014).

Actually, high volatility, high intensity jumps, and bullish and bearish periods make it difficult for investors to track the CO price trend and for process designers/managers to perform a sound economic assessment in feasibility studies. What is the driving force of these instabilities? First of all, it is better to speak about more *driving forces*.

Shale gas, shale oil, international crises, embargos, available infrastructures, industrial and transport accidents, natural calamities, and weather variability are further examples of exogenous variables that play a major role in the fluctuations of quotations even over short time periods with further influence coming from geopolitical backgrounds (Rasello and Manca 2014).

These events may create uncertainty about future supply or demand, which on their turn can lead to higher volatility of prices. Table 9.6 covered the main political/historical/economic contingencies that affected CO prices over the last decade, from the 2008 financial crisis to the more recent European and Chinese upsets. The same market uncertainties might repeat in the future. This is why both DCD and PCD methods lay their foundations on the probabilistic curves or *fan-chart/river of blood* obtained from suitable distributions of price trajectories, which represent the expected economic scenarios. By identifying the possible future trajectories of CO, it is then possible to assess the corresponding scenarios of both derived commodities and utilities.

9.2.4.1.1 Statistical Results

Market uncertainties and fluctuations can be quantified and modeled with the help of time series analysis, whose description is given in Paragraph 7.1. The aim of this paragraph is to provide the reader with suitable means for the analysis of price time series, some of which are used to carry out and assess the econometric model of Paragraph 7.2. The statistical tools used in the literature to highlight and study the volatility of prices are time plots, measures of skewness and kurtosis, and correlation and autocorrelation indexes. The reference unit of CO price is the *shock*, which is the term that signifies the price variation from the previous quotation to the following one on a daily, weekly, monthly, quarterly, or yearly basis. Manca (2015) showed that, since 2000, most of the CO weekly shocks had a maximum ±10% spread and followed a Gaussian-like distribution. As reported by Salisu and Fasanya (2013), Brent is more volatile than WTI in terms of standard deviation of quotations. Conversely, the same volatility dominance of Brent over WTI cannot be observed when both absolute and relative oil price shocks are considered. As occurs frequently in financial markets, the kurtosis of the shocks is greater than 3, thus the density function is characterized by a fatter tail when compared to the standard Gaussian distribution. Conversely, the coefficient of skewness is either positive or negative and rather near zero in different cases for both Brent and WTI prices and shocks. This point shows that there is an acceptable symmetry of the probability distribution of prices and

TABLE 9.7   Descriptive Statistics Results of CO Prices (USD/bbl)

	2009–2014		2005–2014		2000–2014	
	BRENT	WTI	BRENT	WTI	BRENT	WTI
Mean	96.0682	87.3774	86.1997	81.6763	66.7031	64.3951
Median	107.62	90.755	83.745	83.275	63.485	64.065
Mode	43.32	94.51	39.95	94.51	25.62	94.51
Min	43.32	39.09	39.95	39.09	18.71	19.39
Max	125.45	109.53	132.72	133.88	132.72	133.88
Standard deviation	21.1076	15.7918	24.579	20.0744	33.8915	29.3389
Skewness	−0.9125	−1.161	−0.0749	0.0422	0.1986	0.1257
Kurtosis	2.776	4.1511	1.6913	2.5265	1.6274	1.8112

TABLE 9.8   Descriptive Statistics Results of CO Shocks (USD/bbl)

	2009–2014		2005–2014		2000–2014	
	BRENT	WTI	BRENT	WTI	BRENT	WTI
Mean	0.8554	0.8068	0.4922	0.4284	0.4324	0.3936
Median	0.81	1.02	1.095	1.075	0.93	0.995
Mode	2.61	−12.35	−1.22	1.4	−2.78	1.4
Min	−15.18	−12.35	−25.65	−27.5	−25.65	−27.5
Max	11.31	14.18	13.73	14.18	13.73	14.18
Standard deviation	5.245	5.3651	6.5798	6.533	5.5891	5.5151
Skewness	−0.3628	−0.147	−1.1581	−1.1345	−1.2394	−1.2365
Kurtosis	3.2484	3.1093	5.2889	5.731	6.6776	7.419

shocks. The econometric models should account for this feature of shocks distribution. Tables 9.7 and 9.8 collect the statistical features of the time series of CO prices and shocks. According to Jarque–Bera's test, the distribution of CO prices does not result to be distributed normally, while the distributions of absolute shocks and relative shocks are normal at the 5% significance level. These results are important to estimate the local drift (i.e., the short-term trend) in the econometric models.

9.2.4.2 Beyond the Supply and Demand Law: Quota Policy

Besides volatility and uncertainties, other characteristic features of the CO market are the supply and demand law and the policy of producer countries. As highlighted in previous paragraphs, CO quotation trend is a quite frangible equilibrium that goes beyond the traditional supply and demand law, which is the typical approach to determine prices in a competitive market. In this situation, the good price will vary according to the quantity transacted until it settles at a point where the quantity demanded will equal the quantity supplied at that price. CO market followed this natural behavior of goods until the last quarter of 2014, when Middle East producers decided to maintain their production competitive with the other participants, such as the United States, even if the collapse of CO quotations can hurt the national budget of these producer and exporter countries, which traditionally pay public administration subsidies by means of their CO business revenues. To better understand the current situation of CO market fundamentals, it is worth going back in the past history of CO trading.

As already discussed in Paragraph 5, the Organization of Petroleum Exporting Countries (OPEC) was formed in 1960, when the international oil market was largely dominated by a group of multinational companies known as the *Seven Sisters*, which tried to eliminate competitors and control the global oil resources. In the following decades, the dominance of these companies and their successors declined because of the increasing influence of OPEC cartel and of the emerging market economies. For instance, the expression *New Seven Sisters* was coined by the *Financial Times* (Hoyos 2007) and indicates the group of most influential national oil and gas companies that are based in countries outside OECD (Organization for Economic Cooperation and Development) and comprises OPEC and BRIC (i.e., Brazil, Russia, India, and China) countries, such as Saudi Arabia, Iran, Venezuela, China, Russia, and Brazil. Almost 72% of the worlds proven oil reserves are located in OPEC countries, with the bulk of CO reserves (66%) in the Middle East. Saudi Arabia reserves are about 20% of the worldwide oil reserves and the resulting production amounts to 40% of the proven reserves (EIA 2014; OPEC 2015; BP 2015). Since the 1970s of the last century, OPEC members have discovered that their oil could be used as a political and economic weapon against other countries. For instance, in October 1973, OPEC declared an oil embargo in response to the USA and Western Europe support to Israel in the Yom Kippur War and CO prices quadrupled from 3 USD/bbl in October 1973 to 12 USD/bbl in March 1974. In response to the high oil prices of the 1970s, industrial nations tried to reduce their dependence on CO. The energy sources used to produce both civil and industrial utilities switched to coal, natural gas, and nuclear power, whereas national governments launched multibillion dollar research programs to develop alternatives to oil. During that period, demand for CO dropped by 5 Mbbl/d and the percentage of oil produced by OPEC fell from 50% to 29% (EIA 2016). The result was a 6-year price decline that culminated with a 46% price drop in 1986. In order to avoid such an imbalance between supply and demand, OPEC self-imposed to fix an overall production ceiling with individual quotas for each member in accordance with each country output capacity. That mechanism was set up by OPEC to adjust production levels if prices moved out of a certain range for a previously defined period. In the case of high prices (i.e., above the range), a production increase is needed, while low prices and cut in production have the opposite effect on the market. As an example of the disposition to keep prices in line, when CO prices began climbing to the upper limit of the band at the end of September 2002, OPEC increased production by 760,000 bbl/d. By the end of October of the same year, CO prices had fallen more than 10%, returning to their target range (Dahmani and Al-Osaimy 2001; Sandrea 2003; Horn 2004). The production ceiling has been frequently adjusted in a systematic and timely way according to the prevailing market conditions, but a moderate level of overproduction (up to an average 7%) has been the norm for OPEC over the last 20 years, because OPEC has no way to enforce compliance by its members with the agreed quotas. Nowadays the OPEC behavior has changed and the quota system has been used more as a political weapon that an economic instrument. New scenarios are emerging thanks to the American production of shale oil. These conditions are bringing up the political instabilities in the Middle East, the reentry of Iran in CO market, the ghost of the reminded Cold War, and the global economic turmoil. Are there any other solutions for Saudi Arabia except the decision to not cut the production quota and go below its own breakeven point? The economic models presented in Paragraph 7.3 account for this policy behind the CO market.

9.2.5 CO Price Forecast

In preparing the *Annual Energy Outlook 2015*, EIA (2015a) provided a summary of six cases with the major assumptions underlying the projections of energy prices based on GDP, OPEC, and OECD market share and amount of conventional and unconventional oil and gas resources. This fact is particularly representative of the importance of forecast models creating several distinct scenarios of CO/commodity/utility prices, which correspond to the different historical or technological situations. The same scenarios are needed also to answer the typical question of PSE and CAPE applications about the feasibility and sustainability of products and processes (Manca 2015).

While past literature studies mainly focused on one-step-ahead models, which allow predicting the variable of interest for just the following time step, PSE/CAPE applications (except for real-time optimizations) are more complex and usually based on short-, medium-, and long-term fully predictive time horizons. Frequently, PSE/CAPE applications involve time intervals of up to 5 years, because in the fields of Chemical Engineering and DCD/PCD the time scale considered is rather large as a plant runs for several years (e.g.,15–20 years). Examples of applications of price/cost models over long-term horizons are the hydro-de-alkylation (HAD) process and the styrene monomer plant, largely discussed in Manca (2013) and Barzaghi et al. (2016), respectively. The problems that cover short- and medium-term horizons (i.e., scheduling and planning) deal with manufacturing management processes, by which raw materials and production capacity are optimally allocated to meet demand. This approach is conceptually simple but quite challenging, as a suitable model should really adapt to the changes in demand, resource capacity, and material availability that affect each operational and decisional level of the chemical supply chain.

9.2.5.1 Literature Overview and Models Classification

The scientific literature has shown a significant attention to the investigation of the dynamics of oil prices and their volatilities over different time horizons, by means of a plethora of econometric and economic models described in Table 9.9.

As far as the recent literature is concerned, several articles focus on the CO price dynamics in the 2000–2014 period that were characterized by high volatility, high intensity jump, strong upward drift, and was concurrent with underlying fundamentals of oil markets and world economy. How can the process designer/manager use the best model for their technical purposes?

Most of the published manuscripts have a financial background and focus on the forecasting capability to trade the CO by means of futures, selling/buying options, and other financial tools that have a much shorter time horizon than the one typically used in PSE/CAPE applications (Manca 2013). Financial modeling distinguishes between futures and spot prices. Futures denote a contract between two parties to buy or sell an asset for a price agreed upon today with delivery and payment occurring at the delivery date. Spot prices specify the settlement price of a contract of buying or selling a commodity or currency, which is normally two business days after the trade date.

The large set of CO price models calls for a classification proposal according to three taxonomic criteria:

- Time horizon of the forecast
- Time granularity of data
- Mechanism used to generate the simulated prices and explain their trend

The first classification criterion involves the length of the time horizon that can be used reliably to forecast the CO prices. Depending on the specific problem, the time horizon taken for the economic assessment can cover a short-, medium-, and long-term period. Short-term horizon models cover periods

TABLE 9.9 Classification of the Literature CO Models according to Different Taxonomic Criteria

Authors	Time Horizon	Time Granularity	Model Typology
Chen (2014)	Short	Monthly	Economic
Chevallier (2014)	Long	Weekly	Economic
Cifarelli and Paladino (2010)	Long	Weekly	Econometric
Dées et al. (2007)	Long	Quarterly	Economic
Dvir and Rogoff (2014)	Medium	Monthly	Econometric
Ghaffari and Zare (2009)	Short	Daily	Econometric
Kang and Yoon (2013)	Medium	Daily	Econometric
Kaufmann et al. (2004)	Medium	Quarterly	Economic
Kaufmann and Ullman (2009)	Short	Daily	Econometric
Kilian (2009)	Medium	Monthly	Economic
Rasello and Manca (2014)	Long	Monthly	Econometric
Salisu and Fasanya (2013)	Medium	Daily	Econometric
Ye et al. (2009)	Short	Monthly	Economic
Zagaglia (2010)	Medium	Monthly	Econometric

from a few days to a few weeks and are usually related to financial activities and sell and buy purposes. Medium- and long-term horizon problems are challenging due to the need for forecasting the different variables involved in the market background (e.g., the levels of supply, demand, production, and capacity storage) for intervals from a few months to some years, which are the horizons required by PSE/CAPE targets related to either CD or planning of chemical processes and plants.

The second classification criterion is based on the time granularity or discretization of the dataset of input variables, which can be gathered daily, weekly, monthly, quarterly, or yearly. This concept is closely related to the extent of the forecasting horizon, according to the number of steps that the model is able to forecast consistently.

The third classification criterion is based on the intrinsic nature of the forecasting models that can be divided into economic and econometric models. The first category simulates the CO market trend by means of the physical, economic, and financial features involved in the supply and demand law. The second category does not take into account the forces that cause price fluctuations, but simulates market uncertainties and possible price evolutions by the statistical analysis of past price shocks. The reliability of economic models is more challenging as the time horizon becomes longer and longer and such models call for the long-term forecasts of the variables that affect the price trend over long periods.

Recent studies of CO market cover a number of different areas and issues and examine the characteristics of these markets according to various aspects. Chen (2014) forecast the CO price fluctuations by means of oil-sensitive stock indices with 1-month ahead predictions. The novelty of his study consists in suggesting a new and valuable predictor (i.e., American Stock Exchange oil index) that reflected timely market information and was readily available since it was a price-weighted index of the leading companies involved in the exploration, production, and development of petroleum. Other studies focused on the growing presence of other financial operators in the oil markets and led to the diffusion of trading techniques based on expectations (Kaufmann and Ullman 2009; Cifarelli and Paladino 2010). Zagaglia (2010) studied the dynamics of NYMEX oil futures by using a large dataset that included global macroeconomic indicators, financial market indices, and quantities and prices of energy products.

A number of papers suggest that the supply-and-demand law plays a major role on oil prices. For instance, Kilian (2009) decomposed the real price of CO into supply shocks, shocks to the global demand for industrial commodities, and demand shocks. Similarly, Dvir and Rogoff (2014) presented evidence of connections among four variables in the American and global oil markets: (i) oil production, (ii) stocks of CO, (iii) real price of oil, and (iv) incomes.

Kaufmann et al. (2004) applied statistical models to estimate the causal relationship between CO prices and several factors, such as capacity utilization, production quotas, and production levels. Kaufmann, Karadeloglou, and Di Mauro (2008) investigated the factors that might have contributed to the increase of CO price in the first quarter of 2008. They revised a model of CO prices to include refinery utilization rates, a nonlinear effect of OPEC capacity utilization, and conditions in futures markets, and found that their model performed relatively well when used for forecasting purposes.

Other studies tribute a great importance to oil price volatility. Fluctuations of oil prices have often had great impacts on the economy. For instance, most of the American post World War II recession was preceded by sharp increases of CO prices (Hamilton 1983). Quite a few authors have used the standard deviation of price differences (i.e., price shocks) as a measure of volatility of commodity prices (Kang and Yoon 2013; Salisu and Fasanya 2013). These papers introduce plausible models to measure the oil price volatility and compare the performance of such models to Brent and WTI markets. As extensively discussed in Manca (2013), the distribution of shocks is nonlinear and cannot be modeled by an averaged trend line. Indeed, high volatility, high intensity jumps, and bullish and bearish periods make it difficult for investors to track the CO price trend and for process designers/managers to attain a sound economic assessment in feasibility studies. Kang and Yoon (2013) worked on more sophisticated econometric techniques. They discussed the superior performance of the generalized autoregressive conditional heteroskedastic models (GARCH) and its modifications (e.g., IGARCH, TGARCH, EGARCH) when high-frequency time series in financial markets are involved.

The scientific literature questioned also on the role of CO to forecast the price/cost of other raw materials (e.g., Manca 2013; Rasello and Manca 2014). Mazzetto et al. (2013) used CO as the reference component for econometric models in bioprocesses and showed a functional dependency of both raw biomaterials and final bio-products from the CO market. They evaluated also autoregressive distributed lag, fully stochastic, and time series decomposition models to remove the limiting assumption of keeping constant, for long-term horizons, the price/cost of commodities and utilities (as it happens in conventional CD techniques (Douglas 1988)).

9.2.5.2 Econometric Models

The identification of CO econometric models requires the acquisition of historical price data and the statistical analysis of both price series and shock series. Indeed, if one performs an analysis of the relative price variations between a time unit (e.g., 1 month) and the next one, a set of values that are representative of the price volatility is obtained. Manca, Fini, and Oliosi (2011) demonstrated that the shock series are normally distributed. A time series is a collection of observations made sequentially through time. Several examples occur in a variety of fields, ranging from economics (e.g., CO quotations, commodity prices, utility costs) to engineering (e.g., failure temperatures, yield of a batch process, and rate of equipment rupture). Methods for the time series analysis constitute an important area of statistics and are mainly concerned with decomposing the variation of a series into components that represent trend, seasonal variation, other incidental cyclic changes, and irregular fluctuations (Smirnov 2010; Chatfield 2000, 2013; Commandeur and Koopman 2007). The study of time series allows to both interpret a phenomenon, by identifying possible trends, seasonal variations, other periodic variations, and stochasticity, and predict its future performance. If future values can be predicted exactly from past values, then a series is said to be deterministic. However, most series are stochastic, or random, as the future can only be partially determined by past values. The stochastic variability is the difference between the true value and trend with seasonality; it can be treated as a stationary stochastic process, i.e., a random series with mean zero and a suitable distribution. Real-world economic processes such as CO quotations are nonstationary because their characteristics are not constant over an observation interval. If one does not pretend to get a precise and unique forecast of future states, a probabilistic approach is traditionally used. Based on such time series, the forecasting models allow investors and process designers/managers to forecast possible future trends based on known past observations. Econometric models can simulate the future quotations of the two benchmarks (i.e., Brent and WTI) and provide a probabilistic distribution of prices, which defines the likely domain where prices can range. Table 9.10 shows a list of econometric models that are widely used in the forecast of price time series, which for the sake of simplicity is called in the general form Y_t.

A general mixed autoregressive moving average model (ARMA) model consists of two parts (AutoRegressive AR and Moving Average MA) that describe CO quotation as a weighted average of past observations with the addition of a white noise error, where $\varphi_0, \varphi_1, \ldots, \varphi_p, \theta_1, \ldots, \theta_q$ are real coefficients, p and q are the orders achieved by plotting the partial autocorrelation functions, and ε_t is a white noise process. The main advantage of this models lies on its mathematical tractability to approximate general stationary processes, and the relatively simple procedure to compute the parameters. However, as it was shown in the literature (Chatfield 2000; Chatfield 2013), ARMA models are not suitable to evaluate the entire distribution of nonlinear processes because they are sensitive to outliers and do not take into account the conditional heteroskedasticity since the variance of the dependent variable (e.g., price, cost, quotation) is constant over time. While the AR models imply the unconditional variance being constant, changes in the variance are very important to understand financial markets and it is worth considering that conditional variance may demonstrate a different behavior and significant changes over time. The main idea behind the autoregressive conditional heteroskedasticity (ARCH) models, which were proposed by Engle (1982), is the following: the forecast based on the past information is presented as a conditional expectation depending upon the values of past observations. Therefore, the variance of such a forecast depends on past information and may be a random variable h_t, while v_t is a random variable

TABLE 9.10 List of Econometric Models Available in Literature to Forecast CO Prices with Their Formulation and Original References

Model	Formulation	References
AR	$Y_t = \mu + \varepsilon_t + \theta_1 \cdot \varepsilon_{t-1} + \theta_2 \cdot \varepsilon_{t-2} + \cdots + \theta_q \cdot \varepsilon_{t-q}$	Box and Jenkins (1976)
MA	$Y_t = \varphi_0 + \varphi_1 \cdot Y_{t-1} + \varphi_2 \cdot Y_{t-2} + \ldots + \varphi_p \cdot Y_{t-p}$	Box and Jenkins (1976)
ARMA	$Y_t = \varphi_0 + \varphi_1 \cdot Y_{t-1} + \varphi_2 \cdot Y_{t-2} + \ldots + \varphi_p \cdot Y_{t-p} - \varepsilon_t - \theta_1 \cdot \varepsilon_{t-1} - \theta_2 \cdot \varepsilon_{t-2} - \ldots - \theta_q \cdot \varepsilon_{t-q}$	Box and Jenkins (1976)
ARCH	$Y_t = \beta_0 + \sum_{i=1}^{p} \left(\beta_i \cdot Y_{t-i} \right) + \varepsilon_t$ $\varepsilon_t = v_t \sqrt{h_t}$ $h_t = \alpha_0 + \sum_{i=1}^{p} \left(\alpha_i \cdot \varepsilon_{t-i}^2 \right)$	Engle (1982)
GARCH	$Y_t = \alpha_0 + \sum_{i=1}^{p} \left(\alpha_i \cdot Y_{t-i} \right) + \varepsilon_t$ $\varepsilon_t = v_t \sqrt{h_t}$ $h_t = \alpha_0 + \sum_{i=1}^{p} \left(\alpha_i \cdot \varepsilon_{t-i}^2 \right) + \sum_{j=1}^{q} \left(\beta_j \cdot h_{t-i} \right)$	Bollerslev (1986)
EGARCH	$\varepsilon_t = v_t \sqrt{h_t}$ $\log(h_t) = \alpha_0 + \sum_{j=1}^{q} \left[\beta_j \cdot \log(h_{t-i}) \right] + \sum_{i=1}^{p} \left[\theta_k \cdot \left(\frac{\varepsilon_{t-k}}{\sqrt{h_{t-k}}} \right) + \gamma_k \cdot \left(\frac{\varepsilon_{t-k}}{\sqrt{h_{t-k}}} \right) - \sqrt{\frac{2}{\pi}} \right]$	Nelson (1991)
GJR-GARCH	$\varepsilon_t = v_t \sqrt{h_t}$ $h_t = \alpha_0 + \sum_{i=1}^{p} \left(\alpha_i \cdot \varepsilon_{t-i}^2 \right) + \sum_{j=1}^{q} \left(\beta_j \cdot h_{t-i} \right) + \sum_{i=1}^{p} \left(\gamma_i \cdot \varepsilon_{t-i}^2 \cdot I_{\{\varepsilon_{t-i} \geq 0\}} \right)$	Glosten et al. (1993)
APARCH	$\varepsilon_t = v_t \sqrt{h_t}$ $h_t^\delta = \alpha_0 + \sum_{i=1}^{p} \alpha_i \left(\varepsilon_{t-i} - \gamma_i \cdot \varepsilon_{t-i} \right)^\delta + \sum_{j=1}^{q} \left(\beta_j \cdot h_{t-i}^\delta \right)$	Ding et al. (1993)

distribution or white noise with mean zero and variance 1, independent of the past error term. The advantage of ARCH models is that they account for the conditional variance being substantially affected by the squared residual term in any of the previous periods. GARCH models, which were elaborated by Bollerslev (1986), state that the conditional variance is an ARMA process. However, ARCH and GARCH models assume that positive and negative shocks have the same effects on volatility because the volatility depends on the square of the previous shocks. In practice, the price of a financial asset such as CO responds differently to positive and negative shocks. In order to overcome the cons of GARCH models, Nelson (1991) proposed the exponential GARCH (EGARCH) model, where the logarithmic conditional variance $\log(h_{t-i})$ is used to relax the positive constraint of model coefficients. Moreover, Glosten et al. (1993) proposed a nonnegativity condition that models asymmetric consequences of positive and negative shocks (i.e., $I_{\{\varepsilon_{t-i} \geq 0\}}$ is 1 if the shock is nonnegative, and zero elsewhere). Finally, the asymmetric power ARCH (APARCH) model of Ding et al. (1993) has the capability to estimate a power coefficient δ, which was assumed equal to 2 in all previous models, that provides a higher flexibility in terms of identification potential. For PSE/CAPE purposes, the most recent econometric model developed in Barzaghi et al. (2016) belongs to the AR model family and comes from the shock analysis of the moving-averaged CO price series. As aforementioned, Manca et al. (2011) demonstrated that price variations are normally distributed and can be assimilated to a stochastic variable by means of the autocorrelogram analysis. They showed that these attributes characterize a typical Markovian process (whose status does not depend from previous historical quotations but varies stochastically starting from the last price). On the contrary, Barzaghi et al. (2016) arrived to opposite deductions. Indeed, they demonstrated that the moving average trend of CO prices is not a Markovian process. As a matter of fact, Manca et al. (2011) worked on real spot values of CO prices, while Barzaghi et al. (2016) used moving-averaged values (with 4-month time spans). By doing so, they eliminated most of the high-frequency fluctuations of CO prices that can be assimilated to a background noise, which, on its turn, is responsible for the Markovian nature of a time series. The corresponding econometric model works with weekly or monthly prices and features a moving-average approach to quotations. It can be formulated as follows:

$$\text{Price}_t = (A + B \cdot \text{Price}_{t-1} + C \cdot \text{Price}_{t-2})(1 + \text{RAND} \cdot \sigma_{\text{CO}} + \bar{X}_{\text{CO}}), \tag{9.1}$$

where A, B, and C are adaptive coefficients that are calculated by means of a linear regression procedure, Price_t, Price_{t-1}^-, and Price_{t-2}^- are the CO moving-averaged prices, σ_{CO} and $\bar{X}_{\text{CO}}$ are the standard deviation and the mean of CO price shock over a suitable time interval, and RAND is a function that returns a random number, normally distributed, with average value 0 and standard deviation 1. The main power of this econometric model is that (1) it catches the trend followed by the prices, (2) allows studying possible future developments based on historical shock volatility, and (3) allows representing probabilistic forecast bounds thanks to the so-called fan-chart representation. Indeed, the time-discrete model of Equation 9.1 has to be used with a step-by-step approach over the prediction horizon of the problem to be solved in order to forecast possible price trajectories (i.e., scenarios), as shown in Figure 9.28. As highlighted in Rasello and Manca (2014), the econometric model should be

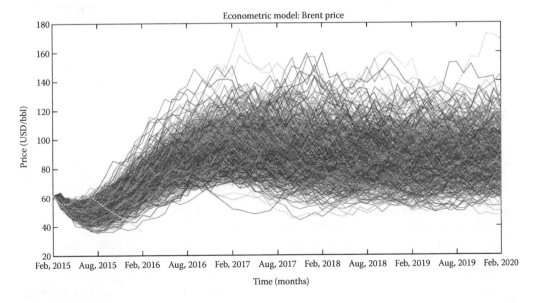

FIGURE 9.28 Fully predictive scenarios of Brent monthly quotations according to the econometric model of Equation 9.1 from February 2015 to February 2020 (500 simulations over 4 years). The yellow dotted line is the real price from February 2015 to February 2016. (From EIA, Energy Information Administration 2016. http://www.eia.gov, 2016.)

applied to the specific CO market that influences the economic assessment under consideration and, with respect to other model typologies, it can be used longer, i.e., up to 5 years.

9.2.5.3 Economic Models

The economic models involve purely economic variables (e.g., GDP, USD exchange rate, inventories, and production capacity) and simulate the fluctuations of spot and future markets with the supply and demand law. Most of these models provide some insights into economic fundamentals behind the market behavior, but they are not practically useful, as they require too much expertise and specific data to be easily implemented. The complexity of economic models can be attributed to the diversity of factors that determine an economic process such as the quotation of CO. These issues include resource limitations (e.g., nonrenewable oil resources), environmental, and geographical constraints (e.g., stocking capacity of Cushing reservoir in the United States), institutional and legal requirements (e.g., Russian sanctions, Iranian embargos), and purely random fluctuations (often due to the mood of investors originated by rumors and other subjective performance indexes). In finance, these predictive models have been used since the 1980s for trading and investment purposes, but to our knowledge, they have not been used in the PSE/CAPE domains, even if economic models can provide a valid contribution to scheduling and planning. The use of economic models on long-term horizons is somehow problematic and difficult to carry out, as there is need for long-term forecasts of the levels of supply, demand, capacity storage, technical advances in exploration and production of hydrocarbons, the political dynamics or changes in collective behaviors, and both national and international legislations. Instead, process designers and plant managers look for reliable models characterized by a simple structure and reduced number of parameters. However, the importance of the economic literature lies on the fact that it acknowledges the impact of physical variables on CO quotations. Indeed, it often discusses the role of excess production capacity, capacity utilization rate (Ye et al. 2009), and spare capacity (Chevallier 2014) on markets in general and on petroleum quotations in particular. The role of OPEC in the worldwide oil market—examined by both the press and the academic community—gave birth to a family of *OPEC behavior models*. These models are based on the evidenced hypothesis that OPEC decisions about quota and capacity utilization have a significant impact on oil prices (Dées et al. 2007; Cooper 2003; Kaufmann et al. 2004; Hamilton 2005). A revised form of the model of Kaufmann and coworkers is proposed in Depetri and Manca (2016):

$$\text{Price}_t = \alpha_0 + \alpha_1 \cdot \text{Days}_t + \alpha_2 \cdot \text{Days}_t + \alpha_3 \cdot \text{Cheat}_t + \alpha_4 \cdot \text{Caputil}_t + \alpha_5 \cdot \text{Delta}_t, \tag{9.2}$$

where

- Price_t is the CO quarterly price at time t (USD/bbl).
- Days_t is the number of days needed to consume current OECD CO stocks.
- Quotas_t is the OPEC production quota (Mbbl/day).
- Cheat_t is the difference between OPEC CO production and OPEC quotas (Mbbl/day).
- Caputil_t is the capacity utilization by OPEC, which is calculated by dividing OPEC production (Mbbl/d) by OPEC capacity of production (Mbbl/day).
- Delta_t is the oversupply of OPEC with respect to the US production in (Mbbl/day), which has fallen under 22 Mbbl/day since the fourth quarter of 2013.
- α_i is the model adaptive parameters whose values are calculated by means of a linear regression procedure and whose signs are representative of the supply and demand considerations of Paragraph 4.5. For instance, the regression coefficient associated with *Days* is negative because an increase in stocks reduces the price of CO by diminishing demand or increasing production. Conversely, the sign of the regression coefficient associated with *Delta* is positive because a decrease in the difference between OPEC and the US production causes the decrease of the CO quotations, as we have seen in recent quarters.

Price and input data have a quarterly (i.e., seasonal) granularity because this is the frequency of political decision about OPEC quotas and availability of supply and demand variables in the most renowned databanks (EIA 2016; IEA 2015). Although CO demand conditions are modeled correctly in Cooper (2003), the supply modeling for PSE/CAPE purposes is extremely difficult as oil markets reflect and translate the complex background of production conditions and OPEC behavior. Table 9.11 introduces the additional models needed to forecast the CO quotations by Equation 9.2 for OECD inventories, demand, OPEC production, production capacity, and the USA production, while OPEC quota deserves further considerations. The decision by OPEC of not cutting production quotas in December 2014 instead of inducing a CO price collapse and going back to the breakeven point of oil countries led to two alternative uses of the proposed model: a *OPEC-based model* and a *politically based model*.

The OPEC-based approach uses quota as an economic strategy and considers that the model, according to the OPEC behavior that has influenced the CO market, acts as the swing producer. Indeed, a forecast equation for OPEC quotas is not proposed, but if the prices decreased during a prolonged period (e.g., two or more quarters), the quotas would automatically change from 30 Mbbl/d to 27 Mbbl/d. On its turn, the decrease in CO production would make its quotations increase in the following few quarters. Eventually, this bullish trend would produce also a return of the quotas to the initial value of 30 Mbbl/d. On the contrary, in case the prices increased for two consecutive quarters the quotas would automatically change from 30 Mbbl/d to 33 Mbbl/d, in order to reequilibrate the prices.

Conversely, the politically-based approach is more concerned to equilibrate the production spread, which means both economic and political standing, than to reequilibrate CO prices. In this case, a general rule for OPEC quota is not tailored, but it contributes to the random features of modern international marketplaces. The cartel tries to produce less than its capacities allow so as to be able to intervene in case of sudden variations in CO markets, but allows also dropping below the breakeven point of member countries without intervening in a predetermined way. As far as price/cost fluctuations are concerned, the deterministic approach of the traditional economic models must be supplemented with stochastic elements, such as the frequency of input variable variations of both supply and demand terms, and major/minor increases/decreases of CO prices, which are described in Manca, Depetri, and Boisard (2015b). The model provides a distribution of supply and demand scenarios, which can be manipulated to create bullish, bearish, and conservative trend setups, and catch the effect of yearly global GDP, which is representative of the degree of economic activity in a country, region, and in the world (Depetri and Manca 2016).

Figure 9.29 clarifies the forecast trend scenarios obtained by both OPEC-based and politically-based models, with the OPEC-based one unable to predict the price crash of the second semester of 2015. As discussed in Depetri and Manca (2016), the probability of scenarios characterized by very high prices (never experienced before) is rather low and consequently has a reduced impact on the distribution of possible future highly bullish trends.

TABLE 9.11 List of Models to Forecast the Input Variables in Equation 9.2 Based on OPEC Behavior

Variable	Model
OECD demand	$\text{Demand}_{t+1}^{OECD} = \beta_0 \text{GDP}_{t+1} + \beta_1 \text{Price}_t + \beta_2$
OECD inventories	$\text{Inventory}_{t+1}^{OECD} = \gamma_0 + \gamma_1 \text{Capacity}_t^{OPEC} + \gamma_2 \text{Demand}_{t+1}^{OECD}$
OPEC production	$\text{Production}_{t+1}^{OPEC} = \xi_0 + \xi_1 \text{Capacity}_{t+1}^{OPEC} + \xi_2 \text{Price}_t$
OPEC production capacity	$\text{Capacity}_{t+1}^{OPEC} = \varepsilon_0 + \varepsilon_1 \text{Capacity}_t^{OPEC} + \varepsilon_2 \text{Production}_t^{OPEC}$
USA production	$\text{Production}_{t+1}^{USA} = \omega_0 + \omega_1 \text{Production}_t^{USA} + \omega_2 \text{Price}_t$

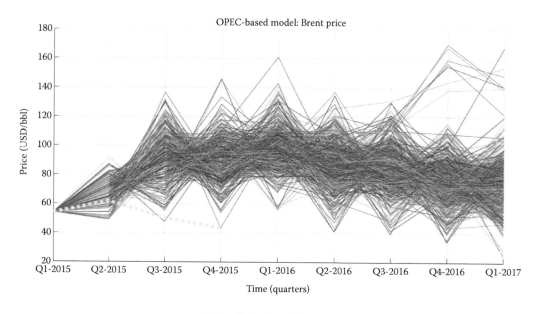

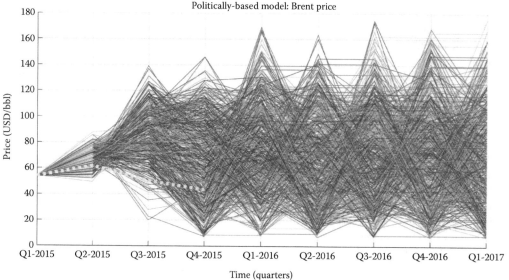

FIGURE 9.29 Fully predictive scenarios of Brent quarterly quotations according to (top) OPEC-based model and (bottom) politically based model from Q1-2015 to Q1-2017 (500 simulations over 8 quarters). The scenarios are obtained by Equation 9.2 and by considering the volatility terms as in Manca et al. (2015b). The yellow dotted line is the real price from Q1-2015 to Q4-2015. (From EIA, Energy Information Administration 2016. http://www.eia. gov, 2016.)

9.2.5.4 A Hybrid Model

The term *hybrid* designates something that is obtained from mixing two or more different components. In this context, hybrid signifies a model that derives from the combination of econometric models with economic ones, and that can simulate the trend originated by supply and demand law in combination with the stochastic fluctuations of CO quotations. The creation of a new hybrid model calls for two taxonomic classification criteria. First of all, the hybrid model combines two econometric and economic models, which are different for both the structure/mechanism used to predict the prices and the type of fundamentals provided about their trend (i.e., physical, economic, and financial features for the economic model; statistical analysis of past price shocks for the econometric model). In particular, the main advantage of economic models (such as the OPEC-based one) is that they involve supply and demand variables, which can be manipulated in order to create future scenarios, with an overall bullish or bearish trend, and simulate possible demand crisis, situations of oversupply, or economic and technological developments. These characteristics make the OPEC/politically-based model interesting also for commercial purposes because it can be implemented for scheduling, planning, and feasibility studies under market uncertainties and fluctuating factors. The feature of the OPEC/politically-based model, which can be improved, is the time granularity of the input variables and consequently of the predicted scenarios. Both price and input data of the OPEC/politically-based model have a quarterly time granularity. Conversely, econometric models may feature daily, weekly, or monthly discretizations. Even if they are not intended to follow the forces that cause price fluctuations, the econometric models can catch the oscillations that characterize CO quotations. The price scenarios obtained by an economic model do not have the swinging/noisy trend that the reader can observe on common trading websites. Indeed, the OPEC/politically-based models usually provide scenarios with frequent and abrupt key reversal points for adjacent quarters, but at the same time, their added value is to predict data that are based on real, objective, and economic, financial, political indicators. This new forecast data set is fed to the econometric model, which can reproduce also the fluctuations inside the quarters, with either a weekly or more commonly monthly granularity. At our knowledge, this procedure is innovative and there are not any hybrid models in the literature capable of combining the stochastic fluctuations to the supply and demand fluctuations. The difference of time granularity between the economic and econometric models is overcome by linear interpolation of the data set provided by the economic model. The quotations might also be interpolated by means of parabolas or cubic splines, but such curves would raise the question of which is the best representative of the price trend. The pseudo-real data provided by the OPEC/politically-based model and linear interpolation are fed to the econometric model that allows determining the adaptive parameters of Equation 9.2. Figure 9.30 shows the comparison between the real Brent price trend and

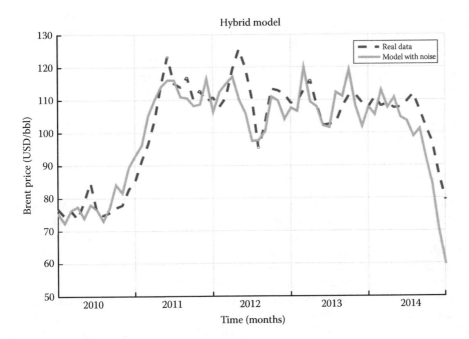

FIGURE 9.30 One-step-ahead simulation of Brent monthly by the hybrid model and comparison with real quotations from January 2010 to January 2015. (From EIA, Energy Information Administration 2016. http://www.eia.gov, 2016.)

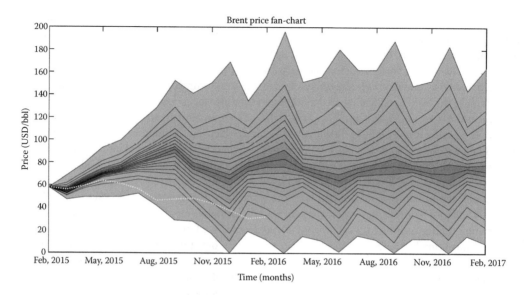

FIGURE 9.31 Fan-chart of Brent monthly prices from February 2015 to February 2017 (500 different simulations).

the hybrid model predictions featuring the background noise. Similarly to the procedure adopted for the econometric model and the economic one, the hybrid model can be adopted to deploy a distribution of fully predictive scenarios over a time horizon of a couple of years in the future for either scheduling or planning purposes under market uncertainties about supply, demand, and historical unbalances. Figure 9.31 shows the *river of blood* diagram obtained from February 2015 to February 2017 and allows observing that the price scenarios of the hybrid model have acquired a background noise that was originally missing in the quarterly data of the OPEC/politically-based economic model. The considerations on the probability distribution of scenarios are the same as ones made for the OPEC/politically-based models. In addition, it is worth observing that future prices belong to the 20–90 USD/bbl interval with a 95% confidence interval. Since the new hybrid model allows creating different price scenarios with a background noise that characterizes market prices, this is a rather promising tool for the optimal management of production sites as a function of the real market demand, global supply, market uncertainties, and historical balances/imbalances. For practical purposes, a probabilistic approach should be preferred than assuming a standard range of prices because it makes the user aware of the probability associated with a specific price range and with the different historical, economical, and technological situations behind that trend.

9.2.6 Economics of Derivatives

The scientific literature questioned on the role played by CO to forecast the price/cost of other raw materials and showed that it affects a number of industrial commodities (i.e., distillates and derived petrochemical commodities) and utilities (e.g., electric energy, hot water, steam) for the economic assessment and feasibility study of chemical plants (Manca 2013, 2016). In order to remove the limiting assumption of keeping constant the price/cost of commodities and utilities for long-term horizons (as it happens when conventional CD techniques are involved (Douglas 1988)), a few econometric models for industrial commodities (reactants, products, and byproducts of chemical processes) and utilities are proposed in this paragraph. Indeed, commodities and utilities contribute to the Operative Expenditures (OPEX) of chemical plants and more in general of the supply chain (Manca 2013). For instance, Rasello and Manca (2014) discussed the application of econometric models to the supply chain management of oil refineries. Mazzetto et al. (2013) used CO as the reference component for econometric models in bioprocesses and showed a functional dependency of both raw biomaterials and final bioproducts from CO markets.

The distinction between *commodity* and *utility* is not always clear because fuel oils distilled from CO are commodities that can be used also as fuels in the furnaces of chemical plants (i.e., utilities).

9.2.6.1 Economics of Commodities

This paragraph deals with the econometric models of commodity prices bound to those of the CO used for their production. Among the models proposed in Mazzetto et al. (2013), the autoregressive with distributed lag (ADL) model occupies the main role thanks to its simplicity and reduced number of adaptive parameters. ADL models allow modeling past quotations and determining future prices of commodities as a function of the previous ones. They are based on the general structure of Equation 9.3:

$$\text{ADL}(p,q) = c_0 + a_0 x_0 + a_1 x_1 + \cdots a_q x_q + b_1 y_1 + b_2 y_2 + \cdots b_p y_p, \tag{9.3}$$

where

- x_i is the price of the reference component, i.e., CO, at time $t–i$.
- y_i is the price of the specific commodity at time $t–i$.
- q and p are the orders of ADL model for the independent and dependent variables, respectively.
- c_0, a_i, and b_i are the adaptive parameters that are evaluated by a multidimensional unconstrained optimization procedure that minimizes the sum of squared errors between real quotations and model values.

A correlogram/autocorrelogram analysis allows determining the number of elements that the specific ADL model should include (as extensively discussed in Manca, Fini, and Oliosi (2011) and Manca (2013)). By applying this procedure, Manca (2013) and Barzaghi et al. (2016), respectively, provided the econometric models of the components involved in the HDA process (i.e., benzene and toluene) and in the styrene monomer plant (i.e., benzene, ethylene, ethylbenzene, and styrene). In particular, the toluene price depends on its previous value and the contemporary price of CO, while benzene price model includes also the dependence of two previous quotations of toluene, which is the main precursor in the hydro-dealkylation reaction. Ethylbenzene is the main reactant of the styrene production process and its main precursors are benzene and ethylene. Since almost all the ethylbenzene produced worldwide is converted to styrene monomer, the quotations of ethylbenzene are not easily available and the ethylbenzene price is estimated as the sum of its precursor prices and a term that accounts for both OPEX and Capital Expenditures (CAPEX) of the ethylbenzene plant as discussed in MacDonald, Roda, and Beresford (2005). The model of ethylene price works with the previous value of its own price and the contemporary value of the CO quotation. The product (i.e., styrene) cost proved to be correlated to its previous price and the contemporary quotations of ethylene and benzene. Similar remarks can be applied to gasoline with 90 and 93 RON and to diesel. Rasello and Manca (2014) pointed out that for these three distillates the model formulation works with one previous value of the corresponding commodity price and the present value of CO quotation. They reported different scenarios over a 15-week time horizon for scheduling purposes in the supply chain of fossil fuels. Table 9.12 reports the econometric models presented above.

Once the econometric models have been identified and validated over a suitable past time interval, they can be extrapolated over different time horizons (from weeks to years) to forecast a distribution of future price scenarios, as described in Paragraph 7.2 for the CO quotations, by accounting for a stochastic term that depends on the standard deviations of prices. Figure 9.32 shows some of these scenarios, which are forecast by the econometric models of commodities. Further scenarios for the styrene plant are reported in Manca, Conte, and Barzaghi (2015a) and Barzaghi et al. (2016).

9.2.6.2 Economics of Utilities

Even though there are several utilities that are used in chemical plants, such as electric energy (EE), steam, fuel oil, cooling water, and diathermic or cryogenic fluids, for the sake of brevity this paragraph

TABLE 9.12 List of Econometric Models Available in the Literature to Forecast Commodity Prices/Costs Correlated to CO Quotations

Commodity	Model	References
Benzene	$P_{Bz}(t) = A + B\ P_{CO}(t) + C\ P_{Bz}(t-1) + D\ P_{Tol}(t-1) + E\ P_{Tol}(t-2)$	Manca et al. (2015a) and Barzaghi et al. (2016)
Ethylene	$P_{Et}(t) = A + B\ P_{Et}(t-1) + C\ P_{CO}(t)$	Barzaghi et al. (2016)
Ethylbenzene	$P_{EtBz}(t) = P_{Et}(t) + P_{Bz}(t) + \text{production costs}$	Barzaghi et al. (2016)
Styrene	$P_{Sty}(t) = A + B\ P_{Sty}(t-1) + C\ P_{Et}(t) + D\ P_{Bz}(t)$	Barzaghi et al. (2016)
Toluene	$P_{Tol}(t) = A + B\ P_{CO}(t) + C\ P_{Tol}(t-1)$	Manca (2013)
Gasoline 90	$P_{G90} = A + B \cdot P_{CO}(t) + C \cdot P_{G90}(t-1)$	Rasello and Manca (2014)
Gasoline 93	$P_{G93} = A + B \cdot P_{CO}(t) + C \cdot P_{G93}(t-1)$	Rasello and Manca (2014)
Diesel	$P_{Diesel} = A + B \cdot P_{CO}(t) + C \cdot P_{Diesel}(t-1)$	Rasello and Manca (2014)

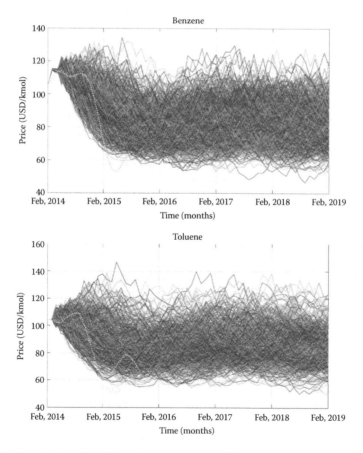

FIGURE 9.32 Five hundred possible fully predictive scenarios of (Top) benzene and (Bottom) toluene monthly prices from February 2014 to February 2019. The yellow dotted line is the real moving averaged price from February 2014 to July 2015 for benzene and from February 2014 to October 2015 for toluene. (ICIS, Trusted market intelligence for the global chemical, energy, and fertilizer 2016, http://www.icis.com/.)

TABLE 9.13 List of Econometric Models Available in the Literature to Forecast Utility Prices/Costs Correlated to CO Quotations

Utility	Model	References
Fuel oil	$P_{Fuel\,oil} = 0.008329\,P_{CO}$	Barzaghi et al. (2016)
Steam	$P_{Steam} = 0.009762\,P_{CO}$	Barzaghi et al. (2016)
Electric energy	Equation 9.4	Manca (2016)

focuses on the ones involved in HDA and styrene production processes ((Manca 2013); (Barzaghi et al. 2016)). Gasoline and diesel commodities can be considered also as material utilities used for heating purposes in chemical plants and their prices have already been modeled in Paragraph 8.1. The main utilities used are the fuel oil in the furnaces in both the plants and the inlet low-pressure steam in the styrene monomer plant (Luyben 2011). Both their prices are assumed proportional to that of CO with the proportionality constants shown in Table 9.13 with $P_{Fuel\,oil}$ in USD/kg, P_{Steam} in USD/kmol, and P_{CO} in USD/bbl.

9.2.6.2.1 Electric Energy

Conversely, EE follows a more complicated behavior with respect to CO quotations, which means that there is not a direct proportionality as it happens for fuel oil and steam (as shown in Table 9.13). EE prices are primarily affected by the contract typology signed between the end user and the producer/seller. Manca (2016) analyzed in detail how the EE quotations can oscillate periodically with different characteristic times, i.e., from days to weeks, and even up to seasons. In addition, for the Italian market, he found a significant correlation and time delay between EE and CO prices, with the time delay approximating an average value of 15 weeks that are nearly 3 months, i.e., a quarter. Manca (2016) proposed Equation 9.4 to forecast the EE prices as a function of CO quotations:

$$P_{EE,i,j} = a_j + b_j \cdot P_{CO,i-t_d} + c_j \cdot \sin\left(\frac{2\pi \cdot i}{T_j} + \varphi_j\right), \tag{9.4}$$

where

- $P_{EE,i,j}$ is the EE price (€/MWh) at the ith discrete time for the jth interval of the day (i.e., morning, afternoon, evening, night).
- a and b are the constants of the linear model that try to catch in a very simplified way the long-term dependency of EE from CO prices.
- c is the proportional constant of the periodic term based on T period and φ phase. Interestingly, when the model is identified by means of a nonlinear regression routine, which minimizes the distance between real past quotations and model predictions, T results usually equal to 52 weeks, which is 1 year. This means an annual periodicity of EE prices.
- t_a is the time delay of the EE quotations with respect to those of P_{CO}, which is the CO price (US$/bbl). For consistency reasons, T and t_a have the same time units of the discretized quotations of CO and EE.

Apparently, Equation 9.4 does not account for any stochastic contribution. Conversely, the stochastic term is embedded in the CO quotation, as outlined in Equation 9.1 and discussed in Paragraph 7.2.

Related to EE prices, the optimization of utilities allocation and consumption is an important application of PSE/CAPE problems when dynamic revamping and retrofitting are concerned. Focusing again on the HDA process, we can highlight a significant enthalpic content of the outlet stream from the reactor that can be exploited to partially preheat the inlet stream in a feed effluent heat exchanger (FEHE) or to produce EE in a different equipment setup called energy production section (EPS). Opting for an alternative operation, one can suppose that the EPS is operated when the revenues from selling the EE are high, while it could be more economically viable operating the FEHE when the price of EE is low, thus saving money by burning a reduced amount of fuel oil in the furnace (Manca and Grana 2010). Without having specific price models of commodities and utilities, these PSE/CAPE applications would not be possible.

9.2.7 Applications Discussion

This chapter focused on the economic elements and forecast of CO markets and prices. The proposed models that account for the time series features and the economic elements behind oil trading can be applied to a plethora of PSE/CAPE applications. A branch of these applications contributes to the optimal design and operation of industrial processes and plants from short- to long-term horizons. Since the late 1980s, process industry has used the CD approach summarized by Douglas (1988). CD is based on a hierarchical approach to plant/process economics structured into five decision levels. CD calls for the evaluation of four economic indexes (aka Economic Potentials) that depend on the OPEX, CAPEX, and revenues of the plants. The prices of commodities and utilities that characterize both OPEX terms and revenues are assumed fixed, even if demand and offer fluctuations and historical uncertainties have always shown the contrary. Some improvements in the solution of PSE/CAPE problems are available in Grana, Manenti, and Manca (2009), Manca and Grana (2010), Manca, Fini, and Oliosi (2011), Manca (2015), and Barzaghi et al. (2016). These papers provide some examples of DCD and PCD procedures. DCD optimizes the design of a plant by considering the historical price time series and by finding a single optimal configuration. Instead, PCD uses the econometric/economic/hybrid models of CO, commodities, and utilities (discussed in this chapter) to devise a set of possible future scenarios of the prices/costs of both process components and utilities, and finds an optimal plant design configuration for each scenario. Based on the future price/cost scenarios, PCD moves from a deterministic approach to a probabilistic approach that delivers a distribution of the optimal values of the design variables and economic indexes. These optimal design configurations proved to be few in case of the styrene monomer plant (Barzaghi et al. 2016).

Prices/costs volatility and demand fluctuations are operational uncertainties that affect not only the design procedure but also the supply-chain operations and production strategies. As price/cost scenarios are based on a stochastic contribution, it is important to understand possible causes of random variations at each temporal scale. With regard to short- and medium-term applications, the economic model (either OPEC-based or politically-based) can be used to create several distinct scenarios that correspond to the different historical or technological situations. According to the geographical position of the chemical plant/refinery, a process designer/manager can either schedule or plan the plant/process/production/resources by analyzing the probability distributions of the quotation scenarios.

9.2.8 Conclusions

The aim of this chapter was to present and discuss the economics of CO and derivatives for PSE/CAPE applications, such as planning, scheduling, and feasibility studies of chemical plants, under market and historical uncertainties. Starting from a number of econometric and economic models proposed in literature and the analysis of prices volatility, the manuscript described the dynamic models capable of forecasting the evolution of CO quotations over different time horizons. CO was chosen as the cornerstone of economic/econometric models because it is the reference component of both the chemical and oil and gas industries. A number of historical events influenced the CO quotations in recent periods. For instance, the increasing concern about shale oil spread in the United States, the Greek crisis, the withdrawal of the embargo to Iran, and the Chinese crisis of stock exchanges are just a few examples. Furthermore, also the economic background is changed and is more complex than the one of 5–10 years ago, as OPEC and OECD do not include the so-called BRIC countries and other emerging markets such as Argentina and South Africa. The proposed price models (i.e., econometric, economic, and hybrid models) can be used to simulate a number of future scenarios based on stochastic contributions from CO shocks and supply and demand variable distributions. These features lead to a probabilistic approach to scheduling, planning, DCD, and other PSE/CAPE problems in general.

List of Abbreviations

ADL	autoregressive distributed lag model
AMEX	American stock exchange
APARCH	asymmetric power autoregressive conditional heteroskedasticity model
AR	autoregressive model
ARCH	autoregressive conditional heteroskedasticity model
ARMA	mixed autoregressive moving average model
BFOE	Brent, Forties, Oseberg, and Ekofisk oil fields in the North Sea
BRIC	Brazil, Russia, India, China
CAPE	computer aided process engineering
CAPEX	capital expenditures
CO	crude oil
CD	conceptual design
DCD	dynamic conceptual design
EE	electric energy
EGARCH	exponential generalized autoregressive conditional heteroskedastic model
EIA	Energy Information Administration
EPS	energy production section
EROI	energy return on energy investment
FEHE	feed effluent heat exchanger
GARCH	generalized autoregressive conditional heteroskedastic models
GDP	gross domestic product
GJR-GARCH	Glosten–Jagannathan–Runkle GARCH
HDA	hydro-de-alkylation (process)
IEA	International Energy Agency
MA	moving average model
NYMEX	New York Mercantile Exchange
OECD	Organization for Economic Cooperation and Development
OPEC	Organization of Petroleum Exporting Countries
OPEX	operative expenditures
PCD	predictive conceptual design
PSE	process systems engineering
USD	United States dollar
WTI	West Texas Intermediate

References

Agnoli, S. 2014. Petrolio sotto quota 90: la reazione di Eni & Co. *Corriere della Sera*, November 4.

Bardi, U. 2009. Peak oil: The four stages of a new idea. *Energy* 34 (3):323–326. doi:10.1016/j.energy.2008.08.015.

Bardi, U., and A. Lavacchi. 2009. A simple interpretation of Hubbert's model of resource exploitation. *Energies* 2 (3):646–661. doi:10.3390/en20300646.

Barzaghi, R., A. Conte, P. Sepiacci, and D. Manca. 2016. Optimal design of a styrene monomer plant under market volatility. In *Computer Aided Process Engineering*. Submitted.

Bassett, E. 2014. Few in US oil, gas realm very concerned about possible oil price collapse: Survey. In *Platts.com*. Houston, TX: McGraw Hill Financial.

Bollerslev, T. 1986. Generalized autoregressive conditional heteroskedasticity. *Journal of Econometrics* 31 (3):307–327. doi:10.1016/0304-4076(86)90063-1.

Box, G.E.P., and G.M. Jenkins. 1976. Time series analysis: Forecasting and control. In *Holden-Day Series in Time Series Analysis*. Holden-Day.

BP. 2014. Statistical review of world energy June 2014.

BP. 2015. Statistical review of world energy June 2015.

Chatfield, C. 2000. *Time-Series Forecasting*. Boca Raton, FL: CRC Press.

Chatfield, C. 2013. *The Analysis of Time Series: An Introduction*. Boca Raton, FL: CRC Press.

Chen, S.S. 2014. Forecasting crude oil price movements with oil-sensitive stocks. *Economic Inquiry* 52 (2):830–844. doi:10.1111/ecin.12053.

Chevallier, J. 2014. The 2008 oil price swing or the quest for a 'smoking gun'. In *Crude Oils: Production, Environmental Impacts and Global Market Challenges*, 39–69.

Cifarelli, G., and G. Paladino. 2010. Oil price dynamics and speculation: A multivariate financial approach. *Energy Economics* 32 (2):363–372. doi:10.1016/j.eneco.2009.08.014.

Commandeur, J.J., and S.J. Koopman. 2007. *An Introduction to State Space Time Series Analysis*. Oxford, UK: Oxford University Press.

Cooper, J.C. 2003. Price elasticity of demand for crude oil: Estimates for 23 countries. *OPEC Review* 27 (1):1–8.

Crooks, E. 2011. Oil sands: Stable source seen as good investment. *Financial Times* October 10.

Dahmani, A., and M.H. Al-Osaimy. 2001. OPEC oil production and market fundamentals: A causality relationship. *OPEC Review* 25 (4):315–337.

Davis, N., and N. Fleming. 2014. The oil price crash and what it means for the petrochemical industry, ICIS.

Dées, S., P. Karadeloglou, R.K. Kaufmann, and M. Sánchez. 2007. Modelling the world oil market: Assessment of a quarterly econometric model. *Energy Policy* 35 (1):178–191. doi:10.1016/j.enpol.2005.10.017.

Depetri, V., and D. Manca. 2016. The impact of OPEC and OECD fundamentals on crude oil prices: A revised economic model. *Energy Policy* Submitted.

Ding, Z., C.W.J. Granger, and R.F. Engle. 1993. A long memory property of stock market returns and a new model. *Journal of Empirical Finance* 1 (1):83–106. doi:10.1016/0927-5398(93)90006-D.

Douglas, J.M. 1988. *Conceptual Design of Chemical Processes*. New York: McGraw-Hill.

Dowling, M., M. Cummins, and B.M. Lucey. 2016. Psychological barriers in oil futures markets. *Energy Economics* 53:293–304. doi:10.1016/j.eneco.2014.03.022.

Dvir, E., and K. Rogoff. 2014. Demand effects and speculation in oil markets: Theory and evidence. *Journal of International Money and Finance* 42:113–128. doi:10.1016/j.jimonfin.2013.08.007.

EIA. 2014. International Energy Outlook. Washington, DC: Energy Information Administration.

EIA. 2015a. Annual Energy Outlook 2015. US Energy Information Administration.

EIA. 2015b. Petroleum and other liquids prices 2015b. http://www.eia.gov.

EIA. 2016. Energy Information Administration 2016. http://www.eia.gov.

Engle, R.F. 1982. Autoregressive conditional heteroscedasticity with estimates of the variance of United Kingdom inflation. *Econometrica: Journal of the Econometric Society* 987–1007.

Fadhil, M. 2015. Iran nuclear deal to alter Mideast polymer trade flows. https://www.icis.com/resources/news/2015/07/15/9903538/iran-nuclear-deal-to-alter-mideast-polymer-trade-flows/.

Fadhil, M., and J. Dennis. 2015. Iran nuclear deal reached: Crude oversupply concerns heighten. http://www.hellenicshippingnews.com/iran-nuclear-deal-reached-crude-oversupply-concerns-heighten/.

Ghaffari, A., and S. Zare. 2009. A novel algorithm for prediction of crude oil price variation based on soft computing. *Energy Economics* 31 (4):531–536. doi:10.1016/j.eneco.2009.01.006.

Glosten, L.R., R. Jagannathan, and D.E. Runkle. 1993. On the relation between the expected value and the volatility of the nominal excess return on stocks. *The Journal of Finance* 48 (5):1779–1801.

Goldman Sachs. 2014. Unlocking the economic potential of North America's energy resources. New York: Global Markets Institute, Goldman Sachs Global Investment Research.

Grana, R., F. Manenti, and D. Manca. 2009. Towards dynamic conceptual design. *Computer Aided Chemical Engineering* 26:665–670. doi:10.1016/S1570-7946(09)70111-7.

Hamilton, J.D. 1983. Oil and the macroeconomy since World War II. *The Journal of Political Economy* 91 (2):228–248.

Hamilton, J.D. 2005. Oil and the macroeconomy. In *The New Palgrave Dictionary of Economics*, 201–228. London: Palgrave Macmillan.

Horn, M. 2004. OPEC's optimal crude oil price. *Energy Policy* 32 (2):269–280. doi:10.1016/S0301-4215(02)00289-6.

Hoyos, C. 2007. The new seven sisters: Oil and gas giants dwarf western rivals. *Financial Times*, March 12.

Hubbert, M.K. 1956. Nuclear energy and the fossil fuel. Paper read at drilling and production practice.

ICIS. Trusted market intelligence for the global chemical, energy, and fertilizer 2016. http://www.icis.com/.

IEA. International Energy Agency 2015. http://www.iea.org.

Kang, S.H., and S.M. Yoon. 2013. Modeling and forecasting the volatility of petroleum futures prices. *Energy Economics* 36:354–362. doi:10.1016/j.eneco.2012.09.010.

Kao, C.W., and J.Y. Wan. 2012. Price discount, inventories and the distortion of WTI benchmark. *Energy Economics* 34 (1):117–124. doi:10.1016/j.eneco.2011.03.004.

Kaufmann, R.K., S. Dees, P. Karadeloglou, and M. Sánchez. 2004. Does OPEC matter? An econometric analysis of oil prices. *Energy Journal* 25 (4):67–90.

Kaufmann, R., P. Karadeloglou, and F. Di Mauro. 2008. Will oil prices decline over the long run? *ECB Occasional Paper*, 98.

Kaufmann, R.K., and B. Ullman. 2009. Oil prices, speculation, and fundamentals: Interpreting causal relations among spot and futures prices. *Energy Economics* 31 (4):550–558. doi:10.1016/j.eneco.2009.01.013.

Kilian, L. 2009. Not all oil price shocks are alike: Disentangling demand and supply shocks in the crude oil market. *American Economic Review* 99 (3):1053–1069. doi: 10.1257/aer.99.3.1053.

Li, L. 2014. Market review: China oil market, ICIS.

Liu, W.M., E. Schultz, and J. Swieringa. 2015. Price dynamics in global crude oil markets. *Journal of Futures Markets* 35 (2):148–162. doi:10.1002/fut.21658.

Luyben, W.L. 2011. Design and control of the styrene process. *Industrial and Engineering Chemistry Research* 50 (3):1231–1246. doi:10.1021/ie100023s.

MacDonald, J., R. Roda, and M. Beresford. 2005. *Liquid Phase Alkylation of Benzene with Ethylene*. Halifax, Canada: Dalhousie University.

Manca, D. 2013. Modeling the commodity fluctuations of OPEX terms. *Computers and Chemical Engineering* 57:3–9. doi:10.1016/j.compchemeng.2013.04.018.

Manca, D. 2015. Economic sustainability of products and processes, Chapter 25. In *Sustainability of Products, Processes and Supply Chains: Theory and Applications*, Fengqi You (Ed.), Elsevier, Amsterdam, Netherlands, Computer Aided Chemical Engineering 36:615–642. doi:10.1016/B978-0-444-63472-6.00025-2.

Manca, D. 2016. Price model of electrical energy for PSE applications. *Computers and Chemical Engineering* 84:208–216. doi:10.1016/j.compchemeng.2015.08.013.

Manca, D., R. Barzaghi, and A. Conte. 2016. Optimal design of a styrene monomer plant under market uncertainty. *Industrial and Engineering Chemistry Research* Submitted.

Manca, D., A. Conte, and R. Barzaghi. 2015a. How to account for market volatility in the conceptual design of chemical processes. *Chemical Engineering Transactions* 43:1333–1338. doi:10.3303/CET1543223.

Manca, D., V. Depetri, and C. Boisard. 2015b. A crude oil economic model for PSE applications. *Computer Aided Chemical Engineering* 37:491–496. doi:10.1016/B978-0-444-63578-5.50077-3.

Manca, D., A. Fini, and M. Oliosi. 2011. Dynamic conceptual design under market uncertainty and price volatility. In *Computer Aided Chemical Engineering* 29:336–340. doi:10.1016/B978-0-444-53711-9.50068-7.

Manca, D., and R. Grana. 2010. Dynamic conceptual design of industrial processes. *Computers and Chemical Engineering* 34 (5):656–667. doi:10.1016/j.compchemeng.2010.01.004.

Maugeri, L. 2012. *Oil: The Next Revolution*. edited by Belfer Center for Science and International Affairs. Cambridge, MA: Harvard Kennedy School.

Mazzetto, F., R.A. Ortiz-Gutiérrez, D. Manca, and F. Bezzo. 2013. Strategic design of bioethanol supply chains including commodity market dynamics. *Industrial and Engineering Chemistry Research* 52 (30):10305–10316. doi:10.1021/ie401226w.

Miller, R.G., and S.R. Sorrell. 2014. The future of oil supply. *Philosophical Transactions of the Royal Society A: Mathematical, Physical and Engineering Sciences* 372 (2006). doi:10.1098/rsta.2013.0179.

Nelson, D.B. 1991. Conditional heteroskedasticity in asset returns: A new approach. *Econometrica: Journal of the Econometric Society* 347–370.

Oil and Gas Journal. 2015. US petroleum product exports rise for 13th consecutive year.

OPEC. 2015. OPEC monthly oil market report. Vienna: Organization of Petroleum Exporting Countries.

Rasello, R., and D. Manca. 2014. Stochastic price/cost models for supply chain management of refineries. *Computer Aided Chemical Engineering* 33:433–438. doi:10.1016/B9 78-0-444-63456-6.50073-9.

Salisu, A.A., and I.O. Fasanya. 2013. Modelling oil price volatility with structural breaks. *Energy Policy* 52:554–562. doi:10.1016/j.enpol.2012.10.003.

Sandrea, R. 2003. OPEC's challenge: Rethinking its quota system. *Oil and Gas Journal* 101 (29):31–36.

Sen, A. 2012. Oil benchmarks in international trade. Paper read at Oxford Energy Forum.

Smirnov, D.A. 2010. *Extracting Knowledge From Time Series*. Berlin and Heidelberg: Springer-Verlag.

Suratman, N. 2015. OPEC to welcome back Indonesia as member after seven-year absence. https://www.icis.com/resources/news/2015/06/05/9892321/opec-to-welcome-back-indonesia-as-member-after-seven-year-absence/.

Yao, K. 2014. China third-quarter GDP growth seen at five-year low of 7.3 percent, more stimuli expected. *Reuters*, October 10.

Ye, M., J. Zyren, C.J. Blumberg, and J. Shore. 2009. A short-run crude oil price forecast model with ratchet effect. *Atlantic Economic Journal* 37 (1):37–50.

Zagaglia, P. 2010. Macroeconomic factors and oil futures prices: A data-rich model. *Energy Economics* 32 (2):409–417. doi:10.1016/j.eneco.2009.11.003.

Zhang, X., K.K. Lai, and S.Y. Wang. 2008. A new approach for crude oil price analysis based on empirical mode decomposition. *Energy Economics* 30 (3):905–918. doi:10.1016/j.eneco.2007.02.012.

10

Bio Fuel

Sumaiya Zainal
Abidin
Universiti Malaysia Pahang

Basudeb Saha
*London South Bank
University*

Raj Patel,
Amir Khan, and
I. M. Mujtaba
University of Bradford

Richard Butterfield
Swansea University

Elisabetta Mercuri
and Davide Manca
*Polytechnic University
of Milan*

10.1 Environmentally Benign Biodiesel Production from Renewable Sources

Sumaiya Zainal Abidin and Basudeb Saha

10.1.1 Introduction

Renewable energy has become an important alternative resource in many countries and considered to be a potential substitute to the conventional fossil fuel. In particular, renewable energy in the form of biodiesel is considered to be one of the best available energy resources (Abidin, 2012; Atabani et al., 2012; Liu et al., 2012). As the fuel's feedstock is originated from renewable sources, this type of fuel is well known to be biodegradable and environment friendly (Kaercher et al., 2013). Apart from this, it also owns a good combustion profile, produces less particulates, i.e., unburned hydrocarbon and hazardous gases (i.e., carbon monoxide, sulfur dioxide), has a higher cetane number, higher flash point, and higher lubricity (Lin et al., 2011) compared to conventional diesel. Biodiesel, comprises monoalkyl esters of fatty acids, is derived from renewable lipid feedstocks, such as edible oil (i.e., palm, sunflower, and soybean) non-edible oils (i.e., jatropha and mahua), animal fats (chicken and lard), and algae. The cost of feedstock alone comprises 75%–85% of the overall cost of biodiesel production (Abbaszaadeh et al., 2012; Atabani et al., 2012). Currently, the popular feedstocks for biodiesel production are the edible oils; however, this was restricted due to the higher price of vegetable oil. The use of vegetable oils in biodiesel production also creates controversial issues on the usage of food elements as the source of fuels.

Cheap, nonedible used cooking oil (UCO) has been found to be an effective feedstock to reduce the cost of biodiesel (Stamenković et al., 2011, Pinzi et al., 2014). According to Balat (2011), UCO is 2.5–3.5 times cheaper than virgin vegetable oils and thus can reduce 60%–90% of the total production cost of biodiesel (Talebian-Kiakalaieh et al., 2013). Yaakob et al. (2013) reported that UCO can reduce water pollution and also prevents blockage in water drainage systems. As these feedstocks contain high amount of free fatty acids (FFA) and water, it cannot be directly used in a base catalyzed transesterification reaction. FFA can react with a base catalyst (neutralization reaction) and accelerates the base catalyst consumption. The high FFA content also causes saponification during the base catalyzed transesterification and lowers the yield of biodiesel. Acid catalyst was found to have better tolerance to high water and FFA content, whereas the base catalysts that are very sensitive toward water and FFA, proved to be effective for transesterification of feedstocks with low FFA content. Therefore, high yield could be achieved using a two-step synthesis of biodiesel. A pretreatment stage (esterification process) is used to reduce the amount of FFA in the feedstock before the reaction proceeds with the base catalyzed transesterification.

The use of heterogeneous catalysts simplifies the production and purification processes because they can be easily separated from the reaction mixture, allowing multiple usage of the catalyst through regeneration process. Among various kinds of acid and base catalysts, ion exchange resins are becoming more popular nowadays because this type of catalyst can catalyze the reactions under mild reaction conditions due to their high concentration of acid/base sites (López et al., 2007). It is an attractive alternative because it is easy to separate and recover from the product mixture. Sulfonated cation exchange resin has been found to be one of the most effective catalysts for esterification of FFA (Tesser et al., 2005; Özbay et al., 2008; Russbueldt and Hoelderich, 2009; Talukder et al., 2009). Few researchers have conducted studies to compare the performance of gelular and macroeticular ion exchange resins. Kouzu et al. (2011) conducted a study on the performance of gelular (Amberlyst 31) and macroeticular (Amberlyst 15) matrices for the esterification of soybean oil. It was found that the gelular resin has a higher catalytic activity compared to macroeticular resin as the swelling capacity controls the accessibility of acid sites in the catalyst, and it simultaneously affects the overall reactivity. A contradictory finding was achieved by Feng et al. (2010) when they conducted a study on different matrix types of cation exchange resins. Three types of resins were employed in this study, namely NKC-9 (macroeticular), 001 × 7 (gelular), and D61 (macroeticular). The highest FFA conversion was obtained using NKC-9 and this resin also showed a good conversion in the reusability study. A similar result was reported by Özbay et al. (2008) when they investigated the esterification of waste cooking oil using Amberlyst 15, Amberlyst 35, Amberlyst 16 (macroeticular), and Dowex HCR-W2 (gelular) as catalysts; Amberlyst 15 was found to give the highest FFA conversion.

There were also a few studies of the esterification process that focused on the macroeticular cation exchange resins as catalysts. Bianchi et al. (2009) studied the deacidification of animal fats using several types of macroeticular cation exchange resins as catalysts, i.e., Amberlyst 15Dry, Amberlyst 36Dry, Amberlyst 39Wet, Amberlyst 40Wet, Amberlyst 46Wet, and Amberlyst 70Wet. From this study, more than 90% of the FFA conversion was successfully achieved when Amberlyst 70Wet was used as the catalyst. This catalyst also showed a good conversion rate in the reusability study and performed well even in less severe operating condition (303 K with 1.25 wt% catalyst). Park et al. (2008) studied the performance of two different macroeticular cation exchange resin catalysts, Amberlyst 15 and Amberlyst BD20. They found that the amount of pores of the catalyst played an important role not only in increasing the catalytic activity but also in reducing the inhibition by water in the esterification process. Study on the comparison between strongly acidic macroporous cation exchange resin (Amberlyst 36) and strongly acidic hypercrosslinked resin (Purolite D5081) has been conducted by Abidin et al. (2012). They found that the catalytic performance of the hypercrosslinked sulfonic acid resin was superior to that of the macroporous catalyst due to the presence of high specific surface area. Comparison between Purolite D5081 and Novozyme 435 was also studied by Haigh et al. (2012, 2013). It was found that Purolite D5081 resin gives a higher conversion and negligible side reactions compared to Novozyme 435. Ilgen (2014) studied on the kinetics and mechanism of oleic acid esterification using Amberlyst 46 as a catalyst. The highest conversion was obtained when the reaction was performed at 3:1 molar ratio of methanol to oleic acid, 100°C reaction temperature, and 15% catalyst loading in 2 h reaction time. From the kinetic study, Eley Rideal mechanism was found to give the best match to the experimental data, and thus surface reaction was found to be the rate limiting step. Shibasaki-Kitakawa et al. (2015) studied continuous production of biodiesel from water rice bran acid oil with 95% of FFA content. The experiment was conducted in a packed column using PK208LH ion exchange resin as a catalyst. It was found that the concentration of FFA was close to zero after 90 h of reaction. However, a decrease in fatty acid methyl ester (FAME) concentration was detected after 90 h of reaction due to the accumulation of water formed by esterification within the resin.

Transesterification of biodiesel is normally conducted using anion exchange resins due to the high time consumption and molar ratio requirement when cation exchange resin is used as a catalyst (dos Reis et al., 2005). A comparison study between anion and cation exchange resin has been carried out by Li et al. (2012). Four types of ion exchange resins, namely Amberlyst 15 (cation), Amberlite IRC-72 (cation), Amberlite IRA-900 (anion), and Amberlite IRC-93 (anion), have been assessed in the transesterification of yellow horn (*Xanthoceras sorbifolia* Bunge.) seed oil. Amberlite IRA-900 was reported to have the highest conversion of 96.3%. Amberlyst 15 was also found to give good conversion (83.5%), however, it is still considered a weak catalyst compared to Amberlite IRA-900. Falco et al. (2010) also conducted a study on the transesterification of soybean oil using basic and acidic ion exchange resins as catalysts. In this study, a strongly basic anion exchange resin, BR 1 was reported to give the highest conversion and the selectivity of FAME could potentially reach 100%. Shibasaki-Kitakawa et al. (2007) investigated the potential of anion exchange resin as the heterogeneous catalyst. Several types of anion exchange catalysts have been tested for the transesterification of triolein, namely the Diaion PA308, Diaion PA306, Diaion PA306s, and HPA 25. Anion exchange resin with a lower cross-linking density and a smaller particle size, Diaion PA306s, was proved to give the highest catalytic activity and resulted in approximately 98.8% purity of biodiesel fuel. In their latest research, Shibasaki-Kitakawa et al. (2011) also reported that Diaion PA306s catalyst could act as both catalyst and adsorbent in the transesterification reaction of waste cooking oil with 1% FFA content. He et al. (2015) studied the continuous two-stage esterification–transesterification reaction in acidic oil using a combination of cation (NKC-9) and anion (D261) exchange resins as catalysts. The reaction was found to convert 95.1% of the acidic oil (combination of oleic acid and soybean oil) to biodiesel. Furthermore, the biodiesel product was also found to meet the biodiesel standard requirement of Chinese Standard.

This chapter concentrates on the production of biodiesel from UCO using two-stage esterification–transesterification catalytic reactions with ion exchange resins as catalysts. This work was conducted in collaboration with Purolite International Limited for possible commercialization of novel Purolite ion exchange resins as potential biodiesel production catalysts. The influence of the following reaction parameters, i.e., mass transfer resistance, catalyst loading, reaction temperature, methanol to oil feed mole ratio, and reusability of catalyst, was investigated.

10.1.2 Materials and Method

10.1.2.1 Materials

The UCO was supplied by Greenfuel Oil Company Limited, UK, with the acid value of 12 mg KOH/g oil. Ion exchange resin catalysts (Purolite CT-122, Purolite CT-169, Purolite CT-175, Purolite CT-275, and Purolite D5081, Purolite D5082) were supplied by Purolite International Limited (UK); Diaion PA306s was supplied by Mitsubishi Chemicals (Japan) and Amberlyst 36 was purchased from Sigma Aldrich, UK. All resins were supplied in the wet form. Methanol (>99.5% purity), sodium hydroxide (98+%) pellets, 0.1 M standardized solution hydrochloric acid, 0.1 M standardized solution sodium hydroxide, 0.1 M standardized solution sodium hydroxide in 2-propanol, toluene (99.5%), 2-propanol (99+%), glacial acetic acid (99.85%), chloroform (>99%), sodium chloride, phenolphthalein, iso-octane (>99.5), and acetonitrile (>99.8%) were purchased from Fisher Scientific, UK, and *p*-naphtholbenzein, *n*-hexane, methyl heptadecanoate (>99%), methyl linoleate (>99%), methyl linolenate (>99%), methyl oleate (>99%), methyl palmitate (>99%), and methyl stearate (>99%) were purchased from Sigma Aldrich, UK.

10.1.2.2 Catalyst Preparation

Two types of resins were used in this research work, i.e., cation exchange resin and anion exchange resin. All of these resins were supplied in wet form. Purolite CT-122, Purolite CT-169, Purolite CT-175, Purolite CT-275, Amberlyst 36, Purolite D5081, and Purolite D5082 are classified as strongly acidic cation exchange resins, whereas Diaion PA306s is classified as a strongly basic anion exchange resin, supplied in chloride form. These resins were pretreated before being used as the reaction catalysts. For cation exchange resin, all resins were immersed in methanol overnight and pretreated with methanol in an ultrasonic bath. The process takes a few cycles of rinsing to ensure that all contaminants were removed. The conductivity of the residual solution was recorded and the process continued until the conductivity of the residual solution was approximately the same with the solvent. Finally, the resins were dried in a vacuum oven at 373 K for 6 h to remove any water and methanol. The dried catalyst was kept in a sealed bottle prior to use. The anion exchange resin, Diaion PA306s was prepared prior to use. It was mixed with 1 M of sodium hydroxide (NaOH) to displace the chloride ions to hydroxyl ions. After that, the resin was washed with reverse osmosis (RO) water. During this washing process, the conductivity of the residual solution was recorded and the process continued until the conductivity of the residual solution was approximately the same as the RO water. The catalyst was then rinsed with methanol, filtered, and decanted and left overnight in a closed environment.

10.1.2.3 Catalyst Characterization

Elemental analysis was performed using a Thermoquest EA1110 Elemental Analyser. The sulfur determination was carried out separately using an oxygen flask combustion analysis, followed by a titration. All the results are reported in weight percentage of carbon, hydrogen, nitrogen, and sulfur. Oxygen cannot be measured by elemental analysis and, therefore, the percentage of oxygen content was determined by the difference from the total weight percentage of other elements (i.e., carbon, hydrogen, nitrogen, and sulfur). The true density (ρ_t) was measured using a Micromeritics Helium Pycnometer 1305. The true density of particles was determined using the standard density formula. A Carl Zeiss (Leo) 1530 VP field emission gun-scanning electron microscope (FEG-SEM), scanning electron microscopy (SEM), and scanning electron microscopy–energy dispersive X-ray (SEM–EDX) were used to study the morphology of the catalysts and to determine the elemental composition. Surface area, pore volume, and average pore diameter were determined from adsorption isotherms using a Micromeritics ASAP 2020 surface area analyzer. The samples were degassed using two-stage temperature ramping under a vacuum of <10 mm Hg, followed by sample analysis at 77 K using nitrogen gas. Table 10.1 shows the elemental analysis results for ion exchange resin catalysts. There was unexpected presence of nitrogen in some of the cation exchange resins and the value was <1%. In this case, nitrogen was assumed to be a contaminant in the sample. Table 10.2 shows the summarized chemical and physical properties of catalysts used in transesterification process.

TABLE 10.1   The Elemental Analysis Results for Ion Exchange Resin Catalysts

Catalyst	% C	% H	% N	% S	% O[a]
Amberlyst 36	41.18	4.10	0.10	18.27	35.35
Diaion PA306s	55.59	9.42	4.34	0.00	30.65
Purolite CT-122	51.06	5.68	0.06	15.99	27.22
Purolite CT-169	48.88	5.07	0.06	16.58	29.42
Purolite CT-175	47.35	4.74	0.00	15.75	32.17
Purolite CT-275	44.59	4.61	0.00	16.61	34.20
Purolite D5081	77.04	5.32	0.95	4.09	12.61
Purolite D5082	68.87	4.44	0.13	5.92	20.65

Source: Abidin, S.Z. et al., *Ind. Eng. Chem. Res.*, 51, 14653–14664, 2012. With permission from American Chemical Society.

[a] Oxygen by difference.

TABLE 10.2 Physical and Chemical Properties of Catalysts Used for Esterification and Transesterification Process

Catalyst Properties	Amberlyst 15	Diaion PA306s	Purolite CT-122	Purolite CT-169	Purolite CT-175	Purolite CT-275	Purolite D5081	Purolite D5082
Physical appearance	Opaque, spherical beads	White beads	Golden, spherical beads	Opaque, spherical beads	Opaque, spherical beads	Opaque, spherical beads	Opaque, spherical beads	Opaque, spherical beads
Functional group	Sulfonic acid	Quaternary ammonium	Sulfonic acid	Sulfonic acid	Sulfonic acid	Sulfonic acid	Sulfonic acid	Sulfonic acid
Moisture capacity (%H$^+$)[a]	52–57	66–76	78–82	51–57	50–57	51–59	56.9	56.2
Polymer structure	Macroporous polystyrene, cross-linked DVB	Gelular polystyrene, cross-linked DVB	Gelular polystyrene, cross-linked DVB	Macroporous polystyrene cross-linked DVB	Macroporous polystyrene, cross-linked DVB	Macroporous polystyrene, cross-linked DVB	Macroporous polystyrene, cross-linked DVB	Macroporous polystyrene, cross-linked DVB
Cross-linking level	Medium cross-linked	Low cross-linked	Low cross-linked	Medium cross-linked	Highly cross-linked	Highly cross-linked	Hyper cross-linked	Hyper cross-linked
Temperature limit (K)[a]	393	333	403	393	418	418	393	393
Specific surface area (m^2 g^{-1})	53	b	b	37.97	23.77	20.9	514.18	459.62
Total pore volume (cm^3 g^{-1})	0.4	b	b	0.27	0.108	0.108	0.47	0.36
Average pore diameter (nm)	30	b	b	27.42	17.37	19.6	3.69	3.14
True density (g cm^{-3})	1.027	1.297	1.297	1.297	1.296	1.296	1.309	1.373

Source: Abidin, S.Z. et al., *Ind. Eng. Chem. Res.*, 51, 14653–14664, 2012. With permission from American Chemical Society.

[a] Manufacturer data.

[b] Data could not be measured.

10.1.2.4 Average Molecular Mass Determination: Analysis of Fatty Acid Composition in the UCO

The fatty acids bonded to the glycerine backbone vary depending on the oil type and as a result an average molecular mass is generally determined based on the fatty acid composition of the oil. Average molecular mass was calculated by multiplying the mass fractions of fatty acids present in the oil with the individual molecular mass of each fatty acid involved. The determination of fatty acid composition was done by converting the triglycerides to glycerine and FAME through a methylation or hydrolysis process (David et al., 2005). Derivatization through the methylation process has been widely used to characterize lipid fractions in fats and oil (Dowd, 1998; Knothe and Steidley, 2009). It is a well-accepted characterization method because of the robustness and reproducibility of the chromatographic data. These methods are also cheaper in terms of reagent usage and do not require expensive equipment.

In this study, the sample was prepared using the method obtained from the official journal of European Union (Official EU Journal, 1991), and the results were verified using European standard method, BS EN 1SO 12966-2, 2011 (European standards, 2011). Derivatization process begins by weighing 100 mg of UCO in a 20-mL screw-cap test tube or reaction vial. Then, the UCO was dissolved in 10 mL of *n*-hexane. 100 μL of 2N potassium hydroxide was added to the reaction vial together with 100 mL of methanol. The tube or vial was closed and mixed vigorously for 60 s. The sample was then transferred into a conical bottom tube for centrifugation process. This process takes about 10 min with 16,000 rpm rotational speed. The clear supernatant in the upper layer, was analyzed by gas chromatography–mass spectrometry (GC–MS).

10.1.2.5 Esterification–Transesterification Reaction

The esterification process was carried out in a four-neck 1000 mL cylindrical jacketed-glass reactor, equipped with a mechanical stirrer, sampling outlet, and reflux condenser to prevent the loss of reactant due to vaporization. Heating was achieved by circulating water from a water bath and through the reactor, and a thermocouple was used for temperature monitoring. Figure 10.1 shows the experimental setup of the reaction process and Figure 10.2 shows the reaction scheme of the esterification process.

A specified amount of UCO and methanol was added to the reactor and the stirring and heating of the reaction mixtures were started. When the reactor reached the required temperature, catalyst was added and this point was taken as the zero time for the reaction. The samples were periodically

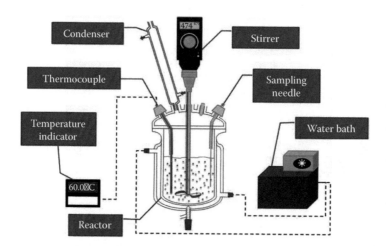

FIGURE 10.1 Experimental setup for the reaction process. (From Abidin, S.Z. et al., *Ind. Eng. Chem. Res.*, 51, 14653–14664, 2012. With permission from American Chemical Society.)

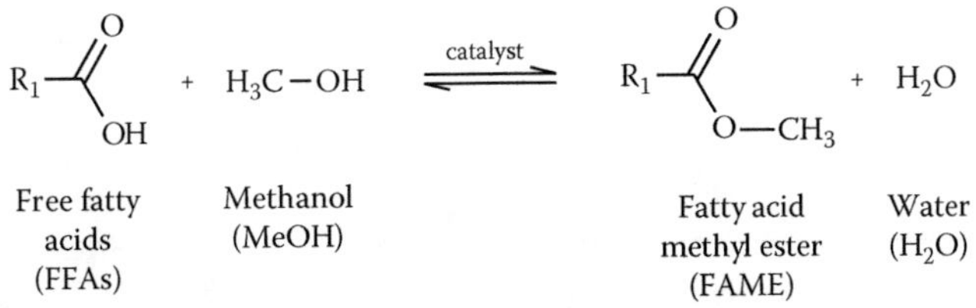

FIGURE 10.2 Reaction scheme for the esterification process: conversion of FFAs to FAME. R_1 represents the fatty acid group. (From Abidin, S.Z. et al., *Ind. Eng. Chem. Res.*, 51, 14653–14664, 2012. With permission from American Chemical Society.)

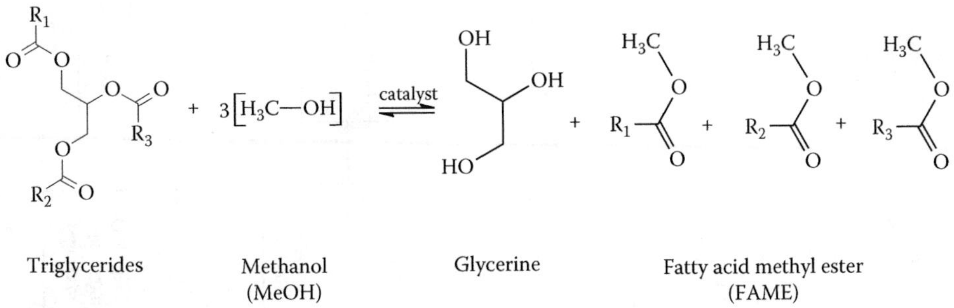

FIGURE 10.3 Reaction scheme for the transesterification process. R_1, R_2, and R_3 represent the fatty acids group attached to the backbone of triglycerides. (From Abidin, S.Z. et al. *Ind. Eng. Chem. Res.*, 51, 14653–14664, 2012. With permission from American Chemical Society.)

taken from the reactor for FFA analysis. After 8 h, the reaction mixture was transferred to a separation funnel and allowed to settle overnight to form two layers; the top layer consisted of excess methanol and its impurities, whereas the bottom layer was mainly unreacted UCO together with traces of methanol, glycerine, esters, and the remaining catalyst. The bottom layer was withdrawn from the separating funnel together with the catalyst and the retained catalyst was washed, dried, and stored for further experimental work. Studies on the mass transfer resistance as well as the effect of methanol to UCO molar ratio, catalyst loading and reaction temperature have been conducted. In addition, a blank run without catalyst has been performed and there was no conversion of FFA after 8 h of reaction time. Therefore, it was concluded that the esterification of FFA occurs only due to the presence of the catalyst. The product from the esterification process is called as pretreated used cooking oil (P-UCO). The P-UCO was used as the raw material for transesterification process. Figure 10.3 shows the reaction scheme of the transesterification process. The experimental setup and procedure for transesterification was similar to the esterification, except that the size for transesterification reactor was smaller, i.e., 250 mL. Studies on the mass transfer resistance as well as the effect of methanol to UCO molar ratio, catalyst loading, and reaction temperature have been conducted. Samples were taken periodically from the reactor for FAME analysis using GC–MS. The results were used to determine triglycerides conversion. Once the experiment was completed, the reaction mixture was separated from the spent catalyst, transferred to a separating funnel, and allowed to settle overnight. The FAME-rich phase (unpurified biodiesel) was withdrawn from the separating funnel and introduced to a rotary evaporator to remove traces of methanol, followed by washing process. Finally, the purified biodiesel was separated from the washing agent and stored for further analysis. In terms of the reproducibility of the experimental data, selected experiments were repeated three times and it was found that there was ±2% difference in the results. Therefore, it was assumed that a similar error applies to all results.

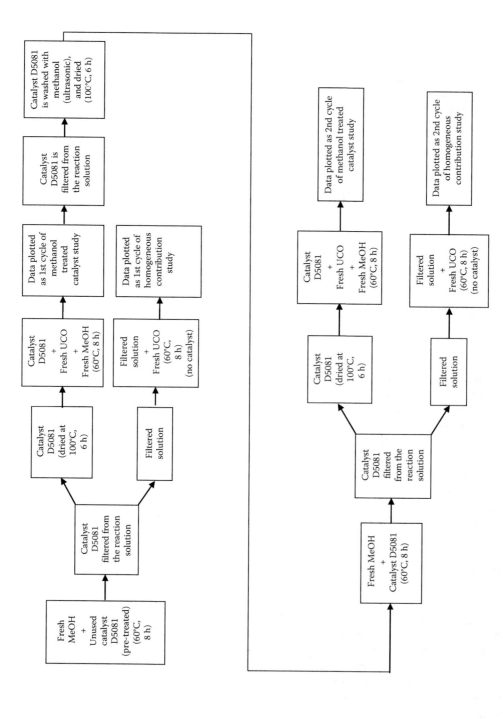

FIGURE 10.4 Process flow diagram for catalyst reusability study. (From Abidin, S.Z. et al., *Ind. Eng. Chem. Res.*, 51, 14653–14664, 2012. With permission from American Chemical Society.)

10.1.2.6 Catalyst Reusability Study

A reusability study was conducted to determine the catalyst life span. For esterification process, used catalyst was washed with methanol with the aid of an ultrasonic bath until there were no traces of oil and colorless solution were obtained. During this washing process, the conductivity of the residual solution was recorded and the process continues until the conductivity of the residual solution is approximately the same with the solvent (methanol). The washed catalyst was filtered and dried in a vacuum oven at 100°C for 6 h. The reusability study was carried out under the optimum process conditions. The fresh and used catalysts were tested for SEM, SEM–EDX, acid capacity, and elemental analyses. An investigation into the homogeneous contribution was carried out to establish if sulfonic acid groups were leached during the reaction process. The process flow diagram for the reusability study is summarized in Figure 10.4. It was done by reacting fresh methanol with unused pretreated catalyst for 8 h at 60°C and 475 rpm stirring speed. After that, the catalyst was filtered and the solution was used in the subsequent reaction with fresh UCO (6:1 (methanol:UCO) molar ratio), in the absence of any catalyst. This is referred to as the first cycle of homogeneous contribution study. The filtered catalyst was dried in the vacuum oven for 6 h (100°C) and this catalyst was then used for an experiment with fresh methanol and fresh UCO at optimum condition to monitor the conversion trend as a result of the catalyst being treated with methanol. This result was plotted as the first cycle of methanol treated catalyst study.

The spent catalyst from the previous methanol treated catalyst experiment was washed thoroughly with methanol using ultrasonic bath until there was no evidence of UCO contamination. The catalyst was filtered and dried in a vacuum oven at 100°C for 6 h. The experimental work proceeded to the second homogeneous cycle, where the reaction took place between the fresh methanol and used catalyst. The used catalyst was the same catalyst used in the first cycle of homogeneous contribution study and the first cycle of methanol treated catalyst study. The filtered solution from the methanol–catalyst reaction was added to fresh UCO [6:1 (methanol:UCO) molar ratio] without the presence of catalyst and the FFA conversion was monitored. The result was plotted as the second cycle of homogeneous contribution study. The used catalyst obtained from the methanol–catalyst reaction was dried (100°C, 6 h) and introduced to fresh UCO and fresh methanol. By using the optimum reaction parameters, the esterification was conducted and the second cycle of methanol treated catalyst data was plotted.

A slightly different approach was carried out for anion exchange resin used in transesterification reaction. The used catalyst was washed with glacial acetic acid in methanol to displace the fatty acid ions. This displacement step was conducted with the aid of an ultrasonic bath until there were no traces of P-UCO, and a colorless solution was obtained. The catalyst was then washed using RO water to remove excess of acetic acid solution. The catalyst was mixed with 1 M NaOH to displace the acetate ions with hydroxyl ions, followed by washing with RO water to remove excess NaOH solution. During this washing process, the conductivity of the residual solution was recorded and the process continued until the conductivity of the residual solution was approximately the same as the RO water. The catalyst was rinsed with methanol, filtered, and decanted overnight in a closed environment.

10.1.2.7 GC–MS Analysis

The FAME content was assayed using a Hewlett Packard GC–MS model HP-6890 and HP5973 (mass selective detector). A DB-WAX (J&W Scientific) capillary column of length 30 m and internal diameter of 0.25×10^{-3} m packed with polyethylene glycol (0.25 μm film thickness) was used. Helium was used as a carrier gas at a constant flow rate of 1.1 mL min^{-1}. The temperature of both the injector and the detector was set at 523 K. The injection volume of 1 μL and a split ratio of 10:1 were used as part of the GC–MS analysis method. The initial oven temperature was held at 343 K for 2 min after the sample injection. The oven temperature was then ramped from 343 to 483 K at a rate of 40 K min^{-1} and from 483 to 503 K at a rate of

7 K min^{-1}. The oven temperature was held at 503 K for 11 min to remove any remaining traces of the sample. The total run time for each sample was approximately 19.5 min. The detailed analysis experimental procedure has been published by Abidin et al. (2013). The determination of monoglycerides, diglycerides, and triglycerides in the UCO was carried out using the method established by Haigh et al. (2014).

10.1.3 Results and Discussions

10.1.3.1 GC–MS Analysis of Derivatized UCO

Figure 10.5 shows a typical chromatogram of the derivatized UCO. From the chromatogram, six different components (including the reference standard) were identified. The retention time for each individual component is as follows: methyl palmitate (C16:0) appeared at 8.324 min retention time, methyl heptadecanoate (C17:0—reference standard) at 8.950 min, methyl stearate (C18:0) at 9.712 min, methyl oleate (C18:1) at 9.914 min, methyl linoleate (C18:2) at 10.349 min, and finally methyl linolenate (C18:3) at 11.005 min. As methylation process converts fatty acids to methyl esters through derivatization method, it could be concluded that there were five main components present in UCO, namely the palmitic acid, stearic acid, oleic acid, linoleic acid, and linolenic acid.

The response factor for each component was determined using the calibration of individual standards. The predetermined response factor was then used to determine the fatty acid composition of the sample. By calculating the mass fractions of each fatty acid, the average molecular mass of the fatty acids could be easily calculated and will be further used to determine the methanol to UCO molar ratio (Zhang and Jiang, 2008). The fatty acids composition is summarized in Table 10.3. These data clearly show that oleic and linoleic acids comprise more than 80% of the total fatty acids in the sample. Using the EU Regulation method, the average molecular mass of fatty acids is 278.11 g/mol. The reliability of

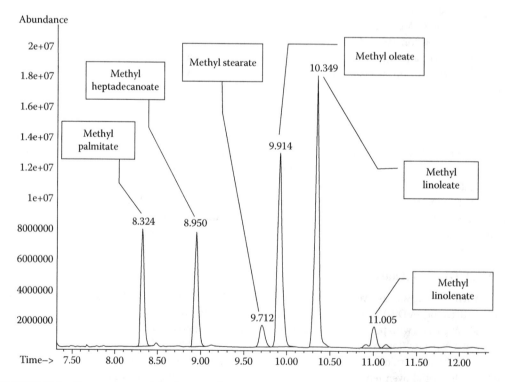

FIGURE 10.5 Chromatogram of the derivatised UCO. (From Abidin, S.Z. et al., *Ind. Eng. Chem. Res.*, 51, 14653–14664, 2012. With permission from American Chemical Society.)

TABLE 10.3   Percentage Composition of Fatty Acid in the UCO

Component	% Composition (w/w)
Palmitic acid (C16:0)	11.34
Stearic acid (C18:0)	3.18
Oleic acid (C18:1)	43.95
Linoleic acid (C18:2)	36.44
Linolenic acid (C18:3)	5.09

Source: Abidin, S.Z. et al., *Ind. Eng. Chem. Res.*, 51, 14653–14664, 2012. With permission from American Chemical Society.

the previous standard method was verified by analyzing the sample according to the British Standard European Norm (BS EN) standard method, EN ISO 12966-2 (2011). It was found that the results were very similar and the average molecular mass of fatty acids obtained using the second method was 277.93 g/mol. Therefore, it can be concluded that both methods are comparable.

10.1.3.2 Esterification Reaction

10.1.3.2.1 Catalysts Screening Study

In order to identify the best of the three ion exchange resins for further experimental work, all three resins, namely Purolite D5081, Purolite D5082, and Amberlyst 36, were evaluated under the same reaction conditions, i.e., at 1% (w/w) of catalyst loading, 6:1 methanol to UCO molar ratio, 60°C reaction temperature, and 350 rpm impeller stirring speed. From Figure 10.6, it can be seen that, after 8 h, Purolite D5081 resin achieved the highest FFA conversion of ~88%, while Purolite D5082 and Amberlyst 36 achieved FFA conversion of ~78% and ~44%, respectively.

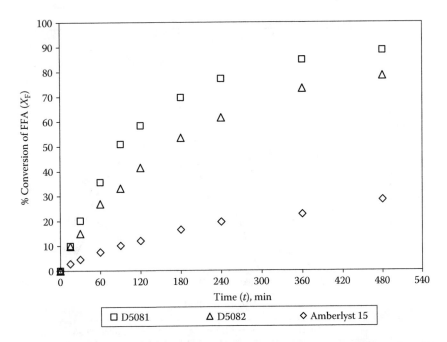

FIGURE 10.6   Effects of different types of ion exchange resins on FFA conversion. Experimental conditions: molar ratio (methanol:UCO), 6:1; catalyst loading, 1% (w/w); stirring speed, 350 rpm; reaction temperature, 60°C. (From Abidin, S.Z. et al., *Ind. Eng. Chem. Res.*, 51, 14653–14664, 2012. With permission from American Chemical Society.)

TABLE 10.4 Elemental Analysis for Fresh and Used Ion Exchange Resins

Catalyst	% C	% H	% N	% S	% O[a]
Fresh D5081	77.04	5.32	0.95	4.09	12.61
Used D5081[b]	77.41	5.69	0.93	3.32	12.66
Fresh D5082	68.87	4.44	0.13	5.92	20.65
Used D5082[b]	69.07	4.53	0.03	5.41	20.97
Fresh Amberlyst 36	42.18	4.10	0.10	18.27	35.35
Used Amberlyst 36[b]	42.18	4.19	0.10	18.17	35.36

Source: Abidin, S.Z. et al., *Ind. Eng. Chem. Res.*, 51, 14653–14664, 2012. With permission from American Chemical Society.

Note: All the percentages are in w/w %.

[a] Oxygen by difference.

[b] Washing after 1st cycle reaction.

TABLE 10.5 Acid Capacity for Fresh and Used Ion Exchange Resins

Catalyst	Fresh Resin (mmol/g)	Used Resin (mmol/g)
D5081	1.59	1.39
D5082	1.79	1.59
Amberlyst 36	5.00	4.99

Source: Abidin, S.Z. et al., *Ind. Eng. Chem. Res.*, 51, 14653–14664, 2012. With permission from American Chemical Society.

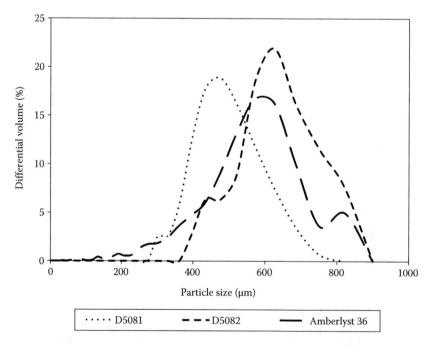

FIGURE 10.7 Particle size distribution of ion exchange resins catalysts. (From Abidin, S.Z. et al., *Ind. Eng. Chem. Res.*, 51, 14653–14664, 2012. With permission from American Chemical Society.)

The differences in the properties of various resins, i.e., surface area measurement, elemental analysis, and acid capacity analysis can be used to explain the differences in catalytic activity (Tables 10.4 and 10.5). These analyses show that even though Purolite D5081 has the lowest sulfur content (Table 10.4), it exhibits the highest specific surface area and total pore volume, which means that there are more accessible active sites for the reaction to occur and hence reaches equilibrium at a faster rate. In addition, the particle size distribution results (Figure 10.7) show that Purolite D5081 has the smallest average particle size, which leads to a larger external surface area compared to the other resins. A larger surface area may contribute to a faster rate of reaction and shorten the time to reach equilibrium; however, these catalysts are highly porous. As a result, this effect is very small when compared to the effect of the pore volume.

Although Amberlyst 36 has the highest sulfur content and the largest average pore diameter, the result from the esterification reaction (Figure 10.6) shows that Amberlyst 36 has the lowest conversion as compared to Purolite D5081 and D5082 resins. This is because Amberlyst 36 has the lowest specific surface area and lowest pore volume compared to the Purolite resins and, therefore, there are less active catalytic sites available for the reaction to occur. The level of divinylbenzene (DVB) cross-linking also contributes significantly to the level of FFA conversion. From Figure 10.6, it could be seen that resins with high DVB cross-linking (Purolite D5081 and D5082) result in higher FFA conversion compared with lower DVB cross-linking resin (Amberlyst 36). The Fourier Transform-Infra Red (FT-IR) analysis has also been carried out and there was no noticeable change of functional groups for all the resin samples, and therefore these results are not presented in the paper. The FT-IR results also indicate that sulfur was present only in sulfonic acid functional group. Since Purolite D5081 showed the best catalytic performance, it was used for further experimental work.

10.1.3.2.2 Investigation on the Effect of Mass Transfer Resistance in Esterification Reaction

There are two types of mass transfer resistances involved in ion exchange catalysis. The first is external mass transfer resistance, which takes place across the solid–liquid interface while the second is the internal mass transfer resistance, associated with the differences in particle size distribution of the catalysts. Mixing is one of the key factors to optimize the production of biodiesel as it increases the interaction between the reactants (methanol and UCO) and the catalyst, predominantly at the early stage of the reaction. However, after the reaction mixture reaches the stage where the reactant and the catalyst are well mixed (e.g., there is sufficient contact between the catalyst and the reactant), there is no additional benefit from increasing the stirring speed. This phenomenon is due to the external mass transfer resistances.

In order to investigate the influence of external mass transfer resistance, three different stirring rates were investigated for the reaction process and the conversion patterns were observed. Figure 10.8 shows the trend for FFA conversion of Purolite D5081 for three levels of impeller agitation speed, i.e., 350, 475, and 600 rpm. The FFA conversion between the selected stirring rates gave almost an identical trend with <2% difference when the agitation speed increased from 350 to 600 rpm. As the stirring speed only has a small impact on the FFA conversion, it was confirmed that there was no evidence of external mass transfer resistance in the esterification reaction.

Internal mass transfer resistance refers to resistance of movement of reactant inside the pores of the catalyst. In order to investigate the occurrence of internal mass transfer resistance, a set of experiments was conducted using various ranges of particle sizes. Figure 10.9 shows the effect of different particle size distribution (PSD) on the FFA conversion. From Figure 10.9, it can be seen that 280–800 μm particle range gives a slightly higher conversion compared to the other two particle ranges. This is because this range, i.e., 280–800 μm, covers a wide range of particles, and therefore the possibility of having smaller particles is higher. This leads to a higher conversion of FFA as smaller particles have shorter pore channels and a larger area of the pores is accessible by the reactant molecules. To support this argument, a comparison was made between 600–710 and 425–500 μm range and it was found that the conversion for 600–710 μm particle size was slightly lower than 425–500 μm particle size in the first 5 h of reaction. Nevertheless, the final conversions of FFA for all three particle ranges were approximately the same at the end of 8 h reaction. This finding proved the previous theory of smaller particles contributing to a

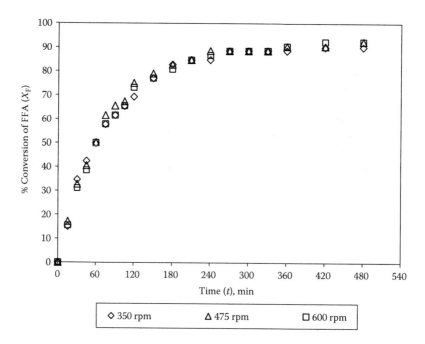

FIGURE 10.8 Effect of different stirring speeds on the FFA conversion—external mass transfer resistance. Experimental conditions: catalyst, Purolite D5081; molar ratio (methanol:UCO), 6:1; catalyst loading, 1.25% (w/w); reaction temperature, 60°C. (From Abidin, S.Z. et al., *Ind. Eng. Chem. Res.*, 51, 14653–14664, 2012. With permission from American Chemical Society.)

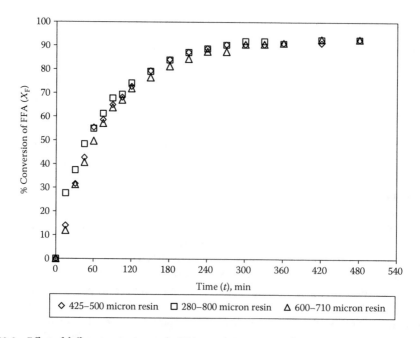

FIGURE 10.9 Effect of different resin size on the FFA conversion—internal mass transfer resistance experimental conditions: catalyst: Purolite D5081; stirring speed: 475 rpm; catalyst loading: 1.25% (w/w); reaction temperature: 60°C; molar ratio (methanol:UCO): 6:1. (From Abidin, S.Z. et al., *Ind. Eng. Chem. Res.*, 51, 14653–14664, 2012. With permission from American Chemical Society.)

larger accessible surface area and thus increasing the FFA conversion. However, the difference between each range was very small (less than ±2%) and the final conversions were approximately the same. Therefore, the effect of internal mass transfer limitation can be eliminated. As a conclusion, stirring speed of 350 rpm and above was selected for subsequent experiments as there was no external mass transfer resistance above this speed limit and the Purolite D5081 resin is used without sieving.

10.1.3.2.3 Effect of Catalyst Loading

The effect of catalyst loading is shown in Figure 10.10. The reaction temperature was set at 60°C with 6:1 methanol to UCO molar ratio, the reaction took approximately 8 h to reach equilibrium. The result shows that the conversion is greatly influenced by the amount of catalyst, especially at the early stage of the esterification reaction. This is attributed to the fact that as the catalyst loading increases, the number of active catalytic sites increases and a shorter time is required to reach equilibrium. Variation in catalyst loading also has an effect on the final conversion of FFA with the difference between the highest and the lowest catalyst loading of about 10%. At a low catalyst concentration [<1% (w/w)], the difference in the conversion was significant. However, as catalyst loading was increased [≥1.25% (w/w)], the trends were less significant. This phenomenon indicates that as the percentage of resin increases, the conversion increases. However, as the number of catalytic sites increases (≥1.25% (w/w) catalyst loading), the benefits are reduced as there are sufficient active catalytic sites in the reactant molecules to catalyze the reaction. In the case of the system investigated, it was found that once the catalyst loading increased to 1.25% (w/w), there were sufficient catalyst sites for this particular reaction. As a result, 1.25% (w/w) loading was chosen as the optimum catalyst loading and was used for all further experimental work.

10.1.3.2.4 Effect of Reaction Temperature

Esterification was investigated at different temperatures (50°C, 55°C, 60°C, 62°C, 65°C) and the results are shown in Figure 10.11. It has been observed that increasing temperatures led to reduction in the

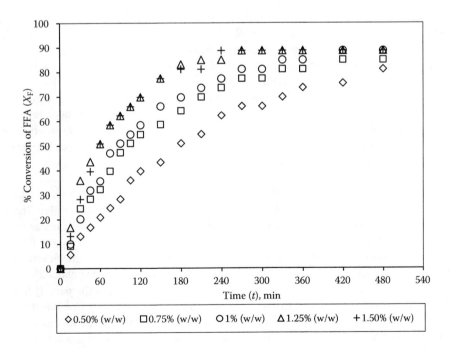

FIGURE 10.10 Effect of different catalyst loading on the FFA conversion. Experimental conditions: catalyst: Purolite D5081; stirring speed: 350 rpm; reaction temperature: 60°C; molar ratio (methanol:UCO): 6:1. (From Abidin, S.Z. et al., *Ind. Eng. Chem. Res.*, 51, 14653–14664, 2012. With permission from American Chemical Society.)

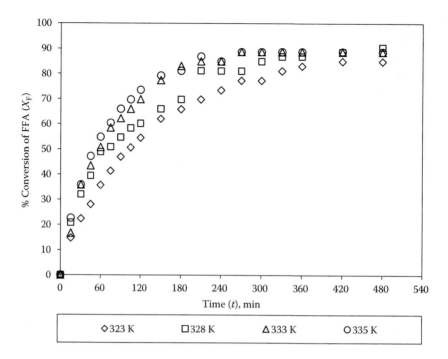

FIGURE 10.11 Effect of different reaction temperatures on FFA conversion. Experimental conditions: catalyst: Purolite D5081; stirring speed: 350 rpm; catalyst loading: 1.25% (w/w); molar ratio (methanol:UCO): 6:1. (From Abidin, S.Z. et al. *Ind. Eng. Chem. Res.*, 51, 14653–14664, 2012. With permission from American Chemical Society.)

viscosity of UCO, which enhances the contact between the methanol and UCO leading to a higher conversion of FFA. The highest conversion was obtained at 65°C. A decrease in the volume of the reaction mixture was observed when the temperature reached the boiling point of methanol, 64.7°C. It was expected as there will be some changes to the system when this temperature is reached, with more methanol present in the headspace of the reactor as vapor. Liu et al. (2009) also claimed that beyond 65°C, methanol started to vaporize rapidly, forming a large number of bubbles to form foam and resulted in decrease in FFA conversion. Generally, in typical biodiesel reaction process, low temperature will result in lower conversion, while higher temperatures lead to excessive methanol loss due to evaporation. After consideration of the safety issues, cost implications, and the conversion trends for each temperature, the optimum reaction temperature was found to be 60°C.

10.1.3.2.5 Effect of Methanol to UCO Molar Ratio

Figure 10.12 shows the effect of methanol to UCO molar ratio on the conversion of FFA. The reaction temperature was set at 60°C, at 350 rpm stirrer speed with 8 h reaction time. The molar ratios were calculated based on the molecular mass of average fatty acid composition. From Figure 10.12 it can be seen that the conversion increased slightly with an increase in methanol to UCO molar ratio. However, the conversion differences between the molar ratios were less noticeable, giving about 2% increments in conversion as the molar ratio increases. In comparison, a ratio of 4:1 has a comparably lower initial conversion as compared to the other three molar ratios. This will be due to insufficient methanol in the mixture to force the reaction, leading to longer reaction time to reach equilibrium. The reaction time can be shortened using a higher methanol to UCO molar ratio; however, the costs will be higher due to higher methanol recovery requirements. A high molar ratio also leads to difficulties in the separation process, and thus hinders separation by gravity (Encinar et al., 2007). Furthermore, this situation, i.e., high molar ratio, could also increase the solubility of glycerine, and thus reduces the yield of FAME as

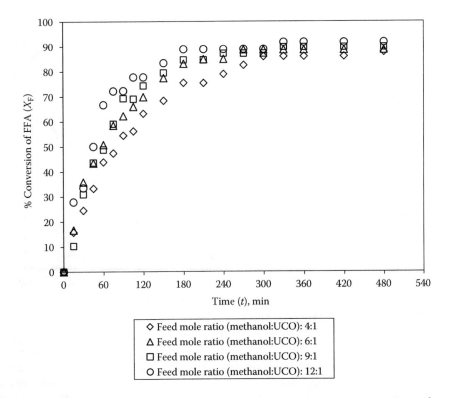

FIGURE 10.12 Effect of different molar ratio (methanol:UCO) on the FFA conversion. Experimental conditions: catalyst: Purolite D5081; catalyst loading: 1.25% (w/w); stirring speed: 350 rpm; reaction temperature: 60°C. (From Abidin, S.Z. et al., *Ind. Eng. Chem. Res.*, 51, 14653–14664, 2012. With permission from American Chemical Society.)

glycerol remains in the pretreated UCO phase. Excess methanol also could drive the combination of methyl ester and glycerine to monoglycerides, which increases the viscosity of the reaction mixture (Liu et al., 2009). Taking into account the safety issues and the capital and operating costs, a molar ratio of 6:1 (methanol:UCO) was chosen as the optimum molar ratio for the esterification reaction.

10.1.3.3 Transesterification Reaction

10.1.3.3.1 Catalyst Screening Study

The transesterification of biodiesel using different types of heterogeneous catalysts has been investigated to select the best catalyst for further optimization work. There are two groups of catalysts involved in this work, which are the cation exchange resins (Purolite CT-122, Purolite CT-169, Purolite CT-175, Purolite CT-275, Purolite D5081) and anion exchange resin (Diaion PA306s), respectively. All catalysts were tested under the same reaction conditions, 1.5% (w/w) of catalyst loading, 333 K reaction temperature, 18:1 methanol to P-UCO feed mole ratio, and 350 rpm impeller stirring speed. The results are collected in Figure 10.13. After 8 h of reaction, the conversion of triglycerides was ca. 50% using Diaion PA306s catalyst, ca. 10% using Purolite CT-275, and ca. 7% using Purolite CT-122 and Purolite CT175. For Purolite D5081 catalyst, there was no measureable formation of FAME. Of all the catalysts investigated, Diaion PA306s gave the highest triglyceride conversion of ca. 50%.

From the results obtained, it was found that the specific surface area of the catalyst did not give any significant impact to the conversion of triglycerides. It was proven by having negligible triglycerides conversion when Purolite D5081, a catalyst with the highest specific surface area was used as the catalyst. The level of cross-linking also gives fluctuate performances where a low cross-linked resin Diaion

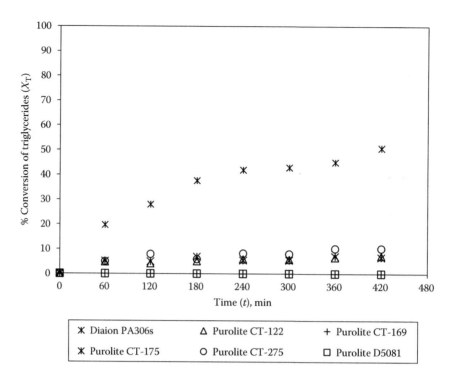

FIGURE 10.13 Effect of different types of catalysts on triglycerides conversion. Experimental conditions: stirring speed: 350 rpm, catalyst loading: 1.5% (w/w); reaction temperature: 333 K; feed mole ratio (methanol:P-UCO): 18:1.

PA306s gives the highest triglycerides conversion (ca. 50%) whereas the other low cross-linked catalyst, Purolite CT-122 was only able to give ca. 7% of triglycerides conversion. In contrast, Purolite CT-275, which has a high degree of cross-linking, gives a slightly higher triglycerides conversion (ca. 10%) as compared to Purolite CT-122.

A huge difference in catalytic performance was observed between Diaion PA306s and the other catalysts and it was expected to be closely related to the acidity and basicity of the catalysts. In this case, Diaion PA306s was classified as a strongly basic anion exchange resin, while the rest of the catalysts are categorized as strongly acidic resins. Few researchers (Mazzotti et al., 1997; Shibasaki-Kitakawa et al., 2007) have also reported that the adsorption strength of the alcohol on the anion exchange resin was much higher compared to cation exchange resins, which results in higher activity for anion exchange resin compared to cation exchange resin. Therefore, it was concluded that the basicity of the catalyst is responsible for its transesterification activity and not specific surface area, particle size distribution, average pore diameter, or the cross-linking level. Since Diaion PA306s showed the best catalytic performance, it was used for the subsequent transesterification reactions.

10.1.3.3.2 Investigation on the Effect of Mass Transfer Resistances in Transesterification Reaction

There are two types of mass transfer resistances involved in ion exchange catalysis, the external mass resistance and internal mass transfer resistance. The external mass transfer resistance, which takes place across the solid–liquid interface, was evaluated using different stirring speeds under the same reaction conditions. Three different agitation speeds were used, 300, 350, and 450 rpm and the result is shown in Figure 10.14. It was found that the stirring speed gives a negligible impact on triglycerides conversion, and therefore it was confirmed that the external mass transfer resistance has negligible effect on the transesterification reaction. The internal mass transfer resistance associated with the differences in particle size of the catalysts can be studied by measuring reaction rates for different average catalyst particle

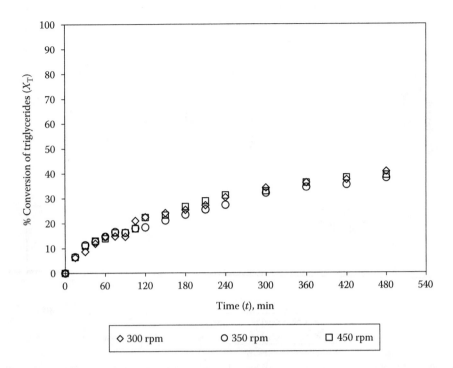

FIGURE 10.14 Effect of stirring speed on triglycerides conversion—External mass transfer resistance. Experimental conditions: catalyst: Diaion PA306s; catalyst loading: 1.5% (w/w); reaction temperature: 323 K; feed mole ratio (methanol:P-UCO): 18:1.

sizes. In this case, absence of internal mass transfer resistance could not be verified because the catalyst was supplied in wet form in the swelling condition. This type of catalyst (anion exchange resin) cannot be heated to more than 333 K, otherwise it affects the stability of the catalyst. As the moisture content cannot be totally removed below 333 K, separation by a sieving will not represent the actual size of catalyst particle. Therefore, PA306s resin was used as received, without sieving for all transesterification reactions.

10.1.3.3.3 Effect of Catalyst Loading

The effect of catalyst concentration on triglycerides conversion was investigated using different catalyst loadings, 1.5% (w/w), 3% (w/w), 4.5% (w/w), 5.5% (w/w), 9% (w/w), and 10% (w/w). Figure 10.15 shows the effect of catalyst loading on the conversion of triglycerides. The reaction temperature was set at 323 K with 18:1 methanol to P-UCO feed mole ratio and 350 rpm stirring speed. As observed from Figure 10.15, increasing the catalyst concentration was found to increase triglycerides conversion. This behavior was expected since with an increase in the number of active catalytic sites, triglycerides conversion increases. As the reaction proceeds, the changes in triglycerides conversion become less significant, indicating that the system is approaching equilibrium. Based on the observation for 9% (w/w) and 10% (w/w) catalyst loading, it could be concluded that a further increase in catalyst concentration would cause negligible increase in the conversion of triglycerides (ca. 64% to 65%). Furthermore, higher catalyst dosage increases the viscosity of the reaction mixtures that increases the mass transfer resistance in the multiphase system. Therefore, using a very high amount of catalyst is unnecessary for this reaction. For all further transesterification study, 9% (w/w) was chosen as the optimum catalyst loading.

10.1.3.3.4 Effect of Reaction Temperature

Figure 10.16 shows the plot of triglycerides conversion over time with the temperature ranging from 313 to 333 K. From the figure, triglycerides conversion was found to increase with an increase in reaction

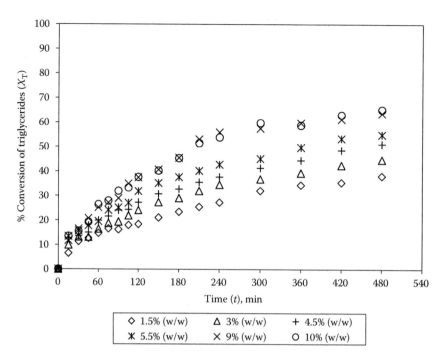

FIGURE 10.15 Effect of catalyst loading on triglycerides conversion. Experimental conditions: catalyst: Diaion PA306s; stirring speed: 350 rpm; reaction temperature: 323 K; feed mole ratio (methanol:P-UCO): 18:1.

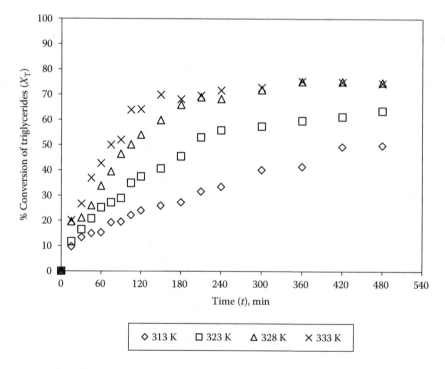

FIGURE 10.16 Effect of reaction temperature on triglycerides conversion. Experimental conditions: catalyst: Diaion PA306s; stirring speed: 350 rpm; catalyst loading: 9% (w/w); feed mole ratio (methanol:P-UCO): 18:1.

temperature. After 8 h of reaction, the final conversion of triglycerides at 313, 325, and 328 K was approximately 50%, 64%, and 75%, respectively. Theoretically, an increase in reaction temperature leads to a reduction in the viscosity of triglycerides, which enhances the contact between methanol and triglycerides. From Figure 10.16, it was also observed that triglycerides conversion for the 328 and 333 K reaction temperatures are similar, although the time for the conversion to reach steady state was faster for 333 K. As the final conversion for 325 and 328 K was approximately the same with ca. 75% conversion, an increase in temperature will only increase the operating cost. Therefore, 325 K was chosen as the optimum reaction temperature and proposed for further transesterification reactions.

10.1.3.3.5 *Effect of Methanol to P-UCO Feed Mole Ratio*

Stoichiometrically, the methanolysis of triglycerides requires three moles of methanol per mole of triglyceride to yield three moles of FAME and one mole of glycerine. Given that the transesterification is a reversible reaction, excess methanol should help the conversion of triglycerides. The molar mass of UCO was determined to be 871.82 g mol^{-1} and this was used to calculate the feed mole ratio of methanol to P-UCO. Figure 10.17 shows the effect of feed mole ratio of methanol to P-UCO on the conversion of triglycerides. As observed from Figure 10.17, the conversion of triglycerides increased with an increase in methanol to P-UCO feed mole ratio from 6:1 to 18:1. The conversion of triglycerides using 6:1, 12:1, and 18:1 methanol to P-UCO feed mole ratio at 8 h is 63%, 69%, and 75%, respectively. From Figure 10.17, it can be seen that a further increase in feed mole ratio of methanol to P-UCO from 18:1 to 24:1 did not result in an increase in conversion of triglycerides and the final triglycerides conversion for both feed mole ratios were approximately same, i.e., 75%. A significantly high feed mole ratio is not preferable in biodiesel

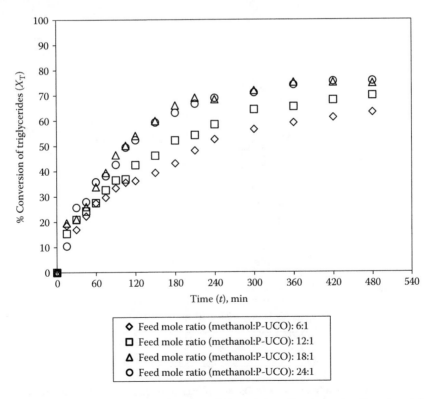

FIGURE 10.17 Effect of feed mole ratio (methanol:P-UCO) on triglycerides conversion. Experimental conditions: catalyst: Diaion PA306s; stirring speed: 350 rpm; catalyst loading: 9% (w/w); reaction temperature: 328 K.

production because it makes the separation process difficult and higher consumption of methanol also requires larger unit operations including reactors, separation column, and methanol recovery equipment that will increase the overall cost of the process. An optimum operating ratio should be selected on the basis of overall economics and equilibrium conversion, and therefore a feed mole ratio of 18:1 methanol to P-UCO was selected as the optimum ratio and used for further transesterification reaction.

10.1.3.4 Catalyst Reusability Study

10.1.3.4.1 Reusability Study on the Spent Esterification Catalyst

A series of reactions, using the same batch of Purolite D5081 catalyst, were carried out at optimum process conditions, i.e., 1.25% (w/w) catalyst loading, 333 K reaction temperature, 6:1 methanol to UCO feed mole ratio, and 475 rpm stirring speed, to determine the catalyst life span. The results are shown in Figure 10.18 and from these data it can be seen that the conversion decreased by approximately 8%–10% per cycle (e.g., fresh catalyst gives ~92% conversion, the first cycle gives ~84% conversion and the second cycle gives ~73% conversion). Potential reasons for the loss of activity include contamination of the external or internal surface of the catalyst and sulfur leaching.

SEM analysis was used to investigate the effect of cleaning regimes on the surface of the ion exchange resin catalysts. Purolite D5081 was cleaned by placing the catalyst in a flask of methanol, which was subsequently placed in an ultrasonic bath, and the method is detailed in Section 10.1.2.6. To determine the effect of ultrasonication, Purolite D5082 was cleaned using methanol but without ultrasonication. A series of SEM images are shown in Figure 10.19, which show the comparison of fresh and used catalyst (after the first run) as well as comparing the effect of cleaning regimes. Figure 10.19a and b shows a sample of Purolite D5081 before and after esterification and it can be seen that there is no trace of UCO

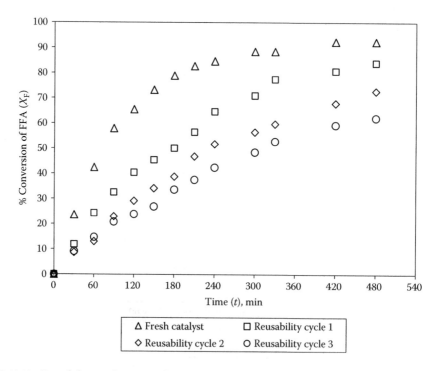

FIGURE 10.18 Reusability study on Purolite D5081 ion exchange resins. Experimental conditions: catalyst: Purolite D5081; stirring speed: 475 rpm; catalyst loading: 1.25% (w/w); reaction temperature: 60°C; molar ratio (methanol:UCO): 6:1. (From Abidin, S.Z. et al., *Ind. Eng. Chem. Res.*, 51, 14653–14664, 2012. With permission from American Chemical Society.)

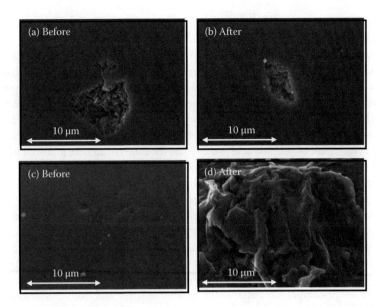

FIGURE 10.19 SEM analysis of Purolite catalysts taken at 5000× magnification: (a, b) is Purolite D5081 before and after esterification process and (c, d) is Purolite D5082 before and after esterification process. (From Abidin, S.Z. et al., *Ind. Eng. Chem. Res.*, 51, 14653–14664, 2012. With permission from American Chemical Society.)

on the surface of the used catalyst in Figure 10.19b. Figure 10.19c and d shows a sample of Purolite D5082 before and after esterification and in this case it can be seen that there is UCO present on the surface of the catalyst, indicating incomplete washing of the catalyst. This observation fits with the results of the elemental analysis for used D5082 (Table 10.4), which shows that there is a slight increase in residual carbon. On this basis, it was decided that ultrasonication was needed as part of the cleaning process in order to ensure UCO was removed from the surface of the catalyst.

Contamination of the internal surface of the catalyst could possibly result from the resin pore blockage, either by the presence of metal ions in the UCO or the blockage by the UCO residue itself. An energy-dispersive X-ray spectroscopy (EDX) analysis of fresh and used Purolite D5081 catalysts has been conducted and trace amount of metal was found on the catalyst. The amount of sodium and iron content was very similar for both fresh and used Purolite D5081 catalysts (~0.20% and ~0.15%). In the used Purolite D5081 catalyst, trace amount of calcium ions was found (approximately 0.012%–0.10%). Given that only trace amount of metal was found on the surface of the catalyst, it has been concluded that the metal ion impurities in UCO do not contribute to the deactivation of Purolite D5081 catalyst.

It was found that the mass of catalyst increased slightly after each reaction cycle and it has been assumed that this was due to the presence of triglycerides, proteins, trace amount of phospholipids, and other impurities in UCO that could potentially foul the Purolite D5081 catalyst. The Brunauer–Emmett–Teller (BET) analysis showed that Purolite D5081 resin has the highest total pore volume of 0.47 cm³/g, which was slightly higher than other conventional ion exchange resins. This means it is possible for the triglycerides, proteins, and trace amount of phospholipids molecules to be retained within the resin catalyst and may contribute to pore blockage. This may also reduce the accessibility of the active catalyst sites and may lead to higher internal mass transfer resistance and finally contribute to decreasing catalytic activity.

Sulfur leaching occurs due to the detachment of sulfonic acid group from the polymer matrix. Water is one of the products formed during the esterification reaction and in theory it could hydrolyze sulfonic acid groups to form homogenous sulfuric acid. An elemental analysis was carried out to determine the level of sulfur within various resin samples and the results are shown in Table 10.4. Fresh D5081 resin had a sulfur content of 4.1%, and after the first reaction cycle this decreased by nearly 20%–33%. Ion exchange capacity results in Table 10.5 show a decrease of acid capacity for fresh and used catalyst of approximately 13%. This indicates that sulfur leaching contributes to the reduction in catalytic activity of Purolite D5081 catalyst.

To further investigate the leaching of homogeneous species in the reaction mixture and UCO blockage within the resin pores, several sets of experiments have been carried out. The experimental work and the process flow diagram have been detailed in Section 10.1.2.6 and Figure 10.4, respectively. Figure 10.20 shows the conversion of FFA during the uncatalyzed reaction between the treated methanol solution and UCO, and this result clearly shows that there was a significant FFA conversion for the first cycle of homogeneous contribution study. This confirms the occurrence of sulfonic acid group leaching, believed to be due to the detachment of sulfonic acid group from the catalyst surface, followed by the hydrolysis of sulfonic acid species with water to form homogeneous species. Data for the second cycle show after the first homogenous contribution cycle that conversion is very low indicating that there is no further leaching of the sulfonic acid group into methanol in subsequent cycles.

In addition, the catalyst deactivation has also been investigated using the catalyst that was previously filtered from the methanol solution. The results were summarized in Figure 10.21. From the figure, it could be seen that the conversion for the second cycle was slightly lower compared to the first cycle. This was solely due to the blockage of large molecules of UCO molecules as the data from Figure 10.20 show that the leaching of sulfonic acid was negligible during the second cycle of the reaction.

Figure 10.22 shows the comparison between the reusability study and methanol treated catalyst study. The difference in FFA conversion between the fresh catalyst and first methanol-treated catalyst cycle is believed to be due to leaching of sulfonic acid groups. This is because both experiments were conducted at the same experimental parameters, the only difference was the condition of the catalyst, where the former experiment used fresh catalyst and the latter one used methanol-treated catalyst. This finding was verified using the homogeneous contribution study shown in Figure 10.20.

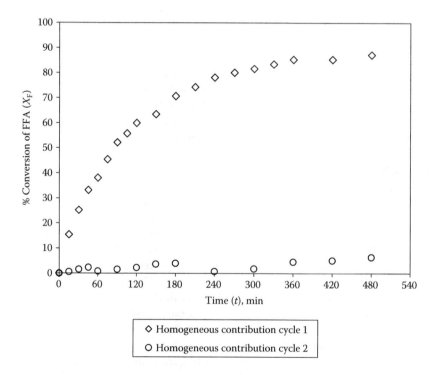

FIGURE 10.20 Study on the homogeneous contribution of Purolite D5081 ion exchange resins. Experimental conditions: catalyst: Purolite D5081; stirring speed: 475 rpm; reaction temperature: 60°C. (From Abidin, S.Z. et al., *Ind. Eng. Chem. Res.*, 51, 14653–14664, 2012. With permission from American Chemical Society.)

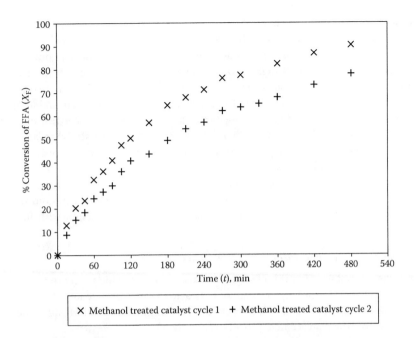

FIGURE 10.21 Study on the methanol treated catalyst deactivation. Experimental conditions: catalyst: Purolite D5081; catalyst loading; 1.25% (w/w), stirring speed: 475 rpm; reaction temperature 60°C; molar ratio (methanol:UCO): 6:1. (From Abidin, S.Z. et al., *Ind. Eng. Chem. Res.*, 51, 14653–14664, 2012. With permission from American Chemical Society.)

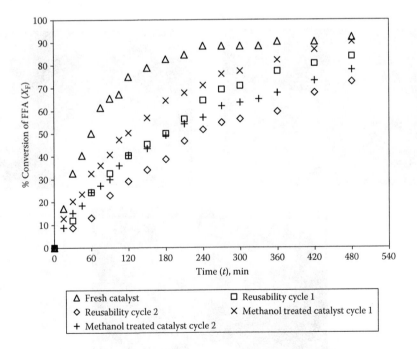

FIGURE 10.22 Comparison between the reusability study and the methanol treated catalyst study. Experimental conditions: catalyst: Purolite D5081; stirring speed: 475 rpm; catalyst loading: 1.25% (w/w); reaction temperature: 60°C; molar ratio (methanol:UCO): 6:1. (From Abidin, S.Z. et al., *Ind. Eng. Chem. Res.*, 51, 14653–14664, 2012. With permission from American Chemical Society.)

On the other hand, the first cycle of methanol treated catalyst study gives a conversion close to the first cycle from the reusability study. For the former experiment, the resin itself was not introduced to any UCO mixture, so there was no possibility of any triglyceride blockage during the reaction and this means the conversion was slightly higher. So, it could be inferred that the 6% reduction in conversion was solely due to blockage of large UCO molecules in the catalysts' pores. The same trend was observed during the second cycle for both cases where the calculated conversion difference was about the same. Therefore, it could be concluded that the reduction of FFA conversion in subsequent cycles (either for the reusability study or the methanol treated catalyst study) is largely due to the progressive blockage of the large triglycerides molecules in the pores of the resin since there is a steady reduction in the FFA conversion for each cycle. Furthermore, the possibility of sulfonic acid groups leaching throughout the process was small. When both conditions (reusability study and methanol treated catalyst study) were evaluated individually, both of them showed sulfur leaching and pore blockage simultaneously and show a reduction in FFA conversion with increasing cycle number.

10.1.3.4.2 Reusability Study on the Spent Transesterification Catalyst

Reusability of the catalyst is an important step as it reduces the cost of biodiesel production. During the preparation of used catalyst, the displacement of fatty acid ion with acetate ion was investigated using acetic acid concentrations of 17.5 and 1 M, respectively. Two analyses (FEG-SEM and elemental analysis) were conducted before the displacement process was finalized.

The FEG-SEM analysis was carried out to observe any changes on the surface of the catalysts after being treated with acetic acid. Figure 10.23 compares the FEG-SEM analysis for (1) fresh Diaion PA306s, (2) used Diaion PA306s (1 M acetic acid treatment), and (3) used Diaion PA306s (17.5 M acetic acid treatment) catalysts captured at 500× magnification.

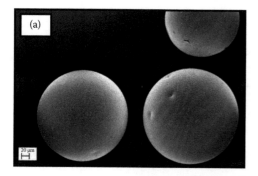
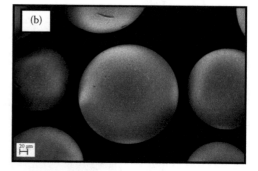
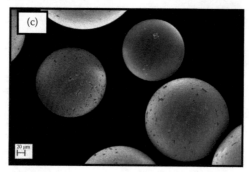

FIGURE 10.23 The FEG-SEM images of Diaion PA306s catalysts, taken at 500× magnification: (a) fresh Diaion PA306s, (b) used Diaion PA306s (1 M acetic acid treatment) and (c) used Diaion PA306s (17.5 M acetic acid treatment).

TABLE 10.6 Elemental Analysis of Fresh and Used Diaion PA306s

| | Elemental Analysis | | | |
Catalysts	% C	% H	% N	% O[a]
Fresh Diaion PA306s	55.59	9.42	4.34	30.65
Used Diaion PA306s (1 M acetic acid)	55.44	9.20	4.31	31.05
Used Diaion PA306s (17.5 M acetic acid)	54.51	8.84	4.35	32.30

[a] Oxygen by difference.

From Figure 10.23, it can be seen that the surface morphology of fresh and used Diaion PA306s (1 M of acetic acid treatment) catalysts appears as a smooth surface, whereas a noticeable deterioration of the surface was found when Diaion PA306s catalyst was treated with 17.5 M acetic acid. This suggests that the concentration of acetic acid is too high. Table 10.6 shows the results of the elemental analysis. From Table 10.6, it can be seen that the used Diaion PA306s catalysts treated with 17.5 and 1 M acetic acid resulted in slightly lower carbon and hydrogen values as compared to the fresh Diaion PA306s. The reduction of carbon and hydrogen values in used Diaion PA306s catalyst treated with 17.5 M of acetic acid was also found to be slightly higher than the used Diaion PA306s catalyst treated with 1 M of acetic acid. This indicates that there are some changes or damage to the structure of Diaion PA306s catalyst when higher acetic acid concentration was used and this could contribute to a loss in catalytic activity. Therefore, the acid displacement method using 1 M acetic acid solution was selected as the displacement process throughout the study.

The reusability study was carried out under the optimum reaction conditions, 9% (w/w) catalyst loading, 328 K reaction temperature, 18:1 methanol to P-UCO feed mole ratio, and 350 rpm stirring speed. The result of reusability study was compared with the optimum result obtained using fresh Diaion PA306s catalyst and shown in Figure 10.24. It was observed that the Diaion PA306s catalyst gave a similar conversion of triglycerides using fresh and used catalysts. The conversion of triglycerides for both catalysts after 8 h of reaction time was approximately 75%. It was concluded that the catalyst can be used several times without losing catalytic activity.

10.1.3.5 Separation and Purification Process

Once the transesterification reaction was completed, the reaction mixture was allowed to cool to room temperature. The reaction mixture was separated from the catalyst and transferred to a separating funnel. The reaction mixture was allowed to settle overnight to form FAME-rich phase and glycerine-rich phase layers. The layers were sequentially withdrawn from the separating funnel and introduced to the washing process. Two types of washing techniques were examined, the conventional wet washing techniques using water and the dry washing techniques using Purolite PD206 as an adsorbent. Table 10.7 shows the purity of FAME using wet and dry washing and the results were compared with the unpurified biodiesel. Results in Table 10.7 show a similar percentage of FAME for all purified and unpurified samples. The dry washing treatment using PD 206 gave the highest percentage of FAME purity of ca. 75%. By having the advantage of being water-free, there is less production of wastewater; purified biodiesel from the dry washing treatment was selected for further testing.

The purified biodiesel from the dry washing process was tested for monoglycerides, diglycerides, triglycerides, and glycerine content. The same analyses were conducted on unpurified biodiesel and the results are presented in Table 10.8. It can be seen from Table 10.8 that biodiesel from the dry washing technique shows a lower percentage of glycerides and glycerine content as compared to the unpurified biodiesel. The finding contradicts with the findings by Shibasaki-Kitakawa et al. (2011) as they claimed that all the impurities such as the residual oil, FFA, water, and dark brown pigment can be removed from the product by adsorption on Diaion PA306s catalyst.

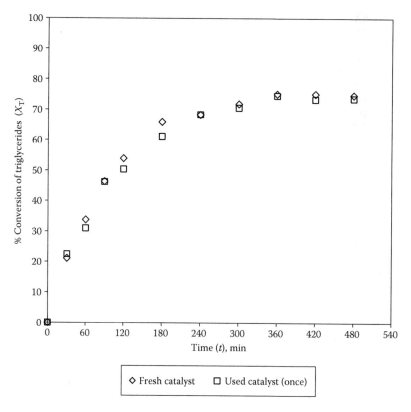

FIGURE 10.24 Effect of catalyst reusability on the conversion of triglycerides. Experimental conditions: catalyst: Diaion PA306s; stirring speed: 350 rpm; catalyst loading: 9% (w/w); reaction temperature: 328 K; feed mole ratio (methanol:UCO): 18:1.

TABLE 10.7 Purity of FAME Using Different Treatment Processes

Treatment Processes	Purity of FAME, %
Ion exchange resin (PD 206) treatment	75.4 ± 1
Water treatment	72.3 ± 2
Unpurified biodiesel	71.6 ± 0.5

Note: (1) Ion exchange treatment, (2) water treatment, and (3) unpurified biodiesel.

TABLE 10.8 Analysis of Monoglycerides, Diglycerides, Triglycerides, and Glycerine Content (Total and Free Glycerine)

Component	Ion Exchange Resin (PD 206) Treatment% (m/m)	Unpurified Biodiesel %(m/m)
Monoglycerides	0.85	1.35
Diglycerides	0.1	2.74
Triglycerides	0.47	1.91
Free glycerine	0.03	0.05
Total glycerine	0.33	0.9

10.1.4 Conclusions

Esterification pretreatment of UCO using various types of ion exchange resins has been investigated. Among the catalysts investigated, Purolite D5081 resin showed the best catalytic performance as compared to other two resins, Purolite D5082 and Amberlyst 36. This is probably due to the catalytic properties of this resin, as it has the highest specific surface area and the largest total pore volume. At the optimum reaction condition of 60°C reaction temperature, 6:1 methanol to UCO molar ratio, 1.25% (w/w) catalyst loading, and 475 rpm stirring speed, Purolite D5081 achieved an FFA conversion of 92%. During the reusability study, the conversion of catalyst dropped by 8%–10% after each cycle. Several experiments have been conducted through the homogeneous contribution study, and the results confirmed that resin pore blockage and sulfur leaching were two dominant factors that decrease the catalytic performance of the polymeric resin. For the transesterification reaction, Diaion PA306s catalyst showed the best catalytic performance and the reason was due to the high basicity of the catalyst. As a result, Diaion PA306s was selected for the optimization study. At the optimum reaction conditions of 9% (w/w) catalyst loading, 328 K reaction temperature, 18:1 methanol to P-UCO feed mole ratio, and 350 rpm stirring speed, triglycerides conversion was ca. 75%. The remaining 25% was predicted to be the unreacted triglycerides. During the reusability study, the Diaion PA306s catalyst gave similar triglycerides conversion after being reused once (at the same conditions). Therefore, it was concluded that the catalyst can be reused several times without losing its catalytic activity. Two types of washing techniques, wet and dry washing, were carried out during the purification process and the results for purified biodiesels were compared with unpurified biodiesel. Biodiesel produced from the dry washing technique shows the lowest percentage of glycerides and glycerine content and therefore was chosen as the best treatment for the purification of biodiesel.

Acknowledgments

We gratefully acknowledge Purolite International Ltd. (the late Dr. Jim Dale and Mr. Brian Windsor) for kindly supplying the catalysts for this research work, Mitsubishi Chemicals for kindly supplying the Diaion catalyst, and GreenFuel Oil Co. Ltd. for supplying the UCO. We would like to thank Universiti Malaysia Pahang and Malaysian Government for the Ph.D. scholarship to S.Z.A.

List of Abbreviations

BET	Brunauer–Emmett–Teller
BSI	British Standard Institution
DVB	Divinylbenzene
FAME	Fatty acid methyl ester
FEG-SEM	Field emission gun-scanning electron microscopy
FFA	Free fatty acids
FT-IR	Fourier transform-infrared
GC–MS	Gas chromatography–mass spectrometry
PSD	Particle size distribution
P-UCO	Pretreated used cooking oil
RO	Reverse osmosis
SEM	Scanning electron microscopy
SEM–EDX	Scanning electron microscopy–energy dispersive X-ray
UCO	Used cooking oil

References

Abbaszaadeh, A., B. Ghobadian, M. R. Omidkhah, and G. Najafi. 2012. Current biodiesel production technologies: A comparative review. *Energy Conversion and Management* 63:138–148.

Abidin, S. Z. (Supervisors: B. Saha and G. T. Vladisavljević). 2012. Production of biodiesel from used cooking oil (UCO) using ion exchange resins as catalysts. PhD Thesis, Loughborough University.

Abidin, S. Z., K. F. Haigh, and B. Saha. 2012. Esterification of free fatty acids in used cooking oil using ion-exchange resins as catalysts: An efficient pretreatment method for biodiesel feedstock. *Industrial and Engineering Chemistry Research* 51:14653–14664.

Abidin, S. Z., D. Patel, and B. Saha. 2013. Quantitative analysis of fatty acids composition in the used cooking oil (UCO) by gas chromatography-mass spectrometry (GC-MS). *The Canadian Journal of Chemical Engineering* 91:1896–1903.

Atabani, A. E., A. S. Silitonga, I. A. Badruddin, T. M. I. Mahlia, H. H. Masjuki, and S. Mekhilef. 2012. A comprehensive review on biodiesel as an alternative energy resource and its characteristics. *Renewable and Sustainable Energy Reviews* 16:2070–2093.

Balat, M. 2011. Potential alternatives to edible oils for biodiesel production—A review of current work. *Energy Conversion and Management* 52:1479–1492.

Bianchi, C. L., D. C. Boffito, C. Pirola, and V. Ragaini. 2009. Low temperature de-acidification process of animal fat as a pre-step to biodiesel production. *Catalysis Letters* 134:179–183.

David, F., P. Sandra, and A. K. Vickers. 2005. Column selection for the analysis of fatty acid methyl esters. *Agilent Application Notes* 1:5989–3760. Retrieved from http://www.agilent.com/cs/library/applications/5989-3760EN.pdf (accessed on 12 May 2017).

dos Reis, S. C. M., E. R. Lachter, R. S. V. Nascimento, J. A. Rodrigues Jr., and M. G. Reid. 2005. Transesterification of Brazilian vegetable oils with methanol over ion-exchange resins. *Journal of the American Oil Chemists' Society* 82:661–665.

Dowd, M. K. 1998. Gas chromatographic characterization of soapstocks from vegetable oil refining. *Journal of Chromatography A* 816:185–193.

European Standards. 2011. BS EN ISO 12966-2. 2011. Animal and vegetable fats and oils–gas chromatography of fatty acid methyl esters—Part 2: Preparation of methyl esters of fatty acids. Retrieved from https://www.cen.eu (accessed on 12 May 2017).

Encinar, J. M., J. F. González, and A. Rodríquez-Reinares. 2007. Ethanolysis of used frying oil. Biodiesel preparation and characterization. *Fuel Processing Technology* 88:513–522.

Falco, M. G., C. D. Córdoba, M. R. Capeletti, and U. Sedran. 2010. Basic ion exchange resins as heterogeneous catalysts for biodiesel synthesis. *Advanced Materials Research* 132:220–227.

Feng, Y., B. He, Y. Cao, J. Li, M. Liu, F. Yan, and X. Liang. 2010. Biodiesel production using cation-exchange resin as heterogeneous catalyst. *Bioresource Technology* 101:1518–1521.

Haigh, K., S. Z. Abidin, B. Saha, and G. Vladisavljević. 2012. Pretreatment of used cooking oil for the preparation of biodiesel using heterogeneous catalysis. *Progress in Colloid and Polymer Science* 139:19–23.

Haigh, K., S. Z. Abidin, G. Vladisavljević, and B. Saha. 2013. Comparison of Novozyme 435 and Purolite D5081 as heterogeneous catalysts for the pretreatment of used cooking oil for biodiesel production. *Fuel* 111:186–193.

Haigh, K., G. Vladisavljević, J. C. Reynolds, Z. Nagy, and B. Saha. 2014. Kinetics of the pre-treatment of used cooking oil using Novozyme 435 for biodiesel production. *Chemical Engineering Research and Design* 92:713–719.

He, B., Y. Shao, Y. Ren, J. Li, and Y. Cheng. 2015. Continuous biodiesel production from acidic oil using a combination of cation- and anion-exchange resins. *Fuel Processing Technology* 130:1–6.

Helfferich, F. 1962. *Ion Exchange*. McGraw-Hill: New York.

Ilgen, O. 2014. Investigation of reaction parameters, kinetics and mechanism of oleic acid esterification with methanol by using Amberlyst 46 as a catalyst. *Fuel Processing Technology* 124:134–139.

Kaercher, J. A., R. D. C. de Souza Schneider, R. A. Klamt, W. L. T. da Silva, W. L. Schmatz, M. da Silva Szarblewski, and E. L. Machado. 2013. Optimization of biodiesel production for self-consumption: Considering its environmental impacts. *Journal of Cleaner Production* 46:74–82.

Knothe, G., and K. R. Steidley. 2009. A comparison of used cooking oils: A very heterogeneous feedstock for biodiesel. *Bioresource Technology* 100:5796–801.

Kouzu, M., A. Nakagaito, and J.-S. Hidaka. 2011. Pre-esterification of FFA in plant oil transesterified into biodiesel with the help of solid acid catalysis of sulfonated cation-exchange resin. *Applied Catalysis A: General* 405:36–44.

Li, J., Y. J. Fu, X. J. Qu, W. Wang, M. Luo, C.-J. Zhao, and Y.-G. Zu. 2012. Biodiesel production from yellow horn (*Xanthoceras sorbifolia* Bunge.) seed oil using ion exchange resin as heterogeneous catalyst. *Bioresource Technology* 108:112–118.

Lin, L., Z. Cunshan, S. Vittayapadung, S. Xiangqian, and D. Mingdong. 2011. Opportunities and challenges for biodiesel fuel. *Applied Energy* 88:1020–1031.

Liu, X., M. Ye, B. Pu, and Z. Tang. 2012. Risk management for jatropha curcas based biodiesel industry of Panzhihua Prefecture in Southwest China. *Renewable and Sustainable Energy Reviews* 16:1721–1734.

Liu, Y., L. Wang, and Y. Yan. 2009. Biodiesel synthesis combining pre-esterification with alkali catalyzed process from rapeseed oil deodorizer distillate. *Fuel Processing Technology* 90:857–862.

López, D. E., J. G. Goodwin Jr., and D. A. Bruce. 2007. Transesterification of triacetin with methanol on Nafion® acid resins. *Journal of Catalysis* 245:381–391.

Mazzotti, M., B. Neri, D. Gelosa, A. Kruglov, and M. Morbidelli. 1997. Kinetics of liquid-phase esterification catalyzed by acidic resins. *Industrial and Engineering Chemistry Research* 36:3–10.

Official EU Journal. 1991. The characteristics of olive oil and olive-residue oil and on the relevant methods of analysis. Commission Regulation (EEC) No. 2568/91. Retrieved from http://data.europa.eu/eli/reg/1991/2568/2015-10-16 (accessed on 12 May 2017).

Özbay, N., N. Oktar, and N. A. Tapan. 2008. Esterification of free fatty acids in waste cooking oils (WCO): Role of ion-exchange resins. *Fuel* 87:1789–1798.

Park, J.-Y., D.-K. Kim, Z.-M. Wang, J.-P. Lee, S.-C. Park, and J.-S. Lee. 2008. Production of biodiesel from soapstock using an ion-exchange resin catalyst. *Korean Journal of Chemical Engineering* 25:1350–1354.

Pinzi, S., D. Leiva-Candia, I. López-García, M. D. Redel-Macías, and M. P. Dorado. 2014. Latest trends in feedstocks for biodiesel production. *Biofuels, Bioproducts and Biorefining* 8:126–143.

Russbueldt, B. M. E., and W. F. Hoelderich. 2009. New sulfonic acid ion-exchange resins for the pre-esterification of different oils and fats with high content of free fatty acids. *Applied Catalysis A: General* 362:47–57.

Shibasaki-Kitakawa, N., K. Hiromori, T. Ihara, K. Nakashima, and T. Yonemoto. 2015. Production of high quality biodiesel from waste acid oil obtained during edible oil refining using ion-exchange resin catalysts. *Fuel* 139:11–17.

Shibasaki-Kitakawa, N., H. Honda, H. Kuribayashi, T. Toda, T. Fukumura, and T. Yonemoto. 2007. Biodiesel production using anionic ion-exchange resin as heterogeneous catalyst. *Bioresource Technology* 98:416–421.

Shibasaki-Kitakawa, N., T. Tsuji, M. Kubo, and T. Yonemoto. 2011. Biodiesel production from waste cooking oil using anion-exchange resin as both catalyst and adsorbent. *BioEnergy Research* 4:287–293.

Stamenković, O. S., A. V. Veličković, and V. B. Veljković. 2011. The production of biodiesel from vegetable oils by ethanolysis: Current state and perspectives. *Fuel* 90:3141–3155.

Talebian-Kiakalaieh, A., N. A. S. Amin, and H. Mazaheri. 2013. A review on novel processes of biodiesel production from waste cooking oil. *Applied Energy* 104:683–710.

Talukder, M. M. R., J. C. Wu, S. K. Lau, L. C. Cui, G. Shimin, and A. Lim. 2009. Comparison of Novozym 435 and Amberlyst 15 as heterogeneous catalyst for production of biodiesel from palm fatty acid distillate. *Energy and Fuels* 23:1–4.

Tesser, R., M. Di Serio, M. Guida, M. Nastasi, and E. Santacesaria, 2005. Kinetics of oleic acid esterification with methanol in the presence of triglycerides. *Industrial and Engineering Chemistry Research* 44:7978–7982.

Yaakob, Z., M. Mohammad, M. Alherbawi, Z. Alam, and K. Sopian. 2013. Overview of the production of biodiesel from waste cooking oil. *Renewable and Sustainable Energy Reviews* 18:184–193.

Yadav, G. D., and H. B. Kulkarni. 2000. Ion-exchange resin catalysis in the synthesis of isopropyl lactate. *Reactive and Functional Polymers* 44:153–165.

Zhang, J., and L. Jiang. 2008. Acid-catalyzed esterification of *Zanthoxylum bungeanum* seed oil with high free fatty acids for biodiesel production. *Bioresource Technology* 99:8995–8998.

10.2 Process for Synthesis of Biodiesel from Used Cooking Oil: Feasibility and Experimental Studies

Raj Patel, Amir Khan, I. M. Mujtaba, Richard Butterfield, Elisabetta Mercuri, and Davide Manca

10.2.1 Introduction

With the current global drive to reduce the reliance on fossil fuels, in particular, which are used as automotive fuels, the search for alternate and renewable sources of fuel is now more prevalent than ever. One of the most commonly cited and researched is biodiesel, a liquid fuel derived from virgin vegetable oils or waste cooking oils that have undergone a transesterification reaction with an alcohol, typically methanol, in presence of a sodium or potassium hydroxide catalyst as reported by the works of Pelly (2003), Demirbas (2005), and Ferdous et al. (2012). There is a plethora of oils that have the ability to be used in the manufacturing. A noninclusive list of oils is shown in Marchetti et al. (2005) and Ma and Hanna (1999), e.g., babassu, coconut, corn, cottonseed, crambe, lard, palm oil, peanut, rapeseed, soybean, sunflower, tallow, canola oil. Each oil is differentiated by the amount and location of double bonds on the unsaturated triglyceride. There are seven major types of fatty acids—lauric, myristic, palmitic, stearic, oleic, linoleic, and linolenic (Marchetti et al., 2005) with the properties of fatty acids affecting the properties of the biodiesel product.

Reaction kinetics has been extensively studied and the rate constants were determined for various oils. The kinetics for Soyabean oil has been reported by Noureddini and Zhu (1997), Freedman et al. (1986), and Mahajan et al. (2006b). The kinetics for Palm oil was reported by Darnoko and Cheryan (2000), and the kinetics for Sesame oil was reported by Ferdous et al. (2012). A study by Benavides and Diwekar (2012a) and continued in Benavides and Diwekar (2012b) describes a batch reactor model for the optimal control of the production of biodiesel, using deterministic and then stochastic control, and based upon the reaction kinetics determined by Noureddini and Zhu (1997). This study was developed further by Benavides and Diwekar (2013) in an effort to accommodate the possible variability in available feedstock and optimize biodiesel production using various performance indices. There is, however, little experimental data in the literature.

Moreover, there are many ways to characterize the final biodiesel product such as viscosity, cold flow (i.e., cloud point, pour point, freezing point), density, acid number, lubricity. These properties were discussed extensively by Knothe (2005), Demirbas (2003, 2005), Singh and Padhi (2009), Akers et al. (2006), Mahajan et al. (2006a), and US DOE (2005). The characterization methods are typically based on offline measurements that cannot be performed *in situ*, and therefore, give little indication about the reaction process. A recent study by Clark et al. (2013) demonstrated that pH could be used to monitor the progress of the transesterification reaction *in situ* and provide real time monitoring of biodiesel production in pure canola oil. The study suggested that the measured pH change is related to the dilution of OH⁻ ions as the oil is converted to products and not from the depletion of OH⁻ ions due to the reaction. As the reaction proceeds, there is a noticeable change in viscosity of the biodiesel compared to the initial vegetable oil. Ultrasound measurements can thus be used to monitor such changes, as the ultrasound wave will pass through the lower viscosity biodiesel much faster than in the higher viscosity untreated oil. Such measurements have only been made offline with waste vegetable oil (WVO) and the biodiesel and the results are promising. If the ultrasound measurements could be correlated to other *in situ* measurements of conversion, such as pH, this would provide a novel and not invasive technique to monitor progress of the transesterification reaction.

In this work, pH measurements have been used to continuously monitor the progression of the reaction in pure vegetable oil. The pH measurements are correlated with reaction conversion and compared with theoretical results obtained from a kinetic model using the gPROMs process simulator. Ultrasound measurements, also taken continuously *in situ*, are correlated with the pH measurements and hence the reaction conversion.

10.2.2 Methods

10.2.2.1 Reaction Kinetics and Modeling

Biodiesel can be produced using a transesterification process whereby a triglyceride (vegetable oil) is reacted with either methanol or ethanol. The reaction also uses a base catalyst usually sodium hydroxide or potassium hydroxide to aid the reaction process. The process involves reacting triglyceride with methanol to produce methyl ester (biodiesel) and a by-product, glycerol. In the study by Noureddini and Zhu (1997), the reaction is considered to consist of three stepwise and reversible reactions, where triglycerides (TG) are converted to diglycerides (DG), diglycerides to monoglycerides (MG), and monoglycerides into glycerol, as presented in Equations 10.1 through 10.4.

$$TG + CH_3OH \overset{k_1,k_2}{\Leftrightarrow} DG + R_1COOCH_3 \tag{10.1}$$

$$DG + CH_3OH \overset{k_3,k_4}{\Leftrightarrow} MG + R_2COOCH_3 \tag{10.2}$$

$$MG + CH_3OH \overset{k_5,k_6}{\Leftrightarrow} GL + R_3COOCH_3 \tag{10.3}$$

The overall reaction is denoted as follows:

$$TG + 3CH_3OH \leftrightarrow GL + 3RCOOCH_3 \tag{10.4}$$

The mathematical model for the biodiesel production in a batch reactor is governed by the following ordinary differential equations, Equations 10.5 through 10.10, which have been derived from the mass balance of a batch reactor and first presented by Noureddini and Zhu (1997):

$$\frac{dC_{TG}}{dt} = -k_1 C_{TG} C_A + k_2 C_{DG} C_E, \tag{10.5}$$

$$\frac{dC_{DG}}{dt} = k_1 C_{TG} C_A - k_2 C_{DG} C_E - k_3 C_{DG} C_A + k_4 C_{MG} C_E, \tag{10.6}$$

$$\frac{dC_{MG}}{dt} = k_3 C_{DG} C_A - k_4 C_{MG} C_E - k_5 C_{MG} C_A + k_6 C_{CE} C_E, \tag{10.7}$$

$$\frac{dC_E}{dt} = k_1 C_{TG} C_A - k_2 C_{DG} C_E + k_3 C_{DG} C_A - k_4 C_{MG} C_E + k_5 C_{MG} C_A - k_6 C_{GL} C_E, \tag{10.8}$$

$$\frac{dC_A}{dt} = -\frac{dC_E}{dt}, \tag{10.9}$$

$$\frac{dC_{GL}}{dt} = k_5 C_{MG} C_A - k_6 C_{GL} C_E, \tag{10.10}$$

where C_{TG}, C_{DG}, C_{MG}, C_E, C_A, C_{GL}, are concentrations of triglycerides, diglycerides, monoglycerides, methyl ester, methanol and glycerol, respectively. The reaction rate constants, k_1–k_6, can be expressed by the Arrhenius equation and the constants are presented by Noureddini and Zhu (1997).

Equations 10.5 through 10.10 were used to model our system using gPROMS, a process simulator, which is appropriate for this type of modeling. The change in concentration of the various components, during the reaction, was calculated from data obtained via pH and ultrasound measurements techniques. Aracil et al. (2005) published activation energies and pre-exponential factors for the transesterification of sunflower oil into biodiesel, which allow for the effect of temperature to be considered. The rate constant k_6 was found to be so small that it could be neglected with respect to the other reactions.

Clark et al. (2013) published a correlation where the relationship between pH and the reaction conversion, X, is quantified by Equation 10.11:

$$X(t) = \frac{\left(10^{-(14-\text{peak pH})} - 10^{-(14-\text{pH at } t)}\right)}{\left(10^{-(14-\text{peak pH})} - 10^{-(14-\text{final expected pH})}\right)}. \tag{10.11}$$

Clark et al. (2013) determined that when the required amount of KOH was dissolved in a methanol volume equivalent to the total batch volume being processed, a similar pH value to that of the final value obtained in the biodiesel reaction for each temperature was achieved. This provided the expected pH values for the correlation. Clark et al. (2013) also showed that the maximum pH values observed in the biodiesel runs were similar to the pH observed for the same amount of KOH dissolved in the methanol volume required for the reaction.

10.2.2.2 Experimental Equipment

Methanol, isopropyl alcohol, potassium hydroxide (KOH), sodium hydroxide (NaOH), distilled water and food grade vegetable oil were used in the conversion process. A Pyrex 1 L jacketed reactor was used with an overhead stirrer to ensure good mixing. Constant temperature was maintained using a Grant GD120 circulating water bath, which pumps the water through the reactor jacket. A DrDAQ data logger connected via USB to a PC was used to monitor and record the process temperature and the pH of the mixture. Ultrasound probes (multicomp—transceivers, centre frequency 200 kHz, 19 mm in diameter, directivity ±7°, distance of detection 0.1–2 m, operating temperature range −20°C to +80°C) were used to measure the time taken for the sound waves to travel through the reaction mixture; the speed of the sound waves could thus be calculated. One probe acts as a transmitter and is connected to a Thandor TG105 5 Hz to 5 MHz pulse generator (settings: 2 ms pulse period, pulse width 10 μs, DC output of 10 V). The other probe is used as a receiver and is connected to a Picoscope 2204 sampler (resolution 8 bits @

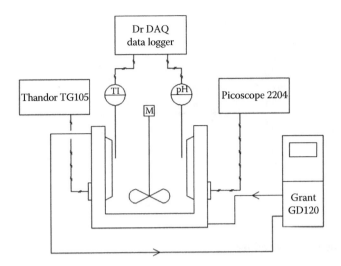

FIGURE 10.25 Diagram of the experimental set up.

100 Ms/s, Bandwith 10 MHz), which is connected via USB to a PC to record the waveforms using the PicoScope 6.0 software.

The time of flight is the time needed for an ultrasonic wave to travel a certain distance. For instance from a transmitter to a target and then, after reflection, back to the receiver located near the transmitter. A pulse was used for that purpose and the ultrasonic system used the time of flight for distance estimation. The distance between the probes can be calculated using Equation 10.12.

$$l = \frac{v\,(\text{ToF})}{2}, \tag{10.12}$$

where v represents the sound propagation velocity, l is the distance between the probes, measured using a ruler and ToF is the time delay. It can be seen that the range of the measurement accuracy depends on the ToF and the accuracy in measuring the sound velocity v.

Figure 10.25 shows a schematic diagram of the set up used. The ultrasound probes are placed outside the reactor vessel, flush to the jacketed wall, making it a nonintrusive technique. The effect of the vessel wall will be a constant factor and thus can be accounted for. As the reaction proceeds, the viscosity of the mixture decreases and the ultrasound wave velocity is expected to increase. This change in velocity of the ultrasound waves can be used to determine the concentration of the reactants and the products.

10.2.2.3 Reaction Conditions

A sample from each vegetable oil used undergoes a titration as described by Pelly (2003) to determine the required amount of potassium hydroxide (or sodium hydroxide) to be used. The amount of methanol used is 20% by mass of the vegetable oil undergoing reaction, as described by Pelly (2003). Experiments conducted with pure sunflower oil were performed as per the method described by Clark et al. (2013), with a 6:1 M ratio of methanol to oil with a potassium hydroxide catalyst content of 0.49 g KOH per 100 g of sunflower oil.

The transesterification of the oil is conducted under the following conditions: 800 rpm stirring, atmospheric pressure and at three different temperatures of 25°C, 35°C, and 45°C. The oil and alcohol are mixed and are thermally conditioned at the reaction temperature before being charged into the reactor.

A typical ultrasound trace is shown in Figure 10.26. The time taken for the pulse to travel across the reactor vessel (peak to peak) can be measured and the velocity can be calculated.

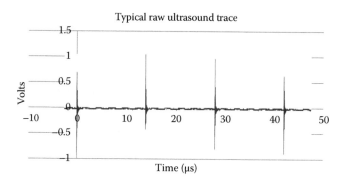

FIGURE 10.26 Typical ultrasound trace obtained from the trials.

10.2.2.4 Experimental Procedure

Half of the methanol required for the reaction is mixed with the oil. This was to establish small droplets of methanol dispersed within the oil to increase the contact surface area and hence the rate of reaction. The required KOH is then dissolved within the remainder of the methanol. The reaction is initiated by adding the mixture to the 1 L reactor and the pH recorded over time for at least 40 min. The time taken for the ultrasound waves to travel between the transmitter and receiver is also measured at 2 min intervals and at a sampling frequency of 1 MHz. The glycerine is separated from the oil by decanting and the oil washed with water. The water–oil mixture is then allowed to settle to form two layers, which are then separated to give the final biodiesel product. Characterization of the biodiesel, although not a topic for discussion here, is also carried out to measure properties such as rheology and calorific value. Between each run, the reactor is cleaned thoroughly using acetone and allowed to dry.

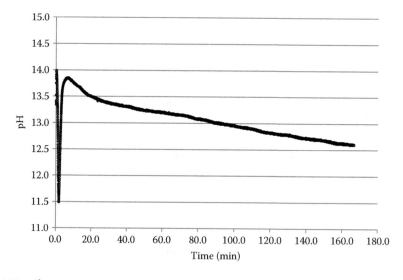

FIGURE 10.27 Change in pH as transesterification of pure sunflower oil proceeds over time at 25°C.

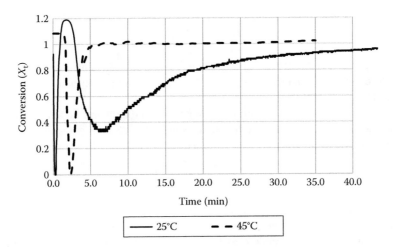

FIGURE 10.28 pH values represented in terms of conversion for transesterification of sunflower oil at 25°C and 45°C.

10.2.3 Results and Discussion

The pH measurements are interpreted by the relationship outlined in Clark et al. (2013), allowing the values for pH to be presented in terms of conversion. Figure 10.27 shows the initial pH measurements for biodiesel production using pure sunflower oil at 25°C.

The corresponding conversion chart is presented in Figure 10.28. Over the initial first 5 min, the pH value drops, showing an s-shape behavior, which Clark et al. (2013) suggest to be an indication of an initial mixing/mass transfer-limited stage.

The reason for this behavior was suggested to be due to an initial mixing/mass transfer limited stage, which diminishes as temperature increases, in agreement with low temperature methanolysis studies conducted by Stamenković et al. (2008). Clark et al. (2013) observed that duration of this mass transfer limiting stage was approximately 3.8 min for a total batch size of 375 mL. At 25°C, Figure 10.28 shows that the duration of this mass transfer limiting stage is approximately 6.5 min for a total batch volume of 582 mL. This suggests that the increased scale of the experiment has an effect on the duration of this

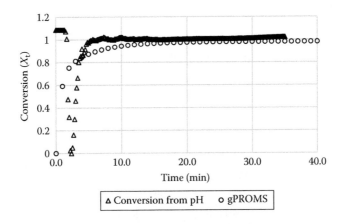

FIGURE 10.29 Comparison of gPROMS simulation results to experimental data for the conversion of sunflower oil at 45°C.

mass transfer limiting stage. Also, shown in Figure 10.28 is the conversion at 45°C. It can be seen that as the reaction temperature is increased, this peak on the conversion curve shifts to the left.

Aracil et al. (2005) published rate constants for the transesterification of sunflower oil into biodiesel. A gPROMS simulation was developed utilizing the published rate constants in Equations 10.5 through 10.10, the results of which are shown in Figure 10.29 together with the conversion obtained experimentally from the pH data.

From Figure 10.29 it can be seen that the agreement between the theoretical and experimental data is quite good beyond about 2 min. There is a discrepancy up to about 2 min because of the erratic pH measurements due to initial mixing/mass transfer limited stage. However, the pH measurement is a "spot-measurement" and in a large reactor it could lead to erroneous data. The use of ultrasounds to monitor the reaction may be more accurate since it is a bulk measurement.

Figure 10.30 shows the change in the velocity of ultrasound wave propagating through bulk mixture as the reaction proceeds with time for the transesterification of sunflower oil at 45°C. It is clear from this data that as the reaction proceeds, the viscosity of the mixture decreases and hence, the velocity increases and levels off when the reaction reaches completion. In order to model the sound velocity to

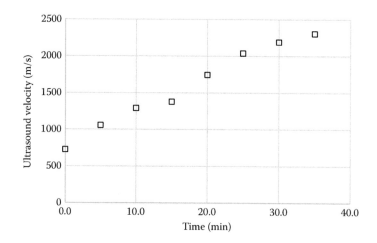

FIGURE 10.30 Ultrasound velocity for transesterification reaction at 45°C.

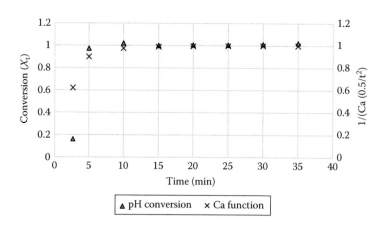

FIGURE 10.31 Comparison of pH conversion and capillary number function for 45°C.

the conversion of sunflower oil, dimensional analysis is used to find a suitable dimensionless number. The factors affecting the ultrasound velocity are viscosity, density, and surface tension of the liquid. Thus with four variables and three dimensions, there should be one dimensionless number that could be used to model the conversion. The Capillary number ($Ca = \mu V/\sigma$ where μ is the viscosity of the liquid, V the ultrasound velocity and σ the surface tension of the liquid) is the candidate dimensionless number. Ellis et al. (2008) have measured the viscosity of the mixture for the transesterification reaction at about 45°C and we have used their data in modeling the conversion as a function of the capillary number. Figure 10.31 shows our experimentally obtained conversion data at 45°C and on the secondary axis the function $1/Ca^{(0.5/t^2)}$.

There is a fair agreement between the two datasets, although about 100% conversion is achieved after the first 10 min of the experiment and for most of the data points. This indicates that it would be possible to use ultrasound as a means of monitoring the transesterification reaction. However, viscosity data would be required to calculate the capillary number for all the reaction temperatures under consideration. Further works need to be carried out to increase the frequency of data collection, using the ultrasound probes, during the first 2–3 min of the reaction. Also, the peculiarity of the s-shaped behavior, at the initial stage of the reaction, needs to be understood better in order to obtain more realistic pH measurements in the first few minutes.

10.2.4 Conclusion

In this ongoing study, initial measurements of biodiesel concentration using ultrasounds have been made and there is a clear difference in the propagation velocities as the viscosity of the oil decreases. A detailed study of ultrasound and pH measurements during the reaction will enable to monitor the concentrations and hence the rate of the reactions. Data on reaction rates make it possible to calculate the efficiency of the process as a function of the operating parameters. Modeling of the reaction process is conducted to corroborate experimental data and enable scale-up studies to be carried out. Use of the ultrasound measurement technique has shown that it is possible to monitor the extent of the reaction. The use of this technique to measure reaction rates is novel and to our knowledge has not been reported elsewhere.

References

Akers, S.M., Conkle, J.L., Thomas, S.N. and Rider, K.B. (2006). Determination of the heat of combustion of biodiesel using bomb calorimetry. *Journal of Chemical Education*, 83(2), 260–263.

Aracil, J., Esteban, A., Martinez, M. and Vincente, G. (2005). Kinetics of sunflower oil methanolysis. *Industrial and Engineering Chemistry Research*, 44, 5447–5454.

Benavides, P.T. and Diwekar, U. (2012a). Optimal control of biodiesel production in a batch reactor. Part I: Deterministic control. *Fuel*, 94, 211–217.

Benavides, P.T. and Diwekar, U. (2012b). Optimal control of biodiesel production in a batch reactor. Part II: Stochastic control. *Fuel*, 94, 218–226.

Benavides, P.T. and Diwekar, U. (2013). Studying various optimal problems in biodiesel production in a batch reactor under uncertainty. *Fuel*, 103, 585–592.

Clark, W.M., Medeiros, N.J., Boyd, D.J. and Snell, J.R. (2013). Biodiesel transesterification kinetics monitored by pH measurement. *Bioresource Technology*, 136, 771–774.

Darnoko, D. and Cheryan, M. (2000). Kinetics of palm oil transesterification in a batch reactor. *Journal of American Oil Chemist Society*, 77(12), 1263–1267.

Demirbas, A. (2003). Biodiesel fuels from vegetable oils via catalytic and non-catalytic supercritical alcohol transesterifications and other methods: A survey. *Energy Conversion and Management*, 44, 2093–2109.

Demirbas, A. (2005). Biodiesel production from vegetable oils via catalytic and non-catalytic supercritical methanol transesterification methods. *Progress in Energy and Combustion Science*, 31, 466–487.

Ellis, N., Guan, F., Chen, T. and Poon, C. (2008). Monitoring biodiesel production using in situ viscometer. *Chenmical Engineering Journal*, 138, 200–206.

Ferdous, K., Uddin, M.R., Khan, M.R. and Islam, M.A. (2012). Biodiesel from sesame oil: Base catalyzed transesterification. *International Journal of Engineering and Technology*, 1(4), 420–431.

Freedman, B., Butterfield, R.O. and Pryde, E.H. (1986). Transesterification kinetics of soybean oil. *Journal of the American Oil Chemists' Society*, 63(10), 1375–1380.

Knothe, G. (2005). Dependence of biodiesel fuel properties on the structure of fatty acid alkyl esters. *Fuel Processing Technology*, 86, 1059–1070.

Ma, F. and Hanna, M.A. (1999). Biodiesel production: A review. *Bioresource Technology*, 70, 1–15.

Mahajan, S., Konar, S.K. and Boocock, D.G.B. (2006a). Determining the acid number of biodiesel. *Journal of the American Oil Chemists' Society*, 83(6), 567–570.

Mahajan, S., Konar, S.K. and Boocock, D.G.B. (2006b). Standard biodiesel from soybean oil by a single chemical reaction. *Journal of the American Oil Chemists' Society*, 83(7), 641–644.

Marchetti, J.M., Miguel, V.U. and Errazu, A.F. (2005). Possible methods for biodiesel production. *Renewable and Sustainable Energy Reviews*, 11, 1300–1311.

Noureddini, H., Zhu, D. (1997). Kinetic of transesterification of soybean oil. *Journal of American Oil Chemistry Society*, 74(11), 1457–1463.

Pelly, M. (2003). Biodiesel from used kitchen grease or waste vegetable oil. http://journeytoforever.org/biodiesel_mike.html

Singh, R.K. and Padhi, S.K. (2009). Characterization of jatropha oil for the preparation of biodiesel. *Natural Product Radiance*, 8(2), 127–132.

Stamenković, O.S., Todorović, Z.B., Lazić, M.L., Veljković, V.B. and Skala, D.U. (2008). Kinetics of sunflower oil methanolysis at low temperatures. *Bioresource Technology*, 99, 1131–1140.

U.S Department of Energy. (2005). Characterization of biodiesel oxidation and oxidation products: CRC Project No. AVFL-2b. http://www.nrel.gov/docs/fy06osti/39096.pdf

11

Synthetic Fuel and Renewable Energy

N. O. Elbashir,
Wajdi Ahmed, Hanif
A. Choudhury,
Nasr Mohammad
Nasr, Naila
Mahdi, Kumaran
Kannaiyan,
Reza Sadr, Laial
Bani Nassr, and
Mohammed Ghouri
*Texas A&M University
at Qatar*

Rajasekhar Batchu,
Kalpesh Joshi, and
Naran M. Pindoriya
*Indian Institute of
Technology Gandhinagar*

Haile-Selassie
Rajamani
*University of
Wollongong Dubai*

Laial Bani Nassr and
Mohamed Ghouri
*Texas A&M University
at Qatar*

11.1 Gas-to-Liquid (GTL)-Derived Synthetic Fuels: Role of Additives in GTL-Derived Diesel Fuels

N. O. Elbashir, Wajdi Ahmed, Hanif A. Choudhury,
Nasr Mohammad Nasr, and Naila Mahdi

11.1.1 Introduction

The global fossil oil reserve, which is depleting rapidly due to its extensive use, has been estimated at approximately 1.47 trillion barrels. Today, the transportation industry is considered the major consumer of fossil fuel, the current consumption rate of fossil fuel has been estimated at approximately 84 million barrels per day. Overall, the global energy consumption rate is increasing at approximately 1.4% every year [1]. This rate is due to the rapid expansion of the economy of BRIC nations like Brazil, India, China, and Russia along with other developed nations like Japan, Germany, and the United States. The global natural gas reserve on the other hand has been estimated at approximately 6.6 quadrillion cubic feet and has not been utilized extensively as compared to fossil oil. The natural gas reserves mainly comprise methane that needs further processing in order to be utilized as a commercial fuel for the transportation industry. Natural gas can be processed to liquid fuel either by physical transformation route, like liquefaction of natural gas, or chemical transformation to liquid hydrocarbons, using gas-to-liquid (GTL) technology.

The State of Qatar has articulated an impressive vision of becoming the "gas capital of the world" while embracing the principles of sustainable development. With its abundant reserves of natural gas, Qatar is in a unique position to build the world's largest liquefied natural gas (LNG) and GTL technology plants. The latter technology produces a wide variety of alternative and cleaner fuels as well as value-added chemicals. The GTL technology starts with the reforming of the natural gas to obtain synthesis gas or syngas (primarily a carbon monoxide and hydrogen mixture), followed by the synthesis of hydrocarbons and oxygenates from syngas via Fischer–Tropsch synthesis (FTS) technology, and finally the upgrading of liquid hydrocarbons (known as syncrude) via hydrocracking, isomerization, and/or hydroisomerization processes to produce synthetic fuel fractions. Such fuels and chemicals are attractive alternatives to crude oil products due to the large supply coming onstream in the near future and because of their reduced environmental impact resulting from the lack of aromatic and sulfur species [2]. Many of these fuel fractions have already been used for many years [2] with others still waiting for their certification (e.g., synthetic aviation fuels).

During the past 20 years, tremendous investments from the major players in the energy market have been directed toward the GTL technology. For example, Sasol in collaboration with Qatar Petroleum built a US$950 million GTL plant in Ras Laffan, while Shell very recently had started the production at the largest GTL plant in the world, the Pearl GTL Plant. This plant has a capacity to reach 145,000 bbl/day and represents an investment in excess of US$ 20 billion. Typical synthetic products from these two plants include liquefied petroleum gas (LPG), gasoline, different naphtha blends, middle distillate hydrocarbons (diesel and paraffinic kerosene for the aviation industry), base oil, and value-added chemicals. All synthetic fuel fractions from GTL enjoy significant environmental advantages over fuel cuts obtained from crude oil because of their lack of aromatics coupled with their extremely low sulfur content. Furthermore, synthetic fuels obtained from GTL have equivalent energy densities to those obtained from crude oil [3]. GTL synthetic fuels have very high energy densities, on both a mass and volumetric basis, when compared to LNG or using natural gas as a source of energy. Considering all of the aforementioned advantages of GTL-derived synthetic fuels and the political, geological, and environmental complexity of oil exploration and production, these synthetic products have now come sharply into focus as future ultraclean fuels and a valuable source of chemicals (e.g., alpha olefins, alcohols, etc.) [4].

Nevertheless, there are still many challenges facing the design, formulation, and/or the certification of synthetic fuels and chemicals as per the current standards of the energy market. An example is synthetic diesel, which is still awaiting full certification by the aviation industry. Even though the lack of aromatics represents a major advantage of synthetic diesel fuels (because of clean combustion), it is also considered as a disadvantage for jet fuels. This is because the aromatics play a major role in boosting certain physical characteristics (i.e., density, lubricity, etc.) that are crucial for aviation fuels. Also, they help to improve the stability of the fuel tank's sealing (referred to as elastomer swelling behavior). Due to these concerns synthetic jet fuels are normally blended with conventional jet fuels that are obtained from crude oil (referred to as Jet A-1). Qatar Airways was the first airline to fly a "commercial flight" from London Gatwick airport to Doha in October 2009 with a 50/50 blend of a GTL synthetic jet fuel and Jet A-1 [5]. Texas A&M University at Qatar (TAMUQ), in collaboration with industry and academia, was focused on developing a suitable GTL fuel through the use of certain additives, which will be fully compatible with aviation requirements [2]. For this purpose, TAMUQ has built an advanced fuel characterization lab (FCL) that is equipped with sophisticated analytical instruments to measure a number of physical and chemical characteristics of the fuels. The FCL is also capable of determining the detailed hydrocarbon structure of the fuels. This setup and the knowledge garnered from the aforementioned project will be used to fulfill the objective mentioned above.

11.1.2 Synthetic Fuels Composition and Properties

The conventional design and formulation methods of the synthetic fuels are based on the hydrocarbon structure and the properties of equivalent fuel cuts from petroleum oil. Typical protocol for certifying a synthetic fuel is built on two major steps: First, the fuel must meet the specification requirements of a recognized fuel in the energy market (e.g., for aviation fuel specification, ASTM D 1655 is used); second, the fuel must demonstrate that it is "fit for purpose" by having other defined properties and characteristics that fall within the range of experience with conventional, petroleum-derived fuel [2]. Even though such an approach simplifies the design and formulation of synthetic fuels and may lead to the approval of a candidate fuel, it limits the fuel manufacturers' ability to design fuels of superior properties in their performance and properties relative to the existing petroleum-derived fuels.

The synthetic GTL fuels are produced from the heart of the GTL process, which is the Fischer–Tropsch (FT) reactor that contains either a cobalt-based or iron-based catalyst. In the first step of the GTL process, the natural gas is converted to synthesis gas with stoichiometry of $CO:H_2$ (1:2) for cobalt-based catalyst using various reforming technologies available. The conventional steam reforming technology is an endothermic process and therefore a highly energy intensive process [6]. The FT reaction produces a mixture of products that comprise gaseous hydrocarbon, n-alkanes, α-olefins, oxygenates, and wax, which may be referred to as syncrude. Similar to conventional fossil fuels, syncrude also need downstream purification to be used as a synthetic fuel.

The refining of FT products differs from crude oil refining in terms of feed composition, refining focus, and heat management. The design of a FT refinery depends on the type of the feed that must be processed. It has an impact on the heteroatom constraints, compatibility with fuel requirements, and carbon number distribution. Due to the limited scope of this chapter we are not going into much detailed discussion on this topic.

Distillation profile of low-temperature FT syncrude varies from 0°C to >500°C and suggest that hydroprocessing could be an effective approach to refine them to valuable middle distillate cuts like diesel and synthetic paraffinic kerosene. Hydroprocessing usually includes multiple steps such as hydrocracking, hydrogenation, hydroisomerization, and hydrodesulfurization. However, FT products are free from sulfur, and hence, hydrodesulfurization is not required for the refining of syncrude. Thus, synthetic fuels from FT synthesis do not produce any SO_x and NO_x during combustion since all sulfur and nitrogen compounds are removed during the syngas production. These synthetic fuels are very attractive in terms of environmental aspects.

High-temperature FT can be configured to naphtha by using oligomerization technology.

11.1.3 A Method to Design Effective Additives for GTL Diesel Fuels

ORYX GTL, a joint venture between Sasol (49%) and Qatar Petroleum (51%), was the Qatar's first GTL facility. This plant was commissioned in 2007 and located at Ras Laffan Industrial City. With a capacity of 32,400 barrels per day (bpd), the ORYX GTL facility produces a range of GTL products, including diesel, naphtha, and LPG, with diesel as making up the bulk of output [7]. GTL diesel promotes the formation of better fuels than conventional diesel oil because of its low (almost absent) sulfur content and extremely low aromatic content. Therefore, the GTL diesel fuel has been classified as an ultraclean fuel because of its lower emissions of carbon monoxide, particulates, and nitrate oxides upon combustion. Nevertheless, the lack of sulfur in the diesel fuels negatively impacts certain important physical characteristics such as lubricity and density. The lubricity issue is significant for GTL diesel fuels, as required by regulations in the United States, Europe, and elsewhere.

This study focused on identifying the effects of blending biodiesel with GTL diesel and evaluating lubrication ability of blended fuels without compromising any other fuel properties. The study consists of an experimental testing campaign that utilized advanced analytical techniques. This campaign was to be followed by statistical analysis to identify correlations between biodiesel-blended GTL diesel fuel properties in comparison to nonblended GTL diesel and conventional diesel fuel. Methyl and ethyl esters proved to cause significant improvements in the lubricity as well as density of diesel fuel and were thus chosen as biodiesel additives [8–12]. Properties like lubricity, density, viscosity, flash point, heat content, cetane index, cloud and pour point were analyzed. This project is an extension to a series of studies that have been carried out, at the FCL at TAMUQ, with the identification of GTL fuel blends (such as jet fuels and diesel fuels) while comparing them to conventional fuels. Results of this work benefit the local GTL diesel producing companies to market their products globally with an ultraclean diesel by blending them with renewable resources like biodiesel by affording all benefits of a cleaner fuel.

Research conducted on GTL diesel and other synthetic fuels in the FCL at Texas A&M Qatar reflects the collaboration between academia and industry that in Qatar. Projects performed on diesel fuel include respective industry collaborators such as Qatar Shell and Oryx GTL.

The objective of this research work was to optimize GTL diesel properties through blending with biodiesel additives, which are fatty acid esters. Methyl and ethyl esters proved to cause significant improvements in the lubricity as well as in the density of diesel fuel and thus were chosen as biodiesel additives. Complete understanding of the nature of biodiesel and its effect on fuel enhancement was to be investigated. The goal was to develop composition property relationships based on this study to help identify optimum biodiesel concentrations to be blended with GTL diesel to solve the lubricity issue and meet ASTM D975 and D7467 standards.

11.1.4 Experimental Verification of GTL Diesel Fuels Characteristics

The experimental stage of this work included preparation of the GTL and biodiesel blends to be characterized. The GTL diesel is mixed with either *ethyl* or methyl-ester of different volume percentages by following ASTM D5854. Subsequently, all the blends were tested according to the ASTM standards outlined in Table 11.1 and the specification for diesel is given in Table 11.2 according to ASTM D975. This study was comprised of 28 blends of GTL diesel and biodiesel as shown in Table 11.3.

11.1.4.1 Testing Physical Properties

11.1.4.1.1 Flash Point

The flash point of a fuel is the lowest temperature at which application of an ignition source causes the vapors of the fuel to ignite under the conditions of the test. Testing the flash point of a fuel is a crucial step toward safety. Through it, a liquid is identified to be combustible or flammable and its predisposition to that is predicted. Flammable liquids are liquids that contain a flash point below 38.7 and are more dangerous than combustible liquids. On the other hand, combustible liquids are liquids that have a flash point at or above 38.7. Therefore, flammable liquids tend to be more dangerous and need to be identified to accordingly adapt the shipping and safety regulations and procedures. The test was done for the ASTM D93 method, using the Seta PM 93 apparatus.

TABLE 11.1 Physical Properties Analyzed in This Campaign

Property	Testing Method	Reason of Selection
Density	ASTM D4052	Indirect measure of average fuel molecular weight
Flash point	ASTM D93	Important safety measure for the fuel Identifies the tendency to flammability
Cloud point	D5773	Cold flow property of fuel and an estimation of crystallization of fuel in cold conditions
Pour point	D5949	Flow characteristics, like pour point, can be critical for the correct operation of lubricating oil systems, fuel systems, and pipeline operations
Vapor pressure	D6378	This test method can be applied in online applications in which an air saturation procedure prior to the measurement cannot be performed
Cetane index (calculated)	D4737	The cetane number provides a measure of the ignition characteristics of diesel fuel oil in compression ignition engines
Viscosity	ASTM D7042	Measure of flow properties of fuel
Distillation curves	ASTM D86	The distillation (volatility) characteristics of hydrocarbons have an important effect on their safety and performance, especially in the case of fuels and solvents. The boiling range gives information on the composition, the properties, and the behavior of the fuel during storage and use. Volatility is the major determinant of the tendency of a hydrocarbon mixture to produce potentially explosive vapors
Lubricity	ASTM D6078	Evaluation of lubricity of diesel fuel using a scuffing load ball-on-cylinder lubricity evaluator

TABLE 11.2 Diesel Fuel Specification

Property	Diesel	GTL Diesel	ASTM D975 (Gr. 1)
Density at 15°C (kg/L)	0.829	0.768	—
Flash point (°C)	76	57	38 (min.)
Cetane number (cal.)	58	73	40 (min.)
Kinematic viscosity at 40°C (cSt)	3.103	2.008	1.3–2.4
Lower heating value (MJ/kg)	43.174	44.109	—
Total sulfur (ppm)	199.5	0	15 ppm (max.)
Total aromatics (Vol%)	22.9	0.88	35
Cloud Point (°C)	−1.5	−2.5	—
Pour Point (°C)	−9	−12	—
Distillation Temperature			
Initial boiling point (IBP) (°C)	193.8	163.1	—
T10 (°C)	234.3	185.3	—
T50 (°C)	275.7	246.3	—
T90 (°C)	335.4	324.5	288 (max.)
End boiling point (°C)	368.2	346.8	—

11.1.4.1.2 Heat Content

The energy content of the fuel can be measured using its net heat of combustion, which is an important property for any fuel. Here, testing was done according to the ASTM D240 standard to find the heat of combustion of each fuel blend using a Parr 6200EF high precision bomb calorimeter. This method measures the heat of combustion of liquid hydrocarbon fuels ranging in volatility from light distillates to residual fuels. Heat of combustion is important because the higher the heat of combustion, the more

TABLE 11.3   List of Biodiesel Blends

	GTL Diesel	Methyl Oleate	Ethyl Oleate	Methyl Stearate	Ethyl Stearate
			Vol%		
UREP-17-001	95	5	—	—	—
UREP-17-002	90	10	—	—	—
UREP-17-003	85	15	—	—	—
UREP-17-004	80	20	—	—	—
UREP-17-005	95	—	5	—	—
UREP-17-006	90	—	10	—	—
UREP-17-007	85	—	15	—	—
UREP-17-008	80	—	20	—	—
UREP-17-009	95	—	—	5	—
UREP-17-010	90	—	—	10	—
UREP-17-011	85	—	—	15	—
UREP-17-012	80	—	—	20	—
UREP-17-013	95	—	—	—	5
UREP-17-014	90	—	—	—	10
UREP-17-015	85	—	—	—	15
UREP-17-016	80	—	—	—	20
UREP-17-017	90	5	—	5	—
UREP-17-018	85	5	—	10	—
UREP-17-019	80	5	—	15	—
UREP-17-020	85	10	—	5	—
UREP-17-021	80	10	—	10	—
UREP-17-022	80	15	—	5	—
UREP-17-023	90	—	5	—	5
UREP-17-024	85	—	5	—	10
UREP-17-025	80	—	5	—	15
UREP-17-026	85	—	10	—	5
UREP-17-027	80	—	10	—	10
UREP-17-028	80	—	15	—	5

energy will be released upon burning the chemical, which means it will raise the temperature faster than another chemical of lower heat of combustion.

11.1.4.1.3 Density

Fuel density is one of the most important physical properties to measure for a fuel, it gives an estimate of how much fuel can be stored in a given tank volume and how much fuel can be pumped with each pump. Density can also be used with the aid of other physical properties to characterize both light and heavy fractions of petroleum and its products. In this project, density was measured according to the test method ASTM D4052 conducted using an Anton Paar DMA4100 high precision density meter at 20.02°C. Density at 15.0°C was also calculated using an excel converter based on ASTM D1250 for petroleum measurement tables.

11.1.4.1.4 Viscosity

There are two types of viscosity measured with the Viscometer SVM 3000 machine: Dynamic viscosity and kinematic viscosity, which is the dynamic viscosity divided by the density of the liquid, also measured by the same machine. The dynamic viscosity is the ratio of applied stress and the rate of shear of the liquid. The standard method to be used is ASTM D7042. Measurements of viscosity are important because viscosity determines how well a petroleum product can be used as a lubricant, and viscosity is important for determining the optimum storage and handling of petroleum fuels.

11.1.4.1.5 Lubricity

Lubricity is one of the fuel's most critical properties that must be controlled and kept in the certified international standards to be able to run efficiently and safely in engines all around the world. Measuring fuel lubricity basically measures its ability to minimize the friction and damage between different mechanical parts in their motion inside an engine [6]. Lubricity is also one of the biggest problems with GTL diesel, where previous literature has shown that GTL's lubricity is significantly lower than conventional diesel. The main focus of this paper is to identify additives that improve lubricity of GTL diesel without compromising other properties and keeping them at levels that satisfy international standards. The lubricity was tested using a scuffing load ball-on-cylinder lubricity evaluator that follows the method of ASTM D6078.

11.1.4.1.6 Cloud and Pour Points

Cloud and pour point accounts for cold flow properties of fuel [12]. Temperature below cloud point wax in diesel or biowax in biodiesels forms a cloudy appearance, may lead to clogging of the fuel line, and eventually stops the engine. Temperature at which liquid becomes semisolid and loses its flow characteristics is known as pour point. These two properties ensure free flow of fuel during cold weather conditions. All the blends were tested for their cloud and pour point by PSA-70XI equipment.

11.1.4.1.7 Vapor Pressure

Vapor pressure is an important physical property to test for fuels. It gives an idea of how the fuel will run under different operating conditions. Vapor pressure can give an indication on whether the fuel will make vapor locks at high temperatures/altitudes or if it will start at low temperatures. Vapor pressure of a liquid is by definition the equilibrium pressure of the vapor above its liquid state. The method used to calculate the vapor pressure of the GTL diesel follows ASTM D6378 method and it can be applied in online applications where the air saturation procedure prior to its measurement cannot be taken.

11.1.4.1.8 Cetane Index

The Cetane number or Cetane index of a fuel is a property that cannot be overlooked and must be characterized and tested as it is an indication of how fast the diesel combusts. The Cetane number gives a measurement of the quality of the diesel as it captures the readiness of the fuel to ignite when pumped inside an engine [13]. The Cetane index tests the ignition properties of the fuel in comparison to Cetane as an index, and the higher this number is, the easier it is to start a regular direct injection diesel engine. Calculated Cetane index was done following ASTM D4737.

11.1.4.2 Quality Management Services (QMS)

The testing was carried out in the state-of-the-art FCL at TAMUQ. The testing facility abides by a rigorous safety and quality management system developed in collaboration with industrial partners. The laboratory has been certified as ISO9001-2008 and consists of efficient equipment with strict safety standards by personnel. Having a Quality Management System ensures that all methods of data acquisition and reporting are at par with the industry standards for those processes. Folders for individual equipment were developed at the laboratory that specify their use and state the ASTM testing standard to be employed. The QMS also involved developing work instructions, maintenance schedule, qualified operators, inventories for equipment, chemicals needed, glassware, and equipment's spare parts.

TABLE 11.4 ASTM Methods and Equipment for Testing

Property	Equipment	Test Method
Flash point	Seta PM 93	ASTM D93
Heat content	Parr 6200EF	ASTM D240
Density	Anton Paar DMA 4100	ASTM D4052
Viscosity	Anton Paar SVM 3000	ASTM D4052
Lubricity	PCS Instruments AIMS Company	ASTM D6078
Distillation	Petrotest Instrument GmbH	ASTM D86
Cloud and pour points	PSA-70XI	ASTM D5773
		ASTM D5949
Vapor pressure	Grabner Instruments	ASTM D6378

One of the major focuses in the quality control system was to develop a safety plan as well as a lab protocol that insures safe operation in the first place. All lab personnel have been trained to follow the lab protocol. The testing for this project was carried out following the appropriate ASTM standards for each test using the equipment given in Table 11.4 above.

11.1.5 Results and Discussion

The analysis of the experimental data of the critical physical properties of the biodiesel-blended GTL diesel was used to identify the role of the selected additives in boosting certain properties of impotence for diesel engine performance. The outcome of this analysis is discussed as follows.

11.1.5.1 Flash Point

Flash point of diesel fuel ensures safe operation and storage. A higher flash point is always desired for diesel fuel and its blends. Flash point in the present work was measured by Seta PM 93. Figure 11.1 below displays the results obtained. The minimum flash point for Grade1 diesel is 38°C, according to ASTM D975. All the biodiesel-blended GTL diesel have flash point higher than Gr. 1 diesel, but slightly lower than conventional diesel, which is 76°C. The trend of flash point of biodiesel blends is given in Figure 11.1. The increase in the flash point of the blends can be accounted for by the higher flash point of the additive used. Nevertheless, all the GTL blends meet the ASTM specification for Gr. 1 and safe for use in the diesel engine [14].

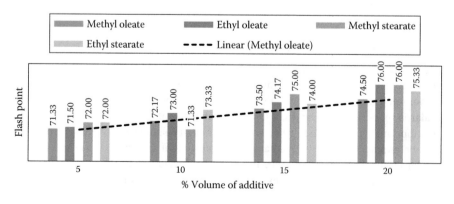

FIGURE 11.1 Flash point as a function of additive concentration.

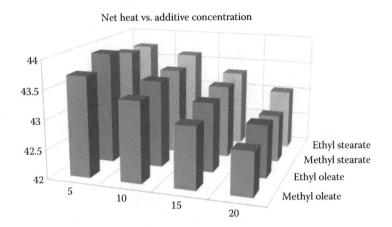

FIGURE 11.2 Heat content as a function of additive concentration.

11.1.5.2 Heat Content

Higher heat of combustion ensures more energy released upon burning the chemical, which in turn raises the temperature of the system faster. A fuel with higher heat of combustion is thus desired. ASTM D975 does not specify any heating value for diesel fuels. The conventional diesel minimum heating value is 43.17 MJ/kg and that of GTL diesel 44/10 MJ/kg. The trend in heating value of biodiesel blends are given in Figure 11.2. In this research, a negative trend in the heating value was observed with the increase in the biodiesel additives. For the blend comprising 20 vol% of additives concentration, there was a higher drop of heat value than the other blends and net heat value of these blends fell below the conventional diesel. The decrease in heating value with the biodiesel additive could be due to the lower heating value of the biodiesel additives. Biodiesel, in general, has lower heating value due to lower hydrogen to carbon ratio (H/C) ratio compared to conventional diesel H/C ratio since ester moiety is present in the molecule. Thus there will be slightly higher fuel consumption that is expected with the use of biodiesel-blended GTL diesel. Nevertheless, this shortcoming of the biodiesel-blended GTL diesel can be overcome if the diesel engines are calibrated with biodiesel-blended diesels [8].

11.1.5.3 Density

Density of the diesel fuel shows an increasing trend with an increase in concentration of biodiesel additives. This trend is consistent with all types of biodiesel additives used in this study. The density values of biodiesel blends are given in Figure 11.3. It is worth mentioning that methyl oleate showed greater impact on the density enhancement than other esters used. Regarding biodiesel additives, the following sequence in descending order is observed to contribute toward the density enhancement of GTL diesel: methyl oleate > ethyl oleate > methyl stearate > ethyl stearate. The unsaturation present in the methyl and ethyl oleate could be the main reason since oleate has a higher density than the stearate. There is no defined speciation for density in the ASTM D975; however, the market specification for density of diesel fuel is 0.88–0.82 kg/L at 15°C. An optimum density of a fuel is required as very high density is also not suitable since it will lead to high dynamic viscosity of the fuel. This in turn has a negative impact on the fuel's spray atomization efficiency resulting in poor combustion with more emissions [15, 16]. The density of all the blends except 5% ethyl oleate blend was found to have lower density than the desired market specifications.

11.1.5.4 Kinematic Viscosity

Viscosity is also an important parameter in determining fuel flow property. A highly viscous fuel will increase the pumping requirement leading to poor spray atomization but also causes an increment in

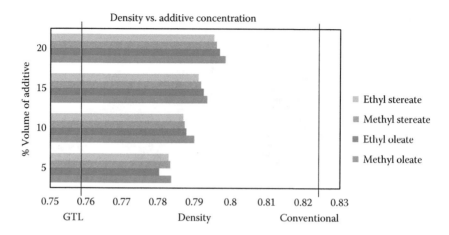

FIGURE 11.3 Densities of blends with different additive concentrations.

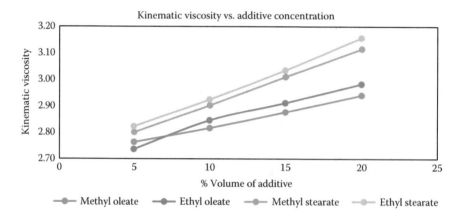

FIGURE 11.4 Kinematic viscosity of blends with different additive concentrations.

fuel consumption. A fuel with a low viscosity will over fill the combustion chamber leading to incomplete combustion. Typically, for Grade 2 diesel fuel, the value of viscosity requirement at 40°C lies between 1.9 and 4.1 cSt according to ASTM D975. The Kinematic Viscosity values of biodiesel blends are given in Figure 11.4. Viscosity of all the blended fuels meets ASTM D975 for Grade 2 diesel. An increase in the viscosity trend was observed for all the additive concentration, which was due to the higher viscosity value of the additives used in this study [9, 10].

11.1.5.5 Lubricity

Lubricity of fuel is a crucial property for proper performance in diesel engines. Lubricity measures the fuels ability to minimize the friction and damage between different mechanical parts in their motion inside an engine [17–19]. According to ASTM D975, lubricity measurement should be done at 60°C to mimic the actual situation of a running engine. The ASTM D975 uses a high-frequency reciprocating rig to determine the lubricity of the diesel fuel; however, in our study, we used scuffing load ball-on-cylinder lubricity evaluator according to ASTM D6078, and according to this method, the maximum frictional coefficient of fuel was found to be less than 0.175 on a single load test done at 25°C.

Lubricity of prepared blends was tested using a scuffing load ball-on-cylinder machine by PCS instruments AIMS Company evaluator that follows ASTM D6078 standard for lubricity measurement.

TABLE 11.5 Lubricity Results of 5% Additive-Blended GTL Diesel

Blends	% Additives	Failing Load (g)	Friction Coefficient
UREP-000	—	2400	0.281
UREP-005	5% EO	4500	0.286
UREP-009	5% MS	4200	0.292
UREP-013	5% ES	4500	0.297
UREP-017	5% MS + 5% MO	5100	0.275
UREP-023	5% ES + 5% EO	5100	0.281

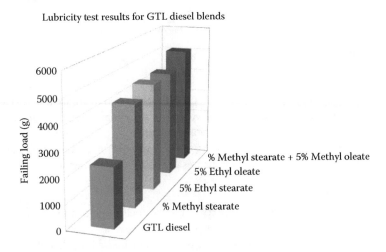

FIGURE 11.5 Lubricity of 5% additive biodiesel-blended GTL diesel.

Figure 11.5 above shows the results for some of the tested blends and the loads at which the test fails, indicating the lubricity of the blend.

The additives clearly enhanced the lubricity of the GTL diesel blends. When testing was done for the lowest concentration of biodiesel blends (5 vol%), the fuel was able to handle approximately double the load handled by pure GTL diesel. It is worth noting here that in all the blended fuel, slightly higher frictional coefficient was observed at double the load of that of pure GTL diesel. This clearly suggests that a significant improvement in lubricity of GTL diesel has been achieved even with the lowest concentration (5 vol%) of biodiesel additive. It was observed that both ethyl esters resulted in a greater improvement to the lubricity than the methyl ester additives. However, no significant change was noticed when using saturated and unsaturated ethyl esters. Table 11.5 below displays the lubricity test results for the blends displayed on Figure 11.5 above. We did not carry out lubricity measurement for higher additive concentrations (10–20 vol%) due to tedious nature of the experiment as well as high cost involved in testing. However, 5 vol% additive was sufficient to cause the desired lubricity enhancement in the GTL diesel and is reported in this study.

11.1.5.6 Distillation

The distillation (volatility) characteristics of hydrocarbons have an important effect on their safety and performance, especially in the case of fuels and solvents. The boiling range gives information on the composition, the properties, and the behavior of the fuel during storage and use. Volatility is the major determinant of the tendency of a hydrocarbon mixture to produce potentially explosive vapors. Distillation profile for all the completed blends follows a similar trend and representative distillation

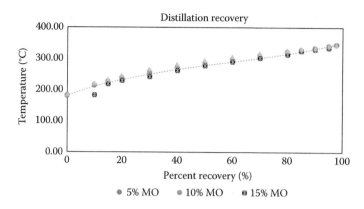

FIGURE 11.6   Distillation recovery profile of biodiesel blends.

recovery profile for methyl oleate additive-blended GTL diesel is given in Figure 11.6. ASTM D7467 limits recovery of biodiesel-blended diesel to 90% by volume at a maximum temperature of 343°C. In this study, all the blends meet ASTM D7467 specification for distillation of biodiesel blend diesel fuel. With the increase of additive concentration, a decreasing trend in volatility observed for all the tested blends. This is a desired property for diesel fuel as it improves the flash point characteristics [17].

11.1.5.7 Vapor Pressure

Vapor pressure is an important physical property to test for fuels. It gives an idea of how the fuel will run under different operating conditions. Vapor pressure can give an indication on whether the fuel will make vapor locks at high temperatures. Vapor pressure of a liquid is by definition the equilibrium pressure of the vapor above its liquid state. For diesel fuel, vapor pressure is not a key factor since it is not volatile and has a very low vapor pressure. For that reason, there is no specification for vapor pressure in diesel fuel ASTM D975. The method used to calculate the vapor pressure of the GTL diesel follows ASTM D6378 standard. Figure 11.7 displays the results obtained from blend tests.

The maximum value noted for vapor pressure was 0.3 kPa. The vapor pressure was expected to decrease based on the trend noticed on the distillation. The IBP increased as additive concentration increased. Therefore, vapor pressure was expected to decrease as additive concentration increased. However, this trend was noted only methyl stearate blends. Other blends showed no observable trends. All values of vapor pressure were between 0.1 and 0.3 kPa.

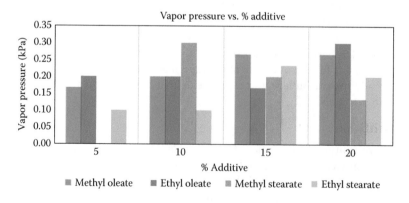

FIGURE 11.7   Vapor pressure of blends with different additive concentrations.

11.1.5.8 Cloud and Pour Point

Cloud and pour point account for cold flow properties of fuel. All the blends were tested for their cloud point and pour point by PSA-70XI equipment according to ASTM D5773 and 5949, respectively. Cloud point and pour point of the samples are given in Figures 11.8 and 11.9. With the increase in the oleate additives concentration, both cloud point and pour point for all the blends were improved. There is no strict guideline for cloud point and pour point for diesel fuel blends mentioned in the ASTM D975; however, a lower value of cloud point and pour point is always desired for colder countries in order to avoid clogging of fuel during winter. In our study, we have only analyzed oleate-blended diesels that showed significant improvement in cloud and pour point values. Very ambiguous trend for both cloud point and point was observed with the stearate-blended diesel due to the formation of crystals during the measurements in the instrument. This could be accounted for the higher melting point of striate additives used (37°C–41°C). The stearate-blended biodiesel starts forming crystals the temperature ramp goes toward a lower temperature value during analysis. Therefore, cloud point and pour point for stearate were not recorded for any sample and are not reported in this study. It is worth mentioning here that stearate additives dissolute when subzero temperature is reached and cannot be used as diesel blend for colder country applications but can be used in hot regions like in the Middle East and Africa.

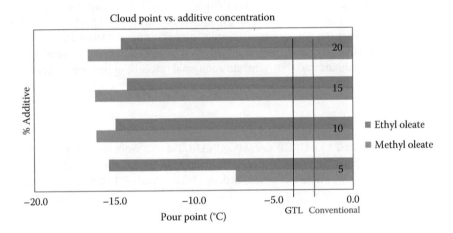

FIGURE 11.8 Cloud point of diesel with oleate additive concentrations.

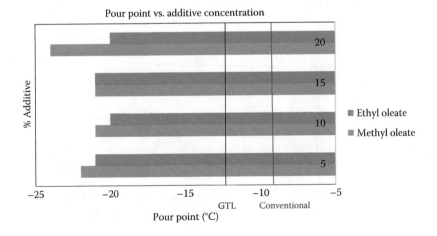

FIGURE 11.9 Pour point of diesel with oleate additive concentrations.

11.1.6 Conclusion

GTL diesel is an ultraclean fuel that is environmentally benign. However, it lacks intrinsic properties that are significant for its performance in diesel engines. The major property drawbacks are lubricity and density. This study aimed at enhancing those intrinsic properties through the use of biodiesel additives. This objective was to be done without compromising any other physical properties of diesel as well as affording benefits of the ultraclean GTL diesel.

Optimization of GTL diesel lubricity and density was achieved without compromising the desired ASTM D975 specifications using methyl and ethyl esters. Ethyl esters' longer chain lengths showed greater enhancement in flash point, viscosity, and lubricity of the GTL blends. However, methyl esters showed better density enhancement than ethyl esters additives. Unsaturated esters such as stearates showed better enhancement of density, viscosity, and flash point. Saturated and unsaturated esters had approximately similar effect on lubricity of GTL diesel. Adding biodiesel additives negatively affected the heat content of the fuel blends. The desired optimization of GTL diesel lubricity was achieved through this study; however, more comprehensive study should be done in order to validate the findings of this study.

Furthermore, the investigation on engine performance with biodiesel-blended GTL diesel will be an interesting study. An engine study would shed light on the real-time performance of the biodiesel-blended GTL diesel in terms of fuel efficiency, torque power, and emission of particulate materials and carbon monoxide in atmosphere. An optimization study on engine performance and emission control with biodiesel-blended GTL diesel would help monetizing GTL diesel, which is environmentally benign, help in diversify product stream, and thus generate additional resource of revenue for state sponsoring GTL technologies.

Acknowledgment

This report was made possible by a UREP award [UREP17-172-2-052] from the Qatar National Research Fund (a member of The Qatar Foundation). The statements made herein are solely the responsibility of the author[s]. The authors also would like to acknowledge Shell's Pearl GTL Plant for providing the GTL fuel samples.

References

1. M. M. A. Shirazi, A. Kargari, M. Tabatabaei, B. Mostafaeid, M. Akia, M. Barkhi, M. J. A. Shirazi, Acceleration of biodiesel–glycerol decantation through NaCl-assisted gravitational settling: A strategy to economize biodiesel production, *Bioresour. Technol.*, vol. 134, pp. 401–406, 2013.
2. E. E. Elmalik, B. Raza, S. Warrag, H. Ramadhan, E. Alborzi, N. O. Elbashir, Role of hydrocarbon building blocks on gas-to-liquid derived synthetic jet fuel characteristics, *Ind. Eng. Chem. Res.*, vol. 53, pp. 1856–1865, 2014.
3. S. Blakey, L. Rye, C. W. Wilson, Aviation gas turbine alternative fuels: A review. *Proc. Combust. Inst.*, vol. 33, pp. 2863–2885.
4. R. Chedid, M. Kobrosly, R. Ghajar, The potential of gas-to-liquid technology in the energy market: The case of Qatar, *Energy Policy*, vol. 35, pp. 4799–4811, 2007.
5. Qatar Airways, World's first commercial passenger flight powered by fuel made from natural gas lands in Qatar. http://www.qatarairways.com/global/en/press-release.page?pr_id=PressRelease_12Oct09_2&locale_id=en_gl.
6. M. M. B. Noureldin, N. O. Elbashir, M. M. El-Halwagi, Optimization and selection of reforming approaches for syngas generation from natural/shale gas, *Ind. Eng. Chem. Res.*, vol. 53, pp. 1841–1855, 2014.

7. E. E. Elmalik, B. Raza, S. Warrag, H. Ramadhan, E. Alborzi, N. O. Elbashir, Role of hydrocarbon building blocks on gas-to-liquid derived synthetic jet fuel characteristics, *Ind. Eng. Chem. Res.*, vol. 53, no. 5, pp. 1856–1865, 2014.

8. M. M. Musthafa, Synthetic lubrication oil influences on performance and emission characteristic of coated diesel engine fuelled by biodiesel blends, *Appl. Therm. Eng.*, vol. 96, pp. 607–612, 2016.

9. M. A. Hazrat, M. G. Rasul, M. M. K. Khan, Lubricity improvement of the ultra-low sulfur diesel fuel with the biodiesel, *Energy Proc.*, vol. 75, pp. 111–117, 2015.

10. D. Stanica-Ezeanu, V. Frangulea, Study regarding biodiesel influence on lubricity of diesel fuel, *Petroleum-Gas Univ. Ploiesti Bull., Tech. Ser.*, vol. LXII, no. 3, pp. 101–104, 2010.

11. H. Liu, S. Jiang, J. Wang, C. Yang, H. Guo, X. Wang, S. Han, Fatty acid esters: A potential cetane number improver for diesel from direct coal liquefaction, *Fuel*, vol. 153, pp. 78–84, 2015.

12. S. Wang, J. Shen, M. J. T. Reaney, Lubricity-enhancing low-temperature diesel fuel additives, *JAOCS, J. Am. Oil Chem. Soc.*, vol. 89, no. 3, pp. 513–522, 2012.

13. H. Sajjad, H. H. Masjuki, M. Varman, M. A. Kalam, M. I. Arbab, S. Imtenan, A. M. Ashraful, Influence of gas-to-liquid (GTL) fuel in the blends of *Calophyllum inophyllum* biodiesel and diesel: An analysis of combustion-performance-emission characteristics, *Energy Convers. Manag.*, vol. 97, pp. 42–45, 2015.

14. H. Sajjad, H. H. Masjuki, M. Varman, M. A. Kalam, M. I. Arbab, S. Imtenan, S. M. A. Rahman, Engine combustion, performance and emission characteristics of gas to liquid (GTL) fuels and its blends with diesel and bio-diesel, *Renew. Sustain. Energy Rev.*, vol. 30, pp. 961–986, 2014.

15. M. Matzke, U. Litzow, A. Jess, R. Caprotti, G. Balfour, Diesel lubricity requirements of future fuel injection equipment, *SAE Int. J. Fuels Lubr.*, vol. 2, no. 1, pp. 273–286, 2015.

16. S. H. Park, K. B. Choi, M. Y. Kim, C. S. Lee, Experimental investigation and prediction of density and viscosity of GTL , GTL–biodiesel , and GTL–diesel blends as a function of temperature, *Energy and Fuels*, vol. 27, pp. 56–65, 2012.

17. P. Y. Hsieh, T. J. Bruno, A perspective on the origin of lubricity in petroleum distillate motor fuels, *Fuel Process. Technol.*, vol. 129, pp. 52–60, 2015.

18. D. Claydon, The use of lubricity additives to maintain fuel quality in low sulphur, *Goriva i maziva*, vol. 53, pp. 342–353.

19. L. Petraru, F. Novotny-farkas, Influence of biodiesel fuels on lubricity of passenger car diesel, *Goriva i maziva*, vol. 51, pp. 157–165, 2010.

11.2 The Role of Alternative Aviation Fuels on Reducing the Carbon Footprint

Kumaran Kannaiyan and Reza Sadr

11.2.1 Introduction

Air and land transportation sectors consume almost half of the global fuel production and account for about 60% of the global greenhouse gas emissions [1]. The aviation industry is also expected to grow at a faster rate in the coming years with conventional fossil fuel as its main source of energy for a foreseeable future [1]. On the other hand, dwindling oil resources, increase in energy demand, and stringent emission norms serve as the major driving forces to identify a sustainable alternative fuel for both modes of transport. While a lot of research is underway to address the above concerns by focusing on the energy efficiency of fossil fuel production and using them in an environmentally friendly way, these efforts may not be adequate to achieve a sustainable solution. This places further emphasis on finding a sustainable alternative fuel to address these concerns. Of the several alternative fuels available, this chapter focuses on gas-to-liquid (GTL) fuel, a liquid fuel derived from natural gas using Fischer–Tropsch synthesis and specifically tailored for aviation gas turbine engines. Since the focus of this work is more on the application side of the GTL jet fuel, the details about the GTL fuel production methodology and the relevant technology advancements will not be discussed here.

In the recent years, there has been a considerable increase in the application of GTL jet fuels in gas turbine engines as it offers significant advantages over other alternative jet fuels from a performance perspective [2–6]. It is worth pointing out that GTL fuel has cleaner combustion and emission performances than its counterparts mainly due to the difference in its chemical composition and absence of aromatics, respectively. However, in addition to the difference in the chemical properties, the physical and volatilization properties of GTL fuels are also different from those of the conventional jet and other alternative fuels. Any change in the physical properties of a liquid fuel will have an effect on its atomization, evaporation, and mixing processes. These processes play a crucial role in determining the fuel placement inside the combustor, which subsequently has an influence on the combustion and emission performances of the combustor. Therefore, it is essential to study the precombustion processes in detail in order to fully understand GTL fuel combustion and emission characteristics.

The present study was started as part of a consortium research work under the auspices of Qatar Science and Technology Park (QSTP) to evaluate the feasibility of using GTL jet fuel as an alternative fuel in aviation engines [7]. The consortium research was broadly categorized into three subresearch groups, namely, properties, combustion, and performance, to evaluate the effect of GTL jet fuel on different aspects of jet engine operation. In all the categories, the GTL fuel performances were compared with those of the Jet A-1 fuel. The properties team was primarily involved in studying the chemical and physical properties of different GTL fuel mixtures and identified the most appropriate GTL fuel chemical composition that meets the ASTM standards for jet fuels. The combustion team was involved in evaluating the fundamental combustion properties like flammability limits [4,8], flame speed [9], chemical kinetics [10], emission, and spray aspects of GTL fuel [11,12]. The mandate of the performance team was to study the effects of GTL jet fuels on the overall aircraft performance.

This chapter presents part of the effort of the combustion team that was carried out in Texas A&M University at Qatar (TAMUQ) to look into spray characteristics of the GTL fuel. The microscopic spray features, i.e., droplet size and velocity, of GTL fuels at different operating conditions are presented and discussed. The spray features of GTL fuels were investigated using optical diagnostic techniques at global (i.e., planar) and local (i.e., point) level using global sizing velocimetry (GSV) and phase doppler anemometry (PDA), respectively, and compared with those of the conventional Jet A-1 fuel. All the investigations were carried out at atmospheric ambient conditions. Although the actual gas turbine combustor ambient conditions are different from those used in this study, the present results will help to decouple the hydrodynamic effects from those related to chemical characteristics of the fuel.

11.2.2 Experimental Setup, Methodology, and Measurement Techniques

In this section, details of the experimental facility, measurement techniques, and fuel properties are presented and discussed. An optically accessible experimental facility was developed to study the spray characteristics of the GTL and Jet A-1 fuels at different operating conditions. Figure 11.10 shows the schematic of the spray experimental facility, which consists of three modules, namely fuel supply module, spray module, and optics module. The primary role of fuel supply module was to deliver the liquid fuel to the spray module at a desired test pressure. As shown in the figure, the fuel was circulated in the fuel module until the line pressure reached the test pressure. The fuel was then supplied to the nozzle located inside the spray module with the help of a three-way solenoid valve. The continuous line in Figure 11.10 represents the fuel supply line and the dashed line denotes the bidirectional signal communication between the computer and control devices. The spray module was continuously purged with nitrogen gas to maintain an inert environment. In the spray module, the fuel nozzle was mounted on a two-dimensional traverse system to enable measuring droplet characteristics at different sections of the spray. Since the spray was symmetric about the nozzle axis, the droplet characteristics were measured

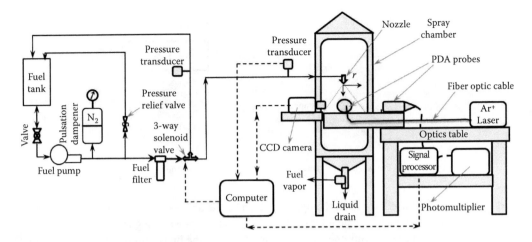

FIGURE 11.10 Schematic of the spray experimental facility with components description.

TABLE 11.6 Chemical and Physical Properties of GTL and Jet A-1 Fuels at 20°C and 1 atm [13]

Properties	GTL-1	GTL-2	Jet A-1
Density (kg/m³)	746.9	751	788.1
Kinematic viscosity (mm²/s)	1.10	1.02	1.66
Surface tension (mN/m)	23.8	24.1	26.9
Distillation Characteristics			
T50-T10 (°C)	8.4	8.3	22.2
T90-T10 (°C)	20.7	21.2	67.1

only on one side of the spray. A charge-coupled device (CCD) camera and PDA transmitter and receiver probes were integrated with the spray module for laser diagnostic measurements. More details about the experimental facility were reported elsewhere [11, 12].

Pressure swirl type of nozzle, typically used as a pilot nozzle in actual jet engine combustors for flame stabilization and high altitude reignition purposes, was used in this work. The fuel injection pressures used in this study represent a range of pressure differentials encountered at different stages of an aircraft engine cycle.

In this study, five samples of GTL fuel with different chemical compositions were investigated at different nozzle operating pressures. However, for the sake of brevity, the spray characteristics of only two GTL samples that are having maximum differences will be presented and compared with those of the reference Jet A-1 fuel. The key fuel properties that are relevant to this study are presented in Table 11.6. All the fuels and their relevant property details were provided by Shell Plc and detailed information about their preparation and chemical composition was reported by Bauldreay et al. [13].

11.2.3 Measurement Technique: Global (Planar) Level

In this work, spray droplet diameter and velocity were obtained at planar level of the spray using GSV. The operating principle of GSV [14] is based on a generalized scattering imaging approach developed by Calabria and Massoli [15]. In this study, GSV measurements were performed by illuminating a region of the spray with the help of a 532 nm Nd:YAG pulse laser sheet with a thickness of about 700 μm. The CCD camera integrated in the spray module was equipped with a macro lens in conjunction with a 532 nm

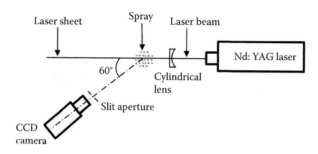

FIGURE 11.11   Top-view of the laser sheet–camera–spray arrangement for the GSV technique.

band-pass filter that is used to collect the light scattered by the droplets in the illuminated region. The laser sheet–camera–spray arrangement for the GSV technique is shown in Figure 11.11.

Prior to the droplet size measurements, the scaling factor of the images to the physical coordinates was obtained using a 2-D calibration target placed in the region of interest. Later, the camera was defocused by moving it away without disturbing the lens arrangement. The defocus distance was measured as the distance between the camera's initial position and its final position. A geometric correction factor was applied by the image processing tool to account for the nonuniform magnification factor across the image. The geometric correction factor was derived based on the angular position, defocus distance, magnification, and the Scheimpflug angle as reported by Pan et al. [14]. The camera collects the scattered light from the defocused droplets as angular oscillations (i.e., fringe patterns) [15], which were then processed to obtain the droplet size information. The fringe patterns were less sensitive to the refractive index of the fuel by positioning the CCD camera at an angle of 60° to the laser sheet. In order to avoid the fringe overlap, a slit was also placed in front of the camera lens to capture only a small portion of the fringe pattern. Furthermore, the slit will enable this technique to be employed in moderately dense sprays, where droplet concentrations can go as high as several thousand droplets/cubic centimeter [16]. The droplet size range that can be measured using GSV depends on several factors, such as camera defocus distance, camera sensor resolution, lens aperture size, and the magnification factor. The droplet velocity information was obtained by capturing two consecutive images separated by a short time interval (frame straddling imaging) and processing the images using a particle tracking technique. More details about the optical settings and other parameters relevant to this technique were reported elsewhere [11]. About 800 images (i.e., 400 image pairs) were captured and processed at each location to obtain a statistically independent data.

11.2.4 Measurement Technique: Local (Point) Level

As shown in Figure 11.10, the 2-D PDA system [17] employed in this study was also integrated to the experimental facility. The droplet size and velocity information at a particular spatial location of the spray were measured simultaneously using the PDA system. An argon ion multiline laser beam with a power of 240 mW was used as light source. The beam was then separated into green ($\lambda = 532$ nm) and blue ($\lambda = 488$ nm) laser beams each with a power of about 40 mW and emitted by the transmitter probe. The intersection of the laser beams created the measurement volume at a distance equivalent to the focal length of the transmitter lens. The green and blue wavelengths were used to measure the axial and radial components of the velocity, respectively. A receiver probe was positioned at an angle of 42° to the transmitter axis to collect the first-order refraction Doppler signal emitted by the droplet crossing the measurement volume. The scattered light received at the receiver probe was then processed by a signal-processing unit to determine the droplet characteristics.

In PDA measurement technique, the droplet diameter is determined from the difference in the phase shift between two Doppler burst signals emitted by the droplet crossing the measurement volume, whereas the droplet velocity is determined from the Doppler burst frequency. Further details regarding the operating principle and methodology of the PDA technique are described elsewhere [18]. The measurements performed in this study were carried out at different axial and radial locations downstream of the nozzle exit as detailed by Kumaran and Sadr [12]. At each location, measurements were conducted to collect either a maximum of 10,000 droplet samples or 15 s of experiment to obtain a statistically independent droplet measurement. Five trials were conducted at each location.

11.2.5 Calibration of Measurement Techniques

Both measurement techniques were calibrated using a monodisperse droplet generator that can produce a stream of uniform droplet diameter. The diameter of the generated droplets depends on the medium flow rate, frequency applied on the piezoelectric transducer, and the orifice plate dimension. More details about the working principle and operational methodology are reported in the literature [19]. In this study, the droplet generator was placed in the field of interest and calibration was performed independently for GSV and PDA systems. In both cases, the measured droplet diameter was reported to be within 15% of the theoretical value provided for the generator with a measurement uncertainty of less than ±3% between the trials [20]. These tests served as a measure of confidence of the accuracy of the measurement methodologies.

11.2.6 Results and Discussion

In this section, the measured spray characteristics of the GTL and Jet A-1 fuels at different operating conditions are presented and discussed. The results obtained at planar level (using GSV) are presented first and followed by the results of the pointwise measurement (using PDA) for comparison and discussion. As mentioned earlier, although the experiments were performed at three operating pressures for five fuel samples, here, the results obtained at two injection pressures and only for two GTL fuels will be presented. In all the cases, the GTL fuel results will be compared with those of the reference Jet A-1 fuel.

Figure 11.12 shows the comparison of the droplet size distribution measured for the two GTL fuels and Jet A-1 fuel at 0.3 and 0.9 MPa injection pressures. The measured region is about 105 mm × 41 mm, located at a distance of 41 mm downstream of the nozzle exit along the spray axis. The measurements along the radial direction (i.e., 105 mm) are carried out in two steps, each covering a radial distance of 55 mm. Furthermore, the number of droplets measured in a particular droplet size bin is presented as a percentage of the total measured droplet sizes. The total number of droplets measured by GSV varies between 12,000 and 18,000 droplets depending on the injection pressure and spatial location. It must

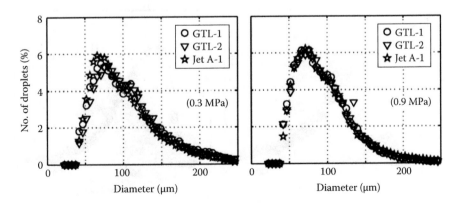

FIGURE 11.12 Comparison of droplet size distribution measured using GSV for different fuels at 41 mm downstream of the nozzle exit for 0.3 (left) and 0.9 MPa (right) injection pressures.

be noted that the droplet size distributions are shown only up to 300 µm as it is difficult to discern the difference between the fuels beyond that range. On the lower range, the minimum droplet that could be measured by the system is limited to 30 µm due to the experimental facility constraints and consequently the optical settings. As can be seen from Figure 11.12, the overall shape and distribution of GTL fuels are similar to that of the conventional Jet A-1 fuel at 0.3 injection pressure. The distribution peak (i.e., maximum number of droplets) for Jet A-1 fuel is found to be at a lower diameter than those of the GTL fuels. The trends of GTL-2 and Jet A-1 fuels are very close to each other owing to the similarities in their fuel properties. At both the injection pressures, the difference in size distribution between the fuels is within the measurement accuracy of the system. Similarly, the droplet size distribution measured at a region further downstream of the nozzle exit (i.e., at 82 mm from the nozzle exit) is shown in Figure 11.13. Even at this location, the size distributions are not very different from each other.

Based on the GSV results, it can be interpreted that the difference in size distributions between the fuels is insignificant. This observation is a positive outcome from a fuel deployment perspective because the combustors may not require any major modifications while using GTL fuels to create a similar fuel placement as that of the Jet A-1 fuel. However, the earlier results are obtained at atmospheric ambient conditions that do not represent the actual combustor conditions. This is an important aspect as the volatilization nature of the fuels is very different as reported in Table 11.6. To further ascertain the earlier inference, the droplet characteristics are also measured at local (i.e., point) level of the spray using PDA diagnostic technique.

The PDA technique is used to measure the droplet size and velocity at different axial and radial locations downstream of the nozzle exit. It must be noted that although the measurements are performed at axial locations of 20, 40, 60, 80, and 100 mm downstream of the nozzle exit, the PDA results will be reported only for 20, 60, and 100 mm axial locations for the sake of brevity. The data rate is a measure of number of droplets sampled per second through the measurement volume. The measurement data rate can be affected by the PDA system settings. Since, the settings are maintained the same across the fuels, any difference in data rate could be an implication of the fuel properties. Furthermore, it is used as a metric to decide on the farthest radial measurement location from the spray axis where the presence of droplets becomes insignificant. Here, the droplet diameter profiles are compared across fuels for a given axial location to facilitate the comparison.

In Figure 11.14, closer to the nozzle axis, the data rate profiles are only marginally different between the GTL and Jet A-1 fuels. With an increase in axial distance, GTL-1 fuel has a higher data rate than the other fuels except in the spray boundary region. With an increase in injection pressure, the data

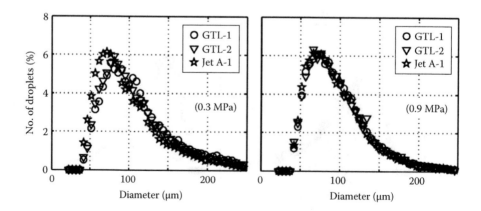

FIGURE 11.13 Comparison of droplet size distribution measured using GSV for different fuels at 82 mm downstream of the nozzle exit for 0.3 (left) and 0.9 MPa (right) injection pressures.

rate profiles for GTL-1 are again higher than those of GTL-2 and Jet A-1 fuels at all axial locations. It is important to note that the *y*-axis scales are different between Figures 11.14 and 11.15. The data rate profiles indicate that more droplets are generated at a faster rate with GTL-1 fuel than the other fuels for a given operating condition.

The mean droplet diameter profiles measured for different fuels at three axial locations $x = 20, 60$, and 100 mm downstream of the nozzle exit are shown in Figures 11.16 and 11.17 for 0.3 and 0.9 MPa injection pressures, respectively.

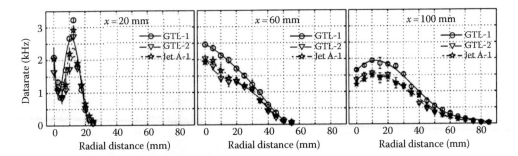

FIGURE 11.14 Data rate profiles of GTL-1, GTL-2, and Jet A-1 fuels measured at different axial locations for an operating pressure of 0.3 MPa.

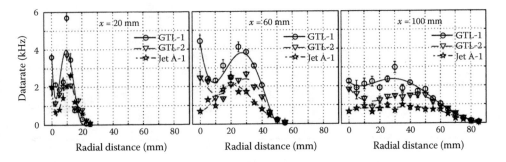

FIGURE 11.15 Data rate profiles of GTL-1, GTL-2, and Jet A-1 fuels measured at different axial locations for an operating pressure of 0.9 MPa.

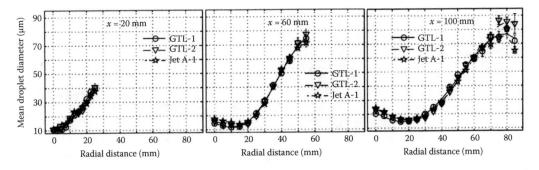

FIGURE 11.16 Comparison of arithmetic mean droplet diameter (d_{10}) between the fuels at three different axial locations, $x = 20, 60$, and 100 mm downstream of the nozzle exit for an operating pressure of 0.3 MPa.

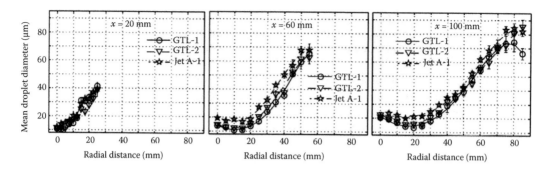

FIGURE 11.17 Comparison of arithmetic mean droplet diameter (d_{10}) between the fuels at three different axial locations, $x = 20$, 60, and 100 mm downstream of the nozzle exit for an operating pressure of 0.9 MPa.

At lower injection pressure, the profiles are almost similar between the fuels. Whereas, with an increase in injection pressure, the droplet diameters of GTL fuels are lower when compared to those of Jet A-1 fuel as close as $x = 60$ mm downstream of the nozzle exit (Figure 11.17).

Furthermore, the mean droplet axial velocities are also compared between the fuels at 0.3 and 0.9 MPa injection pressures in Figures 11.18 and 11.19, respectively. As noticed for droplet diameters, the droplet mean axial velocity measured at 0.3 MPa injection pressure shows only a marginal difference at all three axial locations.

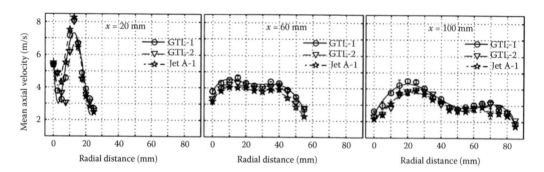

FIGURE 11.18 Comparison of droplet mean axial velocity measured using PDA for different fuels at axial locations $x = 20$, 60, and 100 mm, downstream of the nozzle exit for an operating pressure of 0.3 MPa.

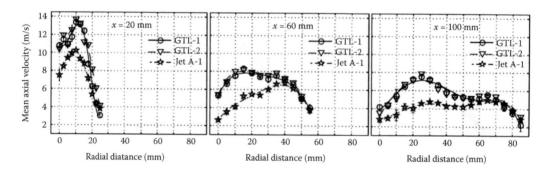

FIGURE 11.19 Comparison of droplet mean axial velocity measured using PDA for different fuels at axial locations, $x = 20$, 60, and 100 mm, downstream of the nozzle exit for an operating pressure of 0.9 MPa.

With an increase in injection pressure to 0.9 MPa, the difference in droplet mean axial velocities is observed as close as $x = 20$ mm from the nozzle exit. The difference in axial velocity between the fuels is seen to widen with an increase in axial distance. This highlights the influence of fuel properties on the droplet characteristics in the near nozzle region. Interestingly, the axial velocities near the spray boundary are observed to be similar between the fuels. This could be attributed to the dominance of the momentum exchange between the spray and ambient.

All the PDA measurements are also pointing to a similar conclusion that the overall spray characteristics are similar between the fuels; however, more insights were gained in the near nozzle region through PDA measurements. This is further shown in Figure 11.20, where the results obtained from GSV and PDA are presented together to highlight the complementary approach followed in this study. Figure 11.20 shows the spatial distribution of droplet diameters detected using GSV for (1) GTL-1 and (2) Jet A-1 fuels at an injection pressure of 0.3 MPa. Here, the GSV results are shown up to 120 mm downstream of the nozzle exit tip. This will give an overall (*global*) perspective of the spray. It must be noted that the region of interest (i.e., $x = 0$ to 120 mm) is covered in three steps using GSV due to the limitations of the field of view settings.

As seen in Figure 11.20, GSV techniques clearly captured the droplets in the spray boundary region, however, the number of droplets detected in the spray core (center) region is less. Furthermore, the transition from one axial station to next (for example, from $0 \leq x \leq 40$ to $40 \leq x \leq 80$) is discrete. The key reasons behind these observations are as follows: First, in a pressure swirl nozzle, smaller droplets are entrained by the swirl motion and stay in the spray core region, while the bigger droplets are pushed toward the spray boundary. Consequently, the average droplet size increases with an increase in the

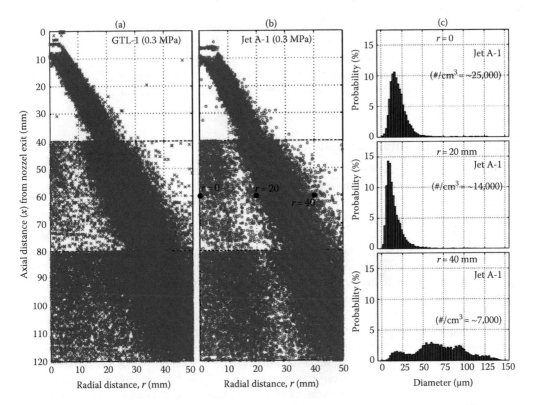

FIGURE 11.20 Spatial distribution of droplets detected using GSV for (a) GTL-1, (b) Jet A-1 at 0.3 MPa injection pressures, and (c) Droplet diameter distribution measured using PDA at an axial location of 60 mm from the nozzle exit along the radial direction at $r = 0$ (top), 20 mm (middle), and 40 mm (bottom).

radial distance from the nozzle axis as shown in Figures 11.18 and 11.19, similar to those reported in the literature [21,22]. The reduction of the smaller droplets detected in the spray core region may be attributed to both the limitation of the GSV technique in detecting small droplets due to optical settings and to the light interference/blocking effect of the bigger/denser droplets in the spray boundary region.

Second, the light intensity of the laser sheet is varying along the axial direction (i.e., higher near the top region of the field of view and lower near the bottom). As a consequence of this nozzle/laser-sheet/camera arrangement, the quality of fringe patterns imaged near the bottom region of the field of view (FOV) is lower than near the top region of the FOV. This variation in fringe quality contributed to more number of droplets discarded near the bottom region when compared to those near the top region. This has caused a difference between the number of droplets detected near the top and bottom region of the FOV as seen in Figure 11.20, which has been the reason behind the discrete change in the number of droplets detected while moving from one axial location to another.

To demonstrate the complementary approach, three radial points, namely $r = 0$, 20, and 40 mm, are shown in Figure 11.20b at an axial location of $x = 60$ mm. At these three radial locations, the droplet diameter size distribution measured using PDA is shown in Figure 11.20c. The droplet diameter densities (i.e., number of droplets per unit volume, #/cm³) measured at these locations are also shown. The droplet diameter histograms and the droplet density obtained using PDA clearly show that more number of smaller droplets exist in the spray core region ($0 \le x \le 20$) than those that were detected by GSV in that region. Along the radial distance in the outward direction, the droplet density decreases and the droplet diameter size shifts toward the higher end. In other words, the diameter distribution is wide (i.e., spread over a broad range of droplet diameter sizes) at the spray boundary when compared to the spray core region.

All the above results are not only consistent but also complementing each other in gaining additional insights on the spray characteristics at global as well as local levels of the spray. The results from this study also emphasize that at atmospheric ambient conditions, the spray characteristics of GTL and Jet A-1 fuels are similar from an overall perspective. At the same time, the influence of fuel physical properties on their spray characteristics is evident in the near nozzle region even at atmospheric ambient conditions.

11.2.7 Summary and Conclusions

An optically accessible spray experimental facility was developed and used to study the spray characteristics of several GTL fuels and the conventional Jet A-1 fuel. The spray characteristics of these fuels were investigated as it plays a major role in subsequent processes of evaporation, mixing, combustion, and emission in a gas turbine combustor. The spray performance was studied using two different laser diagnostic techniques, one at the global level of the spray (GSV) and the other at a local level (PDA). The spray performances were studied for a pilot-scale pressure swirl nozzle, a representative nozzle that is used in actual aviation gas turbine engines for ignition purposes, at two fuel injection pressures. The key objective of this study was to investigate the influence of change in fuel physical properties of GTL fuels on their spray performance and compare them with those of the conventional Jet A-1 fuel. This study was performed at atmospheric ambient conditions.

The results of GSV measurement imply that although the physical properties of GTL fuels differ from those of Jet A-1 by a maximum of 38%, the overall (i.e., global) spray performance of GTL fuels is not significantly different from that of Jet A-1. Although the GSV results show that the GTL and Jet A-1 fuels have similar global spray parameters, it is essential to gain more insights on the spray performance of these fuels where the GSV measurements had limitations. To this end, the PDA technique was used to gather additional information at local level of the spray. The profiles of mean droplet diameter and droplet mean axial velocity profiles were similar between the fuels. However, in the near nozzle region, the influence of fuel properties was evident even at atmospheric ambient conditions. The spray characteristics of the GTL fuels suggest that its lower kinematic viscosity and surface tension assist in faster disintegration and dispersion of the droplets in the core region of the spray when compared to that of the Jet A-1 fuel.

Results obtained from both the techniques clearly emphasize the fact that the spray characteristics of GTL fuels are similar to that of the Jet A-1 fuel at atmospheric conditions. This in turn leads to the support of the hypothesis that GTL fuels could be used in an existing fuel system and combustor without compromising the spray quality and, at the same time, improve the combustion and emission characteristics of the combustor. However, these results must be interpreted with caution as they do not fully take into account the difference in the volatilization characteristics of the tested fuels. Therefore, it is essential to investigate the spray characteristics of the fuels under gas turbine combustor conditions to understand the true benefits of the GTL fuel for such applications. Despite the differences in the ambient conditions of these experiments and the actual combustor condition, the results of this study can help in decoupling the effect of ambient-assisted evaporation from that of atomization.

Path Forward

Currently, in Micro Scale Thermo-Fluids (MSTF) laboratory at TAMUQ, the fuel spray research efforts focus on the understanding of different aspects of the fuel spray affecting combustor performance. The following are some of the studies that are underway:

- As highlighted in the concluding section, it is essential to study the spray performance of GTL fuels at actual aviation gas turbine combustor conditions to understand their true potential. To perform this study, an optically accessible pressure vessel is under construction and the fuels will be tested at different pressures and temperatures that are encountered at different stages of the aircraft engine cycle.
- On a different note, there are recent reports suggesting that the addition of high energetic metal particles at nanoscale to liquid jet fuel enhances combustion and emission characteristics of the fluid. All those studies were performed at high volume percentages of fuel additives. It is well known that addition of nanoparticles at such high concentrations will significantly affect the liquids thermo-physical properties. Therefore, at MSTF lab, the spray performance of nanofuels (GTL/Jet A-1+nanoparticles) under different operating conditions is investigated.

Acknowledgment

This study was made possible by the funding from QSTP GTL consortium program and National Priorities Research Program (NPRP) grant NPRP-7-1499-2-523. The authors would like to acknowledge the support by the project partners at Rolls-Royce (UK), Sheffield University (UK), German Aerospace Center (DLR-Germany), Shell Plc (UK and Qatar), and the Gas and Fuels Research Center (TAMUQ).

References

1. Agarwal, R. K. 2011. Environmentally responsible air and ground transportation. *Proceedings of 49th AIAA Aerospace Sciences Meeting*, Orlando, FL, 4–7 January: AIAA-2011-0965.
2. Corporan, E., DeWitt, M. J., Belovich, V., Pawlik, R., Lynch, A. C., Gord, J. R., and Meyer, T. R. 2007. Emission characteristics of a turbine engine and research combustor burning a Fischer–Tropsch jet fuel. *Energy Fuels* 21 (5): 2615– 2626.
3. Blakey, S., Rye, L., and Wilson, C. W. 2010. Aviation gas turbine alternative fuels: A review. *Proceedings of the Combustion Institute* 33 (2): 2863–2885.
4. Fyffe, D., Moran, J., Kumaran, K., Sadr, R., and Al-Sharshani, A. 2011. Effect of GTL-like jet fuel composition on GT engine altitude ignition performance Part I: Combustor operability. *Proceedings of ASME Turbo Expo: Power for Land, Sea and Air*, Vol. 2, 485–494, Vancouver, BC, Canada, 6–10 June.

5. Lobo, P., Rye, L., Williams, P. I., Christie, S., Uryga-Bugajska, I., Wilson, C. W., Hagen, D. E., Whitefield, P. D., Blakey, S., Coe, H., Raper, D., and Pourkashanian, M. 2012. Impact of alternative fuels on emission characteristics of a gas turbine engine—Part 1: Gaseous and particulate matter emissions. *Environmental Science and Technology* 46 (19): 10805–10811.

6. Badami, M., Nuccio, P., Pastrone, D., and Signoretto, A. 2014. Performance of a small-scale turbojet engine fed with traditional and alternative fuels. *Energy Conversion and Management* 82: 219–228.

7. Shell Qatar media releases. 2009. http://www.shell.com.qa/en/aboutshell/media-centre/news-and-media-releascs/2009/fuel-research-agreement13102009.html (accessed September 2015).

8. Mosbach, T., Gebel, G. C., Le Clercq, P., Sadr, R., Kumaran, K., and Al-Sharshani, A. 2011. Investigation of GTL-like jet fuel composition on GT engine altitude ignition and combustion performance Part II: Detailed diagnostics. *Proceedings of ASME Turbo Expo 2011: Power for Land, Sea and Air*, Vancouver, BC, Canada, 6–10 June: GT2011-45510.

9. Kick, T., Herbst, J., Kathrotia, T., Marquetand, M., Braun-Unkoff, M., Naumann, C., and Riedel, U. 2012. An experimental and modelling study of burning velocities of possible future synthetic jet fuels. *Energy* 43: 111–123.

10. Slavinskaya, N., Riedel, U., Saibov, E., and Kumaran, K. 2012. Surrogate model design for GTL Kerosene. *Proceedings of the 50th AIAA Aerospace Sciences Meeting*, Tennessee, 9–12 Jan: AIAA-2012-0977.

11. Kumaran, K. and Sadr, R. 2014. Effect of fuel properties on spray characteristics of alternative jet fuels using global sizing velocimetry. *Atomization and Sprays* 24 (7): 575–597.

12. Kumaran, K. and Sadr, R. 2014. Experimental investigation of spray characteristics of alternative aviation fuels. *Energy Conversion Management* 88: 1060–1069.

13. Bauldreay, J. M., Bogers, P. F., and Al-Sharshani, A. 2011. Use of surrogate blends to explore combustion-composition links for synthetic paraffinic kerosines. *Proceedings of the 12th International Conference on Stability, Handling and Use of Liquid Fuels* 2: 891–927.

14. Pan, G., Shakal, J., Lai, W., Calabria, R., and Massoli, P. 2006. Spray features measured by GSV: A new planar laser technique. *Proceedings of the 10th International Conference on Liquid Atomization and Spray Systems,* Kyoto; ICLASS06-283.

15. Calabria, R. and Massoli, P. 2007. Generalised scattering imaging laser technique for 2-D characterization of non-isothermal sprays. *Experiments in Thermal and Fluid Sciences* 31: 445–451.

16. Calabria, R., Casaburi, A., and Massoli, P. 2003. Improved GSI Out-of-Focus Technique for Application to Dense Sprays and PIV Measurements. *Proceedings of 9th International Conference on Liquid Atomization and Spray Systems,* Sorrento, Italy, July 13–17, paper 10–6.

17. PDA Reference Manual. 2012. Dantec dynamics. http://www.dantecdynamics.com.

18. Albrecht, H. E., Borys, M., Damaschke, N., Tropea, C. 2003. *Laser Doppler and Phase Doppler Measurement Techniques*, Springer–Verlag, Berlin.

19. MDG-100 Monosize droplet generator, operation and service manual, TSI Inc., USA, 2010.

20. Kumaran, K. and Sadr, R. 2013. Spray characteristics of Fischer-Tropsch alternative jet fuels. *Proceedings of ASME Turbo Expo 2013*, San Antonio, TX, 3–7 June: GT2013-95761.

21. Santolaya, J. L., Aisa, L. A., Calvo, E., Garcia, I., and Cerecedo, L. M. 2007. Experimental study of near-field flow structure in hollow cone pressure swirl sprays. *Journal of Propulsion and Power* 23 (2): 382–389.

22. Li, T., Nishida, K., and Hiroyasu, H. 2011. Droplet size distribution and evaporation characteristics of fuel spray by a swirl type atomizer. *Fuel* 90: 2367–2376.

11.3 Integration of Distributed Renewable Energy Generation with Customer-End Energy Management System for Effective Smart Distribution Grid Operation

Rajasekhar Batchu, Kalpesh Joshi, and Naran M. Pindoriya

11.3.1 Introduction

11.3.1.1 Background

With the advancement of human civilizations and as the energy-intensive lifestyle accelerates, the access of clean energy is a strong prerequisite for the socioeconomic development across the world, particularly to India being the world's fastest growing economy. India poses a huge energy gap, around 400 million Indians, representing one-third of the country's population, either do not have access to electricity at all or have access to limited and/or unreliable electricity. Rising to this challenge, governments across the world have been encouraging and promoting renewable sources of energy for electricity generation, with reasonable degree of success. The emerging trend of bottom-up approach with distributed energy resources (DERs) is the potential solution to meet the electricity demand while simultaneously leveraging clean energy resources as part of the energy mix as well. Rooftop Solar photovoltaic (PV) generation and wind energy generation are the most techno-economically viable options among the DERs and therefore, there is a lot development in effective hybrid solar PV and battery storage systems at residential and small-scale commercial customer's level. In addition, the practicing of demand response (DR) strategies in the form of incentives and dynamic feed-in tariffs for customer end energy management program has presented new opportunities for energy users to reduce their electricity bill by optimizing or scheduling their consumption/generation/storage operations.

The residential DR and energy storage are the key for effective integration of DER in a smart home; which is done by a home energy management system (HEMS). Although independent operations of home energy management system (HEM) modules are beneficial at individual customers, it leads to peak rebounds and poor peak-to-average ratio (P.A.R) at aggregated level. Hence, various collaborative and decentralized scheduling techniques have been proposed for residential smart grid to address peak rebound and data privacy issues. Hierarchical and distributed multiagent-based model has been identified as effective strategy. This has been done in multiple hierarchical levels: (1) demand management of elastic loads and energy management of DERs is performed at the device/residential level acting in response to network and market price signals, which is the bottom layer and is fully distributed; (2) next, the clusters are aggregated in several layers, with collaboration (community level and area level); (3) the top layer being the central control entity, e.g., the distribution system operator. Also it is done in different time scales: day-ahead scheduling followed by a real-time scheduling approach for demand management to tackle the problems of coordinating with electricity markets and to mitigate supply and demand uncertainties.

The key aspects to consider for effective integration of DERs at home level and community level through DR programs for reliable and economic operation of an active distribution network are highlighted in Batchu and Pindoriya (2015b) and those are as folllows:

1. *Prediction and uncertainty:* As the distributed renewable generation is highly intermittent in nature, the uncertainty associated with its prediction is essential. Although appliance jobs are entered from the users, they may be subject to change (e.g., interrupting electric vehicle charging job) and some are entered at the time of execution (e.g., turning on TV, light bulb). Electricity price forecasting is also an important element to consider for effective energy management.
2. *Load demand and renewable energy generation models for scheduling and real time DR strategies:* Models needed for a scheduling and control of appliances and energy dispatch scheduling. The optimizer should be able to recognize the various types of loads and their DR potential.

3. *Multicriteria optimization:* HEMS should consider various objectives of the customer viz. energy cost minimization, inconvenience minimization, appliance priorities, and their trade-offs as well. Also it should be able to coordinate with community level energy management system for DR strategies.

4. *Effect on load demand profile:* Local DER disrupts the P.A.R ratio of the customers demand as well as the aggregated load demand of the utility level without optimal dispatch scheduling that limits the benefit of having a local generation from DER. Also, under dynamic feed-in tariff without optimal dispatch scheduling, the economic benefits are limited.

5. *Coordination with bottom-to-top approaches in multi-timescale:* The utility agent or aggregator is responsible for procuring energy on the day-ahead retail electricity market based on forecasted optimized demand profile and also to participate in DR schemes by aggregating demand flexibility of customers. Hence, a coordination is required between day-ahead scheduling and real-time management to mitigate supply and demand uncertainties and computational complexity.

11.3.2 Customer Level Home Energy Management

11.3.2.1 Understanding the Problem Scenario

Among the local DERs, a rooftop solar PV system is the most economically viable to partially or fully satisfy the demand at the customer end. Typical rating of rooftop solar PV system varies from 1 to 10 kWp and the energy demand of a residential home can range from 10 to 100 kWh/day depending upon the geographical location, house size, and customer's income level. A typical residential load profile will have two peak consumption periods—typically one in the morning period and other in the evening to midnight period. The two main classifications of PV system configurations are (1) grid-connected or utility-interactive mode and (2) stand-alone mode.

Figure 11.21a shows a traditional grid-tied solar PV that can meet electrical loads in a home when solar radiation levels enable adequate generation of electricity. This mode of configuration helps to reduce electricity consumption, but the impact on demand reduction has not been significant. This is largely due to the fact that major residential demand lies between evening and midnight during weekdays; a time when solar PV generation is not always available. Grid connected solar PV plus battery storage

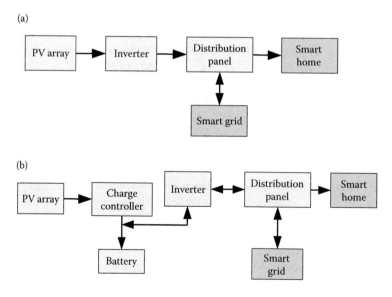

FIGURE 11.21   (a) Grid-connected PV system and (b) Grid-connected PV plus battery storage system.

configuration, as shown in Figure 11.21b, is popularly used because battery system that can be used to serve as backup power, storing excess renewable generation for self use, and grid electricity for bidirectional energy trading under dynamic pricing tariff. The selection of suitable solar PV and battery storage system ratings for a home will depend on techno-economic benefits (Erdinc et al. 2015). In this case, optimal dispatch scheduling of PV power output to minimize the energy cost is possible based on electricity pricing, forecasted PV power output, and load demand. Based on optimal generation dispatch scheduling, the power flows among PV, battery, load, and grid are controlled by sending control signal to the inverter by a HEMS. Here, the focus is on the integration of rooftop solar PV system in conjunction with energy storage technologies.

Recently, there has been a significant effort by researchers on energy management algorithms for HEMS, which considers both DR and DERs integration. Guo et al. (2012) presented a stochastic optimization problem to minimize energy costs considering the stochastic nature of DER output. Zhang et al. (2015) proposed a multiobjective optimization-based real-time load management algorithm taking load uncertainty into account. A multi-timescale model predictive approach based on stochastic modeling is presented by Chen et al. (2013).

The problem of local energy resources and DR management by home energy management becomes more challenging if users manage their demand individually without proper coordination. For example, if a large group of customers start using their washing machines, PEV's charging, etc., at low price period, this will result in a new peak demand. Hence, a proper coordination among the customers is required. This section addresses the question—how to effectively make use of local DER generation to meet the demand and design an effective DR strategy for customer-end energy management.

11.3.2.2 Architecture of Smart Home Integrated with Solar PV+ Battery Storage

Figure 11.22 shows a typical smart home consisting of critical, noncritical and comfort-based appliances, a rooftop solar PV with a battery storage integrated to grid via a smart hybrid inverter, and electric vehicle linked to a HEMS via communication network. It is also to be noted that a HEMS will have machine learning, prediction capabilities, multicommunication capabilities, and interface with the user (Hu and Li 2013) and DR aggregator/community energy management system (CEMS).

The loads in a home are of different types, viz. space cooling, space heating, water heating, cloth drying, cooking, refrigeration, freezer, lighting, and others. Based on their demand management potential and controllability, these can be classified into four major types: fixed/critical, time-shiftable, power-shiftable, comfort based, and local renewable energy generation and storage devices as described in Table 11.7. The simplified load models and detailed physical-based appliance load modelings are provided in literature by Shao et al. (2013), Batchu and Pindoriya (2015b), and Dang and Ringland (2012). The power distribution relationships among PV, battery, loads, and grid are shown in Figure 11.23.

The possible power distributions are as follows: battery can be charged by power from grid $\left(p_{t,n}^{G2B} \right)$ and PV $\left(p_{t,n}^{P2B} \right)$ and it can also supply power loads $\left(p_{t,n}^{B2L} \right)$ and grid $\left(p_{t,n}^{B2G} \right)$. Renewable power output can flow to battery, home or grid and is denoted as $p_{t,n}^{P2G}$, $p_{t,n}^{P2L}$ and $p_{t,n}^{P2B}$, respectively. These flows are scheduled using the optimization algorithm.

A solar PV module output depends upon solar irradiation and module temperature and the module material. The PV power forecast model is developed based on manufacturer specifications, system location, and orientation and forecasted insolation and temperature models. First regression-based solar power forecasting model based on American Society of Heating, Refrigerating and Air-Conditioning Engineers (ASHRAE) models for insolation and temperature was by Bakirci (2009), which was developed by considering historical data from weather sensors and local weather station is considered as shown in Equation 11.1:

$$P_{PV} = \left[a_1 + a_2 \cdot \text{Gpoa} + a_3 \cdot \ln(\text{Gpoa}) \right] \times \left[1 \times \alpha_p \left(T_C - 25 \right) \right], \tag{11.1}$$

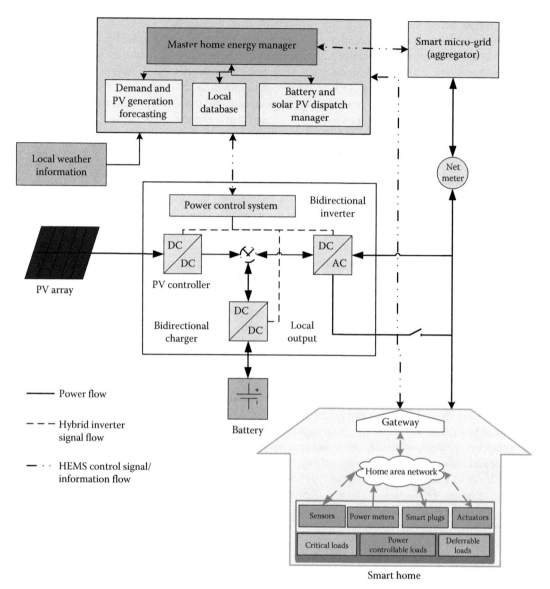

FIGURE 11.22 Block diagram of smart home with gird connected solar PV plus battery storage.

where α_p is temperature coefficient of power, Gpoa is plane of array incidence (W/m²), and a_1, a_2, a_3 are regression coefficients. The generic battery storage system charging and discharging model is given in Equation 11.2.

$$\mathrm{EE}_t^{\mathrm{Bat}} = \mathrm{EE}_{t-1}^{\mathrm{Bat}} - \frac{\delta \cdot p_t^{\mathrm{bat,dis}}}{\eta^{\mathrm{EE}}}$$

$$\mathrm{EE}_t^{\mathrm{Bat}} = \mathrm{EE}_{t-1}^{\mathrm{Bat}} + \frac{\delta \cdot p_t^{\mathrm{bat,dis}}}{\eta^{\mathrm{EE}}_{\mathrm{bat,ch}}},$$

(11.2)

where $\mathrm{EE}_t^{\mathrm{Bat}}$ is the total energy and $p_t^{\mathrm{bat,ch}}$ and $p_t^{\mathrm{bat,dis}}$ represent the amount of charging and discharging power at time interval t, $\eta^{\mathrm{EE}}_{\mathrm{bat,ch}}$ is the charging efficiency, and $\eta^{\mathrm{EE}}_{\mathrm{bat,dis}}$ is the discharging efficiency. As the

TABLE 11.7 Classification of Home Appliances

Critical/Fixed Loads	Time Shiftable Loads	Power Shiftable Load	Comfort-Based Loads	Local RE Generation and Energy Storage Devices
Entertainment loads, lighting, kitchen appliances, electronic devices (~0.5–1 kW)	Clothes dryer (4–5.6 kW) Dish washer (1–2 kW) Clothes washer (0.5–1 kW)	PHEV (3–16 kW), E-byke (1–2 kW), Pool pump (1–2 kW)	Air-conditioner (1–2 kW) Water heater (1–2 kW)	Rooftop solar PV (1–10 kWp) Battery storage (1–10 kWh)
Appliances which are essential and have their own characteristic power consumption for each time slot over the time horizon	Can only be shifted in time and operates in its own required power consumption pattern	Appliances that have a prescribed energy requirement that has to be met over a set of timeslots	Appliances that control a physical variable that influences the user's comfort	Devices which can generate, store, and dispense the electrical energy
There is no DR potential associated with these as are critical and cannot be controlled	A low to high DR potential is associated with these, as starting time can be postponed	A high DR potential is associated with these, as their operation can be interrupted	A low to medium DR potential is associated with these by a reset of thermostat settings within comfort limits	A high DR potential, since they can be used to respond to DR signal and get benefit of reward tariff
Their power absorption is modeled as inequality constraint with lower and upper bounds depending on ON/OFF status and time of operation	User inconvenience can be modeled as associated delay in starting time from its preferred time	Delay in charging finishing time is a measure of inconvenience experienced by the user	Amount of temperature deviation from set value is a measure of user comfort/discomfort level	Manage the demand profile and benefitted by buy/sell power from/to grid

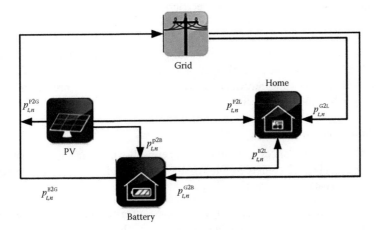

FIGURE 11.23 Power distribution relations of various elements in smart home.

life of a battery will be few thousands of cycles, its depreciation needs to be considered. The normalized hourly cost of cycling per watt-hour $\left(C_{\text{Deg}}^{\text{B}}\right)$ is given in Equation 11.3.

$$C_{\text{Deg}}^{\text{B}} = \left(\frac{\text{Cost of the storage system}}{\text{Cycle life}}\right) \times \frac{1}{2 \times \left(E_{\text{B\_max}}\right)^2}, \tag{11.3}$$

where $E_{\text{B\_max}}$ is the maximum charge or discharge limit of battery.

11.3.2.3 Optimal Scheduling Algorithm under Dynamic Pricing Scheme

This section presents a coordinated multistage optimal scheduling model, where a customer can participate in energy trading and also in DR program. It is assumed that home agents will collaborate with aggregator for further economic benefits. This multistage optimal scheduling strategy combines day-ahead (hourly or 15 min) and real-time control strategy for each minute for effective DR aggregation ensuring home agents and utility interests in a computational and communication efficient way. A set of comfort level indicators power distribution relation among HEMS proposed in Zhang et al. (2015) are adopted here for discomfort modeling. A modified hybrid differential evolution (DE) method is used for solving method to allow integration of DER in an incentive-based HEM system. A brief description of the problem formulation optimization objectives and constraints modeling is explained in subsequent parts of this section. The flow chart for this strategy is shown in Figure 11.24, which could further be summarized as follows:

- In day-ahead, an iterative multiobjective optimization of in-home energy management is performed by participating in cooperative strategies with aggregator/CEMS. The day-ahead scheduling consists of scheduling time-shiftable and power-shiftable appliances, dispatch scheduling of local energy resources, i.e., solar PV power and battery storage considering minimization of net energy cost (i.e., buying minus selling cost), battery utilization cost, peak to average ratio, and user convenience. Heating, ventilation and air conditioning (HVAC) and critical loads demand are considered by a forecasted power consumption profile for them during this stage. The required inputs such as user preferences and load demand are predicted, day-ahead real time pricing is considered from market and the scheduling time interval it can be an hour or few minutes for effective representation of short duration shiftable loads. This multiobjective optimization problem is solved using Non-dominated sorting genetic algorithm (NSGA)-II (Deb et al. 2002).
- The limitations in prediction accuracy and uncertainties in PV generation and demand cause the customers to deviate from the agreed demand limits, which is taking care of rescheduling of DER's dispatch, HVAC load control, and electric vehicle (EV) load.
- The second stage is a short term scheduling cum real-time control of HVAC loads, interruptible loads, and battery storage to maintain the day-ahead scheduled load profile and respond to dynamic DR signal from aggregator, considering thermal comfort preferences, uncertainties and load priorities using DR enabled mathematical load models.
- After completion of one time interval, this second stage is repeated for remaining time period for the day by receding the time interval.

11.3.2.3.1 Day-Ahead Scheduling Strategy

Day-ahead scheduling of appliances operation and energy dispatch can be done either together or individually. Here it has been decomposed into the appliance scheduling from local renewable-energy (RE)-based energy and battery energy scheduling.

11.3.2.3.1.1 Objective Functions   Let the appliance set consists of critical/fixed appliances $(A_{n,F} \hat{I} A_n)$, time shiftable $(A_{n,T} \hat{I} A_n)$, and power shiftable appliances $(A_{n,S} \hat{I} A_n)$. The day-ahead scheduling problem is a multiobjective Mixed integer linear programming (MILP) model with the following different objectives:

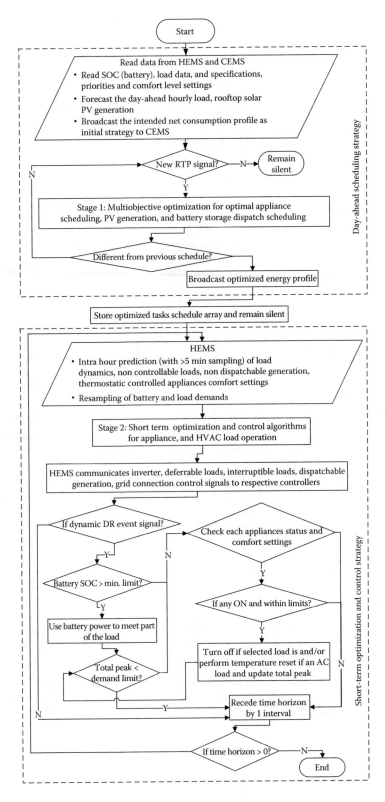

FIGURE 11.24 Multistage home energy management strategy.

1. *Minimization of normalized energy cost:* The overall net electricity cost of customer is formulated as Equation 11.4, which consists of three parts: the first term represents the overall electricity cost of buying from the grid, the second term represents the revenue of selling electricity to the grid, and the third represents degradation cost of the home battery storage.

$$
C_n^{(1)} = \text{Min} \sum_{t=1}^{N_{\text{slot}}} \left\{ \begin{array}{l} C_{t,n}^{\text{buy}} \times \left(p_{t,n}^{\text{G2L}} + p_{t,n}^{\text{G2B}} \right) - C_{t,n}^{\text{sell}} \times \left(p_{t,n}^{\text{P2G}} + p_{t,n}^{\text{B2G}} \right) \\ + C_{\text{Deg},n}^{\text{B}} \left(p_{t,n}^{\text{P2B}} + p_{t,n}^{\text{G2B}} + p_{t,n}^{\text{B2L}} + p_{t,n}^{\text{B2G}} \right) \end{array} \right\},
\tag{11.4}
$$

where $C_{t,n}^{\text{buy}}$ and $C_{t,n}^{\text{sell}}$ are buying and selling electricity costs at time interval t.

2. *Minimization of user discomfort:* Each schedulable task in a home has a preferred time range (PTR) and allowable utilization time range (ATR) for this operation. To quantify the satisfaction/dissatisfaction of the user that depends upon how near the scheduled time range is from the preferable range, inconvenience model described by Zhao et al. (2013) is adopted here. The objective of discomfort minimization is shown in Equation 11.5.

$$
C_n^{(2)} = \text{Min} \left(\frac{\sum_{a \in A} \lambda_{\text{elec},a} \cdot \rho^{I_{n,a}}}{\left(\sum_{a \in A} \lambda_{\text{elec},a} \cdot \rho^{I_{n,a}} \right)_{\max}} \right).
\tag{11.5}
$$

Here, the scaling factor $\lambda_{\text{elec},a}$ denotes an appliance priority, ρ is a delay parameter that is >1. The inconvenience of an appliance a is the waiting time for shiftable appliances and set temperature in the case of thermostatic-controlled appliances given as

$$
I_{n,a} = \frac{t_{n,a} - t_{n,a,s}}{t_{n,a,f} - I_{n,a} - t_{n,a,s}} \quad \forall a \in A_{n,\text{T}}
$$

$$
I_{n,a} = \frac{t_{n,a,f} - (t_{n,a,s} + t_{n,a,\text{operating}} + t_{n,a,\text{delay}})}{t_{n,a,\text{end}} - (t_{n,a,s} + t_{n,a,\text{operating}} + t_{n,a,\text{delay}})} \quad \forall a \in A_{n,s}
$$

$$
I_{n,a} = \frac{T_{\text{set},t,a} - T_{t,a}}{\Delta T_{\max,t,a}} \quad \forall a \in A_{n,\text{HVAC}},
$$

where $I_{n,a}$ is the delay time rate and $t_{n,a,s}$, $t_{n,a,f}$, $I_{n,a}$ indicate starting, ending, and number of time slots, respectively, for operation of appliance $a \in S$. For HVAC, $T_{\text{set},t,a}$ is appliance set temperature, $T_{t,a}$ is actual temperature, and $\Delta T_{\max,t,a}$ is the maximum allowable variation limit.

11.3.2.3.1.2 Appliance Constraints A time shiftable appliance model given by Zhu et al. (2011) is adopted where the scheduling result p_n^s is viewed as one of the cyclic shifts of the pattern $x_{n,a}$, put together in matrix form as and can be described by Equation 11.6 as,

$$
\left[p_n^a \right] = \left[x_{n,a} \right] \cdot \left[b_{n,a} \right],
$$

$$
\sum_{t=1}^{N_{\text{slot}}} b_{n,a,t} = 1 \quad a \in A_{n,\text{T}}.
\tag{11.6}
$$

Suppose a power-shiftable appliance $a \in A_{n,s}$ is required to operate between intervals $t_{n,a,s}$ and $t_{n,a,f}$ and total energy required is $E_{n,a}$. This can be ensured by constraints in Equations 11.7 and 11.8.

$$\sum_{t_{n,a,s}}^{t_{n,a,f}} x_{n,a,t} = E_{n,a} \quad \forall a \in A_{n,s} \tag{11.7}$$

$$\alpha_a \leq x_{n,a,t} \leq \beta_a, \quad \forall t \in [t_{a,s},\ldots,t_{a,f}]$$

$$x_{n,a,t} = 0, \quad \forall t \notin [t_{a,s},\ldots,t_{a,f}]. \tag{11.8}$$

Therefore the total amount of energy required for all the appliances of the consumer n at time t is given by Equation 11.9 as

$$\sum_{a \in A_{n,F}} p_{t,n}^F + \sum_{a \in A_{n,S}} p_{t,n}^S + \sum_{a \in A_{n,T}} p_{t,n}^T = p_{t,n}^L. \tag{11.9}$$

This energy can be supplied either from one or more of the following choices: grid, battery, and renewable source. This constraint is formulated as Equation 11.10.

$$p_{t,n}^L = p_{t,n}^{P2L} + p_{t,n}^{B2L} + p_{t,n}^{G2L}. \tag{11.10}$$

The other local renewable energy dispatch and battery constraints are modeled by Equations 11.11 and 11.12 respectively as

$$p_{t,n}^P = p_{t,n}^{P2L} + p_{t,n}^{P2G} + p_{t,n}^{P2B} \tag{11.11}$$

$$0 \leq p_{t,n}^{P2B} + p_{t,n}^{G2B} \leq p_{max}^{Bat,ch}$$

$$0 \leq p_{t,n}^{B2L} + p_{t,n}^{B2G} \leq p_{max}^{Bat,disch} \tag{11.12}$$

$$E_{Bat,min} \leq \eta_{eff} \cdot \left(p_{t,n}^{P2B} + p_{t,n}^{G2B} - p_{t,n}^{B2L} - p_{t,n}^{B2G} \right) \leq E_{Bat,max}.$$

The overall scheduling problem of each customer n is a constrained multiobjective MILP problem formulated as Equation 11.13:

$$\text{Min} \quad C_n^{(1)}$$

$$\text{Min} \quad C_n^{(2)} \tag{11.13}$$

$$\text{s.t} \quad (6)-(12).$$

The solution to the problem can be solved by using multiobjective evolutionary techniques.

11.3.2.3.2 Real-Time Control Strategy

Using mathematical models for model for power intensive controllable devices, i.e., for air-conditioner (AC), electric vehicle, clothes dryer, refrigeration loads, and prediction-based model for other devices from data gathered, hose detailing's (Shao et al. 2013). Their operating conditions and thermostat settings are controlled with respect to user thermal comfort preferences viz., indoor temperature, for DR based on decision model as shown in Figure 11.24.

11.3.2.4 Case Study: Simulation Results and Discussion

A simulation study has been carried out to verify the approach and effectiveness of evolutionary techniques considered using real-time appliance specifications and preferences. A typical medium income residential home model with a 2 kWp rooftop PV system in conjunction with a 6 kWh battery bank is considered. Assume a typical daily load requirement of 26 kWh, having deferrable loads, viz. 2 hp water pump, 1.2 kW washing machine, 1 hp flour mill, 1 kW dish washer, and thermostatic loads (LG-D292RPJL 258 L refrigerator, 1.5 Ton noninverter type air conditioner along with low power critical home appliances. The power consumption patterns and preferences are given in detail (Batchu and Pindoriya 2015a) and snapshot is shown in Figure 11.25.

The refrigerator was modeled as load with average demand of 30 W and other kitchen and entertainment loads as base load of 90 W. Also assumed an EV and PV+ battery system may exist, and the forecasted power generation for the considered rooftop PV system is shown in Figure 11.27. To simulate the day-ahead real-time pricing (RTP) model, we consider a wholesale electricity market price profile (Power Exchange India Limited 2015) assuming that retail buying and selling electricity price will be scaled and follows the same pattern. Pricing details for the considered case can be seen in Figure 11.27. We considered time interval of $t = 1$ h for simplicity. Two case studies with the same input parameters were done for validating the performance and benefit of HEMS.

1. *Without HEMS: A base case:* In this case, customer uses the appliances at its preferred operating periods without applying any optimal scheduling. Figure 11.26 shows the power flows among PV, battery, and grid to meet the demand.
2. *With HEMS: Optimal scheduling:* The algorithm takes into account forecasted renewable energy generation information, appliances power consumption patterns, user preferences, and day-ahead hourly real time buying and selling price. The multiobjective mixed integer linear programming problem was solved using NSGA II multiobjective genetic algorithm in MATLAB® environment. The average solution time is the order of a few minutes, which can be further improvised by initializing suitable initial population. Figure 11.27 shows the stage 2 optimization output, i.e., the detailed hourly distribution of energy from grid, PV system, and batteries to home as well as among them. It can be seen that depending on the energy price and loads, energy can be bought and sold dynamically and battery system can be charged and discharged to accomplish jobs while minimizing total cost, peak load, and discomfort. Results obtained by applying the proposed optimization approach are provided in Table 11.8.

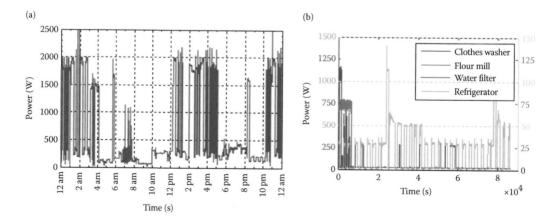

FIGURE 11.25 (a) Typical summer day load profile of the considered home measured using sub circuit power meter and (b) disaggregated load profiles.

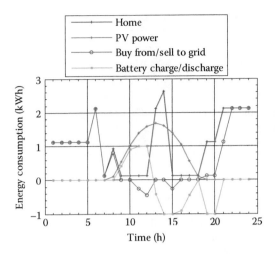

FIGURE 11.26 Without any optimal scheduling: load demand, PV generation, and demand from grid and battery charge/discharge.

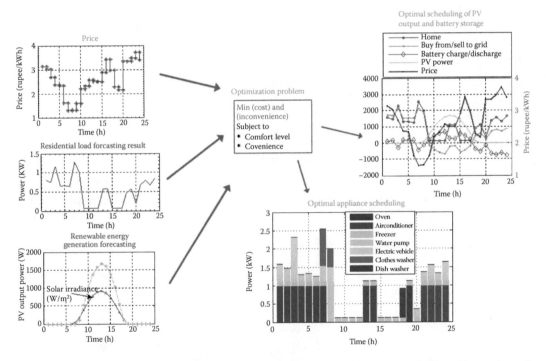

FIGURE 11.27 Optimal appliance scheduling and power flows scheduling among PV, grid, and home obtained using the proposed approach for the considered RTP.

Assuming that there are no uncertainties and operated as per scheduling then, it can be observed that the new reduced peak demand is 2.16 kW—as shown in Table 11.8—which occurs at 7 AM and peak injection to grid is −0.59 kW, which occurs at 12 PM. The negative sign here represents the power injection to grid. In stage 3, HEMS responds to DR signal from aggregator in the form of increased additional peak demand charge (say 1 Rs. per kW) or demand limit (say 2 kW). Based on battery state-of-charge (SOC), appliance priority dynamic thermal comfort levels HEMS reduces demand by interrupting certain appliances based on discharge battery.

TABLE 11.8   Simulation Results Comparison—with/out HEMS

Objective Parameter	Day-Ahead RTP	
	Without HEMS	With HEMS
P.A.R	4.12	3.42
Cost/day (rupees)	63.89	39.24
Discomfort level (%)	0	30
Saving (%)	0	38%
Peak load and duration	2.6 at 2 pm	2.16 at 7 am

11.3.3 Aggregated Customers Level Home Energy Management

Let us consider the scenario of a residential community with one utility company (aggregator) and total of N multiple users and a solar PV plant, wind generation, and battery storage resources at community level, as shown in Figure 11.28. Each user can have essential loads, shiftable loads (both time and power shiftable types), and local renewable generation (rooftop solar PV or wind turbine) in conjunction with battery storage. A CEMS/distribution utility aggregator job is to procure/buy the required energy based on the optimized demand profile of the community from day-ahead energy demand from electricity market and participating in DR programs by aggregating customers demand flexibility. The objectives of CEMS for day-ahead scheduling are minimization of overall cost of energy purchase as shown in Equation 11.14 and minimization of peak to average ratio in Equation 11.15.

$$C_N^{(1)} = \text{Min} \sum_{t=1}^{N_{\text{slot}}} \left\{ \sum_{n=1}^{N_{\text{slot}}} C^t \left(p_{t,n}^L \right) - p_w^t(v) - p_{PV}^t(G, T_a) \pm p_B^t \right\} \tag{11.14}$$

$$C_N^{(2)} = \text{Min} \frac{L_{\text{peak}}}{L_{\text{avg}}}, \tag{11.15}$$

where $p_w^t(v)$ is the community level wind turbine generation power at interval t, which is a function of wind velocity v and $p_{PV}^t(G, T_a)$ is the community level solar PV generation power at time interval

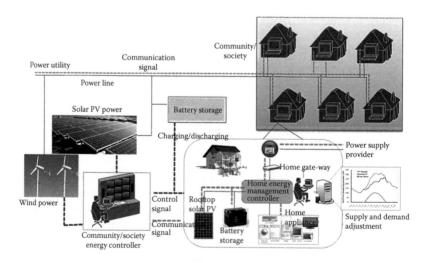

FIGURE 11.28   Aggregator assisted community energy management system.

t, which is a function of irradiance G and ambient temperature T_a. p_B^t is the community level battery storage charging/discharging power. L_{peak} represents the overall peak load demand and L_{avg} represents average load demand at the community level. This problem can be solved as a centralized or decentralized optimization strategy.

To preserve the customer's private information, decentralized game theoretic optimization techniques (Guo et al. 2013; Chavali et al. 2014; Safdarian et al. 2014; Mhanna et al. 2015) are popular compared to centralized optimization techniques (Pedrasa et al. 2010). Liu et al. (2013) proposed a bidirectional power trading mechanism for a residential smart grid, where delay aware consumption schedulings are done in two stages. The key challenges for energy management in residential grid with local DER-based generation and battery storage are

- How to coordinate among the customers and with other community in the residential smart grid?
- How to maximize social welfare and minimize P.A.R ratio of the community?
- How to reach equilibrium solution point with reduced number of iterations and computational complexity?

11.3.4 Impact Investigation of Distributed Renewable Generation on Low Voltage Distribution

Autonomous operation of HEMS and CEM modules, as independent self-interested entities at single and multiple residential customers, although beneficial for individual and societal customers, can threaten the distribution network efficiency. When HEM and CEM modules work selfishly, new peak demands are likely to be created at times when electricity prices are lower. It is, therefore, also important to assess the impact of distributed renewable energy integration in the active distribution network. This section demonstrates the application of sequential time simulations in investigating the influence of DGs on unbalanced operations of active distribution network. The modeling for all major components in active distribution network such as transformers, voltage regulators (VRs), underground (U/G) cables, overhead (O/H) lines, spot, and distributed loads can be referred from Joshi and Pindoriya (2012), wherein the modeling of PV arrays is also included as a small- to medium-sized DER in low to medium voltage (MV) distribution networks. The framework for sequential time simulations using backward forward sweep (BFS) algorithm is outlined as follows.

11.3.4.1 Framework for Sequential Time Simulations (STS)

Recent developments in the analysis of active distribution network suggest that the dynamic nature of load and generators requires a more comprehensive approach than the one that is used for transmission networks. The snapshot power flow studies with extreme demand and generation scenarios are found insufficient for analysis of distribution networks with increasing penetration of distributed resources (Walling et al. 2008). Impact investigation studies (Joshi and Pindoriya 2012, 2014) suggest the need for STS to accurately investigate the effects of intermittent and uncertain characteristics of DERs. It is also observed that time-series simulations are also required where the distribution network has a perceivable presence of Plug-In Hybrid Electric Vehicles (PHEVs) (ElNozahy and Salama 2014; Joshi and Lakum 2014). Furthermore, to optimize the use of energy storage systems (ESS) connected in distribution networks with a high degree of DER penetration also necessitates a small-interval continuous simulation for year-long data sets (Taylor et al. 2011; Joshi and Pindoriya 2015). In view of these changing requirements, a framework for STS is proposed in Figure 11.29.

11.3.4.2 The Approach

The approach consists of three layers namely the data sets, core modules, and intended output variables. As shown in Figure 11.29, blocks 1–5 form the data sets for demand and generation data. These datasets may or may not be available. In case if these datasets are not available, some publically available data for

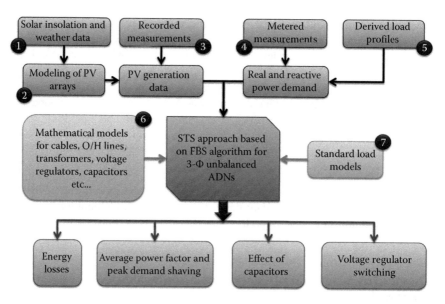

FIGURE 11.29 Framework for STS approach.

intermittent sources and modeling exercise together can produce the year-long reliable data. On the other hand, annual load profiles also can be produced based on some sample load profiles as discussed earlier. The core module includes blocks 6 and 7 in Figure 11.29 as well as a distribution power flow engine—BFS algorithm. Blocks 6 and 7 represent the time invariant models of all the network components including load models. It should, however, be noted that the time-varying load models can also be used in this approach. The distribution power flow engine runs the simulations sequentially for each set of demand generation record available in datasets. The third layer compiles the resulting output variables of interest for yearly analysis. Some typical parameters of interest are shown in Figure 11.29 such as energy losses, power factor, peak demand shaving, capacitor, and voltage regulator switchings.

11.3.4.3 Application of STS and Sample Results and Discussion

Time-series power flow analysis or STS approach finds many applications in steady-state analysis or quasistatic analysis of active distribution network. Several applications including voltage regulating device operations, feeder voltage profiles, voltage flicker analysis, and short circuit analysis are demonstrated with examples in Broderick et al. (2013). This approach is now considered inevitable in the distribution systems' planning exercises as the uniqueness of feeder responses to DER penetrations is captured adequately only by sequential simulations of Smith et al. (2015). It can also be used to propose remedial measures to reduce the impact of DER on the distribution network operations as proposed in Joshi and Pindoriya (2014). The same is also useful for better and effective utilization of network resources such as PHEVs and energy storage in the distribution networks (ElNozahy and Salama 2014; Joshi and Pindoriya 2015). STS approach can provide important insights into the effects of DER on distribution network operations as depicted in sample results shown in following figures. The effect of rooftop PV generator on the daily average power factor of an institutional premise, IIT Gandhinagar campus distribution network is obtained by running distribution power flow (DPF) simulations over a period of 1 year. The effect on power factor—with and without the PV generation—is shown in Figure 11.30. The diurnal variations in real and reactive power demand along with power factor are shown in Figure 11.31 for a typical day. The results of STS approach provide insights at the microlevel and macrolevel simultaneously as shown in Figures 11.31 and 11.32. The influence of PV generators on overall peak demand at the source node is shown in Figure 11.33.

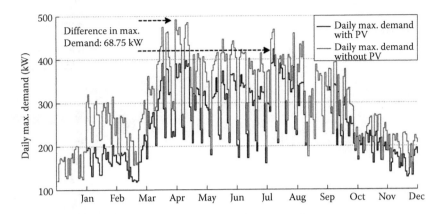

FIGURE 11.30 Diurnal variations in power factor on February 27, 2012 along with real and reactive power demand.

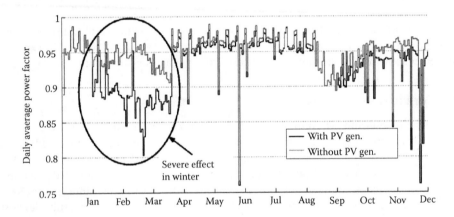

FIGURE 11.31 Daily average power factor—effect of local PV generation is significant during winter days when local demand is substantially less.

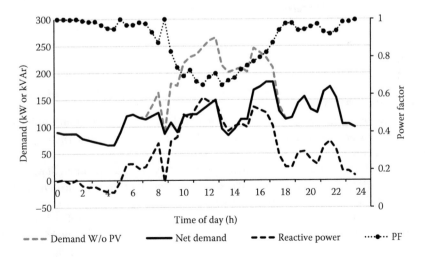

FIGURE 11.32 Diurnal variations in power factor on February 27, 2012 along with real and reactive power demand.

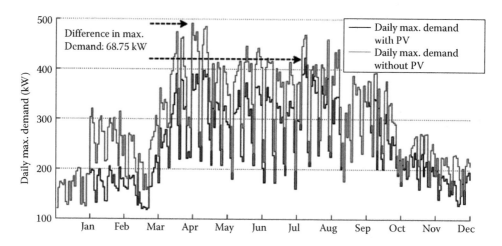

FIGURE 11.33   Effect of local PV generation on daily maximum demand—reduced by 68.75 kW for the year 2012.

These results can also be used for deferring investments on distribution and transmission network upgrades, estimating hosting capacity of a network for DGs, estimating ESS size and its expected benefits, and other such critical issues for distribution system planners and operators.

11.3.5 Conclusion

Implementation of residential DR through HEMS and CEMS has attracted much interest because of its demand flexibility. This chapter presents an optimal HEMS algorithm to integrate local DER and battery storage systems in a smart home environment considering customer level and community level objectives. The HEMS is developed for a day-ahead RTP coupled with critical peak pricing-based DR program, which employs multistage algorithms to optimize the power consumption/generation/storage operations based on signals from an aggregator, customer priority, and preference settings. Simulation results clearly showed effective energy management through HEMS and validated that the customer benefit can be extracted in terms of reducing demand restrike peak and duration. Moreover, the study on impact assessment of solar PV integration in low voltage distribution grid is presented and clearly observed on how the penetration of solar PV generation at the node alters/affects the distribution grid operation.

References

Bakirci, K. 2009. Estimation of solar radiation by using ASHRAE clear-sky model in Erzurum, Turkey. *Energy Sources, Part A: Recovery, Utilization, and Environmental Effects* 31(3): 208–216. doi:10.1080/15567030701522534.

Batchu, R., and N. M. Pindoriya. 2015a. Multi-stage scheduling for a smart home with solar PV and battery energy storage—A case study. In *IEEE Innovative Smart Grid Technologies (ISGT Aisa)* Bangkok, 4-6 Nov. 2015.

Batchu, R., and N. M. Pindoriya. 2015b. Residential demand response algorithms : State-of-the art, key issues and challenges. In *International Workshop on Communication Applications in Smart Grid*, pp. 18–34, Bradford, UK. doi:10.1007/978-3-319-25479-1_2.

Broderick, R. J., J. E. Quiroz, M. J. Reno, A. Ellis, J. Smith, and R. Dugan. 2013. Time series power flow analysis for distribution connected PV generation. Sandia National Laboratory. prod.sandia.gov/techlib/access-control.cgi/2013/130537.pdf.

Chavali, P., P. Yang, and A. Nehorai. 2014. A distributed algorithm of appliance scheduling for home energy management system. *IEEE Transactions on Smart Grid* 5(1): 282–290. doi:10.1109/TSG.2013.2291003.

Chen, C., J. Wang, Y. Heo, and S. Kishore. 2013. MPC-based appliance scheduling for residential building energy management controller. *IEEE Transactions on Smart Grid* 4(3): 1401–1410. doi:10.1109/TSG.2013.2265239.

Dang, T., and K. Ringland. 2012. Optimal load scheduling for residential renewable energy integration. In *2012 IEEE 3rd International Conference on Smart Grid Communications, SmartGridComm 2012*, pp. 516–21, Taiwan, November 5–8. 2012. IEEE. doi:10.1109/SmartGridComm.2012.6486037.

Deb, K., S. Pratab, S. Agarwal, and T. Meyarivan. 2002. A fast and elitist multiobjective genetic algorithm: NGSA-II. *IEEE Transactions on Evolutionary Computing* 6(2): 182–197.

ElNozahy, M. S., and M. M. A. Salama. 2014. A comprehensive study of the impacts of PHEVs on residential distribution networks. *IEEE Transactions on Sustainable Energy* 5(1): 332–342. doi:10.1109/TSTE.2013.2284573.

Erdinc, O., N. G. Paterakis, I. N. Pappi, A. G. Bakirtzis, and J. P. S. Catalão. 2015. A new perspective for sizing of distributed generation and energy storage for smart households under demand response. *Applied Energy* 143: 26–37. doi:10.1016/j.apenergy.2015.01.025.

Guo, Y., M. Pan, and Y. Fang. 2012. Optimal power management of residential customers in the smart grid. *IEEE Transactions on Parallel and Distributed Systems* 23(9): 1593–1606. doi:10.1109/TPDS.2012.25.

Guo, Y., M. Pan, Y. Fang, and P. P. Khargonekar. 2013. Decentralized coordination of energy utilization for residential households in the smart grid. *IEEE Transactions on Smart Grid* 4(3): 1341–1350. doi:10.1109/TSG.2013.2268581.

Hu, Q., and F. Li. 2013. Hardware design of smart home energy management system with dynamic price response. *IEEE Transactions on Smart Grid* 4(4): 1878–1887. doi:10.1109/TSG.2013.2258181.

Joshi, K., and A. Lakum. 2014. Assessing the impact of plug-in hybrid electric vehicles on distribution network operations using time-series distribution power flow analysis. In *2014 IEEE International Conference on Power Electronics, Drives and Energy Systems (PEDES)*, Mumbai, India, December 16–19, pp. 1–6. IEEE. doi:10.1109/PEDES.2014.7042120.

Joshi, K. A., and N. M. Pindoriya. 2012. Impact investigation of rooftop solar PV system: A case study in India. In *2012 3rd IEEE PES Innovative Smart Grid Technologies Europe (ISGT Europe)*, Berlin, October 14–17, 2012, pp.1–8. IEEE. doi:10.1109/ISGTEurope.2012.6465813.

Joshi, K. A., and N. M. Pindoriya. 2014. Reactive Resource Reallocation in DG Integrated Secondary Distribution Networks with Time-Series Distribution Power Flow. In *2014 IEEE International Conference on Power Electronics, Drives and Energy Systems (PEDES)*, Mumbai, India, December 16–19, 2014, pp. 1–6. IEEE. doi:10.1109/PEDES.2014.7042111.

Joshi, K. A., and N. M. Pindoriya. 2015. Day-ahead dispatch of battery energy storage system for peak load shaving and load leveling in low voltage unbalance distribution networks. In *IEEE PES General Meeting*, 1–5. Denver, CO: IEEE.

Liu, Y., N. U. Hassan, S. Huang, and C. Yuen. 2013. Electricity cost minimization for a residential smart grid with distributed generation and bidirectional power transactions. In *2013 IEEE PES Innovative Smart Grid Technologies Conference, ISGT 2013*, Washington, DC, February 24–27. doi:10.1109/ISGT.2013.6497859.

Mhanna, S., G. Verbic, and A. C. Chapman. 2015. A faithful distributed mechanism for demand response aggregation. *IEEE Transactions on Smart Grid*, 7(3):1743–1753. doi:10.1109/TSG.2015.2429152.

Pedrasa, M. A. A., T. D. Spooner, and I. F. MacGill. 2010. Coordinated scheduling of residential distributed energy resources to optimize smart home energy services. *IEEE Transactions on Smart Grid* 1(2): 134–43. doi:10.1109/TSG.2010.2053053.

Power Exchange India Limited. 2015. Day-ahead electricity market. http://www.powerexindia.com/pxil/.

Safdarian, A., M. Fotuhi-Firuzabad, and M. Lehtonen. 2014. A distributed algorithm for managing residential demand response in smart grids. *IEEE Transactions on Industrial Informatics* 10(4): 2385–2393. doi:10.1109/TII.2014.2316639.

Shao, S., M. Pipattanasomporn, and S. Rahman. 2013. Development of physical-based demand response-enabled residential load models. *IEEE Transactions on Power Systems* 28(2): 607–614. doi:10.1109/TPWRS.2012.2208232.

Smith, J., M. Rylander, L. Rogers, and R. Dugan. 2015. It's all in the plans: Maximizing the benefits and minimizing the impacts of DERs in an integrated grid. *IEEE Power and Energy Magazine* 13(2): 20–29. doi:10.1109/MPE.2014.2379855.

Taylor, J., J. W. Smith, and R. Dugan. 2011. Distribution modeling requirements for integration of PV, PEV, and storage in a smart grid environment. *IEEE Power and Energy Society General Meeting* doi:10.1109/PES.2011.6038952.

Walling, R. A., R. Saint, R. C. Dugan, J. Burke, and L. A. Kojovic. 2008. Summary of distributed resources impact on power delivery systems. *IEEE Transactions on Power Delivery* 23(3): 1636–1644. doi:10.1109/TPWRD.2007.909115.

Zhang, Y., P. Zeng, S. Li, C. Zang, and H. Li. 2015. A novel multiobjective optimization algorithm for home energy management system in smart grid. *Mathematical Problems in Engineering* 2015: 1–19. doi:10.1155/2015/807527.

Zhao, Z., W. C. Lee, Y. Shin, and K. B. Song. 2013. An optimal power scheduling method for demand response in home energy management system. *IEEE Transactions on Smart Grid* 4(3): 1391–1400. doi:10.1109/TSG.2013.2251018.

Zhu, Z., J. Tang, S. Lambotharan, W. H. Chin, and Z. Fan. 2011. An integer linear programming and game theory based optimization for demand-side management in smart grid. In *2011 IEEE GLOBECOM Workshops (GC Wkshps)*, pp. 1205–10, Houston, TX, December 5–9. IEEE. doi:10.1109/GLOCOMW.2011.6162372.

11.4 Evaluation and Modeling of Demand and Generation at Distribution Level for Smart Grid Implementation

Haile-Selassie Rajamani

11.4.1 The Smart Grid

Urban areas with high economic activity generally have a good electricity energy infrastructure to facilitate economic growth. However, with rapidly rising energy demands and the desire for low carbon technologies, there is a need to maximize the utilization of the current infrastructure and all its resources to ensure that electricity supply is reliable and sustainable. The forecasted rise in electrical vehicles and in heat pumps across Europe is a challenge to the power grid. These technologies can lead to the collapse of the power grid if measures are not put in place to try and manage both load and demand in a more efficient way.

Advances in power electronics, monitoring systems, sensor technology, communication systems, data analytics, smart metering, embedded generation, and storage have led to the evolution of the smart electricity grid where the resources and demand can be better managed. Various roadmaps on the evolution of the smart grid have been published by governments showing both the benefits and the technology requirements [1,2].

Figure 11.34 shows the various linkages between the components of a modern power system that now has the markets, the service provider, and the customer as key participants in the grid. These new players are linked to the traditional power, and the idea is that they all become more active in the grid, bringing in more private investment, particularly at the customer end where renewable energy technologies, storage technologies, and techniques to change loads are being implemented.

The drivers behind the evolution to the smart grid are as follows:

1. The deregulation of the electricity industry, moving away from single large utilities to a market-driven industry with competition.

2. The need for energy security by spreading supply to a wider number of participants.
3. The need to permit new approaches to reducing CO_2 emissions.
4. The rising cost of fossil fuels is making other forms of energy attractive that could be utilized at distribution levels.
5. The rapid technology developments in the area of power electronics, control systems, communication systems, and renewable energy conversion technologies.
6. The need to bring in private investment to upgrade existing infrastructure and to develop new infrastructure.

The smart grid covers a broad area, ranging from the control of the large power stations to ensure stability of supply, to the control of demand. In this chapter, we will focus more closely on the consumer side, known as the demand side of the smart grid. Figure 11.35 shows the UK total energy consumption over

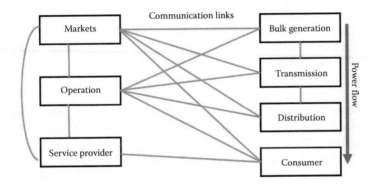

FIGURE 11.34 The smart grid.

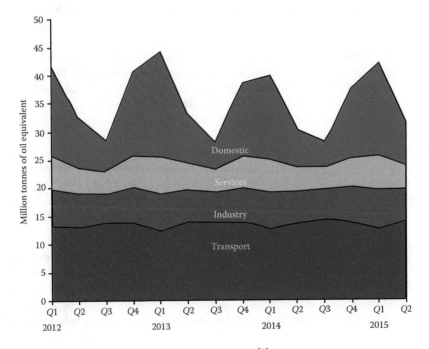

FIGURE 11.35 United Kingdom energy consumption over time [3].

the last few years in quarterly intervals by sector. It is clear that the domestic sector shows a significant seasonal variation. Naturally, this means that the supply of energy is stressed to meet the peak demand and so must be designed to cater for the peak. This additional peak demand aspect of energy usage is critical and gives the opportunity for techniques to reduce it. In this paper, we will focus on the electrical form of energy where the smart grid is likely to first have impact.

11.4.2 Electrical Distribution Side Aspects

The electrical distribution side forms the demand side of the smart grid. Changes in the distribution side are occurring because of the following:

1. *Embedding of renewable energy and energy storage:* There has been significant effort by many countries to introduce technologies such as solar panels and wind turbines, and more recently storage, to the residential and industrial sectors. This changes the way in which energy is billed as consumers with such energy generating capacity would be able to supply energy back into the grid for which they would want some financial reward. Governments are keen on this as such an uptake of renewable energy is often a privately funded scheme with some support from the government. Effectively, the government is engaging private finance to support the grid. This has made the consumer into a prosumer.

2. Purchasing schemes, such as buying your energy from a green source, or simply getting the cheapest price available, that enable competition. This involves better metering and billing systems to permit competition. The UK government has a 2020 target to have smart meters installed in all houses. It is expected that competition in the retail market will be enhanced.

3. *Demand side response techniques:* These are techniques where the actual load itself is modified. A reduction in peak demand is of significant benefit to the supplier as the cost of electricity at peak times is high. This is because normally the most efficient generators are run first to keep costs low. As demand increases, less efficient generators are turned on or energy is imported from more costly sources including from other countries. Since the cost can be very high, there is a benefit to reducing demand at peak times for which the consumer can demand reward. This is the essence behind demand side response techniques.

4. *Rise in electrical vehicles:* While electrical vehicles have in theory been around for many years, there is now a renewed interest in them. Of particular interest is the very likely introduction of driverless cars in the United Kingdom over the next year or two. Electrical vehicles will increase the demand placed on the already stretched grid. The demand patterns will change due to this. In addition, there are ideas of using the batteries in electrical vehicles as energy storage to meet peak demand [4].

5. Increase in heat pump usage, both for heating and cooling across the globe. This significant change from gas boilers and fans is because it is envisaged that electricity is likely to be the best way to transmit low carbon energy. This will increase electricity demand [5].

6. Maximizing infrastructure is the main aim of advanced countries that have already invested heavily. While emerging economies would more likely be able to add new infrastructure, advanced countries have more difficulty replacing old infrastructure.

The actual demand from a customer changes for a number of reasons. The behavior of the consumer may change over a period. For example, a consumer may stop drinking lots of coffee, which means the electrical kettle is not being used. There are also demographic factors such as the employment status, size of household, and age of occupants of a particular household. For example, if two people in a household work, then it is likely that the energy demand during the daytime will be low, but that in the mornings and evenings there may be a peak spike in demand. Other factors are the climate, summer and winter patterns, and the daily weather changes. A more recent factor is the impact of the renewable energy supplies on the demand seen by the supplier. For example, if the photovoltaic (PV) panels provide most of the energy during the daytime, then the household will be seen as having no load or even negative load by the distribution network.

11.4.3 Load Shifting Initiatives

To change the load, various methods have been used [6]. One technique used is to have peak and off peak pricing schemes. The issue with this is whether it would actually change behavior significantly. Definitely setting a significant difference in price would affect change in behavior but this may not be desirable in a competitive environment where another supplier may offer a better deal. Research has shown that consumers are more likely to change behavior if they are well informed. Another issue is that the critical peaks are often short term and not regular. As such having a fixed pricing scheme may not bring the necessary change that suppliers may be willing to pay for [7]. Another method of changing the demand would be to have some kind of automated shifting of individual loads such as heat pumps. Research has shown that the most impact is derived from having flexible energy units. This could be extended to larger loads such as manufacturing units. However, the difficulty with such schemes is again the pricing issue.

Research on demand side has developed around the areas of simulations, historical and demographical analysis, and business aspects, particularly with respect to government initiatives. Various trail studies have been conducted, mainly in the United States of America, Germany, the United Kingdom, and the Nordic countries, to test various aspects of smart grid implementation. Technology aspects were core to the tests as issues of safety and grid stability are key to success. Financial aspects such as tariff schemes and capital investments were also considered.

11.4.4 Modeling of Community Load

National projections that are based on historical consumption patterns and long-term demographic changes already exist. They are very good for long-term planning purposes. However, they lack detail that is more useful for local community planning over shorter periods of time. Case studies of energy consumption in small communities have been done and the results are available. However, small communities can vary a lot depending on where they are located. For example, the energy consumption pattern of a community in upper class London would be different from a rural community in the north of the United Kingdom. On the other hand, detailed measurements of sample households are a way forward. While these can be very useful, there are many issues on data security, which makes it difficult for local communities to engage with.

To address this, various simulations models have been developed that are published in the literature that give an indication of energy demand. Advanced models include detailed appliance characteristics [8]. However, it is still difficult based on standard models and simulation to understand a local community's energy demand behavior so as to be able to modify it. In this chapter, we report on work done in developing a community demand profile generator that can be adjusted to study particular communities.

11.4.5 Community Demand Profile Generator

To simulate a typical community, key national statistics are used together with some data from the local community that can be obtained by interview of a sample [9]. The UK national composition of households, which is obtained from the English Housing survey [10], is the first key input so as to determine what the composition of the households in a community is in terms of the number of people working, number of children, etc. From the data, it is possible to classify the households into eight groups. There is opportunity to refine this data if actual household details are available.

The second piece of information that is needed is the appliance information data. In this work, the data used was from that published in Reference [11]. Again refinement of data is possible, but requires much more detailed study of individual appliances that the community use, which is not the aim of the work. Once the percentage of appliances is known, the appliances are randomly distributed. It is possible to allocate appliances according to household types, but this would be difficult to do without understanding the link between appliance usage and type of household.

To really make this applicable to the local community, a survey is taken on how appliances are actually used. This gives an idea of the actual behavior of the community and most importantly accounts for the location setting. From the data collected, it is then possible to develop a cumulative distribution function for each appliance usage by household type, which is shown in Figure 11.36. This then is used to allocate time of usage to the households. The overall profile generator scheme is shown in Figure 11.37.

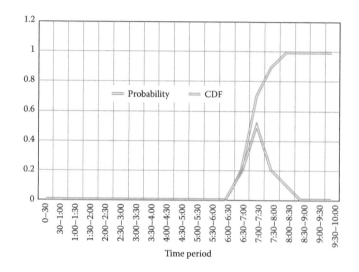

FIGURE 11.36 Probability distribution for an appliance over the morning.

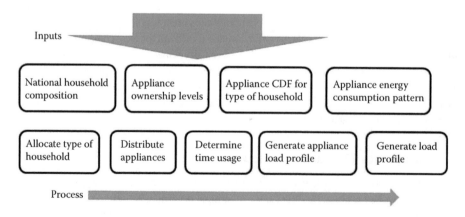

FIGURE 11.37 Framework for modeling a local community.

Typical results from the profile generator for the test community are shown in Figure 11.38 for the single adult type homes in a local community. This shows key details such as when the peaks and lows in energy consumption occur. It also reveals the fact that there are opportunities to modify the demand in energy.

Another aspect is that it would be possible to target particular types of households. For example, the total energy consumed by all the single households could be the same size as that of families. This would give the possibility of trying to modify the behavior of the single person household rather than trying to modify the behavior of the family households.

Figure 11.39 shows the total energy consumed by the particular community. It is clear that the main usage is during the evenings. What is interesting is that the peak demand is actually around 8–9 pm. This is well past the national peak demand that tends to occur around 4–6 pm. It is also worth noticing that the energy consumed during the day is relatively low. As such it maybe, simply by looking at the figure, thought that the best thing would be to adjust the demand in such a way as to reduce the peak consumption at night time. The next sections discuss the possible interventions.

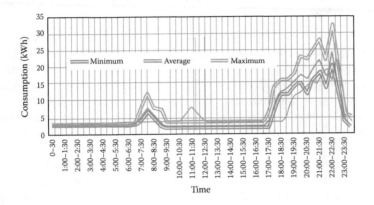

FIGURE 11.38 Energy usage of single adult household types in local community.

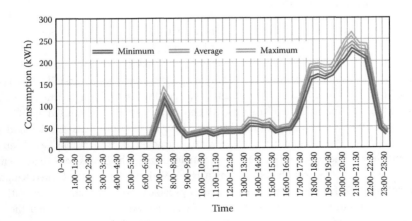

FIGURE 11.39 Total community energy usage.

11.4.6 Interventions

Interventions possible in such a community would include the following:

1. *Renewable energy:* Renewable energy could provide locally part of the energy thereby reducing the load demand. For example, PVs could be used to reduce the load during the day.
2. *Storage options:* As solar energy is available during the day, the demand by the community is actually low, it may be better to store the energy during the day. The energy would then be released during the peak times, thereby reducing the load demand.
3. *Appliance load shifting:* It is possible that some of the appliances such as the washing machine could be shifted automatically to occur at a different time.
4. *Energy efficiency measure:* A significant and effective way of modifying the energy demand is by using more efficient appliances. For example, energy efficient lighting has already reduced night time consumption significantly. Other efficiency aspects such as better insulation mean heating appliances are less used.
5. *Behavior modification/education:* Changing the behavior of people is an essential aspect of demand response. For example, getting people to boil only sufficient water in their kettles for the tea they wish to have reduces energy consumption.

However, we need to be able to determine which type of intervention would actually be useful. Currently, the way of evaluating interventions has been to look at CO_2 emissions reduction, which is based on how much energy is actually used. This has been driven very much by government agenda to meet global CO_2 obligations. Another way to evaluate is to use standard business evaluation methods. This is simply based on how much money you invest and the payback time. This is generally short term and based on current tariffs set by the retailers. However, the weakness with this is that in a volatile energy price environment, the calculations can be misleading. It is very difficult to actually work out what the real price on a particular day is as buyers and suppliers are both transacting business confidentially, and also over different periods of times. For example, coal is bought and sold on the stock market. Furthermore, interventions such as renewable energy can also be affected by the actual generation possible for a given day and also the efficiency of generation tends to fall with age. Economic models are a possibility, but often these are based on lumped models that do not necessarily help a local community with their planning or control of electricity. In the next section, a brief overview of the UK electricity market is given followed by a proposal on how we can evaluate interventions.

11.4.7 UK Electricity Market

The UK electricity market is deregulated. Although electricity is used continuously and the supply will be exactly equal to usage as electricity cannot be stored, it is traded in half hour blocks known as the settlement period. Suppliers and buyers are expected to make contractual purchases ahead of the period of settlement in which electricity is actually used. The base load is normally purchased in this way. However, there may still be a need for changes closer to the actual usage. This can be purchased using power exchanges. In the United Kingdom, all contracts have to be completed up to 1 h before the settlement period. In the half hour itself, suppliers may have wrongly predicted the demand, producers may have been unable to produce the supply they were contracted to, or there may have been a problem in transporting the electricity. Therefore, real-time management of the grid is done so as to ensure a stable supply. This is what the System Operator, known as the National Grid in the United Kingdom, does.

In order for this level of control, suppliers can make available more or less energy generation capacity to the National Grid. Buyers also that can reduce or increase their load may also offer the capacity to change their load to the National Grid. Suppliers and buyers set their bids and offer prices. The System Operator in real time will match supply and demand. Thereafter, the actual metered volumes of electrical energy that were used or generated, that were above or below the contracted amounts for that settlement period, would be settled based on the bids and offers. This is known as the settlement. The buy and sell prices of this market are available and reflect to some extent the cost of additional electricity above the base load. The sell prices for 1 week are shown in Figure 11.40.

The figure shows that the price of electricity transacted varies over the day, work day and weekend, and also over the week. There are also variations over the year. What is interesting to note is that the prices are high during particular periods of the day. Between 16:30 and 20:00 it is particularly high as would be expected based on people returning to homes. It may be assumed that this is due to the national demand increasing significantly. Figure 11.41 shows the demand over 3 days in half hour blocks. It is clear that the demand is highest around 6 pm.

Table 11.9 shows what makes up the price paid by a consumer in the United Kingdom. It is clear that the wholesale cost is ~40% of the bill. Network costs and supplier costs account for 25% and 13%. The environmental and social obligation costs that are key to moving the country toward renewables and higher efficiencies contribute to about 10%. Value added tax (VAT) and profit margins account for the

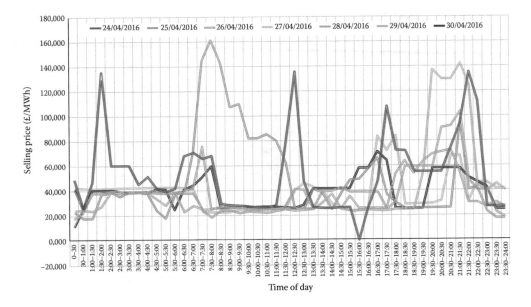

FIGURE 11.40 UK system buy prices for 1 week.

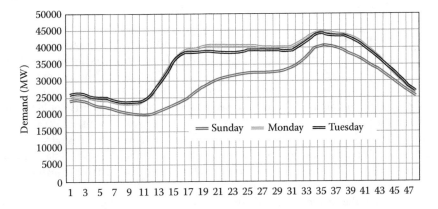

FIGURE 11.41 Demand curves for England and Wales over January 10, 2016–January 12, 2016 [12].

TABLE 11.9 Breakdown of Consumer Electricity Bill [13]

Wholesale costs	Electricity	£ 229
	Unbilled volumes	£ 7
	Electricity imbalance costs	£ 1
	Total	£ 237
Network costs	Electricity network (transmission)	£ 32
	Electricity network (distribution)	£ 116
	Balancing (BSUoS)	£ 6
	Total	£ 154
Environmental and social obligation costs	Renewable Obligation Certificates (ROCs)	£ 38
	Electricity Energy Companies Obligation (ECO)	£ 20
	Feed-in tariffs (FiTs)	£ 10
	Elec. WHD	£ 6
	Government-funded rebate	−£ 12
	Total	£ 62
Supplier operating costs	Operational (inc. meters and smart meters)	£ 75
	Depreciation and amortization	£ 4
	Total	£ 79
	VAT	£ 29
	Total costs	£ 562
	Revenue (average customer bill)	£ 614
	Pretax margin	£ 37
	Pretax profit margin (%)	6%

remainder. This information does not in any way help consumers modify their demand profiles, nor does it help local planning authorities understand the costs of peak demand.

The buy prices, as previously obtained, are only a small portion of the customer's bill. However, it is possible to get an understanding of the difficulty with which the suppliers provide peak energy. To be able to use this information at the distribution level, the data are normalized to be in per unit form. So the base unit for demand would be the average demand, and the base unit for price would be the average price. Figure 11.42 shows the upper quartile, median, and lower quartile curves that can be used as indicative of a high, medium, or low price. As demand increases, the price does not rise linearly but rather exponentially. This would be expected given that the cost of additional generation would normally be much higher given that less efficient generation may be turned on or that energy is imported.

The graph shows that when the demand goes beyond 1 p.u., which is the average demand, the median and the upper quartile do show very steep rises in price. As the demand gets higher, it is clear that the median gets closer to the upper quartile showing that overall high prices. The aim of this paper is to develop a way of evaluating the local community strategies. Figure 11.42 can be used as an indicator of pricing for the local community. The average demand of the local community could be used as the base, and hence, the pricing based on national demand and pricing could be used to evaluate interventions. Figure 11.43 shows the communities total energy cost per unit over a 24 h period for all the pricing curves. The use of a 24 h circular diagram is very helpful in visualizing the daily cycle and where costs actually occur. The figure shows the demand in p.u., and the three total costs are shown in dotted lines based on using the best fit curves for Figure 11.42.

If it is assumed that the current standard tariff schemes work out to 1 p.u., it can then be seen that the high costs mean that the grid is actually losing money. While this is offset by the periods of low

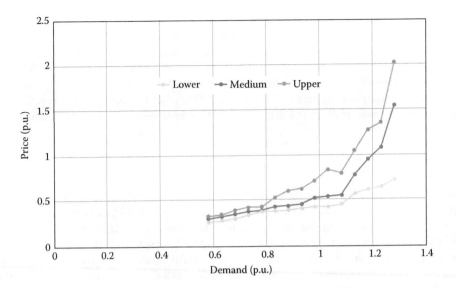

FIGURE 11.42 Upper, lower, and median quartiles for the price versus demand over a year.

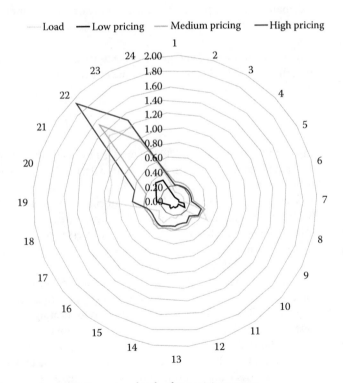

FIGURE 11.43 Community electricity cost under the three pricing ranges.

demand, this may still disadvantage the supplier. For example, the peak cost is 1.9 p.u. at upper quartile. This indicates the grid may actually be willing to pay the consumer to bring demand down.

While this graph only gives indication of cost of electricity supply, it does provide a visual basis for evaluating an intervention. We need this because of the new requirements that the smart grid evolution is bringing. Introduction of more market forces at the consumer levels means that we need to have some way of establishing the relative cost of demand and supply over a day, over weeks, and over months. We also need the service providers to be able to identify consumer behavior, and opportunities in order to develop market instruments. There is a need to manage loads remotely and also to monitor them. Ultimately, the consumer needs to be able to participate in a financial manner, which is only possible if he can engage with the national demand and cost models. This last point is important because consumers are naturally dealing only with big private utilities that are aiming to make profit while supplying services. A consumer needs to feel comfortable that he or she is not being cheated by the private companies.

11.4.8 Conclusions

The evolution of the smart grid is aimed at enabling the consumer to play a more significant role in the energy sector. In order for local communities to engage with the sector, an energy profile generator suitable for the local community has been presented, and results indicate where interventions would be possible, including which type of households to target. However, while the demand is obtainable, it is very difficult to evaluate interventions as currently most countries use standard tariffs and also because the actual bill consists of many components. The split in the consumer's bill leaves the consumer with only one objective which is to minimize usage. However, the real cost of electricity production, transport, and distribution is high due to the uneven demand pattern. Therefore, a method of evaluating price and demand utilizing p.u. values has been presented. The method of presenting information in a visual diagram with both demand and cost on the same diagram is very useful in identifying opportunities.

Electrical energy will be the main form of energy used as electrical vehicles and heat pumps start replacing the last century technologies. It will therefore be more critical to look at how energy balancing can be done so as to ensure reliable energy.

References

1. Department of Energy and Climate Change. (2014). Smart grid vision and routemap. https://www.gov.uk/government/uploads/system/uploads/attachment_data/file/285417/Smart_Grid_Vision_and_RoutemapFINAL.pdf [Accessed November 20, 2015].
2. National Institute of Standards and Technology, US Department of Commerce. (2012). NIST framework and roadmap for smart grid interoperability standards, Release 2.0. http://www.nist.gov/smartgrid/upload/NIST_Framework_Release_2–0_corr.pdf [Accessed November 20, 2015].
3. United Kingdom Department of Energy and Climate Change. (2015). Energy Trends, London, p. 7. https://www.gov.uk/government/collections/energy-trends.
4. Liu, C., Chau, K.T., Wu, D., and Gao, S. (2013). Opportunities and challenges of vehicle-to-home, vehicle-to-vehicle, and vehicle-to-grid technologies. *Proceedings of the IEEE*, 101(11), 2409–2427.
5. Department of Energy and Climate Change. (2012). The future of heating: A strategic framework for a low carbon heat in the UK. https://www.gov.uk/government/uploads/system/uploads/attachment_data/file/48574/4805-future-heating-strategic-framework.pdf [Accessed November 20, 2015].
6. Department of Energy and Climate Change. (2012). Demand side response in the domestic sector— A literature review of major trials final report. https://www.gov.uk/government/uploads/system/uploads/attachment_data/file/48552/5756-demand-side-response-in-the-domestic-sector-a-lit.pdf [Accessed November 20, 2015].
7. Darby, S.A. (2010). Load management at home: Advantages and drawbacks of some 'active demand side' options. *Proceedings of the Institution of Mechanical Engineers, Part A: Journal of Power and Energy*, 227(1), 9–17.
8. Richardson, I., Thomson, M., Infield, D., and Clifford, C. (2010). Domestic electricity use: A high-resolution energy demand model. *Energy and Buildings*, 42(10), 1878–1887.

9. Ihbal, A., Rajamani, H.S., Abd-Alhameed, R.A., and Jalboub, M. (2012). The generation of electric load profiles in the UK domestic buildings through statistical predictions. *Journal of Energy and Power Engineering*, 6(2), 250–258.

10. Department for Communities and Local Government. (2015). English Housing Survey HOUSEHOLDS 2013–14. https://www.gov.uk/government/uploads/system/uploads/attachment_data/file/461439/EHS_Households_2013–14.pdf [Accessed November 20, 2015].

11. Mansouri, I., Newborough, M., and Probert, D. (1996). Energy consumption in the UK households: Impact of domestic electrical appliances. *Applied Energy*, 54(3), 211–285.

12. NationalGrid plc. (2016). Electricity demand data. http://www2.nationalgrid.com/UK/Industry-information/Electricity-transmission-operational-data/Data-Explorer/. [Accessed May 18, 2016].

13. United Kingdom Office of Gas and Electricity Markets. (2015). Supply market indicator—March 2014 to April 2015. www.ofgem.gov.uk/publications-and-updates/supply-market-indicator-march-2014-april-2015.

11.5 A Process to Model Fischer–Tropsch Reactors

N. O. Elbashir, Laial Bani Nassr, and Mohamed Ghouri

11.5.1 Introduction

With the continuous increase in global demand for cleaner energy sources, gas-to-liquid (GTL) technology is receiving significant interest as a viable alternative to conventional energy sources. GTL technology is a chemical process that converts natural gas to ultraclean fuels (i.e., gasoline, jet fuel, diesel, and kerosene) and value-added chemicals through what is known as the Fischer–Tropsch synthesis (FTS). Qatar has the third largest natural gas reserves in the world with a total capacity of 910 tcf (Chedid et al. 2007). This has motivated Qatar to have a long-term vision to establish world-class, commercial-scale GTL facilities. Shell has developed several generations of the FTS fixed-bed reactors that are currently a part of the largest GTL plant in the world, the Pearl GTL Plant in Ras Laffan, Qatar. However, Sasol has developed the slurry bubble column FTS reactors, which is part of several GTL plants including their Oryx GTL plant in Qatar. These unique large-scale GTL plants lead Qatar to be described as the "world capital of GTL."

However, despite this there are currently only three main reactor designs used in commercial-scale FTS plants: fixed bed, fluidized bed, and slurry. Each existing technology has its strengths and weaknesses, and thus, the selection of a particular reactor design represents a trade-off, e.g., fixed-bed reactors (which operate in the gas phase with the formation of heavy waxes leading to a trickle-bed regime) have the advantage of lower capital costs, less system complexity, and easier catalyst recovery, but have challenges of temperature control and catalyst deactivation. On the other hand, slurry reactors (which consist of gaseous reactants bubbled through a liquid slurry phase) have the advantage of greater control over reaction heat and prevention of runaway; however, catalyst recovery, high recycle rates, and low per pass conversions remain a challenge.

FTS is an extremely exothermic process, which is an important characteristic that influences the efficiency of the overall system (Elbashir and Roberts 2005). Consequently, the process development and reactor design were mainly focused on temperature control and effective heat removal (Spath and Dayton 2003). Insufficient heat removal leads to short catalyst lifetime or catalyst deactivation, low conversion, high methane selectivity, and low chain growth probability to produce light hydrocarbon fractions (Huang and Roberts 2003; Mogalicherla and Elbashir 2011). The three types of FTS reactor designs used currently in commercial scale are: (1) multitubular fixed-bed reactor, (2) slurry-bubble column reactor, (3) fluidized-bed reactor (fluidized-bed, circulating, or bubbling) (see Figure 11.44). Comparing these reactor types, FTS that has been commercially operated using fixed-bed reactors (i.e., gas phase) provides unique reactant diffusivity and a high rate of reaction. However, the fixed-bed reactor is subjected to local overheating of the catalyst surface that may lead to deactivation of the catalyst active sites and also to favoring light

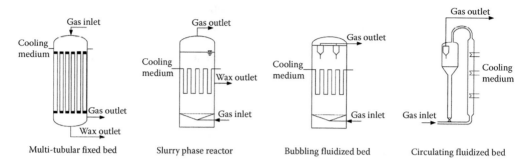

FIGURE 11.44 Commercial FTS reactors. (From Davis, B. H., *Catalysis Today* 71 (3–4):249–300, 2002.)

TABLE 11.10 Comparison of Advantages and Drawbacks for Fixed-Bed Reactor and Slurry-Bubble Reactor

	Fixed-Bed Reactor	Slurry-Bubble Column Reactor
Temperature control	−	+
Product–catalyst separation	+	−
Pressure drop	−	+
Catalyst make-up	−	+
Scale-up	+	−

Source: Sie, S. T. and R. Krishna, *Reviews in Chemical Engineering* 14:203–352, 1998.

hydrocarbon production (i.e., poor temperature control enhances the termination of growing chains and the methanation reaction route) (Fan and Fujimoto 1999). Furthermore, the heavy wax formation inside catalyst pores would limit the accessibility of the reactants (i.e., CO and H_2) to the micropores due to mass transfer and diffusional limitations (Fan et al. 1992). The slurry-bubble column reactor is composed of fine catalyst particles suspended in heavy paraffinic slurry at high boiling point. This technology has been developed to overcome the limitation of the gas-phase FTS in fixed-bed reactor as the liquid phase provides an optimum medium for the highly exothermic reaction due to its heat capacity and density (Fan et al. 1995). Besides its excellent temperature control, this medium facilitates the in situ extraction of heavy liquid hydrocarbons in addition to other advantages (Kölbel and Ralek 1980; Abbaslou et al. 2009). However, the diffusion of the reactants into the catalyst pores is relatively slow in the slurry-phase FTS, such that the overall rate of reaction is considerably lower than that in the gas-phase FTS. Other disadvantages of the slurry-phase FTS are low productivity, low catalyst hold-up, and difficult catalyst separation from the heavy products (Stern et al. 1983; Fan et al. 1992; Mogalicherla et al. 2012) (Table 11.10).

The previous challenges of the commercial FTS reactors directed research efforts toward the application of a reaction media that provided the advantages of both the gas phase (i.e., fixed-bed reactor) and the liquid phase (i.e., slurry-bubble column reactor), while at the same time overcoming their limitations. More importantly, this technology was developed to mitigate the weaknesses of the gas-phase FTS and, at the same time, allow the use of fixed-bed reactor. The supercritical fluid solvents have been suggested as a suitable media for FTS due to the desirable advantage of the existing gas-phase transport properties and the liquid-phase heat capacity and solubility while sustaining a single-phase operation where mass transfer barrier is eliminated (Lang et al. 1995; Elbashir et al. 2010).

11.5.2 Modeling Fischer–Tropsch Reactors

The material in this chapter is part of a study that is focused on advancing the design of GTL processes, specifically those related to the FTS reactors, the heart of the GTL technology. The studies completed in this project cover multiscale investigations that are aimed at better understanding of the mechanism of the reaction, investigating the potentials of the unique characteristics of the supercritical fluids to enhance both the activity and the selectivity of the reaction, and finally to design a novel reactor technology of potentials to provide selective control of the hydrocarbon product distribution toward ultraclean fuels and value-added chemicals.

Texas A&M Qatar takes the lead in forming a global consortium to advance FTS processes and reactor design, whereby this report presents a few examples of these microscale and mesoscale studies aimed at developing models for the kinetics, the thermodynamics and phase behavior (via experimental and modeling studies), and the intraparticle catalyst effectiveness performance. On the other hand, the macroscale investigations covered: (1) identifying the overall (heat/mass/hydrodynamic) profile inside the reactor, (2) selecting an appropriate supercritical solvent, and (3) building a lab-scale reactor unit. The outcome of these studies is that we were able to identify the most applicable solvent(s) while providing an opportunity to develop a detailed techno-economic and safety evaluation of this process.

The data generated from these studies have been used to design a novel FTS reactor technology. A bench scale of this unit has been commissioned at Texas A&M Qatar building and proved to produce typical hydrocarbon products under different operating conditions. This unit will be used to explore the potential of upgrading this technology to a pilot scale and later to a commercial scale.

11.5.2.1 Modeling Supercritical Phase Fischer–Tropsch Reactor

The unique properties of the supercritical fluid (SCF)-FTS reaction media, e.g., liquid-like density and heat capacity coupled with gas-like diffusivity, have resulted in many improvements in terms of product selectivity and catalyst activity (Huang and Roberts 2003). However, limited efforts have been devoted to quantify these improvements either with the bulk fluid (i.e., macroscale) or inside confined catalyst pores (i.e., microscale).

There are several unknowns that need to be addressed to move this technology from lab scale to commercial scale; the following are examples of these questions:

- To what degree could the supercritical fluids media impact the performance of FTS reactions related to the conventional fixed-bed reactor?
- How will we be able to quantify the role of this media to better understand the in situ behavior of the FTS reactor bed?
- What types of modeling tools could help us to quantify the possible improvements in supercritical fluids FTS?
- Is it possible to model the reactor bed in this nonideal reaction media while simultaneously investigating the micro and the macroscale behavior of the reactor bed?
- Will these models provide knowledge about experimentally observed phenomena in SCF-FTS, such as enhancement in the in situ mass and heat transfer processes?

The focus of this study is to utilize modeling techniques to simulate and predict the performance of a fixed-bed reactor under SCF-FTS reaction condition mainly to

1. Develop a mathematical model to simulate the concentration and temperature profiles inside the catalyst pellet ("microlevel assessment") under both SCF- and gas-phase FTS.
2. Structure the required model equations that predict the heat and mass transfer behavior inside the reactor bed ("macrolevel assessment") under the gas-phase and SCF reaction conditions utilizing the catalyst effectiveness factor estimated from microlevel assessment analysis.
3. To investigate the role of the main controlling parameters, such as operating conditions (i.e., pressure and temperature), reaction media (i.e., gas phase and SCF), and catalyst pellet size.
4. To study the effect of the reaction media and particle size on the overall catalyst effectiveness factor.

11.5.2.2 Kinetic and Rate Expressions of the Fischer–Tropsch Reaction

Several kinetic studies of the consumption of syngas on cobalt- and iron-based catalysts have been reported in the literature (see Table 11.11). Bub et al. (1980) and Lox and Froment (1993) fitted an empirical

TABLE 11.11   Power-Law and LHHW Rate Expressions for FTS

Rate Expression	Catalyst	References
$-r_{CO} = k\ P_{H_2}^m\ P_{CO}^n$	Fe	Bub et al. (1980), Lox and Froment (1993)
$-r_{CO} = \dfrac{kP_{CO}^{1/2}P_{H_2}^{1/2}}{\left(1+k_1 P_{H_2}^{1/2}+k_2 P_{CO}^{1/2}+k_3 P_{CO}\right)^2}$	Co/kieselguhr	Sarup and Wojciechowski (1989)
$-r_{CO} = \dfrac{kP_{H_2}P_{CO}}{\left(1+k_1 P_{CO}\right)^2}$	Co/MgO/SiO$_2$	Yates and Satterfield (1991)
$-r_{CO} = \dfrac{kP_{H_2}^{1/2}P_{CO}}{\left(1+k_1 P_{H_2}^{1/2}+k_2 P_{CO}\right)^2}$	Co/kieselguhr	Wojciechowski (1988)
$-r_{CO+H_2} = \dfrac{kP_{H_2}P_{CO}^{1/2}}{\left(1+k_1 P_{CO}^{1/2}\right)^3}$	Co/Al$_2$O$_3$	Rautavuoma and Vanderbaan (1981)

power-law rate expression for the reaction rate of carbon monoxide. While Sarup and Wojciechowski (1989) and others developed the Langmuir–Hinshelwood–Hougen–Watson (LHHW) rate expression for Co-based catalyst. A detailed literature review of FTS kinetics is given by Van Der Laan et al. (1999).

11.5.2.3 Modeling of FTS in a Fixed-Bed Reactor

The simulation of a fixed-bed FTS reactor has been done in several previous studies using different models, such as one- and two-dimensional models. Wang et al. (2003) have used a one-dimensional heterogeneous model to investigate the performance of FTS using fixed-bed reactor. While Jess and Kern (2009) developed a two-dimensional pseudohomogeneous model for FTS in a multitubular fixed-bed reactor. The following Table 11.12 shows a summary of related modeling studies that have been published.

TABLE 11.12   Summary of Related Modeling Studies

Paper Title and Year	Research Work and Main Findings	References
Steady state and dynamic behavior of fixed-bed catalytic reactor for FTS (1999)	• Simulate the fixed-bed FTS reactor packed with Co catalyst using 2-D heterogeneous model. • Consider the mass transfer, pore diffusion, momentum, and pressure drop. • Estimate the chemical and physical parameter from experiments or using equation of state.	Quan-Sheng et al. (1999)
Heterogeneous modeling for fixed-bed FTS reactor model and its applications (2003)	• Simulate the fixed-bed FTS reactor using 1-D heterogeneous reactor model. • Study the effect of tube diameter, recycle ratio, cooling temperature, and pressure on temperature profile.	Wang et al. (2003)
Modeling of multitubular reactors for FTS (2009)	• Model of fixed-bed FTS reactor packed with Fe and Co catalysts using 1-D and 2-D homogeneous model.	Jess and Kern (2009)
FTS in a fixed bed reactor (2011)	• Simulate the fixed-bed FTS reactor packed with Fe-based catalyst. • Study the effect of different process parameters and operating conditions (i.e., syngas feed ratio, pressure, reactor length) on product distribution.	Moutsoglou and Sunkara (2011)
A trickle fixed-bed recycle reactor model for the FTS (2012)	• Simulate the trickle fixed-bed FTS reactor using Co- and Fe-based catalyst. • Validate data by SASOL's Arge reactors.	Brunner et al. (2012)

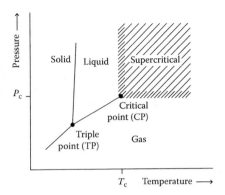

FIGURE 11.45 Definition of the supercritical state of pure components. (With kind permission from Springer Science+Business Media: Gas Extraction: An Introduction to Fundamentals of Supercritical Fluids and the Application to Separation Processes, 1994, Brunner, G.) With kind permission from **Springer Science+Business Media**: [book/journal title, chapter/article title, volume, year of publication, page, name(s) of author(s), figure number(s), and any original (first) copyright notice displayed with material.

TABLE 11.13 Magnitudes of Physical Properties of Gases, Liquids, and Supercritical Solvents

Physical Quantity	Gas	SCF	Liquid
Density (kg/m³)	10^0	10^2	10^3
Viscosity (Pa.s)	10^{-5}	10^{-4}	10^{-3}
Diffusivity (m²/s)	10^{-5}	10^{-7}	10^{-9}–10^{-10}

11.5.2.4 Supercritical Fluids

Supercritical fluids are substances in a thermodynamic state where their pressure and temperature are higher than the critical pressure (P_C) and temperature (T_C) (see Figure 11.45) (Elbashir et al. 2010). At the supercritical condition, the fluid exists as a single phase having unique physical properties (e.g., diffusivity and heat capacity) that are in between those of the gas and liquid phase (Baiker 1999).

Table 11.13 shows the magnitude of supercritical solvents properties compared to the gas and liquid phase. The density of the supercritical fluids is liquid like, while the viscosity and the diffusivity are more gas like.

The utilization of supercritical fluids as media for a chemical reaction provides several advantages for catalytic reactions as shown in the following (Moutsoglou and Sunkara 2011):

1. Improve the diffusion-control liquid phase reaction by eliminating the gas/liquid and liquid/liquid resistance due to the gas-like diffusivity.
2. The reaction environment can be continuously adjusted by a small change in pressure and/or temperature to enhance the reactants and product solubility and to eliminate mass transport resistance.
3. Easy access to the catalyst pores to extract the nonvolatile substances due to the low surface tension of SCF.
4. Increase the catalyst lifetime because nonvolatile substances can be dissolved in the SCF due to liquid-like density.
5. Enhance the mass transfer because of the high diffusivity and the low viscosity and the heat transfer due to the higher thermal conductivity of SCF than the corresponding gas phase.

11.5.2.5 SCF-FTS

A number of papers have attempted to study the reaction performance of the SCF-FTS. Yokota and Fujimoto (1991), the pioneer of utilizing supercritical solvents in FTS, have investigated the SCF-FTS using *n*-hexane as a solvent. They also compared the reaction performance in the gas phase, liquid phase and SCF

TABLE 11.14 Summary of the Main Related Research Work

Paper Title and Year	Operating Condition	Research Work and Main Findings	Ref.
Supercritical phase FTS (1990)	$T = 240°C$, $P = 45$ bar $CO/H_2 = 1/2$ Solvent/syngas = 3.5 Type of solvent: n-hexane	• The overall rate of the reaction in the supercritical phase reaction was lower than the gas-phase reaction. • The diffusion of the reactants was also lower in the gas phase compared to the supercritical phase reaction conditions. • Effective removal of heat generated through the exothermic reaction and heavy waxy products from the catalyst pellet in the SCF than that in the gas-phase reaction.	Yokota et al. (1990)
Supercritical phase FTS reaction 3. Extraction capability of supercritical fluids (1991)	$T = 240°C$, $P = 45$ bar $CO/H_2 = 1/2$ Solvent/syngas = 3.5 Type of solvent: n-hexane, n-heptane, benzene and methanol	• Heavy wax was effectively extracted from the catalyst bed. • Using n-hexane as supercritical solvent gave the highest rate of reaction and highest extraction capability. • High olefin content in the hydrocarbon for SCF reaction. • CO_2 selectivity was lower in the SCF-FTS.	Yokota et al. (1991)
Supercritical phase FTS: Catalyst pore-size effect (1992)	$T = 240°C$, $P = 45$ bar $CO/H_2 = 1/2$ Solvent/syngas = 3.5 Type of solvent: n-hexane	• For the large pore size, the proportion of the heavy hydrocarbon was high. • While for the small pore size catalyst tended to produce lighter hydrocarbons.	Fan et al. (1992)
Enhanced incorporation of α-olefins in the FTS chain-growth process over an alumina-supported Co catalyst in near-critical and supercritical hexane media (2005)	$T = (230–260)°C$, $P = (30–80)$ bar Type of solvent: n-hexane	• They study the product distribution in the SCF-FTS using Co-based catalyst in fixed-bed reactor. • They also measure the critical point of n-Hexane, syngas and products using variable-volume view cell apparatus. • Significant deviation of hydrocarbon distribution from the standard Anderson–Schultz–Flory model.	Elbashir and Roberts (2005)

(Continued)

TABLE 11.14 (*Continued*) Summary of the Main Related Research Work

Paper Title and Year	Operating Condition	Research Work and Main Findings	Ref.
Development of a kinetic model for supercritical fluids FTS (2011)	$T = (230-250)°C$, $P = (35-79)$ bar $CO/H_2 = 1/2$ Solvent/syngas = 3 Type of solvent: *n*-hexane	• Derived fugacity-based kinetic models to account for the nonideal reaction behavior in the gas-phase media and SCF using Co-based catalyst.	Mogalicherla and Elbashir (2011)
Selective FTS over an Al_2O_3 supported Co catalyst in supercritical hexane (2003)	$T = 250°C$, $P = (35, 41, 65$ and 80) bar $CO/H_2 = 1/2$ Type of solvent: *n*-hexane	• The optimum operating conditions to maximize the conversion of carbon monoxide and olefin selectivity are $T = 250°C$ and $P = 65$ bar. • The catalyst bed temperature was well controlled and in SCF-FTS compared to the gas-phase reaction.	Huang and Roberts (2003)
Effect of process conditions on olefin selectivity during conventional and supercritical FTS (1997)	$T = 250°C$, $P = 55$ bar $CO/H_2 = 1/2$ Type of solvent: *n*-propane	• SCF-FTS is attractive for producing a high molecular weight ∝-olefins. • Total olefin content decreases with increasing syngas molar feed ratio. • Olefin selectivity was independent of reaction temperature. • Total olefin content was greater during SCF-FTS. • High diffusivities and desorption rates of ∝-olefins in the SCF-FTS than the liquid hydrocarbon wax produced in the gas-phase reaction.	Bukur et al. (1997)
Impact of Co-based catalyst characteristics on the performance of conventional gas phase and supercritical-phase FTS (2005)	$T = (230-250)°C$, $P = (20-65)$ bar $CO/H_2 = 1/2$ Type of solvent: *n*-hexane	• SCF-FTS minimize methane selectivity even at high syngas conversions. • While in the gas-phase reaction methane selectivity increases as syngas conversion increases. • The selectivity of CO_2 was lower in the SCF-FTS.	Elbashir et al. (2005)

using fixed-bed reactor. These efforts were followed by several researchers to study the effect of supercritical solvents on FTS. Table 11.14 provides the main studies published in open literature in the area of SCF-FTS.

11.5.3 Reactor Modeling Structure

11.5.3.1 Microscale Model

In this study, the microscale model was developed to simulate the diffusion and reaction in a spherical catalyst pellet (Figure 11.46) under both SCF and gas-phase reaction conditions. The steps involved in modeling a chemical system with diffusion and reaction starts with defining the system and all relevant assumptions, writing mole balance in terms of molar flux on a specific species, using Fick's first law for mass transfer to obtain a second-order differential equation in terms of concentration, stating all relevant assumptions, and then solving the resulting differential equation to obtain the concentration profile. The heat balance equation was also performed in the same manner using Fourier's Law.

11.5.3.1.1 Model Main Assumptions

The modeling of FTS in a fixed-bed reactor is a complex task due to the abundant factors to be taken into account in order to obtain a realistic model. The following are the model assumptions:

1. Steady state conditions
2. One-dimensional model in the radial coordinate
3. Spherical catalyst pellet with radius R
4. Pores are filled with supercritical n-hexane in the SCF-FTS and heavy wax (n-$C_{28}H_{58}$) in the gas-phase FTS
5. Single phase operation under SCF reaction condition

11.5.3.1.2 Mass and Heat Balances

The mass balance equation for a spherical catalyst pellet assuming steady state conditions can be expressed by the following second-order differential equation that describes the diffusion and reaction:

$$\frac{D_i^e}{r^2} \frac{d}{dr}\left(r^2 \frac{dC_i}{dr}\right) = (-r_i)\rho_{cat}, \tag{11.16}$$

where i is the species in the reaction mixture, r is the radius of the pellet, r_i is the rate of formation of compound i (mol/g.s), ρ_{cat} is the catalyst density, and D_i^e is the effective diffusivity of i in catalyst pore (cm/s).

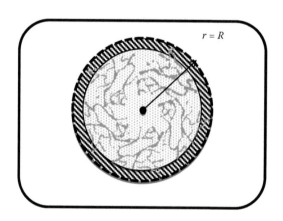

FIGURE 11.46 Spherical catalyst pellet.

The corresponding mass balance for the reactants CO and H_2 are (Fogler 2006)

$$\frac{D_{CO}^e}{r^2} \frac{d}{dr}\left(r^2 \frac{dC_{CO}}{dr}\right) = (-r_{CO})\rho_{cat} \tag{11.17}$$

$$\frac{D_{H_2}^e}{r^2} \frac{d}{dr}\left(r^2 \frac{dC_{H_2}}{dr}\right) = (-r_{H_2})\rho_{cat} \tag{11.18}$$

It is important to mention here that the mass balance equation was derived in terms of fugacity to account for the nonideal reaction mixture in the SCF reaction. The thermodynamic properties of the mixture were calculated using the SRK-EOS.

The corresponding heat balance equation was derived in the same manner as follows:

$$\frac{\lambda^e}{r^2} \frac{d}{dr}\left(r^2 \frac{dT}{dr}\right) = (-r_i)\rho_{cat}(\Delta H)_r, \tag{11.19}$$

where i is the species in the reaction mixture, r is the radius of the pellet, r_i is the rate of formation of compound i (mol/g·s), ρ_{cat} is the catalyst density, λ^e is the pellet effective thermal conductivity (cal/s cm k), and ΔH_r is the heat of reaction (cal/mol CO).

11.5.3.1.3 Boundary Conditions

Two sets of boundary conditions are considered in the present study (Fogler 2006):

Case 1: Catalyst particles have no external mass transfer limitation, where the concentration at the entrance of the catalyst pore is equal to the concentration in the bulk solution and the concentration remains finite at the center of the catalyst pellet.

$$C_{CO} = C_{CO}^s = C_{CO}^b \quad \text{at} \quad r = R_{cat}$$

$$C_{H_2} = C_{H_2}^s = C_{H_2}^b \quad \text{at} \quad r = R_{cat}$$

$$T = T^s = T^b \quad \text{at} \quad r = R_{cat}$$

$$\frac{dC_{CO}}{dr} = 0 \quad \text{at} \quad r = 0$$

$$\frac{dC_{H_2}}{dr} = 0 \quad \text{at} \quad r = 0$$

$$\frac{dT}{dr} = 0 \quad \text{at} \quad r = 0$$

where R_{cat} is the catalyst radius, $C_{CO}^s, C_{H_2}^s$, and T^s are the concentrations of CO and H_2 and the temperature at the surface of the catalyst pellet, $C_{CO}^b, C_{H_2}^b$, and T^b are the concentrations of CO and H_2 and the temperature at the bulk solution (Figure 11.47).

Case 2: Catalyst particles with external mass transfer limitation, where the molar flux to the boundary layer is equal to the convective mass transport across the boundary layer thickness, and the concentration remains finite at the center of the pellet.

$$D_{CO}^e \frac{dC_{CO}}{dr} = k_m\left(C_{CO}^b - C_{CO}^s\right) \quad \text{at} \quad r = R_{cat}$$

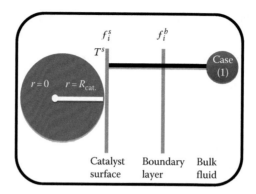

FIGURE 11.47 Graphical representation of the boundary condition of the system represented in Case 1.

$$D_{H_2}^e \frac{dC_{H_2}}{dr} = k_m \left(C_{H_2}^b - C_{H_2}^s \right) \quad \text{at } r = R_{cat}$$

$$\lambda^e \frac{dT}{dr} = h \left(T^b - T^s \right) \quad \text{at } r = R_{cat}$$

$$\frac{dC_{CO}}{dr} = 0 \quad \text{at } r = 0$$

$$\frac{dC_{H_2}}{dr} = 0 \quad \text{at } r = 0$$

$$\frac{dT}{dr} = 0 \quad \text{at } r = 0,$$

where k_m is the internal mass transfer coefficient and h is the internal heat transfer coefficient (Figure 11.48).

11.5.3.1.4 Kinetics

In the present study, a LHHW kinetic model was used to express the reaction behavior in both the conventional gas-phase FTS and the nonconventional SCF-FTS. The kinetic model was developed from the experimental data for an alumina-supported cobalt catalyst (15% Co/Al$_2$O$_3$) in a fixed-bed reactor.

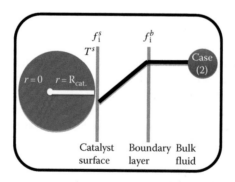

FIGURE 11.48 Graphical representation of the boundary condition of the system represented in Case 2.

TABLE 11.15 Kinetic Parameters for SCF and Gas Phase at $T = 513\,K$

Kinetic Parameter	SCF-FTS	Gas-Phase FTS
K (mol/g$_{cat}$·min·bar)	$4\times10^{-4}\exp\left[-12496.99\left(\dfrac{1}{T}-\dfrac{1}{513}\right)\right]$	$3\times10^{-4}\exp\left[-8960.78\left(\dfrac{1}{T}-\dfrac{1}{513}\right)\right]$
K_1 (bar$^{-0.5}$)	$16.9\times10^{-2}\exp\left[6025.98\left(\dfrac{1}{T}-\dfrac{1}{513}\right)\right]$	$24\times10^{-2}\exp\left[7072\left(\dfrac{1}{T}-\dfrac{1}{513}\right)\right]$
K_2 (bar$^{-0.5}$)	$20\times10^{-2}\exp\left[-17981.77\left(\dfrac{1}{T}-\dfrac{1}{513}\right)\right]$	$16\times10^{-2}\exp\left[-17007.45\left(\dfrac{1}{T}-\dfrac{1}{513}\right)\right]$
K_3 (bar^{-1})	$2\times10^{-4}\exp\left[8996.88\left(\dfrac{1}{T}-\dfrac{1}{513}\right)\right]$	$1\times10^{-4}\exp\left[8995.3\left(\dfrac{1}{T}-\dfrac{1}{513}\right)\right]$

Source: Mogalicherla, A. K. and N. O. Elbashir, *Energy and Fuels* 25 (3):878–889, 2011.

Also the model was derived in terms of fugacity to account for the nonideal behavior reaction mixture under high pressure.

$$-r_{CO} = \frac{K f_{CO}^{1/2} f_{H_2}^{1/2}}{\left(1 + K_1 f_{H_2}^{1/2} + K_2 f_{CO}^{1/2} + K_3 f_{CO}\right)^2}, \tag{11.20}$$

where f_{CO} and f_{H_2} are the fugacities of CO and H$_2$, respectively (bar), r_{CO} is the CO consumption rate (mol/g$_{cat}$ min), K, K_1, K_2, and K_3 are the kinetics parameters.

The kinetic parameters for this model were reported by Mogalicherla and Elbashir (2011) at $T = 513\,K$ for the near critical and SCF-FTS reaction and gas-phase FTS reaction. The following Table 11.15 shows the temperature dependence of the kinetic constants expressed by the Arrhenius equation.

11.5.3.1.5 Operating Conditions and Catalyst Physical Properties

The operating conditions and catalyst physical properties used for simulation are given in the following Table 11.16.

11.5.3.1.6 Effective Diffusion Parameters

11.5.3.1.6.1 For SCF Reaction: Catalyst Pores Filled with Supercritical n-Hexane
Catalyst pores have different cross-sectional areas and the paths are tortuous. It will be hard to describe the diffusion inside each tortuous pathway. Accordingly, the effective diffusivity is used to account for the average diffusion taking place at any position inside the catalyst pellet. The following equation is used to calculate the

TABLE 11.16 Operating Conditions and Catalyst Properties

Temperatures (T)	513 K
Total pressure (P)	80 bar
H$_2$/CO feed ratio (V)	2
n-Hexane/syngas ratio (S)	3
Catalyst type	Co/Al$_2$O$_3$
Pellet shape	Spherical
Pellet diameter (d$_{Pellet}$)	1 mm
Pellet porosity (ε_p)	0.5
Pellet density (ρ_{cat})	1.5 g/cm^3
Pellet tortuosity (τ)	3

effective diffusivity using the binary diffusion coefficients, catalyst porosity (which is the volume of the void divided by the total volume), constriction factor (which accounts for a different cross-sectional area), and tortuosity (which is the actual distance the molecule travels divided by the shortest distance) (Fogler 2006):

$$D_i^e = \frac{D_{21}\ \varepsilon_P\ \sigma_C}{\tau}, \tag{11.21}$$

where D_{21} is the binary diffusion coefficient of solute (2) in solvent (1), ε_P is the catalyst porosity, σ_C is the constriction factor, and τ is the catalyst tortuosity.

The ability to predict the binary diffusion coefficients in SCF is considerably important to the design and efficient operation of SCF-FTS. In this research work, binary diffusion coefficients were estimated using the correlation proposed by He (1997). This correlation determines the binary diffusion coefficient of liquid and solid solutes in supercritical solvents and it was tested for more than 107 solute–solvent systems including *n*-hexane. The correlation required solvent properties (i.e., critical pressure, critical volume, molecular weight, and density), solute properties (i.e., molecular weight), and system temperature is shown below:

$$D_{21} = \left[0.61614 + 3.0902 \times \exp\left(-0.87756\frac{\sqrt{M_1 V_{C_1}}}{P_{C1}}\right) \right] \times 10^{-6} \left(V_1^k - 23\right) \times \sqrt{\frac{T}{M_2}}, \tag{11.22}$$

where the subscript 1 and 2 refers to solvent and solute, respectively, M is the molecular weight (g/mol), V_C is the critical molar volume (cm³/mol), P_C is the critical pressure (bar), V is the molar volume of the solvent (cm³/mol), T is the temperature (K), and the parameter k is a function of solvent reduced density (ρ_r) as the following:

$$k = 1 \quad \text{for} \quad \rho_r > 1.2$$

$$k = 1 + \frac{(\rho_r - 1.2)}{\sqrt{M_1}} \quad \text{for} \quad \rho_r < 1.2$$

In this study, the correlation was used to calculate the diffusivity of reactants (CO and H₂) in supercritical *n*-hexane, which was used as a solvent in several SCF-FTS research studies.

11.5.3.1.6.2 For Gas-Phase Reaction: Catalyst Pores Filled with Heavy Wax
The effective diffusivity for the gas-phase reaction is calculated in the same manner using Equation 11.22 except for the binary diffusion coefficient. In SCF-FTS reaction, the binary diffusion coefficient was calculated assuming that the reactants diffuse through the solvent (*n*-hexane), while in the gas-phase FTS reaction, it is calculated assuming that the reactant diffuses through the heavy wax.

The binary diffusion coefficient for the case when the catalyst pores are filled with liquid wax was calculated using the following correlation proposed by Yong Wang et al. (2001). These correlations assume that the liquid wax is *n*-C₂₈H₅₈ and it was simply derived by fitting the reported diffusivity of CO and H₂ is *n*-C₂₈H₅₈, as shown below in Equations 11.23 and 11.24:

$$D_{CO,Wax} = 5.584 \times 10^{-7} \exp\left(\frac{-1786.29}{T}\right) \left(m^2/s\right) \tag{11.23}$$

$$D_{H_2,Wax} = 1.085 \times 10^{-6} \exp\left(\frac{-1624.63}{T}\right) \left(m^2/s\right) \tag{11.24}$$

11.5.3.1.7 Effective Thermal Conductivity

11.5.3.1.7.1 For SCF Reaction: Catalyst Pores Filled with Supercritical n-Hexane   The effective thermal conductivity of porous catalyst plays a significant role in determining the temperature gradient inside the catalyst pellet, especially for highly exothermic reactions. The following relationship was used to predict an approximation for the effective thermal conductivity as a function of the pellet porosity and the thermal conductivity of both the bulk fluid and the catalyst pellet (Hill 1977):

$$\lambda^e = \lambda_{\text{Pellet}} \left(\frac{\lambda_{\text{fluid}}}{\lambda_{\text{Pellet}}} \right)^{\varepsilon_P}, \tag{11.25}$$

where λ_{Pellet} is the catalyst pellet thermal conductivity (W/m.K) and λ_{fluid} is the bulk fluid (*n*-hexane) thermal conductivity (W/m·K).

In this work, the thermal conductivity of catalyst pellet was used based on the correlation developed by Wu et al. (2010). This correlation was developed for a Co-based catalyst for FTS over the temperature range from 160°C to 255°C. Also, it was derived by fitting the catalyst thermal conductivity into a linear relationship with temperature as follows:

$$\lambda_{\text{Pellet}} = a + bT \quad \left(\text{W/m} \cdot {}^\circ\text{C} \right), \tag{11.26}$$

where the constants *a* and *b* were calculated by linear regression from the experimental data, $a = 0.8652$, $b = 0.00108$.

Equation 11.26 can be rewritten as follows:

$$\lambda_{\text{Pellet}} = 0.8652 + 0.00108 \left(T - 273.15 \right) \quad \left(\text{W/m} \cdot \text{K} \right) \tag{11.27}$$

Calculating the thermal conductivity for the bulk fluid (i.e., *n*-hexane) at supercritical phase using the available correlation is a complex task (Arai et al. 2001). Near the critical point, the liquid solvent behaves somewhat like a dense gas and the thermal conductivity varies significantly with a small change in the pressure or temperatures. The thermal conductivity of *n*-hexane near the critical point can be estimated based on dense gas thermal conductivity correlations developed by Stiel and Thodos as follows (Poling et al. 2001).

In the present research work, the thermal conductivity of *n*-hexane as a function of temperature was obtained using Aspen Plus simulation package in the near critical and supercritical region. Aspen Plus physical properties system was used to calculate *n*-hexane thermal conductivity utilizing SRK-EOS at different temperature and pressure to fit the following polynomial equation for the sake of simulation.

$$\lambda_{\text{fluid}} = a \left(\frac{P}{65} \right)^b \left(\frac{T}{235} \right)^C, \tag{11.28}$$

where the parameter *a*, *b*, and *c* are constants calculated by linear regression using Aspen data, $a = 0.000142$, $b = 2.017551$, and $c = 0.218024$.

Equation 11.28 can be represented as follows:

$$\lambda_{\text{fluid}} = 0.000142 \left(\frac{P}{65} \right)^{2.017551} \left(\frac{T}{235} \right)^{0.218024} \tag{11.29}$$

11.5.3.1.7.2 For Gas-Phase Reaction: Catalyst Pores Filled with Heavy Wax The effective thermal conductivity for the gas phase (i.e., reactants: CO, H_2, and inert N_2) was estimated by two steps (Poling et al. 2001):

1. Estimate the thermal conductivity of the gas mixture at low pressure using Stiel and Thodos correlations (Poling et al. 2001).
2. To account for the influence of high pressure in the system, the thermal conductivity was estimated using Brokaw's empirical method (Poling et al. 2001).

Stiel and Thodos (Poling et al. 2001) correlations (Equation 11.30 through 11.32) are generally used to calculate the thermal conductivity at low pressure knowing the reduced density of the gas mixture.

$$\left(\lambda_m - \lambda_m^\circ\right)\Gamma z_c^5 = 14.0 \times 10^{-8}\left[\exp\left(0.535\,\rho_r\right) - 1\right] \quad \text{for } \rho_r < 0.5 \tag{11.30}$$

$$\left(\lambda_m - \lambda_m^\circ\right)\Gamma z_c^5 = 13.1 \times 10^{-8}\left[\exp\left(0.67\,\rho_r\right) - 1.069\right] \quad \text{for } 0.5 < \rho_r < 2.0 \tag{11.31}$$

$$\left(\lambda_m - \lambda_m^\circ\right)\Gamma z_c^5 = 2.976 \times 10^{-8}\left[\exp\left(1.155\,\rho_r\right) + 2.016\right] \quad \text{for } 2.0 < \rho_r < 2.8, \tag{11.32}$$

where λ_m is the thermal conductivity of the gas mixture (W/m·K), λ_m° is the thermal conductivity calculated from pure component thermal conductivity (W/m·K), z_C is the critical compressibility factor, and Γ is the reduced inverse thermal conductivity (W/m·K)$^{-1}$.

Then, the thermal conductivity for the gas mixture at high pressure assuming two component mixtures (syngas and N_2) was calculated using Brokaw's empirical method (Poling et al. 2001).

$$k_m = q k_{mL} + \left(1 - q\right) k_{mR} \tag{11.33}$$

$$k_{mL} = y_1\lambda_1 + y_2\lambda_2 \quad \text{and} \quad \frac{1}{k_{mR}} = \frac{y_1}{\lambda_1} + \frac{y_2}{\lambda_2}, \tag{11.34}$$

where k_m is the thermal conductivity of the gas mixture at high pressure, q is the Brokaw parameter, y_1 is the mole fraction of the light component, y_2 is the mole fraction of the heavy component, λ_1 is the thermal conductivity of component 1 (syngas) at low pressure, and λ_2 is the thermal conductivity of component 2 (N_2).

11.5.3.1.8 Effectiveness Factor Calculation

For the catalyst pellet simulation, the overall catalyst effectiveness factor is defined here as the ratio of the actual overall rate of reaction to the rate of reaction if the catalyst surface was exposed to the bulk conditions. The overall effectiveness factor can be calculated using the following equation (Fogler 2006):

$$\eta = \frac{3}{R_p^3} \frac{\displaystyle\int_0^{R_p} r_{CO} r^2 \, dr}{r_{CO,\,bulk}} \tag{11.35}$$

11.5.3.2 Macroscale Model

This section considers the FTS taking place in the backed bed of the catalyst pellets rather than one single pellet (i.e., zooming out from microscale to macroscale) to understand the reactor bed behavior in supercritical phase and gas-phase reaction (Figure 11.49).

The advantages of the supercritical-phase compared with the gas-phase reaction media have been proved experimentally by many researchers (Yokota et al. 1990; Huang and Roberts 2003; Mogalicherla et al. 2012). The purpose of macroscale modeling is to capture these advantages from the modeling results.

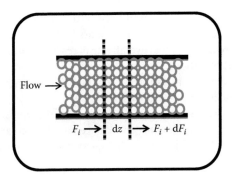

FIGURE 11.49 Illustration of fixed-bed reactor.

A one-dimensional heterogeneous mathematical model of fixed-bed reactor was developed in this work to obtain the concentration and temperature profiles in the reactor bed (see Figure 11.49). The steps involved in modeling an FTS reactor bed at the macroscale are the following:

1. Define the system boundaries
2. State all relevant assumptions
3. Write the reactor bed balance equations for mass, heat, and momentum and explain their physical significance
4. Define initial conditions and their physical interpretation
5. Calculate all necessary variables to solve the balance equations (e.g., superficial velocity, overall heat transfer coefficient, etc.)
6. Use a modeling tool to solve the balance equations
7. Validate the model outcomes with experimental data published in the literature

11.5.3.2.1 Model's Main Assumptions

A number of assumptions were made to simplify the complex phenomena of heat, mass, and momentum into a mathematical model. The main assumptions are

1. Steady-state conditions
2. One-dimensional plug-flow model in the axial direction
3. Constant superficial velocity in the axial direction
4. Catalyst pores are filled with n-hexane in the SCF-FTS and heavy wax (n-$C_{28}H_{58}$) in the gas-phase FTS
5. Single-phase operation under SCF reaction condition

11.5.3.2.2 Mass, Heat and Momentum Balances

The mass balance equation of a tube packed with a solid catalyst (system in Figure 11.49) was developed using a one-dimensional steady-state model in the axial direction as per the following first-order differential equation:

$$-u\frac{dC_i}{dz} = \eta_0 \rho_B (-v_i) r_i,$$ (11.36)

where u is the superficial velocity, C_i is the concentration of species i, η_0 is the overall effectiveness factor, ρ_B is the bed density (mass of catalyst/volume of bed), v_i is the stoichiometric coefficient of species i, and r_i is the rate of reaction over the solid catalyst (mole/mass of catalyst/time).

In Equation 11.36, the rate of reaction per unit mass of the catalyst is multiplied by the bed density to obtain the rate of reaction per unit volume in the mass balance equation. Additionally, the overall effectiveness factor is used to relate the actual overall rate of reaction within the catalyst pellet to the rate that would result in bulk fluid conditions.

The fugacity-based mass balance was estimated for the previous Equation 11.36 to account for the nonideal reaction mixture under high-pressure condition.

An energy balance equation was also developed to account for the temperature gradients for the fixed-bed reactor with heat exchange (i.e., heat is either added or removed) in the axial direction. The energy balance for the reactor bed used in this work is shown in Equation 11.37 (Davis and Davis 2002; Fogler 2006).

$$u_s \rho_g C_p \frac{dT}{dz} = (-\Delta H) \eta r_{CO} \rho_B - \frac{4U}{d_t}(T - T_r) \tag{11.37}$$

To account for the pressure drop through the porous backed bed, a common pressure drop equation was used ("Ergun equation") as per the following (Fogler 2006; Froment et al. 2010):

$$-\frac{dP}{dz} = f \frac{\rho_g u_s^2}{d_p} \tag{11.38}$$

For Equation 11.37, the bed friction factor f was calculated using Hicks's correlation for spherical particles (Froment et al. 2010) as follows:

$$f = 6.8 \frac{(1 - \varepsilon_p)^{1.2}}{\varepsilon_p^3} Re^{-0.2} \tag{11.39}$$

11.5.3.2.3 Initial Conditions

The initial conditions considered in the present study are based on the inlet conditions of the reactor bed entrance ($z = 0$). The inlet conditions used are (1) the inlet concentration, (2) the inlet temperature, and (3) the inlet pressure.

- For mass equation:

$$C_{CO} = C_{CO,o} \quad \text{at} \quad z = 0$$

- For heat equation:

$$T = T_{in} \quad \text{at} \quad z = 0$$

- For momentum equation:

$$P = P_{in} \quad \text{at} \quad z = 0$$

11.5.3.2.4 Kinetics

For the macrolevel assessment, the same rate expression was used (see Section 24.2.1.4) for CO consumption.

11.5.3.2.5 Operating Conditions and Fixed-Bed Reactor Properties

The operating conditions and the fixed-bed reactor properties used for simulation are given in the following Table 11.17. Figure 11.50 shows a schematic representation of the lab-scale fixed-bed reactor dimensions used in the simulation.

TABLE 11.17 Simulation Conditions Employed in the SCF- and Gas-Phase FTS

Temperatures, T	513 K
Total pressure, P	80 bar
Inlet flow rate (std.)	50 cm³/min
H_2/CO feed ratio	2
n-Hexane/syngas ratio, S	3
Tube length	40.64 cm
Bed length	5 cm
Tube internal diameter	1.57 cm
Wall thickness, d	0.8 cm

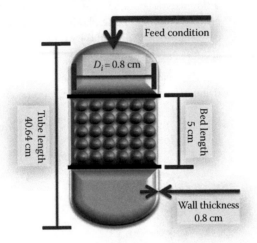

FIGURE 11.50 Schematic diagram of the lab scale fixed-bed reactor dimensions.

11.5.3.2.6 Overall Heat Transfer Calculation

The overall heat transfer coefficient in a fixed-bed reactor was estimated using the following equation (Froment et al. 2010):

$$\frac{1}{U} = \frac{1}{\alpha_i} + \frac{d}{\lambda_{wall}} \frac{A_b}{A_m} + \frac{1}{\alpha_u} \frac{A_b}{A_u}, \tag{11.40}$$

where α_i is the heat transfer coefficient on the bed side, α_u is the heat transfer medium side, λ_{wall} is the heat conductivity of the wall, A_b is the heat exchanging surface areas on the bed side, A_u is the heat transfer medium side, A_m is the log mean of A_b and A_u, and d is the reactor tube wall thickness.

The heat transfer coefficient on the bed side can be found using the following correlation proposed by de Wasch and Froment (1972):

$$\alpha_i = \alpha_i^0 + 0.033 \frac{\lambda_g}{d_p} Pr_g Re_g \tag{11.41}$$

$$\alpha_i^0 = \frac{10.21\lambda_e^0}{d_t^{4/3}}, \tag{11.42}$$

where α_i^0 and λ_e^0 are the static contribution and the static contribution to the effective thermal conductivity, respectively.

11.5.4 Numerical Solution: MATLAB® Implementation

As was mentioned previously, the aim of this study was to develop an appropriate technique to simultaneously evaluate the mass and heat transfer inside the catalyst pellet and reactor bed itself. This, however, considerably complicates the task of finding a numerical solution. MATLAB® is a powerful modeling tool that has the ability to solve a system of ODEs, either boundary-value problem (BVP) or initial-value problem (IVP). Two different cases were developed for the modeling task. The first case is the microlevel modeling in the form of ordinary differential Equations 11.18, 11.19, and 11.20 together with the boundary conditions, lead to a two point BVP. The resulting systems of heat and mass balance equations were solved simultaneously by utilizing bvp4c function from MATLAB®. This case focuses on the concentration and temperature distribution inside the catalyst pellet to study the effect of heat and mass transfer limitations. As a part of the microlevel modeling, the catalyst effectiveness factor was estimated using trapz function that computes an approximation for the integral through the trapezoidal method. While the second case is the macroscale modeling given by Equations 11.32, 11.33, and 11.34 together with the initial conditions lead to an IVP. The obtained equations were solved using ode45 function from MATLAB®. This case is mainly focusing on the overall behavior of the reactor bed itself. The steps involved in the modeling are as per the following:

1. Define the research problem and develop mathematical model for the system.
2. Formulate the mathematical equation and use mathematical simulator to compute the numerical solution.
3. Use the experimental data to verify the numerical results obtained from MATLAB® simulator (see Figure 11.51).

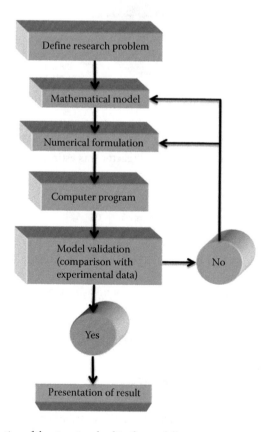

FIGURE 11.51 Representation of the steps involved in the modeling.

11.5.5 Models Results and Discussion

11.5.5.1 Introduction

The following sections present the outcomes of mass, energy, and momentum balances conducted in this study. The first section highlights the main results for the microscale assessment in terms of the concentration profile, temperature profile, and effectiveness factor. It will also investigate the role of the main controlling parameters such as pressure, pellet size, and reaction media. While the second section underlines the results of macroscale assessment that are considered as the principle contribution to the study, showing the impact of supercritical solvents on the temperature profile distribution. This section will also address the impact of the conversion on the overall catalyst effectiveness factor.

11.5.5.2 Microscale Modeling Results

11.5.5.2.1 Diffusivities of H_2 and CO

The calculated results for the binary diffusion coefficient and the effective diffusivity in the conventional gas-phase FTS reaction and the SCF-FTS reaction are listed in the following Table 11.18. The binary diffusivity of the reactants in the supercritical *n*-hexane was estimated using Equation 11.23. The diffusivity of CO was found to be four times less than that of the H_2, under the specified reaction conditions. While the binary diffusion coefficient in the heavy hydrocarbons under the gas-phase reaction conditions was obtained using Equations 11.24 and 11.25. According to the results presented in Table 11.18, it is obvious that the diffusivity of the H_2 and CO is much higher in the SCF-FTS compared to the gas-phase FTS. This is due to the complex mixture of hydrocarbons produced in a typical industrial FTS with fixed-bed reactors. Therefore, the catalyst pores are filled with liquid hydrocarbons, which in turn decrease the reactants accessibility (Madon and Iglesia 1994; Wang et al. 2001). A similar finding has been reported by Yan et al. (1998) at different reaction conditions using *n*-pentane as a supercritical solvent.

11.5.5.2.2 Concentration Profile

The simulated intrapellet concentration profiles for the reactants (CO + H_2) inside catalyst pellet both in the SCF-FTS and gas-phase FTS are shown in the following Figures 11.52 and 11.53. It is clear from the concentration profiles that the diffusivity of the syngas in the heavy waxy product under conventional gas-phase FTS is much slower than in the SCF-FTS. This leads to the significant profiles inside the catalyst pellet. Yan et al. (1998) presented similar findings using power rate law for F-T kinetics and slightly different reaction conditions. As can be noticed from Figure 11.52, CO concentration drops from 5.94 bar to 3.63 bar as it enters the mouth of the pore (when $r/R_P = 1$). It was also noticed that the catalyst pores are rich with H_2 along all the positions (from $r/R_P = 1$ to the center of the pellet $r/R_P = 0$). This is because of the very high effective diffusivity of H_2 relative to CO, even under supercritical condition.

It should be noted here that we simulated the SCF-FTS reactor utilizing experimental data reported in literature for a Co-based catalyst under both conventional gas-phase FTS and near critical and supercritical FTS (Mogalicherla et al. 2012).

TABLE 11.18   Binary Diffusion Coefficient and Effective Diffusivity

Supercritical Phase	Gas Phase
$D_{\text{CO-Hexane}} = 1.03 \times 10^{-3}$ cm^2/s	$D_{\text{CO-Wax}} = 1.72 \times 10^{-4}$ cm^2/s
$D_{\text{H}_2\text{-Hexane}} = 3.86 \times 10^{-3}$ cm^2/s	$D_{\text{H}_2\text{-Wax}} = 4.57 \times 10^{-4}$ cm^2/s
$D^e_{\text{CO-Hexane}} = 1.33 \times 10^{-4}$ cm^2/s	$D^e_{\text{CO-Wax}} = 2.86 \times 10^{-5}$ cm^2/s
$D^e_{\text{H}_2\text{-Hexane}} = 6.43 \times 10^{-4}$ cm^2/s	$D^e_{\text{H}_2\text{-Wax}} = 7.62 \times 10^{-5}$ cm^2/s

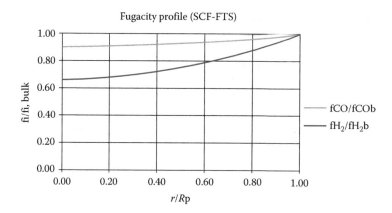

FIGURE 11.52 Dimensionless concentration profiles inside catalyst pellet under SCF-FTS (temperature: 513 K; pressure: 80 bar; syngas ratio (H$_2$/CO): 2:1; solvent/syngas ratio: 3; pellet diameter: 1 mm).

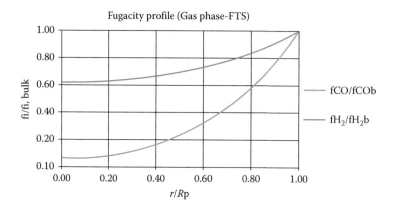

FIGURE 11.53 Dimensionless concentration profiles within catalyst pores under gas-phase FTS (temperature: 513 K; pressure: 80 bar; syngas ratio (H$_2$/CO): 2:1; nitrogen/syngas ratio: 3; pellet diameter: 1mm).

11.5.5.2.3 Pressure-Tuning Effect on SCF-FTS

The following are the concentration profiles inside the catalyst pore under SCF-FTS at different operating pressures (35, 65, and 80 bar) to investigate the effect of pressure tuning in the critical and near the critical phase and also to study the effect of diffusion on the performance of the catalyst pellet.

In Figures 11.54 and 11.55, as the total pressure increases from 35 to 80 bar, the system moves from the gas phase to the liquid phase and then to the SCF by simply tuning the operating pressure. These figures also show the influence of pressure on reactant conversion in the SCF-FTS. It was observed that the reactant conversion decreases with increasing the pressure, since the bulk diffusivity decreases when increasing the pressure. This means that at high pressure, external diffusion limitations control the process.

11.5.5.2.4 Particle Size Effect on SCF-FTS

A simulation of the effect of the catalyst particle size on the overall catalyst effectiveness factor for both reaction media is shown in the following Figure 11.56. In the SCF-FTS, as the diameter of the catalyst pellet increased from 1 to 5 mm, the overall catalyst effectiveness factor decreased from 0.99 to 0.86. While in the gas-phase FTS, the overall effectiveness factor sharply dropped from 0.96 to 0.75.

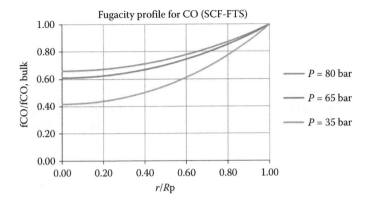

FIGURE 11.54 Carbon monoxide concentration profile inside the catalyst pores under the SCF-FTS conditions under different total pressures (temperature = 250°C, solvent/syngas ratio = 3 and H_2/CO ratio = 2).

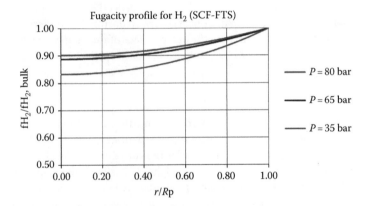

FIGURE 11.55 Hydrogen concentration profile inside the catalyst pores under the SCF-FTS conditions under different total pressures (temperature = 250°C, solvent/syngas ratio = 3 and H_2/CO ratio = 2).

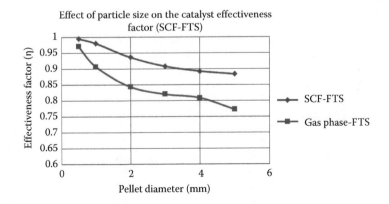

FIGURE 11.56 Modeling of the effect of the catalyst pellet size on the overall effectiveness factor.

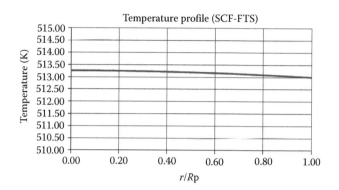

FIGURE 11.57 Temperature profiles within catalyst pores under SCF-FTS (temperature: 513 K; pressure: 80 bar; syngas ratio (H_2/CO): 2:1; solvent/syngas ratio: 3; pellet diameter: 1 mm).

11.5.5.2.5 Temperature Profile

The Figure 11.57 shows the temperature profile inside the catalyst pellet. As can be noticed, there was no significant increase in the temperature inside the catalyst pellet (i.e., the temperature difference was found to be less than 1°C). This indicates that the heat generated during the exothermic reaction is transferred by conduction from the catalyst pore to the outer surface of the catalyst and by convection from the outer catalyst surface to the bulk fluid (i.e., isothermal catalyst pellet).

11.5.5.3 Macroscale Modeling

11.5.5.3.1 Conversion Profile

Figure 11.58 and Table 11.19 present the effects of reaction media on the CO conversion level. Since the reactant mass diffusivity rates are higher in the gas phase than in the SCF-FTS, it might be expected that a higher CO conversion would be obtained under the conventional gas-phase reaction. However, the modeling results presented in Figure 11.58 show higher CO conversions under SCF-FTS (ca. 78%) relative to the gas-phase FTS (ca. 69%) at the same total pressure (80 bar).

In the gas-phase FTS reaction, under the steady-state operation, the catalyst pores are filled with the heavy liquid hydrocarbons in which the reactants must be dissolved and then diffuse to reach the catalyst active sites. While in the SCF-FTS, it is well known that SCF has high solubility that can enhance the in situ extraction of heavy hydrocarbons.

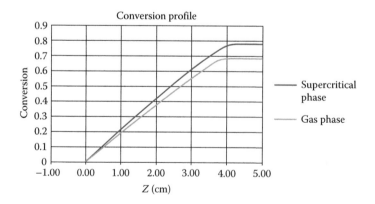

FIGURE 11.58 Conversion profile under SCF phase and gas phase (feed temperature: 513 K; pressure: 80 bar; syngas ratio (H_2/CO): 2:1; nitrogen to syngas ratio: 3:1).

TABLE 11.19 Comparison of the CO Conversion Level for Both SCF- and Gas-Phase FTS

Phase	T (°C)	P (bar)	CO/H$_2$	X_{CO} (%)	Catalyst	Ref.
Gas	210	45	1/2	70	Co/SiO$_2$	Yan et al. (1998)
SCF	210	45	1/2	84		
Gas	250	20	1/2	50	Co-Pt/Al$_2$O$_3$	Huang and Roberts (2003)
SCF	250	80	1/2	70		
Gas	240	55	1/2	54	Co-Ru/Al$_2$O$_3$	Irankhah and Haghtalab (2008)
SCF	240	55	1/2	63		
Gas	240	80	1/2	69	Co/Al$_2$O$_3$	This work
SCF	240	80	1/2	78		

Accordingly, the transportation of the reactants to the catalyst surface is facilitated and the CO conversion is consequently increased (Yan et al. 1998; Subramaniam 2001; Huang and Roberts 2003).

Huang and Roberts (2003), Irankhah and Haghtalab (2008), and Yan et al. (1998) reported similar experimental observations using a Co-based catalyst with different total pressure, as shown in Table 11.19.

11.5.5.3.2 Temperature Profile

The temperature distribution was investigated along the length of the reactor for both SCF-FTS and gas-phase FTS (Figure 11.59). Under a supercritical-phase reaction, the temperature profile is significantly flatter along the reactor compared to the gas-phase reaction. This shows that the supercritical media is more efficient in absorbing and distributing the heat generated by the exothermic reaction. The maximum temperature rise along the catalyst bed in the SCF-FTS is around 5°C, while it is around 10°C in the gas-phase FTS. These results suggest that the heat transfer rate is more effective in the SCF-FTS compared to the gas-phase FTS reaction. This is due to the high heat capacity of the solvent that influences the heat transfer rate. In addition, the thermal conductivity of many gases, including light hydrocarbons, increases significantly by five or six times in the SCF-FTS compared to the conventional gas-phase FTS (TSederberg 1965).

The temperature distribution was done experimentally by (Yokota et al. 1990) using supported silica Co catalyst (Co-La/SiO$_2$). They reported that the temperature rise is 10°C, 13°C, and 18°C in the liquid-, SCF-, and gas-phase FTS reaction, respectively, using different total pressure (45 bar) and solvent/syngas

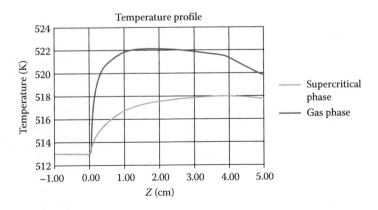

FIGURE 11.59 Temperature distribution under SCF phase and gas phase (feed temperature: 513 K; pressure: 80 bar; syngas ratio (H$_2$/CO): 2:1; solvent to syngas ratio: 3:1).

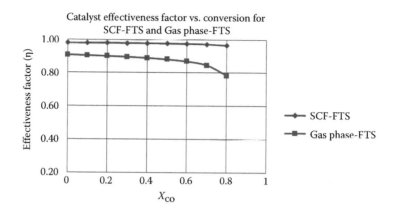

FIGURE 11.60 The catalyst effectiveness factor at different CO conversions for the conventional gas phase and SCF-FTS filled (temperature = 250°C bar, solvent/syngas ratio = 3 and H_2/CO ratio = 2).

ratios (*n*-hexane/syngas = 3). Also, another study by Irankhah and Haghtalab (2008) shows that the maximum temperature rises using Co-based catalyst (Co-Ru/Al_2O_3) were 7°C and 15°C in the SCF-FTS and gas-phase FTS, respectively, using different total pressure (55 bar). Similar results were shown by Huang and Roberts (2003) using a Co-based catalyst (15% Co-0.5% Pt/Al_2O_3) under reaction conditions similar to this study. They stated that the maximum temperature deviation is 5°C in the SCF-FTS compared to 15°C in the gas-phase FTS.

11.5.5.3.3 *Comparison of the Catalyst Effectiveness Factor*

The Figure 11.60 shows the relation between the overall catalyst effectiveness factor and CO conversion for the conventional gas phase and SCF-FTS. The results show high effectiveness factor in both reaction media until a certain conversion (around 50%). However, the overall catalyst effectiveness factor of the gas-phase FTS drops dramatically from 0.9 to 0.75 above a certain conversion (from 50% to 80%). The overall effectiveness factor at the SCF-FTS shows almost a constant pattern at all conversion levels (η is almost equal to 1), which indicates that there is no mass transfer resistance and that the overall effectiveness factor is equal to the internal effectiveness factor. Similar findings were reported by Elbashir et al. (Mogalicherla et al. 2012), assuming that the pores in the gas-phase FTS are filled with the gaseous reactants, while in this work the pores are assumed to be filled with the heavy wax represented by *n*-$C_{28}H_{58}$.

11.5.6 Conclusions

In conclusion, this study provided a framework to understand the performance of SCF-FTS. A one-dimensional heterogeneous model has been developed to simulate temperature and concentration profiles simultaneously in a fixed-bed reactor under both gas phase and SCF-FTS. In the modeling process, a comprehensive catalyst pellet was used to simulate the diffusion and reaction in a typical spherical catalyst pellet. A fugacity-based kinetic model was implemented into the reactor model along with considering the assumption that the catalyst pores would be filled with liquid wax (for the case of conventional gas-phase FTS) and with supercritical *n*-hexane (for the case of SCF-FTS) under realistic operating conditions for the reaction. The thermodynamic properties of the reaction mixture were calculated by using SRK-EOS. The reactor model was validated using the data reported from the experimental investigation available in the literature, and satisfactory agreements were found between the model prediction and experimental results for similar conditions. MATLAB® was used to solve the system of ODEs using ode45 function.

The simulation results provide a prediction for the effect of major variables such as temperature, pressure, and pellet size on the reaction behavior. It was observed that the syngas conversion can be enhanced in the SCF-FTS compared to the gas-phase FTS. Additionally, the simulation in both reaction media indicated that the increase of reaction operating pressure has a significant effect on the increase of CO conversion at certain conditions. A semiflat temperature profile was obtained under SCF-FTS reaction with a temperature rise of 5°C. However, in the gas-phase reaction the temperature profile showed a very sharp increase in temperature (around 10°C) in the first 2 cm of the reactor bed. This result is in agreement with the previous experimental reporting in this regard (Yokota and Fujimoto 1991; Yan et al. 1998; Irankhah and Haghtalab 2008).

The overall catalyst effectiveness factor was higher in the gas-phase FTS compared to the SCF-FTS at the entrance of the reactor bed; however, the effectiveness factor for the SCF become superior at the middle and the bottom of the reactor bed. The decrease of the catalyst effectiveness factor for the gas-phase FTS can be attributed to formation of the wax as result product condensation inside the pores, which result in enhancing the mass transfer limitation. However, in the SCF-FTS the catalyst effectiveness factor shows a small variation as the conversion increased along the bed length because of the in situ extraction of the heavy hydrocarbons by the solvent. Moreover, our findings show that the pellet size had an important effect on the overall catalyst effectiveness factor. The catalyst effectiveness factor showed a clearly decreasing trend as the diameter of the particle increased.

The majority of the research work reported in this chapter has been published as a thesis (Bani Nassr 2013) at Texas A&M University. For any specific details regarding the derivations and the calculations, the readers are referred to this thesis (Bani Nassr 2013).

Future work can be done to improve the research work conducted and presented in this thesis. This model can be extended to other kinetic models, equations of state, catalyst types, or operating conditions. The modeling studies could as well include the overall product distribution of the hydrocarbons obtained from the FTS on both gas phase and the SCF media. Future research could also focus on developing a better represented EOS for the nonideal behavior of the reaction mixture under the high-pressure FTS conditions. Experimental data could be used to validate and further improve the model using our new bench-scale fixed-bed reactor installed recently at TAMUQ. More importantly, the visualization of the reactor in situ behavior will be conducted utilizing advanced MRI and NMR faculties at the University of Cambridge could as well provide accurate measurements of diffusivities and other transport properties.

Acknowledgment

This report was made possible by National Priorities Research Program (NPRP) grant awards [NPRP 08-261-2-082, NPRP 04-1484-2-590, NPRP 07-843-2-312] from the Qatar National Research Fund (a member of the Qatar Foundation). The statements made herein are solely the responsibility of the authors.

Nomenclatures

i	Species in the reaction mixture
λ^e	Pellet effective thermal conductivity
ΔH_r	Heat of reaction
D_i^e	Effective diffusivity of i in catalyst pore
D_{12}	Binary diffusion coefficient, 1 refers to solvent and 2 refers to solute
$D_{i,\,wax}$	Binary diffusion coefficient of reactants in heavy wax
ρ_{cat}	Catalyst density
T	Temperature
P	Pressure
r	Radius of the pellet
r_i	Rate of formation of compound i
R	Universal gas constant
R_{cat}	Catalyst pellet radius

T^s	Temperature at the pellet surface
T^b	Temperature at the bulk fluid
C_i^s	Reactants concentration at the pellet surface
C_i^b	Reactants concentration at the bulk fluid
f_i^s	Reactant concentration at the pellet surface
f_i^b	Reactant concentration at the bulk fluid
U_s	Superficial fluid velocity
ρ_B	Bulk density of the catalyst per unit volume
λ_L	Thermal dispersion coefficient in the axial direction
U	Overall heat transfer coefficient
d_t	Reactor tube internal diameter
T_{wall}	Wall temperature
M	Molecular weight
V_c	Critical molar volume
P_c	Critical pressure
z_c	Critical compressibility factor
V	Molar volume of the solvent
k	Parameter function of reduced density
ρ_r	Reduced density of the solvent
d_P	Catalyst pore diameter
ε_P	Catalyst porosity
τ	Catalyst tortuosity
f_{CO}	Fugacity of CO
f_{H_2}	Fugacity of H_2
r_{CO}	CO consumption rate
K	Rate constant for the rate of CO consumption ($mol/g_{cat}.min.bar$)
K_1, K_2, K_3	Constants for the rate of CO consumption ($1/bar^{0.5}$, $1/bar^{0.5}/1/bar$)
S	Hexane to syngas ratio (feed)
V	H_2 to CO ratio
k_m	Internal mass transfer coefficient
h	Internal heat transfer coefficient
λ_{Pellet}	Pellet thermal conductivity
λ_{fluid}	Bulk fluid (*n*-hexane) thermal conductivity
λ_m	Thermal conductivity of the gas mixture at low pressure
λ_m^0	Thermal conductivity of pure gas components at low pressure
Γ	Reduced inverse thermal conductivity
k_m	Thermal conductivity of the gas mixture at high pressure
Q	Brokaw parameter
y_1	Mole fraction of light component
y_2	Mole fraction of heavy component
R_e	Dimensionless Reynolds number
S_c	Dimensionless Schmidt number

References

Abbaslou, R. M. M., J. S. S. Mohammadzadeh, and A. K. Dalai. 2009. Review on Fischer-Tropsch synthesis in supercritical media. *Fuel Processing Technology* 90:849–856. doi: 10.1016/j.fuproc.2009.03.018.

Arai, Y., T. Sako, and Y. Takebayashi, eds. 2001. *Supercritical Fluid: Molecular Interactions, Physical Properties, and New Applications*. Springer, Berlin.

Baiker, A. 1999. Supercritical fluids in heterogeneous catalysis. *Chemical Reviews* 99 (2):453–473. doi: 10.1021/Cr970090z.

Brunner, G. 1994. *Gas Extraction: An Introduction to Fundamentals of Supercritical Fluids and the Application to Separation Processes.* 1st ed. Jointly published with Springer-Verlag New York; Steinkopff-Verlag Heidelberg.

Brunner, K. M., J. C. Duncan, L. D. Harrison, K. E. Pratt, R. P. S. Peguin, C. H. Bartholomew, and W. C. Hecker. 2012. A trickle fixed-bed recycle reactor model for the Fischer-Tropsch synthesis. *International Journal of Chemical Reactor Engineering* 10:1–21. doi: 10.1515/1542-6580.2840.

Bub, G, M. Baerns, B. Büssemeier, and C. Frohning. 1980. Prediction of the performance of catalytic fixed bed reactors for Fischer-Tropsch synthesis. *Chemical Engineering Science* 35:348–355.

Bukur, D. B., X. Lang, A. Akgerman, and Z. Feng. 1997. Effect of process conditions on olefin selectivity during conventional and supercritical Fischer-Tropsch synthesis. *Industrial and Engineering Chemistry Research* 36:2580–2587. doi: 10.1021/ie960507b.

Chedid, R., M. Kobrosly, and R. Ghajar. 2007. The potential of gas-to-liquid technology in the energy market: The case of Qatar. *Energy Policy* 35 (10):4799–4811. doi: 10.1016/j.enpol.2007.03.017.

Davis, B. H. 2002. Overview of reactors for liquid phase Fischer–Tropsch synthesis. *Catalysis Today* 71 (3–4):249–300. doi: 10.1016/S0920–5861(01)00455-2.

Davis, M. E. and R. J. Davis. 2002. *Fundamentals of Chemical Reaction Engineering,* 1st ed., *McGraw-Hill Chemical Engineering Series,* McGraw-Hill Higher Education, New York.

de Wasch, A. P. and G. F. Froment. 1972. Heat transfer in packed beds. *Chemical Engineering Science* 27 (3):567–576. doi: 10.1016/0009-2509(72)87012-X.

Elbashir, N. O., D. B. Bukur, E. Durham, and C. B. Roberts. 2010. Advancement of Fischer-Tropsch synthesis via utilization of supercritical fluid reaction media. *AIChE Journal* 56:997–1015. doi: 10.1002/aic.12032.

Elbashir, N. O., P. Dutta, A. Manivannan, M. S. Seehra, and C. B. Roberts. 2005. Impact of cobalt-based catalyst characteristics on the performance of conventional gas-phase and supercritical-phase Fischer-Tropsch synthesis. *Applied Catalysis A: General* 285 (1–2):169–180. doi: 10.1016/j.apcata.2005.02.023.

Elbashir, N. O., and C. B. Roberts. 2005. Enhanced incorporation of α-olefins in the Fischer-Tropsch synthesis chain-growth process over an alumina-supported cobalt catalyst in near-critical and supercritical hexane media. *Industrial and Engineering Chemistry Research* 44:505–521. doi: 10.1021/ie0497285.

Fan, L. and K. Fujimoto. 1999. Fischer-Tropsch synthesis in supercritical fluid: Characteristics and application. *Applied Catalysis A* 186:343–354. doi: 10.1016/s0926-860x(99)00153-2.

Fan, L., K. Yokota, and K. Fujimoto. 1992. Supercritical phase Fischer-Tropsch synthesis: Catalyst pore-size effect. *AIChE Journal* 38:1639–48. doi: 10.1002/aic.690381014.

Fan, L., K. Yokota, and K. Fujimoto. 1995. Characterization of mass-transfer in supercritical-phase Fischer-Tropsch synthesis reaction. *Topics in Catalysis* 2 (1–4):267–283.

Fogler, H. S. 2006. *Elements of Chemical Reaction Engineering.* 4th ed. Prentice Hall, Upper Saddle River, NJ.

Froment, G. F., K. B. Bischoff, and J. De Wilde. 2010. *Chemical Reactor Analysis and Design.* 3rd ed. John Wiley & Sons, Inc., New York.

He, C.-H. 1997. Prediction of binary diffusion coefficients of solutes in supercritical solvents. *AIChE Journal* 43 (11):2944–2947. doi: 10.1002/aic.690431107.

Hill, C. G. 1977. *An Introduction to Chemical Engineering Kinetics and Reactor Design.* John Wiley & Sons, New York.

Huang, X. and C. B. Roberts. 2003. Selective Fischer-Tropsch synthesis over an Al_2O_3 supported cobalt catalyst in supercritical hexane. *Fuel Processing Technology* 83:81–99. doi: 10.1016/s0378-3820(03)00060-2.

Irankhah, A. and A. Haghtalab. 2008. Fischer-Tropsch synthesis over Co-Ru/γ-Al_2O_3 catalyst in supercritical media. *Chemical Engineering Technology* 31:525–536. doi: 10.1002/ceat.200700452.

Jess, A. and C. Kern. 2009. Modeling of multi-tubular reactors for Fischer-Tropsch synthesis. *Chemical Engineering and Technology* 32 (8):1164–1175. doi: 10.1002/ceat.200900131.

Kölbel, H. and M. Ralek. 1980. The Fischer-Tropsch synthesis in the liquid phase. *Catalysis Reviews* 21 (2):225–274. doi: 10.1080/03602458008067534.

Lang, X., A. Akgerman, and D. B. Bukur. 1995. Steady state Fischer-Tropsch synthesis in supercritical propane. *Industrial and Engineering Chemical Research* 34:72–7. doi: 10.1021/ie00040a004.

Lox, E. S. and G. F. Froment. 1993. Kinetics of the Fischer-Tropsch reaction on a precipitated promoted iron catalyst. 1. Experimental procedure and results. *Industrial and Engineering Chemistry Research* 32 (1):61–70. doi: 10.1021/ie00013a010.

Madon, R. J. and E. Iglesia. 1994. Hydrogen and CO intrapellet diffusion effects in ruthenium-catalyzed hydrocarbon synthesis. *Journal of Catalysis* 149:428–437.

Mogalicherla, A. K. and N. O. Elbashir. 2011. Development of a kinetic model for supercritical fluids Fischer-Tropsch synthesis. *Energy and Fuels* 25 (3):878–889. doi: 10.1021/Ef101341m.

Mogalicherla, A. K., E. E. Elmalik, and N. O. Elbashir. 2012. Enhancement in the intraparticle diffusion in the supercritical phase Fischer–Tropsch synthesis. *Chemical Engineering and Processing: Process Intensification* 62:59–68. doi: 10.1016/j.cep.2012.09.008.

Moutsoglou, A. and P. P. Sunkara. 2011. Fischer-Tropsch synthesis in a fixed bed reactor. *Energy and Fuels* 25 (5):2242–2257. doi: 10.1021/Ef200160x.

Nassr, L. A. B. 2013. Simulation of Fischer-Tropsch fixed-bed reactor in different reaction media. Master of Science, Chemical Engineering Program, Texas A&M University.

Poling, B. E., J. M. Prausnitz, and J. P. O'Connel. 2001. *The Properties of Gases and Liquids*. 5th ed. McGraw-Hill, New York.

Quan-Sheng, L., Z.-X. Zhang, and J.-L. Zhou. 1999. Steady-state and dynamic behavior of fixed-bed catalytic reactor for Fischer-Tropsch synthesis. *Natural Gas Chemistry* 8:137–150.

Rautavuoma, A. O. I. and H. S. Vanderbaan. 1981. Kinetics and mechanism of the Fischer-Tropsch hydrocarbon synthesis on a cobalt on alumina catalyst. *Applied Catalysis* 1 (5):247–272.

Sarup, B. and B. W. Wojciechowski. 1989. Studies of the Fischer-Tropsch synthesis on a cobalt catalyst II. Kinetics of carbon monoxide conversion to methane and to higher hydrocarbons. *The Canadian Journal of Chemical Engineering* 67 (1):62–74. doi: 10.1002/cjce.5450670110.

Sie, S. T. and R. Krishna. 1998. Process development and scale up: III. Scale-up and scale-down of trickle bed processes. *Reviews in Chemical Engineering* 14:203–352.

Spath, P. L. and D. C. Dayton. 2003. Preliminary screening—Technical and economic assessment of synthesis gas to fuels and chemicals with emphasis on the potential for biomass-derived syngas. National Renewable Energy Laboratory, Golden, CO.

Stern, D., A. T. Bell, and H. Heinemann. 1983. Effects of mass transfer on the performance of slurry reactors used for Fischer-Tropsch synthesis. *Chemical Engineering Science* 38 (4):597–605. doi: 10.1016/0009-2509(83)80119-5.

Subramaniam, B. 2001. Enhancing the stability of porous catalysts with supercritical reaction media. *Applied Catalysis, A* 212:199–213. doi: 10.1016/s0926-860x(00)00848-6.

TSederberg, N. V. 1965. *Thermal Conductivity of Gases and Liquids*. MIT Press, Cambridge, MA.

Van Der Laan, G. P., and A. A. C. M. Beenackers. 1999. Kinetics and selectivity of the Fischer–Tropsch synthesis: A literature review. *Catalysis Reviews* 41 (3–4):255–318. doi: 10.1081/cr-100101170.

Wang, Y. N., Y. Y. Xu, Y. W. Li, Y. L. Zhao, and B. J. Zhang. 2003. Heterogeneous modeling for fixed-bed Fischer-Tropsch synthesis: Reactor model and its applications. *Chemical Engineering Science* 58 (3–6):867–875. doi: 10.1016/S0009-2509(02)00618-8.

Wang, Y. N., Y. Y. Xu, H. W. Xiang, Y. W. Li, and B. J. Zhang. 2001. Modeling of catalyst pellets for Fischer-Tropsch synthesis. *Industrial and Engineering Chemistry Research* 40 (20):4324–4335.

Wojciechowski, B. W. 1988. The kinetics of the Fischer-Tropsch synthesis. *Catalysis Reviews* 30 (4):629–702. doi: 10.1080/01614948808071755.

Wu, J., H. Zhang, W. Ying, and D. Fang. 2010. Thermal conductivity of cobalt-based catalyst for Fischer–Tropsch synthesis. *International Journal of Thermophysics* 31 (3):556–571. doi: 10.1007/s10765-010-0740-x.

Yan, S., L. Fan, Z. Zhang, J. Zhou, and K. Fujimoto. 1998. Supercritical-phase process for selective synthesis of heavy hydrocarbons from syngas on cobalt catalysts. *Applied Catalysis, A* 171:247–254. doi: 10.1016/s0926-860x(98)00049-0.

Yates, I. C. and C. N. Satterfield. 1991. Intrinsic kinetics of the Fischer-Tropsch synthesis on a cobalt catalyst. *Energy and Fuels* 5 (1):168–173. doi: 10.1021/Ef00025a029.

Yokota, K. and K. Fujimoto. 1991. Supercritical-phase Fischer-Tropsch synthesis reaction. 2. The effective diffusion of reactant and products in the supercritical-phase reaction. *Industrial and Engineering Chemistry Research* 30:95–100. doi: 10.1021/ie00049a014.

Yokota, K., Y. Hanakata, and K. Fujimoto. 1990. Supercritical phase Fischer-Tropsch synthesis. *Chemical Engineering Science* 45:2743–50. doi: 10.1016/0009-2509(90)80166-c.

Yokota, K., Y. Hanakata, and K. Fujimoto. 1991. Supercritical phase Fischer-Tropsch synthesis reaction. 3. Extraction capability of supercritical fluids. *Fuel* 70:989–94. doi: 10.1016/0016-2361(91)90056-g.

Atuman S. Joel and
Akeem K. Olaleye
University of Hull

Jonathan G. M. Lee
and KeJun Wu
Newcastle University

DoYeon Kim, Nilay
Shah, C. Kolster,
and N. Mac Dowell
Imperial College London

Meihong Wang,
Colin Ramshaw,
Xiaobo Luo,
Eni Oko, Lin
Ma, Mohamed
Pourkashanian,
and S. Brown
University of Sheffield

Adekola Lawal
*Process Systems
Enterprise Ltd*

H. Mahgerefteh,
R. T. J. Porter,
A. Collard, and
S. Martynov
University College London

M. Fairweather, R.
M. Woolley, and
S. A. E. G. Falle
University of Leeds

G. C. Boulougouris,
Ioannis G.
Economou, D. M.
Tsangaris, and
I. K. Nikolaidis
*National Center for
Scientific Research
"Demokritos"*

A. Beigzadeh, C.
Salvador, and K.
E. Zanganeh
*Canada Centre for Mineral
and Energy Technology*

A. Ceroni, R.
Farret, Y. Flauw, J.
Hébrard, D. Jamois,
and C. Proust
*Institut national
de l'environnement
industriel et des risques*

12

Carbon Capture

S. Y. Chen, J. L.
Yu, and Y. Zhang
*Dalian University
of Technology*

D. Van Hoecke,
R. Hojjati Talemi,
and S. Cooreman
*ArcelorMittal Global
R&D Gent-OCAS NV*

J. Bensabat and
R. Segev
*Environmental and Water
Resources Engineering Ltd*

D. Rebscher
and J. L. Wolf
*Federal Institute
for Geosciences and
Natural Resources*

A. Niemi
Uppsala University

Mert Atilhan
and Ruh Ullah
Qatar University

Cafer T. Yavuz
KAIST

Nejat Rahmanian
University of Bradford

Sina Gilassi
*Petronas University
of Technology*

12.1 Application of Rotating Packed Bed Technology for Intensified Postcombustion CO_2 Capture Based on Chemical Absorption

Atuman S. Joel, Eni Oko, Meihong Wang, Colin Ramshaw, Jonathan G. M. Lee, KeJun Wu, DoYeon Kim, Nilay Shah, Lin Ma, and Mohamed Pourkashanian

12.1.1 Introduction

Rapid increase in emissions of greenhouse gases (GHGs) has become a major concern to the global community. This is associated with the rapid growth in population and corresponding increase in energy demand. Combustion of fossil fuels accounts for the majority of CO_2 emissions. Coal is used mostly for electricity generation, for instance, about 85.5% of coal (produced and imported) in the United

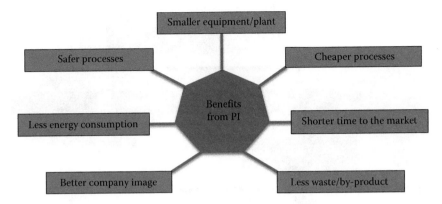

FIGURE 12.1   Main benefits from PI. (Stankiewicz, A.I., Moulijn, J.A., *Chem. Eng. Prog.*, 96, 22–34, 2000.)

Kingdom was used for electricity generation in 2011 [1]. Coal-fired power plants are therefore the largest stationary source of CO_2.

The Intergovernmental Panel on Climate Change (IPCC) has set an ambitious goal to reduce CO_2 emission by 50% in 2050 as compared to the level in 1990. CO_2 capture technology is important for meeting the target. Postcombustion CO_2 capture (PCC) with chemical absorption is the most mature CO_2 capture technology. As such, it is considered a low-risk technology and a promising near-term solution to large-scale CO_2 capture [2].

PCC for coal-fired power plants using packed columns has been reported by many authors. Wang et al. [3] carried out a state-of-the-art review of PCC with chemical absorption. Dugas [4] performed experimental studies of PCC using solvents in the context of fossil fuel-fired power plants. Lawal et al. [5–7] carried out dynamic modeling of CO_2 absorption for postcombustion capture in coal-fired power plants. From Lawal et al. [8], one of the identified challenges to the commercial rollout of the technology has been the large size of the packed columns required for absorption of CO_2 and regeneration of the solvent. This translates to high capital and operating costs and unavoidable increases in the electricity price. Approaches such as heat integration, intercooling, among others, could reduce the operating cost slightly. However, they limit the plant flexibility and will make operation and control more difficult [9]. Process intensification (PI) has the potential to meet these challenges as highlighted in Figure 12.1 [10–13].

12.1.1.1 What Is PI?

Ramshaw [11] defined PI as a strategy for making dramatic reductions in the size of a chemical plant so as to reach a given production objective. According to Stankiewicz and Moulijn [13], Ramshaw's definition is quite narrow, describing PI exclusively in terms of the reduction in plant or equipment size.

Reay et al. [14] defined PI as "Any engineering development that leads to a substantially smaller, cleaner, safer and more energy efficient technology." Reay et al.'s [14] definition is an improvement of Stankiewicz and Moulijn's [13] definition by including the term "safer." Reay et al. [14] view safety as an important driver in motivating businesses to seriously consider PI technologies. There are general approaches to PI with the aim of improving process performance [15] as follows: (1) reducing equipment size using an intensified field (e.g., centrifugal, electrical, and microwave), (2) simplifying processes by integrating multiple process tasks in a single item of equipment, and (3) reducing equipment size by reducing its scale of structure. PI technologies differ in functions and areas of application. Some will be very good at intensifying mass transfer (e.g., rotating packed bed, RPB), while others are good at intensifying heat transfer (e.g., printed circuit heat exchangers).

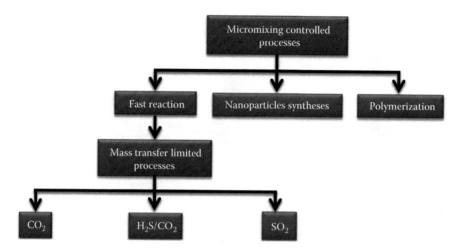

FIGURE 12.2 Summary of why PI for PCC. (From Chen, J.F., The recent developments in the HiGee technology. Presentation at GPE-EPIC, June 14–17, http://inpact.inp-toulouse.fr/GPE-EPIC2009/images/presentation_chen. pdf, 2009.)

12.1.1.2 Why PI for PCC?

It was reported that a $500\,MW_e$ supercritical coal fired power plant operating at 46% efficiency (Lower Heating Value basis) releases over 8000 tons of pure CO_2 per day [16]. PCC using solvents based on the conventional technology (i.e., using packed columns) requires very large columns. Dynamic modeling and simulation studies of a $500\,MW_e$ subcritical coal-fired power plant by Lawal et al. [8] showed that two absorbers of $17\,m$ in packing height and $9\,m$ in diameter will be needed to separate CO_2 from the flue gas. These huge packed columns translate into high capital costs. A significant amount of steam from power plants has to be used for solvent regeneration. This translates into high thermal efficiency penalty. It is reported that 3.2–4.5 MJ energy is required to capture 1 kg of CO_2 using 30 wt% monoethanolamine [8,9,17–19]. On the other hand, through PI techniques there could be significant reductions in equipment sizes and energy consumption, hence lower capital cost [10,20–23] and improved process dynamics.

The reason why PI technology is applicable for the capture of carbon dioxide include (Figure 12.2) (1) enhanced mass transfer due to better gas–liquid interactions. Biliyok et al. [24] reported CO_2 capture using 30 wt% MEA concentration in a conventional packed tower and found that the process is mass transfer controlled [20,25]; (2) improved reaction kinetics due to micromixing in some cases.

12.1.1.3 Success Stories of PI Application in Different Areas

The use of high gravity (HIGEE) mass transfer contactors did not receive serious attention when it was first proposed in 1935 by Podbielniak [26]. Talks about PI potential to achieve an order of magnitude reduction in size of either distillation or absorption column are always greeted with skepticism. Brave researchers such as Podbielniak [26], George [27], Adolph [28], Wilhelm and Wilhelm [29], and Ramshaw and Mallinson [12] continued to push despite the discouragement received.

Regardless of the earlier difficulties, RPB technology has been successfully applied in industrial-scale desulfurization process [30]. The RPB plant has reportedly been in operation for over 2 years as at 2012 [30]. The size reduction compared to a conventional packed tower can be clearly seen in Figure 12.3. Commercial deployment of the RPB for PCC has not been reported anywhere in the open literature, but there are many research efforts from different research groups across the world such as Newcastle University (UK), Beijing University of Chemical Technology (China), Indian University of Technology Kanpur (India), Chang Gung University (Taiwan), National Tsing Hua University (Taiwan), The Delft University of Technology (Netherlands), and the University of Hull (UK).

FIGURE 12.3   Desulfurization site [30] small RPB (left) replaced large packed tower (right).

12.1.2 Current Status of PI for PCC

12.1.2.1 Current Status of Experimental Rigs/Studies in PI for PCC

12.1.2.1.1 PI Experimental Rigs and Studies for PCC: Intensified Absorber

The use of RPB for PCC has been explored by a number of research institutions. Table 12.1 gives the summary of rig specifications for different institutions across the world and their major findings. The rig at Newcastle University, UK is the biggest in size having rig diameter of 1 m and bed thickness of 0.05 m.

12.1.2.1.2 PI Experimental Rigs for PCC: Intensified Stripper

There are only two published studies on intensified stripper using the RPB. Table 12.2 summarizes the rig specifications and their findings (Newcastle University in the UK and National Tsing Hua University in Taiwan). In both groups, the reboiler is the same as that of conventional packed column (i.e., still quite big in size).

12.1.2.2 Current Status of Modeling/Simulation PI for PCC

Process models validated against small scale experimental data can be used to scale up and predict the performance of RPB for PCC at commercial scales.

12.1.2.2.1 Modeling/Simulation of Intensified Absorber

There are some studies in the open literature that discuss the modeling and simulation of an intensified absorber. The group in Taiwan modeled the RPB as a series of continuous stirred tank reactors (CSTRs). Cheng and Tan [38] reported that five CSTRs with a contactor can achieve the set target for a given case through simulation study. Kang et al. [44] reported comparison between packed column and RPB using 3 wt% ammonia solution operating in the lean-to-rich loading range of 0.2–0.4 and using 30 wt% MEA solution operating in the lean-to-rich loading range of 0.35–0.49. They found that intensification effect is more significant in 3 wt% ammonia solution than 30 wt% MEA solution. This is because the loading range for 30 wt% MEA solution is already close to saturation, which means less CO_2 absorption performance; this finding suggests the use of higher MEA concentration for intensified PCC process to be beneficial. The research group at University of Hull, UK reported modeling and simulation of RPB absorber using Aspen Plus and visual FORTRAN [21,23]. Their key findings include (1) the packing volume can be

TABLE 12.1 Summary of Rig Specification and Findings for RPB Absorber

Institutions	Rig Specification	Operating Conditions	Results
Newcastle University, UK	Rig 1 geometry: $d_i = 0.156\,m$ $d_o = 0.398\,m$ $h = 0.025\,m$ Packing type: expamet Surface area: 2132 m^2/m^3; Void fraction: 0.76	Aqueous MEA solutions flowrate: (a) 0.35 kg/s and (b) 0.66 kg/s MEA concentrations of 30, 55, 75, and 100 wt%. Flue gas flow rate 2.86 kmol/h	The effect of lean amine temperature, rotor speed, and MEA concentrations were investigated.
	Rig 2 geometry: $d_i = 0.190\,m$ $d_i = 1.000\,m$ $h = 0.050\,m$ Packing type: Expamet and stainless wire mesh	Water is used as the solvent for hydrodynamic study in the RPB	The following were some of the targets for the research study [31]: 1. Power consumption 2. Pressure drop 3. Liquid distribution
BUCT in China	Rig 1 geometry: $d_i = 0.080\,m$ $d_i = 0.200\,m$ $h = 0.031\,m$ Packing type: stainless wire mesh Surface area: 870 m^2/m^3; void fraction: 0.95	Benfield solution (DEA-promoted hot potassium carbonate). Mass fraction of K_2CO_3 and DEA in aqueous absorbent was 27% and 4%, rotating speed of 1300 rpm, liquid flow rate of 79.70 L h^{-1}, gas flow rate of 0.481 mol min^{-1}, inlet CO_2 concentration of 4.10 mol%, and temperature of 356 K	Yi et al. [32] study effect of rotating speeds, liquid flow rates, gas flow rates and temperatures in RPB, with Benfield solution as the absorbent. End effect was identified in the study which means mass transfer is more at the inlet of the RPB.
	Rig 2 geometry: $d_i = 0.156\,m$ $d_i = 0.306\,m$ $d_i = 0.050\,m$ Packing type: stainless wire mesh Surface area: 500 m^2/m^3; void fraction: 0.96	NaOH solvent The NaOH solution at 20°C–30°C in the range of 40–120 L/h The mixed gas of CO_2 and N_2 in the range of 800–12,000 L/h	Luo et al. [33,34] study effective interfacial area and liquid side mass transfer coefficient in a RPB equipped with blades. Mass transfer rate improve from 8 to 68% compared to conventional RPB without packing.
	Rig 3 geometry: $d_i = 0.020\,m$ $d_i = 0.060\,m$ $h = 0.020\,m$ Packing type: stainless wire mesh Surface area: 850 m^2/m^3; void fraction: 0.90	The ionic liquid (1-*n*-butyl-3-methylimidazolium hexafluorophosphate) gas flow rate, 0.6–1 L/min, liquid flow rate, 29.2–102.2 mL/min	Zhang et al. [35] found that liquid side volumetric mass transfer coefficient for RPB has been improved to around 3.9×10^{-2} s^{-1} compared with 1.9×10^{-3} s^{-1} for the conventional packed column under the same operating conditions

(Continued)

TABLE 12.1 (*Continued*) Summary of Rig Specification and Findings for RPB Absorber

Institutions	Rig Specification	Operating Conditions	Results
Taiwan (National Tsing Hua University, Chang Gung University, and Chung Yuan University)	Rig 1 geometry: $d_i = 0.076\,m$ $d_i = 0.160\,m$ $h = 0.02\,m$ Packing type: stainless wire mesh Surface area: 803 m^2/m^3; void fraction: 0.96	N_2 gas stream containing 10 vol% CO_2. Four different aqueous solutions, DETA, MEA, DETA mixed with PZ, and MEA mixed with PZ, with concentration of 30 wt%. Gas flow rate 30 and 60 L/min, liquid flow rate 50 and 100 mL/min	Yu et al. [36], Tan and Chen [37]. Overall mass transfer coefficient (K_Ga) and HTU corresponding to the most appropriate operating conditions in RPB were found to be higher than 5.8 s^{-1} and lower than 1.0 cm [38]. However, HTU is around 40 cm for conventional PCC process
	Rig 2 geometry: $d_i = 0.048\,m$ $d_o = 0.088\,m$ $h = 0.120, 0.09, 0.06, 0.03\,m$ Packing type: stainless wire mesh Surface area: 855, 873, 879, 830 m^2/m^3; void fraction of 0.95	CO_2–N_2 stream containing 10 vol% CO_2. Gas flow rate was varied from 10 to 70 L/min and the liquid flow rate was varied from 0.2 to 0.5 L/min.	Lin et al. [39] and Lin and Chen [40] reported RPB absorber with flue gas and lean solvent moving in cross flow. Cross flow RPB takes advantage of lower pressure drop and also has mass transfer efficiency comparable to countercurrent-flow RPB
India (India Institute of Technology (IIT) Kanpur)	Rig geometry: $d_i = 0.240\,m$ $d_o = 0.480\,m$ $h = 0.420\,m$ Packing type: Ni–Cr metal foam Specific surface area: 2500 m^2/m^3; void fraction = 0.9	Gas composition 0.98 mol% CH_4; 0.02 mol % CO_2 gas flow rate 2490 kmol h^{-1}, liquid flow rate 3038 kmol h^{-1} (30 wt % DEA)	Agarwal et al. [41], shows that there is good performance in gas phase control processes by enhancing volumetric mass transfer coefficient on the gas side to about 35–280 times compared to those of packed columns, the liquid side volumetric mass transfer coefficient enhances in the range of 25–250 times compared to the packed column [42]
Iran (process Intensification Research Lab, Chemical Engineering Department, Yasouj University)	Rig geometry: $d_i = 0.006\,m$ $d_o = 0.12\,m$ $h = 0.04\,m$ Packing type: (a) Stainless steel wire mesh (b) Aluminum expamet specific surface area: (a) 1800 m^2/m^3; (b) 1300 m^2/m^3; void fraction = (a) 0.9 (b) 0.9	Inlet gas stream contain 5000 ppm CO_2 concentration. Gas and liquid flow rates were from 10 to 40 to 0.2 to 0.8 L/min, respectively. Rotor speed range 400–1600 rpm.	Rahimi and Mosleh [43] study height of transfer unit (HTU) of RPB using two different types of packing. The effect of rotational speed, gas flow rate, liquid flow rate, and MEA concentration on HTU values for CO_2 capture were within 2.4–4 cm depending on the rotational speed, gas and liquid flow rates, and solution concentration

TABLE 12.2 Summary of Rig Specification and Findings for RPB Stripper

Institutions	Rig Specification	Process Variables	Findings
Newcastle University, UK	Rig geometry: $d_i = 0.156$ m $d_i = 0.398$ m $h = 0.025$ m Packing type: expamet Surface area: 2132 m²/m³; void fraction: 0.76	Three rich solvent: 30, 54, and 60 wt% MEA concentration was used. Solution flow rates range from 0.2 to 0.6 kg/s	Conventional packed column was compared with RPB under similar performance and it was found that the height and diameter has reduction factor of 8.4 and 11.3, respectively [20].
Taiwan	Rig geometry: $d_i = 0.076$ m $d_o = 0.160$ m $h = 0.020$ m Packing type: wire mesh Surface area: 803 m²/m³; void fraction: 0.96	Two solvents were used (1) blended 20 wt% DETA + 10 wt% PZ aqueous solution (2) 30 wt% MEA aqueous solution. Rich solvent flow rate 400 mL/min	Back pressure regulator was introduced in order to operate the stripper at higher temperature and pressure. The result shows that specific energy consumption with RPB is less than conventional packed column [22].

reduced 52 times, and the absorber size can be reduced 12 times; (2) there is no temperature bulge, due to the heat released by CO_2 absorption, observed inside the packing [21,23].

12.1.2.2.2 *Modeling/Simulation of Intensified Stripper*

Experimental studies of intensified stripper were only reported in Jassim et al. [20] and Cheng et al. [22]. No modeling and simulation of intensified stripper has been reported in the open literature.

12.1.2.2.3 *Modeling and Simulation of the Whole Plant*

No modeling or simulation studies of the whole intensified PCC process have been published in the open literature at the time of this study.

12.1.3 Challenges of PI for PCC and Potential Solutions

12.1.3.1 Corrosion Issues

Because the contact time between the gas and liquid phases in the RPB absorber is much shorter than in conventional packed columns, the selection of solvent with a fast reaction rate with CO_2 is crucial [12,20,21,23,37,45–47]. This necessitates the use of higher concentration solvent (such as 55 or 75 wt% MEA reported in References [20,21,23]), but this comes with another challenge of increasing corrosion rate [48].

12.1.3.1.1 *Corrosion in PCC Plant with MEA Solvent*

Corrosion in PCC plant with MEA solvent is mostly due to absorbed acid gas (CO_2) and high solution temperature. Kittel and Gonzalez [49] suggested that temperature increase of 10–20 K doubles corrosion rate. Other driving factors for corrosion include MEA concentration and presence of oxygen. On its own, MEA is not corrosive. It only becomes corrosive in the presence of acid gases such as CO_2. Shao and Stangeland [50] concluded that an acceptable level of corrosion for PCC with solvent is approximately 0.25 mm/year (250 μ/year) or less.

Other alkanolamines such as Diethanolamine and Methyl Diethanolamine have significantly lower corrosion rates than MEA [51]. MDEA–MEA blends are less corrosive compared to MEA by itself. MEA–piperazine solvent blends are significantly more corrosive than MEA alone [52].

12.1.3.1.1.1 *MEA Concentration* Corrosion rate increases with MEA concentration (Figure 12.4). Corrosion rate at all concentrations from 30 wt% up to 55 wt% are higher than the acceptable limit (~250 μ/year) suggested by Shao and Stangeland [50].

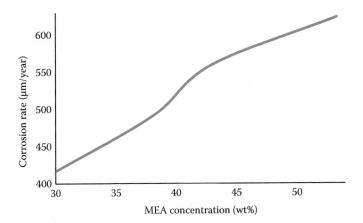

FIGURE 12.4 Corrosion rate of carbon steel versus MEA concentrations at 80°C and 0.2 mol CO_2/mol MEA loading. (From Nainar, M., Veawab, A., *Energy Proc.*, 1, 231–235, 2009.)

12.1.3.1.1.2 CO₂ Loading Dissolved CO_2 (loading) in MEA is the main factor that influences corrosion in PCC plants with MEA solvents [51]. Figure 12.5 shows that corrosion rate increases with CO_2 loading though it decreases between CO_2 loading of 0.3 and 0.45. Corrosion rates are, therefore, expected to be high at the region with high CO_2 loading such as the absorber bottom, rich-side piping, rich side of the cross-heat exchangers, and stripper overhead. There is, however, a strong interplay between the influence of temperature and loading. For instance, the temperature at the absorber bottom is lower than the temperature in the overhead stream from the stripper overhead. So despite the loading in both regions being similar, the rate of corrosion in them is significantly different due to the differences in temperature [53].

Corrosion rate in DuPart et al. [51] is significantly low. More recent experimental results from Soosaiprakasam and Veawab [54], Nainar and Veawab [52], and Kittel et al. [53] suggest higher corrosion rates at about similar conditions. Reported trends in corrosion rate at different conditions such as temperature, CO_2 loading, and MEA concentration are similar across all the studies.

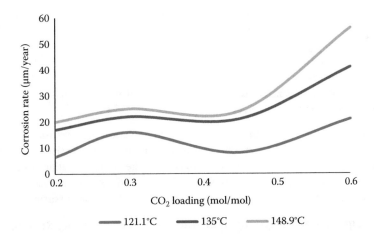

FIGURE 12.5 Corrosion rate for carbon steel versus CO_2 loading at 20 wt% MEA and different temperatures. (From DuPart, M.S. et al., Understanding corrosion in alkanolamines gas treating plants Part 1&2: Proper mechanism diagnosis optimizes amine operations. *Hydrocarbon Processing*, Gulf Publishing Co, Houston, TX, 1993.)

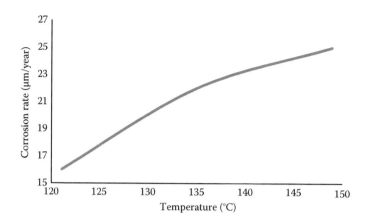

FIGURE 12.6 Corrosion rate of carbon steel versus temperature at 20 wt% MEA and 0.3 mol CO_2/mol MEA loading. (From DuPart, M.S. et al., Understanding corrosion in alkanolamines gas treating plants Part 1&2: Proper mechanism diagnosis optimizes amine operations. *Hydrocarbon Processing*, Gulf Publishing Co, Houston, TX, 1993.)

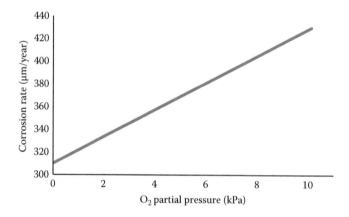

FIGURE 12.7 Corrosion rate of carbon steel versus O_2 partial pressure at 30.7 wt% MEA and 0.2 mol/mol CO_2 loading. (From Soosaiprakasam, I.R., Veawab, A., *Int. J. Greenhouse Gas Control*, 2, 553–562, 2008.)

12.1.3.1.1.3 Solution Temperature   Corrosion rate increases with temperature as shown in Figures 12.5 and 12.6. There is less corrosion at the absorber inlet and outlet area (temperature is lower in these areas). On the other hand, corrosion is more significant at the cross-heat exchanger, stripper inlet and outlet, and the reboiler areas where the temperature is higher.

12.1.3.1.1.4 Presence of Oxygen   The presence of oxygen increases the corrosion rate as shown in Figure 12.7. This is due to degradation products formed when MEA reacts with oxygen [49].

The corrosion problems associated with the use of concentrated amine solutions in carbon steel could be managed by the use of (1) more expensive construction material such as stainless steel rather than the commonly used carbon steel; (2) coating with high-performance polymer on the surface of the carbon steel.

12.1.3.2 New Equipment Design for Intensified Stripper (Including Reboiler)

The experimental data of Jassim et al. [20] and Cheng et al. [22] have shown an RPB stripper would be significantly smaller than a conventional packed tower stripper; however, a conventional reboiler would still be large. In the reboiler, the vapor of the solvent is produced by evaporation with the heat being

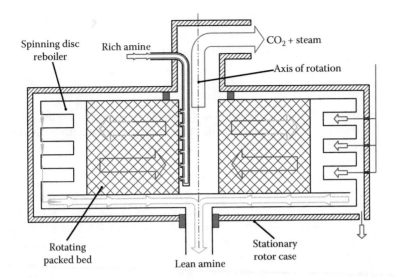

FIGURE 12.8   Section through a RPB amine stripper with integrated spinning disc reboiler.

transferred from the condensation of steam. If these processes are performed on either side of a rotating disc, the heat transfer coefficient will be up to $12\,kW/(m^2\,°C)$ compared to $2–3\,kW/(m^2\,°C)$ for a reboiler of conventional design. This would lead to a significant reduction in the area required for the reboiler. Therefore, to fully intensify the amine regeneration process, the RPB stripper should incorporate a spinning disc reboiler as shown in Figure 12.8. The reboiler comprises of a number of coaxial annular discs joined in a "concertina" configuration placed outboard of the inner stripper packing. Process steam will condense on one side of each disc and liquid solvent for the stripper will boil off on the other as shown in the diagram. In principle, both the reboiling and stripping functions can be performed in one rotating unit.

12.1.3.3 Experimental Studies

12.1.3.3.1 Marangoni Effect

The Marangoni effect occurs when there is a gradient of surface tension at the interface between two phases. In most cases, this is a gas–liquid interface. The surface tension typically changes due to variations in solute concentration, surfactant concentration, and temperature variations along the interface.

According to Sternling and Scriven [55], interfacial turbulence is due to

1. Solute transfer out of the phase of higher viscosity
2. Solute transfer out of the phase in which its diffusivity is lower
3. Large differences in kinematic viscosity and solute diffusivity between the two phases.
4. Steep concentration gradients near the interface.
5. The interfacial tension being highly sensitive to solute concentration.
6. Low viscosities and diffusivities in both phases.
7. An absence of surface active agents.
8. Interfaces of large extent.

With the anticipation of using high solvent concentration for intensified PCC process as highlighted by Ramshaw and Mallinson [12], Jassim et al. [20], Joel et al. [21,23], and Chamber and Walls [56], there is high chance of the presence of Marangoni effect.

The occurrence of the Marangoni effect during gas–liquid mass transfer processes can be detected: (1) indirectly, using the fact that Marangoni convection enhances the transfer rate [57,58]; (2) directly, using visualization methods, such as Schlieren photography [59–61], laser Doppler velocimetry (LDV) [59], and particle image velocimetry (PIV)[62].

Therefore, the major challenge should be how to detect Marangoni effect in RPB either directly or indirectly.

12.1.3.3.2 Physical Properties

12.1.3.3.2.1 Thermodynamic Properties
The design of gas treating processes, whether equilibrium based or rate based, fundamentally depends on the availability of an appropriate thermodynamic model that can accurately interpolate and extrapolate experimental thermodynamic data for the given system [63–65]. If process simulation is to be successful, then the availability of an appropriate thermodynamic model for the system of interest becomes critical and more or less indispensable [64,66].

Phase equilibria is what governs the distribution of molecular species between the vapor and liquid phases in an equilibrium mixture. However, ionic species are normally treated as nonvolatile and they are assumed to be present in the liquid phase only [63,67]. According to Gibbs [68], phase equilibria is said to be attained when the chemical potential of each of the species is same in all the phases present in a given system, in addition to uniform temperature and pressure.

For intensified PCC process, it is expected that high solvent concentration will be used because of the short residence time [12,20,21,23] and most thermodynamic property calculations were done for 30 wt% or less MEA concentration in Aspen plus properties. Vapour Liquid Equilibrium (VLE) data for higher alkanolamines concentrations were presented in Aronu et al. [69] and Mason and Dodge [70]. Since these data are important in calculating thermodynamic properties, getting more dependable data is necessary.

The thermodynamic properties, namely, fugacity, activity coefficient, Henry's constant, enthalpy, and heat of absorption, are calculated using correlations within the eNRTL package in Aspen Plus package.

12.1.3.3.2.2 Transport Properties
The transport properties include density, viscosity, surface tension, thermal conductivity, and binary diffusivity [71]. A summary of some of the models in Aspen Plus that can be used for transport properties calculations is given in Table 12.3. Validation of these models when using them for PI is important; this is because PI is usually operated at higher solvent concentration.

12.1.3.3.3 Mass Transfer Correlations

Table 12.4 summarizes some of the mass transfer correlation that can be used for modeling of intensified absorber or stripper. Some of the mass transfer correlations are derived from the two-film theory, penetration theory, and others such as surface renewal theory. More studies are needed to identify which model of two-film theory or penetration theory or others will be suitable for modeling intensified PCC process.

12.1.3.4 Scale-Up Study

Herzog [83] reported that the challenge for Carbon Capture and Storage (CCS) commercial deployment is to integrate and scale-up the components (absorber, heat exchanger, and the regenerator). Shi et al. [84] and Yang et al. [85] used computational fluid dynamic (CFD) to study fluid flow in RPB to

TABLE 12.3 Summary of the Models in Aspen Plus for Transport Properties Calculations

Property	Gas Phase	Liquid Phase
Density	COSTALD model by Hankinson and Thomson [73]	Clarke density model
Viscosity	Chapman–Enskog model with Wilke approximation	Jones–Dole model
Surface tension	Not applicable	Onsager–Samaras model
Thermal conductivity	Stiel–Thodos model with Wassiljewa–Mason–Saxena mixing rule	Reidel model
Binary diffusivity	Chapman–Enskog Wilke–Lee model	Nernst–Hartley model

Source: AspenTech. Aspen Physical Properties System—Physical Property Methods, http://support.aspentech.com/, 2010.

TABLE 12.4 Mass Transfer Correlations

Correlation	Model	Reference
Liquid-phase mass transfer coefficient	$\dfrac{k_L d_p}{D_L} = 0.919\left(\dfrac{a_t}{a}\right)^{1/3} Sc_L^{1/2} Re_L^{2/3} Gr_L^{1/6}$	Tung and Mah [74]
	$\dfrac{k_L a d_p}{D_L a_t}\left(1 - 0.93\dfrac{V_o}{V_t} - 1.13\dfrac{V_i}{V_t}\right) = 0.35 Sc_L^{0.5} Re_L^{0.17} Gr_L^{0.3} We_L^{0.3}\left(\dfrac{a_t}{a'_p}\right)^{-0.5}\left(\dfrac{\sigma_c}{\sigma_w}\right)^{0.14}$	Chen et al. [75]
	$k_L a = 0.0733 Re_L^{0.3547} Gr_L^{0.2934} Sc_L^{0.5}\left(\dfrac{a_t D_L}{d_p}\right)^{0.8878}$	Rajan et al. [76]
Gas-phase mass transfer coefficient	$k_G = 2.0(a_t D_G) Re_G^{0.7} Sc_G^{\frac{1}{3}}(a_t d_p)^{-2}$	Onda et al. [77]
	$\dfrac{k_G a}{D_G a_t^2}\left(1 - 0.9\dfrac{V_o}{V_t}\right) = 0.023 Re_G^{1.13} Re_L^{0.14} Gr_G^{0.31} We_L^{0.07}\left(\dfrac{a_t}{a'_p}\right)^{1.4}$	Chen [78]
	$k_G a = 0.00738 Re_G^{0.976} Gr_G^{0.132} Sc_G^{0.333}\left(\dfrac{a_t D_G}{d_p}\right)$	Reddy et al. [79]
Total gas–liquid interfacial area	$\dfrac{a}{a_t} = 1 - \exp\left[-1.45\left(\dfrac{\sigma_c}{\sigma}\right)^{0.75} Re_L^{0.1} We_L^{0.2} Fr_L^{-0.05}\right]$	Onda et al. [77]
	$\dfrac{a}{a_t} = 1.05 Re_L^{0.047} We_L^{0.135}\left(\dfrac{\sigma}{\sigma_c}\right)^{-0.206}$	Puranik and Vogelpohl [80]
	$\dfrac{a}{a_t} = 66510 Re_L^{-1.41} Fr_L^{-0.12} We_L^{1.21} \varphi^{-0.74}$	Luo et al. [33]
Liquid holdup	$\varepsilon_L = 0.039\left(\dfrac{g_c}{g_o}\right)^{-0.5}\left(\dfrac{U}{U_o}\right)^{0.6}\left(\dfrac{v}{v_o}\right)^{0.22}$	Burns et al. [81]
	$\varepsilon_L = 3.86 Re_L^{0.545} Ga^{-0.42}\left(\dfrac{a_t d_p}{\varepsilon}\right)^{0.65}\varepsilon$	Lin et al. [82]

understand the hydrodynamics and liquid distribution inside the RPB system. But more studies are required for scale-up of RPB. To be able to carry out the scale-up study of an intensified PCC process, it is recommended to couple process modeling software with CFD software so as to accurately predict the hydraulic behavior and the mass transfer behavior of the RPB. Three-dimensional (3D) CFD simulations can be used to unveil details about the pressure field and velocity distribution, the gross flow patterns, maldistribution [86], and a process modeling software can study the mass transfer behavior.

12.1.4 Modeling of PI for PCC

12.1.4.1 Steady State Modeling

Modeling of the intensified absorber using RPB was reported in Joel et al. [21,23]. The default mass/heat transfer correlations of the Aspen Plus rate-based model were changed with subroutines written in Intel visual FORTRAN. The new model now represents intensified absorber using RPB. Liquid-phase mass transfer coefficient given by Chen et al. [75], gas-phase mass transfer coefficient given by Chen [87], interfacial area correlation estimated by Luo et al. [33], and liquid hold-up correlation given by Burns et al. [81] are all presented in Table 12.4 and also dry pressure drop expression given by Llerena-Chavez and Larachi [86] are the correlations written in Intelvisual FORTRAN for the dynamic linkage with the Aspen Plus model.

12.1.4.1.1 Physical Property

To describe the vapor–liquid equilibrium, the chemical equilibrium, and the physical properties, Joel et al. [21] selected the electrolyte nonrandom-two-liquid (ElecNRTL) activity coefficient model in Aspen Plus. The equilibrium constants for reactions 1–5 are calculated from the standard Gibbs free energy change, the equilibrium reactions are assumed to occur in the liquid film and kinetics reactions equations and parameters are obtained from AspenTech [72].

$$\text{Equilibrium } 2H_2O \leftrightarrow H_3O^+ + OH^- \tag{12.1}$$

$$\text{Equilibrium } CO_2 + 2H_2O \leftrightarrow H_3O^+ + HCO_3^- \tag{12.2}$$

$$\text{Equilibrium } HCO_3^- + H_2O \leftrightarrow H_3O^+ + CO_3^{2-} \tag{12.3}$$

$$\text{Equilibrium } MEAH^+ + H_2O \leftrightarrow MEA + H_3O^+ \tag{12.4}$$

$$\text{Equilibrium } MEACOO^- + H_2O \leftrightarrow MEA + HCO_3^- \tag{12.5}$$

Kinetic reaction used for the calculation is specified by Equations 12.6 through 12.9.

$$\text{Kinetic } CO_2 + OH^- \rightarrow HCO_3^- \tag{12.6}$$

$$\text{Kinetic } HCO_3^- \rightarrow CO_2 + OH^- \tag{12.7}$$

$$\text{Kinetic } MEA + CO_2 + H_2O \rightarrow MEACOO^- + H_3O^+ \tag{12.8}$$

$$\text{Kinetic } MEACOO^- + H_3O^+ \rightarrow MEA + CO_2 + H_2O \tag{12.9}$$

Power law expressions are used for the rate-controlled reactions. The kinetic parameters for reactions in Equations 12.6 through 12.9 are in Table 12.5:

$$r_j = k_j^\circ \exp\left(-\frac{E_j}{R_c}\left[\frac{1}{T} - \frac{1}{298.15}\right]\right)\prod_{i=1}^{N} a_i^{\alpha_{ij}} \tag{12.10}$$

TABLE 12.5 Constants for Power Law Expressions for the Absorption of CO_2 by MEA

Reaction No.	k_j°	E_j (cal/mol)
6	4.32e + 13	13249
7	2.38e + 13	29451
8	9.77e + 13	9855.8
9	2.18e + 13	14138.4

Source: AspenTech. Aspen Physical Properties System—Physical Property Methods, http://support.aspentech.com/, 2010.

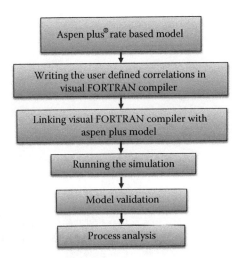

FIGURE 12.9 Methodology employed to model intensified absorber or stripper. (From Joel, A. S. et al., *Int. J. Greenhouse Gas Control*, 21, 91–100, 2014; Joel, A.S. et al., *Appl. Therm. Eng.*, 74, 47–53, 2015.)

12.1.4.1.2 Modeling and Simulation Methodology

The procedure used by Joel et al. [21] to model an intensified RPB absorber is shown in Figure 12.9.

The models were validated based on the experimental data presented by Jassim et al. [20] with relative percentage error of <12%.

12.1.4.1.3 Process Analysis on Intensified Absorber

12.1.4.1.3.1 Effect of Rotor Speed and Temperature on CO_2 Capture Level For this analysis, the model has the same geometry with Jassim et al. [20]. The inner and outer diameters of the RPB are 0.156 and 0.398 m, respectively, and the height of the packing is 0.025 m. Expamet packing with surface area of 2132 m^2/m^3 and void fraction of 0.76 was used. The lean MEA flow rate and flue gas flow rate are 0.66 kg/s and 2.87 kmol/h, respectively. The RPB operates at pressure 1 atm and two lean-MEA temperatures were used at 20.9°C and 39.5°C. The rotor speed was varied from 400 to 1200 rpm.

The effect of rotor speed on CO_2 capture is shown in Figure 12.10. As the rotor speed increases, the capture level increases in all the cases. This is because, increasing the rotor speed leads to increase in mass transfer, since there is more of droplets and film flow in the bed [88]. Cases 1 and 2 have the same operating temperature of 20.9°C, but different MEA concentrations of 56 and 75 wt%, respectively; Cases 3 and 4 have the same temperature of 39.5°C, but different MEA concentrations of 56 and 75 wt%, respectively. Case 2 has a percentage increase in capture level of 8.1% as the rotor speed increases from 400 to 1200 rpm, which is more than twice the percentage increase in the capture level for Case 1 when the rotor speed increases from 400 to 1200 rpm. This is because high viscosity effect of reducing mass transfer performance, which is more significant in Case 2 has been reduced at higher rotor speed leading to more quantity of CO_2 captured. At lower MEA concentrations, reaction rate and mass transfer have influence on the amount of CO_2 capture; while at higher concentrations (e.g., 75 wt%), the absorption process is mass transfer controlled. Figure 12.10 shows that the higher the MEA concentration, the higher the CO_2 capture level, again at higher temperature (39.5°C), CO_2 capture level is also higher.

12.1.4.1.3.2 Comparison between Intensified Absorber and Conventional Absorber To compare intensified and conventional absorber Table 12.6 is used as the input conditions. In both simulation runs, the capture level was fixed at 90%. MEA concentration of the conventional absorber was kept at 30 wt%. In the

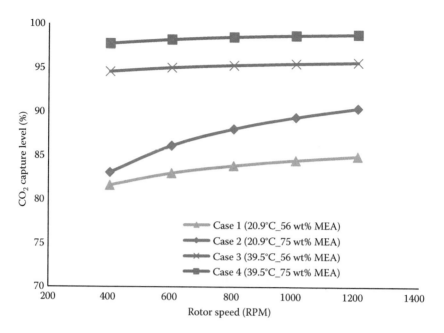

FIGURE 12.10   Effect of rotor speed on CO_2 capture level at 56 and 75 wt% MEA concentrations.

TABLE 12.6   Process Conditions for Conventional and RPB Absorbers

Description	Conventional Absorber		RPB Absorber	
	Flue Gas	Lean-MEA	Flue Gas	Lean-MEA
Temperature (K)	323.15	313.25	323.15	313.25
Pressure (10^5 Pa)	1.186	1.013	1.186	1.013
Total flow (kg/s)	0.0228	0.0454	0.0228	0.0440
L/G (kg/kg)	1.99		1.93	
		Mass Fraction		
H_2O	0.0030	0.6334	0.0030	0.23426
CO_2	0.0666	0.0618	0.0666	0.02574
N_2	0.9304	0	0.9304	0
MEA	0	0.3048	0	0.74000

RPB absorber simulation, MEA concentration of 74 wt% is used. Modeling and simulation of intensified absorber using RPB was done at rotor speed of 1000 rpm.

Keeping the CO_2 capture level at 90%, the simulation results for the conventional absorber using packed column and intensified absorber using RPB are shown in Table 12.7. The study shows a 12 times volume reduction for RPB compared to conventional packed column using the assumption used by Agarwal et al. [41] that the casing volume of RPB is taken as 4.5 times the RPB packing volume.

12.1.5 Technical, Economic, and Environmental Performance Comparison between Intensified Process and Conventional Process

The potential of any new technology to be attractive for businesses is for it to show better performance in terms of its technical, economic, and environmental impact compared to the existing ones.

TABLE 12.7 Comparison between Conventional and RPB Absorber

Description	Conventional Absorber	RPB Absorber
Height of packing (m)	3.85	0.2885 (r_o)
		0.078 (r_i)
Diameter (m)	0.395	0.0377 axial depth
Packing volume (m³)	0.4718	0.0091
Packing volume reduction		52 times
Volume of unit (m³)	0.4718[a]	0.04095[b]
Volume reduction factor		12 times
Specific area (m²/m³)	145	2132
Void fraction	0.79	0.76
Lean-MEA loading (mol CO₂/mol MEA)	0.2814	0.0483
Rich-MEA loading (mol CO₂/mol MEA)	0.4189	0.1069

Note: Where [a]excluding sump [b]using the assumption given by Agarwal et al. [41].

12.1.5.1 Technical Performance

12.1.5.1.1 Size Reduction

Joel et al. [21] carried out a comparative study between conventional absorber and intensified absorber and found the size reduction factor of about 12 times. Cheng et al. [22] stated that a reduction factor around 10 times can be found when using RPB stripper as compared to conventional stripper. Another unit operation in intensified PCC process that needs to be intensified is the heat exchanger. Li et al. [89] reported a reduction factor of four to six times for Printed Circuit Heat Exchanger (PCHE) compared to shell-and-tube heat exchanger operating at the same duty. Therefore, the major units in intensified PCC process have been reduced resulting in lower footprint.

12.1.5.1.2 Specific Energy Consumption

Energy consumption in intensified PCC process includes the electricity (motor) energy consumed to drive the intensified absorber and stripper and also heat energy required to generate steam for solvent regeneration. Therefore, in intensified PCC process, there is additional parasitic energy (i.e., motor energy) for carbon capture but some of this is offset by the reduction in the pumping energy required to get the solvent to the top of a packed tower. Cheng et al. [22] studied the regeneration energy of stripper with RPB and without RPB; their study shows 64% reduction in regeneration energy for the RPB stripper, which means that heat transfer is intensified in the RPB. There is decrease in the amount of vapor lean MEA required from reboiler to RPB due to improved heat transfer zone inside RPB, thereby decreasing its reboiler duty. Only Agarwal et al. [41] studied electricity consumption to drive the motors in RPB absorber using DEA solvent for carbon capture. Generally, the higher the rotating speed, the higher the electricity consumption by the motor. Study by Agarwal et al. [41] indicates that electricity consumption is quite low at 900 rpm, while the number increased significantly at 1500 rpm; this is because power consumption increases as the square of the rotational speed since most of it can be attributed to the kinetic energy gain of the liquid flowing through the RPB. More experimental studies are required to quantify the contribution of electricity used by motors to the overall energy consumption in intensified PCC process.

12.1.5.2 Economic Performance

A preliminary technical and economic analysis for intensified PCC process compared with conventional PCC process was reported by Wang et al. [15]. Wang gave an initial prediction that the capital cost of the intensified PCC process would be reduced by 16.7% for the whole intensified PCC process compared with the same capacity of a conventional PCC process. The reasons behind the capital cost reduction are (1) smaller size of units, average as 12 times; (2) stainless steel to be used instead of carbon steel to avoid corrosion since the process will be operated at higher solvent concentration. This means material price will roughly double, but owing to the smaller volume, the total cost of the unit will be less than that of conventional technology. In terms of energy consumption per unit mass of CO_2 captured, this will be similar for both the intensified PCC process and conventional PCC process [15].

12.1.5.3 Life Cycle Assessment

Detailed life cycle assessment for intensified PCC process should be performed in order to compare it with conventional PCC process.

12.1.6 Conclusions and Recommendations for Future Research

12.1.6.1 Conclusions

Application of PI to PCC process was discussed in this chapter by first defining PI, the motivation for applying PI to PCC and an evaluation of

- The current research status in intensified PCC regarding experimental rigs/studies (for intensified absorber and stripper)
- Challenges of PI for PCC, modeling, and simulation of PI for PCC
- Technical, economic, and environmental performance of PI compared to conventional packed column.

Recommendations were made for improving the intensified PCC process.

12.1.6.2 Recommendations for Future Research

For the application of RPB technology to PCC to be successful, the following recommendations were drawn: (1) It is important to use simple experiments to observe whether interfacial turbulence (i.e., Marangoni effect) exists in RPB absorber. (2) It is vital to develop dynamic models for the whole intensified PCC process for future work in process control. (3) It is necessary to combine Computational Fluid Dynamics (CFD) study and process modeling for scale-up study. (4) More detailed and systematic technical, economic, and environmental performance analysis should be performed.

Acknowledgment

The authors would like to acknowledge financial support from UK Engineering and Physical Sciences Research Council (EPSRC) (Ref: EP/M001458/1). The authors from the University of Hull would like to thank EU International Research Staff Exchange Scheme (IRSES) (Ref: PIRSES-GA-2013-612230).

References

1. Department of Energy and Climate Change (DECC). Solid fuels and derived gases statistics: Data sources and methodologies. 2012; http://webarchive.nationalarchives.gov.uk/20121217150421/ http://decc.gov.uk/en/content/cms/statistics/energy_stats/source/electricity/electricity.aspx (Accessed May 2013).
2. Mac Dowell N, Florin N, Buchard A, Hallett J, Galindo A, Jackson G et al. An overview of CO_2 capture technologies. *Energy and Environmental Science* 2010; 3:1645–69.
3. Wang M, Lawal A, Stephenson P, Sidders J, Ramshaw C. Post-combustion CO_2 capture with chemical absorption: A state-of-the-art review. *Chemical Engineering Research and Design* 2011; 89:1609–24.

4. Dugas RE. Pilot plant study of carbon dioxide capture by aqueous monoethanolamine. MSE Thesis, University of Texas at Austin, 2006.

5. Lawal A, Wang M, Stephenson P, Yeung H. Dynamic modeling and simulation of CO_2 chemical absorption process for coal-fired power plants. *Computer Aided Chemical Engineering* 2009; 27:1725–30.

6. Lawal A, Wang M, Stephenson P, Yeung H. Dynamic modeling of CO_2 absorption for post combustion capture in coal-fired power plants. *Fuel* 2009; 88:2455–62.

7. Lawal A, Wang M, Stephenson P, Koumpouras G, Yeung H. Dynamic modeling and analysis of post-combustion CO_2 chemical absorption process for coal-fired power plants. *Fuel* 2010; 89:2791–801.

8. Lawal A, Wang M, Stephenson P, Obi O. Demonstrating full-scale post-combustion CO_2 capture for coal-fired power plants through dynamic modeling and simulation. *Fuel* 2012; 101:115–28.

9. Kvamsdal HM, Jakobsen JP, Hoff KA. Dynamic modeling and simulation of a CO_2 absorber column for post-combustion CO_2 capture. *Chemical Engineering and Processing: Process Intensification* 2009; 48:135–44.

10. Reay D. The role of process intensification in cutting greenhouse gas emissions. *Applied Thermal Engineering* 2008; 28:2011–9.

11. Ramshaw C. The incentive for process intensification. *Proceedings of 1st International Conference Process Intensification for Chemical Industry*, 18, BHR Group, London, 1, 1995.

12. Ramshaw C, Mallinson RH. Mass transfer process. 1981; US Patent 4,283,255.

13. Stankiewicz AI, Moulijn JA. Process intensification: Transforming chemical engineering. *Chemical Engineering Progress* 2000; 96:22–34.

14. Reay D, Ramshaw C, Harvey A. *Process Intensification: Engineering for Efficiency, Sustainability and Flexibility*. Butterworth-Heinemann, Oxford, 2013.

15. Wang M, Joel AS, Ramshaw C, Eimer D, Musa NM. Process intensification for post-combustion CO_2 capture with chemical absorption: A critical review. *Applied Energy* 2015; 158:275–91.

16. BERR. Advanced power plant using high efficiency boiler/turbine. Report BPB010. BERR, Department for Business Enterprise and Regulatory Reform, 2006; http://webarchive.nationalarchives.gov.uk/20090609003228/http://www.berr.gov.uk/files/file30703.pdf (Accessed August 2015).

17. Mac Dowell N, Shah N. Dynamic modeling and analysis of a coal-fired power plant integrated with a novel split-flow configuration post-combustion CO_2 capture process. *International Journal of Greenhouse Gas Control* 2014; 27:103–19.

18. Agbonghae EO, Best T, Finney KN, Palma CF, Hughes KJ, Pourkashanian M. Experimental and process modeling study of integration of a micro-turbine with an amine plant. *Energy Procedia* 2014; 63:1064–73.

19. Henderson C. Toward zero emission coal-fired power plant. IEA Clean Coal Centre Reports, 2005, www.iea-coal.org.uk/documents/81379/5947/Toward-zero-emission-coal-fired-power-plant%C2%A0 (Accessed February 2015).

20. Jassim MS, Rochelle G, Eimer D, Ramshaw C. Carbon dioxide absorption and desorption in aqueous monoethanolamine solutions in a rotating packed bed. *Industrial and Engineering Chemistry Research* 2007; 46:2823–33.

21. Joel AS, Wang M, Ramshaw C, Oko E. Process analysis of intensified absorber for post-combustion CO_2 capture through modeling and simulation. *International Journal of Greenhouse Gas Control* 2014; 21:91–100.

22. Cheng H, Lai C, Tan C. Thermal regeneration of alkanolamine solutions in a rotating packed bed. *International Journal of Greenhouse Gas Control* 2013; 16:206–16.

23. Joel AS, Wang M, Ramshaw C. Modeling and simulation of intensified absorber for post-combustion CO_2 capture using different mass transfer correlations. *Applied Thermal Engineering* 2015; 74:47–53.

24. Biliyok C, Lawal A, Wang M, Seibert F. Dynamic modeling, validation and analysis of post-combustion chemical absorption CO_2 capture plant. *International Journal of Greenhouse Gas Control* 2012; 9:428–45.

25. Chen JF. The recent developments in the HiGee technology. Presentation at GPE-EPIC, June 14–17, 2009. 2009. http://inpact.inp-toulouse.fr/GPE-EPIC2009/images/presentation_chen.pdf (Accessed February 2014).

26. Podbielniak WJ. Centrifugal countercurrent contact apparatus. US Patent 2,004,011, 1935.

27. George T. Centrifugal countercurrent contacting machine. US Patent 2,176,982, 1939.

28. Adolph P. Process and apparatus for treating liquids with a gaseous medium. US Patent 2,281,616, 1942.

29. Wilhelm PC, Wilhelm DS. Apparatus for intimate contacting of two fluid media having different specific weight. US Patent 2,941,872, 1960.

30. Qian Z, Li Z, Guo K. Industrial applied and modeling research on selective H2S removal using a rotating packed Bed. *Industrial and Engineering Chemistry Research* 2012; 51:8108–16.

31. Lee J, Reay D, Ramshaw C. Post-combustion carbon capture research at Newcastle University, Presentation to PIN, May 2, 2012. 2012. www.pinetwork.org (Accessed June 2013).

32. Yi F, Zou H, Chu G, Shao L, Chen J. Modeling and experimental studies on absorption of CO_2 by Benfield solution in rotating packed bed. *Chemical Engineering Journal* 2009; 145:377–84.

33. Luo Y, Chu G, Zou H, Zhao Z, Dudukovic MP, Chen J. Gas–liquid effective interfacial area in a rotating packed bed. *Industrial and Engineering Chemistry Research* 2012; 51:16320–5.

34. Luo Y, Chu G, Zou H, Xiang Y, Shao L, Chen J. Characteristics of a two-stage counter-current rotating packed bed for continuous distillation. *Chemical Engineering and Processing: Process Intensification* 2012; 52:55–62.

35. Zhang L, Wang J, Xiang Y, Zeng X, Chen J. Absorption of carbon dioxide with ionic liquid in a rotating packed bed contactor: Mass transfer study. *Industrial and Engineering Chemistry Research* 2011; 50:6957–64.

36. Yu C, Cheng H, Tan C. CO_2 capture by alkanolamine solutions containing diethylenetriamine and piperazine in a rotating packed bed. *International Journal of Greenhouse Gas Control* 2012; 9:136–47.

37. Tan C-S, Chen J-E. Absorption of carbon dioxide with piperazine and its mixtures in a rotating packed bed. *Separation and Purification Technology* 2006; 49:174–80.

38. Cheng H, Tan C. Removal of CO_2 from indoor air by alkanolamine in a rotating packed bed. *Separation and Purification Technology* 2011; 82:156–66.

39. Lin C, Lin Y, Tan C. Evaluation of alkanolamine solutions for carbon dioxide removal in cross-flow rotating packed beds. *Journal of Hazardous Materials* 2010; 175:344–51.

40. Lin CC, Chen YW. Performance of a cross-flow rotating packed bed in removing carbon dioxide from gaseous streams by chemical absorption. *International Journal of Greenhouse Gas Control* 2011; 5:668–75.

41. Agarwal L, Pavani V, Rao D, Kaistha N. Process intensification in HiGee absorption and distillation: design procedure and applications. *Industrial and Engineering Chemistry Research* 2010; 49:10046–58.

42. Rajan S, Kumar M, Ansari MJ, Rao D, Kaistha N. Limiting gas liquid flows and mass transfer in a novel rotating packed bed (HiGee). *Industrial and Engineering Chemistry Research* 2010; 50:986–97.

43. Rahimi MR, Mosleh S. CO_2 removal from air in a countercurrent rotating packed bed, experimental determination of height of transfer unit. *Advances in Environmental Technology* 2015; 1:19–24.

44. Kang J, Wong DS, Jang S, Tan C. A comparison between packed beds and rotating packed beds for CO_2 capture using monoethanolamine and dilute aqueous ammonia solutions. *International Journal of Greenhouse Gas Control* 2016; 46:228–39.

45. Cheng H, Tan C. Reduction of CO_2 concentration in a zinc/air battery by absorption in a rotating packed bed. *Journal of Power Sources* 2006; 162:1431–6.

46. Cheng H, Shen J, Tan C. CO_2 capture from hot stove gas in steel making process. *International Journal of Greenhouse Gas Control* 2010; 4:525–31.

47. Chen Y, Liu H, Lin C, Liu W. Micromixing in a rotating packed bed. *Journal of Chemical Engineering of Japan* 2004; 37:1122–8.

48. Barham H, Brahim S, Rozita Y, Mohamed K. Carbon steel corrosion behaviour in aqueous carbonated solution of MEA/bmim DCA. *International Journal of Electrochemical Science* 2011; 6:181–98.

49. Kittel J, Gonzalez S. Corrosion in CO_2 post-combustion capture with alkanolamines—A Review. *Oil and Gas Science and Technology–Revue d'IFP Energies nouvelles* 2014; 69:915–29.

50. Shao R, Stangeland A. Amines used in CO_2 capture-health and environmental impacts. http:// bellona org/filearchive/fil_Bellona_report_September_2009_-_Amines_used_in_CO2_capture pdf [Accessed May 2015] 2009; 49.

51. DuPart, MS, Bacon TR, Edwards, DJ. Understanding corrosion in alkanolamines gas treating plants Part 1&2: Proper mechanism diagnosis optimizes amine operations. *Hydrocarbon Processing*, Gulf Publishing Co, Houston, TX, 1993.

52. Nainar M, Veawab A. Corrosion in CO_2 capture unit using MEA-piperazine blends. *Energy Procedia* 2009; 1:231–5.

53. Kittel J, Idem R, Gelowitz D, Tontiwachwuthikul P, Parrain G, Bonneau A. Corrosion in MEA units for CO_2 capture: Pilot plant studies. *Energy Procedia* 2009; 1:791–7.

54. Soosaiprakasam IR, Veawab A. Corrosion and polarization behavior of carbon steel in MEA-based CO_2 capture process. *International Journal of Greenhouse Gas Control* 2008; 2:553–62.

55. Sternling C, Scriven L. Interfacial turbulence: Hydrodynamic instability and the Marangoni effect. *AIChE Journal* 1959; 5:514–23.

56. Chambers H, Wall MA. Some factors affecting the design of centrifugal gas absorbers. *Transactions of the American Institute of Chemical Engineers* 1954; 32:S96–S107.

57. Buzek J, Podkański J, Warmuziński K. The enhancement of the rate of absorption of CO_2 in amine solutions due to the Marangoni effect. *Energy Conversion and Management* 1997; 38:S69–74.

58. Sobieszuk P, Pohorecki R, Cygański P, Kraut M, Olschewski F. Marangoni effect in a falling film microreactor. *Chemical Engineering Journal* 2010; 164:10–5.

59. Yu L, Zeng A, Yu KT. Effect of interfacial velocity fluctuations on the enhancement of the mass-transfer process in falling-film flow. *Industrial and Engineering Chemistry Research* 2006; 45:1201–10.

60. Sun Z, Yu K, Wang S, Miao Y. Absorption and desorption of carbon dioxide into and from organic solvents: Effects of Rayleigh and Marangoni instability. *Industrial and Engineering Chemistry Research* 2002; 41:1905–13.

61. Okhotsimskii A, Hozawa M. Schlieren visualization of natural convection in binary gas–liquid systems. *Chemical Engineering Science* 1998; 53:2547–73.

62. Chen W, Chen S, Yuan X, Zhang H, Liu B, Yu K. PIV measurement for Rayleigh convection and its effect on mass transfer. *Chinese Journal of Chemical Engineering* 2014; 22:1078–86.

63. Austgen DM, Rochelle GT, Chen CC. Model of vapor-liquid equilibria for aqueous acid gas-alkanolamine systems. 2. Representation of hydrogen sulfide and carbon dioxide solubility in aqueous MDEA and carbon dioxide solubility in aqueous mixtures of MDEA with MEA or DEA. *Industrial and Engineering Chemistry Research* 1991; 30:543–55.

64. Austgen DM, Rochelle GT, Peng X, Chen CC. Model of vapor-liquid equilibria for aqueous acid gas-alkanolamine systems using the electrolyte-NRTL equation. *Industrial and Engineering Chemistry Research* 1989; 28:1060–73.

65. Li Y, Mather AE. Correlation and prediction of the solubility of carbon dioxide in a mixed alkanolamine solution. *Industrial & Engineering Chemistry Research* 1994; 33:2006–15.

66. Liu Y, Zhang L, Watanasiri S. Representing vapor–liquid equilibrium for an aqueous MEA-CO_2 system using the electrolyte nonrandom-two-liquid model. *Industrial and Engineering Chemistry Research* 1999; 38:2080–90.

67. Edwards T, Maurer G, Newman J, Prausnitz J. Vapor-liquid equilibria in multicomponent aqueous solutions of volatile weak electrolytes. *AIChE Journal* 1978; 24:966–76.
68. Gibbs J. *The Scientific Papers of J Willard Gibbs Thermodynamics*, Vol. 1, Dover Publications Inc., New York, 1961.
69. Aronu UE, Gondal S, Hessen ET, Haug-Warberg T, Hartono A, Hoff KA et al. Solubility of CO_2 in 15, 30, 45 and 60 mass% MEA from 40 to 120°C and model representation using the extended UNIQUAC framework. *Chemical Engineering Science* 2011; 66:6393–406.
70. Mason JW, Dodge BF. Equilibrium absorption of carbon dioxide by solutions of the ethanolamines. *Transactions of the American Institute of Chemical Engineers* 1936; 32:27–48.
71. Zhang Y, Chen H, Chen C, Plaza JM, Dugas R, Rochelle GT. Rate-based process modeling study of CO_2 capture with aqueous monoethanolamine solution. *Industrial and Engineering Chemistry Research* 2009; 48:9233–46.
72. AspenTech. Aspen physical properties system—Physical property methods. 2010, http://support.aspentech.com/ (Accessed May 2012).
73. Hankinson RW, Thomson GH. A new correlation for saturated densities of liquids and their mixtures. *AIChE Journal* 1979; 25:653–63.
74. Tung H, Mah RS. Modeling liquid mass transfer in HIGEE separation process. *Chemical Engineering Communications* 1985; 39:147–53.
75. Chen Y, Lin F, Lin C, Tai CY, Liu H. Packing characteristics for mass transfer in a rotating packed bed. *Industrial and Engineering Chemistry Research* 2006; 45:6846–53.
76. Rajan SK. Limiting gas liquid flows and mass transfer in a novel rotating packed bed. M Tech Thesis, Department of Chemical Engineering, Indian Institute of Technology Kanpur, 2008.
77. Onda K, Sada E, Takeuchi H. Gas absorption with chemical reaction in packed columns. *Journal of Chemical Engineering of Japan* 1968; 1:62–6.
78. Cheng H, Tan C. Carbon dioxide capture by blended alkanolamines in rotating packed bed. *Energy Procedia* 2009; 1:925–32.
79. Reddy KJ, Gupta A, Rao DP, Rama OP. Process intensification in a HIGEE with split packing. *Industrial and Engineering Chemistry Research* 2006; 45:4270–7.
80. Puranik SS, Vogelpohl A. Effective interfacial area in irrigated packed columns. *Chemical Engineering Science* 1974; 29:501–7.
81. Burns J, Jamil J, Ramshaw C. Process intensification: Operating characteristics of rotating packed beds—Determination of liquid hold-up for a high-voidage structured packing. *Chemical Engineering Science* 2000; 55:2401–15.
82. Lin C, Chen Y, Liu H. Prediction of liquid holdup in countercurrent-flow rotating packed bed. *Chemical Engineering Research and Design* 2000; 78:397–403.
83. Herzog HJ. Scaling up carbon dioxide capture and storage: From megatons to gigatons. *Energy Economics* 2011; 33:597–604.
84. Shi X, Xiang Y, Wen L, Chen J. CFD analysis of liquid phase flow in a rotating packed bed reactor. *Chemical Engineering Journal* 2013; 228:1040–9.
85. Yang W, Wang Y, Chen J, Fei W. Computational fluid dynamic simulation of fluid flow in a rotating packed bed. *Chemical Engineering Journal* 2010; 156:582–7.
86. Llerena-Chavez H, Larachi F. Analysis of flow in rotating packed beds via CFD simulations—Dry pressure drop and gas flow maldistribution. *Chemical Engineering Science* 2009; 64:2113–26.
87. Chen Y. Correlations of mass transfer coefficients in a rotating packed bed. *Industrial and Engineering Chemistry Research* 2011; 50:1778–85.
88. Burns JR, Ramshaw C. Process intensification: Visual study of liquid maldistribution in rotating packed beds. *Chemical Engineering Science* 1996; 51:1347–52.
89. Li X, Le Pierres R, Dewson SJ. Heat exchangers for the next generation of nuclear reactors. *Proceedings of the 2006 International Congress on Advances in Nuclear Power Plants (ICAPP'06)*, 2006.

12.2 Process Simulation and Integration of Natural Gas Combined Cycle (NGCC) Power Plant Integrated with Chemical Absorption Carbon Capture and Compression

Xiaobo Luo and Meihong Wang

12.2.1 Introduction

Controlling greenhouse gases emission plays an important role to address the global climate change challenges. Intergovernmental Panel on Climate Change (IPCC) (2005) reported that fossil-fired power generation is the single largest source of CO_2 emissions. However, with the advantages of high energy density and high reliability, fossil energy is projected to remain a major source of energy in the near future (Mac Dowell and Shah, 2013).

Natural gas is a major source of electricity generation and currently accounts for around 22% of global electricity generation capacity (BP P.L.C., 2014). Natural gas can be burned more clearly than other fossil fuels such as coal and oil. Another remarkable advantage of natural gas power generation is its high net low heating value (LHV) efficiency close to 60% with the application of combined cycle gas turbine (CCGT) technology. Table 12.8 shows CO_2 emissions from different fuels in the world for 2015 from the United States Energy Information Administration (EIA) (2016). To generate the same amount of electricity, burning natural gas emits about 42% less carbon dioxide than burning coal.

With these advantages, more natural gas combined cycle (NGCC) power plants would be built especially in the developed countries (BBC, 2015). But NGCC power generation is not a zero carbon emission solution. The International Energy Agency (IEA) (2010) reported that NGCC power plants integrated with carbon capture processes would contribute 5% electricity supply in 2050 in order to achieve the target of CO_2 emission control.

Among different carbon capture technologies (see in Figure 12.11) used for fossil fuel fired power plants, solvent-based PCC is widely regarded as the most possible technology to be implemented (IEAGHG, 2012; Wang et al., 2011). Amine solvent such as monoethanolamine (MEA) is often used to absorb the CO_2 from the flue gases. But for the full commercial scale application of this carbon capture technology, the main barrier is the high-cost increment of the electricity including massive capital costs and high thermal energy penalty to the power plants (Luo and Wang, 2015; Luo et al., 2015; Marchioro Ystad et al., 2013). The cost of electricity from NGCC power generation would increase to £144.1 from £66 per MWh when it is integrated with a PCC process, reported by the Department of Energy and Climate Change (DECC) of the United Kingdom (2013).

The main contributors for this energy penalty include (1) steam extracted from the power plant for solvent generation, (2) electricity consumption of the CO_2 compression train, and (3) electricity

TABLE 12.8 CO_2 Emissions from Different Fuels

Fuel	CO_2 Content (kg/GJ)	Heat Rate (kg/kWh)	CO_2 Emission (kg/kWh)
Bituminous coal	98.25	10,644.50	0.94
Subbituminous coal	101.79	10,644.50	0.98
Lignite coal	103.08	10,644.50	0.98
Natural gas	56.03	10,924.09	0.55
Distillate oil (no. 2)	77.23	10,902.99	0.76
Residual oil (no. 6)	83.23	10,902.99	0.82

Source: U.S. EIA. How much carbon dioxide is produced per kilowatthour when generating electricity with fossil fuels? www.eia.gov/tools/faqs/faq.cfm?id=74&t=11, 2016.

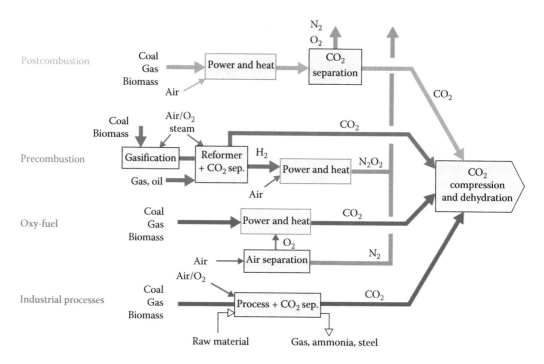

FIGURE 12.11 Processes of CO_2 capture technologies. (From IPCC, Carbon dioxide capture and storage, 2005.)

consumption of the PCC process. Many researches were conducted on how to improve the performance of the PCC process to reduce its energy requirement (Marchioro Ystad et al., 2013). For capturing 1 ton of CO_2 from flue gases from NGCC power plants, the typical range of required thermal energy is from 3.4 GJ to 4.2 GJ (Abu-Zahra et al., 2007b; Canepa and Wang, 2015; Kvamsdal et al., 2007; Mac Dowell and Shah, 2013; Marchioro Ystad et al., 2013). For a certain task, the required thermal energy is determined by several key process control variables such as CO_2 capture level, solvent concentration, and CO_2 loading in lean solvent. The lean solvent CO_2 loading was found to have a major effect on the thermal energy requirement, while its economic range depends on the interreactions with other parameters of the processes (Abu-Zahra et al., 2007a,b; Mac Dowell and Shah, 2013). High operating pressure in the stripper leads to a reduction of energy requirement of both solvent regeneration and the CO_2 compression. But it has an upper limitation because higher pressure would require higher temperature, which results in a higher thermal degradation of the solvent inside the stripper (Davis and Rochelle, 2009).

Normally, the captured CO_2 would be pressurized to 100–150 bar for pipeline transport for geologic storage using a multistage compression train. Except for being very expensive, the CO_2 compression train is also a big electricity power consumer (Jordal et al., 2012; Luo et al., 2014b; Marchioro Ystad et al., 2013; Witkowski et al., 2013) because of a big temperature increment during the compression process (Darby, 2001). One improvement is to recover compression heat using low-temperature heat recovery technologies such as Rankine cycle turbine (Marchioro Ystad et al., 2013). Recently, aiming at addressing the challenge of high total pressure ratio of the CO_2 compression, an advanced supersonic shock wave compression technology was developed (Lawlor, 2009). The special design can achieve very high pressure ratio per stage and only two stages are required for the CO_2 compression train. At the same time, the discharge temperature of each stage is as high as 246–285°C (Witkowski et al., 2013), providing an opportunity for heat integration between the compression train and power plant and PCC process.

12.2.2 Model Development of NGCC Power Plant

12.2.2.1 CCGT Technology

For gas-fired power plant, CCGT is the prevailing technology because of its high thermal efficiency (IEA, 2008). Figure 12.12 presents a typical schematic of a CCGT power plant, which is a dual-cycle process. Air and fuel combusts to generate heat and then gas mixture expands through gas turbine to generate a part of electricity. Exhaust gas enters the heat recovery steam generator (HRSG) by which waste heat of the exhaust gas is recovered to create steam. In the steam cycle, steams at different pressure levels enter multisteam turbines to generate another part of electricity.

12.2.2.2 Model Development

For process simulation of a power plant, some professional software packages such as GateCycle (General Electric, 2014) and GT-Pro (Thermoflow Inc., 2016) provide a good prediction of the plant performance. In this study, considering integrating with the PCC process model, the power plant model was developed in Aspen Plus and was validated with the results of GT Pro in a benchmark report (IEAGHG, 2012).

The process model of this 435 MW$_e$ NGCC power plant consists of a GE 9351FB gas turbine, a triple-level pressures reheat HRSG, and steam turbines. It should be noticed that, in the real process, the gas turbine is integral equipment (see Figure 12.12), but it is separated into air compression, combustion, and expansion as three unit processes in the model in Aspen Plus (See in Figure 12.13). The same philosophy was applied to the model of HRSG. The integrated system was separated into serials of heat exchangers, pumps, and steam tanks with the multisteam turbine.

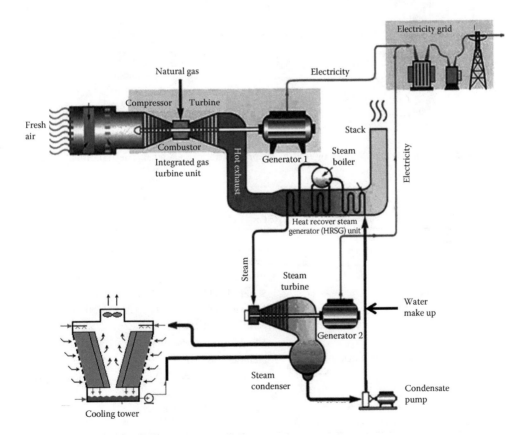

FIGURE 12.12 CCGT power plant schematic. (Adapted from blog.gerbilnow.com, 2012.)

Figure 12.13 displays the model flow sheet of the NGCC power plant. Fresh air is compressed to mix with natural gas to enter the combustion chamber. The exhaust gas generated from the combustion enters the turbine at high temperature and then expands to generate a part of electricity. Flowing through the HRSG, the hot gas exchanges heat to the steam cycle to generate steams at different pressure levels of 170, 40, and 5 bar. These steams are lined to the high pressure steam turbine (HP-ST), the intermediate pressure steam turbine (IP-ST), and the low pressure steam turbine (LP-ST), respectively, to generate another part of electricity. One design feature of this model is that both the temperature and pressure of the steams are higher than what are currently typical (Canepa and Wang, 2015; Marchioro Ystad et al., 2013) as similar steam conditions are considered to be applied in NGCC power plants by 2020 (IEAGHG, 2012). Some key model parameters can be checked in Table 12.9.

12.2.2.3 Model Validation

For full commercial scale NGCC power plant simulations, there is no variable operational or experimental data for model validation purposes. In order to make a brief validation, the simulation results of this model developed in Aspen Plus were compared with the simulation results of the model using another software package, GT Pro (IEAGHG, 2012). The comparison results can be seen in Table 12.10, which appear to be in good agreement.

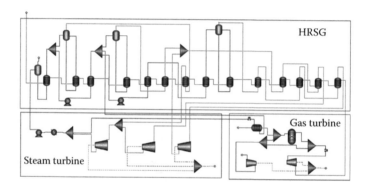

FIGURE 12.13 The flow sheet of NGCC power plant. (Reprinted from Luo, X. et al., *Fuel*, 151, 110–117, 2015.)

TABLE 12.9 Model Parameters of the NGCC Power Plant

Parameters		Value
Gas turbine	Type	GE 9371FB
	Pressure ratio	18.2
Steam turbine	Steam inlet of HP turbine (bar/°C)	172.6/601.7
	Steam inlet of IP turbine (bar/°C)	41.5/601
	Steam inlet of LP turbine (bar/°C)	5.8/293.1
	HP/IP/LP turbine efficiencies (%)	92/94/90
Minimum temperature approach of HRSG	Steam and gas (°C)	25
	Gas and boiling water (°C)	10
	Water liquid and gas (°C)	10
	Approach of economizer (°C)	5
Condenser pressure and temperature (bar/°C)		0.039/29.0

TABLE 12.10 Model Validation Results

Input Conditions	
Parameters	Input Value
Natural gas composition (vol) CH_4, C_2H_6, C_3–C_5, CO_2, N_2	89%, 7%, 1.11%, 2.0%, 0.89%
Flow rate of natural gas (kg/s)	16.62
Flow rate of air (kg/s)	656.94
Temperature of flue gas to HRSG (°C)	638.4

Validation Results		
Parameters	IEAGHG (2012)	This Study
Flow rate of flue gas to HRSG (kg/s)	673.58	673.57
HP turbine inlet pressure and temperature (bar/°C)	172.5/601.7	172.6/601.7
IP turbine inlet pressure and temperature (bar/°C)	41.4/601.5	41.5/601.0
LP turbine inlet pressure and temperature (bar/°C)	5.81/293.3	5.8/293.1
Condenser pressure and temperature (bar/°C)	0.04/29.2	0.039/29.0
Gas turbine power output (MW_e)	295.238	295.03
Steam turbine power output (MW_e)	171.78	170.71
Net plant power output (MW_e)	455.15	453.872
Net plant efficiency (%, LHV)	58.87	58.74

12.2.3 Model Development of PCC Process

12.2.3.1 PCC Process

PCC process is a chemical reactive absorption using chemical solvent to absorb CO_2 from the flue gas from fossil-fired power plants or industrial facilities. The flue gas is treated by a preconditioning process (desulfurizing and cooling) and then enters the absorber, in which the solvent reacts with the CO_2. The scrubbed flue gas is emitted to the atmosphere and the CO_2-rich solvent is discharged from the bottom of the absorber and enters the stripper. The CO_2-rich solvent is regenerated inside the stripper with heat input to the reboiler. The regenerated solvent is cooled and recirculated to the absorber for reuse. Figure 12.14 shows a typical PCC process schematic for a power plant.

12.2.3.2 Model Development

Generally, the accuracy of the performance prediction by process simulations depends on the complexity of the model. Rate-based mass transfer and kinetics controlled reactions offer better accuracy than equilibrium approaches for this reactive absorption process (Kenig et al., 2001). Zhang et al. (2009) presents the details of the rate-based model development of PCC process in Aspen Plus. The study examined several key parameters and operational variables, such as lean solvent loading, rich solvent loading, capture level, and the temperature profiles of the absorber against the University of Texas at Austin (Dugas, 2006) for a validation purpose. The comparison results show good agreements between the simulation results and experimental data. The study proves that the rate-based model using Aspen Plus is capable of providing an acceptable accuracy for the performance prediction of PCC process.

Aqueous phase chemical reactions involved in the MEA–H_2O–CO_2 system can be expressed as below equilibrium reactions and kinetics controlled reactions.

The equilibrium reactions are presented by Equations 12.11 through 12.13.

Water dissociation:

$$2H_2O \leftrightarrow H_3O^+ + OH^-$$ (12.11)

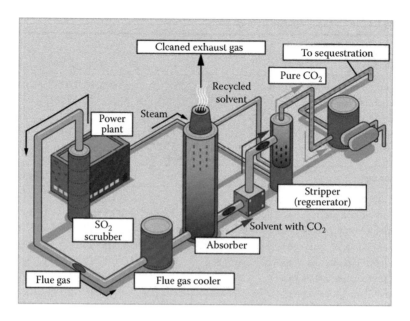

FIGURE 12.14   PCC process schematic. (Courtesy of Clean Air Task Force, 2015.)

Dissociation of carbonate:

$$HCO_3^- + H_2O \leftrightarrow H_3O^+ + CO_3^{2-} \tag{12.12}$$

Dissociation of the protonated amine:

$$MEAH^+ + H_2O \leftrightarrow H_3O^+ + MEA \tag{12.13}$$

The kinetics controlled reactions are presented by Equations 12.14 through 12.17.
   Dissociation of CO_2:

$$CO_2 + 2H_2O \rightarrow H_3O^+ + HCO_3^- \tag{12.14}$$

$$H_3O^+ + HCO_3^- \rightarrow CO_2 + 2H_2O \tag{12.15}$$

Carbonate formation:

$$MEACOO^- + H_2O \rightarrow HCO_3^- + MEA \tag{12.16}$$

$$HCO_3^- + MEA \rightarrow MEACOO^- + H_2O \tag{12.17}$$

Chemical equilibrium constants of those reactions are calculated by Equation (12.18) (Austgen et al., 1989).

$$\ln K_j = C_1 + \frac{C_2}{T} + C_3 \ln T + C_4 T, \tag{12.18}$$

TABLE 12.11   Correlations for Chemical Equilibrium Constants in Equation 12.18

Reaction	C_1	C_2	C_3	C_4	Source
(1.1)	132.899	−13445.90	−22.4773	0	Edwards et al. (1978)
(1.2)	216.049	−12431.70	−35.4819	0	Edwards et al. (1978)
(1.3)	−3.03833	−7008.36	0	−0.0031348	Canepa et al. (2013)

TABLE 12.12   Parameters k and E in Equation 12.19

Reaction	k (kmol/m³/s)	E (J/kmol)
(1.4)	4.32e +·13	5.54709e + 7
(1.5)	2.38e + 17	1.23305e + 8
(1.6)	9.77e + 10	4.12643e + 7
(1.7)	2.18e + 18	5.91947e + 7

Source: Canepa, R. et al., *Proceedings of the Institution of Mechanical Engineers, Part E: Journal of Process Mechanical Engineering,* 227 (2), 89–105, 2013.

where K_j is the chemical equilibrium constant for each equation j, T is system temperature, C_1, C_2, C_3, C_4 are correlations for Henry's constants, whose values are seen in Table 12.11.

Power law expressions are used for the rate-controlled reactions. The kinetic parameters for reactions in Equations 12.14 through 12.17 are in Table 12.12.

$$r = kT^n \exp\left(-\frac{E}{RT}\right) \prod_{i=1}^{N} C_i^{a_i} \tag{12.19}$$

The details of the rate-based model for the absorber and the stripper developed with Aspen Plus can refer to previous studies (Canepa and Wang, 2015; Canepa et al., 2013). Electrolyte-Non-random two-Liquid (NRTL) (Chen and Song, 2004) method is used to describe the thermodynamic and physical properties. Two-film theory (Whitman, 1962) is used to describe the mass transfer of components across the gas phase and the liquid phase inside the columns. The above chemical reactions are assumed to occur only in the liquid film. The close-loop process model was developed at the pilot scale and validated with the experimental data (Canepa and Wang, 2015). Then the model was scaled up (Canepa et al., 2013; Lawal et al., 2012; Liu et al., 2015) to match the NGCC power plant at industry scale, which is described in Section 12.2.2 Figure 12.15.

The design features and process parameter of scale-up PCC process are presented in Table 12.13. The results show that the large sizes of the columns are required because of huge flow rate of the glue gas and normal operating pressure of the PCC process. It is noticed that, in the results, only the cross-sectional areas were given to present a generic sizing of the absorber and the stripper. The main consideration is that

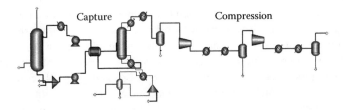

FIGURE 12.15   The flow sheet of PCC process and compression train. (Reprinted from Luo, X. et al., *Fuel,* 151, 110–117, 2015.)

TABLE 12.13 Model Parameters of PCC Process

Parameter	Value
CO_2 concentration in flue gas (mol%)	4.5
CO_2 capture level (%)	90
CO_2 captured (kg/s)	41.11
Columns flooding (%)	65
Lean loading (mol CO_2/mol MEA)	0.32
Rich loading (mol CO_2/mol MEA)	0.461
L/G (mol/mol)	1.79
Reboiler duty (kW)	186,805
Reboiler duty (GJ/ton CO_2)	4.54
Lean solvent MEA concentration (wt%)	32.5
Lean solvent temperature (°C)	30
Absorber pressure (bar)	1.07
Absorber pressure loss (bar)	0.069
Absorber packing	IMTP no. 40
Absorber packing height (m)	25
Absorber columns cross-sectional area (m^2)	307.91
Stripper pressure (bar)	2.1
Stripper pressure loss (bar)	0.01355
Stripper packing	Flexipack 1Y
Stripper packing height (m)	15
Stripper cross-sectional area (m^2)	81.71

Source: Luo, X. et al., *Fuel*, 151, 110–117, 2015.

the upper limit of the diameter of a column will vary with different column internal technologies used by different equipment manufacturers. Even, rather than a cylindrical metal material tower, a concrete rectangular tower with appropriate lining could be used for the absorber (IEAGHG, 2012) in a power plant to get a better economic profile as the operating pressure of the absorber is near the atmosphere pressure.

12.2.4 Model Development of CO_2 Compression Train

For onshore and offshore transport of large volumes of CO_2, pipeline is the preferred method (Zhang et al., 2006). The supercritical phase is regarded as the most energy-efficient condition for the CO_2 pipeline transport due to high density and low viscosity (Luo et al., 2014b). Thus, one important operating practice is to maintain the pressure well above the critical pressure of the CO_2 stream. A compression train is required to pressurize the CO_2 stream from PCC captured plant to reach an entry pressure as high as 110–150 bar (Luo et al., 2014a,b; Roussanaly et al., 2014).

There are several types of compression configurations for CO_2 pipeline transport (McCollum and Ogden, 2006; Zhang et al., 2006). Through compression process, the volume of the CO_2 stream would decrease sharply, which results in a large reduction of the impeller diameter between the first stage and the last stage. And the efficiency decays obviously because of the temperature rise with pressure rise in this adiabatic process. Thus, a conventional compressor normally needs 6–16 stages. In the study of Witkowski et al. (2013), a thermal performance evaluation was carried out for different configurations with 6–12 stages compressor based on centrifugal compressor and integrally geared compressor. Luo et al. (2014b) conducted a comprehensive techno-economic evaluation for different compression configurations with an objective function of minimizing the annual cost including annualized capital cost, operating and maintenance cost, and energy cost. The optimal configuration consists of six stage compressor with intercoolers at an exit temperature of 20°C following a pump. The conventional multistage

TABLE 12.14 Model Parameters of Compression Train

Parameters	Value
Flow rate of CO_2 stream (kg/s)	41.11
Inlet pressure (bar)	1.9
Inlet temperature (°C)	20
Outlet pressure (bar)	136
Stage number	2
Pressure ratio per stage	8.65
Isentropic efficiency (%)	85
1st stage exit temperature (°C)	214.5
2nd stage exit temperature (°C)	230.5
Intercoolers exit temperature (°C)	20
Pressure drop of intercoolers (bar)	0.3
Power consumption (MW_e)	14.8

Source: Luo, X. et al., *Fuel*, 151, 110–117, 2015.

compressor not only has high costs for equipment material, construction, and installation, but also need a large installation space, which results in a great capital investment.

Specifically for the CO_2 compression, a supersonic shock wave compression technology is developed (Ramgen Power Systems, 2008). With a high pressure ratio of single stage, the shock wave compression only needs two stages of compression compared with 6–16 stages for the conventional multistage approach. The potential capital cost saving is up to 50% (Ciferno et al., 2009). Meanwhile, the exit temperature of single stage is as high as 246–285°C (Witkowski et al., 2013) due to high pressure ratio, providing chances for compression heat integration with the NGCC power plant and the PCC process.

For the simulation of this study, the supersonic shock wave compression technology was developed also in Aspen Plus and validated with the published data. And then the inputs and discharge pressure requirement of the compression train were set to be consistent with boundary conditions of the PCC process. Table 12.14 summarizes the model parameters of the compression train.

12.2.5 Integration of NGCC with PCC and CO_2 Compression

When a NGCC power plant is designed or retrofitted to integrate with a PCC process and compression, there are some structure modifications required for basic interfaces, including the following: (1) flue gas is lined from the exit of HRSG of the power plant to the PCC process, (2) low pressure steam is extracted from the steam cycle of the power plant to provide heat for solvent regeneration in the PCC process, (3) steam condensate generated by the PCC process returns to the steam cycle of the power plant, (4) the power plant provides electrical power supply for the PCC process and CO_2 compression.

The process flow diagram can be seen in Figure 12.16. Flue gas leaves the HRSG at a temperature of around 80°C and enters a gas conditioning unit, which consists of a direct contact column (DCC), a water cycling pump and a blower. The flue gas is then cooled down to 40°C–50°C (Kvamsdal et al., 2011) by a spray of water at 25°C, in order to improve the absorption efficiency and to reduce solvent evaporation losses in the absorber (Wang et al., 2011). At the same time, a part of the water is removed from the flue gas due to the condensation. The flue gas is pressurized to a certain pressure by the blower before it feeds into the absorber.

The solvent regeneration process requires a great heat input, which is normally provided by extracting low pressure steam from the steam cycle of the NGCC power plant into the stripper reboiler of the PCC process. The flow rate of the steam extraction is decided by the operating conditions of PCC process and has a large impact to the output of the power plant. Considering that high temperature would result in thermal degradation of the solvent in the reboiler and the stripper, normally, the temperature of the reboiler is maintained between 110°C and 130°C at an operating pressure of 1.6–2.1 bar. There are three

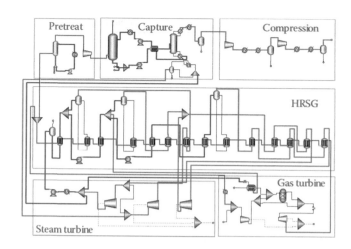

FIGURE 12.16 The flow sheet of NGCC power plant with EGR integrated with PCC process and compression. (Reprinted from Luo, X. et al., *Fuel*, 151, 110–117, 2015.)

potential positions, clutched turbine, throttled turbine, and floating crossover pressure, in the NGCC power plant process for steam extraction (Kang et al., 2011). In this study, the steam is extracted off from the floating Intermediate Pressure (IP)/Low Pressure (LP) crossover, the most feasible solutions for steam extraction (Lucquiaud and Gibbins, 2009), at 5.8 bar and 303°C. Before the steam enters the reboiler, it is cooled down just above saturation temperature with a spray of the condensate circulated from the reboiler, which helps to reduce the requirement of steam to be extracted from the power plant. After heat exchange inside the reboiler, the steam is cooled down to condensate, which then is returned to the deaerator in the HRSG of the power plant for cycling.

12.2.6 Exhaust Gas Recirculation

Compared with coal fired power plants, NGCC power plant emits only half CO_2 per unit power. Consequently, the CO_2 concentration in flue gas from a NGCC power plant is as low as 3–4 mol%, while it is 11–13 mol% for a coal-fired power plant. Low CO_2 concentration causes low absorption efficiency, while large flow rate of inert gas requires large equipment size in PCC capture plant (Biliyok et al., 2013; Jonshagen et al., 2011). Exhaust gas recirculation (EGR) is regarded as an effective solution. The flue gas leaving the HRSG is split into two streams. One is lined to the PCC process and the other is cooled and recirculated to the compressor inlet where it is mixed with fresh air (see Figure 12.16).

The underpinning of EGR is that the O_2 concentration in the flue gas leaving the HRSG is still high (11.41 mol% in this study). Even though EGR is applied, a relatively high oxygen concentration in the combustion air can be ensured with an appropriate recirculation ratio.

Thus the flow rate of fresh air intake reduces greatly. Consequently, the flow rate of flue gas going to be treated by the PCC process would largely decreases, while the CO_2 concentration in the flue gas increase obviously.

EGR ratio is defined as

$$\text{EGR ratio} = \frac{\text{Volume flow of recirculated exhaust gas}}{\text{Volume flow of exhaust gas}} \tag{12.20}$$

The impacts of EGR can be seen in Figures 12.17 and 12.18. In Figure 12.17, exhaust gas composition is shown as a function of the EGR ratio. With the increase of EGR ratio, the concentrations of N_2, H_2O, and CO_2 increase. But O_2 concentration decreases because less oxygen is available in the recirculated stream. Figure 12.18 illustrates the change of O_2 concentration in combustion air when EGR ration varies. The

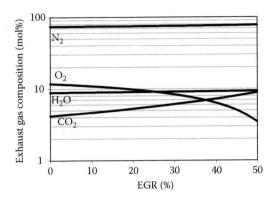

FIGURE 12.17 Composition in exhaust gas as a function of EGR ratio. (From Canepa, R. et al., *Proceedings of the Institution of Mechanical Engineers, Part E: Journal of Process Mechanical Engineering*, 227 (2), 89–105, 2013.)

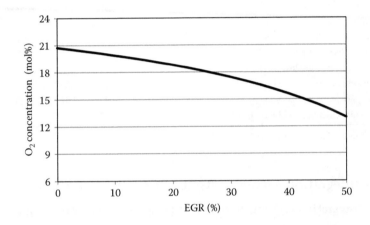

FIGURE 12.18 Oxygen concentration in combustion air as a function of EGR ratio. (From Canepa, R. et al., *Proceedings of the Institution of Mechanical Engineers, Part E: Journal of Process Mechanical Engineering*, 227 (2), 89–105, 2013.)

maximum EGR ratio of flue gas recirculation is limited by combustor-related phenomena. It is believed that the changes in turbomachinery performance may be very small with an oxygen concentration in combustion air of minimum 16–18 mol% (Ulfsnes et al., 2003).

In this study, EGR ratio of 0.38 is selected to ensure the minimum oxygen content of 16 mol%. Table 12.15 presents the comparison results of the integration without EGR and with EGR. With EGR, 4.73 kg/s less steam is extracted for solvent regeneration. So the net power generated from the gas turbine section decreases by 0.39 MW$_e$, but steam turbine section generates more at 4.13 MW$_e$. At the same time, significant equipment size reductions are achieved for both absorber and stripper in the case with EGR. In this case, the flow rate of the flue gas feed reduces to 38%, which results in 28.6% and 7.69% reduction of the cross-sectional area of the absorber and stripper, respectively. The reason for the difference between these values is that the required cross-sectional area of a column is decided by both gas phase and liquid phase loadings inside the column. Although the flow rate of flue gas reduces 38%, the flow rate of lean solvent only decreases 8.1% because higher liquid gas ratio (L/G ratio) is needed for higher CO_2 concentration in the flue gas (it increases from 4.5 to 7.32 mol%). Considering that the absorber is the most expensive equipment accounting for 30% of the total cost of PCC process (Klemeš et al., 2007; Mac Dowell and Shah, 2013), the size reduction of the absorber and stripper contributes to a significant capital cost saving.

TABLE 12.15 The Comparison between Integration without EGR and with EGR

Parameter	Without EGR	With EGR
Flow rate of fresh air intake (kg/s)	656.94	407.45
O_2 concentration in combustion air (mol%)	20.74	16.0
O_2 concentration in gas turbine vent gas (mol%)	11.4	6.45
Gas turbine power output (MW_e)	295.03	294.64
Steam turbine power output (MW_e)	113.56	117.69
Steam extracted for reboiler (kg/s)	76.39	71.06
Flow rate of flue gas to PCC (kg/s)	660.54	408.75
CO_2 concentration in flue gas (mol%)	4.5	7.32
CO_2 captured (kg/s)	41.11	40.92
Lean loading (mol CO_2/mol MEA)	0.32	0.32
Rich loading (mol CO_2/mol MEA)	0.461	0.472
Flow rate of lean solvent (kg/s)	1128.19	1036.81
L/G (mol/mol)	1.79	2.71
Reboiler duty (kW)	186,805	176,227
Reboiler duty (GJ/ton CO_2)	4.54	4.31
Lean solvent MEA concentration (wt%, CO_2 free)	32.5	32.5
Lean solvent temperature (K)	303.15	303.15
Absorber pressure loss (bar)	0.069	0.054
Absorber cross-sectional area (m²)	307.91	216.42
Stripper pressure loss (bar)	0.01355	0.01344
Stripper cross-sectional area (m²)	81.71	75.43

Source: Luo, X. et al., *Fuel*, 151, 110–117, 2015.

12.2.7 Heat Integration between NGCC, PCC, and CO_2 Compression

12.2.7.1 Heat Integration Options Based on Supersonic Shock Wave Compression

In Section 12.2.4, it is discussed that the exit temperature of the CO_2 stream is as high as 214.5°C–230.5°C for each stage of the supersonic shock wave compression. At this temperature, the compression heat could be recovered by integrating the pressurized streams with low-temperature streams in the NGCC power plant and the PCC process. Two options could be justified as follows:

1. The compression heat is integrated into the steam generation cycle of HRSG to generate more steam.

   In the NGCC power plant, the steam coming out of the LP-ST is cooled down to condensate with a temperature of 29.0°C at a pressure of 0.039 bar before it is pressurized to a high pressure by a pump. Then the subcooled water enters the economizer section of HRSG, in which, it is heated to around 158°C by the hot flue gas in normal case. Applying this heat integration option, this subcooled water could be lined to the compression train first as a refrigerant of the intercoolers. With this additional heat recovered from the compression process, more LP steam generation is expected to go to the LP-ST to generate more electricity.

2. The compression heat is integrated into the stripper reboiler of the PCC process for solvent regeneration.

   In the PCC process, the operating temperature range of the stripper reboiler is from 110°C to 125°C, which is much lower than the exit temperature of each stage of the compressor. So the compression heat could be transferred to provide heat to the reboiler. Here, one statement is that the reboiler duty is so high that the compression cannot provide all heat for it. Thus, the steam from the power plant is still required at the same time using a multiple shell kettle reboiler (Shah and Sekulic, 2003).

12.2.7.2 Case Setup

In previous sections, different process integration options were discussed when a NGCC power plant is integrated with a PCC process and CO_2 compression. A case study was conducted for the evaluation of power consumptions and heat requirement of different options for comparison purposes. For the case setup, five scenarios were summarized as below:

1. *Reference case*: NGCC power plant without integration with PCC and compression.
2. *Case 1*: NGCC power plant without EGR integrated with PCC and compression without compression heat integration.
3. *Case 2*: NGCC power plant with EGR integrated with PCC and compression without compression heat integration.
4. *Case 3*: NGCC power plant with EGR integrated with PCC and compression with compression heat integration into the steam cycle of HRSG.
5. *Case 4*: NGCC power plant with EGR integrated with PCC and compression with compression heat integration into the reboiler of the stripper.

12.2.7.3 Results and Discussion

Table 12.16 shows the results of energy and electricity consumptions of each case. By comparing the reference case (NGCC standalone) and Case 1, a total 9.58%-points net power efficiency decrease is observed when the NGCC power plant integrated with the PCC process. This obvious reduction is caused by three main factors: (1) the steam through the LP-ST decreases hugely to lead to a power output reduction because of steam extraction, which contributes 7.40%-points net efficiency decrease, (2) the electricity consumption of CO_2 compression contributes 1.92%-points net efficiency decrease; (3) auxiliary electricity consumption of the blower and solvent circulation pumps accounts for 0.55%-points net efficiency decrease.

TABLE 12.16 Performance Comparison Results of Different Cases

Description	Reference	Case 1	Case 2	Case 3	Case 4
Major process components	NGCC	NGCC + PCC	NGCC + PCC	NGCC + PCC	NGCC + PCC
The application of EGR	Without EGR	Without EGR	With EGR	With EGR	With EGR
Compression heat integration	Without	Without	Without	With HRSG	With reboiler
Gas turbine power output (MW$_e$)	295.03	295.03	294.64	294.64	294.64
Steam turbine power output (MW$_e$)	170.71	113.56	117.69	120.14	121.85
Power island power consumption (MW$_e$)	11.69	9.7	9.7	9.7	9.7
Power consumption in PCC (MW$_e$)	—	4.24	2.035	2.035	2.035
CO_2 compression power consumption (MW$_e$)	—	14.8	14.8	14.8	14.8
Stripper reboiler duty (MW$_{th}$)	—	186.8	176.2	176.2	176.2
Steam extracted for reboiler (kg/s)	—	76.39	71.06	71.06	65.50
CO_2 captured (kg/s)	—	41.11	40.92	40.92	40.92
Specific reboiler duty (MJ$_{th}$/kg CO_2)	—	4.54	4.31	4.31	4.31
Net plant power output (MW$_e$)	453.872	379.85	385.795	388.245	389.955
Net plant LHV efficiency (%)	58.74	49.16	49.93	50.25	50.47
Efficiency decrease (%-points) compared with reference case	—	9.58	8.81	8.49	8.27
Overall efficiency improvement (%) compared with Case 1	—	—	0.77	1.09	1.31

Source: Luo, X. et al., *Fuel*, 151, 110–117, 2015.

The results of Case 2 shows EGR help to achieve a 0.77% efficiency improvement compared with Case 1. The reason is dissected as follow. First, the special reboiler duty decreases from 4.54 MJ_{th}/kg CO_2 to 4.31 MJ_{th}/kg CO_2. The absorption efficiency is improved because of increasing of the CO_2 concentration in the flue gas (from 4.5 to 7.32 mol%, see in Table 12.15), which leads to a higher rich solvent loading and then a lower flow rate of solvent cycling. That results in less reboiler duty for the solvent regeneration. Second, the power consumption of the PCC process reduces from 4.24 MW_e to 2.035 MW_e. With EGR at a ratio of 0.38, the flue gas flow rate decreases significantly, which causes a great reduction of the power consumption of the flue gas blower upstream of the absorber. Meanwhile, the simulation results show that the discharge pressure of the blower also decreases because of the decrease in the whole tower pressure drop of the absorber (see Table 12.15).

Applying compression heat integrations into the main process of NGCC and PCC, Cases 3 and 4 improve the net LHV efficiency of the power plant to 50.25% and 50.47%, respectively. In Case 3, the subcooled water from the feed water pump of the HRSG is lined to the compression train and is heated to around 65°C before entering the economizer of the HRSG. One limitation of this option is that the temperature of the water leaving the economizer should be lower than its boiling temperature, otherwise there would be vapor phase exiting in its downstream pump. In Case 4, the temperature of the stream from compression train is 135°C, which is still higher than expected recoverable temperature of 90°C, after it exchanges heat in the stripper reboiler. Thus, more efficiency improvement could be achieved by combining other low-temperature heat recovery technology. In both Cases 3 and 4, there are no major capital investments for both two heat integration options. So these efficiency improvements are meaningful especially considering the great amount of the total electricity output from the NGCC power plants.

12.2.8 Conclusions

This chapter presents the investigation on thermal performances of different integration options of a 453MW_e NGCC equipped with a PCC process and a CO_2 compression train. The process models of each process were developed using Aspen Plus and were validated with published data or experimental data. Integrated with the PCC process and the compression train, LHV thermal efficiency of the NGCC power plant deceases from 58.74% to 49.16%. This decay includes 7.40% points for steam extraction, 0.55%-points for the PCC power consumption, and 1.92%-points for the compression train power consumption. With the application of EGR for the NGCC power plant at a recirculation ratio of 0.38, the net efficiency increases to 0.77%, while the cross-sectional areas of the absorber and stripper in the PCC process reduce to 28.6% and 7.69%, respectively. The compression heat integration options have been analyzed by applying supersonic shock wave compression technology. Compression heat integration into the steam cycle of HRSG and stripper reboiler achieves 0.32%-points and 0.54%-points net efficiency improvement separately without major capital investment required. The study indicates that EGR technology, supersonic shock wave compression technology, and compression heat integrations could be the future directions of commercial deployment of PCC process for NGCC power plant.

Acknowledgment

The authors would like to acknowledge financial supports from EU IRSES (Ref: PIRSES-GA-2013-612230) and UK NERC (Ref: NE/H013865/1).

References

Abu-Zahra, M. R. M., Niederer, J. P. M., Feron, P. H. M. and Versteeg, G. F. (2007a). CO_2 capture from power plants: Part II. A parametric study of the economical performance based on mono-ethanolamine. *International Journal of Greenhouse Gas Control*, 1, 135–142.

Abu-Zahra, M. R. M., Schneiders, L. H. J., Niederer, J. P. M., Feron, P. H. M. and Versteeg, G. F. (2007b). CO_2 capture from power plants Part I. A parametric study of the technical performance based on monoethanolamine. *International Journal of Greenhouse Gas Control*, 1, 37–46.

Austgen, D. M., Rochelle, G. T., Peng, X. and Chen, C. C. (1989). Model of vapor-liquid equilibria for aqueous acid gas-alkanolamine systems using the electrolyte-NRTL equation. *Industrial and Engineering Chemistry Research*, 28, 1060–1073.

BBC. (2015). UK's coal plants to be phased out within 10 years. www.bbc.co.uk/news/business-34851718 [Accessed April 18, 2016].

Biliyok, C., Canepa, R., Wang, M. and Yeung, H. (2013). Techno-economic analysis of a natural gas combined cycle power plant with CO_2 capture. In: Andrzej, K. and Ilkka, T. (eds) *Computer Aided Chemical Engineering*, Elsevier, Amsterdam.

BP P.L.C. (2014). BP Statistical Review of World Energy 2014. www.bp.com/content/dam/bp-country/de_de/ PDFs/brochures/BP-statistical-review-of-world-energy-2014-full-report.pdf [Accessed April 14, 2016].

Canepa, R. and Wang, M. (2015). Techno-economic analysis of a CO_2 capture plant integrated with a commercial scale combined cycle gas turbine (CCGT) power plant. *Applied Thermal Engineering*, 74, 10–19.

Canepa, R., Wang, M., Biliyok, C. and Satta, A. (2013). Thermodynamic analysis of combined cycle gas turbine power plant with post-combustion CO_2 capture and exhaust gas recirculation. *Proceedings of the Institution of Mechanical Engineers, Part E: Journal of Process Mechanical Engineering*, 227 (2), 89–105.

Chen, C. C. and Song, Y. (2004). Generalized electrolyte-NRTL model for mixed-solvent electrolyte systems. *AIChE Journal*, 50, 1928–1941.

Ciferno, J. P., Fout, T. E., Jones, A. P. and Murphy, J. T. (2009). Capturing carbon from existing coal-fired power plants. *Chemical Engineering Progress*, 105, 33–41.

Clean Air Task Force. (2015). Post-combustion capture. www.fossiltransition.org/pages/post_combustion_capture_/128.php [Accessed April 18, 2016].

Darby, R. (2001). *Chemical Engineering Fluid Mechanics*, CRC Press LLC, Boca Raton, FL.

Davis, J. and Rochelle, G. (2009). Thermal degradation of monoethanolamine at stripper conditions. *Energy Procedia*, 1, 327–333.

DECC of the UK. (2013). CCS cost reduction task force: Final report. www.gov.uk/government/publications/ccs-cost-reduction-task-force-final-report [Accessed April 18, 2016].

Dugas, R. E. (2006). Pilot plant study of carbon dioxide capture by aqueous monoethanolamine. MSE Thesis, University of Texas at Austin.

Edwards, T., Maurer, G., Newman, J. and Prausnitz, J. (1978). Vapor-liquid equilibria in multicomponent aqueous solutions of volatile weak electrolytes. *AIChE Journal*, 24, 966–976.

General Electric. (2014). GateCycle™: Performance/heat balance software for power plant simulation. https://getotalplant.com/GateCycle/docs/GateCycle/index.html [Accessed April 18, 2016].

Gerbilnow. (2012). Why choose between solar and coal when you can use both?. http://blog.gerbilnow.com/2012/11/solar-coal-power.html [Accessed April 18, 2016].

IEA. (2008). Energy technology perspectives 2008: In support of the G8 plan of action, International Energy Agency, Paris.

IEA. (2010). Energy technology perspectives 2010. International Energy Agency, Paris.

IEAGHG. (2012). CO_2 capture at gas fired power plants. International Energy Agency, Paris.

IPCC. (2005). Carbon dioxide capture and storage, Cambridge, UK.

Jonshagen, K., Sipöcz, N. and Genrup, M. (2011). A novel approach of retrofitting a combined cycle with post combustion CO_2 capture. *Journal of Engineering for Gas Turbines and Power*, 133, 011703.

Jordal, K., Ystad, P. A. M., Anantharaman, R., Chikukwa, A. and Bolland, O. (2012). Design-point and part-load considerations for natural gas combined cycle plants with post combustion capture. *International Journal of Greenhouse Gas Control*, 11, 271–282.

Kang, C. A., Brandt, A. R. and Durlofsky, L. J. (2011). Optimal operation of an integrated energy system including fossil fuel power generation, CO_2 capture and wind. *Energy*, 36, 6806–6820.

Kenig, E. Y., Schneider, R. and Górak, A. (2001). Reactive absorption: Optimal process design via optimal modeling. *Chemical Engineering Science*, 56, 343–350.

Klemeš, J., Bulatov, I. and Cockerill, T. (2007). Techno-economic modeling and cost functions of CO_2 capture processes. *Computers and Chemical Engineering*, 31, 445–455.

Kvamsdal, H. M., Haugen, G. and Svendsen, H. F. (2011). Flue-gas cooling in post-combustion capture plants. *Chemical Engineering Research and Design*, 89, 1544–1552.

Kvamsdal, H. M., Jordal, K. and Bolland, O. (2007). A quantitative comparison of gas turbine cycles with CO_2 capture. *Energy*, 32, 10–24.

Lawal, A., Wang, M., Stephenson, P. and Obi, O. (2012). Demonstrating full-scale post-combustion CO_2 capture for coal-fired power plants through dynamic modeling and simulation. *Fuel*, 101, 115–128.

Lawlor, S. (2009). CO_2 Compression using supersonic shock wave technology, Ramgen Power Systems. https://www.netl.doe.gov/File%20library/Events/2012/CO2%20Capture%20Meeting/K-Lupkes-Ramgen-Shockwave-Compression.pdf [Accessed April 18, 2016].

Liu, X., Chen, J., Luo, X., Wang, M. and Meng, H. (2015). Study on heat integration of supercritical coal-fired power plant with post-combustion CO_2 capture process through process simulation. *Fuel*, 158, 625–633.

Lucquiaud, M. and Gibbins, J. (2009). Retrofitting CO_2 capture ready fossil plants with post-combustion capture. Part 1: Requirements for supercritical pulverized coal plants using solvent-based flue gas scrubbing. *Proceedings of the Institution of Mechanical Engineers, Part A: Journal of Power and Energy*, 223, 213–226.

Luo, X., Mistry, K., Okezue, C., Wang, M., Cooper, R., Oko, E. and Field, J. (2014a). Process simulation and analysis for CO_2 transport pipeline design and operation—Case study for the Humber region in the UK. *Computer Aided Chemical Engineering*, 33, 1633–1638.

Luo, X. and Wang, M. (2015). Optimal operation of MEA-based post-combustion carbon capture for natural gas combined cycle power plants under different market conditions. *International Journal of Greenhouse Gas Control*, 48, 312–320.

Luo, X., Wang, M. and Chen, J. (2015). Heat integration of natural gas combined cycle power plant integrated with post-combustion CO_2 capture and compression. *Fuel*, 151, 110–117.

Luo, X., Wang, M., Oko, E. and Okezue, C. (2014b). Simulation-based techno-economic evaluation for optimal design of CO_2 transport pipeline network. *Applied Energy*, 132, 610–620.

Mac Dowell, N. and Shah, N. (2013). Identification of the cost-optimal degree of CO_2 capture: An optimisation study using dynamic process models. *International Journal of Greenhouse Gas Control*, 13, 44–58.

Marchioro Ystad, P. A., Lakew, A. A. and Bolland, O. (2013). Integration of low-temperature transcritical CO_2 Rankine cycle in natural gas-fired combined cycle (NGCC) with post-combustion CO_2 capture. *International Journal of Greenhouse Gas Control*, 12, 213–219.

Mccollum, D. L. and Ogden, J. M. (2006). Techno-economic models for carbon dioxide compression, transport, and storage and correlations for estimating carbon dioxide density and viscosity. https://www.researchgate.net/publication/46440087_Techno Economic_Models_for_Carbon_Dioxide_Compression_Transport_and_Storage_Correlations_for_Estimating_Carbon_Dioxide_Density_and_Viscosity, [Accessed July 07, 2017].

Ramgen Power Systems. (2008). *What is Shock Compression?* www.ramgen.com/tech_shock_compression.html [Accessed April 18, 2016].

Roussanaly, S., Brunsvold, A. L. and Hognes, E. S. (2014). Benchmarking of CO_2 transport technologies: Part II—Offshore pipeline and shipping to an offshore site. *International Journal of Greenhouse Gas Control*, 28, 283–299.

Shah, R. K. and Sekulic, D. P. (2003). *Fundamentals of Heat Exchanger Design*, John Wiley and Sons, Hoboken, NJ.

Thermoflow Inc. (2016). Gas turbine combined cycle design program to create cycle heat balance and physical equipment needed to realize it. www.thermoflow.com/combinedcycle_GTP.html [Accessed April 18, 2016].

U.S. EIA. (2016). How much carbon dioxide is produced per kilowatthour when generating electricity with fossil fuels? www.eia.gov/tools/faqs/faq.cfm?id=74andt=11 [Accessed March 10, 2016].

Ulfsnes, R., Karlsen, G., Jordal, K., Bolland, O. and Kvamsdal, H. M. (2003). Investigation of physical properties of CO_2/H_2O-mixtures for use in semi-closed O_2/CO_2 gas turbine cycle with CO_2-capture. In *Proceedings of the 16th International Conference on Efficiency, Cost, Operation, Simulation, and Environmental Impact of Energy Systems (ECOS'03)*, Copenhagen, Denmark.

Wang, M., Lawal, A., Stephenson, P., Sidders, J. and Ramshaw, C. (2011). Post-combustion CO_2 capture with chemical absorption: A state-of-the-art review. *Chemical Engineering Research and Design*, 89, 1609–1624.

Whitman, W. G. (1962). The two film theory of gas absorption. *International Journal of Heat and Mass Transfer*, 5, 429–433.

Witkowski, A., Rusin, A., Majkut, M., Rulik, S. and Stolecka, K. (2013). Comprehensive analysis of pipeline transportation systems for CO_2 sequestration. Thermodynamics and safety problems. *Energy Conversion and Management*, 76, 665–673.

Zhang, Y., Chen, H., Chen, C.-C., Plaza, J. M., Dugas, R. and Rochelle, G. T. (2009). Rate-based process modeling study of CO_2 capture with aqueous monoethanolamine solution. *Industrial and Engineering Chemistry Research*, 48, 9233–9246.

Zhang, Z. X., Wang, G. X., Massarotto, P. and Rudolph, V. (2006). Optimization of pipeline transport for CO_2 sequestration. *Energy Conversion and Management*, 47, 702–715.

12.3 Postcombustion CO_2 Capture Based on Chemical Absorption in Power Plants

Eni Oko, Meihong Wang, and Adekola Lawal

12.3.1 Introduction

Anthropogenic CO_2 is the main culprit of global warming. At the current rate, global warming is predicted to reach a dangerous level by 2100 except trends in anthropogenic CO_2 emission are reversed (CCC, 2015). As a result, carbon emission reduction targets have been set globally and in many countries (Adam, 2008). Globally, carbon emission is expected to be reduced by more than half by 2050 compared to 2005 levels (IEA, 2008). To meet this target, the International Energy Agency (IEA) has proposed a portfolio of technologies, one of which is carbon capture and storage (CCS). CCS is considered a sustainable and economic option for cutting down CO_2 emissions from fossil fuel power plants and other carbon-intensive industries. Without CCS, cost of CO_2 emission reduction from these sectors may be up to 70% more (CCSA, 2011).

CCS technology involves capturing CO_2 from large stationary sources such as power plants (e.g., coal-fired power plants) and carbon-intensive industries (e.g., refineries, steelworks, and cement plants), and transporting them to underground storage sites, namely saline aquifer and depleted oil and gas reserves, where they are either stored permanently and prevented from entering the atmosphere or used for enhanced oil recovery (EOR) purposes (IPCC, 2005). CCS is implemented through different technology routes, namely postcombustion, precombustion, and oxy-fuel capture (Figure 12.19). Postcombustion capture (PCC) unlike the other options relies on established technologies and also has the capacity to be retrofitted to existing power plants (IPCC, 2005). As a result, PCC is considered the most matured technology for CCS and many first generation CCS projects are expected to be based on PCC (Wang et al., 2011).

However, PCC is very challenging because it involves capturing CO_2 from flue gases, which are usually at low pressure with dilute amounts of CO_2. This results in incredible cost of the CO_2 capture system, about 70% of the entire CCS project cost (Wang et al., 2011). To meet the challenges, different CO_2 separation options can be explored. These include physical absorption, chemical absorption, cryogenic separation, adsorption, membrane separation, and membrane absorption (Wang et al., 2011). Among these options,

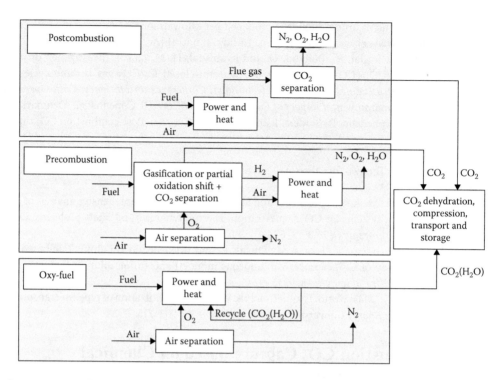

FIGURE 12.19 Technology options for CCS. (From IPCC, IPCC special report on carbon dioxide capture and storage, 2005.)

chemical absorption using solvents has the highest CO_2 selectivity (Yu et al., 2012). As a result, it is most suitable for commercial deployment in power plants where huge volume of flue gas is expected.

12.3.2 Overview of PCC Based on Chemical Absorption in Power Plants

12.3.2.1 Process Description

In PCC process (Figure 12.20), flue gas from a power plant enters an absorber where most of the CO_2 in the flue gas is removed by a solvent in countercurrent flow. The solvent captures CO_2 from the flue gas by reacting with them to form weakly bonded intermediate compounds (Wang et al. 2011). The scrubbed gas is then water washed to remove entrained solvent and thereafter vented to the atmosphere. The lean solvent gradually heats up as it absorbs CO_2 leading to temperature of typically 40°C–60°C inside the absorber. The rich solvent is pumped to the stripper column where the solvent is regenerated through application of heat (100°C–120°C). Before entering the stripper, the rich solvent goes through a cross heat exchanger where it is preheated by regenerated hot lean solvent from the stripper. Regeneration heat is supplied by steam extracted from the steam turbine stages in the power plant steam cycle, although it can also be obtained from auxiliary boilers (Lucquiaud and Gibbins, 2011). Regenerated solvent is pumped back to the absorber, while the recovered CO_2 (up to 99% in purity) is compressed to dense phase conditions (over 100 bar) and transported to storage or EOR sites.

A 30 wt% monoethanolamine (MEA) solution is used commonly as solvent in this process. This is due to their high reactivity with CO_2 and cost effectiveness. However, it has high regeneration energy requirement. The host power plant could be derated by more than one-third of its capacity when integrated to a PCC plant with 30 wt% MEA solution as solvent (Fisher et al., 2007). For a coal-fired power plant, this would mean an increase in electricity cost from about 5 cents/kWh to about 10.7 cents/kWh (Fisher et al., 2007). Other disadvantages of using 30 wt% aqueous MEA as solvent include high

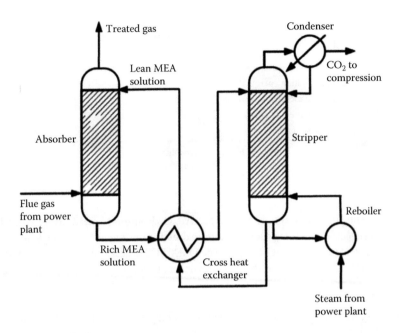

FIGURE 12.20 Conventional PCC with chemical absorption. (From Wang, M. et al., *Applied Energy* 158, 275–291, 2015.)

corrosivity, high potential for chemical and thermal degradation, high circulation rate leading to large equipment sizes, environmental problems due to fugitive emissions, waste water treatment, and surface water contamination among others.

12.3.2.2 Flue Gas Conditioning

Solvents used in PCC process are generally susceptible to degradation by oxygen, SO_x and NO_x are usually present in the flue gas (Eswaran et al., 2010). Particulates in the flue gas can also hinder solvent performance in the columns. Also, depending on the flue gas temperature, there could be solvent loss due to evaporation. As a result, the flue gas is conditioned to meet design specifications for the solvent before entering the PCC plant. For aqueous MEA solvent, the flue gas is recommended to have < 10ppm SO_x, < 1ppm O_2, and temperature of about 45–50°C (Davidson, 2007; Ramezan et al., 2007; Rao et al., 2004). The flue gas must be conditioned to meet this requirement through appropriate equipment installed upstream of the PCC plant (Ramezan et al., 2007; Rao et al., 2004).

12.3.2.3 Improvement of PCC Performance

Globally, PCC commercialization has been stalled by its huge cost and energy penalty, many commercial projects were cancelled in the last decade, only a few were actualized. As a result, new developments in PCC are targeted at reducing the cost and energy penalty through improvements in solvent performance and energy utilization. These developments are summarized below.

12.3.2.3.1 New Solvents

Blend of amine solvents (e.g., MEA/PZ, 2-Amino-2-Methyl-1-Propanal/Piperazine, and proprietary solvents like KS-1) and nonamine solvents such as ammonia and mixture of ionic liquid and amine solvents have shown great promise (Darde et al., 2010; Dubois and Thomas, 2012; Yang et al., 2014). For example, KS-1 solvent, developed jointly by Mitsubishi Heavy Industries Ltd and Kansai Electric Power Co. (Iijima et al., 2011) show significant improvement compared to aqueous MEA (Table 12.17). Different commercial PCC technologies have been developed using different combinations of amine

TABLE 12.17 MEA versus KS-1 Solvents in PCC Process

Performance Parameters	MEA	KS-1
Solvent circulation rate	1	0.6
Regeneration energy	1	0.8
Degradation of the solvent	1	0.1
Solvent loss	1	0.1
Corrosion inhibitor	Yes	No

Source: Iijima, M., Flue gas CO_2 capture.
http://gcep.stanford.edu/pdfs/energy_workshops_04_04/
carbon_iijima.pdf, 2005.

and nonamine solvents, namely CanSolv, PostCap, AAP, CAP, and KM-CDR. More recently, biphasic solvents have been suggested and they show even more drastic improvements in CO_2 loading and regeneration energy requirement (Zhang et al., 2013). The DMX PCC technology developed by IFP Energies Nouvelles (IFPEN) is based on biphasic solvents (Raynal et al., 2013).

12.3.2.3.2 New Process Configurations

New process configurations have been developed by adding extra equipment (e.g., heat exchangers and compressors) to the conventional process (Figure 12.20). Typical examples include configurations involving absorber intercooling (Figure 12.21), multipressure stripping (Figure 12.22), and split flow (Figure 12.23) among others (Ahn et al., 2013; Boot-Handford et al., 2014; Fisher et al., 2007). Thermodynamic analysis of these configurations indicates that some of them are more energy efficient than the conventional process configuration (Ahn et al., 2013). However, due to the extra instrumentation/equipment, they have higher capital cost. Some of them have been incorporated in some commercial PCC processes (e.g., CanSolv PCC that includes absorber intercooling).

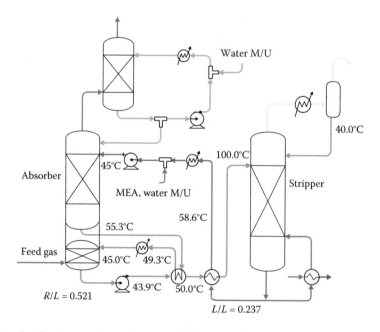

FIGURE 12.21 Absorber intercooler configuration. (Ahn, H. et al., *International Journal of Greenhouse Gas Control* 16, 29–40, 2013.)

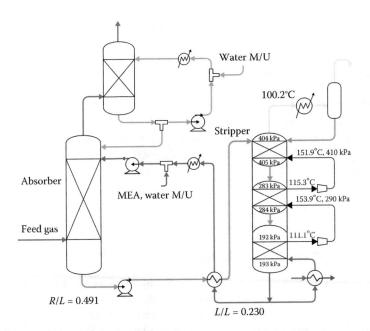

FIGURE 12.22 Multipressure stripping configuration. (Ahn, H. et al., *International Journal of Greenhouse Gas Control* 16, 29–40, 2013.)

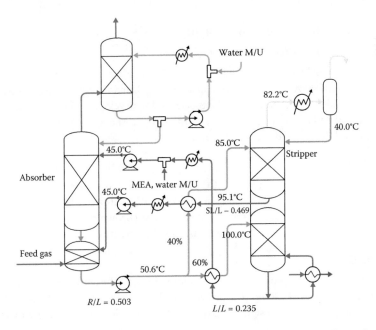

FIGURE 12.23 Split solvent flow configuration. (Ahn, H. et al., *International Journal of Greenhouse Gas Control* 16, 29–40, 2013.)

12.3.2.3.3 *Steam Cycle Options*

Steam for the PCC reboiler is obtained from either the host power plant steam cycle or an auxiliary boiler. Many studies agree that obtaining the steam from the host power plant steam cycle is more efficient (Lucquiaud and Gibbins, 2011). As a result, in commercial PCC projects (e.g., Boundary Dam

CCS), steam is obtained from the host power plant steam cycle. There are different design options for retrofitting the reboiler and the steam cycle (Lucquiaud and Gibbins, 2011). Lucquiaud and Gibbins (2011) showed through thermodynamic analysis of the options that any selected design option could have major impact on energy penalty.

12.3.2.3.4 Heat Integration Options

Different options have been explored: integration of the compressor intercooling with condensates from the PCC (Harkin et al., 2010), integration involving cooling stripper overhead vapor with a portion of cold rich solvent from absorber (Ahn et al., 2013), etc. There is a typical case of heat integration in the Boundary Dam CCS (Just, 2013); vapor from the SO_x stripper is used in the reboiler for the CO_2 stripper column. Generally, heat integration could reduce operating costs by up to 10% (Fisher et al., 2005). However, the process becomes more complex and some plant flexibility may be sacrificed.

12.3.3 Pilot PCC Plant Research Programs

Pilot PCC plants have been commissioned across the world, either as stand-alone plants or integrated to live power plants via flue gas slipstream, in the last decade leading up to commercial/demonstration PCC project development. The outcomes of these projects have helped shape up deployment strategies for the commercial/demonstration projects. Summary of pilot plants integrated to live power plants via a flue gas slipstream will be discussed here, summary of stand-alone pilot PCC plants has been reported in Wang et al. (2011). Most of the information presented in this section were obtained from MIT CCS database (MIT CCS, 2015).

12.3.3.1 Pleasant Prairie Project, Wisconsin

12.3.3.1.1 Project Participants, Objective, and Cost

The project was developed by a consortium involving Alstom, Electric Power Research Institute (EPRI) and We Energies to validate Alstom's Chilled Ammonia PCC technology in live coal-fired power plant. The project was intended to provide needed experience for developing the larger scale Mountaineer PCC project. The initial phase of the project was estimated to cost about US$8.6 million, contributed by DOE's Office of Fossil Energy and the consortium.

12.3.3.1.2 Description

The project involved capture of about 15,000 tCO_2/year from flue gas slipstream (5 MWth equivalent) at about 90% capture level from one of the units in the 1210 MW_e coal-fired Pleasant Prairie power station at Wisconsin using Alstom's Chilled Ammonia PCC process (Brown et al., 2009). The project operated from June 2008 to October 2009. CO_2 captured was vented to the atmosphere.

12.3.3.1.3 Lessons/Challenges

One of the key lessons from the project was strategy for scaling up the chilled ammonia process for large scale PCC projects (Brown et al., 2009). The project was the first attempt to integrate PCC plant to a live power plant. As a result, several redesigns were inevitable at the initial phase of the project (Brown et al., 2009).

12.3.3.2 E.ON Karlshamn Project, Sweden

12.3.3.2.1 Project Participants, Objective, and Cost

The project was developed by a consortium comprised of E.ON Thermal Power and Alstom Power to test Alstom's Chilled Ammonia PCC process in oil/gas-fired power plants and provide needed experience for the development of a larger scale PCC project for a CCGT plant at Malmö, Sweden. Initial project cost was estimated to be about US$15 million and this was provided by the consortium.

12.3.3.2.2 Description

The Chilled Ammonia PCC process is integrated to an auxiliary boiler that combusts high sulfur oil. About 15,000 tCO_2/year (5 MWth equivalent) was captured from a slipstream of the boiler flue gas. The project was commissioned in April 2009. CO_2 captured was vented into the atmosphere.

12.3.3.2.3 Lessons/Challenges

The project demonstrated Chilled Ammonia PCC process for oil-fired boilers. Also, the results revalidated findings from other projects, which showed that chilled ammonia process incur about 10% less regeneration energy than conventional PCC with aqueous MEA solvent. Like in the Pleasant Prairie Project, transitioning to the project scale from laboratory scale presented a major challenge.

12.3.3.3 AEP Mountaineer PCC Project Phase 1, WV

12.3.3.3.1 Project Participants, Objective, and Cost

The project was developed by a consortium involving American Electric Power (AEP), Alstom Power, Rheinisch-Westfälisches Elektrizitätswerk (German), National Energy Technology Laboratory (NETL), and Battelle Memorial Institute. The project aim was to provide larger scale validation of Alstom's Chilled Ammonia PCC Process and to demonstrate the interoperability of PCC and power plant. Total project cost was about US$668 million; 50% of which was paid by DOE and the remaining paid by the consortium.

12.3.3.3.2 Description

The project, commissioned in October, 2009, involved capture of about 100,000 tCO_2/year from a flue gas slipstream (30 MWth equivalent) from 1300 MW_e coal-fired Mountaineer Power Station using PCC with chilled ammonia solvent at 90% capture level. Field experience from the Pleasant Prairie Project was useful for scaling up the Chilled Ammonia Process for the Mountaineer Project. The project was to be extended in the Phase II to capture 1,500,000 tCO_2/year (235 MWth equivalent). However, the Phase II was cancelled due to uncertain climate policies. CO_2 captured was transported and injected at the saline Mount Simon Sandstone.

12.3.3.3.3 Lessons/Challenges

In this project, the differences in the operating philosophy and process dynamics of the PCC and power plant were highlighted (Cerimele, 2012). It was recommended that these differences be recognized and addressed early in the design process of PCC. Waste water management was also identified as a major challenge.

12.3.3.4 Brindisi Project, Brindisi, Italy

12.3.3.4.1 Project Participants, Objective, and Cost

The project was developed by a consortium involving Enel and Eni. It was intended to provide the experience needed for the development of a larger PCC project at Porto Tolle, Italy. The PCC part of the project is estimated to have cost about €20 million, most of that coming from the European Union (EU) as part of the European Energy Programme for Recovery (EEPR).

12.3.3.4.2 Description

In the project, 8000 tCO_2/year was captured from a flue gas slipstream of a 660 MW_e coal-fired power plant unit at Brindisi Power Station, Italy, using PCC with aqueous MEA as solvent. Different solvents, 20 wt% MEA, 30 wt% MEA, etc., were tested. The plant started operations in June 2010. Captured CO_2

was stored briefly in tanks from mid-2011 and later transported and injected in the storage site (deep saline aquifer) at Stogit field, Cortemaggiore in North Italy.

12.3.3.4.3 Lesson/Challenges

The main challenges encountered here include waste water management and reclamation of degraded solvent. Also, due to lack of experience, there was a lot of redesign at the initial stages to fit with emerging findings.

12.3.3.5 Plant Barry PCC Project, Alabama

12.3.3.5.1 Project Participants, Objectives, and Cost

The project was developed by a consortium involving Southern Energy, Mitsubishi Heavy Industries (MHI), Southern Company, US DOE's Southeast Regional Carbon Sequestration Partnership (SECARB), and EPRI. The aim was to demonstrate the technical feasibility of PCC integrated to a power plant and also to validate the KM-CDR™ PCC technology at higher power plant capacity. Actual cost of this project is unknown.

12.3.3.5.2 Description

In the first stage of the project, 150,000 tCO_2/year was captured from the flue gas slipstream (25 MWth equivalent) of the 2657 MW_e coal-fired Plant Barry Power Station at Mobile in Alabama using KM-CDR PCC process. The plant became operational in June 2011. Captured CO_2 was compressed and transported via a 19 km pipeline to the Citronelle Dome oil field for injection into the deep saline reservoir (2865 m underground) for permanent storage. At the time, it was the only pilot CCS project where the complete CCS chain was demonstrated.

12.3.3.5.3 Lesson/Challenges

Fugitive solvent emissions were learnt to increase significantly with small increase in SO_x concentration in the flue gas (Hill, 2014). Some of the components such as the reboiler and column internals were redesigned based on emerging experience.

12.3.3.6 Shidongkou CCS Project, Shanghai, China

12.3.3.6.1 Project Participants, Objectives, and Cost

The project was developed by Huaneng Power Group to validate an in-house PCC process developed by the company. It is estimated to have cost about US$24 million.

12.3.3.6.2 Description

In the project, PCC plant using proprietary amine solvent developed by Huaneng Power is used to capture about 120,000 tCO_2/year from a flue gas slipstream (3%) of the 1320 MW_e coal-fired Shidongkou power station. It became operational in 2011. CO_2 captured was further purified and sold to food and beverage industries.

12.3.3.6.3 Lessons/Challenges

Certain combination of amine solvents was found to give significantly lower capture cost as shown from performance analysis. Cost estimates with the amine solvent combination (proprietary) used in the project is about US$20/$tCO_2$. In comparison, operating cost of about US$100/$tCO_2$ is incurred with 30 wt% MEA solvent (Tollefson, 2011). It is been argued that the lower cost is down to cheaper labor and less regulatory burden in China.

12.3.3.7 Boryeong Station CCS Project, South Korea

12.3.3.7.1 Project Participants, Objectives, and Cost

The PCC project was developed by Korea Electric Power Corporation (KEPCO) to test new amine-based solvent (KoSol-4) developed by the company. The cost of the initial phase of the project is unknown.

However, it cost an estimated US$42 million to upscale the initial phase from the original 0.1 MWth equivalent to the Phase II of 10 MWth equivalent capacity.

12.3.3.7.2 Description

Phase 1 of the project took off around 2010. In this phase, about 2 tCO_2/day (0.1 MWth equivalent) was captured using proprietary amine solvent (KoSol-4) from flue gas slipstream of Unit #8 (500 MW_e) of Boryeong Power Station, South Korea. This was later upscaled in 2013 to capture about 200 tCO_2/day (10 MWth) using the same solvent and from the same power plant. In both phases, 90% capture level was achieved and further upscaling to about 100–500 MWth equivalent capacity is targeted in the future.

12.3.3.7.3 Lessons/Challenges

One of the lessons is that solvent performance combined with process improvements is a key factor which determines energy consumption of PCC process (GCCSI, 2013). The KoSol-4 solvent showed about 14%–23% less regeneration energy compared to aqueous MEA solvent. Also, scale-up is expensive and difficult due to limited data and experience with the process.

12.3.3.8 Wilhelmshaven PCC Project, Bremen, Germany

12.3.3.8.1 Project Participants, Objectives, and Cost

The project was developed by Fluor and E.ON Kraftwerke to test Fluor's Econamine FG Plus PCC technology on a live power plant. The total project cost is unknown.

12.3.3.8.2 Description

The project involved capturing about 70 tCO_2/day from a flue gas slipstream (3.5 MWth equivalent) at the 757 MW_e coal-fired Wilhelmshaven power plant in Bremen, Germany using Econamine FG Plus PCC process at 90% capture level. The plant was commissioned in October 2012. CO_2 captured was vented into the atmosphere.

12.3.3.8.3 Lessons/Challenges

One of the lessons from this project is that meaningful improvement in energy efficiency can be achieved through a combination of energy-saving designs and better solvent capability (Fluor, 2015).

12.3.3.9 Ferrybridge CCSPilot100 + Project, UK

12.3.3.9.1 Project Participants, Objectives, and Cost

The project was developed by Scottish and Southern Energy (SSE), Doosan Babcock and Vattenfall at a cost of £21 million. It was used to demonstrate scale-up of PCC process with reference to the smaller scale Doosan Babcock's Emissions Reduction Test Facility (ERTF) PCC pilot plant and to test a wide range of solvents. The project was funded by the Department of Energy and Climate Change (DECC), Technology Strategy Board (TSB), and Northern Way.

12.3.3.9.2 Description

This project involved the capture of 100 tCO_2/day (5 MW_e equivalent) from a flue gas slipstream of Unit #4 (500 MW_e) of Ferrybridge Coal/biomass-fired power station in the United Kingdom using amine solvents. It operated for 2 years from 2012 to 2013. Captured CO_2 was vented to the atmosphere. The pilot plant was used to carry out extensive parametric and long-term performance testing of the proprietary RS-2 solvent with 30 wt% aqueous MEA solvent as benchmark.

12.3.3.9.3 Lessons/Challenges

There was a lot of redesign at the initial phase of the project based on insights gained as the testing progressed. Also, the RS-2 proprietary solvent has better loading characteristic and regenerability than aqueous MEA solvent.

12.3.3.10 ECO₂ Burger PCC Project, OH

12.3.3.10.1 Project Participants, Objectives, and Cost

The project was developed by a consortium involving First Energy, Powerspan, and Ohio Coal Development Office. The objective was to test Powerspan's ECO_2 PCC process. Total cost of the project is unknown.

12.3.3.10.2 Description

The project involved capture of over 20 tCO_2/day from a flue gas slipstream (1 MW$_e$ equivalent) at the 50 MW$_e$ Burger Plant ECO unit at a capture level of 90% using Powerspan's ECO_2 PCC process. The ECO_2 process uses proprietary solvent made up of a mixture of aqueous amines (GCCS, 2012). The plant operated from December 2008 to December 2010 with all the captured CO_2 vented into the atmosphere.

12.3.3.10.3 Lessons/Challenges

From independent assessment by Worley Parson, it was found that the energy penalty was less compared to conventional PCC with aqueous MEA solvent (GCCSI, 2012).

12.3.3.11 Aberthaw CCS Project Wales, UK

12.3.3.11.1 Project Participants, Objectives, and Cost

The project was developed by a consortium involving RWE nPower and CanSolv Technologies Inc. The objective of the project was to gain more insights into operating mechanisms of CCS technology in the context of coal-fired power plants and also to evaluate the viability of full-scale PCC process integrated to a power plant. Total project cost is unknown.

12.3.3.11.2 Description

In the project, 50 tCO_2/day (3 MWth equivalent) was captured from a flue gas slipstream at the 1500 MW$_e$ Aberthaw Power station in Wales using amine solvents. The plant is based on the CanSolv PCC process. The plant operated from January 2013 to May 2014. Operation of an earlier 1 tCO_2/day capacity PCC plant with aqueous MEA solvent commissioned in 2008 at Didcot, UK, provided the experience needed for the development and operation of the Aberthaw pilot plant.

12.3.3.11.3 Lessons/Challenges

The plant was designed to include a lot of heat integration. Operating data from the plant highlighted the significance of heat integration in reducing energy penalty of the process.

12.3.3.12 Pikes Peak South CCS Project, Saskatchewan, Canada

12.3.3.12.1 Project Participants, Objectives, and Cost

This project was developed by a consortium involving Husky Energy Inc. and CO_2 Solutions to determine the potentials of commercializing the enzyme-catalyzed solvent process developed by CO_2 Solutions. The total project cost is estimated to be around US$12.132 million, part of which were provided by the Climate Change and Emissions Management Corporation (CCEMC) and Government of Canada's ecoENERGY Innovation Initiative (ecoEII) program.

12.3.3.12.2 Description

The plant was designed to capture 15 tCO_2/day from flue gas of a natural gas-fired boiler using enzyme-catalyzed solvent. The plant layout is similar to the conventional process and equipped with HTC's proprietary and patented solvent reclaimer system (SRS). It became operational in May 2015.

12.3.3.12.3 Lessons/Challenges

The SRS™ minimized solvent replacement requirements, lowered energy cost and reduced waste disposal requirements compared to traditional solvent reclaimer systems used in PCC plants. Also, the enzyme-catalyzed solvent showed good promise in CO_2 loading capacity and regenerability.

12.3.3.13 Big Bend PCC Project, FL

12.3.3.13.1 Project Participants, Objectives, and Cost

The project was developed by Tampa Electric and Siemens. The objective of the project is to determine if Siemens PostCap PCC technology can offer lower cost option for CO_2 capture compared to the conventional process using aqueous MEA solvent. Actual cost of the project is unknown, although it is reported to have received a grant of US$8.9 million in July 2010 from US DOE.

12.3.3.13.2 Description

The project involves capture of CO_2 from a 1 MW_e equivalent flue gas slipstream from the 1892 MW_e Big Bend Coal-fired Power Station at Ruskin in Florida at a capture level of 90% using Siemens PostCap process. The technology uses an amino acid salt formulation as solvent. The project became operational in 2013.

12.3.3.13.3 Lessons/Challenges

The solvent showed good thermal and chemical stability as no fugitive emissions were detected and energy requirement for solvent reclamation was comparatively less.

12.3.3.14 The European CO_2 Test Center Mongstad (TCM), Norway

12.3.3.14.1 Project Participants, Objectives, and Cost

The project is owned by a consortium involving Gassnova (On-behalf of the Norwegian Government), Statoil, Sasol, and Shell. TCM is intended to be used for testing Chilled Ammonia and Amine-based PCC process. The estimated project cost is about US$1.02 billion.

12.3.3.14.2 Description

The plant is comprised of a chilled ammonia and an amine-based process either of which can be connected to exhaust gases from a Residue Catalytic Cracker (RCC) or a Natural Gas Combined Heat and Power (CHP) plant within the site. About 74,000–82,000 and 22,000–25,000 tCO_2/year can be captured from the RCC and CHP, respectively. CO_2 captured is released into the atmosphere. The plant became operational in May 2012. Several PCC processes have been tested at TCM, namely Alstom's Chilled Ammonia Process, Aker's Clean Carbon, Shell's CanSolv, and Siemens PostCap among others.

12.3.3.14.3 Lessons/Challenges

Regeneration steam is obtained from waste heat process with some top-up from steam purchased from other sources. This basically isolates the capture plant and power plant economics leading to lower-cost solution. However, the waste heat from the CHP plant is not utilized optimally.

12.3.3.15 EDF Le Havre PCC Project, Le Havre, France

12.3.3.15.1 Project Participants, Objectives, and Cost

The project was developed by a consortium involving EDF, Veolia, and Alstom Power/Dow Chemical for validating key process performance parameters such as CO_2 capture efficiency, solvent degradation, and related emissions among others for the Alstom's advanced amine PCC process. Total project cost is estimated to be about €22 million.

12.3.3.15.2 Description

In the project, about 25 tCO_2/day is captured from flue gas slip stream (5 MWth equivalent) from 580 MW_e coal-fired Le Havre Power Plant in France using Alstom's advanced amine process at 90% capture level. Captured CO_2 is vented into the atmosphere.

12.3.3.15.3 Lesson/Challenges

The plant was designed for diverse operating conditions, posing challenges to operators and equipment (Chopin, 2014). Better regenerability and lower fugitive emissions were reported.

12.3.4 Commercial Deployment of PCC Using Solvents

Many turnkey PCC processes have been developed by different vendors in the past three decades (GCCSI, 2012). They are similar to conventional PCC process albeit with extra unit operations and proprietary process chemistry. The processes have been proven in relevant environments through flue gas slip streams and some of them are currently in operation in commercial CCS projects in different places across the world. A summary of these processes are given in the following sections.

12.3.4.1 Shell CanSolv PCC Process

In this process, the absorbers are equipped with intercoolers and circulated water prescrubber for flue gas cooling is used (Shaw, 2009). The solvent comprises of a blend of aqueous amine, mainly tertiary amines plus promoters and oxidation inhibitors. The solvent is reportedly more resistant to oxidative and thermal degradation than conventional aqueous MEA solvent. Also, it has lower regeneration, faster reaction kinetics with CO_2 and higher loading capacity. In some designs of the process, CO_2 and SO_x are removed using an integrated absorber (Shaw, 2009). The process was tested at TCM in Norway and Aberthaw Pilot PCC Plant in Wales, UK. Currently, it is in use at Boundary Dam CCS Project, the only power plant commercial PCC project in existence. In addition, the technology have been used in industrial CO_2 capture applications involving natural gas and coal-fired boilers, blast furnace, and cement kiln (Shaw, 2009).

12.3.4.2 Aker Clean Carbon's Advanced Carbon Capture Process

The Aker process (Gorset et al., 2014) uses intercooled and interheated absorber and stripper, respectively, and advanced amine blend as solvent. Also, the absorber design includes multiple lean solvent feeds at different heights. The solvent is reported to have lower energy requirement, environment impact, corrosivity, and degradation potential than conventional aqueous MEA solvent. The technology has been tested at TCM, Norway and was selected for the Longannet CCS project in the United Kingdom. Longannet project was however cancelled later.

12.3.4.3 Siemens PostCap Process

The PostCap process (Siemens A.G., 2015) uses lean solvent flash and vapor compression, circulating water prescrubber for flue gas cooling, split feed for the stripper, proprietary reclaimer technology, and an environmentally friendly amino acid salt solvent. The solvent requires lower regeneration energy, about 2.7 GJ/ton of CO_2 compared to about 4.2 GJ/ton CO_2 for 30 wt% MEA solvent. It has good stability against various degradation mechanisms especially oxygen. The solvent has low vapor and solvent losses through evaporation are, therefore, minimized as the treated flue gas does not require to be water washed like in conventional PCC with 30 wt% MEA solvent. It has been tested at TCM, Norway and at the Big Bend PCC Project, Florida. The PostCap technology will be used in the planned ROAD CCS Project in the Netherlands.

12.3.4.4 Fluor's Econamine FG Plus PCC Process

The Econamine FG Plus is a second generation technology of the Econamine FG technology, which was acquired by Fluor from Dow Chemical in 1989 (Herzog et al., 2009). Econamine FG, based on 30 wt% MEA solvent + inhibitors, is already proven commercially (Abu-Zahra et al., 2013). Econamine FG Plus uses a proprietary formulation of aqueous amine solvent and incorporates a number of energy saving features, namely absorber intercooling, heat integration, and high efficiency reclaimer, leading to lower regeneration energy requirement and circulation rate. The technology tolerates high oxygen (up to 20%) in flue gases. It has been tested at TCM, Norway and at the Wilhelmshaven PCC Project in Germany.

12.3.4.5 Kerr McGee/ABB Lummus Crest Process

The Kerr McGee/ABB Lummus Crest process alongside Econamine FG are the oldest PCC technologies in the world (Herzog et al., 2009). The process uses 15–20 wt% MEA + oxidation inhibitors as solvent. The process includes a rich solvent flash before the stripper, which releases some of the CO_2 from the rich solvent (Nsakala et al., 2001). This improves the regenerability of the solvent. The process is the most commercially proven PCC technology. It was first used in 1978 for CO_2 capture from a boiler firing a mix of coal and petroleum coke at Kerr-McGee's soda ash plant in Trona, California (Abu-Zahra et al., 2013). The low solvent concentration is both a plus and a drawback for this technology; low corrosion rates are achieved without corrosion inhibitors but circulation rates are huge leading to relatively large equipment sizes and reboiler duty.

12.3.4.6 Alstom's Advanced Amine Process

This process (Alstom, 2015; Baburao et al., 2014) is developed jointly by Alstom and Dow Chemical Company based on a proprietary Dow amine solvent (DOW UCARSOL FGC 3000). Benefits of the process include higher tolerance for oxygen and trace contaminants, lower solvent degradation, and regeneration energy requirement. The process have been tested at the EDF Power Plant Pilot PCC Project at Le Havre, France, and the Charleston Industrial Pilot PCC Plant at WV. Also, it will be used in the planned Elektownia Belchatow CCS Project in Poland.

12.3.4.7 Alstom's Chilled Ammonia Process (CAP)

CAP (Alstom, 2015) uses chilled ammonia as solvent. Compared to amine solvents, chilled ammonia is more thermally and chemically stable and requires less energy for regeneration. However, the technology includes an energy intensive refrigeration system. The process has been tested at TCM (Norway), Pleasant Prairie PCC Project (Milwaukee), Karlshamn PCC Project (Sweden), and AEP Mountaineer CCS (West Virginia). The process was to be used at AEP Mountaineer CCS Phase II which was later cancelled.

12.3.4.8 KM CDR PCC Process

The process (MHI, 2015) was developed jointly by MHI and Kansai Electric Power Company. It uses a proprietary solvent named KS-1, which is reported to have relatively lower regeneration energy requirement. It has been tested at the pilot PCC plants at Plant Barry, Alabama and Plant Yates at Georgia both in the United States. Also, it will be used in the commercial Petro-Nova CCS due for commissioning later in 2016. The process has also been used in commercial projects in India, the Middle East, Malaysia, etc. for capturing CO_2 from natural gas fired steam reformer flue gases.

12.3.4.9 Other PCC Processes

There are a host of other technologies that are still pretty much at development stages. They include Powerspan's ECO_2™ technology, HTC Purenergy/Doosan Babcock technology, and CO_2 solutions capture technology among others.

12.3.4.9.1 Powerspan ECO₂ Process

ECO_2 process (GCCSI, 2012), developed by Powerspan, uses proprietary amine solvent and incorporates energy saving features, specific regeneration energy is about 2.3 GJ/ton CO_2. The process has been tested at a live power plant at the Burger PCC project, Ohio. Independent assessment by Worley Parsons concluded that the technology is ready for scale-up with a cost of less than US$40/ton of CO_2 captured and compressed.

12.3.4.9.2 HTC Purenergy/Doosan Babcock Process

HTC Purenergy/Doosan Babcock process (GCCSI, 2012) was developed at the University of Regina, Saskatchewan and field tested at the International Test Centre (Pilot PCC Plant) at SaskPower's Boundary Dam power plant. The process uses mixed amine solvent reported to have about 30% less regeneration energy requirement. Further tests and development are ongoing at a newly commissioned RD&D pilot plant in Renfrew, Scotland.

12.3.4.9.3 CO₂ Solutions Capture Process

Finally, the CO_2 solutions capture technology (CO_2 Solutions, 2016) uses enzyme-based solvent for CO_2 capture. The process also includes HTC's proprietary and patented solvent reclaimer system (SRS). There are ongoing tests of the process at Pikes Peak South PCC Project in Saskatchewan, Canada. Early results indicate impressive CO_2 loading capacity and regenerability.

12.3.5 Modeling and Simulation of PCC Based on Chemical Absorption

12.3.5.1 Basic Modeling Theory

CO_2 absorption/stripping in PCC process with chemical absorption involve complicated gas–liquid heat and mass transfer and chemical reactions (R1–R2). In modeling the process, these phenomena must be captured accurately. Interfacial heat and mass transfer can be described using any of surface renewal model (Danckwerts et al., 1963), penetration model (Tobiesen et al., 2007), or two-film theory (Lawal et al., 2010; Oko et al., 2015). However, two-film theory is more commonly used due to its relative simplicity.

$$2H_2O \leftrightarrow H_3O^+ + OH^-$$ (12.21)

$$CO_2 + 2H_2O \leftrightarrow H_3O^+ + HCO_3^-$$ (12.22)

$$HCO_3^- + H_2O \leftrightarrow H_3O^+ + CO_3^{2-}$$ (12.23)

$$RH^+ + H_2O \leftrightarrow R + H_3O^+$$ (12.24)

$$RCOO^- + H_2O \leftrightarrow R + HCO_3^-$$ (12.25)

Two-film theory (Figure 12.24) assumes that the liquid and vapor phases both consist of ideally mixed bulk regions and laminar film regions. Heat and mass transfer resistances are assumed to be restricted to the laminar film regions (Danckwerts, 1970). In describing the mass transfer phenomena, the detailed two-film theory can be applied (rate-based approach) or an equilibrium-based approach can be used (Lawal et al., 2010). In equilibrium-based approach, mass transfer calculations are simplified by the use of efficiency parameters (e.g., Murphree efficiency).

On the other hand, CO_2 reaction kinetics are often neglected and the reactions assumed to reach equilibrium (Lawal et al., 2010). This is reasonable if the solvent–CO_2 system has rapid kinetics, e.g., MEA–H_2O–CO_2 system (Kenig et al., 2001; Lawal et al., 2010). More accurate description of the CO_2 reaction kinetics can be obtained using the actual kinetic model (Zhang et al., 2009). This can be simplified by assuming pseudo first-order reaction and introducing an enhancement factor to account for kinetics (Kucka et al., 2003; Kvamsdal et al., 2009).

From a combination of the methods for describing mass transfer (rate-based or equilibrium-based) and CO_2 reaction kinetics (equilibrium reaction, kinetic model, or enhancement factor), models of PCC processes with chemical reactions can be classified into five levels of complexities as shown in Figure 12.25 (Kenig et al., 2001). Level 5 is considered the most accurate because it adopts a rate-based approach for describing mass transfer and kinetic model for the chemical reactions.

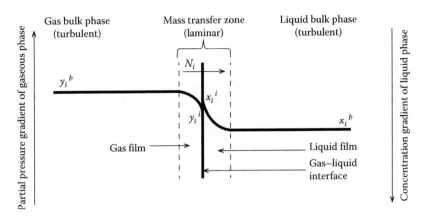

FIGURE 12.24 Schematic illustration of two-film theory. (Lawal, A. et al., *Fuel* 89 (10), 2791–2801, 2010.)

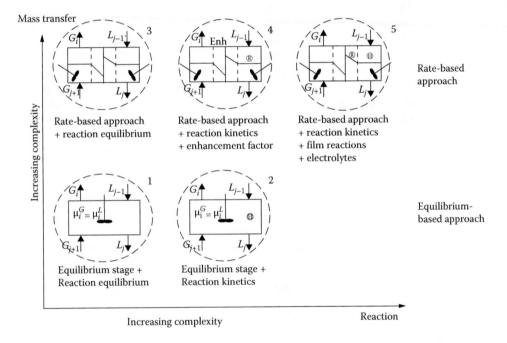

FIGURE 12.25 Complexity levels for reactive absorption/stripping model. (Lawal, A. et al., *Fuel* 89 (10), 2791–2801, 2010.)

12.3.5.2 Thermodynamic Model

The thermodynamic model is used to define the interfacial equilibrium, speciation in liquid phase, and enthalpy of absorption during modeling of CO_2 absorption/stripping with chemical reactions (Boot-Handford et al., 2014). It has a strong impact on the model fidelity and as such thermodynamic model selection is an important phase of the absorption/stripping model development. The CO_2–solvent–H_2O system in the process is a complex electrolyte system made up of molecules and ions. Thermodynamic model of such system is quite complex.

Many thermodynamic models for the system have been reported, the simplest of which are empirical correlations with vapor–liquid chemical equilibria obtained by fitting numerical parameters on experimental data (Gabrielsen et al., 2005). Some other models involve extending existing models for nonelectrolyte systems to address electrolyte species using Debye–Huckel theory (Deshmukh and Mather, 1981; Pitzer, 1973; Weiland et al., 1993). There are also complex activity coefficient models such as electrolyte Non-Random Two-Liquid (eNRTL) (Chen and Evans, 1986; Chen et al., 1982) and extended UNIversal QUAsiChemical (eUNIQUAC) (Thomsen and Rasmussen, 1999). The activity coefficient models have been used extensively in modeling CO_2 absorption/stripping in PCC processes (Faramarzi et al., 2009; Lawal et al., 2010). Recently, SAFT-VR model (Chapman et al., 1989, 1990; Gil-Villegas et al., 1997) that provides implicit treatment for the chemical reactions and ionic speciation in the liquid phase have also been proposed. The SAFT-VR model has been used successfully in modeling absorption/stripping processes (Mac Dowell et al., 2011).

12.3.5.3 Current Status of the Process Modeling

Dynamic models of the absorber (Khan et al., 2011; Kvamsdal and Rochelle, 2008; Kvamsdal et al., 2009; Lawal et al., 2009a; Posch and Haider, 2013) and the stripper (Lawal et al., 2009b; Ziaii et al., 2009), which are the main components of the PCC are available in literature. The two-film theory with detailed mass transfer calculations (rate-based) is used in most of the models to describe mass transfer (Khan et al., 2011; Kvamsdal and Rochelle, 2008; Kvamsdal et al., 2009; Lawal et al., 2009a,b; Ziaii et al., 2009). Equilibrium-based approach involving approximate mass transfer calculations has also been used (Posch and Haider, 2013). Comparative assessments of the two approaches in Lawal et al. (2009a) indicated that rate-based approach gives better predictions.

Reaction kinetics are neglected and the reactions assumed to reach equilibrium in some of the models (Lawal et al., 2009a,b; Ziaii et al., 2009). This is justified by the selection of MEA solvent, which has rapid kinetics with CO_2. Approximate kinetic models obtained by assuming pseudo first-order kinetics (Kvamsdal and Rochelle, 2008; Kvamsdal et al., 2009), more complex second-order kinetics (Khan et al., 2011), and thermomolecular reaction mechanism (Posch and Haider, 2013) have also been reported.

Dynamic model of the complete PCC process with chemical absorption including the absorber and stripper is also available (Gáspár and Cormoş, 2011; Harun et al., 2011; Lawal et al., 2010; Mac Dowell et al., 2013). Mass transfer process in these models were described using rate-based approach. The reaction kinetics are either approximated by assuming that the reactions reach equilibrium (Lawal et al., 2010; Mac Dowell et al., 2013) or by using approximate kinetic models with the assumption of pseudo first-order reaction (Harun et al., 2011; Gáspár and Cormoş, 2011).

In summary, published models of PCC have been developed with varying levels of complexities (see Figure 12.25) as follows:

1. Equilibrium-based mass transfer and approximate kinetic model for chemical reactions (Level 2 complexity)
2. Rate-based mass transfer and chemical equilibrium reactions (Level 3 complexity)
3. Rate-based mass transfer and approximate kinetic model for chemical reactions (Level 4 complexity).

Literature evidences suggest that level 4 complexity models are superior to others but are subject to greater computational requirement (Gáspár and Cormoş, 2011). Also, there are no published Level 5 complexity models, which involves detailed rate-based mass transfer and kinetic model.

12.3.5.4 Validation of the Process Models

There is a general insufficiency of plant data for detailed validation of existing models (Chikukwa et al., 2012). Lawal et al. (2010) and Biliyok et al. (2012), respectively, carried out steady-state and dynamic validations using plant data logs obtained from University of Texas at Austin (Dugas, 2006). Equipment specification for the pilot plant is given in Table 12.18.

12.3.5.4.1 Steady-State Validation

Steady-state validations by Lawal et al. (2010) were based on two experimental cases (32 and 47) selected from the pilot plant data sets. Process conditions for Cases 32 and 47 are given in Table 12.19.

TABLE 12.18 Equipment Specifications

Description	Value
Column inside diameter (absorber and regenerator) (m)	0.427
Height of packing (absorber and regenerator) (m)	6.1
Nominal packing size (absorber and regenerator) (m)	0.0381
Packing specific area (absorber) (m²/m³)	145
Packing specific area (regenerator) (m²/m³)	420
Cross heat exchanger heat transfer area (m²)	7
Reboiler volume (m³)	1
Condenser volume (m³)	2

Source: Dugas, E. R., Pilot plant study of carbon dioxide capture by aqueous monoethanolamine, 2006.

TABLE 12.19 Process Conditions for Cases 32 and 47

Stream ID	Case 32			Case 47		
	Flue Gas	Lean MEA	Rich MEA	Flue Gas	Lean MEA	Rich MEA
Temperature (K)	320	316	358	332	313	356
Total flow (kg/s)	0.130	0.720	0.745	0.158	0.642	0.746
			Mass Fraction			
H_2O	0.0148	0.6334	0.6122	0.0193	0.6334	0.6085
CO	0.2520	0.0618	0.0971	0.2415	0.0618	0.0366
MEA	0.0000	0.3048	0.2901	0.0000	0.3048	0.2943
N_2	0.7732	0.0000	0.0006	0.7392	0.0000	0.0006
Loading (mol CO_2/mol MEA)	—	0.2814	0.4646	—	0.2814	0.4556

Source: Dugas, E. R., Pilot plant study of carbon dioxide capture by aqueous monoethanolamine, 2006.

Models of absorber and stripper only were validated by comparing the temperature profile with the pilot plant absorber and stripper (regenerator) temperature profile for Case 32 (Figure 12.26). Although there is some deviation, it was observed that the prediction of the standalone models generally agrees with data logs of the pilot plant. Temperature bulges are clearly observed in the absorbers (Figure 12.27). Considering the stand-alone regenerator, an abrupt temperature drop of about 20 K is observed around the middle of the column height (Figure 12.27) in both pilot plant and model. The reboiler temperature is the main driver of the regenerator temperature profiles (Lawal et al., 2010).

Similarly, validations of the complete PCC model (i.e., with integrated absorber and stripper) are performed by comparing the temperature profile of the integrated columns and the pilot plant measurements for Case 32 (Figure 12.27). The results revealed an improved temperature profile prediction in Figure 12.28 for the absorber and regenerator when the integrated column is compared to the standalone column.

12.3.5.4.2 Dynamic Validation

Dynamic validations in Biliyok et al. (2012) were performed using Case 25 of the pilot plant data. Initial process conditions for this case are given in Table 12.20.

The model was simulated over a 10 h period subject to disturbances arising from changes in lean solvent flow rate to the absorber, concentration of CO_2 in the inlet flue gas stream to the absorber and temperature of the inlet flue gas stream to the absorber. During the run, other variables, namely reboiler temperature and reboiler liquid level among others were maintained at a constant value. The model predictions were compared with plant measurements taken under a similar scenario (Biliyok et al., 2012). The variables measured include temperature at top, middle, and bottom of the absorber, CO_2 mass fraction

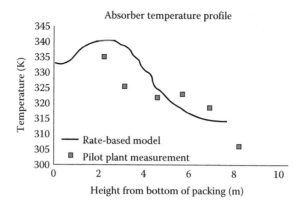

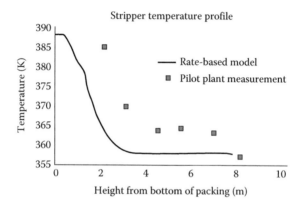

FIGURE 12.26 Stand-alone absorber and stripper temperature profile. (Lawal, A. et al., *Fuel* 89 (10), 2791–2801, 2010.)

in treated gas stream, and reboiler duty. The results (Biliyok et al., 2012) indicated that the model was able to track the data log very well regardless of the disturbances that were imposed throughout the run.

12.3.6 Process Analysis

12.3.6.1 Analysis Pilot Plant Scale

So far, most of the analysis of the process has been done at pilot scale and has covered steady-state and dynamic scenario. The analysis here has been done using the model presented in Lawal et al. (2010). Validation results of this model are satisfactory. As a result, the model can be used to study various characteristics of the process.

12.3.6.1.1 Steady State Analysis

Various parameters such as flue gas flow rate, lean solvent flow rate, and reboiler temperature were varied and the impact on other parameters, namely, CO_2 loading in the lean solvent and rich solvent and capture level, were recorded.

The CO_2 loading (mol CO_2/mol MEA) is given by

$$\frac{[CO_2]+\left[HCO_3^-\right]+\left[CO_3^{2-}\right]+\left[MEACOO^-\right]}{[MEA]+\left[MEA^+\right]+\left[MEACOO^-\right]} \tag{12.26}$$

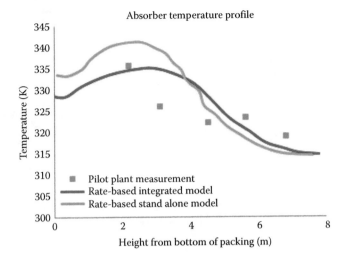

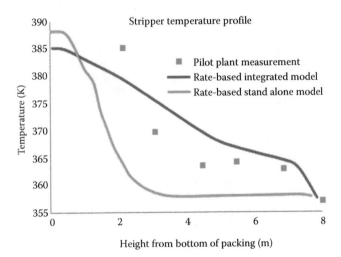

FIGURE 12.27 Absorber and stripper temperature profile. (Lawal, A. et al., *Fuel* 89 (10), 2791–2801, 2010.)

TABLE 12.20 Initial Process Conditions for Dynamic Test Case 25

	Stream ID	
	Flue gas	Lean MEA
Temperature (K)	313	316
Pressure (10^5 Pa)	1.05	1.05
Total flow (kg/s)	0.235	1.840
Mass Fraction		
H_2O	0.015	0.633
CO_2	0.159	0.059
MEA	0.000	0.308
N_2	0.826	0.000

Source: Dugas, E. R., Pilot plant study of carbon dioxide capture by aqueous monoethanolamine, 2006.

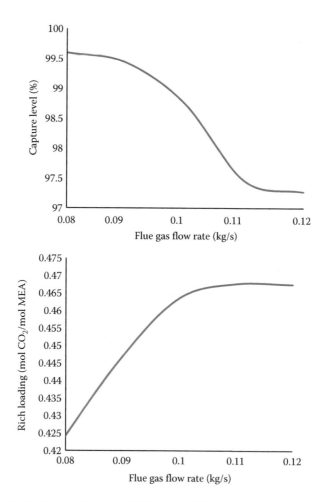

FIGURE 12.28 Sensitivity of capture level and rich loading at different flue gas flow rate.

The CO_2 capture level (%) on the other hand is given by

$$\left(\frac{y_{CO_2,in} - y_{CO_2,out}}{y_{CO_2,in}} \right) \times 100$$

(12.27)

12.3.6.1.1.1 Flue Gas Flow Rate   For this analysis, the flue gas flow rate was varied with all other parameters remaining unchanged. The impact of the change on capture level and CO_2 loading in the rich solvent was recorded (Figure 12.28). The results showed a decrease in the capture level and increase in the rich loading, respectively. This is expected since the solvent flow rate remained the same. Less changes are observed beyond about 0.11 kg/s flue gas flow rate. At this point, the solvent has reached its maximum CO_2 loading capacity. As a result, further increase in flue gas flow rate has little effect on these variables.

12.3.6.1.1.2 Lean Solvent Flow Rate   For this analysis, the lean solvent flow rate was varied with all other parameters remaining unchanged. The impact of the change on capture level and CO_2 loading in the rich solvent was recorded (Figure 12.29). With increasing solvent flow rate, more MEA becomes available to react with CO_2 resulting in higher capture and lower rich loading as seen in the results.

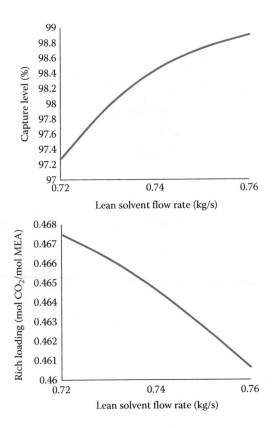

FIGURE 12.29 Sensitivity of capture level and rich loading at different flue gas flow rate.

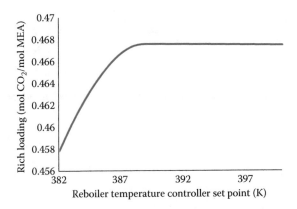

FIGURE 12.30 Sensitivity of rich loading at different reboiler temperature controller set point.

12.3.6.1.1.3 Reboiler Temperature Controller Set Point Reboiler temperature is an important parameter that drives solvent regeneration. As shown in the results (Figure 12.30), capture efficiency improves significantly at higher reboiler temperature as seen in the increasing rich solvent loading. Again, beyond 387 K temperature, the solvent becomes saturated (i.e., reaches its maximum CO_2 loading capacity). Further increase in temperature has little effect as a result. Also, note that increasing reboiler temperature could promote solvent degradation and inevitably increase energy consumption. Design temperature for the reboiler must, therefore, be balanced against these variables.

12.3.6.1.2 Dynamic Analysis

The model (Lawal et al., 2010) is used to simulate the effect of an increase in power plant output over a period of 10 min. It is assumed that this leads to a corresponding increase in flue gas flow rate to the absorber and its composition is maintained. The process was simulated at base-load conditions, Case 32 (Table 12.19), for 2 h and a 10% increase in flue gas flow rate was implemented over a period of 10 min after which the simulation was left to run for just less than 8 h. Two cases were investigated (Figure 12.31):

- Case 1: The solvent circulation rate was kept constant.
- Case 2: The solvent circulation rate was correspondingly increased to maintain the molar Liquid/ Gas ratio.

Figure 12.32 shows a decrease in CO_2 capture levels at the onset of the disturbance for Case 1 and virtually the same capture level maintained for Case 2 after a slight drop. This is, however, at the cost of increased heat requirement for capture (Figure 12.33). The heat required to capture a kg of CO_2 in Case 1 drops because the lean solvent had the capacity to capture more CO_2 at the same circulation rate. The difference between the rich and lean CO_2 loading (i.e., absorption capacity) is estimated and plotted for the two cases (Figure 12.34). In Case 1, the solvent is in contact with significantly more CO_2 in the flue gas and thus absorbs more CO_2 giving higher rich loading. As the lean loading is roughly maintained due to the constant reboiler temperature, CO_2 wt% in the top product increases. With the constant solvent circulation rate, the reboiler duty needed to heat the solvent to the required set point of 387 K does

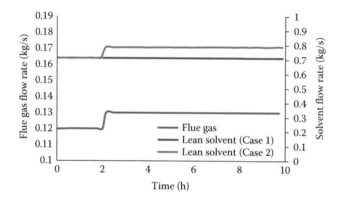

FIGURE 12.31 Change in flue gas flow rate. (Biliyok, C. et al., *International Journal of Greenhouse Gas Control* 9, 428–445, 2012.)

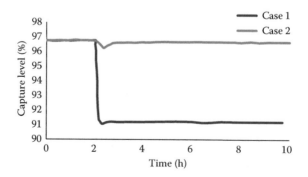

FIGURE 12.32 CO_2 capture level for Case 1 and Case 2. (Biliyok, C. et al., *International Journal of Greenhouse Gas Control* 9, 428–445, 2012.)

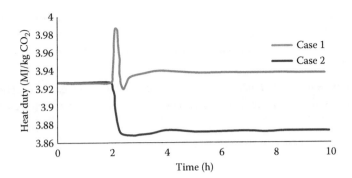

FIGURE 12.33   Heat duty for Case 1 and Case 2. (Biliyok, C. et al., *International Journal of Greenhouse Gas Control* 9, 428–445, 2012.)

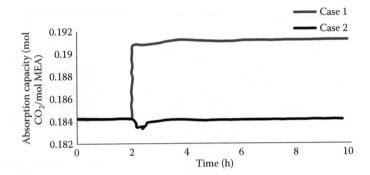

FIGURE 12.34   Absorption capacity for Case 1 and Case 2. (Biliyok, C. et al., *International Journal of Greenhouse Gas Control* 9, 428–445, 2012.)

not increase significantly. As a result, heat duty requirement for Case 1 reduces significantly; heat duty required for Case 2 is higher than that of Case 1.

12.3.6.2 Analysis at Large Scale

Due to the lack of data at large scale scenario, validated pilot-scale model of the process has been scaled up and used to study the process behavior at large-scale scenario (Lawal et al., 2012). Scale-up of the absorber and stripper is complex due to the high nonlinearity of hydrodynamics in the packed column.

12.3.6.2.1 Methodology for Absorber and Stripper Scale-Up

There are different methodologies for process scale-up in Chemical Engineering:

1. Model-based methodology
   Model of a process can be used to perform scale-up calculations. This will usually be computationally intensive because it requires many rounds of iterations before convergence can be reached. Also, in this methodology, the model must be proven through detailed validations.
2. Correlation-based methodology
   This methodology involves the use of equipment sizing correlations from literature. This approach is very simple, but less accurate. Also, this approach does not give optimal sizes but it will give a good starting point for optimization.

Lawal et al. (2012) used the correlation-based approach to scale up the absorber and stripper in a PCC process. This involved conventional hydraulic correlations for sizing the packed column (Sinnot, 2005). Parameters calculated include the diameter and height of the packed column. The correlation-based approach is illustrated here.

12.3.6.2.1.1 *Assumptions and Calculation of Solvent Circulation Rate*   Typical assumptions will include setting flooding limits, usually not more than 80% and also determining packing type, etc. Other information required include

- Large-scale flue gas flow rate and composition
- Pilot-scale solvent type and concentration
- Pilot-scale lean and rich loading
- Economic pressure drop.

The solvent flow rate can be obtained as follows:

$$\text{Solvent flow rate } (\text{kg/s}) = \frac{\dot{m}_{FG} M_{MEA}}{A_c wt_{MEA}}, \tag{12.28}$$

where $\dot{m}_{FG}$ = molar flue gas flow rate (kmol/s), M_{MEA} = molecular weight of the solvent, e.g., MEA (kg/kmol), A_c = absorption capacity (mol CO_2/mol MEA), and wt_{MEA} = solvent (MEA) concentration (wt%).

For the pilot plant (Lawal et al. 2010), the lean and rich loadings are, respectively, 0.28 and 0.48 mol CO_2/mol MEA. Therefore, A_c (i.e., rich loading–lean loading) = 0.20 mol CO_2/mol MEA. For a 500 MW$_e$ coal-fired subcritical power plant, the molar flue gas flow rate is about 2.845 kmol/s. With 30 wt% MEA solvent, the solvent flow rate that will be needed for the flue gas will be 2850 kg/s using Equation 12.28 and the pilot plant (Lawal et al., 2010) lean and rich loading data.

12.3.6.2.1.2 *Packed Column Diameter*   Packed column diameter is subject to tight boundaries defined by flooding and minimum liquid load (Lawal et al., 2012) and good liquid and gas distribution requirements (Sinnot et al., 2005). This must all be considered when estimating the packed column diameter. The criteria are met by designing the column to operate at the highest economical pressure drop. Sinnot et al. (2005) recommended economic pressure drop of 15–50 mmH$_2$O per meter of packing for absorbers and strippers.

To estimate the cross-sectional area and consequently the diameter, the generalized pressure drop correlation (GPDC) chart for packed columns is generally used. The pressure drop lines in the chart are in mmH$_2$O per meter of packing. The GPDC comprise of a flow parameter abscissa (F_{LV}) that represents the ratio of liquid kinetic energy to vapor kinetic energy (Kister et al., 2007). This is calculated as follows:

$$F_{LV} = \frac{L_w^*}{V_w^*} \sqrt{\frac{\rho_V}{\rho_L}} \tag{12.29}$$

The $\dfrac{L_w^*}{V_w^*}$ term is the ratio of liquid to vapor mass flow rate per unit column cross-sectional area. This is similar to L/G ratio (i.e., liquid to vapor mass flow rate ratio). The liquid and vapor densities, ρ_L and ρ_V >, are obtained at pilot-scale conditions. On the other hand, the GPDC ordinate is the capacity parameter (K_4). The K_4 parameter is read off from the GPDC (see Sinnot, 2005) at a defined pressure drop and calculated F_{LV}. As a check, flooding (%) can be calculated from K_4 values at design and flooding pressure drops. As a rule of thumb, flooding should be less than 80% for the design to be acceptable. This is calculated as follows:

$$\text{Flooding } (\%) = \left[\frac{K_4 \text{ at design pressure drop}}{K_4 \text{ at flooding}}\right]^{1/2} \times 100 \tag{12.30}$$

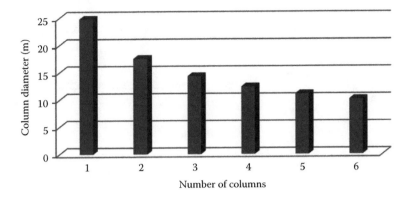

FIGURE 12.35 Required number of packed columns (absorber) and diameters. (Oko, E. Study of power plant, carbon capture and transport network through dynamic modeling and simulation, 2015.)

Thereafter, packed column diameter is calculated:

$$K_4 = \frac{13.1\left(V_w^*\right)^2 F_P \left(\dfrac{\mu_L}{\rho_L}\right)^{0.1}}{\rho_V \left(\rho_L - \rho_V\right)} \tag{12.31}$$

Like the densities, the liquid phase viscosity (μ_L) is obtained at the pilot plant condition. The packing factor (F_p) depends on selected packing type and size. The cross-sectional area (and hence the column diameter) is obtained from the vapor (flue gas) mass flow rate per unit cross sectional area (V_w^*) calculated using Equation 12.31.

Using this methodology, packed column diameter and height for a PCC plant for treating flue gases from a 500 MW$_e$ coal-fired power plant was estimated. Pilot-scale model by Lawal et al. (2010) was used as basis. A 38 mm ceramic Raschig ring packing and economic pressure drop of 42 mmH$_2$O was used. The estimated absorber and stripper diameter following these assumptions are about 25 and 18 m diameter, respectively. Such large columns could result in severe liquid maldistribution leading to lower column efficiency (Ramezan et al., 2007). Multiple columns could be used to avoid this possibility. Based on the criteria for limiting sizes of packed columns (Ramezan et al., 2007), it is found that about four absorber columns will be needed (Figure 12.35).

12.3.6.2.1.3 Packed Column Height To some degree, a trade-off exists between the absorber column packed height and the solvent regeneration energy requirement. As the solvent circulation rate decreases, the rich loading of the solvent achieved in the absorber column increases till it approaches saturation. Toward this point, the system experiences diminishing returns for increases in absorber packed height. An optimization study can be carried out to determine the optimal packed height required.

12.3.6.3 Process Analysis

The validated model was upscaled to treat flue gas from a 500 MW$_e$ coal-fired subcritical power plant (Lawal et al., 2012) as shown above. The model was integrated to a model of the power plant (Oko and Wang, 2014) to represent a scenario of power plant with PCC (Lawal et al., 2012). The entire plant was analyzed under a situation where the power plant load is switched from one level to another. Corresponding deviations in various key variables are shown in Figure 12.36.

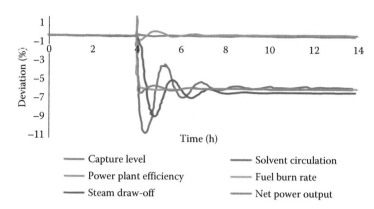

FIGURE 12.36 Deviations in different variables during a load change. (Lawal, A. et al., *Fuel* 101, 115–128, 2012.)

12.3.6.4 Commercial Applications for Modeling

12.3.6.4.1 RADFRAC

RADFRAC is a rigorous multistage separation model developed by Aspen Technology Inc., Texas. The RADFRAC model can be used from commercial applications of Aspen Technologies Inc. such as Aspen Plus and Aspen HYSYS. The model can be used in either rate-based or equilibrium-based mode. There is also option for incorporating user defined subroutines written in FORTRAN. Validation results of PCC models developed using RADFRAC show good agreement (Zhang et al., 2009).

12.3.6.4.2 gCCS

gCCS is a CCS system modeling toolkit that provides high-fidelity modeling capability for all operations from power generation to CO_2 injection. gCCS (PSE, 2015) was built on Process Systems Enterprise's (PSE) Ltd gPROMS advanced process modeling platform with input from E.ON, EDF, Rolls-Royce, and CO_2 Deepstore in a 3-year project funded by the Energy Technologies Institute. gCCS enables users to carry out both steady-state and dynamic (transient) analysis of CCS systems. The current scope contains process model libraries of

- Power generation
  - Conventional: Pulverized-coal, CCGT
  - Nonconventional: Oxy-fueled, IGCC
- Solvent-based CO_2 capture
- CO_2 compression and liquefaction
- CO_2 transportation
- CO_2 injection in subsea storage.

In addition to this, material models have been developed using PSE's proprietary general Statistical Associating Fluid Theory (gSAFT) technology. These are used for prediction of thermodynamic properties of various solvents for chemical and physical absorption of CO_2 capture (Rodríguez et al., 2014) as well as for near-pure CO_2 mixtures.

12.3.7 Challenges Facing PCC Commercial Projects

Commercialization of PCC has been an uphill task, many planned commercial PCC projects have either been cancelled or put on hold (e.g., Trailblazer CCS, Texas, Longannet CCS, Scotland, Getica CCS, Romania, Antelope Valley CCS, North Dakota, and Porto Tolle CCS, Italy) for various reasons in the last couple of years. Some of the challenges contributing to these failures are discussed below.

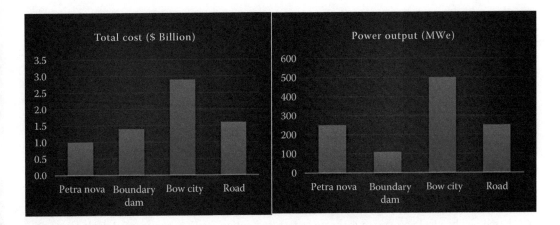

FIGURE 12.37 Cost and plant output of existing/upcoming CCS projects (MIT CCS, 2015).

12.3.7.1 Capital Cost

The cost/MW_e of power plants integrated with PCC is prohibitive as can be seen with some existing/planned projects (Figure 12.37). SaskPower, the developers of Boundary Dam CCS, claims to be able to cut cost by up to 30% in their future projects. Also, with the prospect of relying on cheap coal and revenues from CO_2, fly ash, and sulfuric acid by-product sales, the Boundary Dam CCS is judged to be competitive with CCGT plants.

12.3.7.2 Carbon Pricing

Carbon pricing through carbon tax and/or cap-and-trade is a strong economic driver for CCS commercialization especially in the EU where CO_2 captured from most projects are to be stored permanently underground. In the EU, carbon price under its emission trading system (ETS) took a nosedive from about €16/tCO_2 in 2011 to about €3/tCO_2 in May 2013 (EEX, 2015). This contributed to the cancellation/suspension of many CCS projects in the EU during that time. New data indicate steady rise in carbon price (up to €8/tCO_2 as at 2015) under the EU ETS since mid-2013 (EEX, 2015). In the United States and Canada, CO_2 from many CCS projects is planned to be used for EOR (e.g., Boundary Dam CCS). This will provide some economic incentives for companies to invest in CCS projects.

12.3.7.3 Government Policies

Appropriate government policies that enable the right environment for CCS to thrive is lacking in many countries. In the Netherlands, onshore CO_2 storage was banned in 2011 (MIT CCS, 2015). This will make it difficult for CCS projects expected to be located far away from the coastline. Nuon Magnum CCS (Netherlands) was suspended when this law became effective. Also, permitting process is complicated and rigorous in many countries. The Porto Telle CCS (Italy) was suspended due to permitting and other legislative issues (MIT CCS, 2015).

12.3.7.4 Many Stakeholders

CCS involves many stakeholders with different interests along the entire CCS chain. This constitutes real challenge for effective and optimal design and operation of the CCS network. gCCS modeling toolkit was developed in response to this challenge (PSE, 2015). It is intended to provide a common working platform for different stakeholders along the CCS chain.

12.3.7.5 Water Contamination

Surface water contamination either directly by effluents from the process or indirectly due to wet deposition of effluents released into the air is a real challenge to PCC development (SEPA, 2015). In addition, the process also generates huge volumes of amine contaminated waste water and decontaminating/disposing them safely and economically is challenging. In light of this, it is suggested that environmentally benign and low volatility solvents such as amino acids and ionic liquids should replace amines solvents used commonly in PCC processes (Hasib-ur-Rahman et al., 2010; Siemens A.G., 2015). Amino acid solvents have been successfully commercialized by Siemens in their PostCap process. Ionic liquids on the other hand are still very much at development stage and may not be available commercially for the next couple of years.

12.3.8 Conclusions

PCC based on chemical absorption is the most matured option for CCS commercial deployment. The technology has been widely validated through pilot plant tests. Also, modeling and simulation has been used to gain more in-depth insights of the process dynamics through detailed analysis of its design and operation. Different PCC technologies using solvent have also been commercialized, each of them utilizing different process chemistry and configurations. One of them, CanSolv by Shell, was successfully implemented in a first-of-its-kind power plant CCS project at Boundary Dam Power Plant, Canada, in October, 2014. Regardless of the success stories, PCC commercialization is still beset by a number of challenges (e.g., lack of economic drivers, high capital and operating cost, potential for surface water contamination among others). Unless these challenges are addressed, it is predicted that many planned projects will not be actualized.

Acknowledgment

The authors acknowledge financial supports from EU IRSES (Ref: PIRSES-GA-2013-612230), NERC, UK (Ref: NE/H013865/1), and EPSRC, UK (Ref: EP/M001458/1).

References

Abu-Zahra, M. R .M., Abba, Z., Singh, P. and Feron, P. (2013). Carbon dioxide postcombustion capture: Solvent technologies overview, status and future. In: A. Mendez-Vilas (ed.) *Materials and Process for Energy: Communicating Current Research and Technological Developments.* Formatex Research Centre (ISBN: 978-84-939843-7-3).

Adam, D. (2008). Explainer: Global carbon reduction targets. *The Guardian* Newspaper, October 7, 2008.

Ahn, H., Luberti, M., Liu, Z. and Brandani, S. (2013). Process configuration studies of the amine capture process for coal-fired power plants. *International Journal of Greenhouse Gas Control* 16, 29–40.

Alstom. (2015). Post-combustion capture can easily be retrofitted onto existing power plants. www. alstom.com/products-services/product-catalogue/power-generation/coal-and-oil-power/co2-capture-systems-ccs/co2-capture-post-combustion-ccs/ (Accessed April 22, 2016).

Baburao, B., Bedell, S., Restrepo, P., Schmidt, D., Schubert, C., DeBolt, B., Haji, I. and Chopin, F. (2014). Advanced amine process technology operations and results from demonstration facility at EDF Le Havre. Energy Procedia 63, 6173–6187.

Biliyok, C., Lawal, A., Wang, M. and Seibert, F. (2012). Dynamic modeling, validation and analysis of post-combustion chemical absorption CO_2 capture plant. *International Journal of Greenhouse Gas Control* 9, 428–445.

Boot-Handford, M. E., Abanades, J. C., Anthony, E. J., Blunt, M. J., Brandani, S., Mac Dowell, N., Fernandez, J. R., Ferrari, M.-C., Gross, R., Hallett, J. P., Haszeldine, R. S., Heptonstall, P., Lyngfelt, A., Makuch, Z., Mangano, E., Porter, R. T. J., Pourkashanian, M., Rochelle, G. T., Shah, N., Yao, J. G. and Fennell, P. S. (2014). Carbon capture and storage update. *Energy and Environmental Science* 7, 130.

Brown, T., Perry, C. C. and Manthey, B. (2009). Progress report of Pleasant Prairie carbon capture demonstration project. http://mydocs.epri.com/docs/CorporateDocuments/SectorPages/GEN/CarbonCaptureProject/doc/cc_report.pdf (Accessed May 12, 2017).

Carbon Capture and Storage Association (CCSA). (2011). Strategy for CCS in the UK and beyond. www.ccsassociation.org (Accessed August 30, 2015).

Cerimele, G. L. (2012). CCS lessons learned report American Electric Power Mountaineer CCS II project phase 1. Prepared for The Global CCS Institute Project, January 23, 2012.

Chapman, W. G., Gubbins, K. E., Jackson, G. and Radosz, M. (1989). SAFT: Equation-of-state solution model for associating fluids. *Fluid Phase Equilibrium* 52, 31–38.

Chapman, W. G., Gubbins, K. E., Jackson, G. and Radosz, M. (1990). New reference equation of state for associating liquids. *Industrial and Engineering Chemistry Re*search 29, 1709–1721.

Chen, C.-C. and Evans, L. B. (1986). A local composition model for the excess Gibbs energy of aqueous electrolyte systems. *AIChE Journal* 32, 444–454.

Chen, C.-C., Britt, H. I., Boston, J. F. and Evans, L. B. (1982). Local composition model for excess Gibbs energy of electrolyte systems. Part I: Single solvent, single completely dissociated electrolyte systems. *AIChE Journal* 28, 588–596.

Chikukwa, A., Enaasen, N., Kvamsdal, H. M. and Hillestad, M. (2012). Dynamic modeling of post-combustion CO_2 capture using amines–A review. *Energy Procedia* 23, 82–91.

Chopin, F. (2014). Results of the CO_2 capture demonstration facility at EDF's Le Havre power plant: Status of ALSTOM's advanced amines process. PowerGen Europe 2014, Cologne, Germany, June 3, 2014.

CO_2 Solutions. (2016). Harnessing nature for carbon capture. www.co2solutions.com/ (Accessed April 22, 2016).

The Committee on Climate Change (CCC) UK. (2015). Setting a target for emission reduction. https://www.theccc.org.uk/tackling-climate-change/the-science-of-climate-change/setting-a-target-foremission-reduction/ (Accessed August 30, 2015).

Danckwerts, P. V. (1970). *Gas–Liquid Reactions*. McGraw-Hill, New York.

Danckwerts, P. V., Kennedy, A. M. and Roberts, D. (1963). Kinetics of CO_2 absorption in alkaline solutions—II: Absorption in a packed column and tests of surface-renewal models. *Chemical Engineering Science* 18, 63–72.

Darde, V., Thomsen, K., Van Well, W. J. M. and Stenby, E. H. (2010). Chilled ammonia process for CO_2 capture *International Journal of Greenhouse Gas Control* 4 (2), 131–136.

Davidson, R. M. (2007). Post-combustion carbon capture from coal fired plants—Solvent scrubbing. IEA Clean Coal Centre, CCC/125.

Deshmukh, R. D. and Mather, A. E. (1981). A mathematical model for equilibrium solubility of hydrogen sulfide and carbon dioxide. *Chemical Engineering Science* 36, 355–362.

Dubois, L. and Thomas, D. (2012). Screening of aqueous amine-based solvents for post-combustion CO_2 capture by chemical absorption. *Chemical Engineering Technology* 35 (3), 513–524.

Dugas, E. R. (2006). Pilot plant study of carbon dioxide capture by aqueous monoethanolamine, M Thesis, University of Texas at Austin.

Eswaran, E., Wu, S. and Nicolo, R. (2010). Advanced amine-based CO_2 capture for coal-fired power plants. In: *Coal-Gen 2010*, Pittsburgh, PA, August 10–12, 2010.

European Energy Exchange (EEX). https://www.eex.com/en#/en (Accessed August 30, 2015).

Faramarzi, L., Kontogeorgis, G. M., Thomsen, K. and Stenby, E. H. (2009). Extended UNIQUAC model for thermodynamic modeling of CO_2 absorption in aqueous alkanolamine solutions. *Fluid Phase Equilibria* 282, 121–132.

Fisher, K. S., Beitler, C. M., Rueter, C. O., Searcy, K., Rochelle, G. T. and Jassim, M. (2005). Integrating MEA regeneration with CO_2 compression and peaking to reduce CO_2 capture costs. Final report under DOE Grant DE-FG02-04ER84111, June 9, 2005.

Fisher, K. S., Searcy, K., Rochelle, G. T., Ziaii, S. and Schubert, C. (2007). Advanced amine solvent formulations and process integration for near-term CO_2 capture success. Final Report under DOE Grant DE-FG02-06ER84625, June 28, 2007.

Fluor. (2015). Projects: E.ON Kraftwerke carbon capture technology demonstration plant. www.fluor.com/projects/carbon-capture-plant-design-build (Accessed September 30, 2015).

Gabrielsen, J., Michelsen, M. L., Stenby, E. H. and Kontogeorgis, G. M. (2005). A model for estimating CO_2 solubility in aqueous alkanolamines. *Industrial and Engineering Chemistry Research* 44, 3348–3354.

Gáspár, J. and Cormoş, A. (2011). Dynamic modeling and validation of absorber and desorber columns for post-combustion CO_2 capture. *Computers and Chemical Engineering* 35 (10), 2044–2052.

Gil-Villegas, A., Galindo, A., Whitehead, P. J., Mills, S. J., Jackson, G. and Burgess, A. N. (1997). Statistical associating fluid theory for chain molecules with attractive potentials of variable range. *The Journal of Chemical Physics* 106, 4168–4186.

Global Carbon Capture and Storage Institute (GCCSI). (2012). CO_2 capture technologies: Post-combustion capture. http://hub.globalccsinstitute.com/sites/default/files/publications/29721/co2-capture-technologies-pcc.pdf (Accessed April 22, 2016).

Global CCS Institute (GCCSI). (2013). Capture demonstration at Korea's Boryeong Thermal Power Station. www.globalccsinstitute.com/insights/authors/dennisvanpuyvelde/2013/07/23/capture-demonstration-koreas-boryeong-thermal-power (Accessed September 30, 2015).

Gorset, O., Knudsen, J. N., Bade, O. M. and Askestad, I. (2014). Results from testing of Aker Solutions Advanced Amine Solvents at CO_2 Technology Centre Mongstad. *Energy Procedia* 63, 6267–6280.

Harkin, T., Hoadley, A. and Hooper, B. (2010). Reducing the energy penalty of CO_2 capture and compression using pinch analysis. *Journal of Cleaner Production* 18, 857–866.

Harun, N., Douglas, P. L., Ricardez-Sandoval, L. and Croiset, E. (2011). Dynamic simulation of MEA absorption processes for CO_2 capture from fossil fuel power plant. *Energy Procedia* 4, 1478–1485.

Hasib-ur-Rahman, M., Siaj, M. and Larachi, F. (2010). Ionic liquids for CO_2 capture—Development and progress. *Chemical Engineering and Processing* 49, 313–322.

Herzog, H., Meldon, J. and April, A. (2009). Advanced post-combustion CO_2 capture. Prepared for the clean air task force under a grant from the Doris Duke Foundation. https://mitei.mit.edu/system/files/herzog-meldon-hatton.pdf (Accessed January 20, 2016).

Hill, G. R. (2014). SECARB plant Barry CCS project: Sharing knowledge and learning. CCS Seminar, UKCCSRC, University of Edinburgh, UK, September 2014.

IEA. (2008). Energy technology perspectives: Scenarios and strategies to 2050. www.iea.org/media/etp/etp2008.pdf (Accessed September 1, 2015).

Iijima, M. (2005). Flue gas CO_2 capture. http://gcep.stanford.edu/pdfs/energy_workshops_04_04/carbon_iijima.pdf (Accessed September 25, 2015).

Iijima, M., Nagayasu, T., Kamijyo, T., and Nakatani, S. (2011). MHI's energy efficient flue gas CO_2 capture technology and large scale CCS demonstration test at coal-fired power plants in USA. *Mitsubishi Heavy Industries Technical Review* 48 (1), 26.

IPCC. (2005). IPCC special report on carbon dioxide capture and storage. Prepared by Working Group III of the Intergovernmental Panel on Climate Change, B. Metz, O. Davidson, H. C. de Coninck, M. Loos, and L. A. Meyer (eds). Cambridge University Press, Cambridge, UK and New York, 442 pp.

Just, P.-E. (2013). Shell CanSolv™: Deploying CCS worldwide. http://ieaghg.org/docs/General_Docs/PCCC2/Secured%20pdfs/3_PCCC2-Just-September2013.pdf (Accessed April 24, 2016).

Kenig, E. Y., Schneider, R. and Górak, A. (2001). Reactive absorption: Optimal process design via optimal modeling. *Chemical Engineering Science* 56, 343–350.

Khan, F. M., Krishnamoorthi, V. and Mahmud, T. (2011). Modeling reactive absorption of CO_2 in packed columns for post-combustion carbon capture applications. *Chemical Engineering Research and Design* 89 (9), 1600–1608.

Kister, H. Z. Scherffius, J., Afshar, K. and Abkar, E. (2007). Realistically predict capacity and pressure drop for packed columns. AIChE Spring Meeting, Houston, TX.

Kucka, L., Müller, I., Kenig, E. Y. and Górak, A. (2003). On the modeling and simulation of sour gas absorption by aqueous amine solutions. *Chemical Engineering Science*, 58, 3571–3578.

Kvamsdal, H. M. and Rochelle, G. T. (2008). Effects of the temperature bulge in CO_2 absorption from flue gas by aqueous mono-ethanolamine. *Industrial and* Engineering *Chemistry Research* 47 (3), 867–875.

Kvamsdal, H. M., Jakobsen, J. P. and Hoff, K. A. (2009). Dynamic modeling and simulation of a CO_2 absorber column for post-combustion CO_2 capture. *Chemical. Engineering and Processing: Process Intensification* 48, 135–144.

Lawal, A., Wang, M., Stephenson, P., Koumpouras, G. and Yeung, H. (2010). Dynamic modeling and analysis of post-combustion CO_2 chemical absorption process for coal-fired power plants. *Fuel* 89 (10), 2791–2801.

Lawal, A., Wang, M., Stephenson, P. and Obi, O. (2012). Demonstrating full-scale post-combustion CO_2 capture for coal-fired power plants through dynamic modeling and simulation. *Fuel* 101, 115–128.

Lawal, A., Wang, M., Stephenson, P. and Yeung, H. (2009a). Dynamic modeling of CO_2 absorption for post combustion capture in coal-fired power plants. *Fuel* 88 (12), 2455–2462.

Lawal, A., Wang, M., Stephenson, P. and Yeung, H. (2009b). Dynamic modeling and simulation of CO_2 chemical absorption process for coal-fired power plants. *Computer Aided Chemical Engineering* 27, 1725–1730.

Lucquiaud, M. and Gibbins, J. (2011). Steam cycle options for the retrofit of coal and gas power plants with post-combustion capture. *Energy Procedia* 4, 1812–1819.

Mac Dowell, N., Pereira, F. E., Llovell, F., Blas, F. J., Adjiman, C. S. Jackson, G. and Galindo, A. (2011). Transferable SAFT-VR models for the calculation of the fluid phase equilibria in reactive mixtures of carbon dioxide, water, and n-alkylamines in the context of carbon capture. *The Journal of Physical Chemistry B* 115, 8155–8168.

Mac Dowell, N., Samsatli, N. J. and Shah, N. (2013). Dynamic modeling and analysis of an amine-based post-combustion CO_2 capture absorption column. *International Journal of Greenhouse Gas Control* 12, 247–258.

MHI. (2015). Update of the deployment of the KM CDR process. http://www.ieaghg.org/docs/General_Docs/PCCC3_PDF/1_PCCC3_6_Kamijo.pdf.

MIT CCS. (2015). http://sequestration.mit.edu/index.html (Accessed August 24, 2015).

Nsakala, N., Marion, J., Bozzuto, C., Liljedahl, G., Palkes, M., Vogel, D., Gupta, J. C., Guha, M. and Johnson, H. (2001). Engineering feasibility of CO_2 capture on an existing US coal-fired power plant. First National Conference on Carbon Sequestration May 15–17, 2001, Washington, DC.

Oko, E. (2015). Study of power plant, carbon capture and transport network through dynamic modeling and simulation. PhD Thesis, University of Hull.

Oko, E. and Wang, M. (2014). Dynamic modeling, validation and analysis of coal-fired subcritical power plant *Fuel* 135, 292–300.

Oko, E., Wang, M. and Olaleye, A. K. (2015). Simplification of detailed rate-based model of post-combustion CO_2 capture for full chain CCS integration studies. *Fuel* 142, 87–93.

Pitzer, K. S. (1973). Thermodynamics of electrolytes: I. Theoretical basis and general equations. *The Journal of Physical Chemistry* 77, 268–277.

Posch, S. and Haider, M. (2013). Dynamic modeling of CO_2 absorption from coal-fired power plants into an aqueous monoethanolamine solution. *Chemical Engineering Research and Design* 91 (6), 977–987.

PSE. (2015). www.psenterprise.com/power/ccs/overview.html (Accessed April 23, 2016).

Ramezan, M., Skone, T. J. and Nsakala, N. (2007). Carbon dioxide capture from existing coal-fired power plants. In: DOE/NETL-401/110907.

Rao, A. B., Rubin, E. S. and Berkenpa, M. B. (2004). An integrated modeling framework for carbon management technologies. In: U.S. Department of Energy (DE-FC26-00NT40935).

Raynal, L. La Marca, C., Normand, L. and Broutin, P. (2012). Demonstration of the DMX™ process—description of the Octavius SP3 project. 2nd Post Combustion Capture Conference September 17–20, 2013, Bergen, Norway.

Rodríguez, J., Andrade, A., Lawal, A., Samsatli, N., Calado, M., Ramos, A., Lafitte, T., Fuentes, J., Pantelides, C. C. (2014). An integrated framework for the dynamic modeling of solvent-based CO_2 capture processes. *Energy Procedia* 63, 1206–1217.

Scottish Environmental Protection Agency (SEPA) (2015). Review of amine emissions from carbon capture systems. www.sepa.org.uk/media/155585/review-of-amine-emissions-from-carbon-capture-systems.pdf (Accessed April 23, 2016).

Shaw, D. (2009). CanSolv CO_2 capture: The value of integration. *Energy Procedia* 1, 237–246.

Siemens, A. G. (2015). Post-combustion carbon capture. www.energy.siemens.com/nl/en/fossil-power-generation/power-plants/carbon-capture-solutions/post-combustion-carbon-capture/ (Accessed April 22, 2016).

Sinnot, R. K. (2005). Chemical engineering design. In: J. M. Coulson and J. F. Richardson (Eds) *Chemical Engineering*, Vol. 6, 4th ed., Elsevier Butterworth-Heinemann, Oxford.

Thomsen, K. and Rasmussen, P. (1999). Modeling of vapor–liquid–Solid equilibrium in gas—Aqueous electrolyte systems. *Chemical Engineering Science* 54, 1787–1802.

Tobiesen, F. A., Svenden, H. F. and Juliussen, O. (2007). Experimental validation of a rigorous absorber model for CO_2 post-combustion capture. *AIChE Journal*, 53, 846–865.

Tollefson, J. (2011). Low-cost carbon-capture project sparks interest. *Nature* 469, 276–277.

Wang, M., Joel, A. S., Ramshaw, C., Eimer, D. and Musa, N. M. (2015). Process intensification for post-combustion CO_2 capture with chemical absorption: A critical review. *Applied Energy* 158, 275–291.

Wang, M., Lawal, A., Stephenson, P., Sidders, J. and Ramshaw, C. (2011). Post-combustion CO_2 capture with chemical absorption: A state-of-the-art review. *Chemical Engineering Research Design* 89 (9), 1609–1624.

Weiland, R. H., Chakravarty, T. and Mather, A. E. (1993). Solubility of carbon-dioxide and hydrogen sulfide in aqueous alkanolamines. *Industrial and Engineering Chemistry Research* 32, 1419–1430.

Yang, J., Yu, X., Yan, J. and Tu, S.-T. (2014). CO_2 capture using absorbents of mixed ionic and amine solutions. *Industrial and Engineering Chemistry Research* 53, 2790–2799.

Yu, C.-H., Huang, C.-H. and Tan, C.-S. (2012). A review of CO_2 capture by absorption and adsorption. *Aerosol and Air Quality Research* 12, 745–769.

Ziaii, S., Rochelle, G. T. and Edgar, T. F. (2009). Dynamic modeling to minimize energy use for CO_2 capture in power plants by aqueous monoethanolamine. *Industrial and Engineering Chemistry Research* 48, 6105–6111.

Zhang, J., Qiao, Y., Wang, W., Misch, R., Hussain, K. and Agar, D. W. (2013). Development of an energy-efficient CO_2 capture process using thermomorphic biphasic solvents. *Energy Procedia* 37, 1254–1261.

Zhang, Y., Chen, H., Chen, C., Plaza, J. M., Dugas, R., Rochelle, G. T. (2009). Rate-based process modeling study of CO_2 capture with aqueous monoethanolamine solution. *Industrial and Engineering Chemistry Research* 48, 9233–9246.

12.4 Operation of Supercritical Coal-Fired Power Plant (SCPP) Integrated with CO$_2$ Capture under the UK Grid Code

Akeem K. Olaleye and Meihong Wang

12.4.1 Introduction

12.4.1.1 Background

With the current economic growth of developing countries such as China, India, and Brazil, and expected development in Africa, the overall energy demand is projected to grow by 37% from 2013 to 2040, an average rate of growth of 1.1%. Demand grew faster over the previous decades; the slowdown in demand growth is mainly due to energy efficiency gains and structural changes in the global economy in favor of less energy intensive activities (WEO, 2014). The UK's electricity system will also experience a period of serious investment. 18GW—about a quarter—of power generating capacity is due for decommissioning by 2020. Of this, 8.5GW of coal-fired plants will be decommissioned to meet EU requirements on pollution, and another 2.5GW of oil-fired stations. A further 7GW of nuclear power is scheduled to close by 2020 based on the available lifespan of the plants (Nichos and Maxim, 2008). The impact of these closures on UK's electricity generating capacity is shown in Figure 12.38. In the interim, the demand for electricity is also projected to increase, which will also have to be met by larger plant capacity. A 20% margin in excess of peak demand is the present amount of extra capacity on hand to ensure there are no power cuts when power plants have to be turned off for maintenance and repairs (Nichos and Maxim, 2008).

Worldwide, coal-based power generation technology proffers an immediate optimism for sustaining economic growth and will remain a major contributor to global power supply in the foreseeable future, hence the need for more fossil-fired power plants and development of renewable and nuclear energy sources to bridge the UK energy shortfall created by ageing power plants and projected increase in demand.

12.4.1.2 Power Plant Emission Reduction and Efficiency Improvement

Power generation from coal-fired power plants is the single largest source of CO$_2$ emissions. CO$_2$ is the largest and most important anthropogenic greenhouse gas (GHG) (Freund, 2003). However, coal-fired power plants play a vital role in meeting energy demands. With growing concerns over the increasing

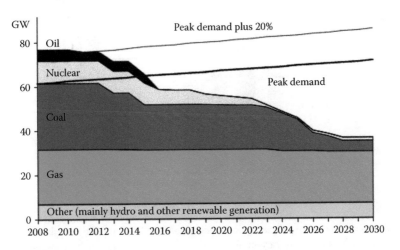

FIGURE 12.38 Predicted UK electricity demand and generation by fuels. (House of Lords, 2008).

atmospheric concentration of anthropogenic GHGs, effective CO_2 emission abatement strategies, such as power plant efficiency improvement, carbon capture and storage (CCS) are required to combat this trend (Wang et al., 2011).

12.4.1.2.1 Power Plant Efficiency Improvement

One of the most promising solutions to combat emissions from power plants is the improvement in design, manufacture, and operation of higher efficiency power plants. As shown in Figure 12.39, every increment in efficiency results in a proportional reduction in emission.

Such improvement lowers the amount of rejected heat; hence the CO_2 that would originally have been emitted per unit of electricity generated. Philibert and Podkanski (2005) also gave useful numeric approximations on the magnitude of the reduction that is achievable in the following quote:

> Coal-fired generating capacity of about 1 TW is installed worldwide. Almost two-thirds of the international coal-fired power plants are over 20 years old, and with average efficiency of 29%, emits almost 4Gt of CO_2 per year. If they are replaced after 40 years with modern plants of 45% efficiency, total GHG emissions will be reduced by about 1.4Gt per year

Philibert and Podkanski, 2005

The main technological evolution that can produce a very high increment in efficiency is the *supercritical pressure technology*.

12.4.1.2.2 Carbon Capture and Storage (CCS)

Excessive accumulation of CO_2 in the atmosphere can have undesirable effects on the climate. Anthropogenic activities mainly due to fossil fuel usage have added to the increase of atmospheric CO_2 concentration from the preindustrial level of 280 ppm to the current value of 400 ppm (WEO, 2014). Under the present carbon emission growth rate, the atmospheric CO_2 concentration could reach 580 ppm within just 50 years (Fan, 2010).

The World Energy Outlook (WEO) report published in 2014 projected that world electricity demand will increase by almost 80% over the period 2012–2040. Fossil fuels (i.e., coal) will continue to dominate the power sector, although their share of generation will decline from 68% in 2012 to 55% in 2040 (WEO, 2014). Global coal demand will grow by 15% from 2013 to 2040, but almost two-thirds of the increase will occur over the next 10 years (WEO, 2014).

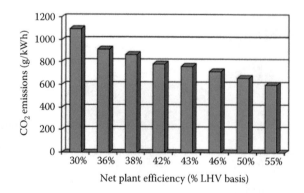

FIGURE 12.39 Effect of increasing efficiency on CO_2 emission. (DTI, Advanced power plant using high efficiency boiler/turbine best practice brochure carbon abatement technologies programme. www.nios.com.cn/common/files/File/BPB010(1).pdf, 2006.)

After several years of disagreement, the IPCC conclusively determined in 2007 that the increase in anthropogenic GHG was responsible for the worldwide temperature rise (Fan, 2010). CO_2 is undoubtedly the most significant anthropogenic GHG. Power generation from fossil fuel is the major source of CO_2 emissions and cause of global warming (IEA GHG, 2002a).

Carbon dioxide levels in flue gases vary depending on the type of fuel used and the excess air level used for optimal combustion conditions. Natural gas-fired power generating plants are typically combined cycle gas turbine (CCGT) power plants which generate flue gases with low CO_2 concentrations, typically 3%–4% by volume. Coal in power generation, on the other hand, is primarily burnt in pulverized-fuel boilers producing an atmospheric pressure flue gas stream with a CO_2 content of up to 15% by volume (IEA GHG, 2002a).

From the foregoing discussion, it is clear that proper CO_2 capture and storage (CCS) is, therefore, imperative. CCS is made up of three main steps, namely CO_2 capture (including separation and compression), transportation, and storage (Fan, 2010). Among the three steps, CO_2 capture step is the most energy intensive. The purpose of CO_2 capture is to produce a concentrated CO_2 stream that can be readily transported for sequestration (IPCC, 2005).

12.4.1.3 Supercritical Coal-Fired Power Plant

Supercritical is a thermodynamic expression describing the state of a substance where there is no clear distinction between the liquid and the gaseous phase (i.e., homogenous fluid). In this chapter referring to power generation, water will be the supercritical fluid under consideration. Water reaches this state at a condition above 373.946°C and 22.064 MPa. The thermal efficiency of a coal-fired power plant shows how much of the energy that is sent into the cycle is converted into electrical energy. The greater the output of electrical energy for a certain amount of energy input, the higher the efficiency. If the energy input to the cycle is kept constant, the output can be increased by selecting elevated pressures and temperatures for the water–steam cycle. Increased thermal efficiency is noticed when the temperature and pressure of the steam is increased. By raising the temperature from 580°C to 760°C and the pressure out of the high-pressure feedwater pump from 33 to 42 MPa (at ultrasupercritical steam condition), the thermal efficiency improves by about 4% (Kjartansson, 2008).

There are a range of operational advantages in the case of supercritical power plants. There are different types of turbine designs available for use in supercritical power plants. These designs do not differ from designs used in subcritical power plants. However, due to the fact that the steam pressure and temperature are higher in supercritical plants, the wall thickness and the materials selected for the high-pressure turbine section requires review. The supercritical plant needs a "once-through" boiler, whereas a "drum" type boiler is required by subcritical power plant.

The performance of a supercritical plant depends on the steam condition. Higher steam temperatures and pressures transform into better efficiency, defined as more electricity produced per British Thermal Unit (BTH) of coal consumed. Naturally, this is attractive to power producers as these increased efficiencies translate into reduced fuel costs and emissions. Nevertheless, fuel flexibility is not compromised in once-through boilers. A wide variety of fuels have already been applied for once-through boilers (Hordeski, 2008). All kinds of coal, oil, and gas have been used.

12.4.1.4 Postcombustion CO_2 Capture Based on Chemical Absorption

Postcombustion CO_2 capture (PCC) based on chemical absorption is one of the strategic technologies identified to reduce emission of GHGs in existing power plants. PCC based on chemical absorption using monoethanol amine (MEA) solvent is the most matured and preferred technology for CO_2 capture from the flue gases in existing power plants. PCC with chemical absorption process has been studied widely in open literature based on experimental studies, steady state, and/or dynamic modeling studies, technoeconomic analysis, and studies based on optimization and energy-saving mechanisms. For more details, please refer to Chapter 10 of this book. In this chapter, experimental data from a CO_2 capture pilot facility are used for validation of the model.

12.4.2 SCPP and the UK Grid Code Compliance

The drum is the key to existing subcritical coal-fired plants delivering the required primary and secondary frequency response. In the absence of the drum, the manufacturers of supercritical coal plants appear unable to offer 10% primary frequency response (Nichos and Maxim, 2008). A value between 3% and 7% appears possible applying a number of techniques used in supercritical plants throughout the world, such as using turbine throttling (Nichos and Maxim, 2008).

However, fast load changes can be achieved with a combination of primary measures using the short-term storage behavior of the power plant. These primary measures include the throttling of live steam of the boiler (i.e., steam throttling) and by interrupting the steam that is usually extracted from the turbines for preheating of condensate referred to as condensate/feedwater stops. These primary measures are vital for the time lag necessary for the boiler to increase the firing rate (Zindler et al., 2008).

Experience of using SCPP in the way the UK Grid Code recommends is nonexistent and there is no practical experience to draw on. The many measures required have not been tested by operational experience. There is a real risk relying on several untried techniques in combination that could result in a shortfall and noncompliant performance. The techniques proposed can lead to temperature and pressure excursions that will reduce the life of the plant. Without operational experience, the effect of this is difficult to predict (Nicholls and Maxim, 2008).

The SCPP integrated with CCS system must also comply with the UK Grid Code. Knowing very well that addition of the capture process will introduce extra design, operational, and controllability issues. It will also result in grid code compliance concerns as it will interact with the plant's frequency response capability (Nicholls and Maxim, 2008). It is expected that the key process variables, such as firing rate, furnace pressure, air–fuel ratio, CO_2 capture efficiency, and overall plant efficiency, be maintained at an optimal value irrespective of variations in load and process disturbances. To achieve this, it is important to understand how these variables interact during operation. It is also important to understand how the entire plant behaves under varying load conditions so that adequate provisions will be made to accommodate such changes so that they do not interfere with safe and efficient operation of the plant.

12.4.2.1 Economics of the SCPP Technology

Fuel costs represent about two-thirds of the total operating costs of a typical SCPP. The main impact of the supercritical plant technology is the increase in overall plant efficiency, thereby reducing the fuel consumption per unit of electricity generated (DTI, 2006). A new supercritical boiler/turbine power plant EPC (Energy Performance Certificate) specific price would be around 800 Euros/kWe gross (around £530/kWe). This is not more expensive than a subcritical plant and is less expensive than an IGCC for which EPC prices are quoted as US$1250/kWe–US$1440/kWe (approx. £700/kWe–£800/kWe; assuming US$1.8/£) for new plants. Investment costs of existing IGCC plants have been between 1500 and 2000 Euros/kWe (£1000/kWe–£1333/kWe) (DTI, 2006).

12.4.2.2 Material Requirements for SCPP

Supercritical plants differ from their subcritical counterparts in terms of operational characteristics due to the higher steam pressures and temperatures, and therefore require more stringent material properties than the subcritical plants. The four key components are high-pressure steam piping and headers, superheater tubing and waterwall tubing (Viswanathan and Bakker, 2000). By increasing the temperature and pressure of the working fluid (steam), the level of corrosion and oxidation to which the tubes and the turbine are exposed become higher. Hence, material requirement is certainly one of the principal challenges facing the supercritical technology (i.e., mechanical and metallurgical problems). Most of the problems are due to the use of austenitic steels with low thermal conductivity and high thermal expansion for components operating at high temperatures, resulting in high thermal stresses and fatigue cracking. Any further improvement in the steam conditions at supercritical condition to meet electricity demand will therefore be based on the manufacture and use of improved steels (DTI, 2006). Intense R&D efforts have been embarked on around the world (i.e., Japan, USA, and Europe) to develop materials suitable for high steam conditions as obtainable in supercritical plants. The Electric Power Research Institute (EPRI), for example, initiated study of development of more economic coal-fired power plant as early as 1978. These studies focused on development of high-temperature-resistant steels for production of materials capable of operating at inlet steam temperatures of up to 650°C.

12.4.2.3 The UK Grid Code: System Frequency and Load Demand under SCPP

The fossil fuel-fired power plants (e.g., SCPP) are being subjected to stringent operational regime due to the influx of renewable resources and the CO_2 emission reduction target. Hence, the fossil fuel power plants will have to be designed and operated flexibly to remain competitive in the energy market. This chapter focuses on modeling and analysis of SCPP integrated with PCC, and its ability to respond to electricity demand changes within the limit of the UK Grid Code.

The UK Grid Code has strict requirements on power generation companies and they are under obligation to comply with the requirements. For grid system operation, it is required that the power generated is continuously matched to demand. One yardstick for this balance is the *system frequency*. The system frequency "is a continuously changing variable that is determined and controlled by the careful balance between system demand and total generation." Table 12.21 shows a summary of the UK Grid Code requirements.

12.4.2.3.1 Base Load and Peak Load

Power generation varies depending on demand regimes such as peak and base load. However, it never goes lower than a minimum referred to as *base load*. Base load is the basis of a sound electrical system (Progress Arkansas, 2010). Peak load generation plants supply electricity at times when power consumption by consumers is highest. Peak load power plants are designed for high responsiveness to changes in power demand. They have a very short start up time and can vary the quantity of power output within minutes. Also, they operate only 10%–15% of the period and are very small compared to base load plants. Nevertheless, because of their size, they are cheaper and easier to construct. They are usually natural gas-fired power plants (Progress Arkansas, 2010).

Renewable sources within the grid (i.e., wind and solar) are intermittent, and hence only produce electricity when the wind is blowing and solar only produces power when the sun is shining. They cannot therefore be trusted with a constant electricity supply requirement. This irregular service often leads to load swing within the grid. As such, the allowance between on-peak and off-peak loads expands, necessitating that existing power plants (i.e., SCPP) operate under more flexible regimes.

TABLE 12.21 The UK Grid Code Requirement for Load/Frequency Control

Parameter	UK Requirement
Grid voltage range (normal)	380...420 kV (nominal 400 kV)
Grid frequency range (normal)	Nominally 50 Hz, normally controlled within 49.5...50.5 Hz. Should maintain constant active power between 49.5 and 50.5 Hz
Load change	Load deviation over 30 min must not exceed 2.5% of registered capacity.
Load control/frequency response	Must be capable of providing frequency response of 10% GRC for a ±0.5 Hz change in frequency. Change of active power to be achieved in 10 s. Response requirement is reduced when operating at high and low ends of registered capacity. Response to high frequency must continue at additional 10% GRC reduction per each additional 0.5 Hz increase in frequency.

Source: National Grid, The grid code, Issue 4, Revision 10. National Grid Electricity Transmission plc, London, UK, 2012.

12.4.2.3.2 Primary and Secondary Frequency

The response on frequency deviation caused by an event in the grid is handled by the frequency control. This is implemented in time ranges. The *Primary Response* capability (P) of a generating unit is the minimum increase in active power output between 10 and 30 s of a ±0.5 Hz drop in system frequency due to an increase in power demand. The *Secondary Response* capability (S) is the minimum increase in active power output between 30 s and 30 min (Figure 12.40) of a ±0.5 Hz drop in system frequency due to increase in power demand (National Grid, 2012).

12.4.2.3.3 High Frequency

The *High Frequency Response* capability (H) of a generating unit is the decrease in active power output provided 10 s after the start of the ramp injection and sustained thereafter (Figure 12.41).

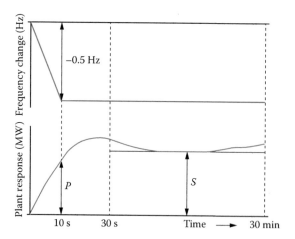

FIGURE 12.40 Interpretation of primary and secondary response values. (National Grid, The grid code, Issue 4, Revision 10. National Grid Electricity Transmission plc, London, UK, 2012.)

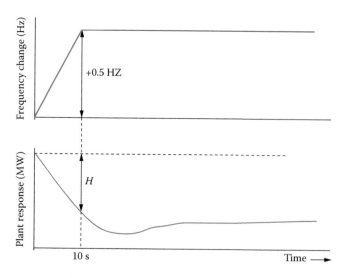

FIGURE 12.41 Interpretation of high frequency response values. (National Grid, National Grid, The grid code, Issue 4, Revision 10. National Grid Electricity Transmission plc, London, UK, 2012.)

12.4.3 Modeling Studies on SCPP

The vast literature of approaches to modeling SCPP can be divided into two distinct classes. In one group, there are sizeable numbers of studies based on first principle modeling, stressing accuracy with the objective of developing numerical simulations (e.g., static and dynamic finite element techniques) as a representation of the complex physical phenomena that characterize the energy transformation process in the power plant (Kitto and Stultz, 2005). The second major group in SCPP modeling refers to the development of linear (transfer function or state space) models produced around the plants' small signal behavior. Without doubt, at given operating conditions, accurately tuned linear models provide the foundation for building an excellent controller, designed around the plant's small signal behavior (Shoureshi and Paynter, 1983). However, the need for safe and efficient adjustment to highly changing demand cycles, for fast response to unexpected demand changes and for automated emergency response, offers a convincing argument for a large signal (and hence nonlinear) design-oriented plant model (Shinohara and Koditschek, 1995).

12.4.3.1 Modeling Purposes

12.4.3.1.1 Modeling for Control

Design and implementation of a good control system requires a model of the real-world system. The model represents the link between the real system and the design of a control system. A good model will help in the understanding of the system to be controlled, which is a vital requirement in control system design. For control, simulation, and analysis of an SCPP, a mathematical model is required. These models represent the physics of the power plants. The trend in power plant control is integrated control of the whole plant concerning the boiler–turbine–generator control. However, the integrated control approach will partly depend on accurate depiction of the power plant dynamics (El-Sayad et al., 1989).

12.4.3.1.2 Modeling for Plant Design and Optimization

Power plant design companies use mathematical models to optimize the design of proprietary and off-the-shelf process equipment. Once an accurate mathematical model of a Greenfield power plant has been developed, it becomes possible to optimize the design using numerical optimization techniques (Marto and Nunn, 1981).

12.4.3.1.3 Modeling for Operational Study

Modeling is also performed in order to simulate the dynamics of power plant under normal and emergency operational conditions. Dynamic models are important tools for studying system response to different operational changes. This is often used for analyzing the dynamics of the plant over its lifetime for investment decision purposes or for operator training. The complexity of the model depends on the end use of the model.

12.4.3.2 Steady-State Modeling and Analysis

12.4.3.2.1 Reference Plant Description

The reference SCPP is a Greenfield plant of 580 MW_e with flue gas desulfurization (FGD) and CO_2 capture equipment described in Woods et al. (2007). The steam turbine conditions correspond to 24.1 MPa/593°C throttle with 593°C at the reheater. The net power, after consideration of the auxiliary power load, is 550 MW_e. The plant operates with an estimated efficiency of 39.1% Higher Heating Value (HHV). The major subsystems of the plant include coal milling system, coal combustion system, ash handling system, FGD, condensate, and feedwater systems. The reference SCPP consists of eight feedwater heaters (including the deaerator); seven were modeled as heat exchangers while the deaerator was modelled as a mixer. The feedwater from the deaerator is pumped into the boiler through the boiler feed pump (turbine driven).

12.4.3.2.2 Aspen Plus Model of SCPP

Steady-state simulation is performed with the aid of Aspen Plus to evaluate the performance of the SCPP. The model was carried out in subsystems (i.e., coal pulverizer, once-through boiler, turbines, feedwater heaters, etc.) to describe the physics of the SCPP operation. The coal milling process reduces the size distribution of the coal by grinding it to a fine powder and thus improves the combustion process in the SCPP boiler by achieving a more uniform distribution of coal in the combustion air.

The SCPP once-through boiler includes coal combustion chamber, air preheater, induced draught (ID) and forced draught (FD) fans, pulverizers, etc. After a pressure increase and slight temperature rise across the FD fan, the combustion air passes through a steam-air heater and then the air preheater. The steam-air heater and air preheater increase the temperature of the incoming combustion air and thereby improve boiler efficiency for the unit. The preheated air then enters the once-through boiler where it is combusted with the pulverized coal.

The work produced by the steam turbine is calculated from the change in enthalpy from the inlet to the outlet in Aspen Plus. The steam turbine for the reference SCPP has Very High Pressure (VHP), High Pressure (HP), Intermediate Pressure (IP), and Lower Pressure (LP) turbines with a total of nine stages that produce electrical power and a tenth turbine stage that powers the boiler feed pump.

The feedwater heating section calculates the increase in heat rate over the feedwater heaters. There are nine feedwater heaters used in the model of the SCPP, eight of these are closed feedwater heaters and one (i.e., the deaerator) is an open feedwater heater. In this study, the feedwater heaters are modelled using countercurrent flow type heat exchangers.

The subsystems (i.e., pulverizer, boiler, turbine, feedwater heaters, etc.) are coupled together to fully model the whole SCPP as shown in Figure 12.42. The process of linking the subsystems is done partly. The pulverizer is linked to the boiler, the boiler to the turbine and the feedwater heaters, etc. Linking the boiler–turbine models together is the most challenging part, and it is accomplished by using multistage heat exchangers (MHEATEX) to model the heat transfer that occurs between the flue gas and the main steam and reheat steam in Aspen Plus.

The coupled SCPP model is then used to determine the performance of the entire power plant. The net power produced by the system is calculated from the net power produced by the turbine cycle less the station service power (obtained by summing up the power required for the pulverizer, ID Fan, FD Fan, and other auxiliary equipment in the power plant).

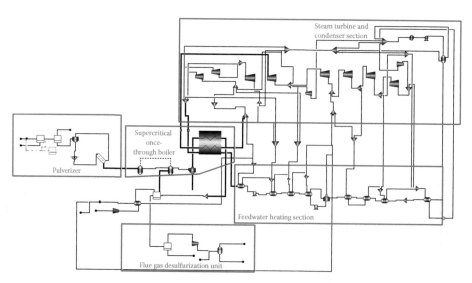

FIGURE 12.42   Aspen Plus model of whole SCPP.

TABLE 12.22 SCPP Performance Summary

Performance Parameters	Reference Plant	Aspen Plus	Relative Error (%)
Total power output (MW$_e$)	580.26	585.39	0.9
Auxiliary load (MW$_e$)	28.28	28.42	0.5
Gross plant power(MW$_e$)	551.98	556.97	0.9
Generator loss (MW)	1.83	1.83	–
Net power output (MW$_e$)	550.15	555.14	0.9
Unit efficiency, HHV (%)	39.1	39.4	0.78

12.4.3.2.3 Model Validation

The SCPP model is made up of subsystems. Each subsystem bases its calculations on inputs from other subsystems, and constants set during the model. The model is validated by comparing the solutions from the model with design data from the reference plant presented in Woods et al. (2007) at full load. For this purpose, selected variables, i.e., unit efficiency, electric power output from the generator, etc., are chosen. The comparison is shown in Table 12.22.

From the results in Table 12.22, it can be seen that the steady-state model is reasonably accurate when comparing the solutions from the model with design data from the reference Greenfield plant. There is consistency between the model output and the data from the reference plant at full load and that gives confidence on the future use of the model to carry out analyses and case studies on energy-efficiency improvement in the SCPP.

12.4.3.3 Dynamic Modeling and Analysis of SCPP

SCPPs were usually operated at base load until recently. The influx of intermittent renewable resources such as wind and solar into the grid, has led to the requirement of coal-fired power plants to be operated more flexibly and reliably to meet the frequently changing dynamics of power demand and supply. Operating an SCPP more flexibly in the face of new grid code challenges requires understanding of the dynamic characteristics of the power plants at various operational modes (i.e., normal or emergency mode due to major or minor disturbances, etc.) and during start-up and shutdown. This section presents the development of the dynamic model suitable for investigating the response of SCPP to changes in the demand under the UK Grid Code requirement.

Development of the SCPP dynamic model was initiated with models of the plant components (i.e., coal pulverizer system, furnace, flue gas–air system, boiler system, turbine–condenser system, the deaerator–feedwater system, etc.). Figure 12.43 shows the modeling structure of the SCPP components. The equation-oriented modeling software gPROMS was used for the component models. Model equations, parameters, correlations and thermodynamic properties for each of the components were obtained from literature, design, and real plant data.

12.4.3.3.1 Reference Plant Description

The reference SCPP selected for the dynamic modeling and analyses in this project is an SCPP with 600MW$_e$ installed capacity at 100% Boiler maximum continuous rating located in Jiangsu, China. The reference plant is a single furnace, single reheater, balanced ventilation, solid slag, steel frame, full suspension structure-type boiler. The process schematic of the 600MW$_e$ SCPP is shown as Figure 12.44. The boiler is capable of generating 1800 t/h of steam at rated conditions of 24.5 MPa and 537°C. The superheated steam from the boiler is expanded in the HP turbines and converted to electric power in the generator. The steam is reheated by a single stage reheater to about 540°C and expanded in the IP and LP turbines. The regenerative feedwater heaters uses steam extracted from various stages of the HP, IP, and LP turbines to preheat the feedwater to improve the plant efficiency.

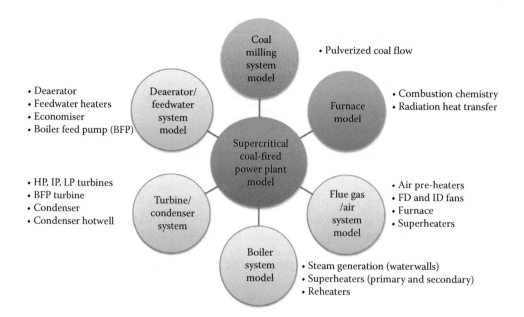

FIGURE 12.43   Component modeling structure of the SCPP dynamics.

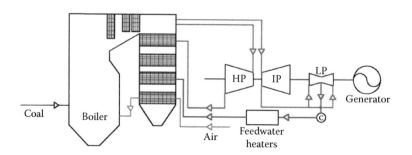

FIGURE 12.44   Process schematic of a typical supercritical coal-fired power plant.

The main parameters of the plant used for model validation are obtained from measurements compiled from weekly performance charts at three load levels: (1) generating at capacities of 520 MW$_e$ (87% BMCR), (2) 450 MW$_e$ (75% BMCR), and (3) 300 MW$_e$ (50% BMCR).

12.4.3.3.2 gPROMS Model of SCPP

12.4.3.3.2.1 Model Development
The dynamic model of SCPP includes models of the plant components such as the coal milling system, furnace, flue gas–air system, boiler system, turbine–condenser system, and the deaerator–feedwater system, etc. The equations describing the dynamic behavior of the once-through boiler are derived from the one-dimensional mass (Equation 12.32), momentum (Equation 12.33), and energy balances (Equation 12.34) for the economizers, the evaporator waterwall tubes, the superheaters, and the reheaters. Other components such as the steam turbines, the pumps, the governing system, the quick-action valves (used for fast response of SCPP during frequency change operation), and the electric generator models are based on lumped parameter models. The equation-oriented modeling software gPROMS was used for the model development. Figure 12.45 shows the model topology of the whole SCPP dynamics.

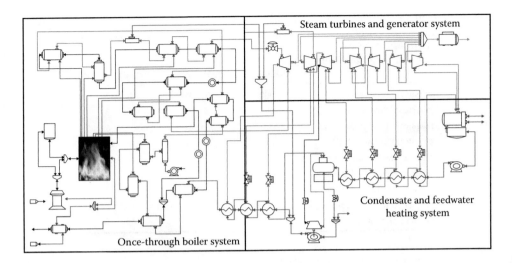

FIGURE 12.45 Whole SCPP dynamic model topology.

The general governing equations are (O'Kelly, 2013) as follows:

Conservation of mass:

$$\frac{\partial(\rho A)}{\partial t} = -\frac{\partial \dot{m}}{\partial z} \tag{12.32}$$

Conservation of momentum:

$$\frac{dp}{dz} = -\frac{dp}{dz}\bigg|_{\text{loss}} - \rho\left(\frac{\partial v}{\partial t} + v\frac{\partial v}{\partial z}\right) - g\rho\sin\beta, \tag{12.33}$$

where

$$\frac{dp}{dz}\bigg|_{\text{loss}} = \frac{1}{2}.\rho v^2$$

Conservation of energy:

$$\frac{1}{A}\frac{dq}{dz} = -\frac{\partial p}{\partial z}\bigg|_{\text{loss}} + \rho\left(\frac{\partial h}{\partial t} + v\frac{\partial h}{\partial z}\right) - \left(\frac{\partial p}{\partial t} + v\frac{\partial p}{\partial z}\right) \tag{12.34}$$

Heat transferred is defined by the supplementary equation

$$\frac{dq}{dz} = \alpha_{hx} A_{hx}\left(T_f - T_{\text{wall}}\right), \tag{12.35}$$

where A_{hx} is the heat transfer area per unit length

The working fluid equation of state:

For single phase (for subcooled water and superheated steam);

$$\rho = \rho\left(p, T_f\right); \quad h = h\left(p, T_f\right) \tag{12.36}$$

TABLE 12.23   Steady-State Validation at 100% BMCR

Performance Data (100% Maximum Continuous Rating (MCR))	Plant Data	gPROMS Model	Relative Error (%)
Net power output (MW$_e$)	600	604.44	−0.32
Main steam flow (kg/s)	513.86	515.78	0.37
Main steam temperature (°C)	538.0	541.42	−0.64
Main steam pressure (MPa)	24.2	24.31	−0.46
Reheater steam temperature (°C)	566.0	568.49	0.65
Reheater steam flow (kg/s)	429.27	431.10	1.28
Reheater steam pressure (MPa)	3.91	4.03	3.06
Flow of condensing steam (kg/s)	229.46	226.61	1.24
Makeup water rate (%)	3.0	3.0	—

For two phase (for water steam two-phase mixture);

$$\rho = \rho(p,x); h = h(p,x) \tag{12.37}$$

12.4.3.3.2 Transport Property Relations   The flue gas/air and the feedwater/steam properties were estimated using Multiflash, a commercial property package embedded in gPROMS. The thermodynamic property of water–steam used in Multiflash is based on the industrial formulation IAPWS-IF97 (Wagner and Kretzschmar, 2008) within the range of pressure and temperature anticipated in the SCPP.

12.4.3.3.3 Model Validation: Steady State and Dynamic

12.4.3.3.3.1 Steady-State Validation   For a model to be considered accurate, it should be able to predict steady state values of variables of interest at different operating levels (or load). A steady-state validation of the model at full load was performed using the reference plant's heat balance data at 100% load. Table 12.23 shows a summary of the main parameters of the plant in comparison with the model. The model shows an average relative error of less than 3%. The model shows an accurate representation of the full load performance of the reference plant at steady state.

12.4.3.3.3.2 Dynamic Validation   Dynamic validation is important for establishing some basis for the capability of the model, and to be able to demonstrate capability for predicting plant behavior over time especially during periods when changes in load (i.e., ramping) are implemented. The model was validated with actual plant operational data for dynamic conditions obtained at transient load ramps of the reference plant. The power output was initially simulated for 100 s at 87% MCR (normal operating load) before initiating the load ramp. The load is then ramped down from 540 to 430 MW$_e$ at 15 MW/min as shown in Figure 12.46a. It is then maintained at these load levels for a further 500 s.

Responses of the main steam flow rate, main steam pressure, and main steam temperatures were compared with the reference plant as shown in Figure 12.46a through d, respectively. The results are agreeable with the trends in the reference plant response to the ramp change.

12.4.4 Study of Efficiency Improvement in SCPP Integrated with PCC

12.4.4.1 Steady-State Model and Integration of SCPP with PCC

The subsystems (i.e., pulverizer, boiler, turbine, feedwater heaters, etc.) were coupled together to fully model the SCPP integrated with PCC, as shown in Figure 12.47. The process of linking the subsystems is done partly. The pulverizer is linked to the boiler, the boiler to the turbine, and the feedwater heaters, etc. Linking the boiler–turbine models together is the most challenging part, and it is accomplished by using multiple heat exchangers to model the heat transfer that occurs between the flue gas and the

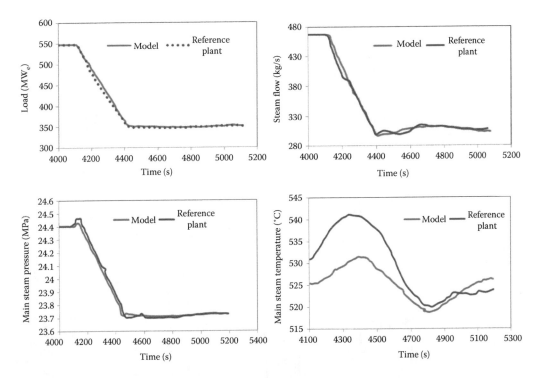

FIGURE 12.46 Comparison of ramping of SCPP model with reference plant.

main steam and the reheat steam in Aspen Plus. Figure 12.47 also shows the connection point of the two models. The coupled SCPP–PCC model is then used to determine the performance of the entire power plant. The net power produced by the system is calculated from the net power produced by the turbine cycle less the station service power (obtained by summing up the power required for the pulverizer, ID Fan, FD Fan, and other auxiliary equipment in the power plant).

12.4.4.2 Exergy Analysis of SCPP Integrated with PCC

12.4.4.2.1 Conventional Exergy Analysis

Integrating SCPP with CO_2 capture incurs serious efficiency and energy penalties due to use of energy for solvent regeneration in the capture process. Reducing the exergy destruction and losses associated with the systems can improve the rational efficiency of the system and thereby reducing energy penalties. Exergy method of analysis defines the maximum possible work potential of a system using the state of the environment as the datum. Conventional exergy analysis identifies the location, magnitude, and sources of thermodynamic inefficiencies in a thermal system. Individual stream exergies were estimated from the Aspen Plus V8 simulation. Aspen Plus V8 contains three new property sets; EXERGYMS, EXERGYML (calculated on mass and molar basis, respectively), and EXERGYFL for estimating exergy of material/energy streams, unit operation, and utilities. The exergies are estimated at a reference environmental temperature and pressure of 25°C and 1atm, respectively. Detailed calculation for physical and chemical exergies for each component of the SCPP and the PCC are estimated using the individual stream flow based on the Aspen Plus EXERGYMS stream calculations.

Thermodynamic reversibility demands that all process driving forces, i.e., temperature, pressure, and chemical potential differences, be zero at all points and times (Leites et al., 2003). Such a theoretical process results in the production of the maximal amount of useful work (exergy), or in the consumption of the minimal amount of work. Unfortunately, a reversible chemical process operates at an infinitesimal rate, and requires an infinitely large plant. It has been generally believed that thermodynamic

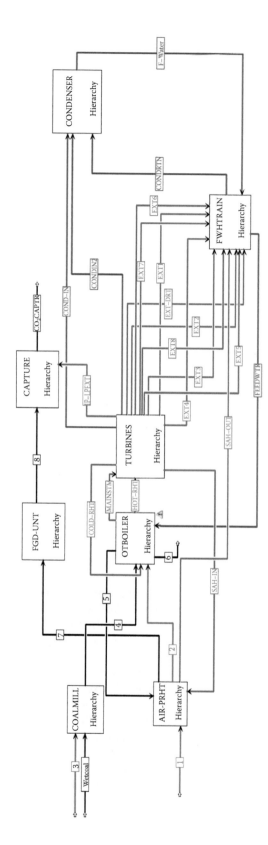

FIGURE 12.47   Aspen Plus hierarchical model of the whole SCPP integrated with PCC. (Olaleye, A. K. et al., *Fuel* Special Issue, 151: 57–72, 2015.)

irreversibility in chemical processes/reactions is almost inevitable and leads to large energy consumption and losses. However, some thermodynamic principles based on the second law of thermodynamics have been investigated and proven effective for lowering energy consumption (Leites et al, 2003). This study uses the driving force method to reduce exergy destruction and hence reduce energy consumption in the PCC process. Three configurations of the MEA-based PCC based on chemical absorption were integrated with the SCPP and analyzed. This includes (1) absorber intercooling (AIC), (2) split flow approach (SF), and (3) combination of both methods (AIC+SF).

12.4.4.2.2 Advanced Exergy Analysis

Conventional exergy analysis cannot determine the interaction among components or the true potential for the improvement of each component. However, an advanced exergy analysis evaluates the interaction among components and the real potential for improving a system component/the overall system (Morosuk et al, 2013). It involves splitting the exergy destruction in system components into endogenous/exogenous and avoidable/unavoidable parts. *Endogenous* exergy destruction is the part of exergy destruction within a component obtained when all other components operate in ideal/reversible condition and the component being considered operates with the same efficiency as in the real system (Morosuk et al., 2013). The *Exogenous* part of the variable is the difference between the value of the variable within the component in the real system and the endogenous part. The *Unavoidable* exergy destruction is the part of exergy destruction that cannot be further reduced or eliminated due to technological limitations such as availability and cost of materials and manufacturing methods. The avoidable part is the difference between the total and the unavoidable exergy destruction. For the SCPP and PCC components, the avoidable exergy destruction is the part that should be considered during the improvement of a system design.

12.4.4.3 Efficiency and Energy Penalties Improvement in SCPP Integrated with PCC

12.4.4.3.1 Conventional Exergy

In the boiler subsystem, the most exergy destruction occurs in the furnace combustion chamber (76%). The steam condenser has the highest exergy destruction (52%) in the turbine subsystem. The HP feedwater heaters accounts for most of the exergy destruction (43.65%) in the feedwater heating subsystem. For the PCC subsystem, the absorber (26%) and the desorber (36%) are the main sources of exergy destruction. The study also investigates the improvement of energy penalties due to exergy destruction in the CO_2 capture subsystem. The results reveal that the absorber (26%) and the desorber (36%) are the main sources of exergy destruction in the PCC subsystem. The feed cooler (18%) and the blower (16.5%) are also contributing strongly. The total exergy destruction is about 203 MW (1.58 MJ/kg CO_2). Process equipment such as the pump, blower, and solvent cooler are minor contributors to the exergy destruction. The exergy loss due to the consumption of MEA was included in the overall exergy destruction. Analysis of the energy consumption of the CO_2 capture system and the overall exergy destruction in the integrated system necessitated the development of several variations of the conventional CO_2 capture (Olaleye et al, 2015). Table 12.24 shows a summary of the performance indicators of the studied cases.

12.4.4.3.2 Advanced Exergy Analysis

Based on the advanced exergy analysis, the fuel saving potential of the turbine subsystem (104 MW) is almost double that of the boiler subsystem (61 MW). The feedwater heater almost has no influence on the fuel consumption. Most of the exergy destruction in the SCPP components is endogenous (over 70%). 30%–50% of exergy destruction in the turbine subsystem is generally avoidable. The real potential for improving a component is not fully revealed by its total exergy destruction but by its avoidable part. Most of the avoidable exergy destructions within the heat exchangers in the boiler subsystems (75%), turbine stages (92%) are endogenous, the improvement measures for these components should be concentrated on the components themselves. Within the feedwater heating subsystem, the exogenous exergy destruction contributes over 70% of the avoidable part.

TABLE 12.24   Performance Indicators

Description	Reference SCPP	SCPP + Conventional	SCPP + (AIC)	SCPP + (SF)	SCPP + (AIC + SF)
Performance Summary					
Total (steam turbine) power (MW$_e$)	580.26	482.28	484.52	486.42	488.58
Auxiliary load (MW)	28.28	52.04	51.95	48.45	42.8
Gross plant power (MW)	551.98	430.24	432.57	437.97	445.78
Generator loss (MW)	1.83	1.83	1.83	1.83	1.83
Net power output (MW$_e$)	550.15	428.41	430.74	436.14	443.95
Unit efficiency, HHV (%)	39.10	30.45	30.61	31.00	31.55
CO$_2$ Capture Performance Summary					
Reboiler duty (MW)	—	528.78	511.81	492.02	466.57
Energy penalty (%)	—	22.13	21.70	20.72	19.30
Efficiency penalty (%)	—	8.65	8.49	8.10	7.55
Exergetic Performance					
Exergy destruction, y_D (%)	52.61	46.27	46.15	45.81	43.19
Exergy losses, E_L (%)	8.34	5.03	4.62	4.37	3.58
Exergetic efficiency, ε (%)	39.05	48.7	49.23	49.82	53.23

Source: Olaleye, A. K. et al., *Fuel* Special Issue, 151: 57–72, 2015.

The stripper and absorber in the PCC system have large absolute endogenous exergy destruction of about 75.5 and 50 MW, respectively. Hence, their performances will be significantly affected by improving the exergy destructions within the components themselves. However, the potential for improvement is governed by the avoidability or unavoidability of the exergy destructions. Majority of the exergy destruction within the PCC components is unavoidable. However, the ratio of the avoidable part of the exergy destruction differs considerably from component to component. For the stripper, about 17% (13.83 MW) of the overall exergy destroyed within it is avoidable. The result also reveals that about 27% (16.23 MW) of the exergy destructions in the absorber are avoidable. In the heat exchanger, the blower, and the cooler, the avoidable exergy destroyed are 41% (5.01 MW), 63% (5.01 MW), and 65% (13.02 MW) respectively. It is important to know the sources (exogenous or endogenous) of the avoidable exergies in the components so as to focus attention on reducing the avoidable exergy destruction of a component based on its source. Most of the avoidable exergy destructions within the stripper (98%), absorber (77%), blower (67%), cooler (78%), and heat exchangers (65%) are endogenous; hence, the improvement measures for these components should be concentrated on the components themselves.

12.4.5 Strategies for Operating SCPP under the UK Grid Code Requirement

Operating the SCPP in the manner required by the UK Grid Code will require some major modifications to the plant's operation and control systems especially its response to changes in load, system frequency (i.e., the primary frequency) and emergency situation. This study presents results of the different strategies proposed to investigate the plant's behavior under typical UK grid scenario (i.e., during primary frequency change). It considered two conventional approaches used worldwide for SCPP load response (i.e., main steam throttling, and Condensate stop), and a novel approach (i.e., steam extraction stop, which is a combination of extraction stops and steam throttling). The strategy also incorporated the use of an indirect firing system (i.e., with pulverized coal silo) to reduce the time required for increasing the firing rate in the boiler as suggested by Mercier and Drenik (2013).

12.4.5.1 Main Steam Throttling

Opening the main steam throttling valve (Figure 12.48) in the HP steam turbine inlet increases the flow of steam into the HP turbine, and consequently leading to increase in the load. This method is, however, not sustainable for meeting primary frequency response within the short time (i.e., 10 s) required by the UK grid. It only produces about 3.3% MCR increase in load within the short time before firing rate is increased.

12.4.5.2 Condensate Stop

In this approach, the steam normally extracted from the last stages of the LP turbine to preheat the condensate is stopped and allowed to undergo further expansion in the turbine to increase the load during frequency change (Figure 12.49).

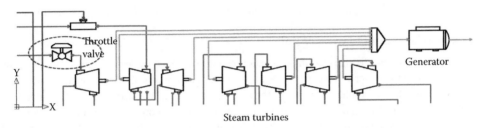

FIGURE 12.48   Main steam throttling.

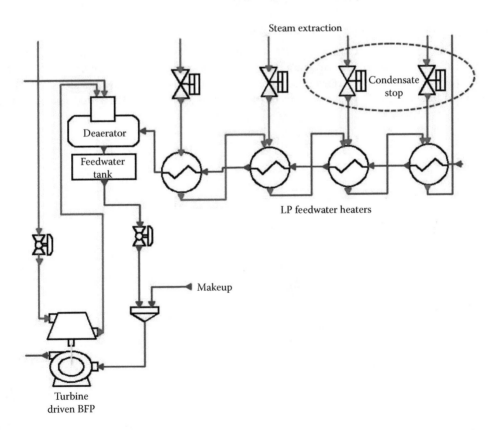

FIGURE 12.49   Condensate stop.

In this study, four LP heaters are installed, and the condensate stop refers to the contribution of the two LP heaters before the condenser hot well. Its contribution to the primary frequency response was found to be about 1.5% MCR increase in load. It is therefore not useful when applied alone. This process is achievable through the use of fast-acting valves.

12.4.5.3 Extraction/Feedwater Stop

In this approach, the steam normally extracted from the IP and LP turbines for preheating of the HP feedwater heaters and part of the LP condensate preheaters, respectively, are stopped (Figure 12.50) and allowed to undergo further expansion in the turbine to increase the steam flow to the turbine and consequently the load during frequency change.

The HP turbine extraction is not stopped during this process to avoid thermal imbalance in the boiler evaporator tubes due to feedwater temperature fluctuations. Extraction stop contributes about 6.8% MCR increase in load due to the higher number of fast-acting valves involved and the extracted steam condition.

12.4.5.4 Partial Indirect Firing of Boiler

The use of partial indirect firing is to enable a faster response of firing increase by considerably reducing the reaction time in the boiler as a secondary response period (within 30 s of frequency drop and sustained for a further 30 min). This is achieved through the use of a pulverized coal storage system (Figure 12.51) to ensure pulverized coal of similar particle size is available to be called upon during primary frequency drop.

The response of steam flow during frequency change is compared for both conventional firing system and partial indirect firing system. The results show that the partial indirect firing method has faster response than the conventional direct firing method (Figure 12.52).

12.4.5.5 Combination of Different Strategies

Since the use of steam throttling, condensate stop, and extraction stop alone was unable to fulfill the 10% MCR requirement of the UK Grid as a response to primary frequency, a combination of any two or all three of the strategies described in Sections 5.1 through 5.3 was investigated to analyze its ability to achieve a 10% increase in generating capacity (MCR) within 10–30s of 0.5 Hz frequency change as

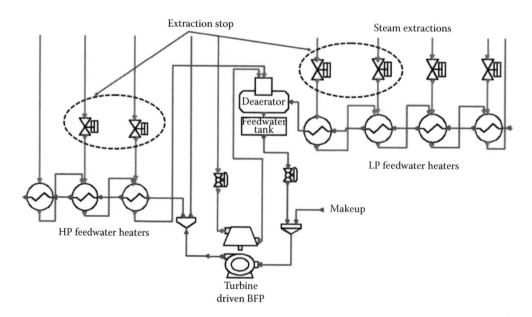

FIGURE 12.50   Extraction/feedwater stop.

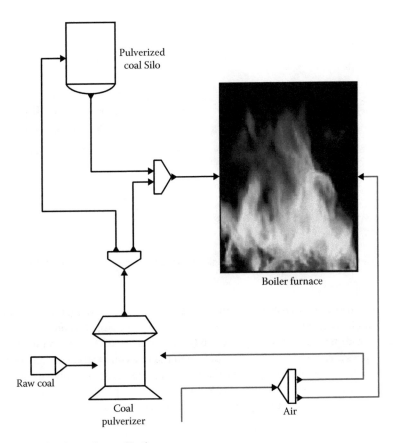

FIGURE 12.51 Partial indirect firing of boiler.

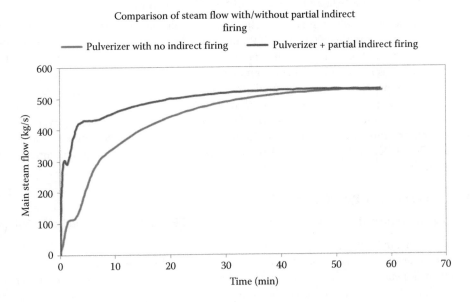

FIGURE 12.52 Comparison of steam flow rate in conventional and partial indirect firing system.

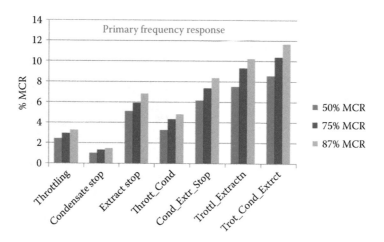

FIGURE 12.53   Contribution of the different strategies to primary frequency response (at 10 s).

required by the UK grid. Four scenarios were investigated at three load levels (50%, 75%, and 87% MCR). These include combinations of: (1) steam throttling and condensate stop, (2) condensate and extraction stops, (3) steam throttling and extraction stop, and (4) combination of the three strategies.

Figure 12.53 shows a summary of the contribution of the individual strategies and combinations of the strategies at 50%, 75%, and 87% MCR. From the simulation result, a combination of steam throttling and extraction stop can achieve 10% MCR increase when operating at above 87% to full load, while a combination of the three strategies is able to achieve the 10% MCR increase at above 75% to full load.

12.4.6 Dynamic Modeling and Process Analysis of SCPP Integrated with PCC

12.4.6.1 Dynamic Modeling and Scale-Up of the PCC System

The PCC dynamic model which was originally based on pilot plant scale was scaled-up to handle flue gas flow rate from the 600 MW$_e$ SCPP unit using the GPDC principle (Sinnot and Towler, 2013). The absorber and the stripper diameters were estimated to be 25.32 and 17.15 m, respectively. Using one column each would be difficult to manage in a real SCPP–PCC integrated plant due to structural limitations. However, for the modeling purpose one column each was retained for the dynamic model due to numerical issues. From the scale-up calculations, it is found that estimated column capacity (i.e., cross-section area of column) is affected by selected packing type and maximum allowable pressure drop in the column. Figure 12.54 shows the gPROMS model topology of the Integrated SCPP–PCC units. Table 12.25 also shows a summary of the key design parameters of the scaled-up system.

12.4.6.2 Integrated SCPP–PCC Model

In the integrated system, three links are included between SCPP and PCC. This includes the following:

- Flue gas stream from the SCPP to the absorber column.
- Steam draw-off from the power plant for solvent regeneration in the absorber.
- Return of spent steam (condensate) from the PCC back to the SCPP low pressure feedwater heater. Details of the integration between the power plant and the PCC regenerator plant are already given in Lawal et al. (2012). The CO$_2$ stream leaving the PCC is approximately 95 wt% CO$_2$ with

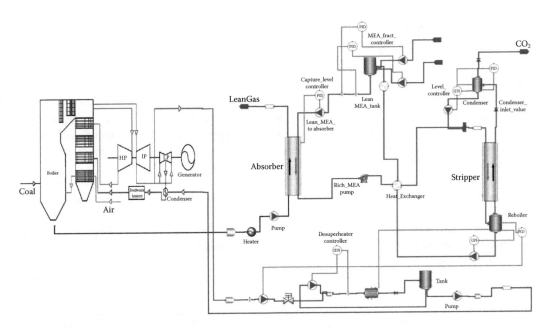

FIGURE 12.54 gPROMS model topology of the integrated SCPP–PCC dynamic model.

TABLE 12.25 Design Parameters from the PCC Scale-Up

Column Parameters	Absorber	Stripper
Packing type	Raschig ring	Raschig ring
Height (m)	30.0	30.0
Diameter (m)	25.32	17.4
Number	1	1
Capture level (%)	90.0	—

TABLE 12.26 SCPP at Full Load (100% MCR)

Variables	Value
Net power output (MW$_e$)	600
Efficiency (%)	40.93
Fuel flow (kg/s)	56.4
CO_2 in flue gas (wt%)	19.46
Flue gas flow rate (kg/s)	612.03

some H_2O and N_2. The SCPP model developed in section 3.3 is used to determine the base case conditions at full load (100% MCR) before the addition of the downstream PCC system. Key variables at the full load condition of SCPP are presented in Table 12.26.

With the addition of the downstream PCC units, there was 24.13% reduction in the net power output of the SCPP to about 455 MW$_e$ (at full load) at about 95% CO_2 capture level. This reduction was due to the steam drawn-off from the SCPP for the reboiler duty in the stripper. This also leads to a drop in efficiency of the SCPP unit to about 34.62% (about 7% efficiency penalty). Figure 12.54 shows the gPROMS model topology of the integrated model.

12.4.6.3 Analysis of the Integrated Model

12.4.6.3.1 *Response of the PCC to Ramp Change in SCPP Load*

During the ramping down of the SCPP load, the flue gas produced reduces correspondingly and hence resulting in reduction in the flow of flue gas to the absorber column. The solvent circulation flow remains unchanged to simulate the impact of higher L/G ratio. The behavior of some variables during these changes was assessed. The SCPP controls were in place during the injection of the load ramp. Also, the reboiler temperature was maintained via the reboiler temperature controller. The reboiler duty needed to regenerate solvent in the stripper thus decreased, therefore a reduced demand for steam drawn-off from the SCPP was observed. The capture level was not controlled in order to monitor its sensitivity to the SCPP load ramp, the lower flue gas flow and the constant solvent circulation (i.e., higher L/G ratio).

The model was initially simulated for 1.5 h at full load with CO_2 capture ($\sim$455 MW_e net power output) before introducing the load ramp. The load is then ramped down from 455 to 350 MW_e for 10 min at 10.5 MW/min. It is then maintained at these load levels for a further 2.0 h. Responses of the pulverized coal flow, flue gas flow, net power plant efficiency, and the CO_2 capture level were monitored as shown in Figure 12.55a through d, respectively.

The results of the response conforms to expected trends in literature. As the load is ramped down from 455 MW_e to 350 MW_e as shown in Figure 12.55a, pulverized coal flow rate decreases from 42.8 to 33.9 kg/s. Also, the flue gas flow rate follows the same trend as shown in Figure 12.55d. The net efficiency of the plant also decreases correspondingly as shown in Figure 12.55b. The CO_2 capture level increases from 95% to 97% (Figure 12.55c) since the solvent flow rate does not change during the load ramp, the L/G ratio increases due to the decrease in flue gas flow rate. It can be concluded from the simulation that the capture level is very sensitive to the solvent-flue gas (L/G) ratio.

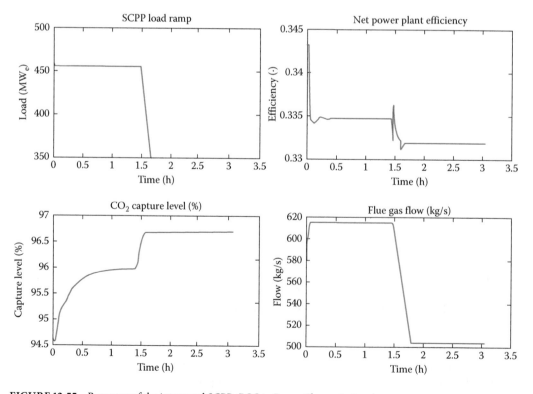

FIGURE 12.55 Response of the integrated SCPP–PCC to Ramp Change in Load.

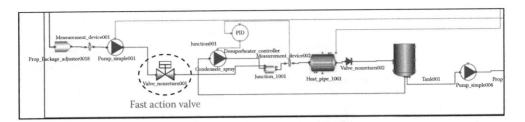

FIGURE 12.56 Stripper stop mechanisms at the steam draw-off point of the PCC model.

12.4.6.4 Operating SCPP Integrated with PCC under the UK Grid Code Requirement

Operating an integrated SCPP with CO_2 capture ability under the UK grid requirement (especially the primary frequency response) presents its own opportunities and challenges. A very important strategy is the use of the PCC link (i.e., the steam draw-off) with the SCPP as a buffer for meeting the 10% MCR requirement during primary frequency response.

12.4.6.4.1 Stripper Stop

As suggested in Haines and Davison (2014), a "stripper stop" mechanism, whereby the steam normally drawn-off from the IP/LP crossover can be reduced or stopped temporarily and the steam redirected to generate more power during primary frequency response (Figure 12.56). This mechanism also requires the use of a fast-action valve to execute the steam reduction or stoppage.

The integrated model was simulated to determine the contribution of the stripper stop approach to the primary frequency response. The results show that the stripper stop mechanism produces about 4.67% MCR (~28 MW_e) increase in the SCPP at full load condition. Other mechanisms i.e., combining stripper stop with condensate/extraction stop etc. can also be investigated to evaluate its ability to meet the UK grid requirement.

12.4.7 Challenges and Prospects of SCPP–PCC Operation under the UK Grid

12.4.7.1 Challenges

12.4.7.1.1 Grid Code Conundrum

The challenges to the SCPP operation in the current UK Grid Code is in the area of primary frequency response. As noted in Section 5, the solution to meeting the primary frequency response requirement of the grid will be a likely use of steam extraction stop technologies. However, steam extraction capability is only guaranteed at high generating load levels (above 75% MCR). Combination of the extraction technologies could also improve the SCPP's capability to meet the grid requirement but this puts the plant on serious uncertainty in its operational integrity over a long-term period. This poses a big challenge to the SCPP's successful integration to the UK Grid.

There have been proposed areas where changes to the grid code could be considered for the SCPP such as revised definition to facilitate averaging, lower requirement at lower load, variable performance requirements, etc. (Nichos and Maxim, 2008). It is believed that such changes to the grid code will solve the SCPP challenge to function successfully within the UK Grid.

12.4.7.1.2 Influx of Low Cost Renewables, Resources, and Advances in Energy Storage

Advances in low cost and relatively available renewable resources, such as solar, wind, etc. has put more challenges on the attractiveness of the fossil-fueled power generating units. The nonstorable nature of electricity is, however, one of the banes of the substantial renewables influx into the electricity grid.

Energy storage on the other hand has emerged as the catalyst for solving the problem of intermittence in renewable resources. If energy storage is deployed at large scale, it has the potential to reduce the intermittency constraints (i.e., cost, physical grid constraints, etc.) that has prevented the large scale penetration of renewables into the UK electricity grid. A massive increase in the progress of these two technologies means even tighter constraints on fossil-fired power generations.

12.4.7.2 Prospects

12.4.7.2.1 Abundance of Clean Coal-Based SCPP Experience Worldwide

The abundant availability of SCPP worldwide, and the evolving advances in the supercritical power plant efficiency improvement (above 50%) means the SCPP has the potential to become the UK's leading power generation source; benefiting from the vast experience of operating SCPP worldwide.

12.4.7.2.2 Size of SCPP Generating Unit

The size of the generating unit goes a long way to reduce the footprint of the power plant. In the UK, the subcritical power plant in existence can only boast of $500\,MW_e$ single unit. This is a well-known limitation of the subcritical units. However, an SCPP can range between 600 and $2000\,MW_e$ single unit. This means that a larger generating capacity can be ensured and at higher efficiency than the current crop of subcritical units. If the UK hopes to meet the shortfall in energy that has been projected due to old subcritical units decommissioning, new advanced supercritical units with higher capacity, higher thermal efficiency, higher availability, and faster response is the power plant for the future.

12.4.7.2.3 SCPP-CCUS Plant for the Future

The SCPP no doubt has the potential to meet the energy demand of the UK. The critical issue of meeting the UK's carbon emission reduction target (CERT)—of reducing emission by 80% to 1990 level by 2050—is also a very important index to be taken into consideration for any power plant of the future. A very important issue to deal with in meeting this target is uninterrupted availability of energy from high efficient and zero emission based generating units at competitive cost. These factors make supercritical generating units integrated with carbon capture, utilization, and storage very attractive choices. The power plant makes the energy available at all times, the carbon capture addresses the issue of emission reduction, and the CO_2 utilization means a more economical way of running the plant at lower cost. Running the integrated plant at a lower cost will result in a lower net cost of electricity to the final consumers.

12.4.7.2.4 Rising Energy Demand and Intermittence of Renewable Resources

The increase in demand for cheap and affordable energy means that the influx of renewables into the power market will come at a cost availability due to their intermittent nature. This is a good opportunity for the emergence of improved, advanced, clean and cheap SCPP technologies. Renewables are not always available all year round, hence the need for a stable supply, means the SCPP will still remain very relevant in the electricity market.

12.4.8 Concluding Remarks

This study addresses the issue of operating a SCPP integrated with PCC under the current UK Grid Code requirement. A steady-state model of the SCPP integrated with PCC was developed with Aspen Plus There was consistency between the model output and the data from the reference plant at full load. Conventional and advanced exergetic analysis of the steady-state model was performed to allow for a consistent and thorough evaluation of energy consumption in the SCPP integrated with CO_2 capture from the thermodynamic point of view.

Dynamic modeling, validation, and analyses of the strategies for operating the SCPP under the UK grid requirement as regards to primary frequency response was performed. The model was used to simulate the flexibility of the SCPP for rapid load changes and variations in system frequency. The simulation results show that using turbine throttling approach, extraction stop or condensate stop individually was not sufficient to meet the grid requirement. On the other hand, the study shows that a combination of turbine throttling, extraction stop and/or condensate stop can achieve a 10% increase in generating capacity (MCR) of a SCPP within 10s to 30s of primary frequency change as required by the UK grid.

The dynamic model of SCPP was integrated with a validated and scaled-up dynamic model of PCC. Analyses of the strategies for operating the SCPP integrated with PCC unit under the UK grid requirement as regards to primary frequency response was undertaken. The integrated model was simulated to determine the contribution of the stripper stop approach to the primary frequency response. The results show that the stripper stop mechanism is not sufficient for the 10% MCR required for the primary response. Further studies and analyses are required to investigate the possibility of combining the "stripper stop" mechanisms with the SCPP's steam throttling or condensate and/or extraction stops.

Acknowledgment

The authors would like to acknowledge the financial support of the Biomass and Fossil Fuel Research Alliance (BF2RA/Ref: 001) and the EU Marie Curie International Research Staff Exchange Scheme (FP7-PEOPLE-2013-IRSES).

References

Department of Trade and Innovation (DTI). (2006). Advanced power plant using high efficiency boiler/turbine best practice brochure carbon abatement technologies programme. Technical report. www.nios.com.cn/common/files/File/BPB010(1).pdf.

Energy Information Administration (EIA). (2010). International energy outlook 2010, Office of Integrated Analysis and Forecasting, U.S. Department of Energy Washington, DC.

Fan, L.-S. (2010). *Chemical Looping Systems for Fossil Energy Conversions*, John Wiley and Sons, Hoboken, NJ.

Freund, P. (2003). Making deep reductions in CO_2 emissions from coal-fired power plant using capture and storage of CO_2, *Proceedings of the Institution of Mechanical Engineers, Part A: Journal of Power and Energy* 217: 1–8.

Haines, M. R. and Davison, J. (2014), Enhancing dynamic response of power plant with post combustion capture using "Stripper stop," *International Journal of Greenhouse Gas Control* 20:49–56.

Hordeski, M. F. (2008). *Alternative Fuels: The Future of Hydrogen*, 2nd ed., The Fairmont Press Inc, Georgia.

House of Lords. (2008). The economics of renewable energy, Select Committee on Economic Affairs, 4th Report of Sessions 2007–2008. Authority of the House of Lords, The Stationery Office Limited, London. Volume I: Report IEA Greenhouse Gas R&D Programme (IEA GHG). 2002a: Building the Cost Curves for CO_2 Storage, Part 1: Sources of CO_2, PH4/9, July, 48 pp.

Intergovernmental panel on climate change (IPCC). (2005). IPCC special report on carbon dioxide capture and storage, Cambridge University Press, Cambridge, UK and New York, USA.

Kitto, J. B. and Stultz, S. C. (2005). *Steam: Its Generation and Use*, 41st ed. The Babcock and Wilcox Company, Barberton, OH.

Kjarttansson, O. (2008). Investigating the potential of ultra-supercritical coal-fired power plants FACE 10-D Project, Institute of Energy Technology, Aalborg University, Denmark.

Lawal, A., Wang, M., Stephenson, P., and Obi, O. (2012). Demonstrating full-scale post combustion CO_2 capture for coal-fired power plants through dynamic modeling and simulation, *Fuel* 101:115–128.

Leites, I. L., Sama, D. A., and Lior, N. (2003). The theory and practice of energy saving in the chemical industry: Some methods for reducing thermodynamic irreversibility in chemical technology processes, *Energy* 28, 55–97.

Marto, P. J. and Nunn, R. H. (1981). *Power Condenser Heat Transfer Technology*, Hemisphere Publishing Corporation, Washington, DC.

Mercier, J. and Drenik, O. (2013). Power plant and method of operating a power plant, US Patent (WO2013005071A1), International Application No: PCT/IB2011/002321.

Morosuk, T., Tsatsaronis, G., and Schult, M. (2013). Conventional and advanced exergy analyses: Theory and application, *Arabian Journal for Science and Engineering* 38:395–404.

National Grid. (2012). The Grid Code, Issue 4, Revision 10. National Grid Electricity Transmission plc, London, UK.

Nicholls, R. J. and Maxim, C. (2008), Supercritical coal-fired plant requirements and the grid code, E.ON UK.

O'Kelly, P. (2013). *Computer Simulation of Thermal Plant Operations*, Springer, London, ISBN 978-1-4614-4255-4.

Olaleye, A. K., Wang, M., and Kelsall, G. (2015). Steady state simulation and exergy analysis of supercritical coal-fired power plant integrated with CO_2 capture, *Fuel* Special Issue, 151: 57–72.

Philibert, C. and Podkanski, J. (2005). International energy technology collaboration and climate change mitigation. Case study 4: Clean coal technologies. Technical report 29 p, OECD/IEA. www.oecd.org/env/cc/34878689.pdf.

Shinohara, W. and Koditschet, D. E., A simplified model of a supercritical power plant, University of Michigan, Control Group reports, CGR-95-08, October 1995.

Shoureshi, R. and Paynter H. M. (1983). Simple model for dynamics and control of heat exchangers, *Proceedings of the American Control Conference*, pp. 1294–1298.

Sinnott, R. K. and Towler, G. (2013). *Chemical Engineering Design—Principles, Practice and Economics of Plant and Process Design*, 2nd ed. Elsevier, Oxford, UK.

Viswanathan, R. and Bakker, W. T. (2000). Materials for boilers in ultra supercritical power plants. *2000 International Joint Power Generation Conference*, ASME, Miami Beach, FL.

Wagner, W. and Kretzschmar, H.-J. (2008). *International Steam Tables. Properties of Water and Steam. Based on the Industrial Formulation IAPWS-IF97*, 2nd ed., Springer, Berlin.

Wang, M., Lawal, A., Stephenson, P., Sidders, J. and Ramshaw, C. (2011). Post-combustion CO_2 capture with chemical absorption: A state-of-the-art review, *Chemical Engineering Research and Design*, 89: 1609–1624.

WEO. (2014). World energy outlook 2014 factsheets, International Energy Agency. www.worldenergy-outlook.org/media/weowebsite/2014/141112_weo_factsheets.pdf.

Woods, M. C., Capicotto, P. J., Halsbeck, J. L., Kuehn, N. J., Matuszewski, M., Pinkerton, L. L., Rutkowski, M. D., Schoff, R. L., and Vaysman V. (2007). Bituminous coal and natural gas to electricity. In: Cost and Performance Baseline for Fossil Energy Plants, Final report DOE/NETL-2007/1281, National Energy Technology Laboratory (NETL), Pittsburgh, PA.

Zindler, H., Walter, H., Hauschke, A., and Leithner, R. (2008). Dynamic simulation of a 800MW$_e$ hard coal once-through supercritical power plant to fulfill the Great Britain grid code, *6th IASME/WSEAS International Conference on Heat Transfer, Thermal Engineering and Environment*, August, 2008, Rhodes, Greece, pp. 184–192.

12.5 Whole System Experimental and Theoretical Modeling Investigation of the Optimal CO$_2$ Stream Composition in the Carbon Capture and Sequestration Chain

H. Mahgerefteh, R. T. J. Porter, S. Brown, S. Martynov, A. Collard,
M. Fairweather, S. A. E. G. Falle, R. M. Woolley, G. C. Boulougouris,
Ioannis G. Economou, D. M. Tsangaris, I. K. Nikolaidis, A. Beigzadeh,
C. Salvador, K. E. Zanganeh, A. Ceroni, R. Farret, Y. Flauw, J. Hébrard,
D. Jamois, C. Proust, S. Y. Chen, J. L. Yu, Y. Zhang, D. Van Hoecke,
R. Hojjati Talemi, S. Cooreman, J. Bensabat, R. Segev, D. Rebscher,
J. L. Wolf, A. Niemi, C. Kolster, N. Mac Dowell, and Nilay. Shah

12.5.1 Introduction

CO$_2$ capture and storage (CCS) refers to a collection of technologies that allow for the continued use of fossil fuels for energy generation and heavy industries while abating their atmospheric emissions of the greenhouse gas (GHG) carbon dioxide (CO$_2$) by capturing it in the given process, and then transporting it to a suitable location for subsurface geological storage. For large-scale applications of CCS, transport of CO$_2$ using pressurized pipelines is found to be the most practical and economic method [1]. However, the CO$_2$ stream captured from fossil fuel power plants or other CO$_2$ intensive industries will contain a range of different types of impurities each having its own impact on the different parts of the CCS chain.

Determining the "optimum composition" of the captured CO$_2$ stream addressing the cost, safety, and environmental concerns is, therefore, fundamentally important in facilitating CCS as a viable technology for addressing the impact of global warming.

This chapter presents the main issues and provides an overview of the experimental and theoretical modeling work carried out as part of the CO$_2$ QUEST European Commission Collaborative Project [2] tasked with addressing this challenge.

12.5.1.1 Classes of CO$_2$ Impurities by Origin

Impurities contained in the CO$_2$ streams from different carbon capture technologies may be classified broadly by origin into three main categories arising from fuel oxidation, excess oxidant/air ingress, and process fluids as shown in Table 12.27 [3]. Water is a major combustion product and is considered as an impurity in the CO$_2$ stream. The elements inherently present in a fuel such as coal include sulfur, chlorine, and mercury and are released upon complete or incomplete combustion and form compounds

TABLE 12.27 Classes of Potential CO$_2$ Impurities by Origin

Coal/Biomass Oxidation Products	
Complete	Partial
H$_2$O, SO$_X$, NO$_X$, HCl, HF	CO, H$_2$S, COS, NH$_3$, HCN
Volatiles	Biomass Alkali Metals
H$_2$, CH$_4$, C$_2$H$_6$, C$_3$+	KCl, NaCl, K$_2$SO$_4$, KOH etc.
Trace Metals	Particulates
Hg (HgCl$_2$), Pb, Se, As etc.	Ash, PAH/soot
Oxidant/Air Ingress	Process Fluids
O$_2$, N$_2$, Ar	Glycol, MEA, Selexol, NH$_3$ etc.

in the gas phase that may remain to some extent as impurities in the CO_2 after it is captured and compressed. The oxidizing agent used for combustion such as air may result in residual impurities of N_2, O_2, and Ar; these same impurities may also result from any air ingress into the process. The materials and chemicals used for the CO_2 separation process such as monoethanolamine (MEA) used for postcombustion capture or Selexol in precombustion capture and their degradation products can be carried over into the CO_2 stream constituting a further class of impurity.

12.5.1.2 Impacts of CO_2 Impurities on CCS Systems

Impurities that arise from CO_2 capture sources can have a number of important impacts on the downstream transport and storage infrastructure and operation. The presence of some impurities in CO_2, such as the air-derived noncondensable species (N_2, O_2, and Ar), can shift the boundaries in the CO_2 phase diagram to higher pressures, meaning that higher operating pressures are needed to keep CO_2 in the dense phase and hence impacting on compression and transport costs. In addition, these species can reduce the CO_2 structural trapping capacity in geological formations to a greater degree than their molar fractions [4] and may therefore change the behavior of the CO_2 plume and the storage capacity. Hydrogen may be present in precombustion capture-derived CO_2 streams and is also believed to impact required pipeline inlet pressures significantly [5]. Enhanced Oil Recovery (EOR) applications require stricter limits, particularly on O_2 due to its promoting microbial growth and reaction with hydrocarbons. Water should be limited in CCS applications in order to mitigate corrosion due to the formation of in situ carbonic acid [6], clathrate formation, and condensation at given operating conditions [1]. On the other hand, water may be of benefit even at high concentrations in storage given its immobilization effect on CO_2.

Sulfur species (H_2S, COS, SO_2, and SO_3) may pose corrosion risk in the presence of water, and there are additional toxicity concerns for H_2S. NO_X species may be present in CO_2 streams as combustion by-products and also pose a corrosion risk due to nitric acid formation [7]. Trace elements such as lead, mercury, and arsenic in the CO_2 stream are of concern for geological storage due to their toxicity and the possibility that they could contaminate groundwater. Among the numerous potential trace metal CO_2 impurities, mercury receives further attention for its corrosive effects on a number of metals. Due to its toxicity, limits are also suggested for carbon monoxide and H_2S. For other components that may be present in CO_2 streams (e.g., HCl, HF, NH_3, MEA, Selexol), not enough information is available to fully understand their downstream impacts on transport and storage and determine maximum allowable amounts. Further work is therefore required to understand the impacts of these species in transport and storage applications and to elucidate potential crossover effects.

12.5.1.3 The QUEST for "Optimal" CO_2 Purity

In CCS systems, as illustrated in Figure 12.57, capture costs can be expected to increase with CO_2 purity when additional process unit operations and increased energy penalty associated with achieving the desired CO_2 purity is taken into consideration. Conversely, transport and storage costs (per tonne of CO_2 transported), as depicted in Figure 12.57, may be expected to decrease with increasing CO_2 purity due to the lower compression requirements to keep CO_2 in the dense phase, lower rates of corrosion, and the relaxation of safety measures needed to deal with hazardous impurities. When capital and operating costs are factored together to calculate a levelized "total cost" for CCS systems, a minima for a given purity and composition range is expected where the system may be assumed to be cost optimized in addition to having the necessary safety precautions taken into consideration. The challenge, therefore, is to find optimum range and concentration of impurities that can be permitted in the CO_2 stream to enable its safe and cost-effective transportation and storage.

The CO_2QUEST project [2] addresses the fundamentally important issues surrounding the impact of typical impurities in the CO_2 stream captured from fossil fuel power plants and other CO_2 intensive industrial emitters on its safe and economic pipeline transportation and storage. This 40 months duration project was funded by the European Commission under the FP7 framework program commenced in February 2013. It involves the collaboration of 10 academic and industry partners across Europe, China, and Canada.

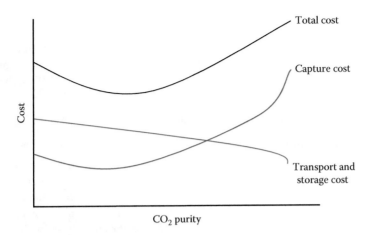

FIGURE 12.57 Cost trade-offs associated with CO_2 purity.

This chapter presents an overview of the project and some of its main findings. Key gaps in knowledge relating to the impact of impurities on the chemical, physical, and transport properties of the captured CO_2 stream under different operating conditions are addressed.

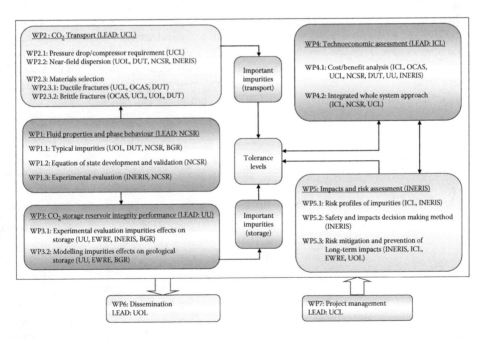

FIGURE 12.58 CO_2QUEST work program structure. Project partner abbreviations: University College London (UCL), University of Leeds (UOL), Dalian University of Technology (DUT), National Center for Scientific Research (NCSR), L'nstitut national de l'environnement industriel et des risques (INERIS), Onderzoekscentrum voor de Aanwending van Staal (OCAS), Bundesanstalt für Geowissenschaften und Rohstoffe (BGR), Uppsala University (UU), Environmental and Water Resources Engineering Ltd (EWRE).

12.5.2 Project Work Packages

Figure 12.58 shows a schematic representation of the CO_2QUEST work programme divided into seven work packages (WPs), encompassing technical (WP1–WP5), dissemination (WP6), and project management (WP7) activities. The following is a description of each of the technical WPs, their interactions with other WPs, and the main findings.

12.5.2.1 WP1: Fluid Properties and Phase Behavior

12.5.2.1.1 WP1.1: Typical Impurities and Cost-Benefit Analysis

The work in this section underpins the other project activities by providing analysis of the ranges and levels of impurities present in CO_2 streams derived from different carbon capture sources, including those from both the power sector and heavy industries based on literature studies. The factors that affect the ranges and levels of impurities for given power generation technologies, such as postcombustion, precombustion, and oxy-fuel combustion capture, have also been identified leading to the ranges as set out in Table 12.28.

A technoeconomic modeling study of power plants with CO_2 capture technologies with focus on process scenarios that deliver different grades of CO_2 product purity was also carried out in this work package. The three leading CO_2 capture technologies for the power sector are considered, namely; oxy-fuel combustion, precombustion, and postcombustion capture. The study uses a combination of process simulation of flue gas cleaning processes, modeling with a power plant cost, and performance calculator and literature values of key performance criteria in order to calculate capital costs operational and maintenance costs, the levelized cost of electricity, and CO_2 product purity of the considered CO_2 capture options.

For oxy-fuel combustion capture, the calculations are based on a 400 MWg retrofitted power station that uses a low sulfur coal and considers three raw CO_2 flue gas processing strategies of compression and dehydration only, double flash system purification, and distillation purification. Analysis of precombustion capture options is based on new build integrated gasification combined cycle plants with one gas turbine and a GE entrained-flow gasifier. Integrated physical solvent systems for capturing CO_2 and sulfur species were considered in three ways: cocapture of sulfur impurities with the CO_2 stream using Selexol solvent, separate capture of CO_2 and sulfur impurities using Selexol, and Rectisol solvent systems for separate capture of sulfur impurities and CO_2. Analysis of postcombustion capture plants was made with and without some conventional pollution control devices.

TABLE 12.28 CO_2 Impurities from Different CO_2 Capture Technologies

| | Oxy-Fuel Combustion | | | | |
	Raw/Dehumidified	Double Flashing	Distillation	Precombustion	Postcombustion
CO_2 % v/v	74.8–85.0	95.84–96.7	99.3–99.9	95–99	99.6–99.8
O_2 % v/v	3.21–6.0	1.05–1.2	0.0003–0.4	0	0.015–0.0035
N_2 % v/v	5.80–16.6	1.6–2.03	0.01–0.2	0.195–1	0.045–0.29
Ar % v/v	2.3–4.47	0.4–0.61	0.01–0.1	0.0001–0.15	0.0011–0.021
NO_x ppmv	100–709	0–150	3–100	400	20–38.8
SO_2 ppmv	50–800	0–4500	1–50	25	0–67.1
SO_3 ppmv	20	–	0.1–20	–	N.I.
H_2O ppmv	100–1000	0	0–100	0.1–600	100–640
CO ppmv	50	–	2–50	0–2000	1.2–10
H_2S/COS ppmv				0.2–34,000	
H_2 ppmv				20–30,000	
CH_4 ppmv				0–112	

Of the different cases considered, precombustion capture with cocapture of impurities and CO_2 using Selexol offered the lowest cost with reasonably high purity of CO_2 at 97.64 mol%, but high estimated levels of H_2S (at 3974 ppm$_v$) in the captured stream. The most expensive system was precombustion capture using Rectisol with separate capture of CO_2 and sulfur impurities, producing a dry 99.51 mol% pure CO_2 stream. The system with the lowest grade of CO_2 was oxy-fuel combustion capture with compression and dehydration of the raw CO_2 stream only, which resulted in 77.69 mol% pure CO_2 and with the second lowest cost. The oxy-fuel plant with distillation purification system and the postcombustion capture plant with conventional pollution control devices had the joint highest CO_2 purity (99.99 mol%), with the postcombustion capture system estimated to be the cheaper of the two. The calculations performed are of use in further analyses of whole chain CCS for the safe and economic capture, transport, and storage of CO_2.

12.5.2.1.2 WP1.2: Equation of State Development and Validation

An integrated CCS process requires the transportation of a CO_2-rich stream from the capture plant to a sequestration site. Process simulation and design require accurate knowledge of the physical properties of the mixtures transported through the pipeline and advanced Equations of State (EoS) can be a very useful tool for the prediction and correlation of these properties [8,9]. Moreover, an important consideration for the design and operation of CCS facilities is the understanding of the phase equilibria of the CO_2 mixtures associated with the process. Studying the Vapor–Liquid Equilibrium (VLE) of these mixtures has attracted much attention both in terms of experimental measurements and modeling, but relatively little work has been performed to understand dry ice formation when other gases are present in the CO_2 stream. CO_2 exhibits a relatively high Joule–Thomson expansion coefficient and during transportation, a sudden pipeline depressurization will lead to rapid cooling so that very low temperatures can be reached [10]. Consequently, solidification of CO_2 may take place and this can affect the safety of CCS facilities during equipment depressurization or other process upsets [11,12].

In this WP, three solid thermodynamic models were applied to model the Solid–Fluid Equilibrium (SFE) of pure CO_2 and CO_2 mixtures with other compounds, typically found in CO_2 streams from industrial sources, namely N_2, H_2, and CH_4. These models are an empirical correlation fitted on experimental data at SFE conditions, a thermodynamic integration model and a solid EoS developed for pure CO_2. The solid models have to be combined with a fluid EoS. In this work, the Peng–Robinson (PR), Soave–Redlich–Kwong (SRK), and Perturbed Chain Statistical Associating Fluid Theory (PC-SAFT) EoS were used. All three fluid EoS are used widely for liquids and gases.

12.5.2.1.2.1 Solid Models

Empirical Correlation Model   SFE of a mixture requires that the chemical potentials of the solid former in the two coexisting phases (S: solid phase, F: fluid phase) are equal at the same temperature and pressure.

$$\mu_i^S(T,P) = \mu_i^F(T,P,\mathbf{x}^F) \tag{12.38}$$

where μ_i^S is the chemical potential of the solid former i in the pure solid phase and μ_i^F is its chemical potential in the coexisting fluid phase of molar composition $\mathbf{x}^F$. If the ideal gas reference state is used to calculate the chemical potential for both phases, Equation 12.38 can be substituted by the equation of fugacities [13]:

$$\hat{f}_i^S(T,P) = \hat{f}_i^F(T,P,\mathbf{x}^F) \tag{12.39}$$

and subsequently to the expression:

$$P_{0i}^{sat}(T)\,\hat{\varphi}_{0i}^{sat}\left(T,P_{0i}^{sat}\right)\exp\left[\frac{v_{0i}^S}{RT}\left(P-P_{0i}^{sat}(T)\right)\right] = x_i^F\,\hat{\varphi}_i^F\left(T,P,\mathbf{x}^F\right)P \tag{12.40}$$

where $P_{0i}^{\text{sat}}(T)$ is the saturation pressure of the pure solid former at temperature T, $\hat{\varphi}_{0i}^{\text{sat}}\left(T, P_{0i}^{\text{sat}}\right)$ is the fugacity coefficient of the pure solid former at temperature T, and pressure P_{0i}^{sat}, $\hat{\varphi}_i^{\text{F}}\left(T, P, \mathbf{x}^{\text{F}}\right)$ is the fugacity coefficient of the solid former in the fluid mixture of molar composition $\mathbf{x}^{\text{F}}$, at temperature T and pressure P, and v_{0i}^{S} is the temperature and pressure independent pure solid molar volume.

Equation 12.40 can be employed to calculate the SFE of a multicomponent mixture with the use of a fluid EoS for the fugacity coefficients and a model that provides the saturation pressure of the solid former at Solid-Liquid Equilibrium (SLE) or Solid-Vapor Equilibrium (SVE) conditions, which can be an empirical correlation fitted to experimental data.

12.5.2.1.2.2 Thermodynamic Integration Model Seiler et al. [14] proposed a different methodology for SFE modeling. In their approach, for SLE calculation, the reference state is the pure, subcooled melt, at system temperature and standard pressure (P^+), whereas P^+ is selected accordingly by taking into account the existence of caloric data at this reference state. The expression that applies to the SLE with this model is as follows:

$$x_i^{\text{L}} = \frac{\varphi_{0i}^{\text{L}^*}}{\varphi_i^{\text{L}}} \cdot \exp\left[-\frac{\left(v_{0i}^{\text{S}} - v_{0i}^{\text{L}^*}\right)\left(P^+ - P\right)}{RT} - \frac{\Delta h_{0i}^{\text{SL}}}{RT}\left(1 - \frac{T}{T_{0i}^{\text{SL}}}\right) + \frac{\Delta c_{P,0i}^{\text{SL}^*}}{RT}\left(T_{0i}^{\text{SL}} - T\right) - \frac{\Delta c_{P,0i}^{\text{SL}^*}}{R}\ln\frac{T_{0i}^{\text{SL}}}{T}\right], \quad (12.41)$$

where $\Delta h_{0i}^{\text{SL}}$ is the enthalpy of melting at melting temperature, T_{0i}^{SL}, v_{0i}^{S}, and $v_{0i}^{\text{L}^*}$ are the pure solid former, solid molar volume, and liquid molar volume at the solid–liquid transition, and $\Delta c_{P,0i}^{\text{SL}^*}$ is the difference of the molar, isobaric heat capacities between the hypothetical subcooled melt and the solid.

12.5.2.1.2.3 Gibbs Free Energy Equation of State for Solid CO_2 Jäger and Span [15] developed an empirical EoS that describes the thermodynamic behavior of solid CO_2, which is explicit in the Gibbs free energy. A fundamental expression for the Gibbs free energy is used and is fitted to the appropriate experimental data of solid CO_2. The Gibbs free energy can be fundamentally written as follows:

$$g(P,T) = h_0 - Ts_0 + \int_{T_0}^{T} c_P(T,P_0)\,\mathrm{d}T - T\int_{T_0}^{T} \frac{c_P(T,P_0)}{T}\,\mathrm{d}T + \int_{P_0}^{P} v(P,T)\,\mathrm{d}P \quad (12.42)$$

By using the appropriate functional forms for the heat capacity, the expansion coefficient and the partial derivative of the molar volume with respect to pressure can be accurately fitted to the experimental data. The resulting expression for the Gibbs free energy is as follows:

$$\frac{g}{RT_0} = g_0 + g_1\Delta\vartheta + g_1\Delta\vartheta^2 + g_3\left\{\ln\left(\frac{\vartheta^2 + g_4^2}{1 + g_4^2}\right) - \frac{2\vartheta}{g_4}\left[\arctan\left(\frac{\vartheta}{g_4}\right) - \arctan\left(\frac{1}{g_4}\right)\right]\right\}$$

$$+ g_5\left\{\ln\left(\frac{\vartheta^2 + g_6^2}{1 + g_6^2}\right) - \frac{2\vartheta}{g_6}\left[\arctan\left(\frac{\vartheta}{g_6}\right) - \arctan\left(\frac{1}{g_6}\right)\right]\right\}$$

$$+ g_7\Delta\pi\left[e^{f_\alpha(\vartheta)} + \mathrm{K}(\vartheta)g_8\right]$$

$$+ g_9\mathrm{K}(\vartheta)\left[(\pi + g_{10})^{(n-1)/n} - (1 + g_{10})^{(n-1)/n}\right], \quad (12.43)$$

where T_0 is a reference temperature set equal to $150\,\text{K}$ and $\vartheta = T/T_0$. Equation 12.43 uses 23 adjustable parameters that are fitted to the experimental data.

12.5.2.1.2.2 Fluid Equations of State Based on the pioneering work of van der Waals, cubic EoS represent an important family of EoS with the most well known being SRK and PR. The equations are used for the calculation of pure component thermodynamic properties and can be extended to mixtures with the introduction of suitable mixing rules. In this work, the van der Waals one fluid theory (vdW1f) mixing rules were applied using one temperature-independent Binary Interaction Parameter (BIP), k_{ij}, which allows for reliable extrapolation over a wide temperature range. The SFE calculation was also performed using the PC-SAFT [16] EoS, which is a model based on rigorous perturbation theory. In PC-SAFT framework, the Helmholtz energy of a fluid is described as the sum of the Helmholtz energy of a simple reference fluid which is known accurately and a perturbation term. This way, PC-SAFT EoS is written as summation of residual Helmholtz energy (A^{res}) terms that contribute to different molecular interactions. In PC-SAFT, mixing rules are needed for the dispersion interactions and the ones derived from the vdW1f theory were used, coupled with the Lorentz–Berthelot combining rules. A temperature-independent BIP was used in this EoS also, in the combining rule for the energy parameter.

12.5.2.1.2.3 Methodology Correlation of the pure CO_2 SLE and SVE was the first step in assessing the performance of every combined model and the agreement between them. Moreover, accurately describing the SFE of pure CO_2 is a prerequisite for successful two-phase and three-phase Solid–Liquid–Gas Equilibrium (SLGE) mixture calculations.

12.5.2.1.2.4 Main Research Outcomes Figure 12.59 presents the modeling results for pure CO_2 SVE with the three different solid models, when PC-SAFT is used for the vapor phase fugacities and properties calculation. The empirical correlation and the Jäger–PC-SAFT models are in excellent agreement with each other, whereas the thermodynamic integration model deviates at higher temperatures. The same calculations were performed for the SLE of pure CO_2 and results are presented in Figure 12.60. In this case, all models are in excellent agreement at low temperatures up to 226 K, but at higher temperatures the Jäger–PC-SAFT model deviates from the other two, which remain in excellent agreement.

Because of the lack of experimental data for the two phase SFE for the CO_2 mixtures of interest, we evaluated the performance of the various combined models on SLGE conditions and compared to experimental data available in the literature [17–19].

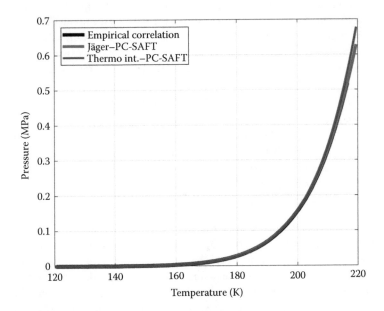

FIGURE 12.59 Comparison of empirical correlation, thermodynamic integration, and Jäger and span EoS models, coupled with PC-SAFT EoS for pure CO_2 SVE.

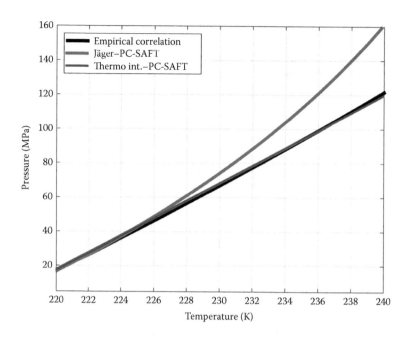

FIGURE 12.60 Comparison of empirical correlation, thermodynamic integration, and Jäger and span EoS models, coupled with PC-SAFT EoS for pure CO_2 SLE.

Application of the combined solid–fluid models on the mixtures of interest in this work revealed that an accurate prediction of the SLG mixture locus requires successful reproduction of the pure CO_2 triple point. In general, when all BIPs are set equal to zero, the thermodynamic integration model and the Jäger and Span EoS are more accurate in predicting the SLGE for the mixtures of CO_2 with N_2 and H_2, when compared to the empirical correlation model. Combining this solid model with PC-SAFT EoS improves the modeling results, because the reproduction of pure CO_2 triple point is also improved. The use of BIPs, regressed from binary VLE data, significantly improves the prediction of the SLG behavior for most models. Very good behavior with all models is observed for the CO_2–CH_4 mixture and the use of BIPs led to very low deviations from experimental data. Representative experimental data and model calculations are shown in Figures 12.61 through 12.63 for the three CO_2 mixtures.

12.5.2.1.3 WP1.3: Experimental Evaluation of Physical Properties of CO_2 Mixtures

In this work package, we identify the gaps in current knowledge and guide the experimental program for physical properties measurements of CO_2 mixtures with impurities at CCS conditions. Experiments have been performed to generate VLE data for the binary, ternary, and multicomponent mixtures of CO_2 with impurities. A further set of experiments has been performed to generate data on the transportation properties of CO_2 with impurities. The experimental data are then used to support the development and validation of the EoS in WP1.2.

12.5.2.1.3.1 VLE and Transport Properties Measurement

Work on the thermophysical properties of CO_2 mixtures has been carried out by CanmetENERGY, located in Regina, Canada. This work supports the development of more accurate EoS for these mixtures. Unfortunately, there is an absence of data for CO_2 mixtures, particularly for quaternary and tertiary mixtures, at the pressure and temperatures typical of CCS processes in the open literature, forcing modelers to use more generic EoS to make predictions about the thermophysical properties of CO_2 mixtures. Experiments have therefore been conducted to generate data and fill some of the gaps in the available thermophysical properties of CO_2 mixtures.

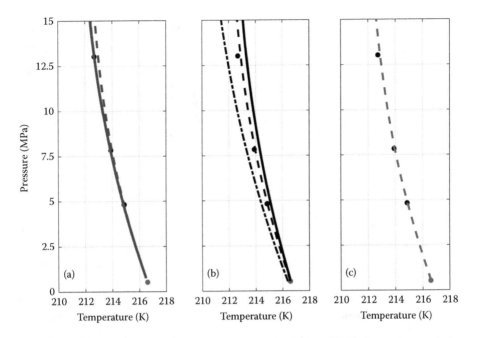

FIGURE 12.61   Prediction of the SLG equilibrium curve of the CO_2–N_2 system with k_{ij} parameters fitted to experimental binary VLE data at low temperature. Results with three different solid models, (a) thermodynamic integration, (b) empirical correlation, and (c) Jäger and Span EoS models, coupled with three fluid EoS. Experimental data [17] are represented by data points and calculations are represented by lines: (–) SRK, (–·–)PR, (– – –) PC-SAFT.

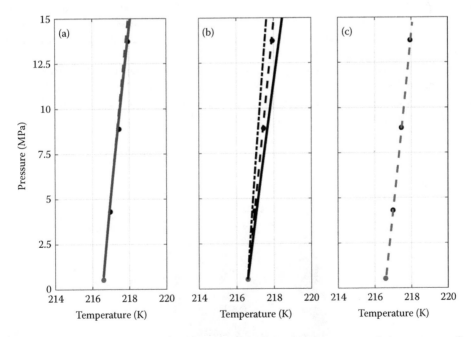

FIGURE 12.62   Prediction of the SLG equilibrium curve of the CO_2–H_2 system with k_{ij} parameters fitted to experimental binary VLE data at low temperature. Results with three different solid models, (a) thermodynamic integration, (b) empirical correlation, and (c) Jäger and Span EoS, coupled with three fluid EoS. Experimental data [13] are represented by data points and calculations are represented by lines: (–) SRK, (–·–)PR, (– – –) PC-SAFT

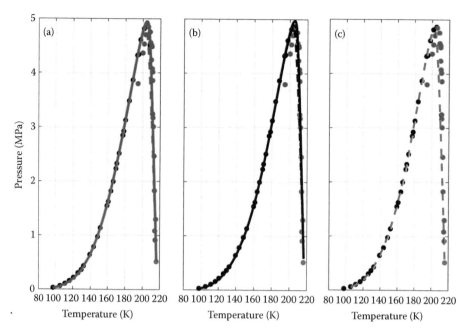

FIGURE 12.63 Prediction of the SLG equilibrium curve of the CO_2–CH_4 system with k_{ij} parameters fitted to experimental binary VLE data from Diamantonis et al. [20]. Results with three different solid models, (a) thermodynamic integration, (b) empirical correlation, and (c) Jäger and span EoS, coupled with three fluid EoS. Experimental data [18,19] are represented by data points and calculations are represented by lines: (–) SRK, (–·–) PR, (– – –) PC-SAFT.

In this context, experiments are carried out using a unique bench-scale CO_2 pressure cell apparatus (0–200 bar and –60°C–150°C). In brief, the pressure cell assembly consisting of a high pressure view chamber, gas mixing and booster pump assembly, syringe pump, recirculating pump, heating/cooling enclosure, density meter, and gas chromatograph (Figure 12.64).

A camera system, situated immediately outside the high pressure view chamber, records the phase change of the fluid within the high-pressure cell in real time. The novel design for the high-pressure view chamber provides the unique opportunity to observe the phase change visually (Figure 12.65) ensuring the correct sampling is made from either the liquid or vapor.

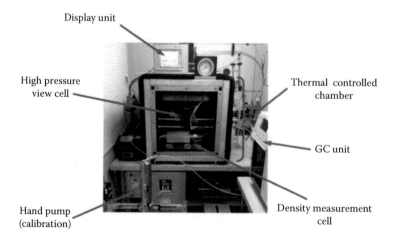

FIGURE 12.64 CanmetENERGY's bench-scale CO_2 pressure cell apparatus.

This experimental investigation is focused on the measurement of density and concentration for pressures from 5 to 80 bar and temperatures from 220 to 300 K, for a single quaternary mixture and three tertiary mixtures. Testing for the quaternary mixture [CO_2 (93%), O_2 (5.4%), N_2 (1.49%), and Ar (649 ppm)] is ongoing with positive preliminary data at 300 K and 280 K, when compared with estimates from Aspen HYSYS. A subsequent snapshot of the density data relative to values from HYSYS is shown in Figure 12.66.

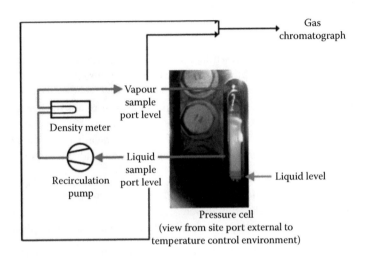

FIGURE 12.65 Two-phase mixture as viewed from observation window.

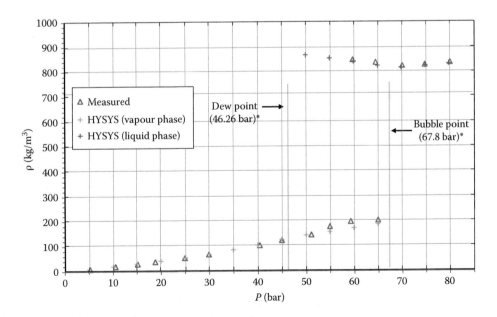

FIGURE 12.66 Experimental and HYSYS calculated densities for a quaternary gas mixture (CO_2 93%, O_2 5.41%, N_2 1.49%, Ar 649 ppm) plotted against mixture pressure.

12.5.2.2 WP2: CO₂ Transport

12.5.2.2.1 WP2.1: Pressure Drop and Compressor Requirement

In this section, nonisothermal steady-state flow has been modeled in order to calculate pressure drop (and hence compressor power requirements) in pipeline networks transporting CO_2 with typical stream impurities using the dedicated EoS developed under WP1. Compression strategies for minimizing compressor power requirements have been developed. We have also performed parametric studies using a nonisothermal steady-state flow model developed in CO₂QUEST to identify the type and composition of stream impurities that have the most adverse impact on the CO_2 pipeline pressure drop, pipeline capacity, fluid phase, and compressor power requirements.

12.5.2.2.1.1 Compressor Power Requirements

Minimizing the pressure drop and avoiding two-phase flows within CO_2 pipeline networks are essential for reducing compressor power requirements. This is critically important given that the compression penalty for CO_2 capture from coal-fired power plants is estimated to be as high as 12% [21].

In order to evaluate the impact of CO_2 impurities on compressor power requirements, a thermodynamic analysis method is applied to CO_2 streams captured using oxy-fuel, precombustion, and postcombustion capture technologies, the compositions of which are detailed in Table 12.29. The analysis is performed for several methods of compression previously recommended for pure CO_2, including the following options [22]:

- Option A: Using the centrifugal integrally-geared multistage compressors.
- Option B: Using supersonic axial compressors.
- Option C: Using compressors combined with liquefaction followed by pumping.

Figure 12.67 shows an example of the calculation of thermodynamic paths of multistage compression combined with intercooling relative to the phase envelopes for pure and impure CO_2 streams, whereas

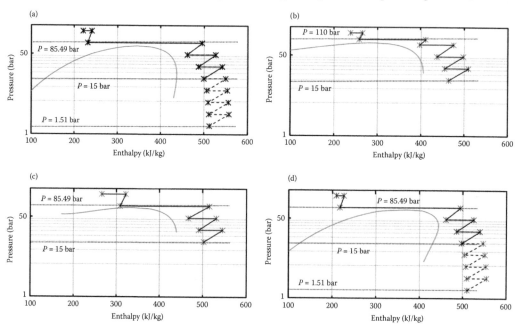

FIGURE 12.67 The thermodynamic paths for compression of pure CO_2, (a) CO_2 mixtures from oxy-fuel (b) precombustion (c) postcombustion, and (d) capture, using compression and pumping with supercritical liquefaction. Note that the compressor inlet pressure is 1.5 bar for pure CO_2 and postcombustion streams, and 15 bar for the precombustion and oxy-fuel streams.

TABLE 12.29 Compositions of CO_2 Mixtures Captured from Oxy-Fuel and Pre- and Postcombustion Technologies

	Oxy-Fuel	Precombustion	Postcombustion
CO_2 (% v/v)	81.344	98.066	99.664
O_2	6.000	–	0.0035
N_2	8.500	0.0200	0.2900
Ar	4.000	0.0180	0.0210
NO_2	609.0	–	38.800
SO_2	800.0	700.00	67.100
H_2O	100.0	150.00	100.00
CO	50.00	1300.0	10.000
H_2S	–	1700.0	–
H_2	–	15000	–
CH_4	–	110.00	–

Source: S. Martynov, S. Brown, CO_2QUEST internal report: A report describing the optimum CO_2 compression strategy, University College, London, 2014.

Figure 12.68 shows the power consumption in compression and interstate cooling for each compression strategy (options A, B, and C) for pure CO_2 and its mixtures with impurities (Table 12.29).

In particular, Figure 12.68 shows that in agreement with the data published in the literature, the integration of the multistage compression with liquefaction and pumping (Option C) reduces (by ca 15% in case of the oxy-fuel stream) the power consumption in compression when compared to conventional gas-phase compression (Option A) and supersonic compression (Option B). Given the relatively small power consumption for the interstage cooling of high-purity CO_2, the option C becomes particularly attractive for compression of almost pure CO_2, when liquefaction can be achieved using utility streams at 20°C for a postcombustion mixture of purity 99.6 vol%, and 8°C for a precombustion mixture (CO_2 purity approximately 98 vol%). At the same time, the cryogenic temperatures needed for liquefaction of

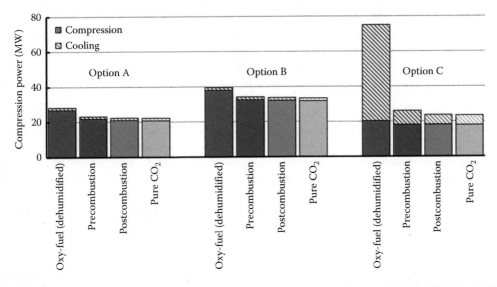

FIGURE 12.68 Power demand for high-pressure multistage compression (options A, B, and C) of pure CO_2, oxy-fuel (raw/dehumidified), pre- and postcombustion streams (Table 12.29) from 15 to 151 bar at a capture rate of 156 kg/s.

FIGURE 12.69   UK CCS network, 2015–2025.

oxy-fuel CO_2 streams carrying 74–85 vol% of impurities, may result in significant increase in the power demand for refrigeration [24]. The data in Figure 12.68 show that when using option C, the liquefaction of the oxy-fuel stream is ca 2.5 times more power demanding than that in compression. Clearly, such information forms the foundation for practical optimization of CO_2 compression, which should be performed in consideration of other processes involved in the CCS chain, such as the CO_2 capture and transport.

12.5.2.2.1.2 Nonisothermal Steady-State Flow Modeling   To evaluate the impact of CO_2 impurities on pressure drop in pipelines, a computer model has been developed that can calculate one-dimensional, transient, compressible, multiphase flows in pipes. This model accounts for both flow and phase-dependent viscous friction, and heat transfer between the transported fluid and the pipeline environment [25–27]. The model has been applied to study the impact of variation in concentration of CO_2 stream impurities at the inlets of a hypothetical pipeline network, upon the pressure and temperature profiles along the pipeline and the delivery composition for given flow rates and temperatures of the feed streams [28].

Figure 12.69 shows an example of the considered realistic pipeline network, transporting CO_2 from Cottam and Drax power stations to the sequestration point at Morecambe South on the East Irish Sea.

For a pipeline network configuration shown in Figure 12.69, the analysis of the steady-state pressure drop and temperature profiles has been performed for CO_2 mixtures carrying various impurities, including water, argon, nitrogen, and oxygen, which are typically present in the oxy-fuel combustion CO_2 stream.

The model developed can be applied to perform sensitivity studies to identify impurities having the most adverse impact on the CO_2 pipeline transport.

12.5.2.2.2 WP2.2: Near-Field Dispersion

In this work package, a computational fluid dynamic (CFD) model capable of predicting the near-field structure of high pressure releases of supercritical, dense phase, and gaseous CO_2 has been developed. The model is capable of handling CO_2 that contains impurities typical of those to be encountered in an integrated CCS chain, due to it incorporating an EoS that covers CO_2 with impurities and models for the formation of liquid droplets and solid particles. CO_2 release experiments were performed in order to support this modeling effort. These included controlled small and medium-scale experiments involving high pressure releases of CO_2 with a range of impurities, with near-field measurements of the dispersing jets and pipe-surface temperatures in the vicinity of punctures representative of typical geometries. Another set of experiments were conducted that comprised controlled large-scale experiments involving high pressure releases of CO_2 with a range of impurities, with near-field measurements of the dispersing jets, and temperature measurements in the vicinity of a predesigned crack geometry. The developed CFD model was then validated against experimental generated as part of the project described above in addition to data available in the literature. The usefulness of the CFD model developed was then demonstrated by interfacing its predictions for a number of realistic release scenarios with existing far-field dispersion models in order to predict hazards at large distances for use in risk assessments.

12.5.2.2.2.1 Development of Near-Field Dispersion Model   Following the puncture or rupture of a CO_2 pipeline, a gas–liquid droplet mixture or gas alone will be released to atmosphere and disperse over large distances. This may be then followed by gas–solid discharge during the latter stages of pipeline depressurization due to the significant degree of cooling taking place. The possibility of releases of three-phase mixtures also exists [11]. This subsection focuses on the detailed mathematical modeling of the near-field characteristics of these complex releases, since predictions of major hazards used in risk assessments are based on the use of near-field source terms that provide input to far-field dispersion models.

Because the pre- to postexpansion pressure ratio resulting from a release will initially be large, the sonic velocity will be reached at the outlet of the pipeline and the resulting free jet will be sonic. This pressure difference leads to a complex shock cell structure within the jet which for the initial highly underexpanded flow will give rise to a flow that contains a Mach disk followed by a series of shock diamonds as it gradually adjusts to ambient conditions. Subsequently, consideration has to be made to the complex physics representing the effects of compressibility upon turbulence generation and destruction. Also, the behavior of a multiphase nonideal system has to be represented by the incorporation of a nonideal EoS to represent mass-transfer between phases at a range of temperatures and pressures.

12.5.2.2.2.1.1 Turbulent Flow Modeling   Descriptions of the numerical approaches to the solution of the fluid dynamics equations applied are presented elsewhere [29], and not repeated here. A modified two-equation turbulence model to represent a compressible system, has also been applied. With respect to two-equation turbulence modeling, however, it is now widely accepted that the main contributor to the structural compressibility effects is the pressure–strain term (Π_{ij}) appearing in the transported Reynolds stress equations [30]. Hence, the use of compressible dissipation models is now considered physically inaccurate, although their performance is good.

Ignoring the rapid part of the pressure–strain correlation, Rotta [31] models the term as Equation 12.44 where ε is the dissipation of turbulence kinetic energy, and b_{ij} is the Reynolds stress anisotropy, as defined below:

$$\Pi_{ij} = -C_1 \varepsilon b_{ij} \tag{12.44}$$

Later, this model was extended [32] to directly incorporate terms arising from compressibility effects as Equation 12.45:

$$\Pi_{ij} = -C_1 \left(1 - \beta M_t^2\right) \varepsilon b_{ij} \tag{12.45}$$

where βM_t is a function of the turbulent Mach number, vanishing in an incompressible flow.

Prior to this development, Jones and Musonge [33] provided a model to account for the "rapid" element of the correlation. Defining a function for the fourth-rank linear tensor in the strain-containing term with the necessary symmetry properties, they obtain,

$$\Pi_{ij} = -C_1 \varepsilon \left(\frac{\overline{\rho u_i'' u_j''}}{k} - \frac{2}{3} \delta_{ij} \overline{\rho} \right) + C_2 \delta_{ij} \overline{\rho u_i'' u_j''} \frac{\partial \tilde{u}_k}{\partial x_l} - C_3 P_{ij} + C_4 \overline{\rho} k \left(\frac{\partial \tilde{u}_i}{\partial x_j} + \frac{\partial \tilde{u}_j}{x_i} \right) + C_5 \overline{\rho u_i'' u_i''} \frac{\partial \mu_l}{\partial x_l}$$

$$+ C_6 \left(\overline{\rho u_k'' u_i''} \frac{\partial}{\partial x_k} (\tilde{u}_k) + \overline{\rho u_k'' u_i''} \frac{\partial}{\partial x_j} (\tilde{u}_k) \right) + C_7 \overline{\rho} k \delta_{ij} \frac{\partial \tilde{u}_l}{\partial x_l} \tag{12.46}$$

where C_1 is the term corresponding to the "slow" part as previously defined as Equation 12.44, is the turbulence kinetic energy, $\overline{\rho}$ is the mean density, δ_{ij} is the Kroncker delta, and $u''u''\tilde{u}$ are the Favre-averaged Reynolds stresses and velocity components, respectively.

Defining the gradient and turbulent Mach numbers as

$$M_g \equiv \frac{Sl}{a} \text{ and } M_t \equiv \frac{\sqrt{2k}}{a} \tag{12.47}$$

Gomez and Girimaji [34] introduce corrections to their derivation of Π_{ij} as Equation 12.45, which is implemented in terms of a modification to Equation 12.46 in the current work:

$$\Pi_{ij} = -C_1 (M_t) b_{ij} + \sum_k C_k (M_g) T_{ij}^k \tag{12.48}$$

The turbulent Mach number is the ratio of the magnitude of the velocity fluctuations to the speed of sound, and the dependence of the "slow" part reflects the degree of influence of dilatational fluctuations. The gradient Mach number characterizes the shear to acoustic time scales and its influence upon the "rapid" part corresponds to the fluctuating pressure field which arises due to the presence of the mean velocity gradient.

Figure 12.70 depicts the application of these turbulence closures to the prediction of centerline axial velocity, normalized by the magnitude at the nozzle exit, for the highly underexpanded air jet studied by Donaldson and Snedeker [35]. Results obtained using the k-ε model and its associated correction attributed to Sarkar, Erlebacher [36] can be seen to conform with observations previously made with respect to the moderately underexpanded air jet. In this case, the initial shock structure is poorly defined by the standard model, and the overpredicted dissipation of these phenomena is considerable by 10 nozzle diameters from the jet outflow. The Sarkar modification to the turbulence dissipation goes some way to reducing the overprediction up to approximately 15 diameters, but the resolution of the initial shock-laden region remains poor, and the solution subsequently becomes overly dissipative.

The Reynolds stress transport model with the closure of the pressure–strain correlation attributed to Rotta [31] notably improves upon the resolution of the shock region and the prediction of the dissipation of turbulence kinetic energy. The introduction of a compressible element to the "slow" part of the model as discussed by Khlifi and Lili [32] effects an additional increase in peak magnitude predictions in the near field, although has little effect upon the subsequent downstream turbulence dissipation. The application of a model for the "rapid" part of the pressure–strain term [33], incorporated with the simple model of Rotta for the "slow" part proves a significant improvement with respect to predictions of both the shock resolution and the turbulence dissipation. This is again improved upon by the introduction of corrections based upon the turbulent and gradient Mach numbers as outlined by Gomez and Girimaji [34].

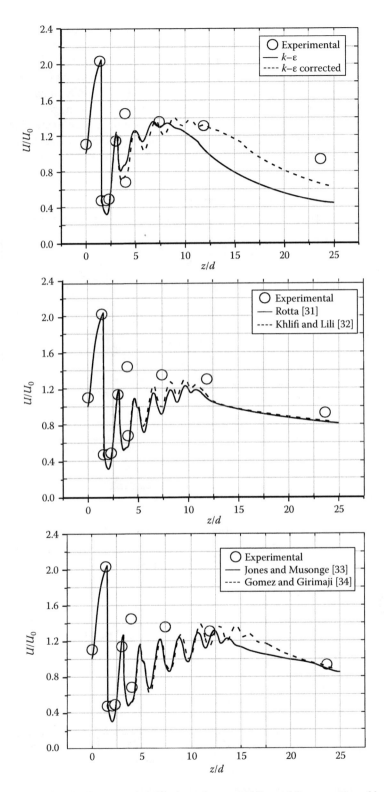

FIGURE 12.70 Normalized velocity predictions (top—$k - \varepsilon$, middle, and bottom—Reynolds stress) plotted against experiment.

12.5.2.2.2.1.2 Thermodynamics modeling   The Peng and Robinson [36] EoS is satisfactory for modeling the gas phase, but when compared to that of Span and Wagner [37], it is not so for the condensed phase, as demonstrated by Wareing et al. [38]. Furthermore, it is not accurate for the gas pressure below the triple point and, in common with any single equation, it does not account for the discontinuity in properties at the triple point. In particular, there is no latent heat of fusion. A number of composite EoS have therefore been constructed by the authors for the purpose of comparing performance in application to practical problems of engineering interest. In these, the gas phase is computed from the Peng and Robinson equation of state, and the liquid phase and saturation pressure are calculated from tabulated data generated with the Span and Wagner EoS, or from an advanced model based upon SAFT [14]. Solid

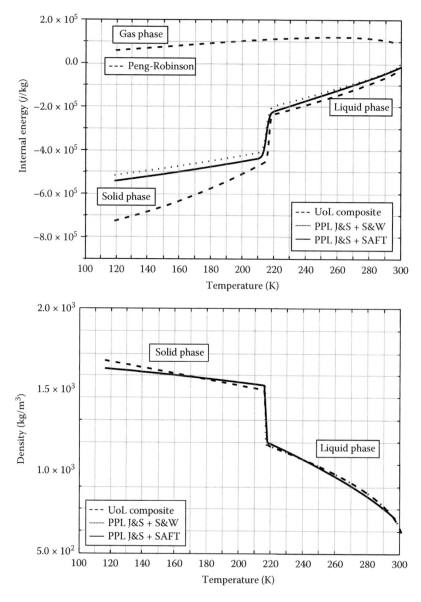

FIGURE 12.71   Saturation line internal energy and density predictions of pure CO_2 using three different equations of state compositions.

phase properties are obtained from an available source of thermodynamic data for CO_2, the Design Institute for Physical Properties (DIPPR) 801 database (www.aiche.org/dippr), or from a recently developed model attributed to Jäger and Span [39].

Under the remit of the CO_2QUEST FP7 project, the development of a Physical Properties Library (PPL), and hence the provision of a "platform" capable of predicting physical and thermodynamic properties of pure CO_2 and its mixtures, has been undertaken. The PPL contains a collection of models which can be applied regardless of the application, and include a number of cubic formulations such as those described by Peng and Robinson, and Soave, and the formulation of Span and Wagner. Also available is the analytical equation of Yokozeki [40], and the advanced molecular models described as SAFT and PC-SAFT by Gross and Sadowski [16]. A comparative study of the performance of these equations of state has been undertaken using experimental data of high-pressure CO_2 releases for validation. Due to space restrictions, a representative sample of these results is presented which include the composite model described above, and the PPL-derived Jäger and Span [39] model coupled with both Span and Wagner [37] and SAFT. Figure 12.71 depicts density and internal energy predictions for both the condensed and the gaseous phases on the saturation line for a pure CO_2 system obtained using these three different approaches. Importantly, it can be seen that all models incorporate the latent heat of fusion which must be considered over the liquid–solid phase boundary. They can also be said to similarly represent the internal energy of the liquid phase, although a discrepancy is observed with the composite model. This can be attributed to the inclusion in that model of a small value to ensure conformity with Span and Wagner [37] in terms of the predicted difference between gas and liquid energies [38]. The major discrepancy in predictions lies in the solid-phase region, in that the composite model is in notable disagreement with the models incorporating Jäger and Span [39]. This can be attributed to the sources of experimental data used in the derivation of the respective models, and the reader is referred to these papers for further information regarding the data sources.

12.5.2.2.2.1.3 Validation against Experimental CO_2 Releases Figure 12.72 presents sample density and temperature predictions of one of the large-scale test cases used for the validation of the code, obtained using the second-moment turbulence closure [34] and three of the composite equations of state. Undertaken by Institut national de l'environment industriel et des risques (INERIS), the experimental parameters were a reservoir pressure of 83 bar, an exit nozzle diameter of 12.0×10^{-3} m, and an observed mass flow rate of 7.7 kg s^{-1}. The predictions are of the very near field of the jet, encompassing the nozzle exit, and in the case of the temperature, extending to 0.5 m downstream. This equates to a distance of approximately 42 nozzle diameters (d), and the Mach disc can clearly be seen as a step change in temperature just before 0.1 m. Temperature predictions are in good quantitative agreement with experimental data within this region, although it is difficult to assess how the model performs qualitatively due to the resolution of the temperature measurements available. The difference in predicted solid-phase properties can be seen to influence the temperature predictions most notably in the region 0.02 m to 0.1 m, where the composite model predicts a warmer jet. Not unexpectedly, this region coincides with the system passing through the triple point at 216.5 K, being notable by the small step-change in the temperature curve of Figure 12.72, and the subsequent freezing of the liquid CO_2.

Density predictions derived from the three models appear to be in good agreement within the very near-field, although closer scrutiny of the predictions reveals the effect of the different solid-phase models. Contrary to intuition, the composite equation predicts a slightly higher density in the first 0.03 m, but conversely predicts a slightly higher temperature in the region bounded by the triple point and the stationary shock. It is considered that this observation is due to two factors. Initially, the predicted density of the liquid release obtained from this model is greater due to the observed differences in the liquid-phase predictions (Figure 12.71). Subsequently, as the system passes through the triple point, the composite model predicts slightly higher temperatures for the solid-phase, which relates to a small correction in the density, bringing them in to line with the PPL-derived models.

Further downstream, and with reference to Figure 12.73, predictions obtained using the Reynolds stress model of Gomez and Girimaji [34] and the composite equation of state, are conforming with

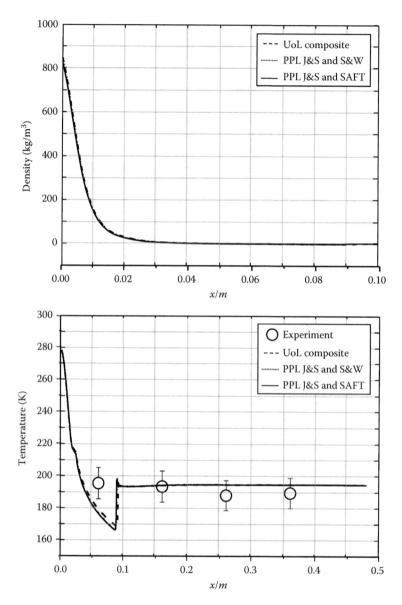

FIGURE 12.72 Mean density and temperature predictions of a large scale, sonic release of CO_2 (test 11) plotted against experimental data. Experimental data are from location $y = 0$ m.

experimental data both qualitatively and quantitatively, although a slight overprediction of temperature can be seen across the width of the jet and also with downstream progression. This is indicative of a marginally over-predicted rate of mixing, in line with previous observations made of the turbulence model validations. It can be said that the second-moment model outperforms the $k - \varepsilon$ approach when coupled with the composite equation of state, in that the rate of mixing is better represented. This is manifest as a notable underprediction of temperatures along the jet axis and also across its width, where the shear layer of the jet is observable as a temperature change of steep gradient, not observed in the Reynolds stress predictions.

Presented are sample results from a number of turbulence and thermodynamics models which have been developed to accurately predict the flow structure and phase behavior of accidental releases of

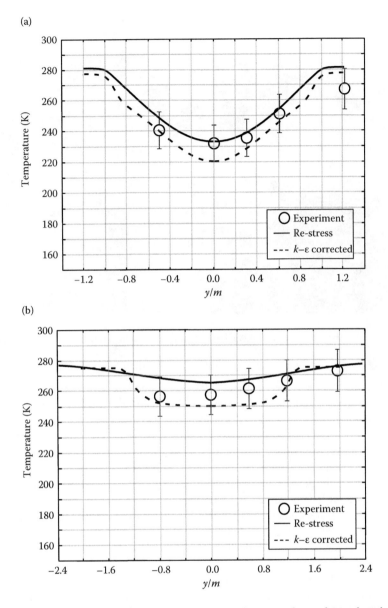

(a)

(b)

FIGURE 12.73 Mean temperature predictions within a large scale, sonic release of CO_2 plotted against experimental data. (a) Test 11, $x = 5$ m, $d = 417$ and (b) test 11, $x = 10$ m, $d = 833$.

high-pressure CO_2 for engineering applications. An excellent level of agreement between modeling approaches and experimental data has been observed.

12.5.2.2.2 Medium Scale CO_2 Release Experimental Investigation Medium-scale experiments involving high-pressure releases of CO_2 containing a range of impurities have been conducted at INERIS, France. An experimental rig and techniques developed during the CO_2PipeHaz project [41,42] were used again in order to establish a better understanding of the influence of the presence of impurities on both the flow inside the pipe and on the external dispersion. Measurements in the near-field and in the pipeline outflow allow the further development of databases and will be used for the validation of mathematical models.

FIGURE 12.74   Left: View of the pipe resting on the weighing masts (blue); right: near field equipment.

The rig and equipment developed in the CO_2PipeHaz project is shown in Figure 12.74 (Left). This includes a 40 m long, 50 mm internal diameter pipeline, with pressure and temperature transducers placed along the tube at intervals of 10 m. The near-field is also instrumented with pressure transducers and thermocouples. Equipment has also been especially developed to be adaptable to the study of the "flash zone" and is shown in Figure 12.74 (Right).

A transparent section is also installed in the center of the pipe, and the inside flow is visualized during the release with a high speed camera. The pipe is filled via a pump, and the transparent section also permits a check upon the level of liquid CO_2 in the pipe.

A particular focus is made upon the mass flow rate measurement. To this end, electronic measuring equipment records data at six weighing points. A weighing mast is visible in Figure 12.74 (Left). Two tests have been run with 100% of CO_2 liquid. The experimental conditions are presented in the next table (Table 12.30):

The evolution of the mass versus time at each measurement point is presented in Figure 12.75 for the two tests. Before the release the total measured mass is not uniform over all measuring points and ranges from 45 to 90 kg. These disparities can be explained by the presence of different equipment on the pipe (measurement, filling, mixing, transparent section).

Figure 12.75 also presents the evolution of the mass released versus the time. The first period (a few hundred milliseconds after the start of the release) corresponds to the short time during which the fluid inside the pipe is nucleating but appears homogenous. The next period (21 s for test 1 and 5 s for test 2) corresponds to the period where a defined two-phase mixture is present in the pipe and the level

TABLE 12.30   Experimental Conditions

Test n	1	2
Pressure (bar)	55	65
Ambient temperature (°C)	18	25
Orifice diameter (mm)	6	12

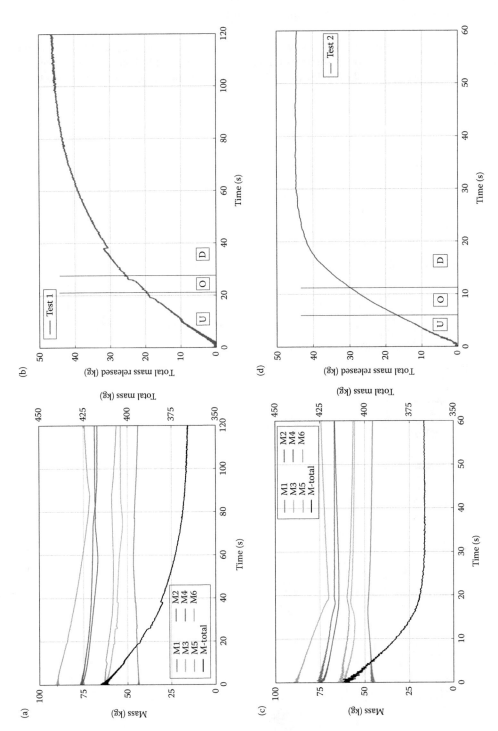

FIGURE 12.75 (a) and (b) Evolution of total inventory mass at each measuring point and of the total mass released respectively for test 1. (c) and (d) Evolution of total inventory mass at each measuring point and of the total mass released respectively for test 2.

of liquid is above the orifice (point U in Figure 12.75). Subsequently, the level of liquid reaches the orifice (O). After 26 s for test 1 and 11 s for test 2, the level of liquid is below the orifice (D). The two-phase flow ends after less than 2 minutes for test 1 and approximately 25 s for test 2 when the fluid becomes vapor.

The mass flow-rate measurement is crucial to the validation of the models [11], and its evolution is calculated by a temporal derivation of the mass released and is presented in Figure 12.76 for the two-phase flow. The different phases of the releases are identified on the figure depending upon the liquid level in the pipe.

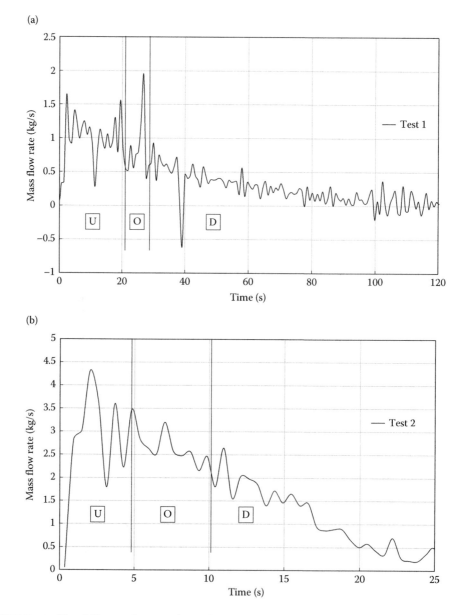

FIGURE 12.76 (a) and (b) Mass flow rates from release experiments 1 and 2 respectively. U = liquid level is above the orifice, O = liquid level with the orifice, D = liquid level below the orifice.

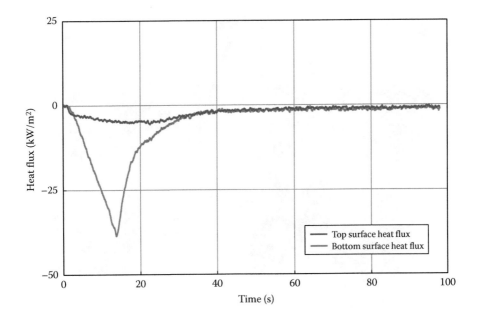

FIGURE 12.77   Heat fluxes measured on the top and on the bottom of the pipe.

Another issue for modelers is the assumption of adiabaticity. INERIS has added heat-flux meters to the skin of the pipe in order to measure the heat exchanges between it and the surroundings. An example of the evolution of the heat fluxes are presented for test 2 in Figure 12.77.

The blue line corresponds to the probe placed on the top of the pipe (gas), and the second probe is located on the bottom (red) corresponding to the liquid.

In addition, insulation has been added to the pipe in order to facilitate the interpretation of the experimental results and assist modelers.

12.5.2.2.2.3 Industrial Scale Pipeline for CO_2 Release Experiments   Three pure CO_2 release experiments, including one full bore rupture release, have been undertaken using the relocated industrial-scale pipeline in Dalian, China. Impure CO_2 release experiments were also performed in October 2014. The fully instrumented pipeline is 256 m long with a 233 mm inner diameter, and CO_2 temperature measurements were obtained at numerous points in the near field. Photos of the facility and a release experiment are shown in Figures 12.78 through 12.84.

The pipeline for industrial scale CO_2 release experiments has a unique design and opening device for controlling CO_2 releases at 80 bar. Full releases are controlled with two disk sections with CO_2 or N_2 present between the gaskets. A data acquisition system (pressure and temperature) is installed in the pipeline and operated using LabVIEW with the ability to make measurements of CO_2 and the dispersion area such as the temperature in the near field from the temperature measurement points shown in Figure 12.84. Over 20 large-scale experimental CO_2 releases including impurities have been conducted using the pipeline providing validation for CFDs modeling studies. Experimental measurements from the pipeline including orifice temperature, pressure drop during releases are shown in Figure 12.80, pipeline temperature at different locations are shown in Figures 12.81 through 12.83. Near-field CO_2 concentration and decompression behavior has been investigated using different orifices. The experimental pipeline aims to assist in establishing design standards for CO_2 transportation.

12.5.2.2.2.4 Ductile and Brittle Pipeline Fractures   In the CO_2QUEST project, ductile fracture propagation behavior in gas and dense phase CO_2 pipelines made from proposed candidate steels transporting

FIGURE 12.78 CO_2 release experimental facility in Dalian, China.

FIGURE 12.79   Full-bore rupture release stills with an inventory of 2.43 tons pure CO_2 under the following conditions: ambient temperature 36°C, pipeline pressure 52.5 bar, humidity 59%, and CO_2 temperature in the pipeline of 20°C–30°C.

typical stream impurities has been investigated by developing a computationally efficient CFD based fluid/structure interaction model. Sensitivity analyses using the developed fracture model were conducted to identify the type and composition of stream impurities that have the most pronounced impact on ductile facture propagation behavior in CO_2 pipelines. Shock tube experiments using the fully instrumented CO_2 pipeline test facility constructed in China were performed validate the fracture model developed in WP2 in addition to pipeline puncture and rupture experiments to validate the outflow model.

A fluid/structure interaction model for brittle fracture propagation in CO_2 pipelines made from the proposed candidate steels transporting typical stream impurities has also been developed. The fracture toughness and crack propagation data of the steel grades have been determined experimentally. The type of impurities and operating conditions that have the most adverse impact on a pipeline's resistance

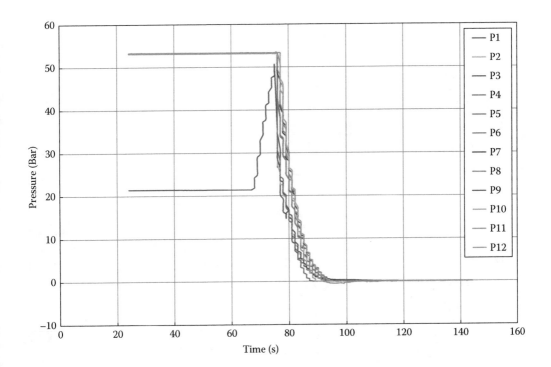

FIGURE 12.80 Pressure drop within the pipeline during the full-bore rupture release. P1 (release end) to P12 (closed end).

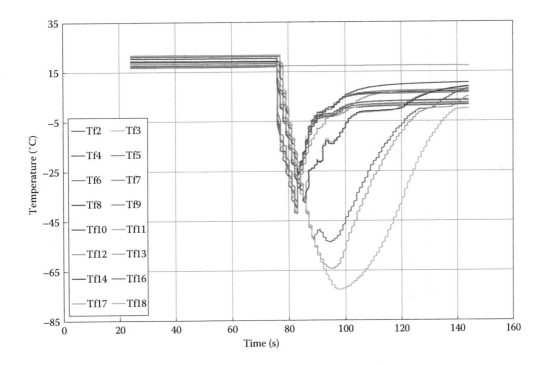

FIGURE 12.81 Temperature change along the pipeline during the full-bore release.

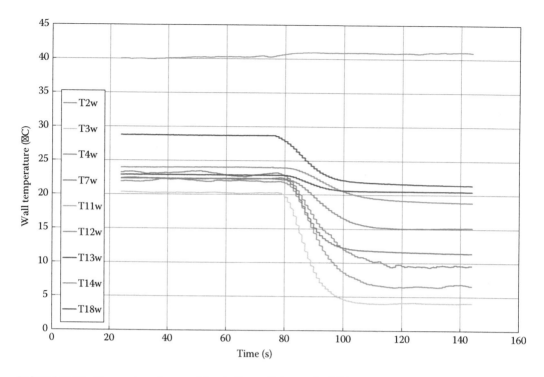

FIGURE 12.82 Temperature change of the pipeline wall during the full-bore release.

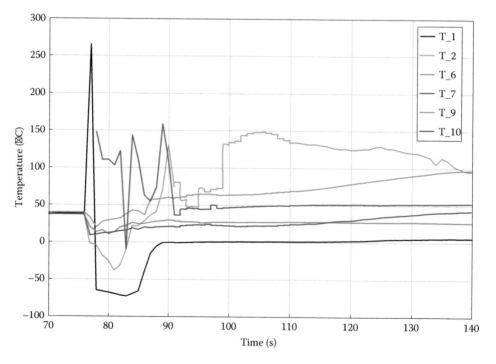

FIGURE 12.83 Temperature change release in the near field during the full-bore release.

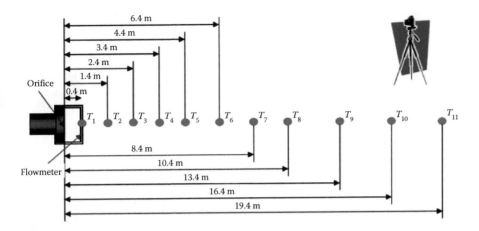

FIGURE 12.84 Temperature measurement points in the near field.

to withstanding long-running fractures for various pipeline steel types have been identified. Prolonged release experiments using the fully instrumented pipeline test facility to validate the model have also been conducted.

Three different steel material grades have been characterized in the CO_2QUEST project, namely X65, X70, and X80. The first two grades are already used in existing CO_2 pipelines, while the third grade (X80) is not yet in service. Investigating this material may provide some insights on its usability for CO_2 pipelines. Important aspects for material selection are

- The response to impurities such as H_2S which may cause sulfide stress cracking.
- Corrosion behavior: The CO_2 stream is likely to contain impurities affecting the corrosion of pipelines and thus pipeline integrity. Samples of the selected steel grades have been prepared and are being exposed in a lab simulator of another project partner, to a CO_2 gas stream containing typical impurities to investigate the corrosion behavior experimentally.
- Low temperature behavior: A leak in a buried CO_2 pipeline can lead to substantial cooling of the surrounding soil lowering the pipe temperature.

To narrow down the material choice and select the most suitable for more detailed investigations, Charpy and Battelle drop weight tear tests were conducted on the three steel grades. The Charpy test results are given in Figure 12.85 which shows that the X70HIC material had the highest Charpy toughness, and if one takes into account the higher thickness of the test X70 material, it has a similar behavior in Battelle testing. Since it is also relatively resistant to (accidental) H_2S exposure, the X70HIC was selected for this project as the preferred choice for more in-depth characterization and validation tests.

The actual behavior of the steel during a CO_2 pipeline failure is not likely to be predictable solely by Charpy and Battelle tests since it also depends on the gas decompression behavior leading to specific fracture conditions for the material (i.e. temperature, stress concentrations, high deformation speeds). To tackle this in a universal approach, activities have concentrated upon

- A numerical strategy to couple the hybrid ductile–brittle damage model used by Onderzoekscentrum voor de Aanwending van Staal (OCAS) to the CFD model of University College London describing the gas decompression [43].
- Improvement of the brittle fracture criterion [44].
- Fine tuning of the parameters of the damage model for the operating conditions typical for CO_2 pipeline failures.

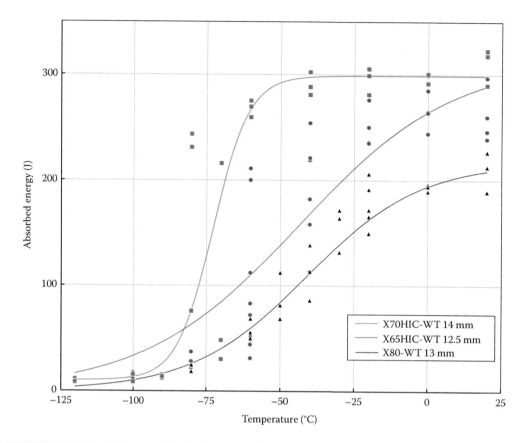

FIGURE 12.85 Transition curves for the three materials using Charpy samples.

The damage model was not optimally calibrated for high deformation speeds. Therefore, so-called Hopkinson bar experiments were carried out to quantify the material response at strain rates above 500 s⁻¹, which can be observed during crack propagation. Figure 12.86 compares the fine tune correction function (red solid line) to the old correction function (black solid line).

The brittle fracture criterion has been implemented in the Abaqus damage model. The coupling of this model with the CFD model will be tested and optimized if necessary. Next to this the material and damage model predictions have been validated using medium scale crack propagation experiments and small and medium scale CO_2 release experiments by other partners on components containing an artificial defect.

12.5.2.3 WP3: CO_2 Storage Reservoir Performance

In this work package, an understanding of the effects of impurities on the performance of the geological storage operation, in terms of fluid/rock interactions and leakage of trace elements has been developed by means of

1. A unique field injection test of water and of supercritical CO_2 (with and without impurities, conducted at the experimental site of Heletz, Israel).
2. Laboratory experiments aimed at determining the impact of the impurities on the mechanical properties of the reservoir and the caprock.

The focus is to use this data to support extensive model development and model application to enhance the understanding of CO_2 geological storage performance in the presence of impurities. The investigation

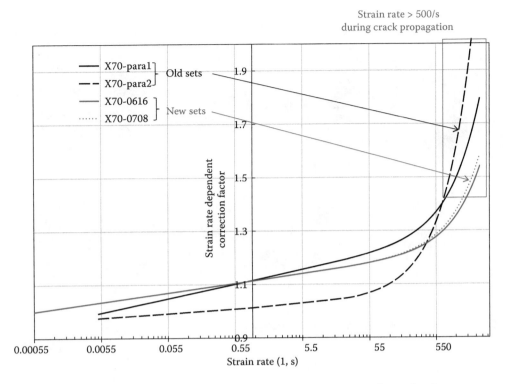

FIGURE 12.86 Strain rate dependency in fracture model before and after new Hopkinson bar data.

of a potential CO_2 leakage will involve the injection of industrial grade CO_2 into a shallow water aquifer in France, followed by the monitoring of the injected fluid, its trace impurities, and the potential mobilization of metallic trace elements from the rock.

12.5.2.3.1 Subsurface Injection Tests of Impure CO_2

In this section, laboratory and field scale investigation of the impact of impurities in the CO_2 stream on subsurface rock properties and CO_2 spreading and trapping behavior and the subsequent effects on storage performance are described. These experiments have provided unique data sets of field data for model validation. The impact on freshwater aquifers due to possible leakage of trace elements from the injected CO_2 stream has also been investigated and appropriate leakage monitoring methods for such overlying aquifers and at the ground surface have been recommended.

The work by Environmental and Water Resource Engineering (EWRE) at the Heletz site, Israel (shown in Figure 12.87), has focused on the configuration of the field experiment aimed at investigating the impact of impurities upon the storage of CO_2. First, major likely impurities emitted with CO_2 from the capture systems were reviewed, and it was suggested to use one with a potential geochemical impact (SO_2) and one with potential physical impact (N_2). Simulations of the behavior of these impurities upon the reservoir were conducted in order to determine suitable mass fractions that induce a detectable change in the reservoir response. Once the impurities and the mass fractions were known, preparations of the experiment in the field were conducted with the actual procedure for the injection of the impurities with the CO_2 determined. Three options were reviewed: (1) purchasing a prepared mixture of CO_2, N_2, and SO_2, that could be injected directly; (2) conducting the mixing at the wellhead of the injection well; and (3) conducting the mixing at a depth of approximately 1000 m, as the injection well has an independent connection from the ground surface to this depth.

FIGURE 12.87 Aerial photograph of the CO_2 injection site at Heletz, Israel.

The field experimental facility consists of two deep wells (1650 m) fully instrumented for monitoring downhole pressure and temperature, continuous temperature sensing via optical fiber, fluid sampling, fluid abstraction, and CO_2 injection. Design and procurement (or rental) of the equipment needed for the injection of the impurities have been undertaken. The structure of one of the wells is shown in Figure 12.88.

12.5.2.3.2 Subsurface Hydrological and Geochemical Modeling

In this section, the theoretical understanding of physical and chemical processes involving impure CO_2 in geological storage is developed and validated by embedding it within numerical models for simulating CO_2 spreading and trapping and the related coupled thermal–hydrological–chemical processes. In particular, through the analysis of laboratory data, in addition to theoretical considerations, the relevant processes and parametric models accounting for the effect of CO_2 impurities CO_2 stream have been further developed and improved upon. Through model application to the region around the injection site as well as on reservoir scale, an understanding of the relevant effects of impurities on storage performance has been obtained.

12.5.2.3.2.1 Geochemical Impact on Impure CO_2 Storage Reservoir Integrity
The presence of impurities within the CO_2 stream might influence physical or chemical properties of a reservoir. In order to enhance the understanding of these changes due to impure CO_2 injection, the critical geochemical impacts on a deep saline aquifer have been modeled using SO_2 as a typical flue gas impurity. Based on geological data of the Heletz saline aquifer situated at a depth of approximately 1600 m [45] a 2-D axially symmetric model was built comprising a total number of cells of 3700 [46]. The grid represents three Lower Cretaceous sandstone layers (thickness from bottom to top 11, 1.5, and 1.5 m, respectively) with two intercalated shale layers (each with 4 m thickness). Hence, the model covers 18 m in the vertical direction and provides a horizontal extent of 1000 m. Using the coupled thermal–hydrological–chemical program TOUGHREACT V3.0-OMP [47] with the fluid property module ECO2N [48], the SO_2 impurity is introduced as a trace gas within the CO_2 stream with a mole fraction of 97 to 3. This coinjection of

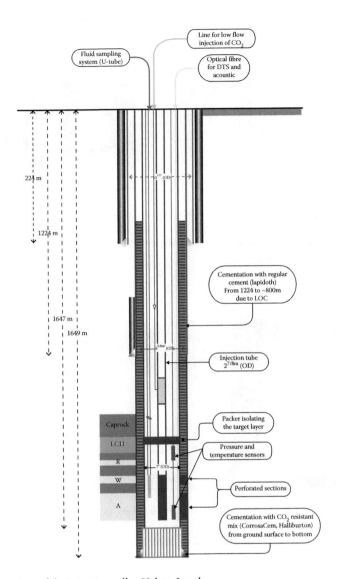

FIGURE 12.88 Structure of the injection well at Heletz, Israel.

CO_2 and SO_2 with a constant rate of 0.28 kg/s takes place at the lower left corner in Figures 12.89 through 12.91. The injection period lasted 100 h, representing the planned push–pull test at the Heletz pilot site.

In response to the injected fluid, the native brine is pushed away from the injection point and a distinct dry-out zone evolves, illustrated in high saturation values, i.e., the red area in the vicinity of the injection point in Figure 12.90. The simulations indicate a preference of flow in the horizontal direction, which is due to a ratio of seven between horizontal and vertical permeability. Both CO_2 and SO_2 injection lower the pH of the initial brine. At the end of the injection period, these changes are detected within horizontal distances of about 18 m in the lowest sandstone layer, whereas in the first overlying shale layer, only the first few meters are affected, see Figure 12.90.

The strongest impact of SO_2 is related to the fast reacting, pH sensitive carbonate mineral ankerite ($CaFe_{0.7}Mg_{0.3}(CO_3)_2$). The dissolution of this primary mineral resulting in a release of Ca^{2+} leads to the precipitation of anhydrite ($CaSO_4$). In addition with the precipitation of pyrite (FeS_2), these three mineral alteration processes change the spatial distribution of the sandstone porosity. The

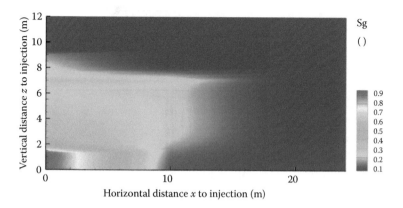

FIGURE 12.89 Gas saturation after 10 days of coinjecting about 100 t CO_2 and SO_2, development of a dry-out zone in the vicinity of the injection point.

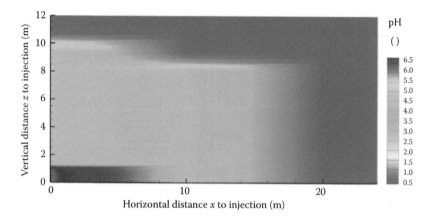

FIGURE 12.90 Distribution of pH value after 10 days coinjecting about 100 tCO_2 and SO_2.

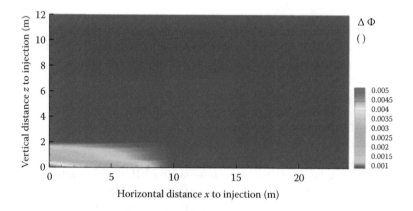

FIGURE 12.91 Changes in porosity after 10 days of coinjecting about 100 tCO_2 and SO_2.

quantitative modifications to the porosity assumed to be constant within each layer at the start of the simulation are depicted in Figure 12.91. For the case presented here, i.e. coinjection of about 100 t CO_2 and SO_2, after about 10 days there is a porosity decrease in the simulated Carboniferous sandstone reservoir of less than 0.005. This change is about 2.5 % of the initial value of 0.2, which seems relatively small, but considering inhomogeneities and potential reservoirs with lower porosities this effect might have a significant impact on injectivity, hence, on pressure management and storage efficiency. Results are highly dependent on initial mineral composition of the reservoir complex. The simulations endorse site-specific modeling based on thorough compilation of input parameter set, including chemical impacts of impurities within the CO_2 stream. Further details can be found in [46].

12.5.2.4 WP4: Technoeconomic Assessment of CCS Systems

In this package, a multiscale whole systems approach is developed that underpins the overall assessments in CO_2QUEST has been carried out.

12.5.2.4.1 Whole-Systems Modeling

Imperial College London has applied a multiscale modeling approach to the analysis of a CO_2 capture and transport system. Focus is upon the development of a series of scale-specific models which interact across a range of length and time scales. This approach is particularly relevant to the CO_2QUEST project as the hour-to-hour behavior of the CO_2 sources within a given decarbonized bubble will dictate the flow-rate and composition of the CO_2 in the transport network. This modeling approach is illustrated in Figure 12.92 below.

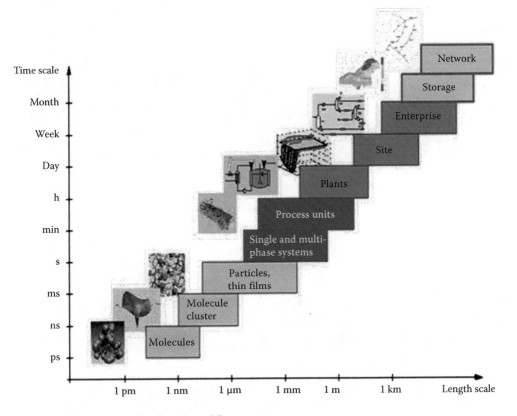

FIGURE 12.92 Illustration of multiscale modeling concept.

As our energy system evolves, incorporating ever greater quantities of intermittent renewable energy, so too will the roles of the various generators which contribute to this system. This is illustrated in Figure 12.93. As can be seen, as we move from the 2030s to the 2050s and beyond, the role of CCS plants within the energy system significantly changes, and the ability to operate in a flexible, load-following manner becomes more valued.

In this work, the theory of Grossmann and Sargent [49] has been applied to dynamic, nonequilibrium models of a decarbonized power plant [50–58] and evaluate three distinct options for flexible operation:

1. Solvent storage—a portion of solvent is stored during periods of peak electricity demand and regenerated during off-peak periods.
2. Exhaust gas venting—a portion of the exhaust gas is vented during periods of peak electricity demand.

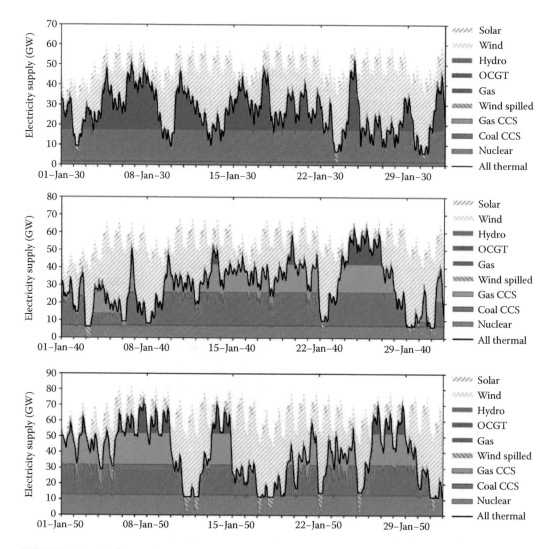

FIGURE 12.93 Evolution of operating patterns of the UK power generation system from 2030 to 2050—output from Multi-scale Energy Systems Modelling Encompassing Renewable, Intermittent, Stored Energy and Carbon Capture and Storage (MESMERISE) project using the ETRC model.

3. Time-varying solvent regeneration—CO_2 is allowed to accumulate in the working solvent during periods of peak electricity demand and the solvent is more thoroughly regenerated during off-peak periods.

In all cases, a load-following power plant with postcombustion CO_2 capture as a reference case is used. We evaluate each option for flexible operation based upon the integrated degree of capture (IDoC) and cumulative profit realized by the power plant over the course of the simulation. Economic considerations are based on both fuel (coal and gas) prices and a cost of CO_2 emission based on the carbon price floor proposed by the UK's Department of Energy and Climate Change. The time period considered is the early 2030s.

In this section, the results of our optimization problem are presented. For all scenarios, the optimization problem solved was the maximization of profit subject to the end point constraint of the IDoC being greater than or equal to 90%.

12.5.2.4.2 Load-Following

This was our benchmark scenario. Here, the capture plant was designed and the operating parameters specified via a steady state optimization as described in our previous paper [57]. We evaluated the effect that a multiperiod dynamic optimization on the end design of the plant. However, as the duration of the period for which the power plant is operating at full load is long relative to the period for which the power plant is operating at part load, the multiperiod design was essentially identical to the steady state design, and the solution period of the steady state design problem was an order of magnitude faster than that of the multiperiod problem. Thus, in this case, all operating parameters in the CO_2 capture plant, i.e., the L/G ratio, lean loading, φ_{lean}, solvent inlet temperature, T_{In}^{Solv} etc., were held constant throughout the simulation.

It can be observed from Figure 12.94 that, as the capacity factor (CF) of the power plant changes, so too does the instantaneous degree of capture (DoC), albeit by a very small amount. This is associated with the changing gas and liquid flow rates in the absorption column leading to a variation in the

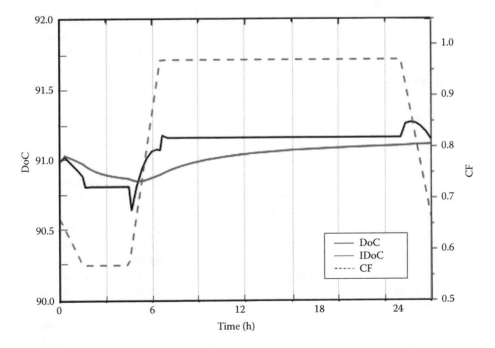

FIGURE 12.94 Degree of CO_2 capture varying with power plant CF.

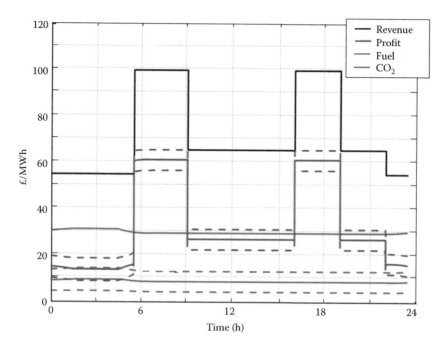

FIGURE 12.95 An analysis of the various financial streams associated with the generation of low-carbon electricity in the case of conventional load following.

effective surface area in the packed section. This implies that if the duration or frequency of periods of dynamic operation are long or great relative to the duration or frequency of periods of steady state operation, the IDoC could be reduced as a result.

As illustrated in Figure 12.95, an evaluation of the various financial streams associated with the power plant has been provided. It can be observed that the fuel cost per MWh is slightly increased during periods of operation at a low load factor in comparison with periods of operation at a high load factor. This is commensurate with the decreased efficiency of the power plant under this mode of operation. The cost associated with CO_2 emission is similarly elevated during this period, commensurate with the slightly reduced DoC as illustrated in Figure 12.87. Finally, as the plant is operating at an essentially constant DoC of approximately 90%, it can be observed that the CO_2 price exerts an important influence on the profitability of the plant.

12.5.2.4.3 Solvent Storage

In this scenario, in order to decouple the operation of the power and capture plants, we consider the option of storing a fraction of the rich solvent during periods of peak electricity prices and subsequent regeneration of the solvent during periods of off-peak electricity prices. Here, the extra parameter to be determined is the quantity of solvent stored or regenerated in each time period. Therefore, this is a multiperiod, piece-wise linear, dynamic optimization problem. As can be observed in Figure 12.96, when electricity prices are lowest, the flow rate of solvent to regeneration is greatest. Similarly, during periods of peak electricity price, rich solvent is directed to storage, bypassing the regeneration process.

The advantage of the solvent storage strategy is that it reduces the quantity of steam required for solvent regeneration during key periods of peak electricity demand, as illustrated in Figure 12.97. However, the availability of steam is a major constraint on the amount of solvent which can be stored during periods of peak electricity demand. It was a constraint in this problem that whatever solvent was stored during periods of peak electricity prices had to be regenerated during off-peak periods. This meant that

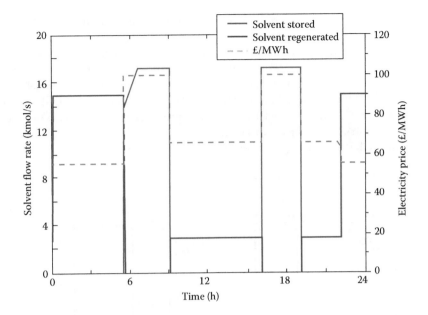

FIGURE 12.96 An analysis of the solvent storage and regeneration is presented here.

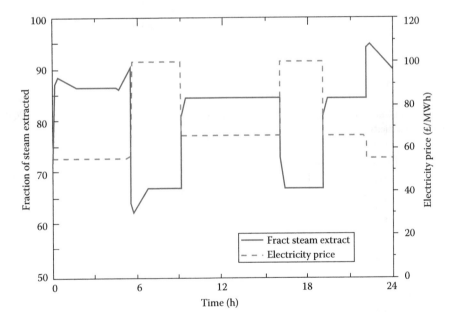

FIGURE 12.97 Steam requirement for solvent regeneration under the solvent storage option.

it was only possible to store approximately 15% of the total solvent flow. This corresponds to a cumulative volume of stored solvent of approximately 11,750 m^3 of solvent over both periods.

12.5.2.4.4 Exhaust Gas Venting

In this scenario, the power plant ramps up and down, but in order to decouple the operation of the power and capture plants, we consider the option of venting a fraction of the exhaust gas during periods of peak electricity prices. Here, the extra parameter to be determined is the quantity of exhaust gas to be vented in each time period. Therefore, this is a multiperiod, piece-wise linear dynamic optimization problem. The key result of this optimization problem is illustrated in Figure 12.98 below.

As can be observed from Figure 12.98, during periods of peak electricity prices, approximately 23% of the total exhaust gas was vented. In this scenario, we departed from the conventional procedure of capturing 90% of the CO_2 at all times. Rather, of the exhaust gas which was introduced into the absorption column, 95% of the CO_2 was captured. However, during the periods of exhaust gas venting, the DoC fell to approximately 74%, increasing the carbon intensity of the electricity generated to approximately 225 kg/MWh. This is illustrated in Figure 12.99. It is noted that the venting of this quantity of CO_2 had

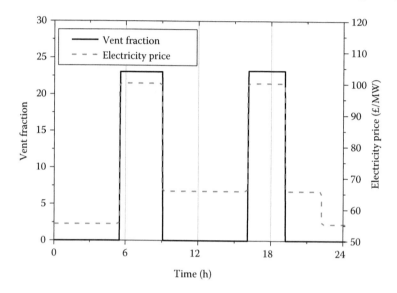

FIGURE 12.98 Exhaust gas venting. In this scenario, approximately 23% of the exhaust gas was vented during periods of peak electricity prices.

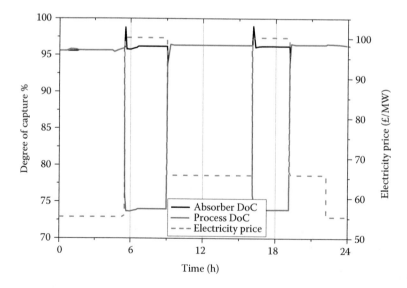

FIGURE 12.99 Variation in DoC with exhaust gas venting. It can be observed here that the absorber DoC remains approximately constant.

the primary consequence of imposing a significant cost penalty on the profitability of the plant and the secondary consequence that it was not possible to solve the optimization problem with the end point constraint of IDoC $\geq$ 90%. Here, this had to be loosened to 89%.

12.5.2.4.5 Time-Varying Solvent Regeneration

In this section, we evaluate the option of using the working solvent as means to provide flexibility to the power plant. This is achieved by allowing the CO_2 to accumulate in the working solvent during hours of peak electricity prices and regenerating the solvent more completely during off peak periods. This is not storing solvent separately to the capture plant; this is storing CO_2 in the solvent which is circulating within the capture plant. This means that the lean loading of the solvent is no longer a time-invariant process parameter. Rather, this control vector is now parameterized such that it is expressed as

$$\theta_{Lean}^k = \alpha^k t^2 + \beta^k t + \gamma^k, \tag{12.49}$$

where θ_{Lean}^k is the function describing the way in which θ_{Lean} varies across a given period k, α^k, β^k, and γ^k describe this function in each period k, and finally t is the time in each period k. Here, the variable t is the time in each period k and which is set to zero at the beginning of each new period. The only additional constraint we have imposed upon the optimization problem by this formulation is the quadratic nature of the parameterization. This could have equally been a cubic or higher order polynomial, however a quadratic polynomial was chosen in the interest of simplicity. Further, we did not constrain the values of the coefficients of this polynomial, i.e., the magnitude of the subsequent behavior was strictly a function of the process response to the time-varying electricity prices. The results of this optimization problem are presented in Figure 12.100.

As can be observed, and as opposed to operating at a constant lean loading (the continuous blue curve), the degree of solvent regeneration varies in sympathy with the prevailing electricity price, as might be expected for the solution of this kind of optimization problem. The solvent is deeply regenerated during periods of low electricity price, whereas CO_2 is allowed to accumulate in the working solvent volume during periods of high electricity price. This has the effect of allowing the plant to direct substantially less steam to solvent regeneration operations when the opportunity cost associated with doing so is high. Obviously, following this kind of operating strategy will have the effect of varying the carbon intensity of the electricity generated.

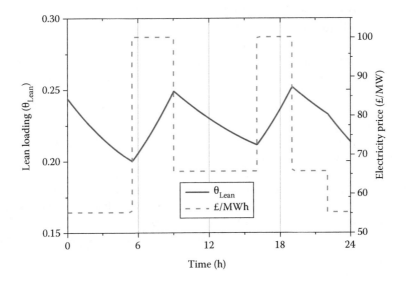

FIGURE 12.100 Solvent regeneration as a function of time and electricity price is shown here.

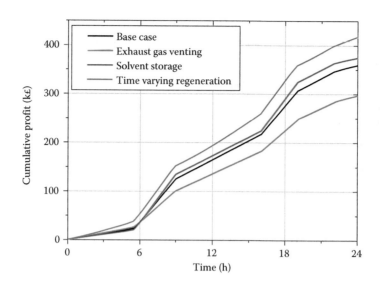

FIGURE 12.101   Cumulative profit available for each option.

Finally, it is instructive to consider the profitability of the power plant, which is illustrated in Figure 12.101. It is apparent that the solvent storage option offers marginally greater profitability (ca. £374k) when compared with the base case (ca. £359.6k), approximately a 4% improvement, whereas the exhaust gas venting option (ca. £298.6k) is approximately 17% less profitable than the base case. Finally, the variable solvent regeneration option (£417.5k) is 16% more profitable than the base case scenario.

12.5.2.5 WP5: Impurity Impacts and Risk Assessment

In this WP, the incremental risks across the CCS chain implied by the presence of impurities have been explored and defined. A decision-making risk assessment method covering both safety and impact on the environment with a particular focus on the role of impurities has been defined. This method has been applied to devise appropriate prevention and mitigation measures for selected risks.

12.5.2.5.1 Risk Analysis and Impact Profiles of Impurities

In a risk analysis, one must consider a number of risk scenarios, this applies also to CCS [59–61]. All scenarios follow a similar pattern, an initial cause creates a transfer in the environment (e.g. pressure transfer and transfer of a substance), that in turn exposes a target through an "impacting phenomenon." In the CO_2QUEST FP7 project, we consider three main targets: performance of storage, humans, and near-surface ecosystem.

After Farret et al. [62], eight categories of impacting phenomena are taken into account. An additional challenge in the CO_2QUEST project was to identify and assess the incremental risk due to the presence of impurities in a CO_2 stream. A first step toward this objective is to undertake an impact-profile risk-analysis for each part of the CCS chain and each common impurity.

An overview of the current knowledge regarding the key effects of impurities along the CCS chain has been carried out. The work undertaken is based on a range of impurities identified in Section 40.1.1 (WP1 of the CO_2QUEST project), and all potential effects were identified, as well as threshold values when they exist. In this respect, key information is given by [63,64].

For each impurity, its effects all along the CCS chain are identified via three categories of impacts:

1. The physical impacts
2. The chemical impacts
3. The toxicological and ecotoxicological impacts

12.5.2.5.2 Physical Impacts

Physical impacts of the impurities contained in captured CO_2 streams are a major concern for two main reasons. First, the impurities responsible for these impacts are very common in the CO_2 streams produced from the main capture technologies (precombustion, postcombustion, and oxy-fuel combustion). Second, the physical behavior of the CO_2 stream has macroscopic impacts all along the CCS chain. In fact, by reducing the overall efficiency of the CCS application or by improving its global cost, physical effects of impurities could have major consequences, which have to be assessed and forecast. Figure 12.102 depicts example impact mechanisms.

Following this detailed study, a limited number of adequate "indicators" were identified; finally, only one indicator was selected to represent the physical effects: the CO_2 stream density. The other possible indicators at stake have a minor effect (e.g. interfacial tension) or are redundant with this one (e.g. CO_2 stream viscosity).

12.5.2.5.3 Chemical Impacts

During transport and storage, a CO_2 stream experiences a number of different steps and physical conditions that can lead to significant chemical alterations. This alteration of CO_2 properties on a chemical level can instigate unwanted subsequent reactions. Figure 12.103 highlights some of these possibilities.

Finally, two indicators were selected to represent the chemical effects in an integrated method: acidity and corrosivity.

12.5.2.5.4 Toxical and Ecotoxical Impacts

Once injected into the storage environment, potential impacts of the CO_2 need to be considered on a long-term scale. Indeed, environmental impacts may still occur in the case of loss of containment. Such a loss can be a result of either an integrity default of the storage (fault through the caprock, existing or well) or too high caprock porosity. Concerning toxicity and environmental sensitivity, even low concentration of impurities may have significant impacts. Figure 12.104 graphically depicts some of the toxicity mechanisms.

Finally, two indicators were selected for this category of effects: one addresses effects for humans and focuses on acute effects (inhalation in case of an accidental release), while the second indicator addresses chronic (long-term) effect on underground water.

12.5.2.5.5 Unified Method

A second key step is to perform a global, integrated, assessment. We chose a semiquantitative approach, that is very common in risk analysis, and that is especially recommended when there is a need to compare a large range of phenomena, such as here.

In order to assess a given CO_2 stream, each impurity is assessed with its associated effects, with a score that ranges from 0 to 4. Score 4 is associated to a nonacceptable level (or nonacceptable effect), given by threshold values from regulation or from literature. The lower scores are then deduced, either with linear scales (e.g. CO_2 stream density) or logarithmic scales (e.g. toxic effects). Essentially, level 0 is dedicated to absence of effect or negligible effect, e.g., when the impurity is hardly detected.

The result for a whole CO_2 stream is the sum of the results for individual impurities. An illustration is given in Figure 12.105 for two different CO_2 streams, as identified in Section 40.1.1. The method displays the result for each individual indicator

The method allows to show the differences for individual indicators: in the example above, we can see that corrosivity is likely to be a greater concern with a postcombustion process than for precombustion. But acute toxic effects are to be managed with greater care in the case of precombustion (because of the possible presence of H_2S).

The method also compares CO_2 streams globally, through an integrated index that is the sum of all the scores for individual impurities. Such a sum is coherent with a semiquantitative assessment and with

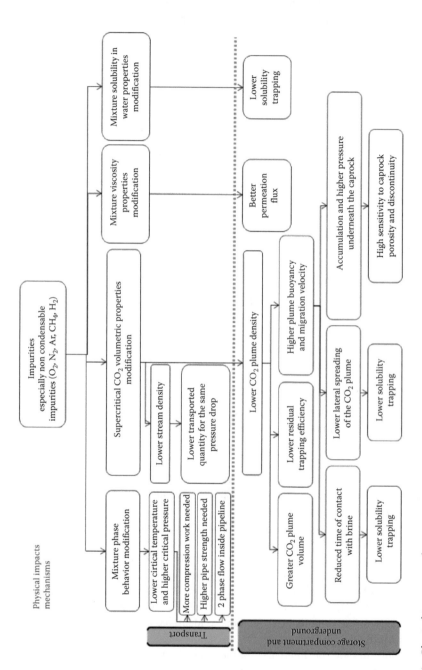

FIGURE 12.102 Physical impacts mechanism.

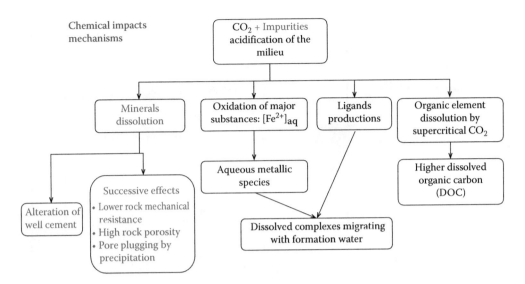

FIGURE 12.103 Chemical impacts mechanisms.

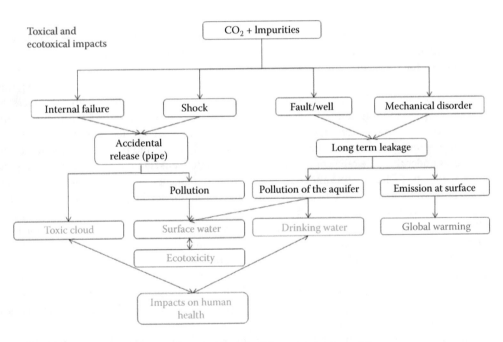

FIGURE 12.104 Toxical and ecotoxical mechanisms for CO_2 containing impurities.

the fact that very few data are available concerning the effect of mixtures of impurities. This index can in turn be combined with other social or economic criteria, e.g., in a cost-benefit analysis.

12.5.3 Summary and Conclusion

During the course of the CO_2QUEST project an extensive campaign of experimental investigations backed by mathematical modeling aimed at determining the optimum CO_2 stream composition based

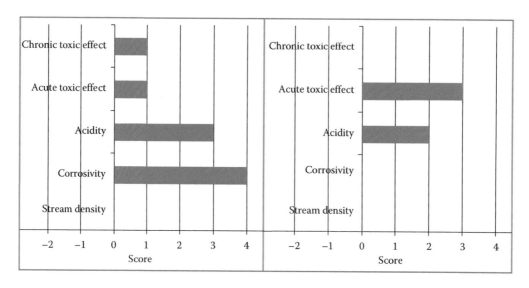

FIGURE 12.105 Display of scores for the five indicators for a CO_2 stream (left: a hypothetical postcombustion stream, global impact index = 9; right: a hypothetical high-purity CO_2 after precombustion, global impact index = 5).

on economical, technical, environmental, and safety considerations was performed. The project work was subdivided into a number of distinct and interlinking work packages (WPs).

WP1 provided the cornerstone to the project in two main ways. First, an extensive literature review was conducted to establish the likely range and level of impurities present in streams arising from power sector and industrial CO_2 capture technologies. This was followed by a cost-benefit analysis to assess the impact of the type of CO_2 capture technology on the overall process costs and the CO_2 product purity. Next, the development of dedicated and accurate computationally efficient Equations of State (EoS) and methods for providing the required phase equilibria, thermodynamic and transport properties of the CO_2 mixtures under the wide range of pressures and temperatures likely to be encountered during the various stages in the CCS chain was undertaken. The EoS development was supported by measurements of pertinent thermophysical properties of CO_2 mixtures, including tertiary and quaternary mixtures using unique experimental facilities.

WP2 focused on addressing the technical, economic, and safety challenges associated with the pipeline transportation of CO_2 mixtures during normal operation and accidental failure scenarios. In the case of the former, the impact of impurities on CO_2 compression duty was investigated using a thermodynamic model utilizing real-fluid EoS leading to the design of optimum multistage compression strategies for minimizing the power requirements. Additionally, a computational steady-state flow model was developed to simulate the mixing of CO_2 streams of various compositions in pipeline networks typical of those in CCS industrial clusters.

In order to provide an in-depth understanding of the types of phenomena that can occur during accidental releases of high-pressure CO_2, medium and large-scale pipeline rupture experiments incorporating extensive real-time monitoring instrumentation were conducted. In particular, these experiments for the first time elucidated the influence of the presence of impurities on the flow characteristics inside the pipe and hence the near-field and far-field dispersion behavior of the released CO_2 cloud. The data recorded using these experiments were in turn employed for the validation of the rigorous heterogeneous flow and dispersion models developed for predicting the impact of the various impurities on the discharge behavior and hence the minimum safe distances.

Investigations of three different steel materials for pipelines were conducted to establish their sensitivities to CO_2 impurities on their corrosion behavior and susceptibility to brittle and ductile fracture propagation. The fracture studies were complemented with the development of fully predictive fluid/ structure models coupling finite element analysis and the CFD flow model to predict the pipeline fracture toughness required in order to resist running fractures.

WP3 was devoted to developing understanding of the impacts of impurities in the subsurface during geological storage of CO_2. The results of CO_2 mixture injection experiments in close proximity to a shallow aquifer indicated no measureable cross-over contamination of the water table by the CO_2 impurities. The deep well CO_2 mixture injection experiments investigating fluid/rock interactions have been highly challenging and are an ongoing area of research.

The geochemical impact of impure CO_2 on storage reservoir integrity in a deep saline aquifer has also been studied through modeling. An axially symmetric 2-D model was constructed to simulate the relevant reactive transport models focusing on the impact of SO_2 on the brine–rock system in sandstone, showing the acidic and reactive properties of this impurity.

Work presented in the WP4 section has focused on flexible operation of power plants integrated with CCS, exhaust gas venting, and time varying solvent regeneration. Such studies can enable full-chain CCS technoeconomic analysis of CCS networks, Leading to optimized overall CAPEX and OPEX of full-chain systems.

WP5 analyzed the downstream impacts of CO_2 impurities on pipeline transport and geological storage by classifying impurity impacts into physical, chemical, and toxic, and by developing a point scoring mechanism to describe their severity.

In conclusion, the CO_2QUEST project has led to the development of rigorous validated theoretical modeling tools [12,27,28,43,56] complemented with the state-of-the-art experimentation [41,42] to determine the optimum level of CO_2 purification required in order to enable the design of safe and economic CCS technologies.

Acknowledgments

The research leading to the results contained in this chapter received funding from the European Commission 7th Framework Programme FP7-ENERGY-2012-1 under grant number 309102 coordinated by Prof. H. Mahgerefteh at UCL. The chapter reflects only the authors' views and the European Commission is not liable for any use that may be made of the information contained therein. A few text excerpts are taken from the CO_2QUEST newsletter Autumn 2014 [65].

References

1. J. Serpa, J. Morbee, E. Tzimas, *Technical and economic characteristics of a CO_2 transmission pipeline infrastructure*, Joint Research Centre Institute for Energy, Petten, 2011.
2. CO_2QUEST. Impact of the quality of CO_2 on storage and transport, CO2QUEST Project Website. www.co2quest.eu/ (Accessed July 01, 2015), 2013.
3. R. T. J. Porter, M. Fairweather, M. Pourkashanian, R. M. Woolley. The range and level of impurities in CO_2 streams from different carbon capture sources. *Int. J. Greenh. Gas Control* 36, 161–174, 2015.
4. J. Wang, D. Ryan, E. J. Anthony, L. Basava-Reddi, N. Wildgust. The effect of impurities in oxyfuel flue gas on CO_2 storage capacity. *Int. J. Greenh. Gas Control* 11, 158–162 2012.
5. B. Wetenhall, J. M. Race, M. J. Downie. The effect of CO_2 purity on the development of pipeline networks for carbon capture and storage schemes. *Int. J. Greenh. Gas Control* 30, 197–211, 2014.
6. I. S. Cole, P. Corrigan, S. Sim, N. Birbilis. Corrosion of pipelines used for CO_2 transport in CCS: Is it a real problem? *Int. J. Greenh. Gas Control* 5, 749–756, 2011.
7. S. Sim, I. S. Cole, F. Bocher, P. Corrigan, R. P. Gamage, N. Ukwattage, N. Birbilis. Investigating the effect of salt and acid impurities in supercritical CO_2 as relevant to the corrosion of carbon capture and storage pipelines. *Int. J. Greenh. Gas Control* 17, 534–541, 2013.

8. S. Brown, S. Martynov, H. Mahgerefteh, D. M. Tsangaris, G. C. Boulougouris, I. G. Economou, N. I. Diamantonis. Impact of equation of state on simulating CO_2 pipeline decompression. *Process Saf. Environ.*, 2015, submitted.

9. S. Brown, L. D. Peristeras, S. Martynov, R. T. J. Porter, H. Mahgerefteh, I. K. Nikolaidis, G. C. Boulougouris, D. M. Tsangaris, I. G. Economou. Thermodynamic interpolation for the simulation of two-phase flow of complex mixtures, *J. Comput. Phys.*, 2016, submitted.

10. R. M. Woolley, M. Fairweather, C. J. Wareing, C. Proust, J. Hebrard, D. Jamois, V. D. Narasimhamurthy, I. E. Storvik, T. Skjold, S. A. E. G. Falle, S. Brown, H. Mahgerefteh, S. Martynov, S. E. Gant, D. M. Tsangaris, I. G. Economou, G. C. Boulougouris, N. I. Diamantonis. An integrated, multi-scale modeling approach for the simulation of multiphase dispersion from accidental CO_2 pipeline releases in realistic terrain. *Int. J. Green. Gas Control* 27 (2014) 221–238.

11. S. Martynov, S. Brown, H. Mahgerefteh, V. Sundra, S. Chen, Y. Zhang. Modeling three-phase releases of carbon dioxide from high-pressure pipelines. *Process Saf. Environ.* 92, 36–46, 2013.

12. S. Martynov, W. Zheng, S. Brown, H. Mahgerefteh. Numerical simulation of CO_2 flows in pipes with phase transition across the triple point. 12th International Conference on Heat Transfer, Fluid Mechanics and Thermodynamics (HEFAT2016), 11–13 July 2016, Spain.

13. J. M. Prausnitz, R. N. Lichtenthaler, E. G. de Azevedo. *Molecular Thermodynamics of Fluid-Phase Equilibria*. 3rd ed., Prentice-Hall, Upper Saddle River, NJ, 1999.

14. M. Seiler, J. Groß, B. Bungert, G. Sadowski, W. Arlt. Modeling of solid/fluid phase equilibria in multicomponent systems at high pressure. *Chem. Eng. Technol.* 24, 607–612, 2001.

15. A. Jäger, R. Span. Equation of state for solid carbon dioxide based on the Gibbs free energy. *J. Chem. Eng. Data* 57, 590–597, 2012.

16. J. Gross, G. Sadowski. Perturbed-chain SAFT: An equation of state based on a perturbation theory for chain molecules. *Ind. Eng. Chem. Res.* 40, 1244–1260, 2001.

17. O. Fandiño, J. P. M. Trusler, D. Vega-Maza. Phase behavior of (CO_2 + H_2) and (CO_2 + N_2) at temperatures between (218.15 and 303.15) K at pressures up to 15 MPa. *Int. J. Greenh. Gas Control* 36, 78–92, 2015.

18. H. G. Donnelly, D. L. Katz. Phase Equilibria in the carbon dioxide–methane system. *Ind. Eng. Chem.* 46, 511–517, 1954.

19. J. A. Davis, N. Rodewald, F. Kurata. Solid-liquid-vapor phase behavior of the methane-carbon dioxide system. *AIChE J.* 8, 537–539, 1962.

20. N. I. Diamantonis, G. C. Boulougouris, E. Mansoor, D. M. Tsangaris, I. G. Economou. Evaluation of cubic, SAFT, and PC-SAFT equations of state for the vapor–liquid equilibrium modeling of CO_2 mixtures with other gases. *Ind. Eng. Chem. Res.* 52, 3933–3942, 2013.

21. J. J. Moore, T. Allison, A. Lerche, J. Pacheco, H. Delgado. Development of advanced centrifugal compressors and pumps for carbon capture and sequestration applications. The Fortieth Turbomachinery Symposium 2011, Houston, TX, 107–120.

22. A. Witkowski, M. Majkut. The impact of CO_2 compression systems on the compressor power required for a pulverized coal-fired power plant in post-combustion carbon dioxide sequestration. *Archive Mech. Eng.* 59, 343–360, 2012.

23. S. Martynov, S. Brown. CO_2QUEST internal report: A report describing the optimum CO_2 compression strategy. Deliverable 2.2, University College, London, UK, 2014.

24. S. Martynov, N. Daud, H. Mahgerefteh, S. Brown, R. T. J. Porter. Impact of stream impurities on compressor power requirements for CO_2 pipeline transportation. *Int. J. Greenh. Gas Control*, 2016, submitted.

25. H. Mahgerefteh, G. Denton, Y. Rykov. A hybrid multiphase flow model. *AIChE J.* 54, 2261–2268, 2008.

26. H. Mahgerefteh, A. Oke, O. Atti. Modeling outflow following rupture in pipeline networks. *Chem. Eng. Sci.* 61, 1811–1818, 2006.

27. S. Martynov, N. Mac Dowell, S. Brown, H. Mahgerefteh. Assessment of thermo-hydraulic models for pipeline transportation of dense-phase and supercritical CO_2. *Ind. Eng. Chem. Res.* 54, 8587–8599, 2015.

28. S. Brown, H. Mahgerefteh, S. Martynov, V. Sundara, N. Mac Dowell. A multi-source flow model for CCS pipeline transportation networks. *Int. J. Greenh. Gas Control.* 43, 108–114, 2015.

29. R. M. Woolley, M. Fairweather, C. J. Wareing, S. A. E. G. Falle, C. Proust, J. Hebrard, D. Jamois. Experimental measurement and Reynolds-averaged navier-stokes modeling of the near-field structure of multi-phase CO_2 jet releases. *Int. J. Greenh. Gas Control* 18, 139–149, 2013.

30. S. Sarkar. The stabilizing effect of compressibility in turbulent shear flow. *J. Fluid Mech.* 282, 163–186, 1995.

31. J. Rotta. Statistische Theorie nichthomogener Turbulenz. 1. Mitteilung. *Zeitschrift fur Physik A-Hadrons and Nuclei* 129, 547–572, 1951.

32. H. Khlifi, T. Lili. A Reynolds stress closure for compressible turbulent flow. *J. Appl. Fluid Mech.* 4, 99–104, 2011.

33. W. P. Jones, P. Musonge. Closure of the Reynolds stress and scalar flux equations. *Phys. Fluids* 31, 3589–3604, 1988.

34. C. A. Gomez, S. S. Girimaji. Toward second-moment closure modeling of compressible shear flows. *J. Fluid Mech.* 733, 325–369, 2013.

35. C. D. Donaldson, R. S. Snedeker. A study of free jet impingement. Part 1. Mean properties of free and impinging jets. *J. Fluid Mech.* 45, 281–319, 1971.

36. D. Y. Peng, D. B. Robinson. A new two-constant equation of state. *Ind. Eng. Chem. Fundam.* 15, 59–64, 1976.

37. R. Span, W. Wagner. A new equation of state for carbon dioxide covering the fluid region from the triple-point temperature to 1100 K at pressures up to 800 MPa. *J. Phys. Chem. Ref. Data* 25, 1509–1596, 1996.

38. C. J. Wareing, R. M. Woolley, M. Fairweather, S. A. E. G. Falle. A composite equation of state for the modeling of sonic carbon dioxide jets. *AIChE J.* 59, 3928–3942, 2013.

39. A. Jäger, R. Span. Equation of state for solid carbon dioxide based on the Gibbs free energy. *J. Chem. Eng. Data* 57, 590–597, 2012.

40. A. Yokozeki. Solid–liquid–vapor phases of water and water–carbon dioxide mixtures using a simple analytical equation of state. *Fluid Phase Equilibr.* 222–223, 55–66, 2004.

41. CO2PipeHaz. Quantitative failure consequence hazard assessment for next generation CO_2 pipelines: The missing link, 2009. Accessed 11 November, 2015; [CO2PipeHaz Project Website]. www.co2pipehaz.eu/.

42. D. Jamois, C. Proust, J. Hebrard. Hardware and instrumentation to investigate massive spills of dense phase CO_2. *Chem. Eng. Trans.* 36, 601–606, 2014.

43. R. Talemi, S. Brown, S. Martynov, H. Mahgerefteh. A hybrid fluid-structure interaction modeling of dynamic brittle fracture in pipeline steel transporting CO_2 streams. *Int. J. Greenh. Gas Control*, 2016, submitted.

44. R. Talemi, S. Brown, S. Martynov, H. Mahgerefteh. Assessment of brittle fractures in CO_2 transportation pipelines: A hybrid fluid-structure interaction model. 21st European Conference on Fracture, ECF21, 20–24 June 2016, Catania, Italy. *Procedia Structural Integrity*, Submitted, April 2016.

45. A. Niemi, J. Bensabat, V. Shtivelman, K. Edlmann, P. Gouze, L. Luquot, F. Hingerl, S. M. Benson, P. A. Pezard, K. Rasmusson, T. Liang, F. Fagerlund, M. Gendler, I. Goldberg, A. Tatomir, T. Lange, M. Sauter, B. Freifeld. Heletz experimental site overview, characterization and data analysis for CO_2 injection and geological storage. *Int. J. Greenh. Gas Control*, 48, 3–23, 2016.

46. D. Rebscher, J. L. Wolf, B. Jung, J. Bensabat, A. Niemi. Numerical simulations of the chemical impact of impurities on geological CO_2 storage—Comparison between TOUGHREACT V2.0 and TOUGHREACT V3.0-OMP, 2015, LBNL-190559, Proceedings, TOUGH Symposium 2015,

Lawrence Berkeley National Laboratory, Berkeley, CA, September 28–30, 2015, 493–500. http://esd1.lbl.gov/files/research/projects/tough/events/symposia/toughsymposium15/Proceedings_TOUGHSymposium2015.pdf.

47. T. Xu, E. Sonnenthal, N. Spycher, L. Zheng. TOUGHREACT V3.0-OMP reference manual: A parallel simulation program for non-isothermal multiphase geochemical reactive transport. LBNL-DRAFT, Lawrence Berkeley National Laboratory, Berkeley, CA, 2014. http://esd1.lbl.gov/files/research/projects/tough/documentation/TOUGHREACT_V3-OMP_RefManual.pdf.

48. L. Pan, N. Spycher, C. Doughty, K. Pruess. ECO2N V2.0: A TOUGH2 fluid property module for mixtures of water, NaCl and CO_2, LBNL-6930E, Lawrence Berkeley National Laboratory, Berkeley, CA, 2015. http://esd1.lbl.gov/files/research/projects/tough/documentation/TOUGH2-ECO2N_V2.0_Users_Guide.pdf.

49. I. E. Grossmann, R. W. H. Sargent. Optimum design of multipurpose chemical plants. *Ind. Eng. Chem. Process Des. Dev.* 18, 343–348, 1979.

50. A. Arce, N. Mac Dowell, N. Shah, L. F. Vega. Flexible operation of solvent regeneration systems for CO_2 capture processes using advanced control techniques: Toward operational cost minimisation. *Int. J. Greenh. Gas Control* 11, 236–250, 2012.

51. N. Mac Dowell. Optimisation of post-combustion CCS for flexible operation, in The 14th Annual APGTF Workshop—'The Role of Fossil Fuel Power Plant in Providing Flexible Generation' 2014, London.

52. N. Mac Dowell, A. Galindo, G. Jackson, C. S. Adjiman. Integrated solvent and process design for the reactive separation of CO_2 from flue gas. *Comput. Aided Chem. Eng.* 28, 1231–1236, 2010.

53. N. Mac Dowell, F. E. Pereira, F. Llovell, F. J. Blas, C. S. Adjiman, G. Jackson, A. Galindo. Transferable SAFT-VR models for the calculation of the fluid phase equilibria in reactive mixtures of carbon dioxide, water, and n-alkylamines in the context of carbon capture. *J. Phys. Chem.* B 115, 8155–8168, 2011.

54. N. Mac Dowell, N. J. Samsatli, N. Shah. Dynamic modeling and analysis of an amine-based post-combustion CO_2 capture absorption column. *Int. J. Greenh. Gas Control* 12, 247–258, 2013.

55. N. Mac Dowell, N. Shah. Dynamic modeling and analysis of a coal-fired power plant integrated with a novel split-flow configuration post-combustion CO_2 capture process. *Int. J. Greenh. Gas Control* 27, 103–119, 2014.

56. N. Mac Dowell, N. Shah. The multi-period operation of an amine-based CO_2 capture process integrated with a supercritical coal-fired power station. *Comput. Chem. Eng.* 74, 169–183, 2015.

57. J. Rodriguez, N. Mac Dowell, F. Llovell, C. S. Adjiman, G. Jackson, A. Galindo. Modeling the fluid phase behaviour of aqueous mixtures of multifunctional alkanolamines and carbon dioxide using transferable parameters with the SAFT-VR approach. *Mol. Phys.* 110, 1325–1348, 2012.

58. N. Mac Dowell, N. Shah. Identification of the cost-optimal degree of CO_2 capture: An optimisation study using dynamic process models. *Int. J. Greenh. Gas Control* 13, 44–58, 2013.

59. J. Condor, D. Unatrakarn, M. Wilson, K. Asghari. A comparative analysis of risk assessment methodologies for the geologic storage of carbon dioxide. *Energy Procedia* 4, 4036–4043, 2011.

60. D. Savage, P. R. Maul, S. Benbow, R. C. Walke. A generic FEP database for the assessment of long-term performance and safety of the geological storage of CO_2. Quintessa Report QRS-1060A-1, 2004.

61. J. Wilday, N. Paltrinieri, R. Farret, J. Hebrard, L. Breedveld. Addressing emerging risks using carbon capture and storage as an example. *Process Saf. Environ.* 89, 463–471, 2011.

62. R. Farret, P. Gombert, F. Lahaie, S. Lafortune, A. Cherkaoui, C. Hulot, O. Bour. Design of fault trees as a practical method for risk analysis of CCS: Application to the different life stages of deep aquifer storage, combining long-term and short-term issues. *Energy Procedia* 4, 4193–4198, 2011.

63. IEAGHG, 2011. Effects of impurities on geological storage of CO_2. Report: 2011/04, June.

64. M. Mohitpour, P. Seevam, K. K. Botros, B. Rothwell, C. Ennis, *Pipeline Transportation of Carbon Dioxide Containing Impurities*, ASME Press, New York, 2011.

65. CO_2QUEST Newsletter Autumn 2014, Impact of the quality of CO_2 stream on storage and transport. www.co2quest.eu/main_download/newsletters/CO2QUEST_newsletter_Autumn_2014.pdf.

12.6 Performance of Porous Covalent Organic Polymers for CO$_2$ Capture at Elevated Pressure

Mert Atilhan, Ruh Ullah, and Cafer T. Yavuz

12.6.1 Introduction

Covalent organic polymers (COPs) are interesting materials, which can be synthesized via simple procedure, and their physical properties can be engineered for gas storage, catalysis, photonics, and water purification through appropriate linkers. COPs are known to have permanent porous structure with various pore sizes ranging from nanometer scale up to macrometer scale depending on whether the molecular arrangement is amorphous or crystalline. The building block monomers in COPs are linked with the help of strong covalent bonds originated mainly from constituents such as H, C, O, and N of the linking agents. Unlike covalent organic frameworks (COFs) and with very few exceptions, most of the COP materials reported so far in the literature are amorphous and possess moderate surface areas and pore volumes as compared to the well-established metal organic frameworks (MOFs). Additionally, COPs have high thermal stability, possess suitable porosity, have tunable affinity for gas uptake, and can be regenerated economically for repeated applications without any degradation in the performance. Highly selective removal and efficient capture of CO$_2$ from flue gases of power plants and cement industries require that the sorbents should be chemically inert and physically robust that can withstand various temperature and pressure conditions. In addition to the surface area, pore volume, and pore size, affinity also plays a key role in steering the adsorption capacity and selectivity of COPs. It is important to note that the affinity of CO$_2$ capture is strongly dependent on the type, extent, and the way the function group is bonded to the building block monomer. The functionality of COPs can be tuned easily by incorporating different function groups such as $-NH_2$, $-OH$, $-CH_3$, $-N=N-$, and $-NO_2$, as a result affinity of materials toward specific gas can be enhanced or minimized. Thus, COPs are fascinating compounds since the linking molecules/monomers in these materials can serve various functions of shaping the COPs chain through linking between two consecutives building block monomers, affecting the affinity of surface, and in pores active sites for very selective capture of different gases and engineering the pore sizes, pore volumes, and surface areas.

CO$_2$ emission is alarmingly increasing globally owing to the excessive numbers of power plant units, cement industries, and massive traffic worldwide. Among various known technologies, adsorption by solid sorbents is also considered to be very effective in reducing the emission of CO$_2$ both on large and small scales as long as regeneration of the materials is concerned. COPs are assumed to be the best candidates for application of selective CO$_2$ capture in both pre- and postcombustion processes, because these materials are applicable to any environmental conditions. The key advantages of COPs over other solid sorbents, particularly MOFs, silicas, activated carbon, nanoclays, and zeolites to be used for CO$_2$ capture and separation, are as follows: (1) both feature and volume of porosity and affinity of COPs can be varied by using different linkers, (2) most of the reported COPs have comparatively lower heat of adsorption, (3) CO$_2$ can be only adsorbed via physical boding on the surface and inside the pore, and almost negligible chemisorption can be observed, (4) the rate of CO$_2$ uptake by COPs is much higher than any other solid sorbent, (5) COPs uptake the largest quantity of adsorbate molecules, and (6) COPs possess the highest adsorption capacity both at low and high pressure to date. COPs were also tested for CO$_2$ capture in water vapor environment to study the industrial applications of these compounds. Based on these advantageous features of COPs, the academics as well as industries are now joining hands to upgrade this emerging technology from bench scale to the pilot plant scale and evaluate the possibility of commercializing these fantastic polymers. In this chapter, we will summarize the preparation, characterization, and applications of COPs for CO$_2$ capture and separation under various pressure and temperature conditions.

12.6.2 COPs Preparation and Their Physical Properties

Selection of synthesis procedures, building block monomers, and linkers is directly influencing the physical properties such as surface area, pore volume, pore size, and morphology of COPs. Many techniques including metal catalyzed (such as Friedel–Crafts arylation) and noncatalyzed (like direct Schiff) reaction pathways have been used to prepare amorphous, crystalline, two- and three-dimensional structures of covalently bonded cross-linked nano-, micro-, and macro-scale polymer chains. Patel et al. reported preparation of an amorphous N_2-phobic organic polymer by direct coupling of amine moieties with aromatic nitro group under basic conditions using tetrakis(4-nitrophenyl)methane (TBM) as building block monomer. Azo-linked COPs have amorphous molecular arrangement, high thermal stability, possess large surface area (792 m^2/g) and average pore volume (0.44 cm^3/g), have irregular morphologies with microporous structures, and exhibited type-I nitrogen isotherms confirming reversible uptake behavior. It is important to note that by coupling TBM as building block monomer via nickel-catalyzed cross-coupling reaction, various multiblock COPs can be synthesized that have much higher surface areas as compared to azo-COPs. The surface area and pore volumes of COPs significantly increased when TBM was copolymerized with rigid monomers of different lengths (C_2, C_3, and C_4) and functionalized by groups such as OH, CH_3 NO_2, and NH_2. Surface area as high as 3624 cm^2/g and pore volume of 3.5 cm^3/g was reported for COP-20 along with higher thermal stability compared to azo-COPs. Amorphous porous aromatic framework (PAF-1) was synthesized from TBM as a building block and has the largest surface area of 5600 m^2/g as compared to all COPs. The largest surface in the case of PAF-1 was attributed to the replacement of two or more C–C bonds in a tetrahedral building unit of TBM by phenyl rings, resulting in less denser and highly porous material. However, incorporating metals like Co, Fe, and Mn within tetrahedral building blocks makes COPs crystallinze with reasonable porosity. Xiang et al. prepared a series of porous organic polymers (POPs) by self-polymerization of three reagents (benzene, triazine, and amine) containing 4-bromophenyl as the core ring of the structures. COP-4 (which is made of triazine-based monomers) is found to have the highest surface area of 2015 m^2/g, COP-2 has the largest pore volume of 1.76 cm^3/g, and COP-1 possesses the lowest surface area and lower pore volume among all four materials. Moreover, triazine-based monomer with carbozole (CBZ) ring has been used to prepare 3D porous organic frames through catalyzed reaction of $FeCl_3$ in the presence of cross-linking agents. The propeller-shaped and crystalline organic polymers have a surface area of 424 m^2/g and a pore volume of 0.358 cm^3/g, which can capture a reasonable quantity of CO_2 under standard temperature and pressure and have very good selectivity for CO_2 over N_2. Charged and ordered COPs with two-dimensional structures have very low surface areas and would not be suitable for gas capturing applications, based on their anticipated lower capacity. It is important to note that, similar to the effect of linkers on the physical properties, synthesis procedure including duration of preparation, heating temperature, and environment of synthesis also have significance for the structure and porosity of materials. Thus, Yuan et al. modified the Yamamoto homo-coupling procedure and conducted synthesis of Porous Polymer Networks (PPN)-3, PPN-4, and PPN-5 at room temperature with tetrahydrofuran mixed solvent, which resulted in exceptionally high surface area of 6461 cm^3/g for PPN-4. Metal-porogen strategy was attempted to tune and control the porosity and surface area of POPs by using metallo-porphyrin building blocks with axial-ligands to serve as a source of porogens scheme, which can be removed after polymerization of POPs having permanent pores in the structure. Cooper and coworkers have employed various techniques and prepared large numbers of nanoporous, microporous, mesoporous, amorphous, and crystalline conjugated POPs with tunable physical properties. It was demonstrated that varying the monomers length, increasing the reaction time, reaction temperature, and changing the reaction methodology can effectively increase the surface area, pore size, and pore volume, and scale down the particles size to the nanorange. The hyper-cross-linked polymers (HCPs) prepared by Cooper and coworkers via copolymerization techniques possessed a surface area of 1970 m^2/g and a pore volume of 1.47 cm^3/g, and they were found to have enough storage capacity both at low and high pressure, suggesting materials suitability for precombustion CO_2 capture. Petal et al. substitute chlorides in cyanuric chloride with nucleophiles piperazine and bipiperidine by gradually increasing the temperature of the mixture in the presence of aprotic base, yielding two amorphous COPs, COP-1, COP-2, which have different physical parameters, i.e., surface areas, pore volumes, and pore sizes. This is a general understanding that capturing capacities of solid materials are mostly dependent on the physical properties of materials, but, COP materials prepared from cyanuric chloride as backbone have average physical parameters and remarkable capacity with excellent selectivity of CO_2 over H_2. The lower surface area and less porosity of COP-1 and COP-2 were caused by the restricted interaction of linkers due to the enhanced temperature of mixture for replacement of all chlorides from building block monomers. Effect of solvents on the properties of aromatic polyamides has been investigated, and results indicated that polymers (PA-1) prepared using *N,N*-dimethylacetamide (DMAc)– 1-methyl-2-pyrrolidone (NMP) have better crystallinity and comparatively better CO_2 capturing capacity than PA-2, which was synthesized using 1,4-dioxane.

Based on the earlier discussion, further investigation is required to explore a direct impact of physical properties on the adsorption capacity and selectivity of COPs. All of the steps including selection of core monomers, catalyst, linking agents, synthesis routes, type of solvents, synthesis temperature, and final cleaning of the samples significantly steer physical parameters of the newly synthesized COPs. Furthermore, numerous findings have shown a strong correlation between adsorption capacity and the physical properties of materials under various environmental conditions. However, there are some exceptional cases where the materials have very poor physical properties, but, possess very high capturing capacity along with better selectivity of CO_2 over N_2. Kinetics such as rate of adsorption, heat of adsorption, affinity, and mass transfer coefficients of gases also have similar importance as the capturing capacity and selectivity. Efforts have been made to correlate the kinetic of adsorption to the material properties, e.g., diffusion of gas molecules into the porous structures has been interrelated to the pore size, and heat of adsorption has been associated with the nature of adsorption, i.e., chemisorption or physical attachment of the adsorbate on the adsorbents exterior or interior.

A comprehensive investigation is still demanded to give logical explanation to certain parameters of the adsorption process and set some fundamental parameters for the physical sorbent particularly for organic polymers. In the next section, a detailed discussion will explore the high-pressure adsorption/desorption of COPs, which might be useful in the selection of materials for further investigation on a large scale.

12.6.3 CO_2 Adsorption and Materials Selectivity

COPs have been recently introduced as promising materials to replace aqueous amine solution for CO_2 capture from flue gases at low pressure and lower temperature. In addition to their applicability for postcombustion CO_2 capture, these compounds can also be suitable for precombustion and high-pressure adsorption processes based on their thermo-physical stability. As mentioned in the earlier sections, these materials can have tunable physical properties such as thermal stability, surface area, pore volume, and different functionalities based on building block monomers and linking agents. Few COPs with different functionalities and different physical parameters have been synthesized using TBM as building block monomer and linkers of different geometrical shapes. Some of the TBM-based materials were also functionalized with single and multiple functional groups to obtain materials with different functionalities for selective CO_2 uptake under various conditions. COP-10, which was prepared with TBM and linked by 1,3,5-tribromobenzene have comparatively better physical properties and can capture the largest quantity of CO_2 at a pressure of 18 bars and 298 K. Importantly, some TBM-based COPs have larger surface area and pore volume and functionalized with different functional groups for the improved performance, but, their CO_2 uptake was lower than COP-10. Coupling TBM as building block monomers with the help of phenyl ring gave diamond-like structure of PAF-1, which has the largest surface area and can uptake as large as 33.5 mmol/g at 40 bars and 298 K. PAF-1 exhibited high uptake capacities for hydrogen and other vapors like benzene and toluene, making this material less effective as long as selectivity of the material is concerned. It was also demonstrated that at high pressure (30 bar), CO_2

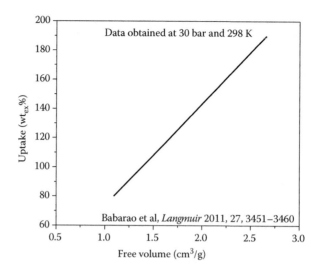

FIGURE 12.106 Dependency of CO_2 adsorption capacity of crystalline porous aromatic framework (PAF-1) and its functionalized analogue on their corresponding free space at 30 bars and 298 K.

adsorption capacities of the three materials were the function of free volume available in the structure. As shown in Figure 12.106, CO_2 uptake was increasing with an increase in the available free space in the crystalline structures of three different materials. It can be deduced from these findings that at low pressure, functionalization of materials dominates the capturing capacity, while at high pressure free space in the structure, or in other words pore volume, and surface area direct the CO_2 adsorption capacity.

Additionally, dimensionality of COFs along with the pore size also contributes to the adsorption capacity of materials. Three groups, 2D structure with one-dimensional small pore size, 2D structure with one-dimensional large pore size, and 3D structure containing three-dimensional medium pore sizes were prepared mainly through condensation techniques. Results indicated that COF-2 and COF-3, which have three-dimensional crystalline structures and three-dimensional pore sizes, can uptake the largest quantity of CO_2 at high pressure of 35 bars and 298K as compared to one- and two-dimensional structures. COF-2 and COF-3, which have almost similar surface areas (3620, 3530 cm^3/g) and similar pore volume can capture about similar quantities of CO_2, i.e., 1200 and 1190 mg/g under similar pressure and temperature conditions. Two benzoxazole-linked and tert-butyldimethylsilyl-functionalized COPs (Box–COP-93, COP-94) were found to have Brunauer -Emmett -Teller (BET) surface areas in the range of 336 cm^3/g and 229 cm^3/g, respectively. Owing to the larger surface of COP-93 than COP-94, the former uptakes more CO_2 and also possesses better selectivity than COP-93, which was reduced with an increase in temperature. It can be seen from Figure 12.107 that at standard temperature and pressure CO_2 adsorption varies almost linearly with an increase in the surface area of azo-modified COPs. Figures 12.106 and 12.107 clearly indicate that the physical properties of materials play a significant role in the capturing capacity of COPs at lower pressure and lower temperature.

Selectivity of CO_2 over N_2 for Azo-COPs was reportedly enhanced by introducing nitrogen phobic moiety, i.e., azo ($-N\equiv N-$) groups within the structure that reject N_2 gas and resulted in very high CO_2/N_2 selectively at 298 K and 1 bar, which was further increased at high temperature. Azo-COPs have almost twice the surface area (729 m^2/g) of COP-93, which adsorb about 76.3 mg/g CO_2 and very nominal N_2 (0.92 m^2/g) under similar conditions. The higher selectivity of CO_2/N_2 was mainly attributed to the increase in the binding affinity of CO_2 toward Azo-COP-2 due the presence of electron-rich environment induced by the $-N\equiv N-$function group.

Rigid triazine-based microporous polymer obtained via direct oxidative coupling (PCBZ) and extensive cross-linking (PCBZL) have surface areas of 341 and 424 m^2/g, respectively, captured 33.8 and 54.6 mg/g

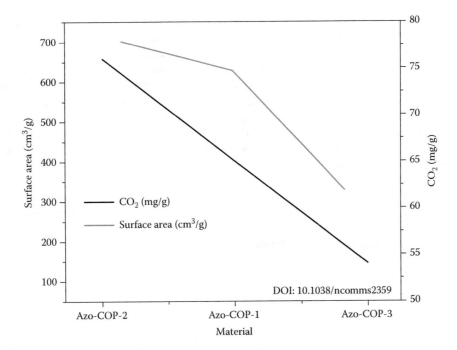

FIGURE 12.107 Relationship of CO_2 uptake with surface area of materials.

CO_2 at 1 bar and 298 K and exhibited very good selectivity for CO_2/N_2. Similar to Azo-COPs, selectivity of microporous polycarbazole materials was also increasing with increasing temperature. COPs with an oxadiazole heterocycle (oz-COPs), which have two nitrogen and single oxygen atom can uptake 61 times more carbon dioxide (1.4 mmol/g) than N_2 (0.15 mmol/g) due to the interaction between CO_2 and heteroatoms in oxadiazole, confirming very high selectivity in mixed gas of $N_2:CO_2$ (85%:15%). Functionalization through various moieties such as OH, CH_2, –N–, and many more have been broadly utilized to explore both the capturing capacity and selectivity of COPs materials that can be subsequently tested/used for large-scale application. Based on the inherent basic behavior of tertiary nitrogen functionality various organic polymers such as TB-COPs, reported by Byun et al., polymers of intrinsic microporosity (PIMs) reported by Carta et al., and crystalline microporous polymers (TB-MOPs) reported by Zhue et al. have been bridged/linked through Troger's base moiety to prepare porous organic materials with adequate physical properties that could be utilized for selective CO_2 capturer subject to various environment conditions. TB-COP-1 possesses very large surface area (1340 m²/g) as compared to other Troger's base linked polymers such as TB-COP-2 (0.94 m²/g), PIM-EA-TB (1028 m²/g), and TB-MOP (694 m²/g). Due its high surface area and comparatively large pore volume, TB-COP-1 was capable of storing 3.16 mmol/g CO_2, which is more than that of TBA-MOP (2.75 mmol/g) along with very high selectivity of 68.9 at ambient conditions. It was suggested that adamantane formed a rigid tetrahedral system by limiting swing of the phenylene legs, ending intramolecular cyclization, and separating the reactive amine sites that consequently insured formation of structure with the largest surface area. The highest capturing capacity of TB-linked polymers can be associated with the significant microporosity and high concentration of N-containing reactive sites, both are essential characteristics for CO_2 separation via adsorption techniques.

Pressure swing adsorption (PSA) has been considered as energy saver and sample strategy for the separation and capture of CO_2 from the exhaust of power-generating plants and other industries operating mainly on coal combustion. Sorption by solid adsorbent is mostly suitable for CO_2 capture at high pressures and different temperatures. Various solid materials such as activated carbon, zeolites, and SBA-15 have been examined for high-pressure CO_2 capture and storage. Costly regeneration, chemisorption, capturing capacity, rate of adsorption, and thermophysical stability of materials are still among the major issues, which hindered practical application of common physical adsorbents. Porous polymers network have been recently familiarized as mechanically robust and thermally stable, which can tolerate temperature of exhaust fumes, and economically suitable candidates and less expensive than other solid materials. COPs that have the benefits of stability, tunable capacity, as well as changeable affinity owing to the insertion of different functional groups within the interior and exterior structures may have potential practicability for high-pressure capture of CO_2. Precombustion (syngas stream) capture, passage through pipelines, and transportation for sequestration of CO_2 require solid adsorbents that are physically strong enough and can withstand pressure as high as 175 bars. Physical adsorption of CO_2 on solid materials involves the van der Waals interaction and pole–ion interaction between the quadrupole of the CO_2 molecules and the polar and/or attractive sites on the solid sorbents. These CO_2 attractive sites within the COPs determine the strength of affinity, which in turn defines the capturing capacity of these materials. COP materials and many other organic polymers have been designed purposely to have robust strength of attraction as well as enough unoccupied space in the vicinity of the attractive sites to accommodate a large number of CO_2 molecules upon pressurizing the system and detach the captured gas after releasing the applied pressure. Sulfur-bridged COPs with triazine moiety as building blocks were found to have surface area of 413 m^2/g and can capture up to 3294 mg/g under similar condition as COP-1 presented by Patel et al.

To investigate the capability of porous polymers, four different hypercross-linked polymers (HCPs) were prepared with different functionalities and subjected to a maximum pressure of 35 bars for the selective adsorption of CO_2 over H_2 under various environment conditions using high-pressure magnetic suspension balance (MSB) (Rubotherm-VTI). Results indicated that among all four materials, HCP-1 can uptake up to the maximum of 13.4 mmol/g at 298 K accompanied by the largest heat of adsorption. The high-pressure CO_2 adsorption of HCP-1 was mainly reasoned to the total micropore volume instead of micropore size of the organic polymers. Cyanuric chloride base, COPs (COP-1 and COP-2) subjected to a pressure of 200 bars at three different temperatures, i.e., 318 K, 328 K, and 338 K after being boiled in water for seven days to evaluate the CO_2 adsorption capacity and stability of these materials for the real application on large scale. Results of high-pressure adsorption experiment conducted with MSB revealed that, COP-1 that possesses larger surface and larger pore volume can uptake 5616 mg/g at 318 K, and COP-2 captured almost half of this quantity under similar conditions. It must be noted that this adsorption capacity of COP-1 is much higher than that reported for all other organic polymers (given in Table 12.31) namely, HCP, PAF-1, COF-103, and even larger than the MOF. In addition to the largest adsorption capacity at high pressure, these materials also showed extremely high selectivity for CO_2 over H_2, which suggested practical applications of COP-1 in syngas technology for precombustion capture. As shown in Figure 12.108, isotherms of COP-1 and COP-2 are almost identical and have type IV characteristic indicating capillary condensation of these porous materials at high pressure. Detailed investigation of isotherms reveals that CO_2 uptake of both the materials is almost linear with pressure from 0 to 71 bars, rapidly increasing above the critical pressure and starts fullness at 125 bars. Beyond this point, saturation starts and uptake capacity is increasing very slowly with an increase in pressure and tends to complete filling of all the available spaces/sites at pressure around 200 bars and above. The very high CO_2 uptake by COP-1 was mainly attributed to the presence of tertiary and aromatic nitrogen along with the large numbers of small pores within the structure, which are not available in COP-2.

Ester-based COPs were prepared with different linkers to investigate the effect of hydroxyl functionality on CO_2 adsorption capacity and selectivity at various temperatures and high pressures. All COPs materials, i.e., COP-35–COP-37 possess moderate surface areas and lower pore volume, have shown strong thermal stability, and perform independently for capturing CO_2, CH_4, and N_2. Unlike other solid sorbents, particularly COPs, the capturing capacity of ester-based COPs may very rarely be related to the porosity (i.e., surface area and pore volume) of these materials. Due to the very low surface area and pore volume, CO_2 adsorption at lower pressure, i.e., up to 1 bar was considerably lower than other counterparts; however, CO_2 adsorption capacity of COP-35 at high pressure was recorded to be the largest

TABLE 12.31 Physical Properties and CO_2 Capturing Capacity of COPs

Name	Surface Area (m²/gm)	Pore Volume (cm³/g)	CO_2 Capacity	Selectivity CO_2/N_2	Author
Azo-COPs	792	0.44	151.3 mg/g (263 K) 67 mg/g (298 K)	288	H. A. Patel
COP-20	3624	3.5	767 mg/g (298 K, 18 bars)	4.5	Z. Xiang
PAF-1	5600		(1300 mg/g) 33.5 mmol/g (298 K, 40 bars)	?	T. Ben
COP-4	2015	1.36	594 mg/g (298 K, 1 bar)	6.1	Z. Xiang
PPN-4	6461		2121 mg/g (295 K, 50 bar)		D. Yuan
PCBZL	424	0.358	54.6 mg/g (298 K, 1 bar)	148	M. Saleh
HCP-1	1646	1.26	13.3 mg/g (298 K, 30 bar)	N/A	C. F. Martin
COP-1	168	0.25	5616 mg/g (218 K, 200 bar) 60 mg/g (298 K, 1 bar)	25	H. A. Patel
PA-1	84.5	0.41	0.09 mmol/g (298 K, 1bar)		S. Zulfiqar
TB-COP-1	1340	0.541	3.16 mmol/g, 1 bar, 298 K	68.9	J. Byun
TB-MOP	694	0.398	2.75 mmol/g, 1 bar 298 K	50.6	X. Zhu

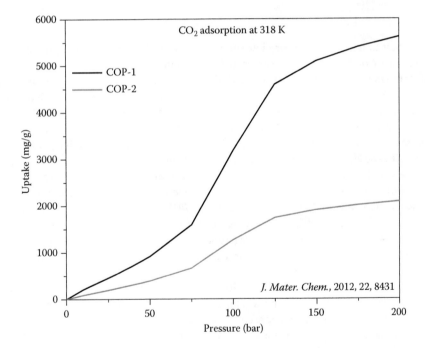

FIGURE 12.108 CO_2 uptake isotherms comparing adsorption capacities of COP-1 and COP-2 under similar conditions of temperature and high pressure.

ever recorded at 308 K and 200 bars. As shown in Figure 12.109, COP-35 adsorbs three times more CO_2 than CH_4 and four times more than nitrogen under similar conditions, indicating very good selectivity at these operating conditions. It can be seen that COP-37 has comparatively lower CO_2 capacity than COP-35, but, its selectivity of CO_2/CH_4 is higher than other materials. COP-36, on the other hand, has comparatively lower performance both for the capturing capacity and selectivity of the materials.

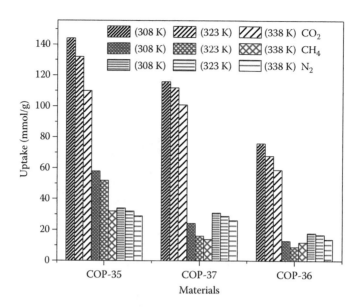

FIGURE 12.109 Gases uptake by ester-based COPs (COP-35, COP-36, and COP-37) at various temperatures and high pressure of 200 bars.

It is important to note that with very low surface area, COP-35 captures an unprecedented amount of CO_2 at a pressure of 200 bars, which can be mainly attributed to the presence of hydroxyl ions on the surface of material. It can be assumed that OH originated from the defect sites of material can effectively interact with CO_2 molecules as compared to the N functionality at high pressure. Herein we assume that at high pressure the defect sites on COP-35 surface become more favorable for CO_2 molecules due to the enhancement in the strength of coulomb forces and Van der Wall forces resulting in significant adsorption capacity. Based on its high CO_2 capacity, comparatively better selectivity, and economically cheap synthesis procedure, COP-35 may have the opportunity to be tested in large-scale applications. Notably, same sample of COP-35 was used repeatedly for capturing of various gases (N_2, CH_4, and CO_2) at different operating conditions without any special regeneration procedure. Each time before adsorption process, the same material was only heated up to 378 K and then subjected to different pressures, temperatures, and different gases to measure the storage capacity. Upon comparing with all materials including COP-1, COP-2, Azo-COP-1, HCP, PAF (porous aromatic framework)-1, and COF-103, it can be concluded that, although COP-35 has lower surface area and very small pore volume, it can capture the largest quantity of CO_2 at elevated pressure.

Upon comparing the adsorption capacity with other solid sorbents such as activated carbon, silica, and organic polymers in the literature, it was found that COP-35 has the largest storage capacity for CO_2 measured at 308 K and 200 bars. It can be seen from Figure 12.110 that the maximum CO_2 uptake capacity presented so far in the literature is for COP-1, which is nearly 100 mmol/g under similar conditions of pressure and temperature. The next largest adsorption capacity of CO_2 shown in Figure 12.110 is for SBA-15 and COP-3 reported by us and our collaborators. Polybenzimidazole, basolites, and MOF have much lower capturing capacity as compared to COP-35 material.

Luo et al. discovered that hyper cross-linked COP-10 could capture 896 mg/g CO_2 at 298 K and 18 bars, whereas the hierarchically porous carbons having surface area of 2734 m^2/g and pore volume of 5.53 cm^3/g possessed the maximum capacity of 27 mmol/g at 27°C and 30 bars. Carbon aerogel modified with melamine has significantly reduced surface area and pore volume; however, CO_2 uptake capacity was considerably large due to the enhanced functionality of the material, which captured 118.77 mg/g at room temperature and 1 bar. It is important to note that, due to the lower physical parameters, COP-35

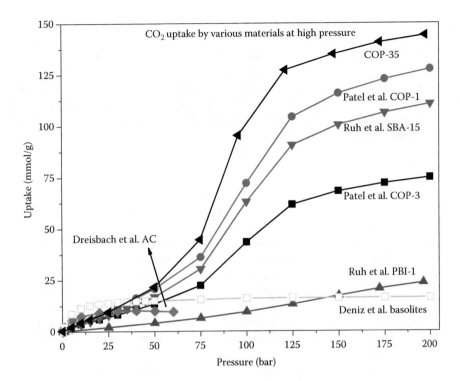

FIGURE 12.110 CO_2 uptake of COP materials and their comparison with other microporous/mesoporous materials.

captures comparatively small quantity of CO_2 at ambient conditions, however, at higher pressure and high temperature, this material has the largest CO_2-capturing capacity reported so far in the field of solid sorbents and organic polymer. Based on the overall outstanding performance of materials, it can be assumed that COPs may have the capability to be considered for the large-scale applications in power generation industries like IGCC. As per the experimental results obtained so far and comparison with other solid sorbents, COPs may possibly lead the CO_2 capture and separation industries in the near future.

12.6.4 Conclusion

In conclusion, it has been demonstrated that COPs are the only solid sorbent materials that accomplish all the requirements including low-cost synthesis, repeated application without performance degradation, thermophysical stability, and cheap regeneration procedure. Based on all these and many other advantages, COPs were suggested to be the best option for both pre- and postcombustion CO_2 capture and separation. COPs are such favorable materials that can be synthesized as per the industrial demand, and their property can be tuned for particular objectives. Capacity and selectivity of COPs can be purposely modified by choosing different monomers as building blocks and different moieties as linking agents. As discussed in detail, COPs have shown excellent performance as long as capturing capacity and selectivity of the materials are concerned. Azo-COPs have shown outstanding selectivity for CO_2/N_2, piperazine-based materials, i.e., COP-1, have extremely high CO_2/H_2 selectivity, and COP-35 can capture the largest amount of CO_2 at elevated pressure and temperature. In many cases, a direct relationship between porosity and capturing capacity at standard temperature and pressure was observed, which may suggest that higher surface area and larger pore volume along with affinity may be beneficial for larger capturing capacity. On the other hand, adsorption capacity at high pressure still requires further investigation, since COP-35 has the least porous structure but captured the largest amount of CO_2 at high pressure. Precombustion CO_2 capture primarily demands materials that can operate under very high pressure and can also separate CO_2 with very high selectivity from syngas. On the other hand, postcombustion process requires sorbents that can remain physically and chemically stable and capture CO_2 from exhaust gas that contains mainly N_2 and water vapors at high temperature. Thermogravimetric analysis, boiling in water before use for a few days, and adsorption capacity at high pressure and high temperature suggested that COPs have the capability to overcome all the technical issues and become the most appropriate materials for CO_2 capture and separation in both pre- and postcombustion processes.

References

Babarao, R., S. Dai, and D. Jiang. Functionalizing porous aromatic frameworks with polar organic groups for high-capacity and selective CO_2 separation: A molecular simulation study. *Langmuir* 27, no. 7 (2011): 3451–60.

Beletskaya, I. P., G. V. Latyshev, A. V. Tsvetkov, and N. V. Lukashev. The nickel-catalyzed Sonogashira–Hagihara reaction. *Tetrahedron Letters* 44, no. 27 (2003): 5011–13.

Ben, T., H. Ren, S. Ma, D. Cao, J. Lan, X. Jing, W. Wang, J. Xu, F. Deng, J. M. Simmons, S. Qiu, G. Zhu. Targeted synthesis of a porous aromatic framework with high stability and exceptionally high surface area. *Angewandte Chemie International Edition* 48, no. 50 (2009): 9457–60.

Carta, M., R. Malpass-Evans, M. Croad, Y. Rogan, J. C. Jansen, P. Bernardo, F. Bazzarelli, and N. B. McKeown. An efficient polymer molecular sieve for membrane gas separations. *Science* 339, no. 6117 (2013): 303–07.

Deniz, E., F. Karadas, H. A. Patel, S. Aparicio, C. T. Yavuz, and M. Atilhan. A combined computational and experimental study of high pressure and supercritical CO_2 adsorption on basolite MOFs. *Microporous and Mesoporous Materials* 175, no. 2013: 34–42.

Furukawa, H. and O. M. Yaghi. Storage of hydrogen, methane, and carbon dioxide in highly porous covalent organic frameworks for clean energy applications. *Journal of the American Chemical Society* 131, no. 25 (2009): 8875–83.

Ko, D., H. A. Patel, and C. T. Yavuz. Synthesis of nanoporous 1,2,4-oxadiazole networks with high CO_2 capture capacity. *Chemical Communications* 51, no. 14 (2015): 2915–17.

Lu, W., J. P. Sculley, D. Yuan, R. Krishna, Z. Wei, and H. C. Zhou. Polyamine-tethered porous polymer networks for carbon dioxide capture from flue gas. *Angewandte Chemie International Edition* 51, no. 30 (2012): 7480–84.

Martin, C. F., E. Stockel, R. Clowes, D. J. Adams, A. I. Cooper, J. J. Pis, F. Rubiera, and C. Pevida. Hypercrosslinked organic polymer networks as potential adsorbents for pre-combustion CO_2 capture. *Journal of Materials Chemistry* 21, no. 14 (2011): 5475–83.

Mu, B., and K. S. Walton. High-pressure adsorption equilibrium of CO_2, CH_4, and CO on an impregnated activated carbon. *Journal of Chemical and Engineering Data* 56, no. 3 (2011): 390–97.

Patel, H. A., S. H. Je, J. Park, D. P. Chen, Y. Jung, C. T. Yavuz, and A. Coskun. Unprecedented high-temperature CO_2 selectivity in N_2-phobic nanoporous covalent organic polymers. *Nature Communications* 4 (2013): 1357.

Patel, H. A., S. H. Je, J. Park, Y. Jung, A. Coskun, and C. T. Yavuz. Directing the structural features of N_2-phobic nanoporous covalent organic polymers for CO_2 capture and separation. *Chemistry—A European Journal* 20, no. 3 (2014): 772–80.

Patel, H. A., F. Karadas, J. Byun, J. Park, E. Deniz, A. Canlier, Y. Jung, M. Atilhan, and C. T. Yavuz. Highly stable nanoporous sulfur-bridged covalent organic polymers for carbon dioxide removal. *Advanced Functional Materials* 23, no. 18 (2013): 2270–76.

Patel, H. A., F. Karadas, A. Canlier, J. Park, E. Deniz, Y. Jung, M. Atilhan, and C. T. Yavuz. High capacity carbon dioxide adsorption by inexpensive covalent organic polymers. *Journal of Materials Chemistry* 22, no. 17 (2012): 8431–37.

Raja, A. A., and C. T. Yavuz. Charge induced formation of crystalline network polymers. *RSC Advances* 4, no. 104 (2014): 59779–84.

Saleh, M., S. B. Baek, H. M. Lee, and K. S. Kim. Triazine-based microporous polymers for selective adsorption of CO_2. *The Journal of Physical Chemistry C* 119, no. 10 (2015): 5395–402.

Suresh, V. M., S. Bonakala, H. S. Atreya, S. Balasubramanian, and T. K. Maji. Amide functionalized microporous organic polymer (Am-MOP) for selective CO_2 sorption and catalysis. *ACS Applied Materials and Interfaces* 6, no. 7 (2014): 4630–37.

Takamura, Y., J. Aoki, S. Uchida, and S. Narita. Application of high-pressure swing adsorption process for improvement of CO_2 recovery system from flue gas. *The Canadian Journal of Chemical Engineering* 79, no. 5 (2001): 812–16.

Ullah, R., M. Atilhan, S. Aparicio, A. Canlier, and C. T. Yavuz. Insights of CO_2 adsorption performance of amine impregnated mesoporous silica (SBA-15) at wide range pressure and temperature conditions. *International Journal of Greenhouse Gas Control* 43 (2015): 22–32.

Xiang, Z., D. Cao, and L. Dai. Well-defined two dimensional covalent organic polymers: Rational design, controlled syntheses, and potential applications. *Polymer Chemistry* 6, no. 11 (2015): 1896–911.

Xiang, Z., R. Mercado, J. M. Huck, H. Wang, Z. Guo, W. Wang, D. Cao, M. Haranczyk, and B. Smit. Systematic tuning and multifunctionalization of covalent organic polymers for enhanced carbon capture. *Journal of the American Chemical Society* 137, no. 41 (2015): 13301–07.

Xiang, Z., X. Zhou, C. Zhou, S. Zhong, X. He, C. Qin, and D. Cao. Covalent-organic polymers for carbon dioxide capture. *Journal of Materials Chemistry* 22, no. 42 (2012): 22663–69.

Xu, Y., S. Jin, H. Xu, A. Nagai, and D. Jiang. Conjugated microporous polymers: Design, synthesis and application. *Chemical Society Reviews* 42, no. 20 (2013): 8012–31.

Zhu, Y., H. Long, and W. Zhang. Imine-linked porous polymer frameworks with high small gas (H_2, CO_2, CH_4, C_2H_2) uptake and CO_2/N_2 selectivity. *Chemistry of Materials* 25, no. 9 (2013): 1630–35.

12.7 Postcombustion Carbon Capture Using Polymeric Membrane

Nejat Rahmanian and Sina Gilassi

One of the most important challenges that human beings are now experiencing is the emission of greenhouse gases (GHGs). The increase in GHG concentration including water vapor (H_2O), carbon dioxide (CO_2), nitrogen oxide (N_2), and ozone (O_3) in the earth's atmosphere causes the phenomenon of global warming. Due to GHG emissions, the climate change dramatically impacts on environment and severely threatens the life on earth. Thus a reduction scheme of the emission is necessary to control and mitigate the emission of such gases. On December 11, 1997, the advent of Kyoto protocol was a beginning for parties to commit plans binding to the reduction of GHG. In this way, the target is to stabilize the concentrations of GHG in the atmosphere and consequently decrease the perilous effect of anthropogenic interference with the climate system on the environment.

Carbon dioxide that is regarded as the main contributor to GHGs is mostly produced by the combustion of fossil fuels (natural gas, oil, and coal) in the field of energy, transportation, and industry. Many industries such as power plants, cement factories, and gas refineries are faced with some difficulties in gas removal processes, for instance, the separation of CO_2, SO_x, and NO_x from flue gas or CO_2 and H_2S from natural gas. Apart from the emission of such gases in the atmosphere, not only the presence of CO_2 and other acidic components reduces the heating value of natural gas, but also compounds a corrosive and acidic stream that is hardly compressed and even damages the pipeline system. At present, the CO_2 capture technologies used in current industries are classified into four processes (1) absorption, (2) adsorption, (3) cryogenic, and (4) membrane. Focusing on the CO_2 removal process, the advantages and disadvantages of the methods are appropriately highlighted. The energy-intensive processes of regeneration, corrosive nature of alkanolamine solutions, solvent degradation, and size of absorber are the main drawbacks in the absorption process. Low CO_2 selectivity and the effect of moisture and SO_2 on separation process and high regeneration temperature all together limit the use of the adsorption process. Based on the separation mechanism, water removal and feed cooling units are installed before a main distillation unit and consequently cryogenic process becomes highly energy intensive.

The membrane is a new technology that is widely used in various industries such as food, biotechnology, and pharmaceuticals. The simple mechanism and promising results have attracted the academia and industry to focus on membrane development to use in the CO_2 separation process. In comparison with the conventional methods, this technology can be regarded as a replacement for CO_2 removal because of low energy consumption, simple design and scale-up, ease of installation, and low capital and maintenance cost. However, there are some limitations of using this technology within which a small number of materials can be chosen for a CO_2 removal process under certain operating conditions. The main drawbacks concern low selectivity of membrane material and a reverse relation between permeability and selectivity. Nevertheless, numerous studies and researches have been carried out to find the difficulties with fabrication, improve the physical and chemical properties, and industrialization over the past years.

The objective of this chapter is to introduce membranes used for CO_2 removal. With respect to the material, the membranes are classified into organic or polymeric and inorganic or ceramic membranes. The mixed matrix membranes (MMMs) are also another type that stands to benefit from the incorporation between organic and inorganic material properties. Moreover, the membrane fabrication and improvement methods for CO_2 separation are described and then experimental results highlight the advantages and disadvantages of using particular membranes under different operating conditions. Finally, the derivation of mathematical models and a CFD procedure are also studied for hollow fiber and dead-end contactors. This source stems from the background and literature of using different types of membrane in CO_2 capture fields and hopefully aims at those who are seeking for the technical information for their research goals and design projects.

12.7.1 Polymeric Membranes

12.7.1.1 Basic Principles and Equations

The solution-diffusion model is suggested to describe the permeation of various gases into the dense polymeric membranes (Wijmans and Baker 1995). The importance of this model is that the transport phenomena across a polymeric film can be explained. Park and Crank (1968) defined the gas transport by the following steps: first, the permeate diffuses through the upstream boundary layer; second, it is dissolved at the membrane surface; third, it diffuses through the material structure of the membrane; and fourth, it is desorbed from the low-pressure side and finally passes through the downstream boundary layer. According to the theory of solution-diffusion model, two gas components in a binary gas mixture are in a competition for passing through the polymeric membrane. In this case, selectivity (S) is defined on the basis of interaction between the molecular structure of a polymeric membrane and kinetic diameters while a gas molecule with lower size passes through the film rather than the larger size. Permeability (P) is also defined on the basis of gas solubility that depends on gas condensability and polymer–penetrant interactions (Li et al. 2011). For a gas molecule (A), the permeability (P_A) in a nonporous polymeric membrane is proportional to the gas solubility (S_A) and diffusivity coefficient (D_A) as follows:

$$P_A = S_A \times D_A. \tag{12.50}$$

And also for a binary gas mixture including A and B components, the selectivity (α_{AB}) is the ratio of the high-permeate gas (A) to the low-permeate gas (B) and expressed as follows:

$$\alpha_{AB} = \frac{P_A}{P_B} = \frac{S_A \times D_A}{S_B \times D_B}. \tag{12.51}$$

The pore flow model is used to define the gas transport in a capillary or microporous membrane. Unlike the solution-diffusion model, the driving force for transport is the gradient of pressure on both the membrane sides. Gas permeates pass through tiny pores and some of them are filtered in the pores due to the molecular size (Richard 2004). Fick's and Darcy's laws justify the gas transport phenomena in two models with concentration and pressure gradients on both membrane sides. The free-volume elements (pores) explain the difference between two models as the polymeric chains are fixed, the gas permeate moves through the pores by pore-flow concept, whereas the polymeric chains fluctuate in a position to randomly make spaces and the gas permeate traverses across the varied-size pores by solution-diffusion concept.

Numerous polymeric membranes have been produced and tested under different operating conditions for the CO_2 capture process. Based on the trade-off between selectivity and permeability (Robeson 1991), most of the researches attempt to improve the physical and chemical properties and find ideal materials structure. The polymeric membrane materials are classified into two groups: glassy and rubbery. The difference in gas–liquid transition temperature, T_g, grants particular properties to each group (Sperling 2015). In general, glassy membranes tend to be less permeable and highly selective, whereas rubbery membranes are highly permeable and less selective (Mark 1996). Therefore, targets of the ongoing research are to fabricate high CO_2 selective and permeable membranes with a capacity for commercialization.

12.7.1.2 Studies on Polymeric Materials

12.7.1.2.1 Polyimides

This section aims to introduce the polymeric membranes and highlight their separation mechanisms for CO_2 capture. The glassy membranes are commonly suggested to purify natural gas from the contaminants including CO_2, H_2S, and H_2O. However, there is still a need to improve CO_2 selectivity of 30–40% and invest in an efficient removal operation compared to the conventional methods (Baker 2001). In this category, polyimide (PI) has shown good thermal stability, chemical resistance, mechanical strength, and high selectivity of CO_2/CH_4 (Cecopieri-Gómez et al. 2007). Moreover, there are some criteria such as reproducibility, resistance to plasticization and swelling, and economical potential in use of polymeric membranes for CO_2 separation process. The PIs that are commonly used in membrane technology include hexafluoroisopropylidene-diphtalic anhydride 6FDA-durene, Lenzing P84, Matrimid, Torlon, Ultem Polyetherimide, and Kapton (Vanherck et al. 2013). The monomers that are linked to the intrinsic PIs play a significant role in obtaining a special characteristic of formed polymers (Liaw et al. 2012).

The new approaches are required to improve the PIs characteristics and also prepare them as desirable membrane materials for CO_2/CH_4 separation. The molecular tailoring and design of a block copolymer system can be considered to grant better properties. There have been other approaches that are more effective and low priced compared to the new material synthesis. The PI membranes can also be modified by using polymer blending, hybrid membrane, thermal annealing, backbone grafting, and cross-linking methods. Staudt-Bickel and Koros (1999) synthesized the cross-linked and uncross-linked PIs and copolyimides to increase the CO_2 selectivity as well as to reduce the swelling effect. Copolyimides with carboxylic acid groups reduce the plasticization due to hydrogen bonding between the carboxylic acid groups and the permeability reduces significantly. The selectivity was increased about 20% by crosslinking of 10% because of the high mobility of polymeric chains. One of the attractive aromatic PIs is 6FDA that is commonly used for gas separation because of the CF3 groups (Xiao et al. 2009). Wind et al. (2004) reported cross-linking of PI membranes derived from 6FDA-DAM: -DABA with 1,4-butylene glycol and 1,4-cyclohexanedimethanol, which increased CO_2 permeability and enhanced the plasticization. Qiu et al. (2013) reported the permeation performance of six 6FDA-based PIs with different chemical structure. They adjusted the packing density by monomer choice and composition, and the permeability of DAM-based PI was found to be higher than other diamines. Cui et al. (2011) studied the effect of DABA incorporation with PIs on CO_2 plasticization, the deficiency was altered by a thermal annealing approach to decarboxylate DABA units, and the cross-linking enhanced the resistance against CO_2 plasticization at high pressure. Kim et al. (2006) correlated permeability of CH_4, N_2, and CO_2 with aging time at 35°C. They noticed a significant reduction in permeability and an increase in selectivity for the crosslinked PIs with DABA units. Bos et al. (1998) utilized the PI Matrimid 5218 as a model polymer, a thermal treatment at 350°C avoided the acceleration of methane permeation due to CO_2 plasticization. Tin et al. (2003) modified the cross-linking of Matrimid 5218 by converting the imide functional groups to amide functional groups during the modification process, and the permeability reduces in the order of $CO_2 > CH_4 > N_2$. Zhang et al. (2008b) combined a microporous metal-organic framework with the Matrimid polymer, Cu–BPY–HFS, which decreased CO_2/CH_4 selectivity from 83 to 35 and favored the CH_4 permeation because of the high affinity with CH_4. Xia et al. (2014) investigated the behavior of aging and CO_2 plasticization of Matrimid in the form of thin film. The result of their work showed that increasinge of CO_2 plasticization was attributed to the reduction in film thickness as well as increasing of the aging time.

12.7.1.2.2 Perfluoropolymers

Other types of glassy polymers are perfluoropolymers (PFs). A combination of outstanding chemical and physical properties draws great interest in using them in a wide range of industrial applications. They are strongly resistant to a variety of hostile chemicals such as acids, alkalis, esters, ethers, ketones, and oxidizers. In addition, the high-energy bonds of C–F (485 kJ/mol) and C–C (360 kJ/mol) make them unique to be used at high temperature (Arcella et al. 1999, 2003). The amorphous PFs commercially used include Hyflon AD, Teflon AF, and Cytop; the polymerization procedure gives them varied glass transition temperatures (T_g), for instance, the polymerization of tetrafluoroethylene (TFE) and 2,2,4-trifluoro-5-trifluoromethoxy-1,3-dioxole (TTD) that produce Hyflon AD 40 and 60 (Arcella et al. 2010). Teflon AF2400 and Cytop are also the most and least permeable materials among fluorinated polymers, respectively (Freeman et al. 2006). Adding TTD aims to enhance the processability, gradually lower the gas permeability, and also improve the selectivity for these polymers. The relatively high gas permeability is normally attributed to the dioxole rings in these materials that hinder the polymer chain packing (Okamoto et al. 2014). The absolute permeability of both thick and thin membrane undergoes the casting process and solvent type, whereas the aging rates are highly dependent on the solvent type rather than the casting type (Tiwari et al. 2014). Moreover, CO_2 plasticization varies with the thickness and also the aging rates. CO_2 sorption for PFs is reduced by the decrease in excess volume and increase in the aging time (Tiwari et al. 2015).

Numerous experimental works have been carried out to modify the structure of these types of membrane and enhance the CO_2 permeability and selectivity. As it was argued earlier, different modification approaches can be used to alter the gas transport and attribute new properties to the primary material. In this case, Macchione et al. (2007) studied the effect of solvent for solution casting of Hyflon AD membrane on the thermal, mechanical, and transport properties. The analysis revealed that the residual solvent plasticized the polymer, reduced the permselectivity, and highly increased the diffusivity coefficient. Scholes et al. (2015) measured the gas permeability for Teflon AF1600 and Hyflon AD60 in the presence of water in a gas mixture including CO_2 and CH_4. The low water permeability was attributed to the low solubility of water in both hydrophobic types that induced reduction of CO_2 and CH_4 due to the water cluster blocking the pathways of gas transport. Furthermore, the gas separation data obtained from literature also revealed that the gas permeability for Hyflon AD copolymers and Teflon AF membrane linearly changed with the glass transition temperature (T_g) (Arcella et al. 1999). Teflon materials as membrane have outstanding advantages such as chemical stability, swelling resistance, and high permeability. Gilassi and Rahmanian (2013) investigated the permeability and selectivity of PTFE membrane for a binary mixture of CH_4 and CO_2. Later, Idris and Rahmanian (2014) also conducted an experiment regarding grafting of various amine solution to an intrinsic Polytetrafluoroethylene (PTFE) membrane film at 80°C and different immersion time. The grafting yield was not increased by adding the higher dose of MDEA grafting solution, whereas an increase was seen for MEA and DEA grafting solutions because of the availability of reactants at the membrane active site.

12.7.1.2.3 Cellulose Acetate

Another polymeric membrane is cellulose acetate (CA) that draws attention of researchers and scientists in reverse osmosis, ultrafiltration, and gas separation field because of unique properties. The product is highly hydrophobic, flexible, and also has very high flux. Having these properties made them very interesting for CO_2 gas separation so that the main usage is in offshore plants where the capital and maintenance cost need to be minimized. Even, the capture plant with CA is as effective as amine absorption for the amount of CO_2 over 20 mol% (Stern et al. 2000). CA is the product of a reaction between cellulose with acetic anhydride and acetic acid in the presence of sulfuric acid (Fischer et al. 2008). With respect to the degree of acetylation, this type of membrane is named CA, cellulose diacetate and cellulose triacetate (Chen et al. 2015). Figure 12.111 shows the chemical structure of CA. As discussed earlier, the mobility of polymer bone is proportional to the permeability and selectivity of species of a gas mixture. The substitution of hydroxyl with acetyl groups affects the chain packing and consequently improves the mobility (Kamide and Saito 1987).

On the other hand, CA membrane is not resistant against the plasticization phenomena aroused by CO_2 and heavy hydrocarbons that highly decline separation efficiency, physical stability, and life time of membrane because of the increase of polymeric chain mobility. Table 12.32 shows the performance of PI membrane in CO_2/CH_4 separation.

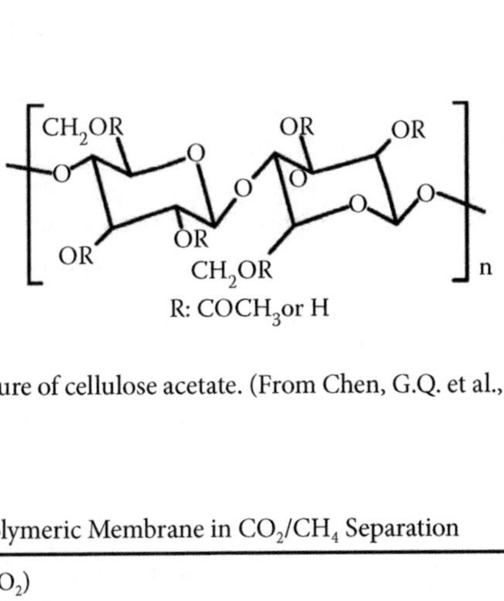

FIGURE 12.111 Chemical structure of cellulose acetate. (From Chen, G.Q. et al., *J. Membr. Sci.*, 487, 249–255, 2015.)

TABLE 12.32 Performance of Polymeric Membrane in CO_2/CH_4 Separation

| Membrane | Permeability (CO_2) | | Selectivity (CO_2/CH_4) | Pressure (bar) | Temperature (°C) | Reference |
	Value	Unit				
6FDA-BAPAF	24.6	GPU	22.78	30	21	Kim et al. (2003)
6FDA-DAP	38.57	GPU	77.82	30	21	Kim et al. (2003)
6FDA-DABA	26.3	GPU	46.96	30	21	Kim et al. (2003)
6FDA-DAT	59	GPU	40	7	20	Cao et al. (2002)
Matrimid 5218	10	Barrer	35.71	1.1	20–25	Vu et al. (2003)
Polycarbonate	2	Barrer	27.2	20	30	Sridhar et al. (2007)
Polyamides	11	Barrer	36.3	2	35	Ghosal et al. (1995)
PSf	80.7	GPU	40.2	5	25	Ismail et al. (1997)
Cellulose acetate	2.5	GPU	20	8	35	Visser et al. (2007)

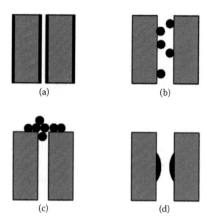

FIGURE 12.112 Schematic drawing of the surface modification of ceramic membrane. (a) Internal deposition of pores by monolayer or multi-layer; (b) pore-plugging of nanoparticles; (c) coating layer on top of the membrane; and (d) constrictions at sites in the top layer. (Mulder 1996).

12.7.2 Ceramic Membranes

The research efforts on development of membrane material are still being conducted to improve the adverse physical and chemical characteristics of the existing products used in the CO_2 separation process. Finding a typical membrane with high-CO_2 selectivity and permeability that is able to resist against acidic gases, be stable at high temperature and also have long and reliable service life is a bit difficult. It seems that ceramic membranes can be more suitable for gas separation process than their polymeric counterparts which fail under that operating condition. They can withstand a higher temperature of 500°C, can be chemically inert, and can also be used in harsh environments. Even, the steam is used to clean them from infections and also the removal of fouling is carried out by backflushing (Noble and Stern 1995). These types of membranes include a support layer with pore size in a range of 5–15 μm and a porosity of 30%–50%, and another layer formed by suspension coating with pore size of 0.2–1 μm. This thin layer is stabilized by slip-coating–sintering procedure to transform the intrinsic pore size to lower nanoscale from 10 to 100 A (Mulder 1996). The sol–gel process is particularly considered as an effective method for membrane modification, the surface and/or active site of the membrane is covered by a catalytic layer or bulk, and this mechanism is highly dependent on the catalyst and membrane material. Figure 12.112 shows schematic drawing of the surface modification of ceramic membranes. More detail about the synthesis of inorganic membrane is also available in literature (Uhlhorn et al. 1991).

Normally, ceramic membranes are divided into two main classes: dense (nonporous) and porous membranes. The porous ceramic membrane consists of distinctive layers which are mainly made of amorphous silica, carbon and zeolite, an intermediate and selective layers as well. The top layer grants the separation mechanism and has pore size smaller than 2 nm, the intermediate layer is like a filler between two porous layers and finally the support layer provides the mechanical strength for the whole membrane structure. Like the polymeric membrane, the selectivity and permeability of the top layer depend on the pore size, porosity, diffusivity, and solubility of the species. Therefore, the molecular size, shape, and affinity to the membrane structure are the important parameters in selection of material layers. The dense ceramic membrane has a thin layer on a support layer and sometimes without a support layer, this layer is also made of perovskites or palladium which enhances gas separation characteristics. The use of this type of membrane is limited and normally applied for the separation of a few gases like hydrogen (H_2).

12.7.2.1 Transport Mechanism

Similar to the polymeric membranes, the ceramic membranes are also classified into porous and non-porous types. Figure 12.113 illustrates various gas transport mechanisms in porous membranes. The gas

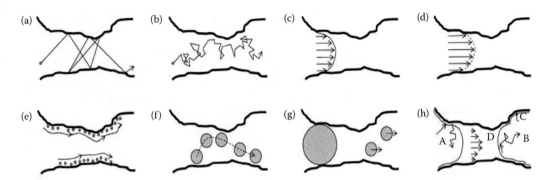

FIGURE 12.113 Gas transport mechanism in porous membranes. (From Drioli, E., Giorno, L., *Comprehensive Membrane Science and Engineering*, Vol. 1, Newnes, 2010.)

transport mechanisms for a porous membrane involve Knudsen diffusion, surface diffusion, capillary condensation, laminar flow, and molecular sieving. Transport of molecular gases through a membrane may be based on an individual or a combination of the mentioned mechanisms. In this case, two variables can be defined to properly describe the gas transport, Knudsen number that is a ratio of mean free path, λ, to average pore size, d_p and also the second index, b_m, is a ration of molecule size, d_m, to the pore size (Civan 2011).

$$Kn = \frac{\lambda}{d_p} \tag{12.52}$$

$$b_m = \frac{d_m}{d_p}. \tag{12.53}$$

In membrane with pore size of more than 10 nm, the viscous flow and surface diffusion take into account the molecular transport through the pores and wall surface. The selected layer with these regions is appropriate for a composite membrane as a support layer that can enhance the flow resistance in a separation system. The flow changes to convective flow and penetrate is transported through the membrane at the flux that is dependent on the mole fraction and also pressure gradient of both sides. This mechanism cannot individually separate a gas mixture and usually conforms to other diffusion mechanisms. The surface diffusion is mainly applied to discuss the gas penetrate through the pore in which the molecules have a tendency to adsorb at the pore walls and create gradient concentration and finally desorb at the gas phase. Knudsen diffusion is defined for a membrane with pore size between 2 and 100 nm, and is the result of a low-pressure condition in which mean free path increases and also small pore size. In this case, penetrate molecules frequently collide with the pore walls even more than other molecules and the direction of molecules changes due to the collisions. This prediction is totally different from molecular diffusion in which the penetrate molecules tend to collide together rather than the pore walls and that is seen for $Kn < 0.01$ when the pressure is about ambient and also the pore size is too large. The molecular sieve mechanism is similar to a filtration process in which a gas mixture is separated based on the size of molecules. The size of membrane pores is between the size of small and large molecules in the gas mixture and the separation process needs cycling in order to obtain acceptable efficiency (Drioli and Giorno 2010).

12.7.2.2 Ceramic Membrane Used for Gas Separation

12.7.2.2.1 Zeolite Membrane

This type of membrane is composed of a polycrystalline zeolite layer deposited on a support inorganic layer. Zeolites are microporous materials constructed from TO_4 tetrahedra, with T being the atom in tetrahedral position (Si, Al, B, Ge, Fe, and P) (Martínez and Corma 2013). In contrast to the polymeric membranes, zeolite membranes are physically and chemically more stable and they keep the initial structure and do not swell and inflame in contact with contaminated species. Moreover, the transport mechanism for some gas species is on the basis of molecular sieving due to the uniform and molecular size pores. Having these unique characteristics make them very attractive for gas separation based on filtration by size or diffusion. For the first time, zeolite membrane was introduced by Suzuki (1987) and then the following research continued to classify the achievement into 13 distinctive types such as Zeolite Socony Mobil -5 (ZSM-5 (five)), Linde type A (LTA), more details are also available in the reference (Bowen et al. 2004; Hilal et al. 2012). The zeolite membrane is also synthesized by hydrothermal and dry gel methods (Nishiyama et al. 1995; Matsufuji et al. 2000; Ismail et al. 2015) in which a gel containing other substitute atoms for tetrahedral framework is gradually crystalized at the top of a support layer such as alumina. There are different types of support layers that have great influence on the crystallization process despite other parameters such as time, temperature, and concentration. Based on the molecular size of gas components, the support layer is chosen, for instance, alumina support (γ-Al_2O_3 and α-Al_2O_3) layer has a pore diameter from 5 to 200 nm. Polymer–zeolite membrane is an accurate combination between physical and chemical properties of two materials, which leads to a MMM. The zeolite crystals are dispersed on dense polymeric film while the zeolite pores become clogged by the polymer matrix and makes a nonselective diffusion pathway (Cejka et al. 2007).

12.7.3 Modeling Approach

The modeling and theoretical study of CO_2 separation processes is an importance of industrial plan as both process efficiency and cost need to be estimated in terms of profitability. The combination of mathematical equations relating to fluid dynamic, thermodynamic, and mass transfer has perfected a reliable way to assure feasibility of a project. In this case, numerous commercial simulators have been released to mathematically show hidden parts of an experiment and also aim to understand a removal process. A new model can be derived based on continuity and momentum equations to use for a particular type of contactor while a series of assumptions needs to be considered to simplify both removal concepts and numerical solutions. The typical contactors used for a gas separation are spiral wound and hollow fiber modules that the configurations are analogous to tube-in-tube and shell-and-tube types of a heat exchanger, respectively. In terms of the flow configuration setup, the membrane contactors are also composed of countercurrent, cocurrent, and cross-flow types in which the separation efficiency is varied. However, other parameters, such as fiber bundle, module size, and membrane material affect the removal performance. When a suggested model is constructed, the next step is to validate modeling result by comparison with an experimental work. The output of developing a new model is either to integrate into a commercial software, for instance, Ansys, Fluent, and COMSOL or introduce new software. The computational fluid dynamic (CFD) is also available to run the model within various research areas and visualize the results while a variable is manipulated. In this case, the gas transport in the contactors can be modeled on the basis of CFD analysis to expose dismal points of design. The following part shows various modeling approaches suggested for CO_2 capture by membrane as well as using numerical solutions in details for those who are involved in CFD and modeling fields.

The hollow fiber membrane contactor (HFMC) is composed of many thin tubular membranes bundled and sealed on end of module with epoxy in a housing. They are commonly used for CO_2 removal compared to other module types, and designed for low-pressure process. First, the feed with low flow rate enters a membrane module with high surface area and low holdup volume, then the penetrants pass through the membrane based on their permeability and the nonpenetrates remain in the residue stream, and finally, the penetrates leaves the module in permeate stream.

Shindo et al. (1985) developed calculation methods for a single-stage permeation of multicomponent gas mixture for different flow patterns. Figure 12.114 shows a schematic diagram of single-stage permeation module with cocurrent flow. The model was derived based on Fick's law and the following

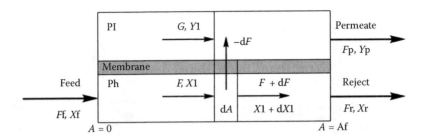

FIGURE 12.114 Schematic diagram of single-stage permeation module with cocurrent. Ff, feed flow rate; Xf, feed composition; Fr, retentate flow rate; Xr, retentate composition; Fp, permeate flow rate; Xp, permeate composition; dA, membrane area; Ph, high pressure; PI, low pressure; F, flow rate; G, flow rate; Y, composition; X, composition. (From Shindo, Y. et al., *Sep. Sci. Technol.*, 20(5–6), 445–459, 1985.)

assumption were made: the permeability of gas components was independent of temperature; there was no pressure drop in both permeate and feed compartments; the thickness of membrane was constant alongside permeator and the concentration gradient in permeation direction was negligible.

The material balance over the differential area d*A* for component *i* can be written as follows:

$$-\mathrm{d}(x_i F) = \mathrm{d}(y_i G) = \mathrm{d}A \frac{Q_i}{\delta}\left(P_h x_i - P_l y_i\right). \tag{12.54}$$

Pan (1986) also developed a mathematical model of multicomponent permeation system for high-flux, asymmetric hollow fiber membrane in which the pressure variation inside the fiber was studied. The governing equation needed solving by an iterative method that was dependent on the initial guess for pressure. Murad Chowdhury et al. (2005) proposed a new numerical method whereby a system of equation was solved with an algorithm for countercurrent flow configuration. The solution started calculating the initial permeate composition by nonlinear equation solver until satisfying the condition related to the unity of summation of all permeate compositions. Later, Khalilpour et al. (2013) presented a new solution to transform at first the ODE of mass balance equation for both cocurrent and countercurrent flow to a simple algebraic equation by backward finite differential equation over all segments. This method had a difficulty in guessing the initial value for residue flow rate and permeate pressure which made the iteration too long as a given tolerance needed to be satisfied for every segment. More details are also available in the reference (Khalilpour et al. 2012; Kundu et al. 2012).

Bansal et al. (1995) developed a new method for modeling of separation of the multicomponent gas mixture using a membrane module. In this model, the flow was assumed to be plug and full mixed for all types of flow configuration. Apart from the material balance over every element, the initial step was to calculate the permeate composition at the module inlet and/or outlet by Newton's iterative procedure. The modeling result also revealed that the separation efficiency reduced from the highest to lowest in the following order countercurrent, cross-flow, cocurrent, and full mixing flow. However, at the lower-pressure ratio, P_r, the separation efficiency reduced and the permeate composition for all types became equal. More theoretical study on the performance of single stage permeation and influence of pressure, membrane area, and flow pattern is also available in the literature (Walawender and Stern 1972).

Thundyil and Koros (1997) presented a new modeling approach regarding mass transfer in permeation by a hollow fiber module. The proposed algorithm benefited the succession of states method (SSM) to separate the module into small size segments in which the mass transfer driving force remained constant. For the cross-flow modeling under isothermal conditions, it was assumed that the pressure variation in the module was neglected in both radial and axial directions but pressure variation in the tube was determined by Hagen–Poiseuille equation (Berman 1953; Pan and Habgood 1978; Pan 1986; Chern et al. 1985). The feed gas concentration was not varied alongside the module and the variation occurred

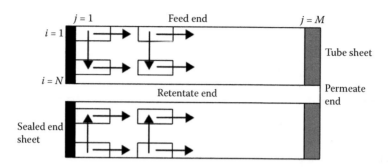

FIGURE 12.115 Schematic representation of flow patterns in radial cross-flow. i, element number; j, element number. (From Lock, S.S.M. et al. *J. Ind. Eng. Chem.*, 21, 542–551, 2015).

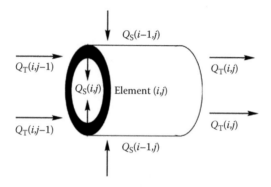

FIGURE 12.116 Elements of mass transfer for radial cross-flow. (From Thundyil, M.J., Koros, W.J., *J. Membr. Sci.*, 125(2):275–291, 1997.)

in the radial direction; this assumption is not valid for long fiber module. Moreover, the permeability coefficient was independent of pressure, temperature, and concentration. Figure 12.115 illustrates the schematic representation of flow patterns in the radial cross-flow.

Figure 12.116 also shows the schematic mass transfer for an element in the radial cross-flow. The mass transfer balance is written for each ring-shaped element with radius r, thickness Δr and Δz and then the input flow, Q_S, is divided into the permeate, Q_T, and the new input flow, Q_S, for the next element. The compositions and flow rates of permeate and residue constitute a system of nonlinear equations that can be solved by a numerical method such as Newton or Bisection methods.

$$Q_S(i,j) = Q_S(i-1,j) - \Delta Q \tag{12.55}$$

$$Q_T(i,j) = Q_T(i,j-1) + \Delta Q \tag{12.56}$$

$$x_1(i,j) = [Q_S(i-1,j)x_1(i-1,j) - \Delta Q_1]/Q_S(i,j) \tag{12.57}$$

$$y_1(i,j) = [Q_T(i,j-1)y_1(i-1,j) + \Delta Q_1]/Q_T(i,j) \tag{12.58}$$

This procedure can be used for cocurrent and countercurrent flows as well, more details on mass balance equations is available in the literature (Thundyil and Koros 1997). In order to validate the suggested model, the removal of carbon dioxide from methane was studied under different conditions. The findings revealed that the cross-flow system was more effective than the other systems that were attributed to the proper feed distribution in this type of flow and also the countercurrent system was appropriate for the module with large bundle size. Later, Ahmad et al. (2013) modified this model to reflect the effect of temperature and pressure on membrane permeation. Unlike the isothermal condition, the experimental work has been shown that the temperature of expanded nonideal gas changes due to Joule Thomson (JT) effect; this phenomenon is inevitable during the CO_2 capture process, while the enthalpy under permeation through the membrane is related to the heat released. Marić (2005, 2007) derived a numerical procedure to calculate the JT coefficient of natural gas, an equation correlated the change in rate of compression factor to the temperature at constant pressure. The general form of JT equation is expressed by Equations 12.59 and 12.60 as follows:

$$\mu_{JT} = \frac{R_g T^2}{\rho C_{m,p}} \left(\frac{\partial Z}{\partial T} \right)_P \tag{12.59}$$

$$T_1 - T_2 = \mu_{JT} \Delta \omega, \tag{12.60}$$

where R_g, T, and Z stand for universal gas constant, temperature, and compression factor, respectively. ρ, $C_{m,p}$, and $\Delta \omega$ are density, molar heat capacity, and pressure drop, respectively.

Coker et al. (1998) presented a model of multicomponent gas separation system using a hollow fiber membrane for all types of flow patterns. The model was also compatible with any change in pressure sweep, permeability coefficient, and pressure gradient on both sides of the membrane. The fiber module was divided into N segments and mass balance is enforced for all of them to have a system of differential equations. But the solution method is complicated as initial guess needs to be provided for the component flow rate on each stage.

Lock et al. (2015) developed a new modeling approach based on Multicomponent Progressive Cell Balance for CO_2 capture process using a hollow fiber membrane. The calculation method was started by an initial guess of first component and then initial stage cut, θ^*, was obtained by Equation 12.61. The flow rate of permeate and retentate were also calculated by initial stage cut and the composition of other permeate components was also determined by the new form of Equation 12.61. Finally, the most accurate value for permeate and retentate composition and flow rates were calculated by a numerical iteration. This method can also be integrated into both cross- and countercurrent flow configurations.

$$y_{T,n}[j]\theta^* Q_S[j-1] = P_n A_m \left\{ \frac{P_h}{1-\theta^*} (x_{S,n}[j-1] - \theta^* y_{T,n}[j]) - p_l y_{T,n}[j] \right\}. \tag{12.61}$$

As mentioned earlier, the model is derived on the basis of mass balance for i component over particular segments. The newer approach is to develop a model based on continuity and momentum equations. Gilassi and Rahmanian (2015b) proposed a numerical model for CO_2 separation using a flat-sheet membrane in a dead-end permeation cell. The new term (Equations 12.61 and 12.62) related to the slip condition (Nield 2009) at the membrane surface was added to the velocity boundaries, then the gas concentration was determined in permeate and retentate chambers.

$$u_x = \frac{Pe^{1/2}}{\alpha_{BJ}} \frac{\partial u_x}{\partial y} \tag{12.62}$$

$$u_y = \frac{Pe^{1/2}}{\alpha_{BJ}} \frac{\partial u_y}{\partial y}. \tag{12.63}$$

The numerical solution is on the basis of finite volume method (FVM) and finite difference method (FDM). At first, the momentum equations are solved by a numerical method to determine the velocity profile inside the feed permeation chamber. In this way, the SIMPLE algorithm (Patankar and Spalding 1972) is used to specify the pressure and velocity in each node. The gas concentration is then calculated for each node with the aim of known value for velocity. The subroutine of solution algorithm with both continuity and momentum equations can be written in mathematical software such as MATLAB®. Other methods for the solution of momentum equations were completely explained in literature (Koukou et al. 1996, 1998, 1999). More information about the SIMPLE algorithm for diffusion–convection equation is also available in Versteeg and Malalasekera (2007).

As discussed earlier, hollow fiber contactors are used to separate gas components based on the molecular size, diffusivity, and solubility of a typical gas mixture. A selected membrane also undergoes some reinforced process to equalize the gas selectivity and permeability in an accepted region of upper bound curve suggested by Robeson. When the tube side is filled with an absorbent which is capable of reacting with specific gas component, for instance, the amine solutions are highly preferable to use in CO_2 capture process because of having good affinity with CO_2. However, high-rich CO_2 is required to purify the regeneration process that is highly energy intensive. In fact, the presence of a liquid phase flowing in tube side is to remove penetrates by a reaction and impede passing of nonreactive penetrates. The membrane is a physical barrier between liquid and gas phases and prevents the liquid penetration. Figure 12.117 shows a schematic diagram of hollow fiber membrane contactor. A mathematical model can be developed for the contactor of CO_2 separation including shell, membrane, and tube. Numerical software such as COMSOL is able to find a solution for a system of stiff and nonstiff differential equations.

The assembling concept of hollow fiber membrane is fairly similar to heat exchangers. As mentioned before, a bundle of fibers inside the module has no support and also the tube is surrounded by the fluid. With respect to the scale of fiber module and numbers, an imaginary shell is assumed over each tube and the fluid flow has no influence on the permeation beyond that boundary. For the first time, Happel (1959) suggested a free-surface model for the fluid passing axially over a cylindrical tube. It was seen that the fluid velocity had no change out of the boundary for a fluid with no slipping at the wall. Thus this model can be employed for the hollow fiber membrane to define static and dynamic region over the tubes and also calculate the fluid velocity inside the shell as follows:

$$r_3 = \left(\frac{1}{1-\phi} \right)^{1/2} r_2$$

$$1-\phi = \frac{n r_2^2}{R^2}$$

$$V_{z-shell} = 2u \left[1 - \left(\frac{r_2}{r_3} \right)^2 \right] \times \frac{(r/r_3)^2 - (r_2/r_3)^2 + 2\ln(r_2/r)}{3 + (r_2/r_3)^4 - 4(r_2/r_3)^2 + 4\ln(r_2/r_3)},$$

where u, r_3, r_2 denote the average velocity (m/s), radius of the free surface (m), and fiber outer radius (m), respectively. R is the module inner radius (m), φ is the volume fraction of the void, and n is the number of fibers. A typical model for CO_2 absorption by an absorbent consists of three type of equations for shell,

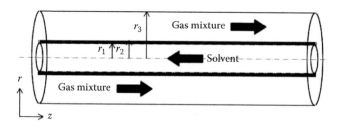

FIGURE 12.117 Schematic diagram of hollow fiber membrane contactor. (From Gilassi, S., Rahmanian, N., *International Conference on Membrane Science and Technology (MST), Kuala Lumpur, Malaysia, 2013.*)

membrane, and tube that solve simultaneously with appropriate boundary condition, details about the equations and model development is available in reference (Wang et al. 2004; Faiz and Al-Marzouqi 2009; Luis et al. 2010; Rezakazemi et al. 2011; Ghasem et al. 2013). Normally, the feed gas fills the pores and channels of polymeric membrane to impede the liquid penetration. Sohrabi et al. (2011) carried out CFD analysis on CO_2 absorption by hollow fiber contactor using different amine solutions and K_2CO_3. The result revealed that CO_2 removal performance with all types of amine solutions was better than K_2CO_3 solution. Furthermore, the proposed model was run to disclose the effect of liquid flow rate on CO_2 removal performance. Figure 12.118 shows the distribution of CO_2 concentration in the membrane contactor.

Increasing the liquid flow rate caused enhanced CO_2 mass transfer rate into the liquid phase. This is because more reactant was available at the gas–liquid boundary layer and this caused to form a high CO_2 gradient between two sides of the membrane. The result also showed that CO_2 removal performance with MEA was the highest among the amine and K_2CO_3 solutions due to high CO_2 reactivity and solubility. Eslami et al. (2011) also developed a model for CO_2 absorption by potassium glycinate (PG) in a hollow fiber contactor and investigated the effect of operating parameters on mass transfer rate and removal efficiency. Increasing the PG concentration caused an increase in the removal efficiency due to the improvement of the active layer at the membrane surface. Increasing the gas flow rate had a negative effect on CO_2 removal efficiency. The reason was that the detention time of CO_2 reduced in the shell, and thus, lower CO_2 was available at the active absorption layer of membrane. Gilassi and Rahmanian (2015a) also developed a mathematical model for CO_2 capture to investigate the performance of three types of amine solutions. The results showed that the MEA solution had better performance than DEA and MDEA because of the high reactivity with CO_2, Figure 12.119 shows the CFD results of CO_2 absorption using three amine solutions in the membrane contactor under same operating condition.

Wang et al. (2005) developed a mathematical model for CO_2 absorption with water under two different operating conditions of wetted and nonwetted modes. The result showed that CO_2 absorption rate for nonwetted mode was about six times higher than the wetted mode. The liquid phase which filled the membrane pores, increases the mass transfer resistance in the membrane and consequently decrease the absorption performance. Zhang et al. (2008a) carried out an experiment and modeling regarding the effect of the wetted membrane on CO_2 chemical and physical absorptions using water and DEA aqueous solution. The theoretical result disclosed that partial and complete wetted membrane increase the membrane mass transfer resistance from 5% to 90% for water but this proportion of membrane resistance immediately increased from 10% to 70% for only 10% of the wetted membrane length. It is expected that the model has new boundary condition for the membrane part filled with the liquid phase and that is the main difference between wetted and nonwetted modes. Figure 12.120 shows hollow fiber membrane in wetted and nonwetted modes.

$$D_{\text{membrane-liquid}}\left(\frac{\partial^2 C_i}{\partial r^2} + \frac{1}{r}\frac{\partial C_i}{\partial r} + \frac{\partial^2 C_i}{\partial z^2}\right) = R_i' \tag{12.64}$$

$$r = r_2 \quad C_{\text{membrane-liquid}} = C_{\text{shell-gas}} \times H \tag{12.65}$$

$$r = r_1 \quad C_{\text{membrane-liquid}} = C_{\text{tube}}, \tag{12.66}$$

where C, r, and H are concentration, radius and Henry's constant, respectively.

All the models discussed in this section have a series of assumptions which simplify the model development and also numerical solution. For example, CO_2 capture using a hollow fiber membrane that is typical of a contactor used for gas separation can turn into a more sophisticated system when the momentum equation for gas and liquid in both shell and tube sides are varied under nonsteady state condition. In this case, it is essential to introduce eddy terms and equations in the model as to the

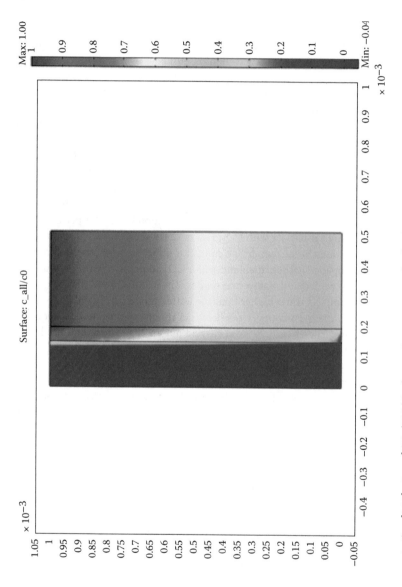

FIGURE 12.118 The concentration distribution of CO_2 (C/C_0) in the membrane contactor for the absorption of CO_2 in MEA. (From Sohrabi, M.R. et al., *Appl. Math. Model.*, 35(1), 174–188, 2011.)

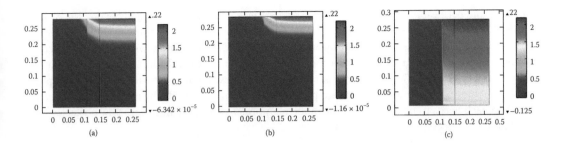

FIGURE 12.119 (a) Concentration of CO_2 in HFMC, $C_{DEA} = 3\,mol/m^3$, $C_{CO2} = 20\%$ vol, $U_g = 5.5\,m/s$, $U_l = 1.5\,m/s$. (b) Concentration of CO_2 in HFMC, $C_{MEA} = 3\,mol/m^3$, $C_{CO2} = 20\%$ vol, $U_g = 5.5\,m/s$, $U_l = 1.5\,m/s$. (C) Concentration of CO_2 in HFMC, $C_{MDEA} = 3\,mol/m^3$, $C_{CO2} = 20\%$ vol, $U_g = 5.5\,m/s$, $U_l = 1.5\,m/s$.

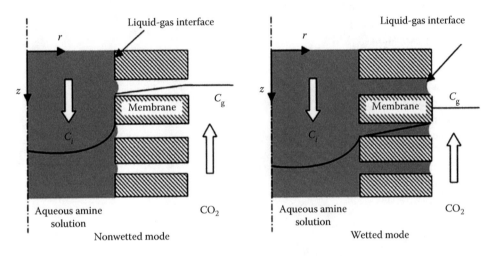

FIGURE 12.120 Hollow fiber membrane in wetted and nonwetted modes. (From Zhang, H.Y. et al., *J. Membr. Sci.*, 308(1):162–170, 2008a.)

turbulent effect is presented in the result. This transit from steady state to turbulent condition causes to intensify the complexity of solution. The thermodynamic effects are also ignored in the previous models. In this case, when CO_2 reacts with the amine solutions, the system temperature changes that is opposite to the isothermal condition assumed for the model development. One of the interesting topics is the lifetime of the polymeric membrane in the gas separation that can be formulated to estimate the process efficiency and also maintenance cost. In micro scale, numerous studies have been carried out to interpret the gas transport inside the pores and channel. The transport mechanism of gas molecules through the porous media defined based on the structure and morphology can simplify the selection and modification of polymeric membranes. Finally, it is likely that the current simulation software will improve the methods and calculations with high accuracy and reliability.

12.7.4 Conclusion

Today the use of membrane contactor for CO_2 separation is speeding up due to the simple separation mechanism and also maintenance and capital cost. However, numerous attempts are going on to improve the removal efficiency. The polymeric membranes are an appropriate choice as both the fabrication and modification methods can experimentally be carried out. Various types of the polymeric membrane were discussed and showed that the modification of structure had a significant effect on both selectivity and permeability. The change in morphology was also sophisticated and beyond the content of this chapter. All of the experimental attempts led to a new approach to enhance and improve the accuracy and efficiency of a separation process. Undoubtedly, CFD simulators with engineering principal and numerical laws can solve one aspect of design in the vast area of the industry if they defined as accurate as a typical experiment of gas separation in a lab. These models were defined for CO_2 capture using membrane contactor with some assumptions which affected the modeling result. Hence, the modeling approach needs to be altered and the modeling in microscale may introduce new obstacles for a gas separation process.

Nomenclature

A	Membrane surface area (m^2)
C	Concentration (mol/m^3)
D	Diffusivity coefficient (m^2/s)
J	Molar flux ($mol/m^2 s$)
H	Henry's constant
N	Mass flux ($kg/m^2 s$)
P	Pressure (bars)
Q	Flow rate (m^3/s)
R	Reaction rate ($mol/m^3 s$)
S	Solubility ($kmol/m^3$ bar)
r_1	Inner tube radius (mm)
r_2	Outer tube radius (mm)
r_3	Inner shell radius (mm)
T	Temperature (°C)
U	Velocity (m/s)

Greek Symbols

α_{BJ}	Beavers–Joseph factor
φ	Module volume fraction
η	CO_2 removal efficiency

References

Ahmad, F., K.K. Lau, A.M. Shariff, and Y.F. Yeong. 2013. Temperature and pressure dependence of membrane permeance and its effect on process economics of hollow fiber gas separation system. *Journal of Membrane Science* 430:44–55.

Arcella, V., P. Colaianna, P. Maccone, A. Sanguineti, A. Gordano, G. Clarizia, and E. Drioli. 1999. A study on a perfluoropolymer purification and its application to membrane formation. *Journal of Membrane Science* 163(2):203–209.

Arcella, V., A. Ghielmi, and G. Tommasi. 2003. High performance perfluoropolymer films and membranes. *Annals of the New York Academy of Sciences* 984(1):226–244.

Arcella, V., P. Toniolo, M. Avataneo, A. Ghielmi, and G. Marchionni. 2010. 1.08—Amorphous perfluoropolymer membranes. In *Comprehensive Membrane Science and Engineering*, edited by E. Drioli and L. Giorno, 147–158. Oxford, UK: Elsevier.

Baker, R. 2001. Future directions of membrane gas-separation technology. *Membrane Technology* (138):5–10. doi:10.1016/S0958–2118(01)80332–3.

Bansal, R., V. Jain, and S.K. Gupta. 1995. Analysis of separation of multicomponent mixtures across membranes in a single permeation unit. *Separation Science and Technology* 30(14):2891–2916.

Berman, A.S. 1953. Laminar flow in channels with porous walls. *Journal of Applied Physics* 24(9):1232–1235.

Bos, A., I.G.M. Pünt, M. Wessling, and H. Strathmann. 1998. Plasticization-resistant glassy polyimide membranes for CO_2/CO_4 separations. *Separation and Purification Technology* 14(1):27–39.

Bowen, T.C, R.D. Noble, and J.L. Falconer. 2004. Fundamentals and applications of pervaporation through zeolite membranes. *Journal of Membrane Science* 245(1):1–33.

Cao, C., R. Wang, T.S. Chung, and Y. Liu. 2002. Formation of high-performance 6FDA-2, 6-DAT asymmetric composite hollow fiber membranes for CO_2/CH_4 separation. *Journal of Membrane Science* 209(1):309–319.

Cecopieri-Gómez, M.L., J. Palacios-Alquisira, and J.M. Dominguez. 2007. On the limits of gas separation in CO_2/CH_4, N_2/CH_4 and CO_2/N_2 binary mixtures using polyimide membranes. *Journal of Membrane Science* 293(1):53–65.

Cejka, J., H. Van Bekkum, A. Corma, and F. Schueth. 2007. *Introduction to Zeolite Molecular Sieves*, Vol. 168. Amsterdam, The Netherlands: Elsevier.

Chen, G.Q., S. Kanehashi, C.M. Doherty, A.J. Hill, and S.E. Kentish. 2015. Water vapor permeation through cellulose acetate membranes and its impact upon membrane separation performance for natural gas purification. *Journal of Membrane Science* 487:249–255. doi:10.1016/j.memsci.2015.03.074.

Chern, R.T., W.J. Koros, and P.S. Fedkiw. 1985. Simulation of a hollow-fiber gas separator: The effects of process and design variables. *Industrial and Engineering Chemistry Process Design and Development* 24(4):1015–1022.

Civan, F. 2011. *Porous Media Transport Phenomena*. Hoboken, NJ: John Wiley & Sons.

Coker, D.T., B.D. Freeman, and G.K. Fleming. 1998. Modeling multicomponent gas separation using hollow-fiber membrane contactors. *AIChE Journal* 44(6):1289–1302. doi:10.1002/aic.690440607.

Cui, L., W. Qiu, D.R. Paul, and W.J. Koros. 2011. Responses of 6FDA-based polyimide thin membranes to CO_2 exposure and physical aging as monitored by gas permeability. *Polymer* 52(24):5528–5537.

Drioli, E., and L. Giorno. 2010. *Comprehensive Membrane Science and Engineering*, Vol. 1. Oxford, UK: Newnes.

Eslami, S., S.M. Mousavi, S. Danesh, and H. Banazadeh. 2011. Modeling and simulation of CO_2 removal from power plant flue gas by PG solution in a hollow fiber membrane contactor. *Advances in Engineering Software* 42(8):612–620.

Faiz, R., and M. Al-Marzouqi. 2009. Mathematical modeling for the simultaneous absorption of CO_2 and H_2S using MEA in hollow fiber membrane contactors. *Journal of Membrane Science* 342(1):269–278.

Fischer, S., K. Thümmler, B. Volkert, K. Hettrich, I. Schmidt, and K. Fischer. 2008. Properties and applications of cellulose acetate. *Macromolecular Symposia* 262(1):89–96.

Freeman, B., Y. Yampolskii, and I. Pinnau. 2006. *Materials Science of Membranes for Gas and Vapor Separation*. Chichester, UK: John Wiley & Sons.

Ghasem, N., M. Al-Marzouqi, and N.A. Rahim. 2013. Modeling of CO_2 absorption in a membrane contactor considering solvent evaporation. *Separation and Purification Technology* 110:1–10.

Ghosal, K., B.D. Freeman, R.T. Chern, J.C. Alvarez, J.G. De La Campa, A.E. Lozano, and J. De Abajo. 1995. Gas separation properties of aromatic polyamides with sulfone groups. *Polymer* 36(4):793–800.

Gilassi, S., and N. Rahmanian. 2013. An experimental investigation on permeability and selectivity of PTFE membrane: A mixture of methane and carbon dioxide. *International Conference on Membrane Science and Technology (MST)*, Kuala Lumpur, Malaysia.

Gilassi, S., and N. Rahmanian. 2015a. CFD modeling of a hollow fibre membrane for CO_2 removal by aqueous amine solutions of MEA, DEA and MDEA. *International Journal of Chemical Reactor Engineering* 14(1):53–61.

Gilassi, S., and N. Rahmanian. 2015b. Mathematical modeling and numerical simulation of CO_2/CH_4 separation in a polymeric membrane. *Applied Mathematical Modeling* 39(21):6599–6611. doi:10.1016/j.apm.2015.02.010.

Happel, J. 1959. Viscous flow relative to arrays of cylinders. *AIChE Journal* 5(2):174–177. doi:10.1002/aic.690050211.

Hilal, N., M. Khayet, and C.J. Wright. 2012. *Membrane Modification: Technology and Applications*. Boca Raton, FL: CRC Press.

Idris, A., and N. Rahmanian. 2014. γ-Ray pre-irradiated grafting of polytetrafluoroethylene film membrane. *Jurnal Teknologi* 69(9):47–51.

Ismail, A.F., K. Khulbe, and T. Matsuura. 2015. *Gas Separation Membranes: Polymeric and Inorganic*. New York: Springer.

Ismail, A.F., S.J. Shilton, I.R. Dunkin, and S.L. Gallivan. 1997. Direct measurement of rheologically induced molecular orientation in gas separation hollow fibre membranes and effects on selectivity. *Journal of Membrane Science* 126(1):133–137.

Kamide, K., and M. Saito. 1987. Cellulose and cellulose derivatives: Recent advances in physical chemistry. In *Biopolymers*, 1–56. Berlin and Heidelberg: Springer, Berlin, Heidelberg.

Khalilpour, R., A. Abbas, Z. Lai, and I. Pinnau. 2012. Modeling and parametric analysis of hollow fiber membrane system for carbon capture from multicomponent flue gas. *AIChE Journal* 58(5):1550–1561.

Khalilpour, R., A. Abbas, Z. Lai, and I. Pinnau. 2013. Analysis of hollow fibre membrane systems for multicomponent gas separation. *Chemical Engineering Research and Design* 91(2):332–347.

Kim, J.H., W.J. Koros, and D.R. Paul. 2006. Effects of CO_2 exposure and physical aging on the gas permeability of thin 6FDA-based polyimide membranes: Part 2. With crosslinking. *Journal of Membrane Science* 282(1):32–43.

Kim, K.J., S.H. Park, W.W. So, D.J. Ahn, and S.J. Moon. 2003. CO_2 separation performances of composite membranes of 6FDA-based polyimides with a polar group. *Journal of Membrane Science* 211(1):41–49.

Koukou, M.K., N. Papayannakos, and N.C. Markatos. 1996. Dispersion effects on membrane reactor performance. *AIChE Journal* 42(9):2607–2615.

Koukou, M.K., N. Papayannakos, N.C. Markatos, M. Bracht, and P.T. Alderliesten. 1998. Simulation tools for the design of industrial-scale membrane reactors. *Chemical Engineering Research and Design* 76(8):911–920.

Koukou, M.K., N. Papayannakos, N.C. Markatos, M. Bracht, H.M. Van Veen, and A. Roskam. 1999. Performance of ceramic membranes at elevated pressure and temperature: effect of non-ideal flow conditions in a pilot scale membrane separator. *Journal of Membrane Science* 155(2):241–259.

Kundu, P.K., A. Chakma, and X. Feng. 2012. Simulation of binary gas separation with asymmetric hollow fibre membranes and case studies of air separation. *The Canadian Journal of Chemical Engineering* 90(5):1253–1268.

Li, N.N., A.G. Fane, W.S.W. Ho, and T. Matsuura. 2011. *Advanced Membrane Technology and Applications*. Hoboken, NJ: John Wiley & Sons.

Liaw, D.J., K.L. Wang, Y.C. Huang, K.R. Lee, J.Y. Lai, and C.S. Ha. 2012. Advanced polyimide materials: Syntheses, physical properties and applications. *Progress in Polymer Science* 37(7):907–974.

Lock, S.S.M., K.K. Lau, and A.M. Shariff. 2015. Effect of recycle ratio on the cost of natural gas processing in countercurrent hollow fiber membrane system. *Journal of Industrial and Engineering Chemistry* 21:542–551.

Luis, P., A. Garea, and A. Irabien. 2010. Modeling of a hollow fibre ceramic contactor for SO_2 absorption. *Separation and Purification Technology* 72(2):174–179.

Macchione, M., J.C. Jansen, G. De Luca, E. Tocci, M. Longeri, and E. Drioli. 2007. Experimental analysis and simulation of the gas transport in dense Hyflon® AD60X membranes: Influence of residual solvent. *Polymer* 48(9):2619–2635. doi:10.1016/j.polymer.2007.02.068.

Marić, I. 2005. The Joule–Thomson effect in natural gas flow-rate measurements. *Flow Measurement and Instrumentation* 16(6):387–395.

Marić, I. 2007. A procedure for the calculation of the natural gas molar heat capacity, the isentropic exponent, and the Joule–Thomson coefficient. *Flow Measurement and Instrumentation* 18(1):18–26.

Mark, J.E. 1996. *Physical Properties of Polymers Handbook*. New York: Springer.

Martínez, C., and A. Corma. 2013. 5.05—Zeolites. In *Comprehensive Inorganic Chemistry II*, 2nd Edition, edited by J. Reedijk and K. Poeppelmeier, 103–131. Amsterdam, the Netherlands: Elsevier.

Matsufuji, T., N. Nishiyama, M. Matsukata, and K. Ueyama. 2000. Separation of butane and xylene isomers with MFI-type zeolitic membrane synthesized by a vapor-phase transport method. *Journal of Membrane Science* 178(1):25–34.

Mulder, M. 1996. *Basic Principles of Membrane Technology*. Amsterdam, The Netherlands: Springer.

Murad Chowdhury, M.H., X. Feng, P. Douglas, and E. Croiset. 2005. A new numerical approach for a detailed multicomponent gas separation membrane model and AspenPlus simulation. *Chemical Engineering and Technology* 28(7):773–782.

Nield, D.A. 2009. The Beavers–Joseph boundary condition and related matters: A historical and critical note. *Transport in Porous Media* 78(3):537–540.

Nishiyama, N.K. Ueyama, and M. Matsukata. 1995. A defect-free mordenite membrane synthesized by vapor-phase transport method. *Journal of the Chemical Society, Chemical Communications* (19):1967–1968.

Noble, R.D., and S.A. Stern. 1995. *Membrane Separations Technology: Principles and Applications*, Vol. 2. Amsterdam, The Netherlands: Elsevier.

Okamoto, Y., H. Zhang, F. Mikes, Y. Koike, Z. He, and T.C. Merkel. 2014. New perfluoro-dioxolane-based membranes for gas separations. *Journal of Membrane Science* 471:412–419.

Pan, C.-Y, and H.W. Habgood. 1978. Gas separation by permeation Part II: Effect of permeate pressure drop and choice of permeate pressure. *The Canadian Journal of Chemical Engineering* 56(2):210–217.

Pan, C.Y. 1986. Gas separation by high-flux, asymmetric hollow-fiber membrane. *AIChE Journal* 32(12):2020–2027.

Park, G.S., and J. Crank. 1968. *Diffusion in Polymers*. New York: Academic Press

Patankar, S.V., and D. Brian Spalding. 1972. A calculation procedure for heat, mass and momentum transfer in three-dimensional parabolic flows. *International Journal of Heat and Mass Transfer* 15(10):1787–1806.

Qiu, W., L. Xu, C.C. Chen, D.R. Paul, and W.J. Koros. 2013. Gas separation performance of 6FDA-based polyimides with different chemical structures. *Polymer* 54(22):6226–6235.

Rezakazemi, M., Z. Niazi, M. Mirfendereski, S. Shirazian, T. Mohammadi, and A. Pak. 2011. CFD simulation of natural gas sweetening in a gas–liquid hollow-fiber membrane contactor. *Chemical Engineering Journal* 168(3):1217–1226.

Richard, W.B. 2004. *Membrane Technology and Applications*. UK: John Wiley & Sons.

Robeson, L.M. 1991. Correlation of separation factor versus permeability for polymeric membranes. *Journal of Membrane Science* 62(2):165–185.

Scholes, C.A., S. Kanehashi, G.W. Stevens, and S.E. Kentish. 2015. Water permeability and competitive permeation with CO_2 and CH_4 in perfluorinated polymeric membranes. *Separation and Purification Technology* 147:203–209. doi:10.1016/j.seppur.2015.04.023.

Shindo, Y., T. Hakuta, H. Yoshitome, and H. Inoue. 1985. Calculation methods for multicomponent gas separation by permeation. *Separation Science and Technology* 20(5–6):445–459.

Sohrabi, M.R., A. Marjani, S. Moradi, M. Davallo, and S. Shirazian. 2011. Mathematical modeling and numerical simulation of CO_2 transport through hollow-fiber membranes. *Applied Mathematical Modeling* 35(1):174–188.

Sperling, L.H. 2015. *Introduction to Physical Polymer Science*. New York: John Wiley & Sons.

Sridhar, S, Tejraj M Aminabhavi, and M Ramakrishna. 2007. Separation of binary mixtures of carbon dioxide and methane through sulfonated polycarbonate membranes. *Journal of Applied Polymer Science* 105(4):1749–1756.

Staudt-Bickel, C., and W.J. Koros. 1999. Improvement of CO_2/CH_4 separation characteristics of polyimides by chemical crosslinking. *Journal of Membrane Science* 155(1):145–154.

Stern, S.A., P.A. Rice, and J. Hao. 2000. *Upgrading natural gas via membrane separation processes.* USA: Syracuse University.

Suzuki, H. 1987. Composite membrane having a surface layer of an ultrathin film of cage-shaped zeolite and processes for production there of. US Patent 4699892.

Thundyil, M.J., and W.J. Koros. 1997. Mathematical modeling of gas separation permeators—For radial crossflow, countercurrent, and cocurrent hollow fiber membrane modules. *Journal of Membrane Science* 125(2):275–291.

Tin, P.S., T.S. Chung, Y. Liu, R. Wang, S.L. Liu, and K.P. Pramoda. 2003. Effects of cross-linking modification on gas separation performance of Matrimid membranes. *Journal of Membrane Science* 225(1):77–90.

Tiwari, R.R., Z.P. Smith, H. Lin, B.D. Freeman, and D.R. Paul. 2014. Gas permeation in thin films of high free-volume glassy perfluoropolymers: Part I. Physical aging. *Polymer* 55(22):5788–5800.

Tiwari, R.R., Z.P. Smith, H. Lin, B.D. Freeman, and D.R. Paul. 2015. Gas permeation in thin films of "high free-volume" glassy perfluoropolymers: Part II. CO_2 plasticization and sorption. *Polymer* 61:1–14.

Uhlhorn, R.J.R., A.J. Burggraaf, and R.R. Bhave. 1991. *Inorganic Membranes: Synthesis, Characteristics, and Applications.* New York: Chapman and Hall.

Vanherck, K., G. Koeckelberghs, and I.F.J. Vankelecom. 2013. Crosslinking polyimides for membrane applications: A review. *Progress in Polymer Science* 38(6):874–896.

Versteeg, H.K., and W. Malalasekera. 2007. *An Introduction to Computational Fluid Dynamics: The Finite Volume Method.* New York: Pearson Education.

Visser, T., N. Masetto, and M. Wessling. 2007. Materials dependence of mixed gas plasticization behavior in asymmetric membranes. *Journal of Membrane Science* 306(1):16–28.

Vu, D.Q., W.J. Koros, and S.J. Miller. 2003. Effect of condensable impurities in CO_2/CH_4 gas feeds on carbon molecular sieve hollow-fiber membranes. *Industrial and Engineering Chemistry Research* 42(5):1064–1075.

Walawender, W.P., and S.A. Stern. 1972. Analysis of membrane separation parameters. II. Countercurrent and cocurrent flow in a single permeation stage. *Separation Science* 7(5):553–584.

Wang, R., D.F. Li, and D.T. Liang. 2004. Modeling of CO_2 capture by three typical amine solutions in hollow fiber membrane contactors. *Chemical Engineering and Processing: Process Intensification* 43(7):849–856.

Wang, R., H.Y. Zhang, P.H.M. Feron, and D.T. Liang. 2005. Influence of membrane wetting on CO_2 capture in microporous hollow fiber membrane contactors. *Separation and Purification Technology* 46(1):33–40.

Wijmans, J.G., and R.W. Baker. 1995. The solution-diffusion model: A review. *Journal of Membrane Science* 107(1):1–21.

Wind, J.D., D.R. Paul, and W.J. Koros. 2004. Natural gas permeation in polyimide membranes. *Journal of Membrane Science* 228(2):227–236.

Xia, J., T.S. Chung, and D.R. Paul. 2014. Physical aging and carbon dioxide plasticization of thin polyimide films in mixed gas permeation. *Journal of Membrane Science* 450:457–468.

Xiao, Y., B.T. Low, S.S. Hosseini, T.S. Chung, and D.R. Paul. 2009. The strategies of molecular architecture and modification of polyimide-based membranes for CO_2 removal from natural gas—A review. *Progress in Polymer Science* 34(6):561–580.

Zhang, H.Y., R. Wang, D.T. Liang, and J.H. Tay. 2008a. Theoretical and experimental studies of membrane wetting in the membrane gas–liquid contacting process for CO_2 absorption. *Journal of Membrane Science* 308(1):162–170.

Zhang, Y., I.H. Musselman, J.P. Ferraris, and K.J. Balkus. 2008b. Gas permeability properties of Matrimid® membranes containing the metal-organic framework Cu–BPY–HFS. *Journal of Membrane Science* 313(1):170–181.

IV

Sustainable
Future

13

The Role of Molecular Thermodynamics in Developing Industrial Processes and Novel Products That Meet the Needs for a Sustainable Future

Ioannis G.
Economou
*Texas A&M University
at Qatar and National
Center for Scientific
Research "Demokritos"*

Panagiotis
Krokidas, Vasileios
K. Michalis
*Texas A&M University
at Qatar*

Othonas A. Moultos
*Delft University
of Technology*

Ioannis N.
Tsimpanogiannis
and Niki Vergadou
*National Center for
Scientific Research
"Demokritos"*

13.1 Introduction

The world is facing today a number of challenges related to energy efficiency and security, climate change, water management and supply, and others. Basic and applied research can provide the necessary tools to address effectively these challenges. In most cases, a better understanding and accurate modeling at different levels of the underlying system is necessary. The unprecedented increase of computing power at relatively low price in recent years, the development of efficient computational algorithms and accurate models, and the design of sophisticated experimental techniques allow us today to calculate, model, and measure phenomena at the microscopic (molecular) level and predict the macroscopic properties of complex chemical and physical systems (Maginn, 2009; Theodorou, 2010; Palmer and Debenedetti, 2015). In this way, we can optimize existing or design new chemical processes and advanced materials that aim to have a positive contribution to some of these challenges.

Multiscale hierarchical modeling provides the means to span a very broad range of time and length scales, from the subatomic to macroscopic scale. Hierarchical approaches typically start with quantum mechanics calculations to determine intra- and intermolecular interactions, continue with atomistic simulations (e.g., Monte Carlo, MC and Molecular Dynamics, MD) addressing length scales of the order of tens of nanometers and time scales of the order of tens of nanoseconds, and proceed with mesoscopic methods (e.g., lattice and coarse-grained simulation) to address longer time- and length-scale phenomena. Finally, for the efficient design of novel processes mainly for the oil and gas, chemical, polymer, pharmaceutical, and other industries, accurate macroscopic models, mostly in the form of equations of state (EoS), are developed for phase equilibria and other thermodynamic properties of multicomponent mixtures (Prausnitz et al., 1999). A number of EoS are rooted in statistical mechanics and can be safely extrapolated to conditions where limited or no experimental data exist (Economou, 2002; Dufal et al., 2015). The challenge that we often face is how to integrate efficiently methods and parameters developed at one time or length scale at a higher one without loss of accuracy. In Table 13.1, a list of physical properties that can be calculated accurately using different molecular simulation techniques is provided.

In this chapter, the aim is to highlight achievements in recent years on the use of molecular simulation and molecular thermodynamics for the prediction of properties of complex chemical systems and materials to be used for the development of new industrial processes and products in line with the goal of sustainable development. The first project refers to the development of molecular simulation methodologies and molecular models (force fields) to predict the phase equilibria and transport properties of H_2O–CO_2–NaCl mixture, which is very important for carbon capture and sequestration (CCS) processes. CCS is one of the most promising technologies for the reduction of CO_2 in the atmosphere. The second example refers to the development of a methodology to predict the three phase (solid–liquid–vapor) equilibria of gas hydrates. Hydrates pose a major flow assurance problem for the oil and gas industry, while at the same time they are known to store large amounts of hydrocarbons. Using the direct phase coexistence methodology, predictions of the hydrate equilibrium conditions for the case of CH_4 and CO_2 hydrates will be discussed. For engineering calculations, accurate EoS are preferred over molecular simulations. A brief introduction to recent developments in the use of statistical mechanics-based EoS for hydrate phase equilibria is presented. The next example refers to modeling of metal organic frameworks (MOFs) for the efficient separation of gas mixtures. The focus here is on MOFs suitable for propylene/propane separation. Finally, we provide a concise overview of our work on the implementation of molecular simulation methods for the design of task-specific ionic liquids (ILs) that can be used as green solvents and novel separation media in a wide range of processes.

TABLE 13.1 Physical Properties Calculated by Molecular Simulation for Process Design

Type	Property
Single phase equilibrium properties	Density, isothermal/isobaric compressibility
	Gibbs free energy, Helmholtz free energy, activity coefficient(s)
	Heat capacities, Joule-Thompson coefficient
Transport properties	Viscosity
	Diffusion coefficients (Self-, Maxwell-Stefan, etc.)
	Thermal conductivity
	Surface tension
Phase equilibria	Liquid–liquid equilibria (LLE)
	Solid–fluid (vapor/liquid) equilibria
	Partition coefficients

13.2 Phase Equilibria and Transport Properties of H_2O-CO_2-NaCl Mixtures

13.2.1 Objectives and Motivation

H_2O-CO_2, $H_2O-NaCl$, and H_2O-CO_2-NaCl mixtures have been in the spotlight of international research during the past few decades due to their relevance to geological, environmental, industrial, biophysical, and interplanetary applications (Orozco et al., 2014a). Such applications include enhanced oil recovery, separation and purification processes, food processing, water desalination, and many others. A particular example is the accurate prediction of thermodynamics and transport properties of these mixtures in order to design processes to control and reduce the amounts of CO_2 in the atmosphere, which has been identified as the main cause of the so-called greenhouse effect. This can be achieved by several ways, with one of the most promising being the CCS process (Metz et al., 2005). The basic principle of this process is that CO_2 is captured from various industrial sources (power plants, steel and cement industries, etc.), it is transported and finally injected into deep geological reservoirs. In these geological formations, many impurities are present with NaCl being the most common dissolved salt.

From the physicochemical point of view, the study of aqueous electrolyte mixtures is far from trivial due to the high nonideality of the mixtures. H_2O is a highly polar, hydrogen-bonding molecule, CO_2 possesses a strong quadrupole moment, and NaCl is a strong electrolyte generating significant Coulombic interactions in the solution. Traditionally, in order to study the physical properties of these mixtures macroscopic thermodynamic models and experiments are utilized (Prausnitz et al., 1999). In particular, activity coefficient models and EoS with empirical or statistical mechanics basis have been widely used for modeling the phase equilibria, while mean-field-theory models have been used for transport coefficients (Brokaw, 1969; Pitzer, 1991; Economou et al., 1995; Haghtalab and Mazloumi, 2009). Although such models have shown tremendous capabilities in many cases, they also exhibit severe drawbacks such as the limited range of applicability as they cannot be extrapolated outside the range of experimental data available due to lack of fundamental understanding of the intermolecular interactions (Valderrama, 2003). On the other hand, experimental measurements of thermodynamic and transport properties are relatively limited and scarce for the mixtures of interest because they are often costly and difficult to perform under a wide range of conditions, especially when high pressures and/or temperatures are involved.

An attractive alternative to macroscopic modeling and a useful complement to experimental measurements are molecular simulations (Allen and Tildesley, 1987). Although simulations in the form of MD and MC algorithms were developed in the mid-1950s, their use was limited due to high demands for computing power. Today, after six decades of integrated circuit's evolution according to the famous Moore's Law, molecular simulations can be performed at a drastically reduced computational cost, enabling the computational study of large and/or chemically complex systems. During the same period, along with the increase in computing power, simulation algorithms have greatly been improved and many robust codes can be freely accessed (open-source via free license) and used by researchers worldwide.

Molecular simulations demand accurate and transferable intra- and intermolecular potential functions, also known as force fields. To this extent, a huge amount of effort has been devoted by the scientific community to the development and validation of such force fields for H_2O, CO_2, NaCl, and their mixtures. Recent detailed reviews of the available force fields for these compounds are those by Vega and co-workers (Vega et al., 2009; Vega and Abascal, 2011) and Economou, Panagiotopoulos, and co-workers (Jiang et al., 2015, 2016a). Extensive studies in our group in collaboration with the group of Prof. A.Z. Panagiotopoulos at Princeton University involved several polarizable and nonpolarizable force fields for the prediction of the liquid density, vapor pressure, chemical potential, viscosity, diffusion and mean ionic activity coefficients, vapor–liquid interfacial tension, and vapor–liquid equilibria (VLE) of the $H_2O-NaCl-CO_2$ mixture and its constituent binaries for a wide temperature and pressure range.

13.2.2 Methodology

For the calculation of liquid density, chemical potential, transport and mean ionic activity coefficients, and vapor–liquid interfacial tension, MD simulations were performed using the open-source software packages GROMACS (Hess et al., 2008; Kiss et al., 2014) and LAMMPS (Plimpton, 1995). Phase equilibria were studied with the Gibbs Ensemble MC method using the Cassandra suite (Shah and Maginn, 2011) and an in-house developed code. Simulations were performed in various statistical ensembles (NVE, NVT, and NPT) and the system size was 250–2,000 molecules, depending on the mixture and the property of interest. In all cases different system sizes were tested to ensure that no system-size dependencies exist. Standard long-range corrections to energy and pressure were applied, except for the case of interfacial simulations for which a high enough cutoff distance was chosen beyond which the contribution of the interatomic potentials is negligible. All the physical quantities reported were calculated from sufficiently long runs in order to ensure relatively low statistical uncertainties. Further details of the computational methods employed are available in the literature (Moultos et al. 2014, 2015, 2016; Orozco et al., 2014a,b; Jiang et al., 2015, 2016a; Michalis et al., 2015, 2016).

13.2.3 Results and Discussion

13.2.3.1 H_2O–CO_2 Mixture

Orozco et al. (2014a) optimized the unlike interactions between H_2O and CO_2 in order to obtain improved mutual solubility predictions for a wide range of temperature and pressure. More precisely, it was found that it is possible to represent properties of both phases using a combination of Exp-6 force fields (referring to a Buckingham type of potential where an exponential term for the intermolecular distance is used for the repulsive part and a term to the -6 power for the intermolecular distance is used for the attractive term) with optimized oxygen–oxygen unlike interactions. The results suggested that an even better agreement to experimental data could be achieved if molecular models that explicitly consider the polarizability of the molecules are used (Orozco et al., 2014a).

An exhaustive study of the mutual diffusion coefficients of H_2O–CO_2 mixture was also performed. It was found that diffusion coefficients of infinitely diluted CO_2 in H_2O and H_2O in CO_2 can be accurately predicted in the range of experimental data available up to approx. 373 K with the combination of TIP4P/2005 force field for H_2O with EPM2 or TraPPE for CO_2 (Moultos et al., 2014; Moultos et al., 2015). The same force fields were shown to be also very accurate at extreme conditions of geological interest, with temperature up to 1000 K and pressure up to 1000 MPa, proving that molecular simulations with the appropriate force fields is a valuable tool for the prediction of key physical properties for a plethora of industrial and geological applications. In parallel, molecular simulation data were used to develop a very simple and computationally efficient phenomenological model, in the form of a generalized Speedy–Angel correlation, for the diffusion coefficients of the H_2O–CO_2 mixture (Moultos et al., 2016).

Molecular models that explicitly take into consideration the polarizability of H_2O and CO_2 can potentially be even more accurate than the simple point charge force fields due to the more realistic representation of the intermolecular interactions. To this extent, and following the success of the Drude oscillator-based polarizable H_2O model (BK3) by Kiss and Baranyai (Kiss and Baranyai 2014; Kiss et al., 2014), a new CO_2 force field was developed (Jiang et al., 2016a). In this polarizable model, the electrostatic interactions are represented by Gaussian charges, while the van der Waals interactions by the Buckingham Exp-6 potential. It can accurately predict thermodynamic and transport properties for a wide range of conditions and, in most cases, is superior compared to most of the popular polarizable and nonpolarizable CO_2 models. Additionally, in order to provide a computationally more efficient version of this new model, a Gaussian charge nonpolarizable one was proposed and found to perform almost equally well in most of the properties examined (Jiang et al., 2016a).

13.2.3.2 H_2O–NaCl Mixture

As mentioned earlier, aqueous electrolyte mixtures are crucial for CCS processes. Therefore, the accurate prediction of thermodynamic and transport properties of these systems is needed. Initially, our study focused on several combinations of existing fixed point charge force fields and it was found that various models can predict different properties with variable accuracy (Orozco et al., 2014b). Particularly, the single-point charge (SPC) water with JC ion models can reproduce vapor pressures very accurately; the semi-flexible SPC/E with JC is the best option for liquid densities and interfacial tensions, while the semi-flexible SPC/E with Smith-Dang (SD) reproduces the experimentally measured viscosities.

In an effort to find a unique combination of molecular models to reproduce simultaneously as many properties as possible, Jiang et al. (2015) studied the BK3 polarizable water force field combined with the AH/BK3 ion force field. It was found that liquid densities, mean ionic activity coefficients, interfacial tensions, and viscosities were predicted with much higher accuracy compared to the nonpolarizable force field combinations. Vapor pressure was predicted with almost the same accuracy as with the SPC+SD models, while for the salt solubility and the temperature effect on the mean ionic activity coefficients the BK3 models yielded a less satisfactory behavior, showing that there is still room for the optimization of the pure component models or the cross-interaction parameters. In conclusion, it is clear that a plethora of physical properties can be very accurately predicted by means of molecular simulations when the intermolecular potentials used are accurate. Additionally, these methods can provide fundamental understanding of the molecular structures and intermolecular interactions.

Work is under way to optimize a new H_2O–CO_2 force field that will be able to predict accurately the properties of the H_2O–CO_2–NaCl mixture.

13.2.3.3 H_2O–CO_2–NaCl Mixture

A SAFT-type EoS was recently proposed by Jiang et al. (2016b) for the calculation of phase equilibria and thermodynamic properties of H_2O–CO_2–electrolyte mixtures. Electrostatic interactions were calculated explicitly using an improved mean spherical approximation in the primitive model. The mean ionic activity coefficients and liquid densities of electrolyte solutions containing Na^+, K^+, Ca^{2+}, Mg^{2+}, Cl^-, Br^-, and SO_4^{2-} from 298.15 to 473.15 K were accurately correlated. Interaction parameters tuned to experimental H_2O–salt and H_2O–CO_2 data were used subsequently to predict CO_2 solubilities in mixed electrolyte solutions and synthetic brines in good agreement with experimental data.

13.3 Molecular Dynamics Studies of Gas Hydrates

13.3.1 Objectives and Motivation

Clathrate hydrates are crystalline inclusion compounds formed out of water and low molecular weight compounds (Sloan and Koh, 2008). The water molecules through hydrogen bonding form cages around the "guest" molecules whose presence is necessary for the stabilization of the hydrate structure. There are more than 130 known molecules that can play the role of the guest and their size and interactions with water define the type of the hydrate structure. Hydrates structures of type sI, sII, and sH are the most common ones differing mainly in the size and ratio of formed cages. The study of hydrates draws considerable interest given their importance in a number of technological and natural processes. Gas hydrates, in particular, which consist of guest molecules that are components of natural gas (primarily methane) play an important role in the oil and gas industry from a flow assurance point of view as well as a potential energy source due to the fact that methane hydrate is an abundant naturally occurring substance (Hammerschmidt, 1934; Koh, 2002). Additionally, gas hydrates are being studied as a potential environmental risk (Kvenvolden, 1999; Archer et al., 2009) and also for storage and transport applications (Papadimitriou et al., 2010; Khokhar et al., 1998).

The study of gas hydrates properties is approached either experimentally or theoretically. There are two major theoretical approaches, the first being a thermodynamic one at the continuum scale using the van der Waals-Platteeuw theory (vdWP; van der Waals and Platteeuw, 1959; Holder et al., 1988), and the second approach is molecular simulation. Molecular simulation techniques are constantly proving their value by offering insight on the molecular level mechanisms, thus improving our understanding enabling at the same time the prediction of materials properties. In this section, we present an MD approach for the calculation of phase coexistence conditions of gas hydrates and in particular of methane (Michalis et al., 2015) and carbon dioxide hydrates (Costandy et al., 2015). The emphasis is on the determination of the three-phase coexistence conditions, where solid hydrate coexists with a liquid water-rich phase and a guest-rich phase, either liquid or gas. The prediction of thermodynamic equilibrium conditions is critical to many technological processes and in this particular example, given the existence of experimental measurements for the three-phase coexistence conditions for both methane and carbon dioxide hydrates, it serves the role of validating the predictive accuracy of the MD approach.

13.3.2 Methodology

The present approach follows the direct phase coexistence methodology (Ladd and Woodcock, 1977). All three phases are brought in contact and MD simulations are carried out in the isothermal-isobaric (NPT) ensemble in order to observe the time evolution of the system at a specific pressure and temperature. For a two-component mixture coexisting in three phases there is only one degree of freedom so by fixing for example the pressure and carrying out a series of simulations at different temperatures, the equilibrium point can be determined. Depending on the conditions of the simulation, the hydrate phase of a three-phase system will either dissociate or grow further. Such a system is depicted in Figure 13.1 (Michalis et al., 2015). In this particular configuration, a methane sI hydrate slab is surrounded by two water slabs, which are in conduct with a methane slab. Three snapshots of a typical trajectory under hydrate growth conditions are presented, namely snapshot (a) depicts the initial configuration, snapshot (b) depicts the system at a later time, and at snapshot (c) the whole system has evolved to the hydrate state.

The hydrate slab used in the simulations consists of a $2 \times 2 \times 2$ sI hydrate supercell. The water oxygen positions within an sI unit cell are those reported by McMullan and Jeffrey (1965). The guest occupancy is 100% and the guest molecules (methane and carbon dioxide) are positioned in the center of the cages. The positions of the hydrogen atoms are determined by energy minimization of the system keeping the positions of the oxygen atoms fixed. This approach is equivalent to using a configuration with minimum dipole moment (Sarupria and Debenedetti, 2011) and results in a structure that respects the Bernal and Fowler rules.

Two water force fields are employed, namely the TIP4P/ice (Abascal et al., 2005) and TIP4P/2005 (Abascal and Vega, 2005), while the OPLS-UA model (Jorgensen et al., 1984) is used for the methane molecules and the TraPPE force field (Potoff and Siepmann, 2001) is used for the carbon dioxide. The Lorentz–Berthelot combining rules have been employed, with an exception in some cases of a modified cross-interaction energy parameter between the oxygen in water and the oxygen in carbon dioxide (Costandy et al., 2015).

A critical characteristic of this approach is the stochastic behavior of the system. At conditions very close to the equilibrium line the system can either exhibit hydrate growth or dissociation and thus statistical averaging of the results is necessary in order to increase the accuracy and consistency of the calculated equilibrium conditions. At each pressure five independent temperature scans were performed by employing different random seeds for the generation of initial velocities. The simulation time required for hydrate growth is significantly larger than in the case of dissociation, and depending on the conditions it can be in the range of 500–4000 ns. Taking into consideration the total number of independent runs along with the necessary simulation time, it is clear that given the current computational capabilities, these simulations are computationally demanding.

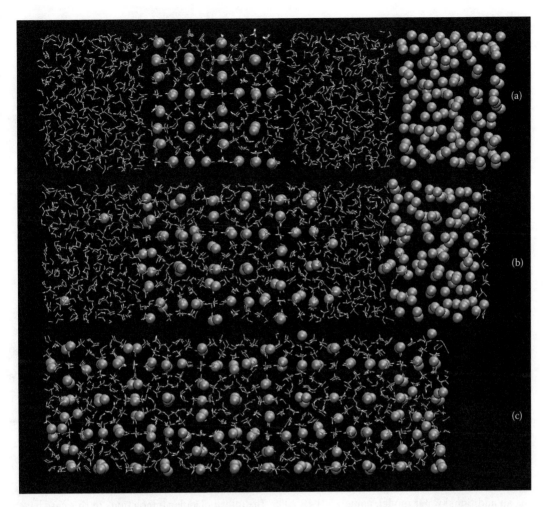

FIGURE 13.1 Three snapshots of a typical trajectory of the methane hydrate system at 294 K and 600 bar: (a) initial configuration at $t = 0$ ns, (b) intermediate step at $t = 600$ ns, and (c) final state at $t = 1500$ ns. The red and white lines represent the methane molecules.

13.3.3 Results and Discussion

The direct phase coexistence methodology along with statistical averaging of the results provides very consistent predictions of the hydrate equilibrium conditions. For the case of methane hydrate, the calculated three-phase coexistence line is presented in Figure 13.2 (Michalis et al., 2015) along with experimental values and results from previous molecular simulation studies from the literature. It can be seen that the equilibrium points predicted by the MD simulations exhibit a consistent trend and have the same slope as that of the experimental data. The observed deviation from the experimental values is −3 K, and the origin of this deviation is the inaccuracy of the TIP4P/ice water force field, used in the case of methane hydrates, in the prediction of melting temperature of ice. The predicted melting temperature of ice by TIP4P/ice is 270(3) K and this average −3 K difference is reflected as well in the case of the determination of the equilibrium temperature of the methane hydrates.

The results of the determination of the three-phase coexistence line for the carbon dioxide hydrate are shown in Figure 13.3 (Costandy et al., 2015). Again, the MD results are compared to experimental values and also other simulation results from the literature. The agreement between the experimental

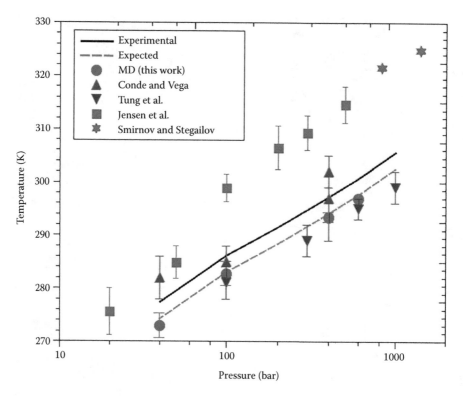

FIGURE 13.2 Experimental and calculated values from this work and from the literature (Berendsen et al., 1984, Conde and Vega, 2010, Tung et al., 2010, Jensen et al., 2010, Smirnov and Stegailov, 2012) for the three-phase coexistence temperature of the methane hydrate system. All authors used TIP4P/ice except Tung et al. who used TIP4P/Ew. The expected values presented in the figure are defined as $T_{3,\text{expected}} = T_{3,\text{expected}} - 3.15\,\text{K}$.

and predicted values is very good again; the consistency of the method is further exhibited by including an additional water model, namely TIP4P/2005. The intermolecular interactions in this case have been optimized in order to take into account the inefficiency of the Lorentz–Berthelot combining rules, for the used pair of force fields, to predict correctly the solubility of carbon dioxide in water. This has been achieved by inserting a modified cross-interaction energy parameter between the oxygen in water and the oxygen in carbon dioxide. This modification of the water–carbon dioxide interaction allows for the correct prediction of the solubility of carbon dioxide in water at hydrate-forming conditions and this is reflected as well in the predicted equilibrium conditions of the carbon dioxide hydrate system. The corrected force field exhibits a consistent deviation from the experimental values exactly as in the case of the methane hydrates, and this deviation is the same as the deviation of the predicted melting temperature of ice of the water force field in use. Thus for the case of TIP4P/ice it is again a −3 K deviation from experimental values, and is consistently constant throughout the wide range of conditions examined (100–5000 bar) that include the area where the equilibrium line exhibits a retrograde behavior.

In conclusion, MD can provide reliable results for the determination of the equilibrium conditions of gas hydrates. The direct phase coexistence methodology has been successfully applied for the case of methane and carbon dioxide hydrates and consistent predictions were provided in both cases. The choice of force fields of course holds a prominent role in the accuracy of the results as they reflect the water–water and water–guest interactions. The examined systems showed that a water force field that can accurately predict the melting temperature of ice is necessary, and it must be used in combination with a force field for the guest molecule that can accurately predict the solubility of the guest in the water

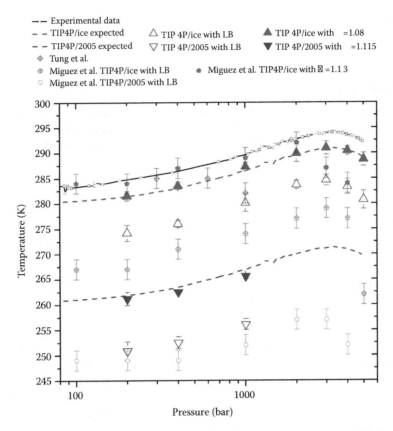

FIGURE 13.3 Experimental (solid line) and calculated values of T_3 for carbon dioxide hydrates from this work (triangles), and from the literature (Costandy et al., 2015). The red up-triangles represent results obtained using the TIP4P/ice force field, and the blue, down-triangles represent results obtained using the TIP4P/2005 force field. Open points indicate the simulations conducted using the LB combining rules, while the filled points indicate that the appropriate modification factor for the force field was applied. The red and blue dashed lines represent the expected values of T_3 for the TIP4P/ice ($T_{3,\text{expected}} = T_{3,\text{experimental}} - 3.15\,\text{K}$) and TIP4P/2005 ($T_{3,\text{expected}} = T_{3,\text{experimental}} - 22.7\,\text{K}$) force fields, respectively.

at hydrate equilibrium conditions. Currently, the methodology is tested for other hydrate systems with different guest molecules and for systems with more than two components and for hydrate systems with inhibitors or promoters.

13.4 EoS for Hydrate Phase Equilibria

13.4.1 Objectives and Motivation

In the previous section, we examined the three-phase coexistence conditions for pure methane and carbon dioxide hydrates using molecular simulations. While such an approach can provide significant insight for systems where no experimental data are available, it is computationally very intensive and therefore cannot be used in routine-type calculations of process design and optimization. Such calculations are traditionally performed at the continuum scale using the vdWP theory (van der Waals and Platteeuw, 1959; Sloan and Koh, 2008) coupled with an EoS. In a recent study (El Meragawi et al., 2016), we examined the case of several hydrate-forming gases of industrial interest using the perturbed-chain statistical associating fluid theory (PC-SAFT) (Gross and Sadowski, 2002), and the Peng–Robinson (PR)

(Peng and Robinson, 1976) cubic EoS. In the current section, we outline the methodology for three-phase equilibrium calculations and provide a brief discussion of the obtained results for hydrates of pure gases, as well as hydrates of some characteristic binary, ternary, and quaternary mixtures.

13.4.2 Methodology

Under three-phase equilibria between a hydrate phase (H), a vapor phase (V), and an aqueous phase (π), the following conditions must be satisfied for the temperature, T, pressure, P, and the chemical potential of each component in all phases (subscript W denotes water and g denotes the hydrate-forming guest):

$$T^H = T^\pi = T^V \tag{13.1a}$$

$$P^H = P^\pi = P^V \tag{13.1b}$$

$$\mu_W^H = \mu_W^\pi = \mu_W^V \tag{13.1c}$$

$$\mu_g^H = \mu_g^\pi = \mu_g^V \tag{13.1d}$$

The aqueous phase can be either liquid water (i.e., for temperatures above 273.15 K) or ice (i.e., for temperatures below 273.15 K, for the case of pure water), denoted with superscripts L or α respectively. Following the notation of Parrish and Prausnitz (1972) and Holder et al. (1988), the ice water phase is known also as the α-phase, while the liquid water phase as the L-phase.

The original formulation of the vdWP theory focused on the equality of the chemical potentials of water μ_W in the hydrate phase and in the liquid phase. A reference state is required in order to proceed with the calculations of the chemical potentials. To this purpose the hypothetical empty hydrate lattice is used as the reference state for the calculation of the chemical potential of water in the hydrate (van der Waals and Platteuw, 1959; Holder et al., 1988). The empty hydrate lattice is denoted as the β-phase, and is a metastable phase. Following the vdWP theory:

$$\frac{\Delta \mu_W^H}{RT} = \frac{\mu_W^\beta - \mu_W^H}{RT} = -\sum_i v_i \ln\left(1 - \sum_j \theta_{ij}\right) \tag{13.2}$$

where μ_W^β is the chemical potential of water in the hypothetical empty hydrate lattice, R is the gas constant, v_i is the number of cavities of type i per water molecule, and θ_{ij} is the occupancy of the cavity of type i by the guest component j. Index j can be equal to one (case of pure hydrate), two (case of binary hydrate), or higher (case of mixed hydrates). Index i takes into account the different type of cavities for each hydrate structure. The occupancy θ_{ij} is given by a Langmuir-type function of the gas fugacity:

$$\theta_{ij} = \frac{C_{ij} f_j}{1 + \sum_j C_{ij} f_j}, \tag{13.3}$$

where C_{ij} is the Langmuir constant of the guest component j in the cavity i and f_j is the fugacity of component j, and can be calculated with an appropriate EoS.

Usually, we assume that the cavity is perfectly spherical and that the water molecules which form the cavity are smeared evenly over the surface of the sphere (Holder et al., 1988). In that case, the Langmuir constants can be calculated from the following simplified configurational integral:

$$C_{ij} = \frac{4\pi}{k_B T} \int_0^\infty \exp\left(-\frac{W(r)}{k_B T}\right) r^2 dr, \tag{13.4}$$

where k_B is the Boltzmann constant, and $W(r)$ is the smoothed-cell potential function along the cavity radius due to the interactions between the guest molecule and the cavity. This potential is derived from the summation of the pair-potentials between the guest molecule and each one of the water molecules of the cavity (McKoy and Sinanoglu, 1963). Usually, the Kihara hard-core spherical potential is used for the water–guest interactions and is given as a function of the energy well depth, ε, the collision diameter for the interaction between the gas and the water molecules, σ, and the radius of the hard core, α. Therefore, by selecting a cell potential such as the Kihara potential, the Langmuir constants can be calculated. Note that the parameters σ and ε still need to be fitted to hydrate-equilibrium data in order to have adequate accuracy to the hydrate-equilibrium calculations. Usually, the obtained values for σ and ε can differ significantly from those values obtained when fitting experimental data of viscosity and second virial coefficients.

The right-hand side (RHS) of Equation (13.1c) can be calculated using the seminal work by Parrish and Prausnitz (1972), and subsequently simplified by Holder et al. (1980). The approach is based on classical thermodynamics, taking again the empty hydrate lattice (β-phase) as the reference state. The approach of Holder et al. (1980) is given as follows, for the case that liquid water is present in the system ($T > 273.15\,\text{K}$). In case there is ice present in the system ($T \leq 273.15\,\text{K}$), Equation 13.1c still holds, however the superscript ($\beta - L$) should be replaced by ($\beta - \alpha$). According to the analysis that was presented by Holder et al. [1988], the RHS of Equation 13.1c can be calculated as

$$\frac{\Delta \mu_W^L}{RT} = \frac{\mu_W^\beta - \mu_W^L}{RT} = \frac{\Delta \mu_W^0 (T_0, 0)}{RT_0} - \int_{T_0}^{T} \frac{\Delta h_w^{\beta-L}(T)}{RT^2} dT + \int_{P_0}^{P} \frac{\Delta v_w^{\beta-L}}{RT} dP - \ln(x_w \gamma_w), \tag{13.5}$$

where $\Delta \mu_W^0 (T_0, 0)$ is the chemical potential difference between water in the pure α-phase (ice) and water in the empty hydrate (metastable β-phase), at the reference conditions T_0 and P_0 (usually taken to be $T_0 = 273.15\,\text{K}$ and $P_0 = 0.0\,\text{MPa}$). $\Delta \mu_W^0 (T_0, 0)$ is an experimentally determined quantity (Dharmawardhana et al., 1980; Holder et al., 1984). The second and third terms on the RHS of Equation 13.5 describe the temperature and pressure dependence of the chemical potential respectively. In particular, $\Delta h_w^{\beta-L}$ is the molar enthalpy difference between the empty hydrate lattice and liquid water at zero pressure, which can be calculated by the following equation:

$$\Delta h_w^{\beta-L} = \Delta h_w^0 (T_0) + \Delta h_w^{\alpha-L}(T_0) + \int_{T_0}^{T} \Delta C_{P_w} \, dT \tag{13.6}$$

$\Delta h_W^0 (T_0)$ is the molar enthalpy difference between the empty hydrate lattice and ice at the reference conditions T_0 and P_0, and is an experimentally determined quantity (Dharmawardhana et al., 1980; Holder et al., 1984). $\Delta h_W^{\alpha-L}(T_0)$ is the molar enthalpy difference between ice and liquid water at the reference conditions. ΔC_{P_w} is the isobaric heat capacity difference between the empty hydrate and the pure water phase. It is calculated by the following equation:

$$\Delta C_{P_w}(T) = \Delta C_{P_w}^0 + \beta(T - T_0) \tag{13.7}$$

where $\Delta C_{P_w}^0$ is the isobaric heat capacity difference between the empty hydrate and ice at the reference conditions, and β is a constant that is fitted to the experimental data. $\Delta C_{P_w}^0$ is an experimentally determined quantity (Dharmawardhana et al., 1980; Holder et al., 1984). Depending on whether the water phase is in liquid or ice form, different values for the parameters need to be used. Values for all the reference properties can be found in recent studies (Tsimpanogiannis et al. 2014; El Meragawi et al., 2016).

The final term in Equation 13.5 accounts for the solubility of gas of the type i in water, x_w, and is calculated from the EoS using the flash calculation routine. This is the condition that ensures that the hydrate phase is in equilibrium with both the liquid and vapor phases and the activity coefficient, γ_w, modifies the calculated equilibrium pressure when hydrates are formed in the presence of inhibitors. However, for the purpose of this work, the activity coefficient has been set to unity.

The calculation of the HL_wV equilibrium in the manner explained in this methodology relies on accurate Kihara parameters. These parameters are fitted to three-phase equilibrium experimental data. The initial values for the Kihara parameters were obtained from the literature and were used to calculate the equilibrium pressure of all data points according to the previously explained methodology.

13.4.3 Results and Discussion

The performance of the PC-SAFT and PR EoS in predicting the hydrate equilibrium pressure for pure and mixed hydrate systems of interest is compared against the commercially available code CSMGem (Sloan and Koh, 2008). The results for the calculation of the three-phase equilibrium pressures of nine pure hydrate components along with some typical binary, ternary, and quaternary mixtures of them are presented as bar charts in Figure 13.4. The results that are presented in this section are from the work of El Meragawi et al. (2016) and are based on the best fit of the Kihara ε (i.e., that produces the minimum overall error) for each case of the PR and PC-SAFT EoS. Figure 13.4 shows the overall performance in the

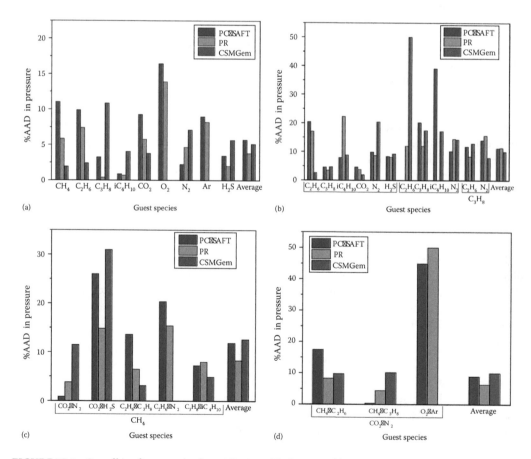

FIGURE 13.4 Overall performance in the prediction of hydrate equilibrium pressure of different hydrate systems: (a) pure hydrates, (b) binary hydrates, (c) ternary hydrates, and (d) quaternary hydrates.

correlation of hydrate equilibrium pressure of different hydrate systems for pure hydrates (Figure 13.4a), and the corresponding predictions for binary hydrates (Figure 13.4b), ternary hydrates (Figure 13.4c), and quaternary hydrates (Figure 13.4d). It is observed that PR performs better than PC-SAFT on the collective average for pure, ternary, and quaternary systems considered, while it performs comparably for the case of binary systems. In addition, for the particular systems examined, PR performs better than CSMGem for pure, ternary, and quaternary systems, while performs comparably for the binary systems.

According to Sloan and Koh (2008), the expected accuracy for the prediction of the incipient hydrate temperature and pressure should be within 0.65 K and 10% of the pressure. These are the expected values for the experimental accuracy of the measurements. For the particular cases examined we observe that improvements of various degrees in the theory are required for certain systems, while for most of them the examined models perform adequately.

In order to improve the accuracy of the calculations, the Kihara ε parameter has been optimized using hydrate equilibrium experimental data for the pure components. Subsequently, the obtained values have been used for the calculation of the mixtures. We observe that reasonable agreement between the experimental and calculated values is obtained for the case of the mixtures, given the large sensitivity of the results on the modified parameters.

13.5 Molecular Simulations in ZIFs for Gas Separation

13.5.1 Introduction

Metal-organic frameworks (MOFs) are a family of nanoporous materials which have pulled the increasing interest of the research community in the last ten years (Furukawa et al., 2013). They exhibit exceptional porosity, thermal and chemical stability, and can be easily modified by replacing small building units without affecting the underlying topology of the framework. They are studied for various applications which take advantage of their unique gas adsorption properties such as hydrogen and methane storage for alternative fuels (Furukawa et al., 2010; Clark, 2011), carbon dioxide capture (Millward and Yaghi, 2005), and separation processes (Chen et al., 2006).

One of the most investigated MOFs for separation of gas mixtures is ZIF-8 (Park et al., 2006) of the subfamily of Zeolitic imidazolate frameworks (ZIFs). A rather unique behavior has been reported which makes this framework very promising for specific separations: molecules larger than the aperture leading to the pores (3.4Å), from methane up to iso-butane (Figure 13.5), can diffuse in ZIF-8 (Bux et al., 2011; Gücüyener et al., 2010; Diestel et al., 2012) taking advantage of a swelling motion of the linkages which has yet to be fully explained. Moreover, a modification of ZIF-8 structure by replacement of Zn metal atom by Co, namely ZIF-67 (Barerjee et al., 2008) has placed this material at the top of the candidates for the industrially demanding separation of propylene/propane mixture. Recent computational studies on ZIF-8 have succeeded in reproducing the environment of ZIF-8 in atomistic level, yielding satisfactory agreement with experiments of gas diffusion reported in literature (Seehamart et al., 2009; Hertag et al., 2011; Pantatosaki et al., 2012; Zheng et al., 2012). Nevertheless, there is still ground to be covered toward the understanding of the underlying mechanisms and the full extent of this framework's capabilities. Topics such as the flexibility of the framework as well as recent advances on the modification of the structure toward the improvement of the separation performance have inspired and motivated the present work.

13.5.2 Methodology

Problems of gas sorption and diffusion in materials with pores in the order of Å are solved in the molecular level, employing molecular mechanics approaches, such as MC and MD techniques. The first step for constructing the desired model is the reconstruction of the unit cell. ZIF-8 and ZIF-67 unit cells were

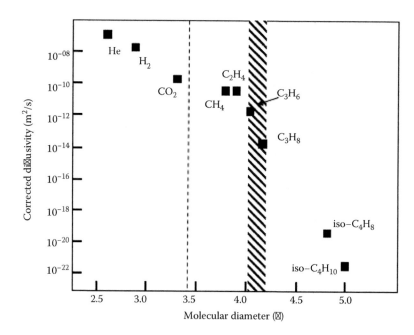

FIGURE 13.5 Corrected diffusivities of various gas molecules in ZIF-8, as a function of their molecular diameter. Dotted line depicts the experimentally observed aperture width giving access to the cavities. The grey ribbon is a critical molecular size beyond which diffusivity drops dramatically.

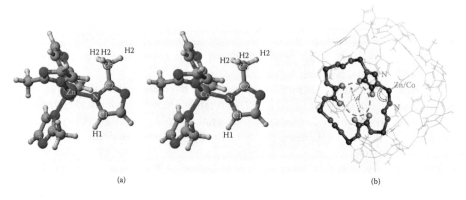

FIGURE 13.6 (a) The basic tetrahedral unit of ZIF-8/ZIF-67 framework (b) and the aperture giving access to the cavity (lighter colored) of the framework.

reconstructed according to the experimentally solved structures by Park et al. (2006) and Banerjee et al. (2008), respectively. Similar tetrahedral units shown in Figure 13.6a are being built and the resulting periodical framework is a sodalite topology of the $I\,\overline{4}3m$ space group. Co atom forms tighter bondings with its surrounding atoms which results in a slightly shrunk unit cell (16.96 Å) when compared with ZIF-8 (16.991 Å) and a smaller aperture size leading to the main cavity as shown in Figure 13.6b.

The forces created between atoms of the structure (host) are described by a force field developed by Krokidas et al. (2015, 2016) which reproduces successfully the structure and major structural properties of ZIF-8 and ZIF-67 frameworks, such as bond lengths and angles. The interactions of diffusing hydrocarbon molecules (guest) were described by the transferable potential for phase equilibria (TraPPE) force field (Martin and Siepmann, 1998). Host–guest interactions were calculated by employing the Lorentz–Berthelot combining rules. Extended 2×2×2 cells (super cells) of ZIF-8 and ZIF-67 were loaded with the desirable amount of hydrocarbon species and MD simulations were carried in the NVT ensemble. The resulted trajectories were processed for the calculation of the corrected diffusivity, D_0, of the diffusing species, which operates in the Darken equation (Darken, 1948):

$$D_t(c) = D_0(c)\left(\frac{\partial \ln f}{\partial \ln c}\right)_T \qquad (13.8)$$

connecting the transport diffusivity, $D_t(c)$, with the ratio at the right side, called the thermodynamic factor at constant temperature, T (f and c are the fugacity and concentration of adsorbed species, respectively). A methodology which was developed by Theodorou et al. (1996) and has been successfully used in gas diffusion in zeolites in the past was employed here. It gives the displacement of center of mass of the swarm of diffusing molecules through the following equation (Skoulidas and Sholl, 2002):

$$D_0 = \frac{1}{6N} \lim_{t \to \infty} \frac{1}{t} \left\langle \left| \sum_{i=1}^{N} (\vec{r}_i(t) - \vec{r}_i(t_0)) \right|^2 \right\rangle \qquad (13.9)$$

Although Equation 13.9 resembles the mean square displacement which provides the self-diffusivity, D_s, the corrected diffusivity is subject to larger error and thus much more computational time is needed for such calculations.

13.5.3 Results and Discussion

MD in the empty frameworks of ZIF-8 and ZIF-67 structures gave average values for characteristic bonds and angles which agree excellently with literature experimental data (Krokidas et al., 2016). Moreover, the simulations managed to reproduce the slight shrinkage of the aperture in the case of ZIF-67, as shown in Table 13.2, along with experimentally measured values. The values were extracted at two different simulation conditions: firstly, at the same temperature as this of the experiments in which apertures were measured originally, for the sake of comparison; secondly, at room temperature in which the computational diffusion of propane and propylene are to be compared with experimentally measured gas diffusion in the two frameworks.

The accurate description of the framework verifies the performance of the proposed force field. The ability to reproduce subtle changes in the structure is of critical importance for separations which depend on such small differences in the size of the diffusing species, such as the propane/propylene mixture (Radius$_{C3H8}$ − Radius$_{C3H6}$≈0.2 Å). This is further proven by calculations of the activation energies of diffusion and of the isosteric heats of adsorption of both propane and propylene, in ZIF-8. Table 13.3 shows that the two species share the same isosteric heat of adsorption, meaning that they interact in the same manner with the atoms of the pores surrounding them. In the same time, the difference in the activation energies of the two species indicates an unhindered propagation of propylene against

TABLE 13.2 Aperture Diameter of ZIF-8 and ZIF-67 as Measured in the MD Simulations, Compared with Relevant Experimentally Measured Values

	$d_{ZIF\text{-}8}$ (Å)	$d_{ZIF\text{-}67}$ (Å)
Experimental data	~3.4	~3.3
MD simulations corresponding to experimental temperature (258K for ZIF-8; 153K for ZIF-67)	3.42	3.31
MD simulations at 295 K	3.44	3.35

Note: Experimental data are from Park et al. (2006) and Banerjee et al. (2008) and MD simulations are from Krokidas et al. (2016).

TABLE 13.3 Isosteric Heats of Adsorption and Activation Energies of Diffusion of Propane and Propylene in ZIF-8

	Isosteric Heat of Adsorption (kJ/mol)		Activation Energy of Diffusion (kJ/mol)	
	C_3H_6	C_3H_8	C_3H_6	C_3H_8
Expt. data	18.4	18.9	38.8	12.7
	34	30	26.6	15.1
MD simulations	26.3	25.9	30.0 ± 1.0	18.0 ± 1.1

Note: Experimental data are from Furukawa et al. (2010, 2013), Liu et al. (2014), Li et al. (2009), and Pan et al. (2015). MD simulations are from Krokidas et al. (2015).

TABLE 13.4 Comparison of D_0 of Propane and Propylene in ZIF-8 and ZIF-67 from MD Simulations and Experiments

		D_0 (m²/s)		$D_{0, C3H6}/D_{0, C3H8}$
		C_3H_8	C_3H_6	
ZIF-8 from author's work	Simulations	$(4.0 \pm 1.5) \times 10^{-14}$	$(1.8 \pm 0.1) \times 10^{-12}$	45 ± 3
	Expt. data	$(2.22 \pm 0.29) \times 10^{-14}$	$(1.26 \pm 0.17) \times 10^{-12}$	57 ± 5
		$(2.77 \pm 0.74) \times 10^{-14}$	$(1.23 \pm 0.04) \times 10^{-12}$	46 ± 10
		$(1.07 \pm 0.21) \times 10^{-15}$	$(0.85 \pm 0.04) \times 10^{-13}$	80 ± 13
ZIF-8 literature expt.		3.70×10^{-14}	1.21×10^{-12}	32.7
		1.1×10^{-14}	1.2×10^{-12}	109
ZIF-67 from author's work	Simulations	$(1.5 \pm 0.5) \times 10^{-15}$	$(2.8 \pm 0.1) \times 10^{-13}$	190 ± 5
	Expt. data	$(0.76 \pm 0.15) \times 10^{-15}$	$(1.49 \pm 0.05) \times 10^{-12}$	200 ± 38

Note: ZIF-8 values are from Li et al. (2009), Furukawa et al. (2013), and Krokidas et al. (2015); ZIF-67 values are from Krokidas et al. (2016).

propane. The comparison with experiments reported in literature justifies this statement. This behavior is attributed to the propylene's size, which being smaller than propane, moves easier through the narrow aperture and diffuse in the material faster.

The smaller aperture of the modified ZIF framework (ZIF-67) takes advantage of the size selectivity of this mixture. Results from MD simulations of propane and propylene diffusion in both ZIF-8 and ZIF-67 are compared in Table 13.4. The ratio of the propylene/propane-corrected diffusivities in ZIF-67 is larger than ZIF-8, yielding a value of 190. This finding agrees with the ratio coming from experimental D_0 values (Krokidas et al., 2015) placing ZIF-67 at the top among the candidates for propylene/propane separation.

Both propane's and propylene's size (4.2 and 4.0 Å) exceed the aperture size (~3.4 and 3.3 Å, for ZIF-8 and ZIF-67, respectively), reported from these simulations and from experiments of literature. This is due to flexibility of the framework, which is accurately described by our forcefield. The motion of the aperture [as shown in Figure 13.5b] in MD simulation were measured and the results are shown in Figure 13.7.

The motion of the ligands, mentioned in the introduction, creates a pseudo-breathing mode under which the aperture oscillates between minimum and maximum value. The average value of the oscillation matches the measured values reported from experiments. This finding agrees with the work by Kolokov who observed experimentally an oscillation of the ZIF-8 ligands, calling it "saloon door" motion (Hara et al., 2014; Kolokov et al. 2015). The observed maximum values justify the mobility of propane and propylene molecules through the apertures of the framework. ZIF-67, having the stiffer bonding of Co affecting the framework, exhibits a shifted aperture oscillation toward lower values. Propane molecular motion is hindered from the narrower aperture of ZIF-67, while propylene motion takes advantage of its slightly smaller size and passes through. The simulations did not reveal any effect of the presence of adsorbed

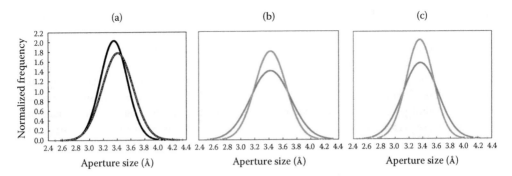

FIGURE 13.7   Distribution of the aperture size as observed in MD simulations. (a) Filled black line (ZIF-67) and open-spaced line (ZIF-8) for 295 K. The blue and orange diagrams show the aperture distribution at 295 and 530 K, respectively, for (b) ZIF-8 and (c) ZIF-67.

molecules on the aperture size. However, the amplitude of the motion is clearly affected by temperature change taking higher values as the system is heated, as shown in Figures 13.7b and c.

13.6 ILs: Advanced Materials and Novel Green Designer Solvents

13.6.1 Introduction

ILs are a rapidly growing class of materials with a broad spectrum of applications that range from green chemistry, catalysis, biotechnology, food, and medical industry to separation technologies and environmental engineering. ILs are composed by ions and are at the liquid state at room temperature and by convention below 100°C (Wasserscheid and Welton, 2008; Weingärtner, 2008). They are usually consisted of an asymmetric organic cation and an organic or inorganic anion, exhibiting a remarkable combination of properties such as almost negligible vapor pressures, nonflammability, wide liquid range, high thermal stability, very good solvation properties, and high conductivity. The set of macroscopic properties of an IL is directly dependent on the chemical structure of the ions involved and therefore one of the ILs unique properties is their chemical tunability. There is a vast number of anions and cations that can be combined to form an IL and although, only a very small fraction of them has been explored so far experimentally or theoretically/computationally, ILs are currently used in the core of many cutting-edge applications and investigated to be utilized in novel technologies that fall within (but not limited to) the Water-Food-Energy nexus that forms the basis of a sustainable future.

ILs exhibit very good extractable capabilities for water purification and wastewater treatments, fulfilling at the same time important safety and environmental issues. Examples of such applications are the use of ILs that can be efficiently regenerated for seawater desalination (Cai et al., 2015), the removal of organic compounds such as phenols from water and wastewater from pharmaceutical and chemical industries (Pilli et al., 2014) using supported IL membranes (SILMs). IL-based systems have been identified as highly efficient for the monitoring and extraction of endocrine disrupting chemicals like ethunylestradiol in wastewater (Dinis et al., 2015). Another important source of pollution is the contamination of water with toxic metals. ILs-based systems of various types (ILs, poly-ILs, IL-grafted nanoparticles, etc.) are investigated and used for the separation of the metal compounds from water (Hu et al., 2008; Poursaberi and Hassanisadi, 2013). In the same spirit, many green ILs have been used in food analysis for the determination and (micro)extraction of metals and organic compounds (Martinis et al., 2010; Martin-Calero et al., 2011) and for the manufacturing of electronic food analysis sensors (Toniolo et al., 2013).

ILs also have the potential to play a crucial role in energy-related applications (Wishart, 2009; McFarlane et al., 2014; Watanabe et al., 2017). Their contribution in energy generation involves for example their use as electrolytes in dye-sensitized solar cells, in biodiesel production, or in biofuel production since ILs have the ability to dissolve natural biomolecules such as cellulose. In parallel, ILs conductivity and electrochemical properties enable their use in energy storage like in batteries and supercapacitors. At the same time, ILs have been identified as being very efficient in gas separations with great environmental impact, especially in novel carbon capture and utilization (CCU) technologies.

The wide range of applications that can benefit from the use of ILs and their indisputable advantage of chemical tunability necessitate the fundamental understanding of the underlying mechanisms that govern the end-use performance of ILs. The unfolding of the relation between chemical structure and macroscopic behavior is essential for the design of task-specific ILs with tuned properties that maximize their performance. Therefore, these systems have been a topic of vivid interest for the molecular simulation community in the recent years.

13.6.2 Methodology and Results

ILs are rather complex systems and there are quite some challenges in simulating their behavior (Salanne 2015). Very strong electrostatic interactions are present in these systems and polarizability and charge transfer effects have to be taken into account. The optimal method for calculation of partial charges is still under debate (Dommert et al., 2012; Hunt et al., 2006), while the chemical diversity in the structure of the plethora of ions that can form an IL hinders the development of general force fields. The predictive power of molecular simulation relies on the accuracy of the interaction potential in use and therefore, especially for the case of ILs, force fields should be validated through direct comparison with experimental data. In Figure 13.8, molecular simulation predictions are shown for the density of 1-alkyl-3-methylimidazolium tricyanomethanide ([$C_n$mim$^+$] [TCM$^-$], $n = 2, 4, 6, 8$) ILs in comparison with experimental measurements. MD simulations of 200 ionic pairs have been performed using NAMD in various ensembles and in a wide temperature range. The force field used for the simulation of these systems has been optimized (Vergadou et al., 2016) for [C_4mim$^+$] [TCM$^-$] IL (Figure 13.9) at 298 K, while all other points are pure predictions. The deviation of the MD predicted densities from the experimental ones is in all cases less than 1.2%.

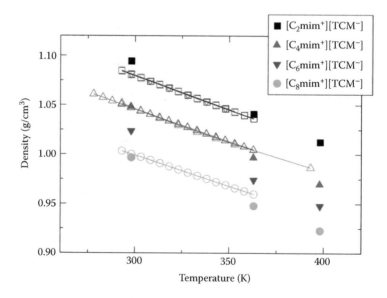

FIGURE 13.8 MD predictions (full points) of density of [$C_n$mim$^+$][TCM$^-$], $n = 2, 4, 6, 8$ (Vergadou et al., 2016) and experimental data (open points) for [C_2mim$^+$][TCM$^-$] (Labropoulos et al., 2013; Larriba et al., 2013; Królikowski et al., 2013), [C_4mim$^+$][TCM$^-$] (Labropoulos et al., 2013), and [C_8mim$^+$][TCM$^-$] (Romanos et al., 2013). Lines are fit to the experimental data.

This IL family was studied as it was identified as a very promising candidate for CO_2 capture from post combustion flue gases within the framework of the project entitled "Novel Ionic Liquid and Supported Ionic Liquid Solvents for Reversible Capture of CO_2," (IOLICAP) funded by the European Commission under the seventh Framework Programme for Research, Technology and Development. IOLICAP aimed at the development, optimization, and evaluation of task-specific ILs and SILMs for the reversible CO_2 capture and separation from postcombustion flue gases.

ILs are heterogeneous fluids and due to the ionic interactions they form polar and nonpolar domains. Structural order is retained at much longer distances compared to ordinary liquids and is clearly depicted in the radial distribution function of the ions' center of mass with the anion–cation interaction being stronger than the other two interactions (Androulaki et al., 2012; Androulaki et al., 2014). The complex dynamical behavior of ILs renders the calculation of their transport properties a very demanding task that requires very long MD simulations. Anisotropy phenomena in the ions diffusion have been observed in these systems, while heterogeneities in the dynamics are present at low temperatures (Androulaki et al., 2014), resembling the dynamics of super-cooled liquids. For ILs, the prediction of diffusivity and viscosity is a determinant factor on the ability of the force field to realistically represent the interactions among the ions. In Figure 13.10, self-diffusion coefficients are shown for [C_8mim$^+$][TCM$^-$] IL (Vergadou et al., 2016) in comparison with NMR experimental data,

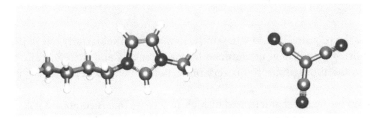

FIGURE 13.9 Chemical structure of the [C_4mim$^+$] cation and [TCM$^-$] anion.

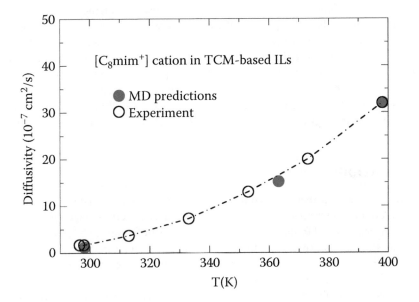

FIGURE 13.10 Self-diffusion coefficients of the cation (squares) as a function of temperature for [C_8mim$^+$] [TCM$^-$] (Vergadou et al., 2016). The open points correspond to experimental results (Papavassiliou and Fardis, 2015). The error bars in the MD predictions are within the symbol size.

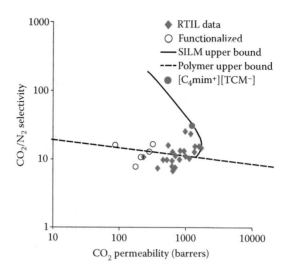

FIGURE 13.11 Estimation of the permeability/selectivity performance of [C₄mim⁺][TCM⁻] in a semi-permeable barrier consideration using a Robeson plot. (Adapted from Scovazzo, P., *J. Membr. Sc.*, 343, 199–211, 2009.)

verifying field for the [C$_n$mim⁺][TCM⁻] IL family. In many technologically relevant applications, ILs are used as supported membranes or confined in porous media and therefore also the effect of surfaces/interfaces in the above properties is very important to be considered (Sha et al., 2012; Kritikos et al., 2016).

Novel separation media are characterized by high selectivity that is combined with sufficient permeability. Gas permeability can be determined as the product of gas diffusivity and solubility. IL permeability and selectivity to gases (Vergadou et al., 2017a,b) has been extracted by performing additional very long MD simulations for the prediction of gas diffusivity, whereas gas solubility has been calculated in the infinite dilution regime using the Widom test particle insertion method (Widom, 1963). The various gases studied appear to have comparable diffusivities in each IL system at hand studied and for that, the solubility is expected to control the selectivity properties of these ILs, a fact that is also supported by experimental results (Finotello et al., 2007). Within a semipermeable barrier consideration of ILs, the permeability and selectivity results of the [C₄mim⁺] [TCM⁻] are well above the upper limit for polymers in a Robeson plot (Robeson, 1991; Scovazzo, 2009) as shown in Figure 13.11. From this analysis, it can be concluded that [TCM⁻] ILs are highly selective (Vergadou et al., 2017a,b) with the ones consisted of the intermediate alkyl chain length cations being the most selective for the critical in CCS processes and CO_2/N_2 separation.

13.7 Conclusions

The critical role of molecular thermodynamics in the design of new sustainable processes and products is undisputed. A number of representative examples were illustrated from ongoing research projects in Texas A&M University at Qatar and National Center for Scientific Research "Demokritos" in Athens, Greece. Despite the fact that molecular thermodynamics and molecular simulation are used widely today in chemical engineering, a number of challenges remain. Development of more accurate, but at the same time more computationally demanding, force fields for the intermolecular and intramolecular interactions based on a combination of *ab initio* quantum mechanics calculations and tuning to experimental data is necessary. In parallel, development of systematic coarse grained molecular models will allow simulation of systems over longer time scales, approaching macroscopic times.

Acknowledgments

This publication was made possible by NPRP grant numbers 6-1157-2-471, 6-1547-2-632 and 7-042-2-021 from the Qatar National Research Fund (a member of the Qatar Foundation). Financial support from the seventh European Framework Programme for Research and Technological Development for the project "Novel Ionic Liquid and Supported Ionic Liquid Solvents for Reversible Capture of CO_2" (IOLICAP Project 283077) is gratefully acknowledged. The statements made herein are solely the responsibility of the authors. We are grateful to the High Performance Computing Center of Texas A&M University at Qatar for generous resource allocation.

References

Abascal, J.L.F., Sanz, E., García Fernández, R. and C. Vega. 2005. A potential model for the study of ices and amorphous water: TIP4P/ice. *J. Chem. Phys.* 122:234511.

Abascal, J.L.F. and C. Vega. 2005. A general purpose model for the condensed phases of water: TIP4P/2005. *J. Chem. Phys.* 123:234505.

Allen, M.P. and D.J. Tildesley. 1987. *Computer Simulation of Liquids.* Oxford, UK: Oxford University Press.

Androulaki, E., Vergadou, N. and I.G. Economou. 2014. Analysis of the heterogeneous dynamics of imidazolium-based [Tf$_2$N$^-$] ionic liquids using molecular simulation. *Mol. Phys.* 112(20):2694–2706.

Androulaki, E., Vergadou, N., Ramos, J. and I.G. Economou 2012. Structure, thermodynamic and transport properties of imidazolium-based bis(trifluoromethylsulfonyl)imide ionic liquids from molecular dynamics simulations. *Mol. Phys.* 110(11–12):1139–1152.

Archer, D., Buffett, B. and V. Brovkin. 2009. Ocean methane hydrates as a slow tipping point in the global carbon cycle. *Proc. Natl. Acad. Sci. USA* 106:20596.

Banerjee, R., Phan, A., Wang, B., Knobler, C., Furukawa, H., O'Keefee, M. and O.M. Yaghi. 2008. High-throughput synthesis of zeolitic imidazolate frameworks and application to CO_2 capture. *Science* 319:939–943.

Berendsen, H.J.C., Postma, J.P.M., Vangunsteren, W.F., Dinola, A. and J.R. Haak. 1984. Molecular-Dynamics with coupling to an external bath. *J. Chem. Phys.* 81:3684–3690.

Brokaw, R.S. 1969. Predicting transport properties of dilute gases. *Ind. Eng. Chem. Process Des. Dev.* 8:240–253.

Bux, H., Chmelik, C., Krishna, R. and J. Caro. 2011. Ethene/ethane separation by the MOF membrane ZIF-8: Molecular correlation of permeation, adsorption, diffusion. *J. Membr. Sci.* 369:284–289.

Cai, Y., Shen, W., Wei, J., Chong, T.H., Wang, R., Krantz, W.B., Fane, A.G. and X. Hu. 2015. Energy-efficient desalination by forward osmosis using responsive ionic liquid draw solutes. *Environ. Sci.* 1:341–347

Chen, B., Liang, C., Yang, J., Contrearas, D.S., Clancy, Y.L., Lobkovsky, E.B., Yaghi, O.M. and S. Dai. 2006. A microporous metal-organic framework for gas-chromatographic separation of alkanes. *Angew. Chem. Int. Ed.* 45:1390–1393.

Clark, J. 2011. Mercedes-Benz F125! Research vehicle technology. http://www.emercedesbenz.com/autos/mercedes-benz/concept-vehicles/mercedes-benz-f125-research-vehicle-technology/.

Conde, M.M. and C. Vega. 2010. Determining the three-phase coexistence line in methane hydrates using computer simulations. *J. Chem. Phys.* 133:064507.

Costandy, J., Michalis, V.K., Tsimpanogiannis, I.N., Stubos, A.K., and I.G. Economou. 2015. The role of intermolecular interactions in the prediction of the phase equilibria of carbon dioxide hydrates. *J. Chem. Phys.* 143:094506.

Darken, L.S. 1948. Diffusion, mobility and their interrelation through free energy in binary metallic systems, *Trans. AIME* 175:184–201.

Dharmawardhana, P.B., Parish, W.R. and E.D. Sloan. 1980. Experimental thermodynamic parameters for the prediction of natural-gas hydrate dissociation conditions. *Ind. Eng. Chem. Fundam.* 19:410–414.

Diestel, L., Bux, H., Wachsmuth, D. and J. Caro. 2012. Pervaporation studies of *n*-hexane, benzene, mesitylene and their mixtures on zeolitic imidazolate framework-8 membranes. *Micropor. Mesopor. Mater.* 164:288–293.

Dinis, T.B.V., Passos, H., Lima, D.L.D., Esteves, V.I., Coutinho, J.A.P and M.G. Freire. 2015. One-step extraction and concentration of estrogens for an adequate monitoring of wastewater using ionic-liquid-based aqueous biphasic systems. *Green Chem.* 17(4):2570–2579.

Dommert, F., Wendler, K., Berger, R., Delle Site, L. and C. Holm. 2012. Force fields for studying structure and dynamics of imidazolium based ionic liquids: A critical review of recent developments. *ChemPhysChem* 13:1625–1637.

Dufal, S., Lafitte, T., Haslam, A.J., Galindo, A., Clark, G.N.I., Vega, C. and G. Jackson. 2015. The A in SAFT: Developing the contribution of associating to the Helmholtz free energy within a Wertheim TPT1 treatment of generic Mie fluids. *Mol. Phys.* 113:948–984.

Economou, I.G. 2002. Statistical associating fluid theory: A successful model for the calculation of thermodynamic and phase equilibrium properties of complex fluid mixtures. *Ind. Eng. Chem. Res.* 41:953–962.

Economou, I.G., Peters, C.J. and J. de Swaan Arons. 1995. Water-salt phase equilibria at elevated temperatures and pressures: Model development and mixture predictions. *J. Phys. Chem.* 99:6182–6193.

El Meragawi, S., Diamantonis, N.I., Tsimpanogiannis, I.N., and I.G. Economou. 2016. Hydrate–fluid phase equilibria modeling using PC-SAFT and Peng–Robinson equations of state. *Fluid Phase Equilib.* 413:209–219. doi:1016/j.fluid.2015.12.003.

Finotello, A., Bara, J.E., Camper, D. and R.D. Noble. 2007. Room-temperature ionic liquids: Temperature dependence of gas solubility selectivity. *Ind. Eng. Chem. Res.* 47:3453–3459.

Furukawa, H., Corcdova, K.E., O'Keefe, M. and O.M. Yaghi. 2013. The chemistry and applications of metal-organic frameworks. *Science* 341:12304441–123044412.

Furukawa, H., Ko, N., Go, Y.B., Aratani, N., Choi, S.B., Choi, E., Yazaydin, A.O., Snurr, R.Q., O' Keeffe, M., Kim, J. and O.M. Yaghi. 2010. Ultrahigh porosity in metal-organic frameworks. *Science* 329:424–428.

Gross, J. and G. Sadowski. 2002. Application of the perturbed-chain SAFT equation of state to associating systems. *Ind. Eng. Chem. Res.* 41:5510–5515.

Gücüyener, C., Van Den Bergh, J., Gascon, J. and F. Kapteijn. 2010. Ethane/ethene separation turned on its head: Selective ethane adsorption on the metal–organic framework ZIF-7 through a gate-opening mechanism. *J. Am. Chem. Soc.* 132:17704–17706.

Haghtalab, A. and S.H. Mazloumi. 2009. A square-well equation of state for aqueous strong electrolyte solutions. *Fluid Phase Equilib.* 285:96–104.

Hammerschmidt, E.G. 1934. Formation of gas hydrates in natural gas transmission Lines. *Ind. Eng. Chem.* 26:851.

Hara, N., Yoshimune, M., Negishi, H., Haraya, K., Hara, S. and T. Yamaguchi. 2014. Diffusive separation of propylene/propane with ZIF-8 membranes. *J. Membr. Sci.* 450:215–223.

Hertag, L., Bux, H., Caro, J., Chmelik, C., Remsungen, T., Knauth, M. and S. Fritzsche. 2011. Diffusion of CH_4 and H_2 in ZIF-8. *J. Membr. Sci.* 377:36–41.

Hess, B., Kutzner, C., der Spoel, D.V. and E. Lindahl. 2008. GROMACS 4: Algorithms for highly efficient, load-balanced, and scalable molecular simulation. *J. Chem. Theory Comput.* 4:435–447.

Holder, G.D., Corbin, G. and K.D. Papadopoulos. 1980. Thermodynamic and molecular properties of gas hydrates from mixtures containing methane, argon, and krypton. *Ind. Eng. Chem. Fund.* 19:282–286.

Holder, G.D., Malekar, S.T. and E.D. Sloan. 1984. Determination of hydrate thermodynamic reference properties from experimental hydrate composition data. *Ind. Eng. Chem. Fund.* 23:123–126.

Holder, G.D., Zetts, S.P. and N. Pradhan. 1988. Phase behavior in systems containing clathrate hydrates: A review. *Rev. Chem. Eng.* 5:1–70.

Hu, J.S., Zhong, L.S., Song, W.G. and L.J. Wan. 2008. Synthesis of hierarchically structured metal oxides and their application in heavy metal ion removal. *Adv. Mater.* 20:2977–2982.

Hunt, P.A., Kirchner, B. and T. Welton. 2006. Characterising the electronic structure of ionic liquids: An examination of the 1-butyl-3-methylimidazolium chloride ion pair. *Chem. Eur. J.*, 12:6762–6775.

Jensen, L., Thomsen, K., von Solms, N., Wierzchowski, S., Walsh, M.R., Koh, C.A., Sloan, E.D., Wu, D.T. and A.K. Sum. 2010. Calculation of liquid water–hydrate–methane vapor phase equilibria from molecular simulations. *J. Phys. Chem. B* 114:5775.

Jiang, H., Mester, Z., Moultos, O.A., Economou, I.G. and A.Z. Panagiotopoulos. 2015. Thermodynamic and transport properties of $H_2O + NaCl$ from polarizable force fields. *J. Chem. Theory Comput.* 11:3802–3810.

Jiang, H., Moultos, O.A., Economou, I.G. and A.Z. Panagiotopoulos. 2016a. Gaussian-charge polarizable and non-polarizable models for CO_2. *J. Phys. Chem. B* 120:984–994. doi:10.1021/acs.jpcb.5b11701.

Jiang, H., Panagiotopoulos, A.Z. and I.G. Economou. 2016b. Modeling of CO_2 solubility in single and mixed electrolyte solutions using statistical associating fluid theory. *Geochim. Cosmochim. Acta* 176:185–197.

Jorgensen, W.L., Madura, J.D. and C.J. Swenson. 1984. Development and testing of the OPLS all-atom force field on conformational energetics and properties of organic liquids. *J. Am. Chem. Soc.* 106:6638.

Khokhar, A.A., Gudmundsson, J.S., and E.D. Sloan. 1998. Gas storage in structure H hydrates. *Fluid Phase Equilib.* 150:383.

Kiss, P.T. and A. Baranyai. 2014. Anomalous properties of water predicted by the BK3 model. *J. Chem. Phys.* 140:154505.

Kiss, P.T., Sega, M., and A. Baranyai. 2014. Efficient handling of Gaussian charge distributions: An application to polarizable molecular models. *J. Chem. Theory Comput.* 10:5513–5519.

Koh, C.A. 2002. Towards a fundamental understanding of natural gas hydrates. *Chem. Soc. Rev.* 31:157–167.

Kolokov, D.I., Stepanov, A.G. and H. Jobic. 2015. Mobility of the 2-methylimidazolate linkers in ZIF-8 probed by ^2H NMR: Saloon doors for the guests. *J. Phys. Chem. C.* 119:27512–27520.

Kritikos, G., Vergadou, N. and I.G. Economou. 2016. Molecular dynamics simulation of highly confined glassy ionic liquids. *J. Phys. Chem. C.* 120:1013–1024.

Krokidas, P., Castier, M., Moncho, S., Brothers, E. and I.G. Economou. 2015. Molecular simulation studies of the diffusion of methane, ethane, propane, and propylene in ZIF-8. *J. Phys. Chem. C* 119:27028–27037.

Krokidas, P., Castier, M., Moncho, S., Sredojevic, D., Brothers, E., Kwon, H.T., Jeong, H-K and I.G. Economou. 2016. ZIF-67 framework: A promising new candidate for propylene/propane separations – experimental data and molecular simulations. *J. Phys. Chem. C* 120:8116–8124.

Królikowski, M., Walczak, K. and U. Domańska. 2013. Solvent extraction of aromatic sulfur compounds from *n*-heptane using the 1-ethyl-3-methylimidazolium tricyanomethanide ionic liquid. *J. Chem. Thermodyn.* 65:168–173.

Kvenvolden, K.A. 1999. Potential effects of gas hydrate on human welfare. *Proc. Natl. Acad. Sci. USA* 96:3420–3426.

Labropoulos, A.I., Romanos, G.E., Kouvelos, E., Falaras, P., Likodimos, V., Francisco, M., Kroon, M.C., Iliev, B., Adamova, G. and T.J.S. Schubert. 2013. Alkyl-methylimidazolium tricyanomethanide ionic liquids under extreme confinement onto nanoporous ceramic membranes. *J. Phys. Chem. C* 117:10114–10127.

Ladd, A.J.C. and L.V. Woodcock. 1977. Triple-point coexistence properties of the Lennard-Jones system. *Chem. Phys. Lett.* 51:155–159.

Larriba, M., Navarro, P., García, J. and F. Rodríguez. 2013. Liquid-liquid extraction of toluene from heptane using [emim][DCA], [bmim][DCA], and [emim][TCM] ionic liquids. *Ind. Eng. Chem. Res.* 52:2714–2720.

Li, K., Olson, H. D., Seidel, J., Emge, J.T., Gong, H., Zeng, H. and J. Li. 2009. Zeolitic imidazolate frameworks for kinetic separation of propane and propene. *J. Am. Chem. Soc.* 131:10368–10369.

Liu, D., Xiaoli, M., Hongxia, X. and Y.S. Lin. 2014. Gas transport properties and propylene/propane separation characteristics of ZIF-8 membranes. *J. Membr. Sci.* 451:85–93.

MacFarlane, D.R., Tachikawa, N, Forsyth, M., Pringle, J.M., Howlett, P.C., Elliott, G.D., Davis, J.H., Watanabe, M., Simon, P. and C.A. Angell. 2014. Energy applications of ionic liquids. *Energy Environ. Sci.* 7: 232–250.

Maginn, E.J. 2009. From discovery to data: What must happen for molecular simulation to become a mainstream chemical engineering tool. *AIChE J.* 55: 1304–1310.

Martin, M. and J.I. Siepmann, 1998. Transferable potentials for phase equilibria. 1. United-atom description of *n*-alkanes. *J. Phys. Chem. B* 102:2569–2577.

Martín-Calero, A., Pino, V. and A.M. Afonso. 2011. Ionic liquids as a tool for determination of metals and organic compounds in food analysis. *Trends Anal. Chem.* 30(10):1598–1619.

Martinis, E.M., Berton, P., Monasterio, R.P. and R.G. Wuilloud. 2010. Emerging ionic liquid-based techniques for total-metal and metal-speciation analysis. *TrAC Trends Anal. Chem.* 29(10):1184–1201.

McKoy, V. and O. Sinanoglu. 1963. Theory of dissociation pressures of some gas hydrates. *J. Chem. Phys.* 38:2946–2956.

McMullan, R. K. and G.A. Jeffrey. 1965. Polyhedral clathrate hydrates. IX. Structure of ethylene oxide hydrate. *J. Chem. Phys.* 42:2725–2732.

Metz, B., Davidson, O., de Concinck, H., Loos, M. and L. Meyer. 2005. *Carbon Dioxide Capture and Storage: Special Report of the Intergovernmental Panel on Climate Change*. Cambridge, MA: Cambridge University Press.

Michalis, V.K., Costandy, J., Tsimpanogiannis, I.N., Stubos, A.K., and I.G. Economou. 2015. Prediction of the phase equilibria of methane hydrates using the direct phase coexistence methodology. *J. Chem. Phys.* 142:044501.

Michalis, V.K., Moultos, O.A., Tsimpanogiannis, I.N. and I.G. Economou. 2016. Molecular dynamics simulations of the diffusion coefficients of light *n*-alkanes in water over a wide range of temperature and pressure. *Fluid Phase Equilib.* 407:236–242.

Millward, A.R. and O.M. Yaghi. 2005. Metal-organic frameworks with exceptionally high capacity for storage of carbon dioxide at room temperature. *J. Am. Chem. Soc.* 127:17998–17999.

Moultos, O.A., Orozco, G.A., Tsimpanogiannis, I.N., Panagiotopoulos, A.Z. and I.G. Economou. 2015. Atomistic molecular dynamics simulations of H_2O diffusivity in liquid and supercritical CO_2. *Mol. Phys.* 113:2805–2814.

Moultos, O.A., Tsimpanogiannis, I.N., Panagiotopoulos, A.Z. and I.G. Economou. 2014. Atomistic molecular dynamics simulations of CO_2 diffusivity in H_2O for a wide range of temperatures and pressures. *J. Phys. Chem. B* 118:5532–5541.

Moultos, O.A., Tsimpanogiannis, I.N., Panagiotopoulos, A.Z. and I.G. Economou. 2016. Self-diffusion coefficients of the binary $(H_2O + CO_2)$ mixture at high temperatures and pressures. *J. Chem. Thermodyn.* 93:424–429.

Orozco, G.A., Economou, I.G. and A.Z. Panagiotopoulos. 2014a. Optimization of intermolecular potential parameters for the CO_2/H_2O system. *J. Phys. Chem. B* 118:11504–11511.

Orozco, G.A., Moultos, O.A., Jiang, H., Economou, I.G. and A.Z. Panagiotopoulos. 2014b. Molecular simulation of thermodynamic and transport properties for the H2O + NaCl system. *J. Chem. Phys.* 141:234507.

Palmer, J.C. and P.G. Debenedetti. 2015. Recent advances in molecular simulation: A chemical engineering perspective. *AIChE J.* 61: 370–383.

Pan, Y., Liu, W., Zhao, Y., Wang, C. and Z. Lai. 2015. Improved ZIF-8 membrane: Effect of activation procedure and determination of diffusivities of light hydrocarbons. *J. Membr. Sci.* 493:88–69.

Pantatosaki, E., Megariotis, G., Pusch, A.K., Chmelik, C., Stallmach, F. and G.K. Papadopoulos. 2012. On the impact of sorbent mobility on the sorbed phase equilibria and dynamics: A study of methane and carbon dioxide within the zeolite imidazolate framework-8. *J. Phys. Chem. C* 116:201–207.

Papadimitriou, N.I., Tsimpanogiannis, I.N. and A. K. Stubos. 2010. Computational approach to study hydrogen storage in clathrate hydrate. *Colloids Surf. A.* 357:67–73.

Papavassiliou, G. and M. Fardis, Nuclear Magnetic Resonance Laboratory, Institute of Nanoscience and Nanotechnology, NCSR "Demokritos," Greece—Unpublished results.

Park, K.S., Ni, Z., Cote, A.P., Choi, J.Y., Huang, R., Uribe-Romo, F.J., Chae, H.K., O'Keeffe, M. and O.M. Yaghi. 2006. Exceptional chemical and thermal stability of zeolitic imidazolate frameworks. *Proc. Natl. Acad. Sci. USA* 103:10186–10911.

Parrish, W.R. and J.M. Prausnitz. 1972. Dissociation pressures of gas hydrates formed by gas mixtures. *Ind. Eng. Chem. Proc. Des. Dev.* 11:26–35.

Peng, D.Y. and D.B. Robinson. 1976. A new two-constant equation of state. *Ind. Eng. Chem. Fundam.* 15:59–64.

Pilli, S.R., Banerjee, T. and K. Mohanty. 2014. Performance of different ionic liquids to remove phenol from aqueous solutions using supported liquid membrane. *Desalin. Water Treat.* 54: 3062–3072.

Pitzer, K.S. 1991. *Acitvity Coefficient in Electrolyte Solutions*. Boca Raton, FL: CRC Press.

Plimpton, S. 1995. Fast parallel algorithms for short-range molecular dynamics. *J. Comput. Phys.* 117:1–19.

Potoff, J.J. and J.I. Siepmann. 2001. Vapor–liquid equilibria of mixtures containing alkanes, carbon dioxide, and nitrogen. *AIChE J.* 47:1676–1682.

Poursaberi, T. and M. Hassanisadi. 2013. Magnetic removal of reactive black 5 from wastewater using ionic liquid grafted-magnetic nanoparticles. *Clean-Soil Air Water.* 41:1208–1215.

Prausnitz, J.M., Lichtenthaler, R.N. and E. Gomes de Azevedo. 1999. *Molecular Thermodynamics of Fluid-Phase Equilibria. 3rd Edition,* Prentice Hall International, Upper Sadle River, NJ

Robeson L.M. 1991. Correlation of separation factor versus permeability for polymeric membranes. *J. Membr. Sci.* 62:165-185.

Romanos, G.E., Zubeir, L.F., Likodimos, V., Falaras, P., Kroon, M.C., Iliev, B., Adamova, G. and T.J.S. Schubert. 2013. Enhanced CO_2 capture in binary mixtures of 1-alkyl-3-methylimidazolium tricyanomethanide ionic liquids with water. *J. Phys. Chem. B* 117:12234–12251.

Salanne, M. 2015. Simulations of room temperature ionic liquids: From polarizable to coarse-grained force fields. *Phys. Chem. Chem. Phys.* 17:14270–14279.

Sarupria, S. and P.G. Debenedetti. 2011. Homogeneous nucleation of methane hydrate in microsecond Molecular Dynamics simulations. *J. Phys. Chem. A.* 115:6102–6111.

Scovazzo P. 2009. Determination of the upper limits, benchmarks, and critical properties for gas separations using stabilized room temperature ionic liquid membranes (SILMs) for the purpose of guiding future research. *J. Membr. Sc.* 343:199-211.

Seehamart, K., Nanok, T., Krishna, R., van Baten, J. M., Remsungnen, T. and S.A. Fritzsche. 2009. A molecular dynamics investigation of the influence of framework flexibility on self-diffusivity of ethane in Zn(tbip) frameworks. *Micropor. Mesopor. Mater.* 125:97–100.

Sha, M., Dou, Q. and G. Wu. 2012. Molecular dynamics simulation of ionic liquids adsorbed onto a solid surface and confined in nanospace. In *Chemical Modelling: Applications and Theory*, Vol. 9. Cambridge, UK: The Royal Society of Chemistry, pp. 186–217.

Shah, J., and E. Maginn. 2011. A general and efficient Monte Carlo method for sampling intramolecular degrees of freedom of branched and cyclic molecules. *J. Chem. Phys.* 135:134121.

Skoulidas, A.I. and D.S. Sholl. 2002. Transport diffusivities of CH_4, CF_4, He, Ne, Ar, Xe, and SF6 in silicalite from atomistic simulations. *J. Phys. Chem. B* 106:5058–5067.

Sloan, E.D. and C.A. Koh, 2008. *Clathrate Hydrates of Natural Gases*, 3rd ed. Boca Raton, FL: CRC Press.

Smirnov, G.S. and V.V. Stegailov. 2012. Melting and superheating of sI methane hydrate: Molecular dynamics study. *J. Chem. Phys.* 136:044523.

Theodorou, D.N., 2010. Progress and outlook in Monte Carlo simulations. *Ind. Eng. Chem. Res.* 49:3047–3058.

Theodorou, D.N., Snurr, R.Q. and A.T. Bell. 1996. *Comprehensive Supramolecular Chemistry*. Pergamon Press: New York, Vol. 7, pp. 507–548.

Toniolo, R., Pizzariello, A., Dossi, N., Lorenzon, S., Abollino, O. and G. Bontempelli. 2013. Room temperature ionic liquids as useful overlayers for estimating food quality from their odor analysis by quartz crystal microbalance measurements. *Anal. Chem.* 85:7241–7247.

Tsimpanogiannis, I.N., Economou, I.G. and A.K. Stubos. 2014. Methane solubility in aqueous solutions under two-phase (H-Lw) hydrate equilibrium conditions. *Fluid Phase Equilib.* 371:106.

Tung, Y.T., Chen, L.J., Chen, Y.P. and S.T. Lin. 2010. The growth of structure I methane hydrate from molecular dynamics simulation. *J. Phys. Chem. B* 114:10804.

Valderrama, J.O. 2003. The state of the cubic equations of state.*Ind. Eng. Chem. Res.* 42:1603–1618.

van der Waals, J.H. and J.C. Platteeuw. 1959. Clathrate solutions. *Adv. Chem. Phys.* 2:1–57.

Vega, C., and J.L.F. Abascal. 2011. Simulating water with rigid non-polarizable models: A general perspective. *Phys. Chem. Chem. Phys.* 13:19663–19688.

Vega, C., Abascal, J.L.F., Conde, M.M. and J.L. Aragones. 2009. What ice can teach us about water interactions: A critical comparison of the performance of different water models. *Faraday Discuss.* 141:251–276.

Vergadou, N., Androulaki E. and I.G. Economou. 2017. Molecular simulation methods for CO_2 capture and gas separation with emphasis on ionic liquids. *Process Systems and Materials for CO_2 Capture: Modelling, Design, Control and Integration*, John Wiley & Sons, Ltd, Chichester, UK. doi: 10.1002/9781119106418.ch3.

Vergadou, N., Androulaki, E. and I.G. Economou. 2017. in preparation.

Vergadou, N., Androulaki, E., Hill, J-R. and I.G. Economou. 2016. Molecular simulation of imidazolium-based tricyanomethanide ionic liquids using an optimized classical force field. *Phys. Chem. Chem. Phys.* 18: 6850–6860. doi:10.1039/C5CP05892A

Wasserscheid, P. and T. Welton. 2008. *Ionic Liquids in Synthesis*, 2nd Edition. Weinheim, Germany: Wiley-VCH.

Watanabe, M., Thomas, M.L., Zhang, S., Ueno, K., Yasuda, T. and K. Dokko. 2017. Application of ionic liquids to energy storage and conversion materials and devices. *Chemical Reviews*. doi: 10.1021/acs.chemrev.6b00504.

Weingärtner, H. 2008. Understanding ionic liquids at the molecular level: Facts, problems and controversies. *Angew. Chem. Int. Ed.* 47:654–670.

Widom, B. 1963. Some topics in the theory of fluids. *J. Chem. Phys.* 39:2808–2812.

Wishart, J.F. 2009. Energy applications of ionic liquids. *Energy Environ. Sci.* 2:956–961.

Zheng, B., Sant, M, Demontis, P. and G.B. Suffritti. 2012. Force field for molecular dynamics computations in flexible ZIF-8 framework. *J. Phys. Chem. C* 116:933–938.

Zubeir, L.F., Rocha, M.A.A., Vergadou, N., Weggemans, W.M.A., Peristeras, L.D., Schulz, P.S., Economou, I.G. and M.C. Kroon. 2016. Thermophysical properties of imidazolium tricyanomethanide ionic liquids: experiments and molecular simulation. *Phys. Chem. Chem. Phys.* 18:23121-23138.

14

Green Engineering in Process Systems: Case Study of Chloromethanes Manufacturing

Chintan Savla and
Ravindra Gudi
Indian Institute of
Technology Bombay

14.1 Introduction

Resource depletion, air, water, and land pollution are examples of environmental problems, which have resulted due to industrial processes. One of the main problems associated with these activities is that they have global impact. For instance, emission of greenhouse gases can occur locally, but resulting greenhouse effect will have a global character. Therefore pressures on improving process performance are rising, and hence, chemical and process industries are constantly under scrutiny of various environmental organizations. This demands more environmentally acceptable processes, products, and practices that can be achieved through ideas of "waste minimization" and "zero emission." Initially, most industries worldwide followed end-of-pipe treatment (i.e., treatment and disposal of waste in nonhazardous form) as an approach toward dealing with process waste. But as environmental regulations have become more strict, increase in cost associated with waste treatment have led to a shift toward reduction of waste at the source or its reuse as more cost-effective waste management methods. Extrapolation of current needs leads to a picture of an unsustainable world since there is a steady decline in the natural resources. So design/retrofit of chemical processes has to be done through the concept of sustainability, which is promoted by green engineering (GE).

14.1.1 Concept of Green Engineering

Green engineering broadly encompasses (1) design, commercialization, and use of products and processes that are feasible and (2) economical while minimizing risk to human health, environment and elimination, and reduction or reuse of wastes at source by effective utilization of material and energy. This takes us to the question whether green engineering is a new discipline created from scratch? The answer to this question is no. Just as we know the "principles" of chemical engineering are basic unit operations, heat and mass tansfer, fluid flow, reactors and kinetics, even GE has its own basic principles. These principles involve applying basic chemical engineering principles in an innovative way to promote sustainability (García-Serna et al., 2007). In the present world, the goal of sustainability would not be attractive if it is not advantageous. Thus, sustainability has to imply "all in one," i.e., economic, social, and environmental profit policy. It has been a myth that investments in green chemistry and engineering increase industrial cost burden. But in many cases it has been proved that costs can be substantially lower due to diminished use of raw material, energy or water, the avoidance of effluent treatment or safety equipment and procedures. Also increase in industrial pollution regulations will inevitably increase waste treatment and disposal costs and in many situations forbid present practices altogether.

14.1.2 Principles of Green Chemistry and Engineering

Principles of green chemistry and engineering have been laid to guide designing of chemical processes and products. Those of green chemistry focus toward the reaction path through atom economy (molecular weight of product/total molecular weight of all reactants), E-factor (kg of waste/kg of desired product), whereas green engineering emphasizes on chemical process as a whole.

The objectives of this work were (1) to showcase the way in which new synthesis routes can be designed to ensure that the developed chemical process is greener as well as economical; (2) develop a simpler methodology to redesign existing processes using process simulation and optimization for waste minimization, recycle, and reuse; and (3) apply the methodology developed to an existing process for the production of chloromethanes by using an economic objective function with cost of waste treatment associated with it. In particular, one of the indices of green chemistry, viz atom economy has been explicitly considered and maximized in the steps of redesigning of the manufacturing route.

14.1.3 Review of Literature Survey

The hierarchy defined by the United States Environmental Protection Agency (U.S. EPA) in handling hazardous/nonhazardous waste is source reduction, recycle, reuse, reclamation, treatment, and disposal. Thus, researchers worldwide started developing process synthesis routes for waste minimization. Process flow sheet synthesis concepts are based either on heuristic generation, evolutionary modification, or on superstructure optimization. While the former two concepts are purely driven by designer experience and creativity throughout the synthesis, the latter is based on mathematical programming. Different approaches for process synthesis for waste minimization applied are as follows:

1. *Hierarchical decision approach*—This procedure developed by Douglas for flow sheet synthesis can be used with minor modifications such as understanding pollution problems that would arise from decisions made at various levels of detail. Here, the best alternative is selected at each level of the process (Douglas, 1992). This approach cannot guarantee to obtain the best solution as it does not take into account the interactions between the different levels of decomposition.

2. *Superstructure-based approach*—Here different alternatives for each level are shown in the form of a superstructure. This approach requires usage of Mixed integer nonlinear programming (MINLP)

algorithms for solving the superstructure. This method is applicable for problems of limited size as solving it becomes difficult. Such MINLP-based work has been done with equation-based simulators (Diwekar and Rubin, 1993; Kravanja and Grossmann, 1996), using ASPEN PLUS by Dantus and High (1996) and Diwekar et al. (1992). Also MINLP problems were solved external to process simulator by usage of different solvers through an activeX client-server application where the whole process was controlled by MATLAB® (Caballero et al., 2007). Creating model, i.e., input–output relationship of flow sheet operations and then solving on GAMS has also been attempted (Kocis and Grossmann, 1987, 1989).

3. *Multilevel-hierarchical approach to MINLP process synthesis*—Involves solving MINLP problems iteratively at various levels using integer cuts (Kravanja and Grossmann, 1997). Thus, bigger flow sheets can be considered but would lead to combinatorial complexity and hence lack of global optimality.

4. *Life cycle assessment (LCA)* as a tool for process selection has been applied which involves doing cradle to cradle analysis, i.e., the environmental impact from raw material exploration, treatment to final product disposal is considered. It is important in cases where uncertainty remains on use of hazardous materials in the future (Kniel et al., 1996).

Chlorination of methane (saturated hydrocarbons) can be carried out by either photochemical or thermal means. The reaction of methane and their chlorination substitution products with chlorine is a chain reaction. Thermal method is preferred over photochemical due to some of the advantages such as complete chlorine conversion, high production capacity and no inhibitor affect. Alternative chlorination methods have also been tried such as reacting methanol and HCl to give methyl chloride (CH_3Cl) which can be further chlorinated. Combination of both the chemistries have also been attempted with a view to have the net production of HCl to be zero as the latter needs to be recycled to the methanol reaction step. Also oxychlorination process has been developed, but has many difficulties such as combustion of methane at the reaction temperature, formation of hot spots in the catalyst bed due to improper temperature control, reduced yield and purity of the product, and increased purity of the raw materials (Stauffer, 1992).

The conventional method of producing chlorinated methanes involves reacting methane with chlorine gas. For each substitution of hydrogen of methane with chlorine atom, one molecule of hydrogen chloride (HCl) is produced. Thus, twice the amount of chlorine is consumed when compared with the quantity that gets incorporated into chlorinated hydrocarbons. So the process gives a chlorine efficiency of 50% (Stauffer, 1992). Also, the process has other disadvantages such as the need to treat HCl waste, deal with hazards of storage and transportation of chlorine. In this work, we have integrated the thermal chlorination process with chlorine production process obtained by catalytic air oxidation of recovered HCl of the thermal process. This process integration has led to several advantages—theoretical chlorine efficiency of the overall process to 100%, increased yield of the product, no end-of-pipe treatment for HCl waste, reduced hazards with respect to chlorine storage and transportation. Moreover, it has also been proved that the integrated process is beneficial from economic as well as environmental point of view.

14.2 Production of Chloromethanes

Chloromethanes include all the four compounds, i.e., CH_3Cl, methylene chloride (CH_2Cl_2), chloroform ($CHCl_3$), and carbon tetrachloride (CCl_4). Of these, CH_3Cl has largest market since it is used in the manufacture of silicones, in methylation reactions, preparation of quaternary amines, and methyl cellulose. Other chloromethanes are used as solvents for extraction or in reaction, but their use is declining due to strict environmental regulations and due to availability of other greener solvents. But still commercial chloromethanes plants are produced to cater to market need, and hence, the need arises to redesign the existing process. The manufacturing process involves many waste streams being produced so an attempt was made to minimize or eliminate these waste streams (DeForest, 1979; Ullmann, 1993). This work showcases novel synthesis process route for chloromethanes production with improved economic and environmental profit.

14.2.1 Methodology for Greener Design

From literature study, we infer that for smaller synthesis processes it was better to compare different alternatives by solving each flow sheet separately followed by further optimizing and selecting best flow sheet. This approach was applied to the selected case study. In this work for designing new synthesis routes our approach involved designing activities along:

1. Development of a base case model.
2. Generation of process alternatives.
3. Evaluation and optimization of different process alternatives.
4. Selection of best alternative.

The developed methodology thus helps in considering interactions between different blocks within a flow sheet and thus the obtained result would be on the basis of overall outcome of the flow sheet. Such interactions are absent in a hierarchical design approach where decisions are made at each level irrespective of its future affect. Any modification of an existing process requires an incentive. Considering green engineering principles, the incentive in addition to economical is having a minimum environmental impact. Thus, a base case model is used to evaluate the current performance of process which serves as guide to analyze different alternatives. The process model developed using an ASPEN PLUS flow sheet simulator would serve as an experimental tool for the various process alternatives in terms of the process performance.

14.2.2 Base Case Process and Reaction Chemistry

The base process involves thermal chlorination of methane. The feed to the reactor is a mixture of chlorine and methane (with minimum amount of hydrocarbon impurities, 100 ppm excluding nitrogen) (DeForest, 1979) which react to produce CH_3Cl and then the subsequent chlorination gives di-, tri-, and tetrachloromethanes. Here, a substantial amount of HCl is produced as by-product. The reaction is irreversible and carried out at about 450°C and at low pressure. Being an exothermic reaction with heat of reaction of 100 kJ/mol of chlorine, reactor explosion could result if the heat of reaction is not removed. Thus, accordingly feed ratio (Cl_2/CH_4) was taken so as to have required product distribution with CH_3Cl production being maximum and also excess methane helps in diluting the reaction mixture, which otherwise would have to be diluted with inert gases, which can further lead to separation problems. The reaction chemistry is as follows with the kinetic data given in Table 14.1:

$$CH_4 + Cl_2 \rightarrow CH_3Cl + HCl$$

$$CH_3Cl + Cl_2 \rightarrow CH_2Cl_2 + HCl$$

$$CH_2Cl_2 + Cl_2 \rightarrow CHCl_3 + HCl$$

$$CHCl_3 + Cl_2 \rightarrow CCl_4 + HCl$$

TABLE 14.1 Kinetic Data for Thermal Chlorination of Methane

Reaction No.	Rate Equation	Preexponential Factor $(s \cdot kmol/m^3)^{-1}$	Activation Energy (kJ/kmol)
1	$k_1[CH_4][Cl_2]$	2.56×10^8	82,000
2	$k_2[CH_3Cl][Cl_2]$	6.28×10^7	71,100
3	$k_3[CH_2Cl_2][Cl_2]$	2.56×10^8	82,000
4	$k_4[CHCl_3][Cl_2]$	2.93×10^8	87,200

Source: Dantus, M., High, K., *Ind. Eng. Chem. Res.*, 35, 4566–4578, 1996.

The reaction is carried out in an isothermal continuously stirred tank reactor (CSTR) which is followed by cooling of reactor effluent, absorption of HCl in water followed by removal of water using sulfuric acid as drying agent and then compressed, condensed flashed to recycle excess methane and liquid product taken to train of distillation columns where each product is separated.

14.2.3 Development of Flow Sheet on ASPEN PLUS

ASPEN PLUS, a sequential modular simulator, was used to develop flow sheets. Each operating unit of the flow sheet has been represented by the corresponding blocks available in the flow sheet simulator as shown in Table 14.2. An important part of any simulation is the selection of thermodynamic model. If the selected thermodynamic model is improper, the simulated results would not be accurate. As complete plant data are not available to compare the simulation results, an appropriate selection was made through using thermodynamic property tree which gives models on the basis of polarity, electrolytic property, operating pressure, availability of interaction parameters, and liquid–liquid split of involved substances (Aspen Plus User Guide, 2004a). Thus, different property methods were used for different blocks and results were compared with available experimental data wherever possible. The selected physical property methods were Non-random two liquid (NRTL) (for nonelectrolytes), Electrolytes – Non-Random-Two-Liquid (for electrolytes), and Soave Redlich–POLAR (SR-POLAR) (for high pressure applications).

Table 14.3 shows the two outlet streams from the absorber—liquid and vapor stream for two simulations using ELEC-NRTL with and without data package H_2OHCl, which takes into account dissociation

TABLE 14.2　Summary of Base Case Blocks

Operating Unit	Block Used
Reactor	RCSTR
Heating, cooling system	HEATER
Absorber	RADFRAC
Dryer	SEPERATOR
Distillation columns	RADFRAC

TABLE 14.3　HCl Absorption at $T = 25\,C$, $P = 1$ atm in Water (226.8 kmol/h)

	Feed	Water	Using ELEC-NRTL		ELEC-NRTL (H_2OHCl Package)	
			Liquid	Vapor	Liquid	Vapor
Temperature (°C)	25	25	26	31.8	60.3	66.2
Mole Flow (kmol/h)						
H_2O	0	226.796	219.524	7.273	183.013	43.783
HCl	38.842	0	1.641	37.201	38.842	0
Methane	101.468	0	2.374	99.094	0.268	101.2
Chlorine	0.181	0	0.035	0.146	0.008	0.173
CH_3Cl	21.048	0	5.337	15.71	0.412	20.636
CH_2Cl_2	6.856	0	6.851	0.005	0.827	6.029
$CHCl_3$	1.671	0	1.671	0	0.556	1.114
CCl_4	0.222	0	0.222	0	0.202	0.02
Mass Fraction						
H_2O		—	0.769	0.034	0.672	0
HCl		—	0.012	0.349	0.289	0.192

of HCl in water using the electrolyte NRTL and Henry's law models. The interaction parameters those of water with di-, tri-, and tetrachloromethanes were missing, so these were obtained from literature and used (Dantus and High, 1996).

All the waste streams obtained in the base case have chlorinated hydrocarbons, which are toxic and can lead to destruction of ozone layer when exposed to atmosphere, and therefore, these have to be treated. Also absorption of HCl in water gives HCl acid, but contaminated with chloromethanes. Thus, new alternatives need to focus toward preventing this waste.

14.3 Generation of Process Alternatives

Two types of process alternatives involve making topological changes in the flow sheet or manipulating the flow sheet variables that have considerable impact on the process. For preventing HCl acid waste, topological changes were required and thus were carried out. Different alternatives that were explored were as follows:

1. For the same chemistry, involved changes would be made in the type of reactor, absorption, and separation step:
   a. Instead of CSTR, isothermal PFR was introduced: It showed different product distribution.
   b. Going for separation step before absorption of HCl in water: Here, the reactor outlet stream of the base process was flashed, the vapor stream returned to the reactor, and liquid stream continued through the separation sequence. The advantage obtained was the reduction of the amount of product exposed to water and thus higher yield were obtained, since this approach prevented loss occurring of products due to solubility in water. However, the problem of HCl acid waste still persisted (Dantus and High, 1996).
   c. Absorption of HCl in part of chloromethane mixture produced in the process: Here, water was substituted with chloromethane mixture as solvent for absorption (Forlano, 1974). This eliminated use of water or other external solvent which could further act as a waste source. Moreover, it became now possible to recover HCl gas, which otherwise when absorbed in water was difficult as it required extractive distillation for recovery due to maximum boiling azeotrope between HCl and H_2O at $T=108.6°C$, $P=1$ bar and HCl concentration of 20.2% w/w.

2. Different reaction chemistry (i.e., hydrochlorination of methanol) to give CH_3Cl which is then to be further chlorinated to give other chloromethanes.
3. Combination of both reaction schemes as recovering HCl gas from methane chlorination can be used for hydrochlorination process.
4. Carrying out oxychlorination of methane using HCl gas (Table 14.4 and Figures 14.1 and 14.2).

Of the above-mentioned alternatives, only some of the promising flow sheets that were evaluated in detail have been shown whereas for others it was qualitatively clear that the required objective of reduction in HCl acid waste would not be met. The second alternative would still lead to generation of HCl acid waste unless recovery is done of anhydrous HCl. Also the drawbacks with the other reaction chemistry is that it yields only CH_3Cl, requires high investment costs, and costlier raw material methanol compared to methane (Table 14.5).

TABLE 14.4 Stream Results of Base Case Process Using RCSTR

	CH_4	CH_4 Recycle	Cl_2	MCDD	DCDD	CHDD	CTDD	W-1	W-2	W-3	W-4
Temperature (°C)	25	−50	25	−2.7	74.5	61	79.5	60.3	170.3	−2.7	97.4
Pressure (bar)	11.86	10.13	6.53	8.11	3.04	1.01	1.01	1.01	1.01	8.11	1.01
Vapor fraction	1	1	1	0	0	0	0	0	0	1	0
Mole flow (kmol/h)	37.195	94.094	39	18.206	6.078	0.973	0.018	224.129	111.796	9.803	0.025

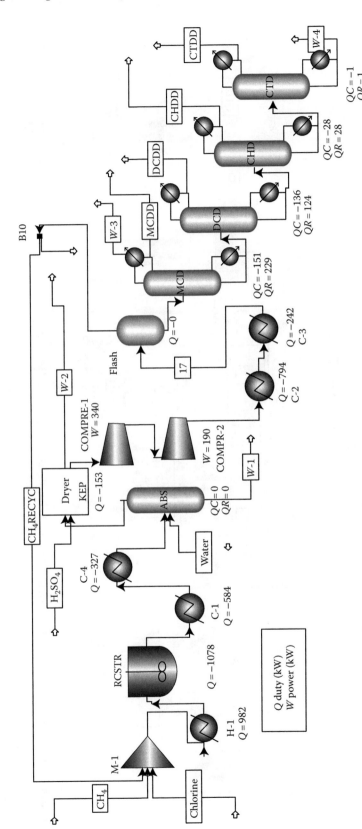

FIGURE 14.1 Preliminary flow sheet of base case process using RCSTR.

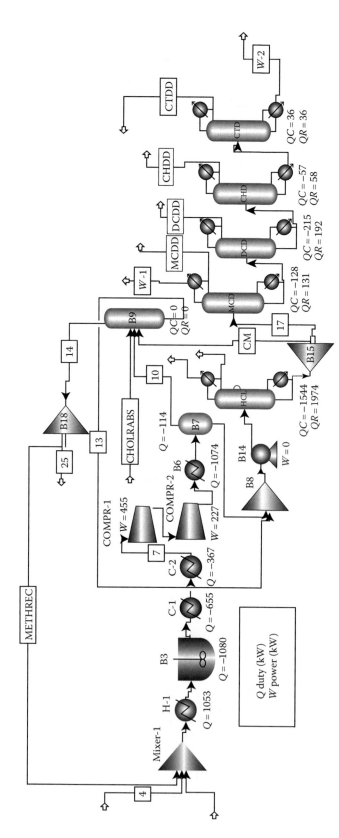

FIGURE 14.2 Preliminary flow sheet of absorption of HCl in chloromethane mixture.

TABLE 14.5 Stream Results of Absorption of HCl in Chloromethane Mixture

	CH$_4$	METHREC	Cl$_2$	HCLD	MCDD	DCDD	CHDD	CTDD	W-1	W-2
Temperature (°C)	25	−4.4	25	−60.3	45.8	36	54.5	67.3	45.8	76.7
Pressure (bar)	11.86	15.2	6.53	14.19	10.13	1.01	1.01	1.01	10.13	1.01
Vapor fraction	1	0.983	1	0	0	0	0	0	1	0
Mole flow (kmol/h)	37.195	112.078	39	39.272	17.478	5.398	1.275	0.985	0.212	0.498

The fourth alternative involves in situ generation of chlorine through catalytic oxidation of HCl. But the drawbacks of this reaction are high temperature and side reactions, which lead to combustion and pyrolysis products. Since separation of desired products is difficult in view of lower yields, and also because the product purity is less, this alternative was not evaluated further (Ullman, 1993).

14.4 Cost Estimation and Optimization

14.4.1 Capital Cost Estimation of Developed Flow Sheets

For any process cost of manufacturing (COM) per year involves both capital cost (annualized) and operating cost. So in order to get an estimate of capital cost, ASPEN IPE (Icarus Process Evaluator) now known as Process Economic Analyzer was used. Aspen IPE is the industry standard for project and process evaluation. Unlike other approaches, the ASPEN-based approach does not rely on capacity-factored curves for equipment pricing, nor does it rely on factors to estimate installation quantities and installed cost from bare equipment. It follows a unique approach where equipment, with associated plant bulks, is represented by comprehensive design-based installation models. So the flow sheet developed on ASPEN PLUS along with necessary stream properties was evaluated by ASPEN IPE, where the simulation blocks are mapped into equipments of desired type. In addition to this project type, design specifications, material of construction (MOC) (i.e., choosing Ni and Ni alloys at high temperature for HCl gas and carbon steel cladded with Ni for chlorinated hydrocarbons) and sizing information needs to be checked. The project type was selected as Plant addition (suppressed infrastructure) to avoid outside battery limit (OSBL) costing. The basis of costing was chosen 1 Q 04 (of ASPEN IPE 14.00) (Aspen Icarus Process Evaluator 2004.1 User Guide). The COM was obtained using the following formula:

$$COM = 0.304 \times FCI + 1.23$$

$$\times (U + RM + WT) (\text{cost of operating labor neglected assuming same for all alternatives}).$$

The profit function was formulated as

Profit function = Product Revenue (PR) − COM (Turton et al., 1998).

The cost of raw materials and products of the chloromethanes manufacturing was obtained from the indicative chemical prices provided by ICIS (formerly called Chemical Marketing Reporter) services for students and academics, which was accessed through website (www.icis.com). The cost of utilities (with charges delivered to the battery limit of a process) and waste treatment were taken for calculations (obtained from Turton et al., 1998) and escalated wherever required.

The results given in Table 14.6 are based on just preliminary analysis of different flow sheets, and therefore, the negative values of profit do not indicate that process is not economically feasible. So on the basis of above results the best alternative was one involving HCl gas absorption in chloromethanes with recovery of HCl as the product. So further optimization of this flow sheet was carried out.

TABLE 14.6 Comparison of Base Case and Different Alternatives

Component Flow (ton/year)	Base Case (cstr)	Base Case (pfr)	Absorption of HCl in Chloromethanes
CH_3Cl	7150	9475	7710
Dichloromethane	4325	4010	3880
$CHCl_3$	965	470	1300
CCl_4	20	—	1200
Total (ton/year)	12460	13955	13780
HCl (ton/year)	—	—	10900
Waste (ton /year)	45090	42480	2520
Utility cost, U (k$/year)	2375	2747	5190
Equipment cost (k$)	2280	2540	3475
Total capital cost, FCI (k$)	5300	4840	6150
Raw material cost, RM (k$)	7896	8025	7895
Waste treatment cost, WT (k$)	1600	1590	310
Product revenue (k$), PR	11124	12186	18490
COM (k$)	16212	16675	18345
Profit, P(k$)	−5088	−4491	145

14.4.2 Heat Integration

Following the principles of green engineering, it is necessary to reduce the energy requirement of the process to the extent possible. This can only be achieved by carrying out heat integration between various process streams subjected to ΔT_{min}. One of the possible cases is shown in Figure 14.3.

Thus, the saving of about $ 2,56,625/year is seen by heat integration of streams around the reactor by using feed effluent heat exchanger as shown in Table 14.7. Further improvements in heat integration can be brought about by utilizing exothermic heat of reaction, carrying out distillation operations at such pressures that condenser of high pressure distillation column can provide heat to the reboiler of other low pressure distillation column.

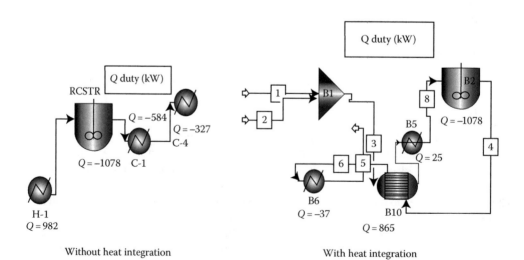

Without heat integration

With heat integration

FIGURE 14.3 Part of flow sheet with and without heat integration.

TABLE 14.7 Results of Heat Integration with Feed Effluent Heat Exchanger

	Without Heat Integration	With Heat Integration
Fixed capital investment, FCI ($)	640,000	1,756,564
Cost of utilities, U ($/year)	641,195	45,155
Equivalent annualized operating cost (EAOC = 0.304 × FCI + U) ($/year)	835,415	579,150

14.4.3 Parametric Optimization

Thus, having fixed the topology of flow sheet it became necessary to find key decision variables to reduce the computational effort and time and make the problem tractable for performing optimization. The flow sheet of absorption of HCl in chloromethane mixture involves two recycle streams—one with methane recycle and other stream consisting of chloromethane mixture that serves as solvent for HCl absorption from vapor stream, with constant molar flow rate obtained by splitting bottom product of HCl distillation column.

Before solving the optimization block on ASPEN PLUS, a sensitivity analysis was done with the selected decision variables of the flow sheet. The sensitivity analysis convergence results were found to be changing with changes in the step size of decision variable value. Thus, an alternative path found was to have separate input of chloromethanes mixture as solvent for HCl absorption column. Molar flow rates for solvent stream and the chloromethanes recycle stream were kept the same. The molar flow rate of the solvent stream was selected after having done the sensitivity analysis to ensure that more than 90% of HCl is absorbed from vapor stream against minimum amount of methane, chlorine being absorbed (<2% by mass fraction).

Thus, the flow sheet with two recycle streams was converted to one with a single recycle stream. This change was validated by monitoring the effect of change in decision variable values on the molar flow rate of bottom product of the HCl recovery distillation column which divides into two streams with one part going for product recovery whereas other part equal to the solvent molar flow rate goes for HCl absorption.

The decision variables and their upper and lower bounds were selected after having done the sensitivity analysis. The selected variables are as follows:

1. Reaction temperature with bounds as $350 < T < 450$ (°C): Reaction temperature was selected because it would affect rate of reaction and hence product distribution, conversion of methane so recycle rate and hence utility requirements. For the reaction to be initiated the reaction temperature has to be >300°C; further since pyrolysis reactions could result in explosion at higher temperatures, an upper limit of 500°C was chosen.
2. Compressor pressure to carry out isothermal flash with bounds as $10 < P_f < 15$ (in atm): Pressure was selected because it would lead to changes in split fraction of HCl in vapor and liquid stream, which would then affect the mole flow of chloromethanes mixture used for absorption and hence total product flow rate. At lower pressures HCl fraction in the liquid stream increases.
3. Absorber pressure for HCl absorption with bounds as $15 < P_{abs} < 20$ (in atm): Absorber pressure affects the percentage of HCl absorbed which increases with increase in P_{abs}. But higher pressures could also result in increased utility requirements.

ASPEN PLUS optimization block was used for solving with objective function as profit, i.e., $O = PR - (RM + U + WT)$ as capital cost remained same within the bounds of decision variables.

The objective function was also evaluated outside the bounds of decision variable, i.e., at lower pressure limit for absorber and flash to be around 5 atm whereas upper bounds to be 20 atm for flash and 25 atm for absorber. It was found that for values in the region of new increased bound, convergence problems were seen, whereas at the new lower bounds of all the decision variables objective function

TABLE 14.8   Decision Variables and Constraints

Decision Variables	Variable Range
Reaction temperature (°C)	350–450
Compressor outlet pressure for flash (atm)	10–15
Absorber pressure (atm)	15–20
Constraints	**Bounds**
Chlorine conversion in reactor	≥99% (Tolerance ± 0.001)
Mass fraction of HCl in input stream to reactor	≤0.008 (Tolerance ± 0.001)
Purity of all the products (mass fraction)	≥95% (Tolerance ± 0.01)

TABLE 14.9   Optimization Results

List of	Final Value
Decision Variables	
Reaction temperature (°C)	450
Compressor outlet pressure for flash (atm)	15
Absorber pressure (atm)	20
Constraints	
Chlorine conversion in reactor	0.9919
Mass fraction of HCl in input stream to reactor	0.0088
Purity of all products obtained (mass fraction)	≥95%
Raw-Material and Product Mass Flow rates	
Methane fresh feed (kg/h)	623
Chlorine fresh feed (kg/h)	2769
CH_3Cl (ton/year)	7598
Dichloromethane (ton/year)	4557
$CHCl_3$ (ton/year)	1697
CCl_4 (ton/year)	1286
Recovered HCl (ton/year)	11204
Cost and Revenue	
Raw material cost, R (k$)	7947
Utility cost, U (k$)	5726
Waste treatment cost, W (k$)	289
Product Revenue, PR (k$)	19357
Objective function, O (k$)	5395

was maximum. But the obtained solution led to change in product distribution, i.e., amount of dichloromethane produced is more than CH_3Cl because of increased recycle of product CH_3Cl to the reactor which leads to further chlorination. Also purity of product streams decreased more than required. Hence, further the objective function, O, was maximized subjected to constraints of the following:

1. Conversion of chlorine in reactor to be greater than 99%.
2. The mass fraction of HCl in the input stream to the reactor to be less than 0.009.
3. The mass fraction of products in all the product streams to be greater than 95%.

These constraints were selected so as to carry out the process safely (no recycle of chlorine), with the required product distribution (i.e., CH_3Cl as major product) and higher purity of product streams. Table 14.8 describes the bounds for decision variables subjected to above-mentioned constraints.

The optimization block was run with initial guess set at the lower bounds for all the decision variables. The bounds of decision variables were narrowed down after having run for some initial guesses as the constraints were violated. It was found that optimization convergence criteria were satisfied but the constraints were violated so the convergence block was reinitialized with the new initial guess being the last converged value of the previous run. This was continued till the constraints were satisfied and optimization convergence achieved. Thus, the problem of requiring very good initial guesses for the optimization routine to convergence was tackled. The optimization results obtained (shown in Table 14.9) produced a local optimum for the problem within the specified bounds and the constraints.

14.5 Process Integration with Chlorine Production

14.5.1 Catalytic Oxidation of Recovered HCl

The flow sheet with HCl recovery was the best alternative so far. The optimization was carried out by considering price of anhydrous HCl gas in the product revenue. But the demand for anhydrous HCl gas is far less, so some alternative route to use recovered HCl was required.

There are a large number of processes capable of producing chlorine from HCl. These include (1) electrolysis of HCl with production of Cl_2 and H_2, (2) direct oxidation of HCl with various inorganic agents, (3) oxidation of HCl by air or O_2 in the presence of a catalyst, e.g., Deacon-type processes. Deacon-type processes have been well known because of their simplicity, low thermal and electrical power requirements when compared with other approaches. The only limitation was incomplete conversion of HCl which in presence of water lead to corrosion problems. So researches then tried developing improved catalysts and equipment and some of those processes have been commercialized. These include the Shell chlorine process, Kel-chlor process, catalytic carrier process, etc. Having studied the advantages and limitations of all the different known processes, the selected process was a two-stage oxidation process using Cu-based solid reactant materials that also acts as a carrier for chlorine and energy.

The process is called as modified Deacon process, where complete conversion of HCl can be obtained. It involves two stages: chlorination (180°C–260°C, exothermic, spontaneous with complete conversion of HCl) and oxidation (300°C–360°C, endothermic, reversible, with 76% conversion of $CuCl_2$) with solid CuO–$CuCl_2$ as catalytic carrier. The solid acts as storage of heat and chlorine. Overall the process is exothermic. It can be carried out in fixed and fluidized bed reactors (Pan et al., 1994; Reddy et al., 2008). The reactions involved are as follows:

$$\text{Chlorination:} \quad 2HCl + CuO \rightarrow CuCl_2 + H_2O$$

$$\text{Dechlorination:} \quad CuCl_2 + \frac{1}{2}O_2 \rightarrow CuO + Cl_2$$

$$\text{Overall:} \quad 2HCl + \frac{1}{2}O_2 \rightarrow Cl_2 + H_2O$$

Having calculated its economic potential at zeroth stage, it is not profitable to carry out this reaction profitably as cost of HCl/kg is far higher than the product chlorine.

14.5.2 Advantages of Process Integration

But this reaction if carried out using recovered HCl, with extensive heat integration and recovered chlorine used on the site itself would lead to production cost of chlorine to be one-third of that obtained from well-known chlor-alkali process (Pan et al. 1994). The chlor-alkali process by itself has issues

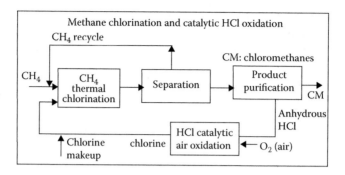

FIGURE 14.4   Cyclic process integration.

TABLE 14.10   Comparison of Chloromethanes Synthesis Process with and without Process Integration

	Without Process Integration	With Process Integration
Chlorine feed cost (k$)	6913	3072
Recovered HCl revenue (k$)	5582	—
Product revenue, PR (k$)	19357	13775
Objective function, O (k$)	5395	2885

related to (1) involves high energy consumption, (2) generates NaOH as by-product which has to be dealt with as demand for NaOH is less compared to that of chlorine, and (3) involves dealing with mercury (in Mercury cell process) a toxic element when discharged. The proposed Deacon process leads to increased economic savings in addition to recovery of chlorine and minimization of waste by-products.

Also when the process is integrated as shown in Figure 14.4 with the chloromethanes synthesis process it reduces raw material cost, dependence on external source (though some makeup of chlorine to be provided due to loss of HCl in upstream process, reversible nature of oxidation reaction of HCl), and also hazards involved with storage and transportation of chlorine which faces increased environmental regulatory restrictions. Table 14.10 compares the objective function values with and without process integration. Though on process integration the objective function value has decreased but it is still profitable and feasible solution. Overall the process integration had been able to make the process greener along with making it profitable when compared to traditional synthesis routes. It can also be seen that this process integration facilitated the atom economy with respect to chlorine to be close to 100%.

14.6 Conclusion

For any process to be made greener and apply it on industrial scale, it is important that there has to be incentive with respect to cost in addition to environmental and safety benefits. Only in this way the promotion of green engineering can be done. Different flow sheet alternatives for chloromethanes manufacturing were considered and the best alternative obtained was the one involving absorption of HCl from vapor stream in chloromethanes mixture integrated with process of production of chlorine from recovered anhydrous HCl. The best alternative was further optimized using optimization block of ASPEN PLUS process simulator which would provide the feed (anhydrous HCl) for the production of chlorine. Thus, an attempt was made to make the synthesis route greener with a check on the economic objective function and positive results were obtained. The chloromethanes synthesis case study was considered due to scope involved in applying green engineering principles such as waste minimization, reuse, and to show the impact of the modified process on the economics.

Abbreviations

HCl	Hydrogen chloride
GE	Green engineering
U.S. EPA	United States Environmental Protection Agency
MINLP	Mixed integer nonlinear programming
CSTR	Continuously stirred tank reactor
PFR	Plug flow reactor
COM	Cost of manufacturing
PR	Product revenue
U	Utility cost
RM	Raw material cost
WT	Waste treatment

References

Aspen Technology Inc., 2004a. Aspen Plus User Guide.

Aspen Technology Inc., 2004b. Aspen Icarus Process Evaluator 2004.1 User Guide.

Caballero, J. A., Odjo, A., and Grossmann, I. E., 2007. Flowsheet optimization with complex cost and size functions using process simulators. *AIChE J.*, 53, 2351–2366 http://onlinelibrary.wiley.com/doi/10.1002/aic.11262/abstract.

Dantus, M. and High, K., 1996. Economic evaluation for the retrofit of chemical processes through waste minimization and process integration. *Ind. Eng. Chem. Res.*, 35, 4566–4578.

DeForest, E. 1979. *Chloromethanes. Encyclopedia of Chemical Processing Design*, edited by J. McKetta and W. Cunningham. Marcel Dekker, New York, Vol. 8, p. 214.

Diwekar, U. M., Grossmann, I. E. and Rubin, E. S., 1992. MINLP process synthesizer for a sequential modular simulator. *Ind. Eng. Chem. Res.*, 31, 313–322.

Diwekar, U. M. and Rubin, E. S., 1993. Efficient handling of implicit constraints problem for the ASPEN MINLP synthesizer. *Ind. Eng. Chem. Res.*, 32, 2006–2011.

Douglas, J. M., 1992. Process synthesis for waste minimization. *Ind. Eng. Chem. Res.*, 31, 238–243.

Forlano, L. Production of methylene chloride. U.S. Patent No. 3,848,007, May 12, 1974.

García-Serna, J., Pérez-Barrigón, L., and Cocero, M. J., 2007. New trends for design towards sustainability in chemical engineering: Green engineering. *Chem. Eng. J.*, 133(1), 7–30.

Kniel, G. E., Delmarco, K., and Petrie, J. G., 1996. Life cycle assessment applied to process design: Environmental and economic analysis and optimization of a nitric acid plant. *Environ. Prog.*, 15(4), 221–228.

Kocis, G. R. and Grossmann, I. E., 1987. Relaxation strategy for the structural optimization of process flow sheets. *Ind. Eng. Chem. Res.*, 26, 1869.

Kocis, G. R. and Grossmann, I. E., 1989. A modeling and decomposition strategy for the MINLP optimization of process flowsheets. *Comput. Chem. Eng.*, 13, 797.

Kravanja, Z. and Grossmann, I. E., 1996. A computational approach for the modelling/decomposition strategy in the MINLP optimization of process flowsheets with implicit models. *Ind. Eng. Chem. Res.*, 35, 2065–2070.

Kravanja, Z. and Grossmann, I. E., 1997. Multilevel-hierarchical MINLP synthesis of Process flowsheets. *Comput. Chem. Eng.*, 21, suppl., S421–S426.

Pan, H. Y., Minet, R. G., Benson, S. W., and Tsotsis, T. T., 1994. Process for converting hydrogen chloride to chlorine. *Ind. Eng. Chem. Res.*, 33(12), 2996–3003.

Reddy, U. V., Cheedipudi, V. L., Bankupalli, S., and Kotra, V., 2008, Recovery of chlorine from anhydrous hydrogen chloride. *Int. J. Chem. React. Eng.*, 6, A97.

Stauffer, J. E., Process for the chlorination of methane. U.S. Patent No. 5,099,084, March 24, 1992.

Turton, R., Bailie, R. C. and Whiting, W. B., 1998, *Analysis, Synthesis, and Design of Chemical Processes*, Prentice Hall, Hoboken, NJ.

Ullman's Encyclopaedia, 1993. Chlorinated hydrocarbons, Vol. A6, 263–384.

Web Source: http://www.icis.com. Accessed October 20, 2010.

15

Fundamental Aspect of Photoelectrochemical Water Splitting

Gyan Prakash
Sharma, Arun
Prakash Upadhyay,
Dilip Kumar Behara,
Sri Sivakumar, and
Raj Ganesh S. Pala
*Indian Institute
of Technology Kanpur*

15.1 Introduction

Photoelectrochemical (PEC) cells are special classes of electrochemical cells, wherein at least one of the electrodes is photoresponsive. More broadly, PEC cell is defined as an electrochemical cell that can convert light energy into a more useful energy product through light-induced electrochemical processes. In PEC cells, water splits into hydrogen and oxygen with the aid of photocatalyst, which absorbs the photon energy and converts it to electrochemical energy (in the form of chemical bonds of hydrogen and oxygen) [1–3]. This is one of the most prominent ways of producing clean and cost-effective hydrogen by taking into account all photons that continuously strike the earth's surface.

The process of splitting water with the aid of electrical energy is called electrolysis of water and if it is aided by photon energy of sunlight it is called photocatalysis of water splitting. PEC water splitting utilizes both solar and electrical energy to split water into H_2 and O_2. PEC water splitting is highly preferred and is a promising field of study for many research communities, since electrolysis alone accounts for a lot of electrical energy to meet the thermodynamic potential for splitting water molecule. As mentioned

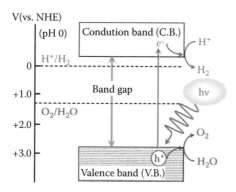

FIGURE 15.1 Band energetic for true water splitting reaction. (Reprinted from Maeda, K. et al., *J. Phys. Chem. C.*, 111, 7851–7861, 2007. With permission. Copyright © 2007, American Chemical Society.)

above, PEC water splitting requires an external energy (electrical power) input as per thermodynamic splitting of water splitting reaction 15.1 [1,3,4]:

$$H_2O \rightarrow \frac{1}{2}\,O_2 + H_2,\ \Delta G = 237\ kJ/mol. \tag{15.1}$$

The PEC cell consists of a working electrode (called as photoanode) and counter electrode (usually Pt) in an electrolyte solution of water. As shown in Figure 15.1, upon photoexcitation within the semiconductor photocatalyst, electrons and holes are generated [5]. These photo-induced electrons reduce water/H$^+$ to form H$_2$ at Pt electrode whereas photo-generated holes oxidize water/OH$^-$ to form O$_2$ at photoanodes under the influence of applied external bias. In principle, the redox potentials of photo-generated electrons and holes should be suitable for performing the redox reactions of water splitting reaction (i.e., oxygen evolution reaction, OER and hydrogen evolution reaction, HER). Since, the thermodynamic potential for water splitting reaction is 1.23 V, the anodic and cathodic reactions will proceed only if the band edges are staggered appropriately w.r.t water redox potentials. As shown in Figure 15.1, the bottom level of the conduction band (CB) has to be more negative than the redox potential of H$^+$/H$_2$ [0 V vs. Reversible Hydrogen Electrode (RHE)], and the top level of the valence band (VB) has to be more positive than the redox potential of O$_2$/H$_2$O (1.23 V vs. RHE) for effective water splitting reaction. Therefore, the primary condition for choosing any semiconductor as photocatalyst is that it should have a minimum band gap of 1.23 eV, with suitable edges of CB and VB for water splitting reaction [4,6]. There are three steps involved in PEC water splitting reaction: (1) absorption of photons followed by generation of charge carriers (mainly governed by band gap and mid-gaps states), (2) charge carriers separation, and (3) migration toward surface reaction sites to perform water redox reaction. The important factors that influence the second step are crystallinity, crystal structure, and particle size of the material under consideration. For example, a photo material with higher crystallinity possesses better charge separation and migration due to less structural defects (which act as recombination centers). Similarly material with less particle size has higher surface area (i.e., more active sites) and shortens the distance that the charge carriers should migrate before they become available to surface redox reactions 15.3. It includes surface electrochemistry, wherein the charge carriers on surface participate in redox reactions (OER and HER) to yield products. This step is more crucial than the first two steps, because even after successful operation/exploitation of former steps, if the material fails to perform surface redox reactions then there is no production of H$_2$ and O$_2$. Generally, addition of co-catalysts improves the surface electrochemistry part of the PEC processes [2, 6–8].

15.2 Basic Thermodynamics of Photoelectrochemical Cells

Many review articles [9–11] and published books/book chapters [12–14] have discussed the fundamental principles of PEC. Nevertheless, a brief summary of thermodynamic principles/basic operations

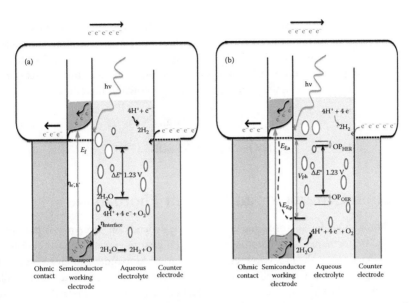

FIGURE 15.2 Schematic representing the band alignments and major factors influencing PEC performance of a photo material (e.g., photon incidence/absorption, charge carrier formation/separation, and electrochemistry) in explaining the interfacial processes of (a) *n*-type photoanode water splitting device and (b) more detailed schematic representing the energetic requirements associated with thermodynamic potential of water splitting along with overpotentials for both OER, HER reactions, and photovoltage. (From Chen, Z. et al., *Photoelectrochemical Water Splitting: Standards, Experimental Methods, and Protocols*, Springer, New York, 2013.)

of PEC device is appropriate here for completeness. The absorbed photons on photo-anode generates electrons (e^-) and holes (h^+), which are separated in the semiconductor and migrate to counter ends to participate in surface redox reactions. The holes drive the OER at the surface of the semiconductor whereas electrons are transferred through external wire to the counter electrode to drive the HER. A detailed demonstration of the energetic requirements for a PEC device is shown in Figure 15.2. In addition to thermodynamic requirements, overpotentials are associated with both anodic (OER) and cathodic (HER) sides, which severely influence the kinetics of the overall reaction system. Therefore, development of an efficient catalyst is essential to drive each half-reaction of water splitting reaction at lower activation energy. It is to be noted that entropic losses are also associated with the photo-generated holes and electrons which hamper the overall performance of the PEC water splitting reaction. Further, it is worthwhile to note that the actual driving force for true water splitting reaction is the photovoltage that includes few losses (e.g., spontaneous emission, incomplete light trapping, and nonradiative recombination) [15]. The sum of the abovementioned losses has to be always less than band gap of the semiconductor. The difference between the quasi-Fermi levels of electrons ($E_{F,n}$) and holes ($E_{F,p}$) under illumination is considered to be the photovoltage responsible for PEC water splitting. The photons from sunlight impinged on semiconductor surface (photoanode) create excitations, which are also called as electron–hole pairs. The photo-generated holes travel to the semiconductor surface of photoanode where they oxidize water/hydroxyl ions in the electrolyte and generate protons along with oxygen (O_2) as per the following:

$$2h^+ + H_2O \rightarrow \frac{1}{2}O_2 + 2H^+, \tag{15.2}$$

$$2h^+ + 2OH^- \rightarrow \frac{1}{2}O_2 + H_2O. \tag{15.3}$$

Further, the photo-generated electrons move to the cathode through the circuit (via externally connected wire) and react with the protons/water (at counter electrode/electrolyte interface) and gets reduced to H_2 according to Equations 15.4 and 15.5. Therefore, two photons are required to produce two electron-hole pairs, which in turn result in one mole of hydrogen and half mole of oxygen as per the equation representing overall PEC water splitting reaction, as shown in Equation 15.1:

$$2H^+ + 2e^- \rightarrow H_2(g), \tag{15.4}$$

$$H_2O + 2e^- \rightarrow H_2(g) + OH^-. \tag{15.5}$$

15.3 Selection Criteria for an Efficient Photoanode

The choice of photoanodes for solar hydrogen production depends on three major parameters which influence the PEC performance: (1) photon absorption, (2) charge carrier separation, and (3) surface electrochemistry [3,11]. Since, the first two events occur in the photoanode, it plays a crucial role in the above design strategies of solar water splitting reaction and the selection of semiconductor photoanode is essential in efficient PEC operation. It is known that metal oxides are suitable materials owing to their high stability in various electrolytic conditions, extensive research is focused on finding various ways of reducing the overpotential associated on metal oxide surfaces which is necessary to drive the water splitting reaction [6,16]. In addition, there is one other important key point to be considered while selecting the material, i.e., to have a balance between the following two quantities: (1) the carrier mobility (an important property related to the material) which is essential for efficient extraction of minority carriers from the semiconductor. (2) Increased bulk conductivity (such as dopants addition/introduction) which minimizes/reduces the voltage drop associated with the material. Since, the increased majority carrier concentration decreases the width of the space charge layer which facilitates charge separation in PEC operation, we need to understand the charge carrier dynamics. Therefore, it is essential to know the interplay of crystal structure, light absorption, and charge recombination pathways before selecting metal oxides for PEC water splitting reaction.

15.4 PEC Characterization Techniques

Several important characterization tools are required to analyze and interpret PEC data in a precise manner. As per PEC group working under Department of Energy (DOE), USA, following are the list of characterization techniques used for analysis of PEC data: (1) UV–Vis spectroscopy for band gap measurement of photo material, (2) illuminated open circuit potential (OCP), (3) Mott–Schottky plot for measurement of flat band potential, (4) dark, light and Chopped I–V curve, (5) photocurrent onset measurement, (6) incident photon conversion efficiency (IPCE) or equivalent efficiency, (7) photocurrent spectroscopy coupled with electrochemical impedance spectroscopy, (8) two-electrode short circuit current density and J–V [solar to hydrogen (STH) efficiency], (9) hydrogen detection (STH efficiency), and (10) photocurrent density vs. time stability. Here, we mainly focus on efficiency-related part of PEC measurement and/or characterization in forthcoming sections [17].

15.5 PEC Measurements/Efficiency Metrics

The most important parameter to characterize a PEC device is to estimate the extent of solar energy that has been used to generate hydrogen and the broad metric indicative of this is the STH conversion efficiency. It is to be noted that, with this single number; all photo materials/devices can be ranked against one another [6]. Unfortunately, existing or published reports on STH efficiency in both photocatalytic (PC)/PEC fields contain confusing information and are more prone to misinterpretation. For example, most of journal articles report PEC data which include either invalid mathematical expressions, or improper experimental methods to obtain efficiency values [17,18]. Further, most articles report efficiencies other than STH without clear distinction. The probable reason for existence of so much pluralism in describing efficiency is existence of different measures of efficiency and each has its own significance. The broader classification of efficiencies that is mainly used to measure the performance of PEC water splitting reaction is given below.

15.5.1 STH Conversion Efficiency

The overall efficiency of a PEC water splitting reaction or device when exposed to solar irradiance (1 Sun, Air Mass (AM) 1.5 Global conditions) under zero bias conditions can be called as STH efficiency. Zero bias conditions correspond to absence of bias between working and counter electrodes and solar energy is the only source for water splitting reaction [17,19,20]. For correct measurement of STH efficiency, both working and counter electrodes should be immersed in same pH solutions and the electrolyte should not contain any sacrificial donors or acceptors (Figure 15.3). Therefore, STH efficiency is defined as the ratio of chemical energy of the hydrogen produced to photon (solar) energy input from sunlight that incident on the process, as shown in the following equation:

$$\text{STH}_{H_2 \text{ production}} = \left[\frac{(\text{mmol } H_2/s) \times (237 \text{ kJ/mol})}{(P_{\text{total}})(\text{mW/cm}^2) \times \text{Area } (\text{cm}^2)} \right]_{AM\ 1.5\ G} \times 100. \qquad (15.6)$$

The chemical energy of hydrogen produced in the numerator can be calculated by multiplying Gibbs free energy per mole of H_2 ($\Delta G° = 237$ kJ/mol at 25°C) to the amount of hydrogen produced per second (mmol/s). In the above equation, the illumination source should closely match the intensity of the Air Mass 1.5 Global (AM 1.5 G, means 1.5 atmosphere thickness with a solar zenith angle of ~48.2° corresponds to the solar spectrum at mid-latitudes where the world's major population centers are) standard as set forth by the American Society of Testing and Materials (ASTM) [17]. The above equation is used for STH calculation based on hydrogen production (monitored by analytical methods such as mass spectrometry or gas chromatography). However, there is an alternative way to calculate STH based on current as shown in Equation 15.7, wherein, the short circuit photocurrent density (j_{SC}, mA/cm^2) is taken into consideration. It is to be noted that the short circuit refers to zero voltage in the external circuit, similar to a PV device. It should be noted that both equations are valid only if the evolution of gases is in stoichiometry proportion (i.e., one mole of water should produce one mole of hydrogen and half mole of oxygen) without any sacrificial (electron donor or acceptor) agent.

$$\text{STH}_{jsc} = \left[\frac{\left(j_{SC} \text{ mA/cm}^2 \right) \times (1.23 \text{ V}) \times \eta_F}{(P_{\text{total}})(\text{mW/cm}^2)} \right]_{AM\ 1.5\ G} \times 100, \qquad (15.7)$$

where j_{SC} is the short circuit current, and η_F is the faradaic efficiency. Figure 15.3 shows STH efficiencies of some important photo materials signifying the importance of this efficiency in material selection for efficient PEC water splitting reaction. Further, it helps to design new photo materials by material tuning/surface structuring and shows the future target to be achieved. It is worth noting that STH efficiency of PEC cells is similar to solar to electric conversion efficiency in photovoltaic (PV) cells.

15.5.2 Applied Bias Photon to Current Efficiency (ABPE)

ABPE is different from STH, wherein, a bias is applied between working and counter electrodes. While bias plays an important role in this definition, when a bias exceeding thermodynamic water splitting potential (1.23 V) is applied, the relevance of PEC becomes limited. In addition, ABPE measurement

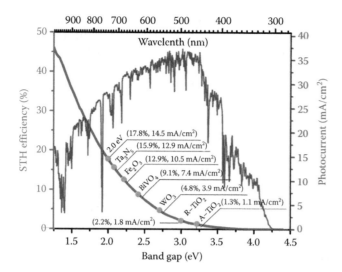

FIGURE 15.3 STH efficiency vs. material bandgap plot of important photo materials: theoretical maximal photo-current (j_{max}) shown on right axis and STH efficiency shown on left axis are plotted against material bandgap (E_g). (Reprinted from Li, J., Wu, N., *Catal. Sci. Technol.*, 5, 1360–1384, 2014. With permission. Copyright © 2014, Royal Society of Chemistry.)

is not truly a STH efficiency measurement, but only serves as a diagnostic efficiency measurement for new material designs and developments. The following is the equation of measurement, where j_{ph} is the photocurrent density obtained under an applied bias V_b:

$$\eta_{ABPE} = \frac{j_{ph} \times (1.23 - V_b)}{P_{solar}}. \tag{15.8}$$

Similar to existence of pitfalls in STH measurement, there are a few misleading perceptions in using and reporting ABPE definition and parameters used. Thus, for the correct measurement of ABPE of PEC cell the following points should be taken care of: (1) value of V_b should be measured between working and counter electrode, (2) ABPE should be measured without sacrificial donor or acceptor in the electrolyte, and (3) two compartment cell with electrolytes have to be at the same pH [17,18].

15.5.3 Incident Photon to Current Efficiency (IPCE)

IPCE is another important diagnostic efficiency measurement of PEC systems, wherein the photocurrent collected per incident photon flux as a function of illumination wavelength is described. Generally an integration of IPCE data over the entire solar spectrum estimates the maximum possible STH efficiency, provided the data are collected at zero bias (2-electrode, short-circuit) conditions. It is worthwhile to note that IPCE efficiency under applied bias is a useful diagnostic tool to know the material properties used in PEC devices.

$$IPCE = \frac{\text{Electrons/cm}^2\text{s}}{\text{Photons/cm}^2\text{s}} = \frac{j_{ph}\left(\text{mA/cm}^2\right) \times 1239.9(\text{V} \times \text{nm})}{P_{mono}\left(\text{mW/cm}^2\right) \times \lambda(\text{nm})}, \tag{15.9}$$

where P_{mono} is the monochromatic illuminated power intensity (mW/cm²) and λ is wavelength in nm. IPCE is an extremely important number which takes into account the spectral variation of incident photons, considered at each energy. Further, it is a realistic number which takes into account "electrons generated out per photon in" rather than "power output per power in," in efficiency measurements [17].

15.5.4 Absorbed Photon to Current Efficiency (APCE)

The inherent performance of material does not account the unavoidable losses (e.g., photons reflected and transmitted) associated in any PEC device. Since STH implicitly includes these losses and results in underestimation of true efficiency, it is advantageous to subtract the losses and measure the efficiency based on specific photons absorbed. APCE is measured as the ratio of photocurrent collected per incident photons absorbed, as shown in the following equation:

$$\text{APCE} = \frac{j_{ph}\left(\text{mA/cm}^2\right) \times 1239.8(\text{V} \times \text{nm})}{P_{mono}\left(\text{mW/cm}^2\right) \times \lambda(\text{nm}) \times \left(1 - 10^{-A}\right)}, \tag{15.10}$$

where A is the absorbance of the sample estimated from Beer's law (relates A to logarithmic ratio of measured output light intensity [I] to initial light intensity [I_0]). This is a very useful number for PEC devices, especially efficiency measurements of thin films [17].

15.5.5 Intrinsic Solar to Chemical Conversion (ISTC) Efficiency

Measurement of intrinsic solar to chemical conversion efficiency (ISTC) is a different approach to estimate the PEC efficiency. It evaluates the electrode materials in 3-electrode mode:

$$\text{ISTC} = \frac{\eta_F \times 1.23 \ (V)}{V_{dark}(V)} \times \frac{j_{photo}\left(\text{mA/cm}^2\right) \times V_{photo}(V)}{P_{solar}\left(\text{mW/cm}^2\right)}. \tag{15.11}$$

Here, V_{photo} is the photovoltage which is often a function of current density and can be estimated as the difference between the voltage required to produce a particular current density in the dark and the voltage required to produce the same current density under illumination. This efficiency is also limited as the $I-V$ behavior of photoelectrode in dark is complex and the dark current generation involves the minority carriers (e.g., holes in n-type semiconductor) which shows typical inversion condition in narrow bandgap materials, but not in wide bandgap materials. Therefore, usage of dark current in PEC efficiencies is not helpful [21].

15.6 Efficiency Losses in PEC

There will always be concentration-dependent and kinetic overpotential losses associated with electron transfer processes at semiconductor/liquid junctions to drive the HER and OER. Therefore, the energy required for water photolysis will be approximately 1.6–2.4 eV per electron-hole pair generated after accounting for the losses. Practically, we should address the following possible losses that severely influence the efficiency: (1) Reflection losses: the transparent collector where PEC reaction occurs should pass all the solar photons without reflecting any fraction. However, ~4–5% reflection losses will be present for untreated glass at normal incidence (more at the higher angles of incidence) [22]. These losses can be reduced using antireflection coating, which leads to the additional cost to the system. (2) Quantum-yield losses: as it is difficult to achieve optimum values of quantum yields, we always assume quantum yields of more than 90% of ideal values. (3) Absorption losses: we always assume that all the sunlight with $\lambda \leq \lambda_g$ (photons) will be absorbed and used to drive the reactions. However, the absorption coefficient varies with wavelength and some fraction of sunlight is absorbed by inactive components which again is a loss to the system. (4) Collection losses: the collection of products from water splitting reaction has significant inherent losses associated in efficiency loss calculations. It is optimistic to assume a collection and storage efficiency of ~90% [23]. There will always be additional losses associated with any PEC system due to the nonideal placement of the band edges w.r.t water redox potentials for charge carrier transport. Sometimes due to poor alignment of band edges the half-cell reactions may not occur at all. Only based on photon absorption (or bandgap of the material), there are inherent losses associated with any solar energy conversion processes [24,25]. The losses can be absorption losses. Only the energy $E \geq E_g$ (or the photons with wavelength $\lambda \leq \lambda_g$ where λ_g is the wavelength corresponding to the bandgap) is absorbed and the rest is lost. Further, out of the absorbed energy, the excess energy $(E - E_g)$ is lost as heat during the relaxation of the absorber to the level of E_g. Apart from these, only 75% of the absorbed energy is utilized to convert into electrical energy or work due to spontaneous emissions such as fluorescence which also contributes to the efficiency limitations.

15.7 Engineering Design

On the basis of technical and economical evaluation, Jaramillo et al. categorized PEC reactors into four designs covering a range of complete practical systems [26]. All the four types of reactors consist of solar photon absorbing material with appropriate band edges for HER and OER and H_2 collection system. Each category has been described in following sections:

15.7.1 Type 1 Reactor: Single Particle Bed Suspension

Type 1 reactor is the simplest design involving big horizontal transparent sealed plastic bags, known as "baggie" (Figure 15.4a). Baggie are made up of high-density polyethylene with upper transparent and bottom as opaque portion. These baggie are designed to contain photoactive particles in 0.1 M KOH electrolyte and evolved H_2 gas. The size of baggie is 323 m × 12.2 m limited by largest possible manufacturing facility available till date. The photon absorbing material consists of 5 nm layer of photocathodic

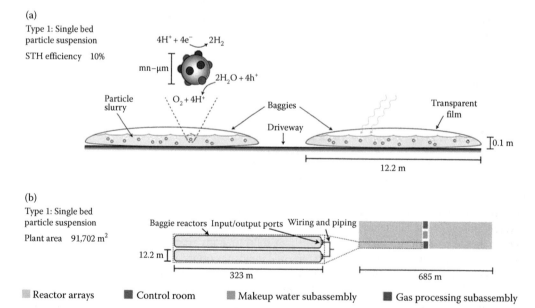

FIGURE 15.4 (a) Schematic representation of Type 1 reactor and (b) design of plant module for 1 TPD H_2 production. (Reprinted from Pinaud, B.A. et al., *Energy Environ. Sci.*, 6, 1983–2002, 2013. With permission. Copyright © 2014, Royal Society of Chemistry.)

and photoanodic material as particle, cluster of particle or thin layer on 40 nm diameter conductive spherical core. The operation under illumination results in generation of both H_2 and O_2 gases which should be purified further. The top layer of long plastic baggies filled with photoactive particle suspension in KOH rests on liquid surface. The photo-generated gases are accommodated within the extra volume of baggie which are drawn off intermittently. Sealing of the baggie prevents evaporative cooling and maintains operation temperature in the range of 60°C which brings down the thermodynamic voltage for water splitting. In spite of having simpler design principle, large size of baggie is more prone to mechanical failure against external damaging parameters such as weather and bird damage. Due to enormous size, even the failure of single baggie results in significant plant capacity loss. In addition, Type 1 reactor shows much lower efficiencies due to greater uncertainty associated with fabrication of photoactive particle as compared to planer electrode. The system is also limited by lack of information available to measure particle density and effective photon capture area for utilizing whole solar spectrum. The particle at the bottom receives lesser solar irradiation as compared to particle at the top which hampers the efficiency of lower region particles thereby reducing overall performance. Based on 5.77 kWh/m² solar irradiation available per day and 10% STH efficiency, Type 1 reactor plant was suggested which contains 18 baggies to produce 1 TPD of H_2. Figure 15.4b illustrates the top view of plant layout. A total of 91,702 m² area is needed considering 30% additional area for other auxiliaries and reactor maintenance access whereby we neglect shading effect.

15.7.2 Type 2 Reactor: Dual Particle Bed Suspension

Type 2 reactor shares most of the feature of Type 1 reactor with only difference of two separate baggie chambers for OER and HER separately which are connected together by a redox mediator (A/A^-) and a porous bridge. Redox mediator is selected on basis of reversible reactivity and large diffusivity in either redox state. Beds are made up of high-density polyethylene with upper transparent and bottom as opaque portion. It is 61.0 m long and 6.1 m wide in dimension (Figure 15.5a). The photon absorbing material consists of single photoactive layer on 40 nm diameter conductive spherical core. The operation under illumination results in generation of both H_2 and O_2 gases in separate beds which eliminates any hazard due to H_2 combustion in presence of O_2. In addition, separate beds provide greater flexibility to opt light absorbing material for effective HER and OER. Effective transport and good mixing of redox mediator is achieved by continuous circulation of slurry through perforated pipes down along the beds. The position of feed through bridges remains below the level of liquid–gas interface and it connects to two adjacent H_2 and O_2 production beds with a porous membrane along the length of bed. It prevents the escape of gas molecules without hampering the transport of redox mediator. Similar to Type 1 design, Type 2 reactors also exhibit lower efficiencies due to greater uncertainty associated with fabrication of photoactive particle, lack of information available to measure particle density and effective photon capture area for utilizing complete solar spectrum. The design is also limited by partial utilization of solar energy by particle at the bottom as compared to top particle. In addition, Type 2 reactor faces the challenges associated with high overpotential across the bridges. Moreover, we cannot assure the presence of pure H_2 and O_2 in separate chambers as the trace amount of these gaseous molecules can pass through the bridge. Under the similar assumption of 5.77 kWh/m² solar irradiation available per day and negligible shading effect, Type 2 reactor plant consists of 347 baggies to produce 1 ton per day (TPD) of H_2 with 5% STH efficiency. Figure 15.5b illustrates the top view of plant layout. Similar to previous design type, addition 30% area is reserved for other auxiliaries and reactor maintenance access. It results in a total of 16×10^6 m² land area for plant operation.

15.7.3 Type 3 Reactor: Fixed Panel Array

Type 3 reactor consists of several photoactive layers inserted between two electrodes to form a planer electrode stack. Upper electrode was chosen as a transparent conductive oxide such as Indium tin oxide

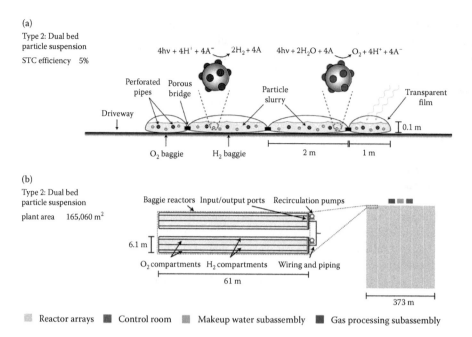

FIGURE 15.5 (a) Schematic representation of Type 2 reactor cross section and (b) design of Plant module for 1 TPD H_2 production. (Reprinted from Pinaud, B.A. et al., *Energy Environ. Sci.*, 6, 1983–2002, 2013. With permission. Copyright © 2014, Royal Society of Chemistry.)

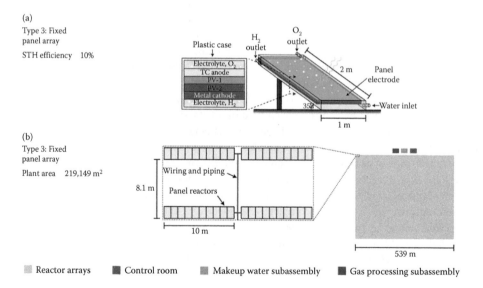

FIGURE 15.6 (a) Schematic representation of Type 3 reactor and (b) design of plant module for 1 TPD H_2 production. (Reprinted from Pinaud, B.A. et al., *Energy Environ. Sci.*, 6, 1983–2002, 2013. With permission. Copyright © 2014, Royal Society of Chemistry.)

(ITO) while the lower electrode was chosen as a metal electrode. Whole system is enclosed in poly (methyl methacrylate) (PMMA) reservoir filled with 0.1 M KOH electrolyte. These cells are assembled together to form a panel of 1 m × 2 m dimensions (Figure 15.6a). These panels are oriented toward the earths equator to maximize the solar photon absorption. Upon illumination, O_2 gas evolves at the top face while H_2 gas evolves at the metal electrode. Both gases are collected separately in a separate chamber. The buoyant

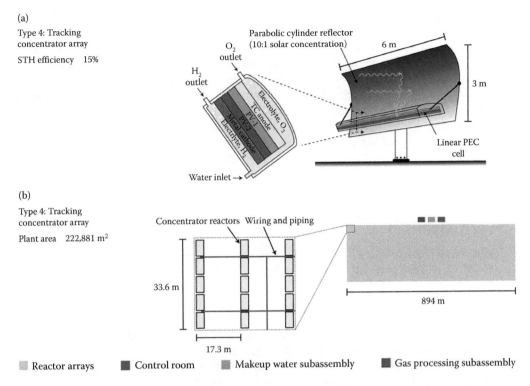

(a)

Type 4: Tracking
concentrator array

STH efficiency   15%

Parabolic cylinder reflector
(10:1 solar concentration)

O_2 outlet

H_2 outlet

6 m

3 m

Linear PEC cell

Electrolyte O_2

TC anode

PV1

PV2

Metal cathode

Electrolyte H_2

Water inlet →

(b)

Type 4: Tracking
concentrator array

Plant area   222,881 m²

Concentrator reactors Wiring and piping

33.6 m

17.3 m

894 m

▨ Reactor arrays   ■ Control room   ▨ Makeup water subassembly   ■ Gas processing subassembly

FIGURE 15.7 (a) Schematic representation of Type 4 reactor and (b) design of plant module for 1 TPD H_2 production. (Reprinted from Pinaud, B.A. et al., *Energy Environ. Sci.*, 6, 1983–2002, 2013. With permission. Copyright © 2014, Royal Society of Chemistry.)

nature of gases helps in collection due to inclination of panel. A design study done at Daggett, California suggests the tilt of 35° to achieve 6.19 kWh/m² solar irradiation per day. To achieve a target production of 1TPD H_2, a total of 26,923 panels are needed with STH efficiency of 10%. Considering negligible shading effect at angle >10°, there should me minimum separation of 8.1 m between adjacent panels. It results in total land area of 2.2×10^5 m². The layout of plant has been illustrated in Figure 15.6b.

15.7.4 Type 4 Reactor: Tracking Concentrator Array

Type 4 reactor has been shown in Figure 15.7a. It consists of an offset parabolic reflector and a solar tracker which maximizes the solar irradiation capture in a smaller area thereby allowing the use of highly efficient and costly materials. This type of reactor allows the installation of photoreactor, electrolyte feed line and H_2 collection line at the reflector base. Additionally, the weight and cost of reactor are also reduced. A typical tracking concentrator array has 6 m wide and 3 m height dimension. Although a concentration ratio up to 400:1 is achievable with this type, the activity of PEC cell is limited by catalyst performance and the current density of PEC system is limited by 1 A/cm². Hence, the concentration ratio of 100:1 has been chosen as upper limit. The design of rector consists of pressurized electrolyte input which evolves H_2 and O_2 gases in separate chambers. The solar tracking system enables the panel pointing toward the direction of sun all over the day. It results in a total of 6.55 kWh/m² solar irradiation per day on annual basis. To produce a target value of 1 TPD H_2 with 15% STH efficiency, a total of 1885 concentrator panels are needed. To prevent shading effect, a spacing of 6.71 m in north–south direction and 17.3 m in east–west direction is essential which results in the total land area of 2.22×10^5 m². Figure 15.7b demonstrates the plant layout for the Type 4 reactor.

15.8 Conclusion

This chapter provides a comprehensive overview on the fundamental aspect of PEC water splitting to generate hydrogen. To accomplish semiconductor-assisted PEC water splitting, semiconductor must have desirable flat band positions to perform the reduction and oxidation reactions. Further, based on above discussions, it can be concluded that material intrinsic properties play an important role in enhancing performance of PEC systems. Moreover, this aspect has to be accounted in material design and fabrication approaches which profoundly influence the performance of PEC systems. Finally, this chapter covers the design of four conceptual systems, i.e., single bed particle suspension (Type 1), dual bed particle suspension (Type 2), fixed panel array (Type 3), and tracking concentrator array (Type 4), which clearly demonstrate that proper technical and plant design advancement is essential to meet the global energy requirement via solar hydrogen.

Acknowledgment

The work was supported by the Department of Science and Technology via the grants DST/TSG/SH/2011/106-G and DST/TM/SERI/2K11/79(G).

References

1. Asahi, R., T. Morikawa, T. Ohwaki, K. Aoki, and Y. Taga, 2001. Visible-light photocatalysis in nitrogen-doped titanium oxides. *Science* 293: 269–271.
2. Hisatomi, T., J. Kubota, and K. Domen, 2014. Recent advances in semiconductors for photocatalytic and photoelectrochemical water splitting. *Chem. Soc. Rev.* 43: 7520–7535.
3. Linic, S., P. Christopher, and D. Ingram, 2011. Plasmonic-metal nanostructures for efficient conversion of solar to chemical energy. *Nat. Mater.* 10: 911–921.
4. Maeda, K. and K. Domen, 2007. New non-oxide photocatalysts designed for overall water splitting under visible light. *J. Phys. Chem. C.* 111: 7851–7861
5. Maeda, K. and K. Domen, 2010. Photocatalytic water splitting: Recent progress and future challenges. *J. Phys. Chem. Lett.* 1: 2655–2661.
6. Walter, M.G., E.L. Warren, J.R. McKone, S.W. Boettcher, Q. Mi, E.A. Santori, and N.S. Lewis, 2010. Solar water splitting cells. *Chem. Soc. Rev.* 110: 6446–6473.
7. Chen, X., S. Shen, L. Guo, and S.S. Mao, 2010. Semiconductor-based photocatalytic hydrogen generation. *Chem. Soc. Rev.* 110: 6503–6570.
8. Zhou, H.L., Y.Q. Qu, T. Zeid, and X.F. Duan, 2012. Towards highly efficient photocatalysts using semiconductor nanoarchitectures. *Energy Environ. Sci.* 5: 6732–6743.
9. Bak, T., J. Nowotny, M. Rekas, and C.C. Sorrell, 2002. Photo-electrochemical properties of the TiO_2-Pt system in aqueous solutions. *Int. J. Hydrogen Energy* 27: 19–26.
10. Gratzel, M., 2001. Photoelectrochemical cells. *Nature* 414: 338–344.
11. Kudo, A. and Y. Miseki, 2009. Heterogeneous photocatalyst materials for water splitting. *Chem. Soc. Rev.* 38: 253–278.
12. Archer, M.D. and A.J. Nozik, 2008. *Nanostructured and Photoelectrochemical Systems for Solar Photon Conversion.* World Scientific, Singapore.
13. Memming, R., 2008. *Semiconductor Electrochemistry.* John Wiley & Sons, New York.
14. Vayssieres, L., 2010. *On Solar Hydrogen and Nanotechnology.* John Wiley & Sons, Singapore.
15. Chen, Z., H. Dinh, and E. Miller, 2013. *Photoelectrochemical Water Splitting: Standards, Experimental Methods, and Protocols.* Springer, New York.
16. Wang, M., L. Chen, and L. Sun, 2012. Recent progress in electrochemical hydrogen production with earth-abundant metal complexes as catalysts. *Energy Environ. Sci.* 5: 6763–6778.

17. Chen, Z., T.F. Jaramillo, T.G. Deutsch, A. Kleiman-Shwarsctein, A.J. Forman, N. Gaillard, R. Garland, K. Takanabe, C. Heske, M. Sunkara, and E.W. McFarland, 2010. Accelerating materials development for photoelectrochemical hydrogen production: Standards for methods, definitions, and reporting protocols. *J. Mater. Res.* 25: 3–16.

18. Hodes, G., 2012. Photoelectrochemical cell measurements: Getting the basics right. *J. Phys. Chem. Lett.* 3: 1208–1213.

19. Li, J. and N. Wu, 2014. Semiconductor-based photocatalysts and photoelectrochemical cells for solar fuel generation: A review. *Catal. Sci. Technol.*, 5: 1360–1384.

20. Smestad, G.P., F.C. Krebs, C.M. Lampert, C.G. Granqvist, K.L. Chopra, X. Mathew, and H. Takakura, 2008. Reporting solar cell efficiencies in solar energy materials and solar cells. *Sol. Energy Mater. Sol. Cells* 92: 371–373.

21. Dotan, H., N. Mathews, T. Hisatomi, M. Graetzel, and A. Rothschild, 2014. On the solar to hydrogen conversion efficiency of photoelectrodes for water splitting. *J. Phys. Chem. Lett.* 5: 3330–3334.

22. Meinel, A.B., 1976. *Applied Solar Energy—An Introduction.* Addison-Wesley Pub. Co, Reading, MA.

23. Bolton, J.R., S.J. Strickler, and J.S. Connolly, 1985. Limiting and realizable efficiencies of solar photolysis of water. *Nature* 316: 495–500.

24. Archer, M.D. and J.R. Bolton, 1990. Requirements for ideal performance of photochemical and photovoltaic solar energy converters. *J. Phys. Chem.* 94: 8028–8036.

25. Bolton, J.R., 1996. Solar photoproduction of hydrogen: A review. *Sol. Energy* 57: 37–50.

26. Pinaud, B.A., J.D. Benck, L.C. Seitz, A. J. Forman, Z. Chen, T.G. Deutsch, B.D. James, K.N. Baum, G.N. Baum, S. Ardo, H. Wang, E. Miller, and T.F. Jaramillo, 2013. Technical and economic feasibility of centralized facilities for solar hydrogen production via photocatalysis and photoelectrochemistry. *Energy Environ. Sci.* 6: 1983–2002.

16

Photoelectrochemical Approaches to Solar-H$_2$ Generation

Gyan Prakash
Sharma, Arun
Prakash Upadhyay,
Dilip Kumar Behara,
Sri Sivakumar, and
Raj Ganesh S. Pala
*Indian Institute of
Technology Kanpur*

16.1 Introduction

The supply of energy for high living standard is one of the eternal and challenging issues for the whole world. According to Energy Information Administration, the projected energy demand for the whole world in 2015 is 572 quadrillion Btu, which will be increased by 43% within a span of next 30 years. Organization for Economic Corporation and Development (OECD) member countries are the top energy consumers [1]. However, such countries are assumed to have slow population and economic growth rate. Therefore, most of the energy demand (~85%) will be required for non-OECD nations due to their emerging population and strong economic growth. For an instance, India and China (non-OECD nations) have the fastest developing economy since 1990. The cumulative energy consumption for both countries has increased from 10% to 24% during 1990–2010 with respect to the total world's energy consumption and is projected to be 34% by 2040.

The supply of energy has been fulfilled either by nonrenewable (e.g., coal, petroleum, natural gas, and nuclear) or renewable (e.g., solar, wind, and geothermal) sources. The exploitation of all such energy sources proliferates with time (Figure 16.1). As projected till 2040, fossil fuel will continue to serve as a primary energy source for worldwide energy consumption. Coal will continue to play an important role for energy supply especially in non-OECD Asia (led by India and China). While the worldwide coal consumption will increase by 1.3% within next 30 years from 2010, non-OECD Asia demand will increase up to 1.9% in the same time horizon. The demand for natural gases will continue to increase by 1.7% per annum till 2040. This demand will be fulfilled especially by shale gas primarily from the United States and Canada. Near future advancement in horizontal drilling and hydraulic fracture technologies will keep its price down as compared to liquid fuel. Among all fossil fuels, petroleum fuels will remain the largest source of energy. However, the petroleum will decrease from 34% in 2010 to 28% in 2040 due to limited reserves. This trend will provide the thrust for the production of alternate fuels.

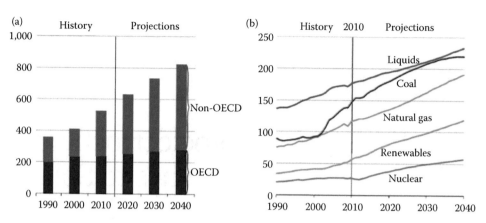

FIGURE 16.1 (a) Global energy demand (quadrillion Btu). (b) Global energy usage by type of fuel (quadrillion Btu). (From EIA, International Energy Outlook, 2013.)

All the above-mentioned fossil fuels will be exhausted in the long run. According to current demand, average life of coal, petroleum, and natural gas reserves will be 150–400, 40–80, and 60–160 years, respectively. Despite being exhaustible reserves, fossil fuels severely impact the environment with the generation of greenhouse gas (e.g., carbon dioxide), which contributes to global warming. The atmospheric CO_2 level has raised from 280 to 394 ppm since the industrial revolution and will continue to increase with a rate of 2 ppm per annum [2]. India ranks third with 5.4% global contribution in CO_2 emission after China and the United States in 2009. As per the information from International Panel on Climate Control (IPCC), 450 ppm CO_2 can lead to global warming above 2°C [2]. Such increase in temperature will deleteriously affect natural climate and ecosystem. Such situation will become unavoidable before year 2020 if we continue to exhaust CO_2 with the same rate in atmosphere [3,4].

To ensure the continuous supply of energy in the long run and to protect the climate from global warming, the world must shift from contemporary primary energy fossil fuels to everlasting renewable fuels [3–5]. Among the available renewable energy sources, solar energy has enormous potential to provide world energy demand, as the earth receives nearly ~1.3×10^5 TW of solar energy, which is almost four times the total world energy consumption (16.9 TW) [5–7]. Thus, effective harvesting and conversion of solar energy into useful form such as chemical fuels and/or electricity can be a panacea to the world energy crisis. In spite of increasing photovoltaic solar cells, storage of electrical energy is economically challenging. As most of the energy demand is based on chemical fuels, direct conversion of solar energy to chemical energy is the more appropriate form of utilizable energy. In this aspect, generation of hydrogen (H_2) via solar water splitting is a very promising technology [5,6,8]. In addition to being most energy dense fuel, the burning of hydrogen results in benign water as a by-product through a reversible cycle. Hence, efficient and cost-effective water splitting technology can pave the path for future hydrogen economy without generation of pollutants and greenhouse gases [5,6,8]. Moreover, hydrogen can be used as renewable reductant for fixing atmospheric CO_2 into useful chemicals (e.g., methanol [9], formic acid [10], and formamide) with both homogeneous and heterogeneous catalysts [9,11–13]. Due to the formation of water as a stable by-product, hydrogenation of CO_2 is a spontaneous reaction. Therefore, hydrogen can act as renewable reductant to fix the greenhouse gas problem at a global scale. Considering that all the CO_2 will be converted to methanol with the following reaction with 100% efficiency, there would be 7.27×10^{14} moles of CO_2 can be converted per year:

$$CO_2 + 3H_2 \rightarrow CH_3OH + H_2O. \tag{16.1}$$

To convert these CO_2 molecules into methanol, 2.18×10^{15} moles of hydrogen is required annually. Therefore, to reduce all the generated CO_2, hydrogen must be produced with the rate of 6.91×10^7 mol/s. Let us take the situation where all the required hydrogen is generated from highly efficient photoelectrochemical water splitting device with 12.3% STH value [14] and current density of 10 mA/cm², a total land area of 133,400 km² will be needed to match the rate of hydrogen consumption of CO_2 reduction. Considering a margin of 20% in the calculated land area, a total of 160,080 km² land area would be required, which approximately covers a little more than the size of Bangladesh.

This chapter provides an overview on the present status of world energy consumption, need of renewable resources, and different pathways of harvesting the solar energy (such as photobiological, solar thermal, and solar light-driven water splitting) to fulfill the world's energy requirement. Among these approaches, a detailed overview on present and future aspects of photoelectrochemical hydrogen generation is discussed here.

16.2 Various Approaches for H₂ Generation

16.2.1 Photobiological Hydrogen Production

A typical photobiological process uses solar light as energy source and photosynthetic bacteria, such as green algae and cyanobacteria, as catalyst to produce H_2 from water as shown in Figure 16.2 [15–20]. Briefly, solar light is absorbed by green algae and cyanobacteria because of the presence of photosystem I and II (PSI and PSII) followed by transfer of absorbed light energy to chlorophyll molecule to facilitate the charge separation and generate oxidants and reductants. Further, H_2-producing organisms utilized PSI-derived electrons to reduce ferredoxin (FD), which plays an important role in generating either NADPH via ferredoxin-NADP oxidoreductase (FNR) or H_2. In such process, hydrogenase prefers to use reductant directly to produce H_2, whereas nitrogenase-based cells use oxidizing carbon compounds to produce oxygen and proton [19–22]. Green algae and cyanobacteria can also use such enzyme under the anaerobic condition where protons are supplied by an organic substance to produce H_2 [19,23]. However, high sensitivity of microorganism toward oxygen

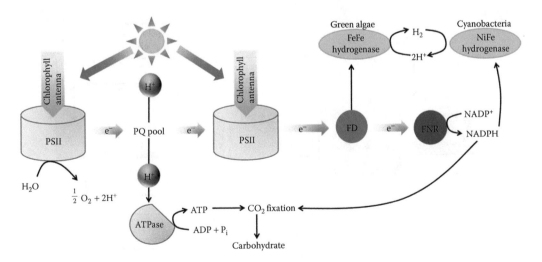

FIGURE 16.2 Photosynthetic pathway of H_2 production in green algae (employing FeFe-hydrogenase) and in cyanobacteria (employing NiFe hydrogenase). PS, photosystem; PQ, plastoquinone; FD, ferredoxin; FNR, ferredoxin-NADP oxidoreductase. (Reproduced from the Nath, K. et al., *Photosynth. Res.*, 126, 237–247, 2015; Ghirardi, M.L. et al., *Chem. Soc. Rev.*, 38, 52–61, 2009.)

and fabrication of catalyst and photosensitizer conjugation to facilitate the generation of carbon-free H_2 using solar light are the major challenges associated with photobiological hydrogen production, which prevents its economical viability.

16.2.2 Solar Thermal Hydrogen Production

Solar thermal hydrogen generation utilizes concentrated solar light as energy source which provides thermal energy to perform the endothermic transformation reaction [24,25]. As described by Tyner and group, a large-scale solar thermal system includes three optical parts, i.e., trough, tower, and dish [25]. Performance of such configuration depends on mean flux concentration ratio over a particular area at the focal plane when normalized w.r.t normal solar irradiation and is given as

$$C = \frac{Q_{solar}}{I \times A},$$
(16.2)

where Q_{solar} is the solar power received by the target and C is the mean flux concentration ratio at solar intensity $I = 1\,kW/m^2$ [25]. There are five possible pathways to utilize the concentrated solar energy to produce hydrogen (Figure 16.3), i.e., water thermolysis, thermochemical route, hybrid thermal processes (i.e., cracking, reforming, and gasification), where fossil fuels are used as reactant and concentrated solar light as energy source [24].

In water thermolysis, concentrated solar light provides the necessary energy required to directly split the water into H_2 and O_2 [24,26–28], whereas thermochemical routes [e.g., sulfur–iodine cycle

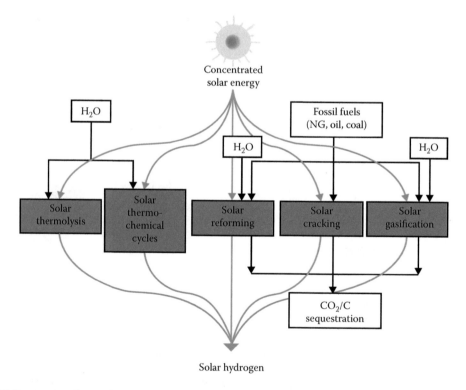

FIGURE 16.3 Five thermochemical routes for the production of solar hydrogen. (Reprinted from Steinfeld, A., *Solar Energy*, 78, 603–615, 2005. With permission. Copyright © 2004 Elsevier Ltd.)

(Figure 16.4), Westinghouse sulfur cycle (Figure 16.5a), and two-step metal oxide reduction and oxidation (Figure 16.5b)] involve the splitting of water through intermediate chemical reactions [24,29,30]. On the other hand, hybrid thermal processes are more favorable in this category because of lesser/no contamination of products, fuel saving which extended the life of fossil fuels [24]. For instance, solar cracking of hydrocarbons (generally methane) pave a pathway to decompose natural gas into hydrogen and carbon which can be directly marketable (Figure 16.6a) [24,31,32]. Further, utilizing the concentrated solar energy in reforming or gasification reaction possesses the advantages of lower temperature, high purity of hydrogen, eliminate the necessity of additional CO to CO$_2$ conversion reactor, and also reduces the emission by 34%–43% (Figure 16.6b) [33–38]. However, requirement of large land area to concentrate the solar light, utilization of fossil fuel to produce H$_2$, and high cost are the major challenges associated with solar thermal-based H$_2$ production and encourage the researcher to search for another economically viable solution.

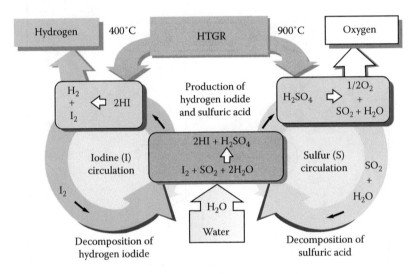

FIGURE 16.4 Scheme of thermochemical water-splitting cycle using iodine and sulfur. (Reprinted from Onuki, K. et al., *Energy Environ. Sci.*, 2, 491–497, 2009. With permission. Copyright © 2009, Royal Society of Chemistry.)

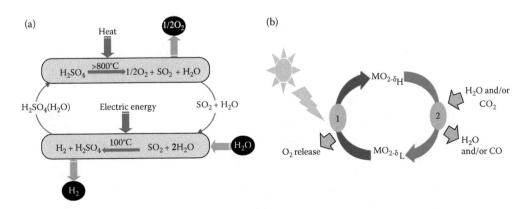

FIGURE 16.5 Schematic representation of (a) Westinghouse sulfur cycle and (b) two-step metal oxide reduction and oxidation.

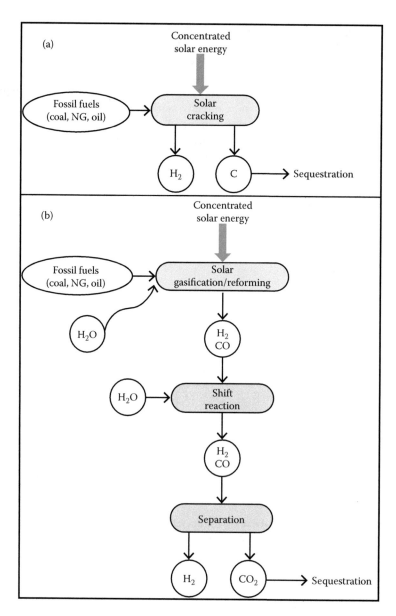

FIGURE 16.6 Schematic of solar thermochemical routes for H_2 production using fossil fuels and H_2O as the chemical source: (a) solar cracking and (b) solar reforming and gasification. (Reproduced from Steinfeld, A., *Solar Energy*, 78, 603–615, 2005. With permission.)

16.2.3 Hydrogen Generation through Water Splitting

As described in previous sections, conversion of solar energy into hydrogen is the most promising alternative to satisfy the world's energy demands in the future. However, development of energy-efficient and economically viable H_2 production processes is the major challenge, which prevents the world from shifting from fossil fuel to hydrogen-based economy [5,6,39,40]. In the recent years, extensive research has been performed and still is going on toward the production of clean hydrogen from water mainly using electrochemical (EC), photocatalytic (PC), and photoelectrochemical (PEC) processes [5,6,40]. Typically, water splitting as shown in Figure 16.7 includes two major

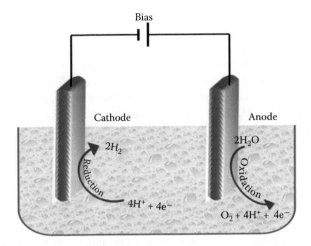

FIGURE 16.7 Schematic representation of electrochemical water splitting.

reactions, i.e., water reduction (for H$_2$ production) and water oxidation (for O$_2$ production) as shown in below reactions:

$$2H_2O + 4h^+ \rightarrow O_2 + 4H^+ \tag{16.3}$$

$$4H^+ + 2e^- \rightarrow 2H_2 \tag{16.4}$$

Hence, development of efficient, stable, and economically viable hydrogen and oxygen evolution catalyst is essential to drive these two half reactions [5,7,40]. However, compared to EC, PC, and PEC water splitting is an environmentally benign strategy to generate clean H$_2$ as it utilizes solar light as energy source. In photocatalytic water splitting configuration, powder semiconductor photocatalyst has been used to drive the water splitting reaction as shown in Figure 16.8 where solar light impinges on the photocatalyst to generate the holes and electron which further take part in the oxygen and hydrogen evolution

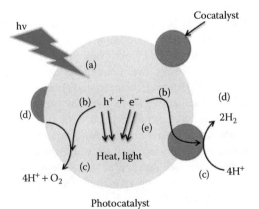

FIGURE 16.8 Reaction processes of water splitting on a heterogeneous photocatalyst. (a) Light absorption, (b) charge transfer, (c) redox reactions, (d) adsorption, desorption, and mass diffusion of chemical species, and (e) charge recombination. (Reproduced from Hisatomi, T. et al., *Chem. Soc. Rev.*, 43, 7520–7535, 2014.)

reaction. Such configurations are also termed as artificial photosynthesis and can be scaled-up to large production [6,7,41]. However, efficient separation of H_2 and O_2 in PC is difficult and prevents its commercialization. To overcome this problem, in the 1970s, Fujishima and Honda described a new configuration where they used titania (TiO_2) semiconductor electrode, which generate holes and electrons upon illumination with UV light. The generated holes drive the oxygen evolution at TiO_2 electrode, whereas electron is transferred to Pt counter electrode via external circuit to produce H_2 [42].

Following this, extensive research has been carried out using various oxide (TiO_2, WO_3, Fe_2O_3, $BiVO_4$, ZnO, SnO_2, NiO, CdO, and PdO) [5,41,43–49] and nonoxide ($CuGaSe_2$, Cu_2ZnSnS_4, Cu (Ga,In)(S,Se), CdS, PbSe, PbSGaN, Ge_3N_4, and TaON) [5,7,41,50–53] semiconductor materials to perform the semiconductor-assisted water splitting (detailed explanation is reported in several review articles [5,8]). However, development of efficient material to produce H_2 and O_2 from water has still not been achieved and research is still continuing in the search for efficient material. The major reasons which prevent the commercialization of hydrogen are (1) low efficiency of photoanode material because of the high band gap of semiconductor and high charge carrier recombination and (2) use of expensive Pt as counter electrode [5,6,8,46]. To circumvent these issues, separate or combination of different possible ways have been explored such as (1) development of electrocatalyst based on earth abundant materials (e.g., Ni- and Co-based electrocatalyst) as a replacement of Pt counter electrode to facilitate the hydrogen evolution reaction [40], (2) tuning of semiconductor band gap to harvest the major fraction of solar spectrum [5,6,8,46], and (3) formation of two or more semiconductor heterostructures to decrease the charge carrier recombination rate [7]. Based on these modifications different configurations have been developed to improve the performance of H_2 production, e.g., (1) PEC device which utilizes light-sensitive materials to fabricate both electrodes and drive water reduction and oxidation using solar light, and (2) production of hydrogen via water electrolysis using efficient and cheaper electrocatalyst where the necessary electric power to drive the water splitting is provided by solar cell [40]. The present chapter will focus mainly on the H_2 production from PEC water splitting using different materials including the basic aspect of PEC water splitting, present and future status PEC devices.

16.2.4 Solar Water Splitting with Semiconductors

As mentioned earlier, the main prerequisite of solar water splitting lies in effective absorption of photon and subsequent charge carrier separation, several important oxides (TiO_2, WO_3, Fe_2O_3, $BiVO_4$, ZnO, SnO_2, NiO, CdO, and PdO) [5,41,43–49] and nonoxides (Si, $CuGaSe_2$, Cu_2ZnSnS_4, Cu(Ga,In)(S,Se), CdS, PbSe, PbSGaN, Ge_3N_4, and TaON) [5,7,41,50–53] semiconductor materials as photo materials have been tested so far. Although all these materials were successfully utilized for water splitting reaction, only a few of them are promising in terms of good efficiency metrics due to requirement of interdependent coupled aspects (photon absorption, charge carrier separation, and surface electrochemistry) of PEC water splitting reaction. Along the line of oxide-based photo-materials, TiO_2 is the most explored photocatalyst for PEC water splitting reaction because of its appropriate band edges for HER and OER, low cost, stability, and nontoxicity [54–58]. However, theoretical STH efficiency of TiO_2 photocatalyst is only 2.2% as large band gap of TiO_2 (rutile ~3.0 eV) limits the solar photon absorption in UV region only [59]. In order to maximize the solar spectrum absorption, several attempts including elemental doping, dye-sensitization, semiconductor sensitization, and hydrogenation have been made [60–72]. In this section, we mainly focus on hydrogenation approach to reduce the band gap of TiO_2 which involves the formation of surface defects (i.e., oxygen vacancy/Ti^{+3} states) to improve the electronic conductivity and photo-responsive property [73,74]. These oxygen vacancies are responsible for formation of shallow donor states below the conduction band with relatively less formation energy and significantly improve the electronic property [75–79]. Hydrogenation of TiO_2 can be achieved by passing hydrogen gas in pressurized chamber with/without high temperature treatment. Synthesis of hydrogenated TiO_2 was first reported by Mao et al. which includes the hydrogen treatment of TiO_2 nanoparticles at 20 bar pressure and 200°C temperature for 5 days [80,81]. Synthesized material appears black in color and has a band gap of 1.0 eV which shows improved performance for methylene blue dye degradation. Additionally, black TiO$_2$ loaded with 0.6 wt% Pt shows improved hydrogen production rate for a test run of 22 h under solar light in a 1:1 methanol–water solution. Similar study has been performed by Zhao et al. where SrTiO$_3$ (STO) was reduced by aluminum reduction method (R-STO), which extends the absorption of R-STO from UV to visible and infrared region. Photocatalytically produced H$_2$ was measured and found to be approximately 2.5 times greater for R-STO as compared to STO due to introduction of oxygen vacancies into the lattice of STO (Figure 16.9a). Cyclic testing curve confirms the stability of R-STO sample up to 21 h (Figure 16.9b).

Liu et al. synthesized different samples of hydrogenated TiO$_2$ with varying degree of hydrogenation by varying the hydrogenation time under constant pressure of 35 bar and ambient temperature [82]. The color of hydrogenated sample changes from pale yellow to dark yellow to gray and ultimately to black. Figure 16.10a shows the digital photograph of hydrogenated samples for different time intervals. The

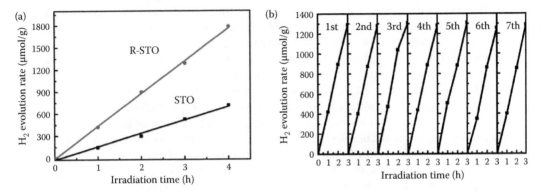

FIGURE 16.9 (a) Photocatalytic H$_2$ generation: STO vs. R-STO activity and (b) repetitive measurements of hydrogen production through direct photocatalytic water splitting using R-STO under simulated solar light. (Reprinted from Zhao, W. et al., *CrystEngComm*, 17, 7528–7534, 2015. With Permission. Copyright © 2015, Royal Society of Chemistry.)

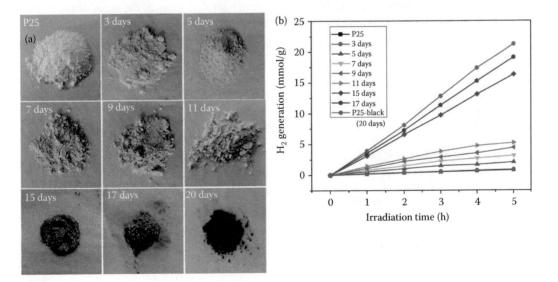

FIGURE 16.10 (a) Digital images of P25 for different hydrogen treatment duration and (b) measurement of H$_2$ generation from various samples in aqueous methanolelectrolyte. (Reprinted from Lu, H. et al., *RSC Adv.*, 4, 1128–1132, 2014. With Permission. Copyright © 2013, Royal Society of Chemistry.)

photochemical activity was analyzed using 20 vol% aqueous solution of methanol (as sacrificial agent) where black TiO_2 shows the highest rate of 3.94 mmol $g^{-1} h^{-1} H_2$ production (Figure 16.10b).

Yu et al. reported the synthesis of hydrogenated TiO_2 that involves the hydrogenation of white TiO_2 nanosheets under ambient pressure with varying time and temperature (500°C–700°C). The color of nanosheets changes from white to blue to gray and finally to black depending on the time of operation. Figure 16.11 demonstrates the color of different samples under varying time of operation and their corresponding UV spectra [83].

In addition to improved optical property, hydrogenated TiO_2 suppresses charge recombination center due to the presence of charge trapping sites and thereby increase the lifetime of holes in bulk TiO_2. High temperature hydrogen treatment is responsible for formation of oxygen vacancies, Ti^{3+} species and hydroxyl groups which ultimately improves the photoactivity of material [84]. Hence, it may be a potential material for photoelectrochemical water splitting in terms of photocatalytic activity as well as electrode stability. Li et al. reported first time the application of hydrogenated TiO_2 in PEC water splitting. Yellow color hydrogenated rutile nanotube was synthesized by hydrogen treatment of pristine TiO_2 nanotubes for 30 min at different temperatures (200°C–550°C) [85]. As shown in Figure 16.12, the photocurrent density is found to be maximum for sample treated at 350°C. Linear sweep studies show that photocurrent density of hydrogenated sample is almost double as compared to pristine TiO_2. Moreover, negative shift in saturation potential suggests better charge separation and transport. The synthesized material shows STH efficiency of 1.1%. The boosted efficiency is attributed to mid-gap states created in TiO_2 during hydrogen treatment, which expands absorption of TiO_2 in visible and NIR region due to transition from valance band to mid-gap state and mid-gap state to conduction band of TiO_2.

Recent studies demonstrated that the hydrogenation of TiO_2 improves its electronic conductivity and enhanced solar spectrum absorption due to controlled introduction of oxygen vacancies. The charge carrier recombination in black TiO_2 is also suppressed due to the presence of carrier trapping site. Hence, it does not only show enhanced photocatalytic activity, but also increases the stability of practical system. The insertion of oxygen vacancies has been extended to WO_3, Fe_2O_3, and ZnO [86]. However, hydrogenation technique may be further extended to materials like $SrTiO_3$ which has suitable band position for water splitting but suffers from large band gap.

In addition to hydrogenation of TiO_2, nonmetal element doping is anticipated to enhance solar spectrum region absorption. Reports are available for enhanced light absorption by the introduction of nonmetal X (X = B, N, C, S) element in crystal lattice of TiO_2 by decrementing electron–hole recombination centers. However, the absorption is not significantly enhanced possibly due to low level of elemental

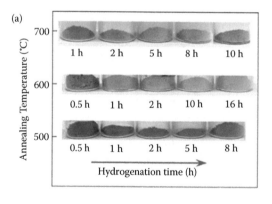
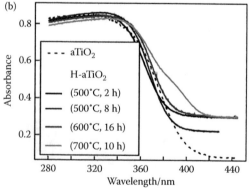

FIGURE 16.11 (a) Digital images of H-aTiO_2 treated at varying hydrogen treatment temperature and duration. (b) Comparison of UV–Vis DR spectra of various samples. (Reprinted from Yu, X. et al., *ACS Catal.*, 3, 2479–2486, 2013. With permission. Copyright © 2013, American Chemical Society.)

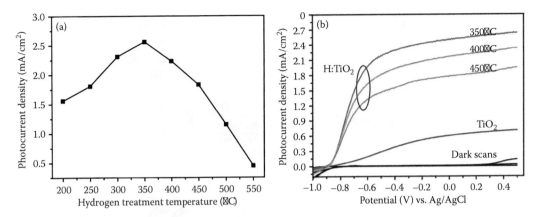

FIGURE 16.12 (a) Measurement of photocurrent density at 1.23V (vs. RHE) of H:TiO$_2$ nanowires treated at different temperatures and (b) linear sweeps voltammetry for various samples under dark and simulated solar light. (Reprinted from Wang, G.M. et al., *Nano Lett.*, 11, 3026–3033, 2011. With permission. Copyright © 2011, American Chemical Society.)

doping [87,88]. The level of elemental doping is proportional to the degree of amorphization at TiO$_2$ surface. The greater the thickness of amorphous layer, the better will be elemental doping in TiO$_2$ lattice [89]. The amorphous layer loses lattice order with generation of oxygen vacancies which assist X atom to accommodate. Further, the diffusion of X atom is slower in crystalline TiO$_2$ as compared to amorphous layer. Additionally, the replacement of more electronegative O atom with lesser electronegative X atom is thermodynamically not feasible.

Lin et al. reported much more thicker amorphous layer of TiO$_2$ by Al reduction method which is responsible for high level of nonmetal elemental doping of X (X = H, N, S, I) as compared to conventional H$_2$ treatment [89]. Schematic of doping has been illustrated in Figure 16.13a. The photocatalytic hydrogen generation of various samples (TiO$_2$–X) was measured for full solar spectrum (Figure 16.13b). It was found to be highest for TiO$_2$–N (15 mmol h^{-1} g^{-1}) which is mainly due to reduced recombination centers. Figure 16.13c demonstrates the photocatalytic H$_2$ production in visible spectrum which confirms the photoactivity of samples in visible region.

Among nonoxide semiconductors, silicon (Si) has been one of the superior choices for photovoltaic application for several decades. Si is one of the cheapest, earth abundant, and nontoxic semiconductor materials which absorbs wide range of solar spectrum due to its smaller band gap (c-Si ~1.1eV), high charge carrier mobility (electrons ~1500 cm^2/V/s and holes ~450 cm^2/V/s) with less charge carrier quenching [90,91]. These advantages make silicon a potential candidate for PEC water splitting. However, silicon is unsuitable for water splitting as valance band maximum (VBM) energy is not energetically high enough to participate in water oxidation. Moreover, silicon does not exhibit catalytic functionality for water splitting reaction. In contrast to transitional metals, silicon lacks reconfigurable electron structure to hold intermediate species. Further, the surface of silicon easily passivated by the formation of SiO$_2$ under PEC operation acts as an insulating layer and hence drops the performance of PEC device drastically. To overcome these limitations, Si is used as a secondary layer in tandem cell configuration where other semiconductor material having suitable band edges are deposited on silicon and exposed to electrolyte. One of the interesting properties of Si is its unique pH response. As most of materials follow Nernstian behavior for change in pH, i.e., flat band potential shifts 0.059 V per pH, silicon behaves differently depending upon the electrolyte. Modou et al. reported that the flat band of silicon shifts −0.03 V per pH in $Fe(CN)_6^{4-}$-based electrolyte [92]. Xhang et al. reported the shift of 0.04 V per pH in NaCl electrolyte [93]. These studies report that the difference in silicon valance band position and oxygen evolution level diminishes at higher pH which suggests

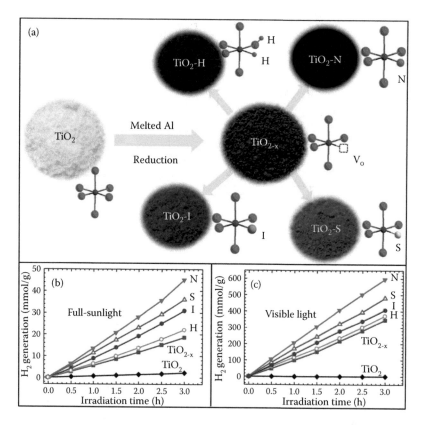

FIGURE 16.13 (a) Schematic representation of pristine titania to oxygen-deficient titania formation and subsequent doping of nonmetal element X (X = H, N, S, I), (b) photocatalytic H_2 generation using various samples under full solar spectrum, and (c) photocatalytic H_2 generation using various samples under visible spectrum. (Reprinted from Lin, T. et al., *Energy Environ. Sci.*, 7, 967–972, 2014. With permission. Copyright © 2013, Royal Society of Chemistry.)

better performance of Si-based electrode. c-Si has been utilized as a photoanode (e.g., Si/TiO$_2$, Si/ TiON, Si/ZnO, and Si/WO$_3$) as well as a photocathode (Si/graphene, Si/Pt, Si/Pd, Si/NiMo) for solar water splitting reaction [94–101]. In contrast to c-Si, hydrogenated amorphous silicon (a-Si:H) exhibits properties that are better suited toward PEC water splitting because of the following advantageous properties: (1) a-Si:H is a direct band gap material which allows thinner absorber layer and hence reduce the cost as compared to c-Si. (2) a-Si has a band gap of 1.7 eV, which is thermodynamically favorable for water splitting reaction [102]. Although the valance band edge (VBE) is a function of extent of hydrogenation, VBE may be tuned to facilitate OER in a hydrogenated sample. Hence, it can be used as front absorber material in tandem cell configuration [96,103–115]. (3) It can be easily scaled-up because its fabrication involves only one-step plasma-enhanced chemical vapor deposition (PECVD) [116,117], which has already been achieved without any difficulty in solar cell industries [118–120]. Further, in the line of amorphous silicon in PEC, multijunction amorphous silicon has gained considerable interest recently. The most common type in this category is a three-junction amorphous cell which can be visualized as thin film cells deposited on top of each other [121]. It can generate voltage up to 2.4 V which is suitable for water splitting reactions. The selection of solar photon absorber is based on "solar spectrum splitting" [122]. The band gap of a-Si is altered by alloying it with different proportion of germanium (Ge). Intrinsic semiconductor is sandwiched between n-type and p-type layer to better separate the electron and hole. All the three individual cells in n-i-p configuration interact together to form a three-junction amorphous cell (Figure 16.14). In such a cell, the top cell harnesses photonic energy corresponding to 300–690 nm wavelengths. Middle cell harnesses

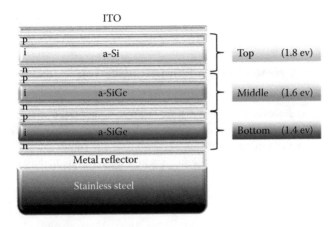

FIGURE 16.14 Schematic representation of a three-junction amorphous silicon solar cell.

in the range of 450–750 nm while the bottom cell harnesses the energy in the range of 550–880 nm [123]. There are three main advantages of such multijunction solar water splitting cells: (1) sufficient water splitting voltage output, (2) high efficiency, and (3) low cost as compared to crystalline multijunction semiconductor cells. Triple-junction amorphous silicon solar cells are exploited for making integrated photovoltaic-electrolysis devices in which the photocurrent generation equivalent to hydrogen generation can easily be measured.

Recently, Reece et al. reported a solar water splitting device, which consists of earth abundant materials such as OER and HER catalyst. The device operates near neutral environment, both with wired and wireless (artificial leaf) configurations [124]. Triple-junction amorphous silicon photovoltaic is interfaced to hydrogen evolving catalyst, made from an alloy of earth abundant cheap metals and oxygen evolving catalysts of cobalt borate (Co-OEC). Both catalysts were interfaced directly through electrodeposition with a commercial triple-junction amorphous silicon solar cell in wired and wireless configuration (Figure 16.15). Fabricated cell consists of three-junction amorphous silicon deposited on a stainless steel substrate and layered with a 70 nm thick ITO [124]. The unoptimized device performance gives 2.5% and 4.7% of solar to fuel efficiency for wireless and wired configuration when driven by 6.2% and 7.7% electricity-efficient solar cell in borate electrolyte of pH ~ 9.2. Wireless configuration was tested

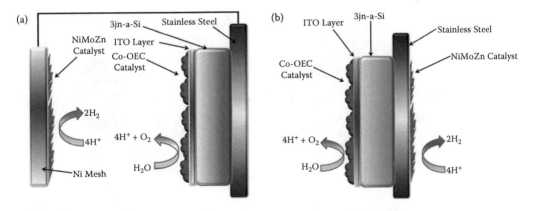

FIGURE 16.15 Schematic representation for (a) wired PEC configuration and (b) wireless PEC configuration. (Reproduced from Reece, S.Y. et al., *Science* 334, 645–648, 2011.)

against stability for 10 h which remains fairly stable. The stability of the cell depends on the material and the method of preparation of the conductive oxide layer [124].

16.2.5 Perovskite: A New Hope in Solar Application

Originally, Perovskite referred to calcium titanium oxide mineral species, having chemical formula of $CaTiO_3$. This mineral was discovered by Gustav Rose in 1839 and is named after the Russian mineralogist Lev Perovskite [125]. Soon after, the "perovskite" described any material having the same crystal structure as that of calcium titanium oxide. There are numerous compounds of the perovskite family found in nature where magnesium silicate perovskite ($MgSiO_3$) has highest abundance in earth's mantle. Chemically, perovskite is defined as formula AMX_3, where "A" and "M" are two different cations of varying sizes, and X is an anion which binds both cations. In an ideal cubic-symmetry structure, perovskite has sixfold coordinated M cation which is bounded by an octahedron of MX_6 and a 12-fold cuboctahedral coordinated A cation (Figure 16.16) [126]. The stability of cubic phase depends on rigidness of relative ion size. Hence, there is slight distortion created in lower-symmetry distorted versions, in which the coordination numbers of either "A" cations or "M" cations or both are reduced. Such distortions are responsible for alteration in many physical properties, e.g., electronic, dielectric, and magnetic properties, which can be harnessed for various applications.

The distortion in an ideal cubic symmetry of perovskite is defined by tolerance factor "t" which measures the deviation from ideal A–X and M–X bond length from ideal cubic structure. Tolerance factor "t" is defined as [126]

$$t = \frac{R_M + R_X}{\sqrt{2}(R_A + R_X)},$$

(16.5)

where R_M = ionic radii of M cation, R_A = ionic radii of A cation, and R_X = ionic radii of X anion. Recently, a set of organic–inorganic halide compounds ($CH_3NH_3PbX_3$, X = Cl, Br, or I) came into picture as an excellent source of light absorbing material. These materials have unique features such as their smaller band gaps, easier synthesis approach, better carrier mobility, and higher extinction coefficient [127–129]. These materials have been greatly utilized for solar cell fabrication since 2009 when Miyasaka et al. reported first perovskite-based solar cell with power conversion efficiency of 3.81% [130]. Nowadays, with great scientific efforts, efficiency has been achieved up to 20.1% [131]. In the era of rapid efficiency growth of such solar cells, their use is limited by stability of these perovskites which are sensitive to oxygen

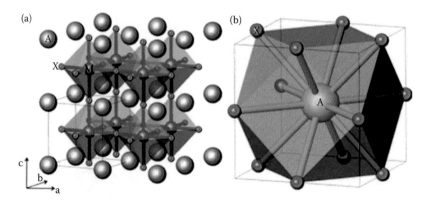

FIGURE 16.16 (a) Schematic of cubic perovskite with formula ABX_3 and (b) 12-fold cuboctahedral coordinated A cation. (Reprinted from Gao, P. et al., *Energy Environ. Sci.*, 7, 2448–2463, 2014. With permission. Copyright © 2014, Royal Society of Chemistry.)

and moisture [129]. Niu et al. reported that under the presence of oxygen, degradation of $CH_3NH_3PbI_3$ occurs due to hydrolysis of the material in presence of moisture under the following steps [132]:

$$CH_3NH_3PbI_3(s) \leftrightarrow PbI_2(s) + CH_3NH_3I \text{ (aq)} \tag{16.6}$$

$$CH_3NH_3I \text{ (aq)} \leftrightarrow CH_3NH_2(aq) + HI \text{ (aq)} \tag{16.7}$$

$$4HI \text{ (aq)} + O_2 \leftrightarrow 2I_2(s) + 2H_2O \text{ (l)} \tag{16.8}$$

$$2HI(aq) \leftrightarrow H_2(g) + I_2(g) \tag{16.9}$$

Reversible reaction 16.7 suggests the coexistence of CH_3NH_3I, CH_3NH_2, and HI under equilibrium. The HI molecule can either be oxidized by oxygen molecule (reaction step 16.8) which is a thermodynamically favorable reaction ($\Delta G = -481.058$ kJ mol⁻¹) or photodegrade under UV exposure (reaction step 16.9) [132]. In another study, Walsh et al. suggested the degradation of $CH_3NH_3PbI_3$ with combination of water molecule as a Lewis base to form $[(CH_3NH_3^+)_{n-1}(CH_3NH_2PbI_3][H_3O^+]$ by removal of one proton from ammonium [133]. Further, this intermediate decomposes to water dissolved HI and CH_3NH_3. It should be noted that the choice of aprotic organic cation, e.g., $(CH_3)_4N^+$, restrict the transfer of proton from ammonium which can make it resistant to moisture. To enhance the moisture stability, Xu et al. reported a new perovskite material $CH_3NH_3Pb(SCN)_2I$ by replacing two iodide ions by two pseudohalidethiocyanate ions in $CH_3NH_3PbI_3$ [134]. The choice of halide ion is an important factor for determining the stability of organic–inorganic halide perovskite as the first degradation step involves the formation of hydrated intermediate containing PbX_6^{4-} octahedral. Stability of the perovskite depends primarily on the binding tightness of halide molecule with lead ion represented as formation constant K. The formation constant for lead and thiocyanate ion is much greater (~7) as compared to that for lead and iodide ion (~3.5) which makes the interaction of thiocyanate with lead ion more intact. Additionally, the linear shape of SCN⁻ stabilizes the frame structure of $CH_3NH_3Pb(SCN)_2I$. The band gap of synthesized material was found to be 1.532 eV which is slightly greater than conventional $CH_3NH_3PbI_3$ (band gap ~1.504). Material was tested under 95% moisture level and found to be stable up to 4 h against $CH_3NH_3PbI_3$ which loses stability within 2.5 h under same conditions (Figure 16.17).

Recently, Karunadasa et al. synthesized a two-dimensional hybrid perovskite with optimized value of $n = 3$ member of series $(PEA)_2(MA)_{n-1}[Pb_nI_{3n+1}]$ (PEA: $C_6H_5(CH_2)_2NH_3^+$; MA: $CH_3NH_3^+$; n: number of Pb-I sheets in each inorganic layer) by mixing (PEA)I, (MA)I, and PbI_2 in 2:2:3 stoichiometric ratio in

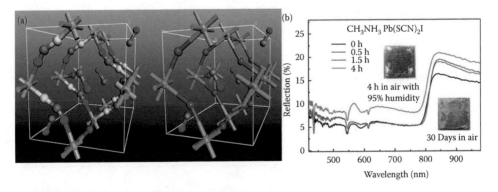

FIGURE 16.17 (a) Crystal structure of $CH_3NH_3Pb(SCN)_2I$ (left) and $CH_3NH_3PbI_3$ (right) perovskite. Gray = C, red = I, pink = Pb, blue = N, yellow = S and (b) UV–Vis spectra of $CH_3NH_3Pb(SCN)_2I$ perovskite for different exposure duration to 95% relative humidity. (Reprinted from Jiang, Q. et al., *Angew. Chem.*, 54, 11006–11006, 2014. With permission. Copyright© 2015 Wiley-VCH Verlag GmbH & Co. KGaA, Weinheim.)

nitromethane/acetone mixture [135]. The optimization of a number of layers was achieved on basis of exciton binding energy which is a difference of band gap and exciton absorption energy. For $n = 1$ and $n = 2$, exciton binding energy was estimated to be 0.22 and 0.17 eV, while for $n = 3$, it was estimated to be 0.04 eV indicating the lesser binding energy closer to MAPbI$_3$. It confirms that the synthesized material is a promising absorber material. Synthesized material was tested under 52% of humidity and found to be stable till 46 days which is confirmed by PXRD analysis. However, MAPbI$_3$ decomposes completely to form PbI$_2$ (Figure 16.18). Additionally, Snaith et al. reported the composite structure of P3HT/SWNT as a hole-transporting layer with resistance toward thermal and moisture degradation of perovskite [136]. They mentioned that hydrophobicity of PMMA (poly(methyl methacrylate)) inhibits the intrusion of moisture into perovskite structure and thus increases its moisture stability while hole transportation is achieved by P3HT assembled over highly conductive SWNTs. The fabricated solar cell based on this approach shows equal performance before and after heat treatment at 80°C for 96 h while cell fabricated with conventional hole transporting material (Li-TFSI doped spiro-OMeTAD (Li-spiro-OMeTAD), P3HT and PTAA) loses their activity after thermal stressing [136].

Yong et al. reports the enhanced stability of CH$_3$NH$_3$PbI$_3$ in solar cell application using surface passivation by hydrophobic coating of Teflon without compromising with light absorption spectrum [137]. This passivation layer acts as a physical barrier for interaction of water with CH$_3$NH$_3$PbI$_3$ as well as chemical barrier as being hydrophobic in nature. The contact angle of unpassivated system was measured to be 53° which increased to 118° upon surface passivation (Figure 16.19). The surface passivation enhances the stability till 30 days of ambient condition storage and 900 s of water immersion. The degradation of perovskite under water immersion was observed to occur at the edges which are inevitably exposed area. However, there was no degradation of perovskite material observed normal to the passivation plane.

In order to harness perovskite in PEC water splitting reaction, Zheng et al. developed multilayer CH$_3$NH$_3$PbI$_3$ photoanode coated with nickel passivation layer [138]. Photoanode was fabricated in a layered manner on which first CH$_3$NH$_3$PbI$_3$ was deposited on FTO using an anti-solvent technique followed by coating of MeOTAD as a hole-transporting material. On top of it, Au was coated followed by final layer

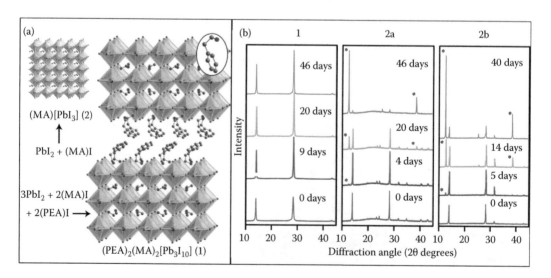

FIGURE 16.18 (a) Schematic representation of the 3D perovskite (MA)[PbI$_3$] and 2D perovskite (PEA)$_2$(MA)$_2$[Pb$_3$I$_{10}$]. Inset image shows a PEA cation and (b) PXRD patterns of various perovskites exposed to 52% relative humidity; 1: (PEA)$_2$(MA)$_2$[Pb$_3$I$_{10}$]; 2a:(MA)[PbI$_3$] formed using PbI$_2$; 2b:(MA)[PbI$_3$] formed using PbCl$_2$. (Reprinted from Smith, I.C. et al., *Angew. Chem.* 53: 11232–11235, 2014. With permission. Copyright© 2014 WILEY-VCH Verlag GmbH & Co. KGaA, Weinheim.)

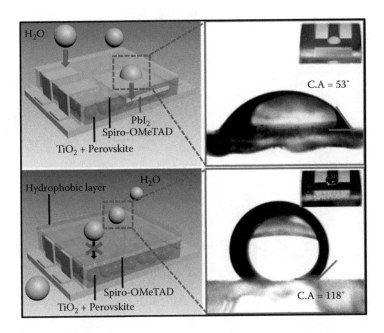

FIGURE 16.19 Non-passivated surface with water contact angle in the range of hydrophilicity (top); Teflon-passivated surface with water contact angle in the range of hydrophobicity (bottom); inset images show the response of surfaces exposed to water droplet, yellow color suggests the dissolution of perovskite in water. (Reprinted from Hwang, I. et al., *ACS Appl. Mater. Interfaces*, 7, 17330–17336, 2015. With permission. Copyright © 2015, American Chemical Society.)

of ultrathin Ni layer using a magnetron sputtering technique (Figure 16.20a). The materials were chosen according to their well-aligned band position shown in Figure 16.20b. The top layer of nickel serves as a physical barrier for electrolyte penetration to perovskite material as well as efficient water oxidation catalyst.

The photoelectrochemical activity of fabricated electrode was measured in a three-electrode cell with $0.1\,M$ Na_2S of pH = 12.8 using platinum wire as counter electrode and Ag/AgCl as reference electrode. The photocurrent of Ni-coated photoanode reaches up to $10\,mA/cm^2$ at $0\,V$ vs Ag/AgCl, whereas non-coated sample exhibit photocurrent $<2\,mA/cm^2$ (Figure 16.21a). Moreover, photocurrent measurement

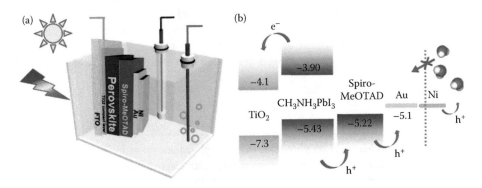

FIGURE 16.20 (a) Three-probe photoelectrochemical cell setup consisting of a Ni-coated perovskite photoanode with back illuminated and (b) schematic of band alignment of TiO_2, $CH_3NH_3PbI_3$, spiro-MeOTAD, Au, and Ni. (Reprinted from Da, P. et al., *Nano Lett.*, 15, 3452–3457, 2015. With permission. Copyright © 2015, American Chemical Society.)

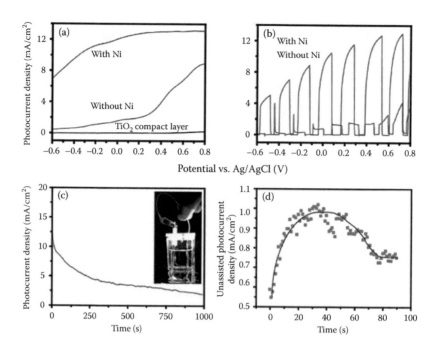

FIGURE 16.21 (a) Comparison of photocurrent densities of the compact TiO₂ layer (black) and CH₃NH₃PbI₃ photoanodes without (blue) and with Ni passivation (red), (b) photocurrent density curves for CH₃NH₃PbI₃ photoanodes (with vs. without Ni passivation) under chopped light, (c) stability analysis of Ni passivized multilayered CH₃NH₃PbI₃ photoanode under continuous back illumination (inset: Digital photograph of photocatalytic setup), and (d) unassisted photocurrent density measurement in two-probe setup using perovskite and Pt electrodes. (Reprinted from Da, P. et al., *Nano Lett.*, 15, 3452–3457, 2015. With permission. Copyright © 2015, American Chemical Society.)

under chopped light shows huge change in current density which confirms light-enabled PEC conversion (Figure 16.21b). The electrode retains its 56% of original activity till first 100 s and produces a photocurrent density of 2 mA/cm² till 15–20 min under continuous PEC measurement in alkaline electrolyte (Figure 16.21c). On the other hand, electrode fabricated without nickel passivation layer instantly degrades or peels off of the electrode surface. In another set of experiments, fabricated electrode was directly connected to Pt wire and tested for photocurrent by directly connecting a digital ampere meter. Under illumination, gas bubble evolution was observed without external bias, which confirms the capability of spontaneous hydrogen generation with photocurrent reaching up to 0.97 mA/cm² (Figure 16.21d).

16.3 Conclusion

The main focus of this chapter is on photoelectrochemical generation of hydrogen wherein the solar photons impinging on a semiconductor electrode generates an electrochemical potential bias sufficient to split water into hydrogen and oxygen. Due to benign nature and abundance of the feed (water) and energy source (sunlight), near-zero pollutants during the operation of the system, this approach to generating hydrogen has been a holy grail of chemical scientists and engineers. Despite extensive research for more than a few decades, significant challenges remain despite enormous depth in understanding the physicochemical principles governing this system as well as extensive exploration of suitable materials, all of which have been outlined in this chapter. While materials like functionalized titania and Si-based systems had shown great potential, none of them have been commercialized. It is hoped that the latest generation of perovskite-based materials will get us closer to the goal of solar hydrogen. In addition to the physicochemical principles, material design, and efficiency metrics, an outline of practical engineering design has also been provided.

Acknowledgment

The work was supported by the Department of Science and Technology via grants DST/TSG/SH/2011/106-G and DST/TM/SERI/2K11/79(G).

References

1. Behravesh, N. and S. Johnson, 2013. The global economy—Bright spots, mixed with worries. IHS Global Insight Global Executive Summary, 2.
2. Allison, I., N.L. Bindoff, R.A. Bindschadler, P.M. Cox, N. de Noblet, M.H. England, J.E. Francis, N. Gruber, A.M. Haywood, D.J. Karoly, G. Kaser, C. Le Quéré, T.M. Lenton, M.E. Mann, B.I. McNeil, A.J. Pitman, S. Rahmstorf, E. Rignot, H.J. Schellnhuber, S.H. Schneider, S.C. Sherwood, R.C.J. Somerville, K. Steffen, E.J. Steig, M. Visbeck, and A.J. Weave, 2009. *The Copenhagen Diagnosis, 2009: Updating the World on the Latest Climate Science.* Elsevier.
3. Schaeffer, M., T. Kram, M. Meinshausen, D.P. van Vuuren, and W.L. Hare, 2008. Near-linear cost increase to reduce climate-change risk. *Proc. Natl. Acad. Sci. USA* 105: 20621–20626.
4. Hansen, J., M. Sato, P. Kharecha, D. Beerling, R. Berner, V. Masson-Delmotte, M. Pagani, M. Raymo, D.L. Royer, and J.C. Zachos 2008. Target atmospheric CO_2: Where should humanity aim? *Open Atmos. Sci. J.* 2: 217–231.
5. Hisatomi, T., J. Kubota, and K. Domen, 2014. Recent advances in semiconductors for photocatalytic and photoelectrochemical water splitting. *Chem. Soc. Rev.* 43: 7520–7535.
6. Kudo, A. and Y. Miseki, 2009. Heterogeneous photocatalyst materials for water splitting. *Chem. Soc. Rev.* 38: 253–278.
7. Yuan, Y.-P., L.-W. Ruan, J. Barber, S.C.J. Loo, and C. Xue, 2014. Hetero-nanostructured suspended photocatalysts for solar-to-fuel conversion. *Energy Environ. Sci.* 7: 3934–3951.
8. Li, Z., W. Luo, M. Zhang, J. Feng, and Z. Zou, 2012. Photoelectrochemical cells for solar hydrogen production: Current state of promising photoelectrodes, methods to improve their properties, and outlook. *Energy Environ. Sci.* 6: 347–370.
9. Olah, G.A., A. Goeppert, and G.K.S. Prakash, 2008. Chemical recycling of carbon dioxide to methanol and dimethyl ether: From greenhouse gas to renewable, environmentally carbon neutral fuels and synthetic hydrocarbons. *J. Org. Chem.* 74: 487–498.
10. Leitner, W., 1995. Carbon dioxide as a raw material: The synthesis of formic acid and its derivatives from CO_2. *Angew. Chem. Int. Ed.* 34: 2207–2221.
11. Saito, M., 1998. R&D activities in Japan on methanol synthesis from CO_2 and H_2. *Catal. Surv. Asia* 2: 175–184.
12. Saito, M. and K. Murata, 2004. Development of high performance Cu/ZnO-based catalysts for methanol synthesis and the water-gas shift reaction. *Catal. Surv. Asia* 8: 285–294.
13. Saito, M., M. Takeuchi, T. Fujitani, J. Toyir, S. Luo, J. Wu, H. Mabuse, K. Ushikoshi, K. Mori, and T. Watanabe 2000. Advances in joint research between Nire and Rite for developing a novel technology for methanol synthesis from CO_2 and H_2. *Appl. Organomet. Chem.* 14: 763–772.
14. Luo, J., J.-H. Im, M.T. Mayer, M. Schreier, M.K. Nazeeruddin, N.G. Park, S.D. Tilley, H.J. Fan, and M. Grätzel 2014. Water photolysis at 12.3% efficiency via perovskite photovoltaics and earth-abundant catalysts. *Science* 345: 1593–1596.

15. Akkerman, I., M. Janssen, J. Rocha, and R.H. Wijffels, 2002. Photobiological hydrogen production: Photochemical efficiency and bioreactor design. *Int. J. Hydrogen Energy* 27: 1195–1208.
16. Chandrasekhar, K., Y.J. Lee, and D.W. Lee, 2015. Biohydrogen production: Strategies to improve process efficiency through microbial routes. *Int. J. Mol. Sci.* 16: 8266–8293.
17. Mersch, D., C.Y. Lee, J.Z. Zhang, K. Brinkert, J.C. Fontecilla-Camps, A.W. Rutherford, and E. Reisner, 2015. Wiring of photosystem II to hydrogenase for photoelectrochemical water splitting. *J. Am. Chem. Soc.* 137: 8541–8549.
18. Nath, K. and D. Das, 2004. Biohydrogen production as a potential energy resource—Present state-of-art. *J. Sci. Ind. Res.* 63: 729–738.
19. Nath, K., M.M. Najafpour, R.A. Voloshin , S.E. Balaghi, E. Tyystjärvi, R. Timilsina, J.J. Eaton-Rye, T. Tomo, H.G. Nam, H. Nishihara, and S. Ramakrishna, 2015. Photobiological hydrogen production and artificial photosynthesis for clean energy: From bio to nanotechnologies. *Photosynth. Res.* 126: 237–247.
20. Sekar, N. and R.P. Ramasamy, 2015. Recent advances in photosynthetic energy conversion. *J. Photochem. Photobiol. C* 22: 19–33.
21. Ghirardi, M.L., A. Dubini, J. Yu, and P.-C. Maness, 2009. Photobiological hydrogen-producing systems. *Chem. Soc. Rev.* 38: 52–61.
22. Rey, F.E., E.K. Heiniger, and C.S. Harwood, 2007. Redirection of metabolism for biological hydrogen production. *Appl. Environ. Microbiol.* 73: 1665–1671.
23. Bandyopadhyay, A., J. Stoeckel, H. Min, L.A. Sherman, and H.B. Pakrasi, 2010. High rates of photobiological H$_2$ production by a cyanobacterium under aerobic conditions. *Nat. Comm.* 1: 139.
24. Steinfeld, A., 2005. Solar thermochemical production of hydrogen: A review. *Solar Energy* 78: 603–615.
25. Tyner, C.E., G.J. Kolb, M. Geyer, and M. Romero. 2001. Concentrating solar power in 2001: An IEA/Solarpaces summary of present status and future prospects. SolarPACES.
26. Bilgen, E., 1984. Solar hydrogen-production by direct water decomposition process: A preliminary engineering assessment. *Int. J. Hydrogen Energy* 9: 53–58.
27. Fletcher, E.A., 1999. Solarthermal and solar quasi-electrolytic processing and separations: Zinc from zinc oxide as an example. *Ind. Eng. Chem. Res.* 38: 2275–2282.
28. Fletcher, E.A. and R.L. Moen, 1977. Hydrogen and oxygen from water. *Science* 197: 1050–1056.
29. Onuki, K., S. Kubo, A. Terada, N. Sakaba, and R. Hino, 2009. Thermochemical water-splitting cycle using iodine and sulfur. *Energy Environ. Sci.* 2: 491–497.
30. Steinfeld, A., P. Kuhn, A. Reller, R. Palumbo, J. Murray, and Y. Tamaura, 1998. Solar-processed metals as clean energy carriers and water-splitters. *Int. J. Hydrogen Energy* 23: 767–774.
31. Meier, A., V.A. Kirillov, G.G. Kuvshinov, Y.I. Mogilnykh, A. Reller, A. Steinfeld, and A. Weidenkaff, 1999. Solar thermal decomposition of hydrocarbons and carbon monoxide for the production of catalytic filamentous carbon. *Chem. Eng. Sci.* 54: 3341–3348.
32. Zedtwitz, P.V., J. Petrasch, D. Trommer, and A. Steinfeld, 2006. Hydrogen production via the solar thermal decarbonization of fossil fuels. *Solar Energy* 80: 1333–1337.
33. Buck, R., M. Abele, H. Bauer, A. Seitz, and R. Tamme, 1994. Development of a volumetric receiver-reactor for solar methane reforming. *J. Sol. Energy Eng* 116: 449.
34. Buck, R., J.F. Muir, R.E. Hogan, and R.D. Skocypec, 1991. Carbon-dioxide reforming of methane in a solar volumetric receiver reactor: The CAESAR project. *Sol. Energy Mater.* 24: 449–463.
35. Dahl, J.K., A.W. Weimer, A. Lewandowski, C. Bingham, F. Bruetsch, and A. Steinfeld, 2004. Dry reforming of methane using a solar-thermal aerosol flow reactor. *Ind. Eng. Chem. Res.* 43: 5489–5495.
36. Gokon, N., Y. Oku, H. Kaneko, and Y. Tamaura, 2002. Methane reforming with CO$_2$ in molten salt using FeO catalyst. *Solar Energy* 72: 243–250.

37. Piatkowski, N., C. Wieckert, A.W. Weimer, and A. Steinfeld, 2011. Solar-driven gasification of carbonaceous feedstock: A review. *Energy Environ. Sci.* 4: 73–82.

38. Von Zedtwitz, P. and A. Steinfeld, 2003. The solar thermal gasification of coal—Energy conversion efficiency and CO$_2$ mitigation potential. *Energy* 28: 441–456.

39. Walter, M.G., E.L. Warren, J.R. McKone, S.W. Boettcher, Q. Mi, E.A. Santori, and N.S. Lewis, 2010. Solar water splitting cells. *Chem. Soc. Rev.* 110: 6446–6473.

40. Wang, M., L. Chen, and L. Sun, 2012. Recent progress in electrochemical hydrogen production with earth-abundant metal complexes as catalysts. *Energy Environ. Sci.* 5: 6763–6778.

41. Maeda, K. and K. Domen, 2010. Photocatalytic water splitting: Recent progress and future challenges. *J. Phys. Chem. Lett.* 1: 2655–2661.

42. Fujishima, A. and K. Honda, 1972. Electrochemical photolysis of water at a semiconductor electrode. *Nature* 238: 37–38.

43. Chen, X.B., C. Li, M. Gratzel, R. Kostecki, and S.S. Mao, 2012. Nanomaterials for renewable energy production and storage. *Chem. Soc. Rev.* 41: 7909–7937.

44. Kubacka, A., M. Fernández-García, and G. Colón, 2012. Advanced nanoarchitectures for solar photocatalytic applications. *Chem. Soc. Rev.* 112: 1555–1614.

45. Mao, S.S., S.H. Shen, and L.J. Guo, 2012. Nanomaterials for renewable hydrogen production, storage and utilization. *Prog. Nat. Sci.* 22: 522–534.

46. Park, Y., K.J. McDonald, and K.S. Choi, 2012. Progress in bismuth vanadate photoanodes for use in solar water oxidation. *Chem. Soc. Rev.* 42: 2321–2337.

47. Tong, H., S.X. Ouyang, Y.P. Bi, N. Umezawa, M. Oshikiri, and J.H. Ye, 2012. Nano-photocatalytic materials: Possibilities and challenges. *Adv. Mater.* 24: 229–251.

48. Yerga, R.M.N., M.C.A. Galvan, F. del Valle, J.A.V. de la Mano, and J.L.G. Fierro, 2009. Water splitting on semiconductor catalysts under visible-light irradiation. *ChemSusChem* 2: 471–485.

49. Zhou, H.L., Y.Q. Qu, T. Zeid, and X.F. Duan, 2012. Towards highly efficient photocatalysts using semiconductor nanoarchitectures. *Energy Environ. Sci.* 5: 6732–6743.

50. Chun, W.-J., A. Ishikawa, H. Fujisawa, T. Takata, J.N. Kondo, M. Hara, M. Kawai, Y. Matsumoto, and K. Domen, 2003. Conduction and valence band positions of Ta$_2$O$_5$, TaON, and Ta$_3$N$_5$ by ups and electrochemical methods. *J. Phys. Chem. B* 107: 1798–1803.

51. Ma, G., T. Minegishi, D. Yokoyama, J. Kubota, and K. Domen, 2011. Photoelectrochemical hydrogen production on Cu$_2$ZnSnS$_4$/Mo-mesh thin-film electrodes prepared by electroplating. *Chem. Phys. Lett.* 501: 619–622.

52. Moriya, M., T. Minegishi, H. Kumagai, M. Katayama, J. Kubota, and K. Domen, 2013. Stable hydrogen evolution from CdS-modified CuGaSe$_2$ photoelectrode under visible-light irradiation. *J. Am. Chem. Soc.* 135: 3733–3735.

53. Yokoyama, D., T. Minegishi, K. Jimbo, T. Hisatomi, G. Ma, M. Katayama, J. Kubota, H. Katagiri, and K. Domen, 2010. H$_2$ evolution from water on modified Cu$_2$ZnSnS$_4$ photoelectrode under solar light. *Appl. Phys. Expr.* 3: 101202.

54. Lin, Y., S. Zhou, X. Liu, S. Sheehan, and D. Wang, 2009. TiO$_2$/TiSi$_2$ heterostructures for high-efficiency photoelectrochemical H$_2$O splitting. *J. Am. Chem. Soc.* 131: 2772–2773.

55. Liu, B. and E.S. Aydil, 2009. Growth of oriented single-crystalline rutile TiO$_2$ nanorods on transparent conducting substrates for dye-sensitized solar cells. *J. Am. Chem. Soc.* 131: 3985–3990.

56. Biswas, S., M.F. Hossain, and T. Takahashi, 2008. Fabrication of gratzel solar cell with TiO$_2$/CdS bilayered photoelectrode. *Thin Solid Films* 517: 1284–1288.

57. Bak, T., J. Nowotny, M. Rekas, and C.C. Sorrell, 2002. Photo-electrochemical properties of the TiO$_2$-Pt system in aqueous solutions. *Int. J. Hydrogen Energy* 27: 19–26.

58. Linsebigler, A.L., G.Q. Lu, and J.T. Yates, 1995. Photocatalysis on TiO$_2$ surfaces—Principles, mechanisms, and selected results. *Chem. Soc. Rev.* 95: 735–758.

59. Murphy, A.B., P.R.F. Barnes, L.K. Randeniya, I.C. Plumb, I.E. Grey, M.D. Horne, and J.A. Glasscock, 2006. Efficiency of solar water splitting using semiconductor electrodes. *Int. J. Hydrogen Energy* 31: 1999–2017.

60. Chen, X.B., S.H. Shen, L.J. Guo, and S.S. Mao, 2010. Semiconductor-based photocatalytic hydrogen generation. *Chem. Soc. Rev.* 110: 6503–6570.

61. Du, H., X. Xie, Q. Zhu et al., 2015. Metallic MoO_2 cocatalyst significantly enhances visible-light photocatalytic hydrogen production over $MoO_2/Zn_{0.5}Cd_{0.5}S$ heterojunction. *Nanoscale* 7: 5752–5759.

62. Gasparotto, A., D. Barreca, D. Bekermann, A. Devi, R.A. Fischer, P. Fornasiero, V. Gombac, O.I. Lebedev, C. Maccato, T. Montini, and G. Van Tendeloo, 2011. F-doped Co_3O_4 photocatalysts for sustainable H_2 generation from water/ethanol. *J. Am. Chem. Soc.* 133: 19362–19365.

63. Jitta, R.R., R. Gundeboina, N.K. Veldurthi, R. Guje, and V. Muga, 2015. Defect pyrochlore oxides: As photocatalyst materials for environmental and energy applications: A review. *J. Chem. Technol. Biotechnol.* 90: 1937–1948.

64. Kim, M., Y.K. Kim, S.K. Lim, S. Kim, and S.-I. Ina, 2015. Efficient visible light-induced H_2 production by Au@ CdS/TiO$_2$ nanofibers: Synergistic effect of core-shell structured Au@ CdS and densely packed TiO$_2$ nanoparticles .*Appl. Catal. B Environ.* 166: 423–431.

65. Li, G., Y. Wang, and L. Mao, 2014. Recent progress in highly efficient Ag-based visible-light photocatalysts. *RSC Adv.* 4: 53649–53661.

66. Mohamed, A.E.R. and S. Rohani, 2011. Modified TiO$_2$ nanotube arrays (TNTAs): Progressive strategies towards visible light responsive photoanode: A review. *Energy Environ. Sci.*, 4: 1065–1086.

67. Ni, M., M.K.H. Leung, D.Y.C. Leung, and K. Sumathy, 2007. A review and recent developments in photocatalytic water-splitting using TiO$_2$ for hydrogen production. *Renew. Sust. Energ. Rev.* 11: 401–425.

68. Shen, M., A. Han, X. Wang, Y.G. Ro, A. Kargar, Y. Lin, H. Guo, P. Du, J. Jiang, J. Zhang, and S.A. Dayeh, 2015. Atomic scale analysis of the enhanced electro- and photo-catalytic activity in high-index faceted porous NiO nanowires. *Sci. Rep.* 5: 1–6.

69. Wang, P., P. Chen, A. Kostka, R. Marschall, and M. Wark, 2013. Control of phase coexistence in calcium tantalate composite photocatalysts for highly efficient hydrogen production. *Chem. Mater.* 25: 4739–4745.

70. Wang, X., J. Chen, X. Guan, and L. Guo, 2015. Enhanced efficiency and stability for visible light driven water splitting hydrogen production over $Cd_{0.5}Zn_{0.5}S/g$-C_3N_4 composite photocatalyst. *Int. J. Hydrogen Energy* 40: 7546–7552.

71. Xu, Y. and R. Xu, 2015. Nickel-based cocatalysts for photocatalytic hydrogen production. *Appl. Surf. Sci.* 351: 779–793.

72. Youngblood, W.J., S.H.A. Lee, K. Maeda, and T.E. Mallouk, 2009. Visible light water splitting using dye-sensitized oxide semiconductors. *Acc. Chem. Res.* 42: 1966–1973.

73. Wang, T., Z. Luo, C. Li, and J. Gong, 2014. Controllable fabrication of nanostructured materials for photoelectrochemical water splitting via atomic layer deposition. *Chem. Soc. Rev.* 43: 7469–7484.

74. Abdi, F.F., L. Han, A.H.M. Smets, M. Zeman, B. Dam, and R. van de Krol, 2013. Efficient solar water splitting by enhanced charge separation in a bismuth vanadate-silicon tandem photoelectrode. *Nat. Commun.* 4: 1–7.

75. Ai, L., 2014. Microwave-assisted synthesis of silver nanocrystals in benzyl alcohol and their subsequent in situ chemical transformation into Ag–AgCl nanohybrids for plasmonic photocatalysis. *Appl. Phys. A* 116: 589–595.

76. Linic, S., P. Christopher, and D.B. Ingram, 2011. Plasmonic-metal nanostructures for efficient conversion of solar to chemical energy. *Nat. Mater.* 10: 911–921.

77. Zhang, X., Y. Liu, and Z. Kang, 2014. 3D branched ZnO nanowire arrays decorated with plasmonic Au nanoparticles for high-performance photoelectrochemical water splitting. *ACS Appl. Mater. Interfaces* 6: 4480–4489.

78. Qu, Y.Q. and X.F. Duan, 2013. Progress, challenge and perspective of heterogeneous photocatalysts. *Chem. Soc. Rev.* 42: 2568–2580.

79. Zhu, H. and T. Lian, 2012. Wave function engineering in quantum confined semiconductor nanoheterostructures for efficient charge separation and solar energy conversion. *Energy Environ. Sci.* 5: 9406–9418.

80. Chen, X., L. Liu, P.Y. Yu, and S.S. Mao, 2011. Increasing solar absorption for photocatalysis with black hydrogenated titanium dioxide nanocrystals. *Science* 331: 746–750.

81. Zhao, W., W. Zhao, G. Zhu, T. Lin, F. Xu, and F. Huang, 2015. Black strontium titanate nanocrystals of enhanced solar absorption for photocatalysis. *CrystEngComm* 17: 7528–7534.

82. Lu, H., B. Zhao, R. Pan et al., 2014. Safe and facile hydrogenation of commercial degussa P25 at room temperature with enhanced photocatalytic activity. *RSC Adv.* 4: 1128–1132.

83. Yu, X., B. Kim, and Y.K. Kim, 2013. Highly enhanced photoactivity of anatase TiO₂ nanocrystals by controlled hydrogenation-induced surface defects. *ACS Catal.* 3: 2479–2486.

84. Harris, L.A. and R. Schumacher, 1980. The influence of preparation on semiconducting rutile (TiO₂). *J. Electrochem. Soc.* 127: 1186–1188.

85. Wang, G.M., H.Y. Wang, Y.C. Ling et al., 2011. Hydrogen-treated TiO₂ nanowire arrays for photoelectrochemical water splitting. *Nano Lett.* 11: 3026–3033.

86. Wang, G.M., Y.C. Ling, and Y. Li, 2012. Oxygen-deficient metal oxide nanostructures for photoelectrochemical water oxidation and other applications. *Nanoscale* 4: 6682–6691.

87. Yang, C., Z. Wang, T. Lin et al., 2013. Core-shell nanostructured "Black" Rutile titania as excellent catalyst for hydrogen production enhanced by sulfur doping. *J. Am. Chem. Soc.* 135: 17831–17838.

88. Di Valentin, C., G. Pacchioni, and A. Selloni, 2009. Reduced and n-type doped TiO₂: Nature of Ti³⁺ species. *J. Phys. Chem. C* 113: 20543–20552.

89. Lin, T., C. Yang, Z. Wang, H. Yin, X. Lü, F. Huang, J. Lin, X. Xie, and M. Jiang, 2014. Effective nonmetal incorporation in black titania with enhanced solar energy utilization. *Energy Environ. Sci.* 7: 967–972.

90. Boettcher, S.W., E.L. Warren, M.C. Putnam, E.A. Santori, D. Turner-Evans, M.D. Kelzenberg, M.G. Walter, J.R. McKone, B.S. Brunschwig, H.A. Atwater, and N.S. Lewis, 2011. Photoelectrochemical hydrogen evolution using Si microwire arrays. *J. Am. Chem. Soc.* 133: 1216–1219.

91. Razeghi, M. 2009. *Fundamentals of Solid State Engineering.* Springer.

92. Madou, M.J., B.H. Loo, K.W. Frese, and S.R. Morrison, 1981. Bulk and surface characterization of the silicon electrode. *Surf. Sci.* 108: 135–152.

93. Zhang, X.G. 2001. *Electrochemistry of Silicon and Its Oxide.* Springer.

94. Betty, C.A., R. Sasikala, O.D. Jayakumar, T. Sakuntala, and A.K. Tyagi, 2011. Photoelectrochemical properties of porous silicon based novel photoelectrodes. *Prog. Photovoltaics* 19: 266–274.

95. Hwang, Y., A. Boukai, and P. Yang, 2009. High density n-Si/n-TiO₂ core/shell nanowire arrays with enhanced photoactivity. *Nano Lett.* 9: 410–415.

96. Lombardi, I., S. Marchionna, G. Zangari, and S. Pizzini, 2007. Effect of Pt particle size and distribution on photoelectrochemical hydrogen evolution by p-Si photocathodes. *Langmuir* 23: 12413–12420.

97. Sun, K., Y. Jing, C. Li et al., 2012. 3D branched nanowire heterojunction photoelectrodes for high-efficiency solar water splitting and H₂ generation. *Nanoscale* 4: 1515–1521.

98. Wang, W., S. Chen, P.-X. Yang, C.-G. Duan, and L.-W. Wang, 2013. Si:WO₃ heterostructure for z-scheme water splitting: An ab initio study. *J. Mater. Chem. A* 1: 1078–1085.

99. Warren, E.L., S.W. Boettcher, J.R. McKone, and N.S. Lewis, 2010. Photoelectrochemical water splitting: Silicon photocathodes for hydrogen evolution. *Inter. Soc. Optics Photo.* 7770: 1–7.

100. Wu, K., W. Quan, H. Yu, H. Zhao, and S. Chen, 2011. Graphene/silicon photoelectrode with high and stable photoelectrochemical response in aqueous solution. *Appl. Surf. Sci.* 257: 7714–7718.

101. Yu, H., S. Chen, X. Quan, H. Zhao, and Y. Zhang, 2009. Silicon nanowire/TiO₂ heterojunction arrays for effective photoelectrocatalysis under simulated solar light irradiation. *Appl. Catal. B-Environ.* 90: 242–248.

102. Hu, S., C. Xiang, S. Haussener, A.D. Berger, and N.S. Lewis, 2013. An analysis of the optimal band gaps of light absorbers in integrated tandem photoelectrochemical water-splitting systems. *Energy Environ. Sci.* 6: 2984–2993.

103. Adler, D., 1987. Electronic-structure of hydrogenated amorphous-silicon. *J. Non-Cryst. Solids* 90: 77–89.

104. Belkouch, S., L. Paquin, A. Deneuville, and E. Gheeraert, 1991. Physicochemical and electronic characterization of the structure platinum + hydrogenated amorphous-silicon + monocrystalline silicon. *Can. J. Phys.* 69: 357–360.

105. Clare, B.W., P.J. Jennings, J.C.L. Cornish, G. Talukder, C.P. Lund, and G.T. Hefter, 1993. Simulation of the electronic and vibrational structure of hydrogenated amorphous-silicon using cluster-models. *J. Comput. Chem.* 14: 1423–1428.

106. Didinchuk, V.A. and A.G. Petukhov, 1993. Calculation of the electronic-structure of hydrogenated amorphous-silicon. *Ukrainskii Fizicheskii Zhurnal* 38: 1797–1802.

107. Holender, J.M., G.J. Morgan, and R. Jones, 1993. Model of hydrogenated amorphous-silicon and its electronic-structure. *Phys. Rev. B* 47: 3991–3994.

108. Kleider, J.P., C. Longeaud, and F. Dayoub, 1999. Electronic properties of bottom gate silicon nitride/hydrogenated amorphous silicon structures. *Thin Solid Films* 337: 208–212.

109. Pollard, W., 1994. Electronic-structure of phosphorus in doped amorphous hydrogenated silicon. *J. Non-Cryst. Solids* 175: 145–154.

110. Senemaud, C. and I. Ardelean, 1990. Electronic-structure of hydrogenated amorphous-silicon germanium alloys studies by x-ray photoelectron-spectroscopy and soft-x-ray spectroscopy. *J. Phys.* 2: 8741–8750.

111. Srinivasan, G., 1990. Thermal equilibration in electronic-structure of hydrogenated amorphous-silicon. *Mater. Sci. Eng., B* 6: 247–255.

112. Srinivasan, G. and A.S. Nigavekar, 1991. Role of dopants in the electronic-structure of hydrogenated amorphous-silicon. *Mater. Sci. Eng., B* 8: 23–37.

113. Korte, L., R. Roessler, and C. Pettenkofer, 2014. Direct determination of the band offset in atomic layer deposited ZnO/hydrogenated amorphous silicon heterojunctions from x-ray photoelectron spectroscopy valence band spectra. *J. Appl. Phys.* 115: 203715-1–203715-7.

114. Vonroedern, B., L. Ley, and M. Cardona, 1977. Photoelectron-spectra of hydrogenated amorphous silicon. *Phys. Rev. Lett.* 39: 1576–1580.

115. Schulze, T.F., L. Korte, F. Ruske, and B. Rech, 2011. Band lineup in amorphous/crystalline silicon heterojunctions and the impact of hydrogen microstructure and topological disorder. *Phys. Rev. B* 83: 165314-1–135314-11.

116. Han, L., I.A. Digdaya, T.W.F. Buijs, F.F. Abdi, Z. Huang, R. Liu, B. Dam, M. Zeman, W.A. Smith, and A.H. Smets, 2015. Gradient dopant profiling and spectral utilization of monolithic thin-film silicon photoelectrochemical tandem devices for solar water splitting. *J. Mater. Chem. A* 3: 4155–4162.

117. Digdaya, I.A., L. Han, T.W.F. Buijs, M. Zeman, B. Dam, A.H. Smets, and W.A. Smith, 2015. Extracting large photovoltages from a-SiC photocathodes with an amorphous TiO₂ front surface field layer for solar hydrogen evolution. *Energy Environ. Sci.* 8: 1585–1593.

118. Guha, S., J. Yang, A. Banerjee, T. Glatfelter, K. Hoffman, S.R. Ovshinsky, M. Izu, H.C. Ovshinsky, and X. Deng, 1994. Amorphous silicon alloy photovoltaic technology—From R&D to production. *MRS Proceedings*, 645.

119. Yang, J., A. Banerjee, and S. Guha, 2003. Amorphous silicon based photovoltaics—From earth to the "Final frontier." *Sol. Energy Mater. Sol. Cells* 78: 597–612.

120. Guha, S., J. Yang, and B. Yan, 2013. High efficiency multi-junction thin film silicon cells incorporating nanocrystalline silicon. *Sol. Energy Mater. Sol. Cells* 119: 1–11.

121. Deng, X.M., X.B. Liao, S.J. Han, H. Povolny, and P. Agarwal, 2000. Amorphous silicon and silicon germanium materials for high-efficiency triple-junction solar cells. *Sol. Energy Mater. Sol. Cells* 62: 89–95.

122. Stern, T.G. and D.M. Peterson, Inherent spectrum-splitting photovoltaic concentrator system. 1982, Google Patents.

123. Kelly, N.A. and T.L. Gibson, 2006. Design and characterization of a robust photoelectrochemical device to generate hydrogen using solar water splitting. *Int. J. Hydrogen Energy* 31: 1658–1673.

124. Reece, S.Y., J.A. Hamel, K. Sung, T.D. Jarvi, A.J. Esswein, J.J. Pijpers, and D.G. Nocera, 2011. Wireless solar water splitting using silicon-based semiconductors and earth-abundant catalysts. *Science* 334: 645–648.

125. De Graef, M. and M.E. McHenry. *Structure of Materials: An Introduction to Crystallography, Diffraction and Symmetry.* Cambridge University Press.

126. Gao, P., M. Gratzel, and M.K. Nazeeruddin, 2014. Organohalide lead perovskites for photovoltaic applications. *Energy Environ. Sci.* 7: 2448–2463.

127. Graetzel, M., 2014. The light and shade of perovskite solar cells. *Nat. Mater.* 13: 838–842.

128. Xing, G., N. Mathews, S. Sun, S.S. Lim, Y.M. Lam, M. Grätzel, S. Mhaisalkar, and T.C. Sum, 2013. Long-range balanced electron- and hole-transport lengths in organic-inorganic $CH_3NH_3PbI_3$. *Science* 342: 344–347.

129. Park, N.-G., 2013. Organometal perovskite light absorbers toward a 20% efficiency low-cost solid-state mesoscopic solar cell. *J. Phys. Chem. Lett.* 4: 2423–2429.

130. Kojima, A., K. Teshima, Y. Shirai, and T. Miyasaka, 2009. Organometal halide perovskites as visible-light sensitizers for photovoltaic cells. *J. Am. Chem. Soc.* 131: 6050–6051.

131. Zhou, H., Q. Chen, G. Li, S. Luo, T.B. Song, H.S. Duan, Z. Hong, J. You, Y. Liu, and Y. Yang, 2014. Interface engineering of highly efficient perovskite solar cells. *Science* 345: 542–546.

132. Niu, G., W. Li, F. Meng, L. Wang, H. Dong, and Y. Qiu, 2013. Study on the stability of $CH_3NH_3PbI_3$ films and the effect of post-modification by aluminum oxide in all-solid-state hybrid solar cells. *J. Mater. Chem. A* 2: 705–710.

133. Frost, J.M., K.T. Butler, F. Brivio, C.H. Hendon, M. van Schilfgaarde, and A. Walsh, 2014. Atomistic origins of high-performance in hybrid halide perovskite solar cells. *Nano Lett.* 14: 2584–2590.

134. Jiang, Q., D. Rebollar, J. Gong, E.L. Piacentino, C. Zheng, and T. Xu, 2014. Pseudohalide-induced moisture-tolerance in perovskite $CH_3NH_3Pb(SCN)$ 2I thin films. *Angew. Chem.* 54: 11006–11006.

135. Smith, I.C., E.T. Hoke, D. Solis-Ibarra, M.D. McGehee, and H.I. Karunadasa, 2014. A layered hybrid perovskite solar-cell absorber with enhanced moisture stability. *Angew. Chem.* 53: 11232–11235.

136. Habisreutinger, S.N., T. Leijtens, G.E. Eperon, S.D. Stranks, R.J. Nicholas, and H.J. Snaith, 2014. Carbon nanotube/polymer composites as a highly stable hole collection layer in perovskite solar cells. *Nano Lett.* 14: 5561–5568.

137. Hwang, I., I. Jeong, J. Lee, M.J. Ko, and K. Yong, 2015. Enhancing stability of perovskite solar cells to moisture by the facile hydrophobic passivation. *ACS Appl. Mater. Interfaces* 7: 17330–17336.

138. Da, P., M. Cha, L. Sun, Y. Wu, Z.-S. Wang, and G. Zheng, 2015. High-performance perovskite photoanode enabled by Ni passivation and catalysis. *Nano Lett.* 15: 3452–3457.

Design and Operating Strategy Innovations for Energy-Efficient Process Operation

Ojasvi and
Nitin Kaistha
*Indian Institute of
Technology Kanpur*

17.1 Introduction

The process industry is an integral part of modern society, manufacturing a vast array of low-margin, high-volume, bulk chemicals such as refinery fuels and petrochemicals. These are typically produced in chemical plants that continuously process appropriate feedstocks to the respective value-added chemicals. For economic reasons of material/energy efficiency, material/energy recycle, also referred to as process integration, has traditionally been employed in the chemical process industry. In more recent decades, long-term sustainability concerns reflected in increasingly stringent product quality and environment discharge norms as well as carbon emissions/energy efficiency targets have pushed the industry toward higher levels of process integration. The objective of the integration is to eliminate waste/side product discharge and minimize energy consumption per kg product. Material/energy integration (recycle) is thus the bedrock of designing sustainable, zero waste discharge, and highly energy-efficient continuous chemical processes.

On the flip side, the multiple energy/material recycle loops introduce positive feedback into the process causing high system nonlinearity. Several recycle paths between the interconnected units also provide multiple paths over which "local" unit-specific disturbances can propagate, potentially disturbing, and in extreme cases, destabilizing the entire plant. Material/energy recycle thus introduces unique control issues that must be effectively addressed by the plantwide control system. This is a prerequisite for realizing the sustainability/economic benefit of highly integrated chemical processes.

Clearly, the effective plantwide control system design is predicated on a thorough and proper understanding of the unique control issues due to recycle. Only such an understanding would allow the proper utilization of the available control degrees of freedom (DOF) in designing an effective plantwide control system. It may also reveal the need for process redesign, e.g., provision for additional control DOF, to ensure robust process operability.

What exactly are the implications of material/energy recycle on plant control [1]? In this chapter, through some very simple examples, we illustrate the control issues introduced by material/energy recycle and develop very simple commonsense guidelines to address them. We then report quantitative case-study results done previously in our research group to demonstrate the significant economic/sustainability benefit of the application of these guidelines. Even as the reported case studies are from the process industry, the developed control system guidelines are readily applicable to the food processing industry as well as water treatment plants for improved energy conservation as well as fresh water utilization. The chapter ends with the customary conclusions that may be drawn from the work.

17.2 An Energy Recycle Example

Consider a low conversion (say 30%) exothermic adiabatic reactor and a feed effluent heat exchanger (FEHE), as shown in Figure 17.1. The hot reactor effluent preheats the cold feed in the FEHE. The preheated feed may be further heated to the desired reactor temperature in the heater. The reactor effluent, postcooling in the FEHE, is further cooled, liquefied, and discharged from a drum. The FEHE creates an energy recycle loop in the process.

For argument's sake, let us say that the process conditions and characteristics are such that the hot reactor effluent can provide all the heat needed (Q_{Tot}) to raise the temperature of the cold feed to the desired reactor inlet temperature. The designer can then fix of FEHE area to extract the necessary heat (Q_{Tot}) and at the steady-state design conditions, the heater duty (Q_{Htr}) is zero. In other words, the steady process operation requires no external heat with the reaction exothermic heat being sufficient to heat the cold feed to the desired reactor inlet temperature. The design thus maximizes the steady-state process energy efficiency. We refer to this flow sheet as Design 1.

Now, let us bring the dynamic operation into the picture. As shown in Figure 17.2a, the feed rate (F; throughput) is flow controlled, the preheater duty valve is fully closed ($Q_{Htr}=0$), the surge drum pressure is controlled by adjusting the cooler duty and the drum liquid level is controlled by the drum outflow. Now, if the cold feed inlet temperature (T_F) rises, the reactor inlet temperature also (T_R) rise. Since the feed to the reactor is hotter, the reaction proceeds faster and more reaction heat gets released per kg throughput. The reactor outlet temperature (T_E) thus increases, which in turn causes T_R to rise further. The self-fuelling cycle pushes all temperatures in the recycle loop to rise. The extent of the temperature rise would be such that all the reactants in the feed get fully converted to products. With $Q_{Htr}=0$ and F fixed, the system thus exhibits a temperature runaway with all temperatures inside the energy recycle loop rising significantly.

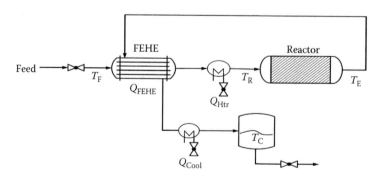

FIGURE 17.1   Reactor-FEHE energy recycle process.

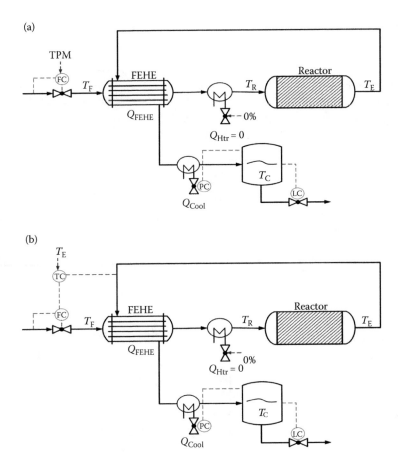

FIGURE 17.2 Control of the energy recycle process. (a) No temperature control inside energy recycle loop. (b) Temperature controlled by adjusting fresh feed.

One may also refer to the above phenomenon as temperature snowballing, where a small increase in the inlet enthalpy in the recycle loop causes a very large increase in the overall temperature inside the energy recycle loop. Clearly, temperature snowballing is unacceptable. The large temperature rise would likely destroy the reactor catalyst. Also the cooler duty valve would saturate at fully open, and the consequent rise in pressure would cause a pressure relief valve to rupture.

If the process is already designed and existing, then the only control solution to prevent the temperature runaway is to manipulate F to hold the temperature inside the energy recycle loop constant. Let us say the T_E–F pairing is used, as shown in Figure 17.2b, then as T_E rises, the cold feed rate, F, is increased. This causes T_R to decrease. Further, the increased F reduces the reactor residence time so that its single pass conversion drops. The reaction heat released per kg throughput then drops. This coupled with the lower T_R brings the rising T_E back down. Conversely, as T_E decreases, F is decreased causing T_E to rise again due to a hotter reactor feed and more reaction heat release per kg throughput. This operating strategy ensures energy-efficient operation ($Q_{Htr} = 0$); however, the throughput is now dependent and cannot be set independently, e.g., by management assessed market demand.

In most operating scenarios, the ability to independently set the process throughput is essential. To provide this flexibility, a process redesign is called for. The simplest option is to redesign the FEHE to have smaller area so that the FEHE provides only say 70% of the total cold feed preheating duty with the feed heater chipping in the remaining 30%. We refer to this as Design 2, as shown in Figure 17.3a. Q_{Htr} can then

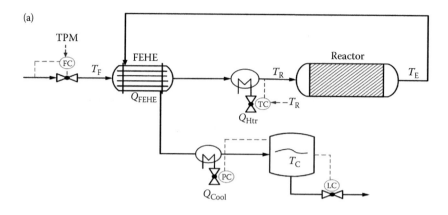

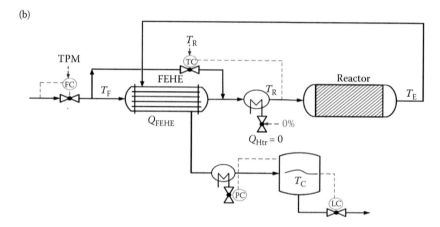

FIGURE 17.3 Reactor-FEHE process design alternative and their control. (a) Small FEHE. (b) Cold side bypass around FEHE.

be manipulated to hold T_R with F being set independently. In this case, we have a smaller FEHE, higher steady-state energy consumption per kg throughput and the throughput being set independently.

A more useful design modification is to provide, say a 20% cold side bypass around the FEHE. We refer to this as Design 3, as shown in Figure 17.3b [2]. Again, for the sake of argument, assume hot and cold stream temperature levels are such that the FEHE with bypass can provide Q_{Tot} without violating the minimum approach temperature constraint. In this case again, $Q_{Htr} = 0$ and the steady process operation is energy efficient. However, since the cold side temperature rise must be higher to achieve T_R postmixing with the cold bypass, the average temperature difference across the FEHE is lower so that the area must be larger than the corresponding FEHE in Design 1. In Design 3, the cold side bypass can be manipulated to hold T_R constant (Figure 17.3b). This loop prevents temperature snowballing. We thus have a more expensive FEHE, zero external energy consumption with the throughput being set independently.

This simple energy recycle example brings out some of the unique control issues attributable to recycle. The positive feedback due to energy recycle in the FEHE coupled with the exponential dependence of reaction rates on temperature creates the nonlinear phenomenon of temperature snowballing. The control solution to mitigate temperature snowballing is to hold a temperature inside the energy recycle loop. In this example, by holding T_E constant, the amount of energy being recycled in the FEHE is not allowed to blow up (snowball). One of the most fundamental questions in control system design is "what to control." This example suggests that in a recycle loop, a process variable that helps to directly regulate the recycle rate (material or energy) should be controlled.

The related next question is "what to manipulate." Since we wish to keep the recycle rate fixed (or within a small band via indirect control), the only reasonable option is to manipulate control DOF that influence what is going in or out of the recycle loop. Thus, if T_F increases (disturbance entering the recycle loop), then by manipulating either F (Figure 17.2b) or Q_{Htr} (Figure 17.3a) or the FEHE bypass rate (Figure 17.3b) to hold T_E, the energy balance disturbance is transformed out of the recycle loop.

These simple qualitative arguments on the very simple process suggest the following control system design guideline for a recycle system:

> Maintain the material/energy recycle rate (or inventory) by manipulating what's going in or out of the recycle loop.

Implicit in the above is the assumption that the design of the process provides at least one control degree of freedom to manipulate either an inflow or an outflow from the recycle loop. If not, a process redesign becomes necessary to provide the control degree of freedom.

17.3 Single Feed Reactor–Separator–Material Recycle Example

Let us now consider the simple chemical plant in Figure 17.4. The irreversible isomerization reaction $A \rightarrow B$ occurs in a continuous stirred tank reactor (CSTR), operating at a given fixed temperature. The reactor effluent is separated in a distillation column into nearly pure B bottoms product with near pure A being recovered up the top as the distillate and recycled back to the reactor. Usually, if the reactant–product separation is not difficult, the optimum design uses a low single pass conversion reactor to reduce the cost of the expensive reactor (including catalyst). For such a typical design, the recycle rate (R) is then noticeably larger than the fresh feed rate (F). Luyben [3] considered such a low conversion design and showed that at a fixed reactor level and temperature, a small change in F causes a disproportionately larger change in R. This was again referred to as snowballing in the material recycle loop.

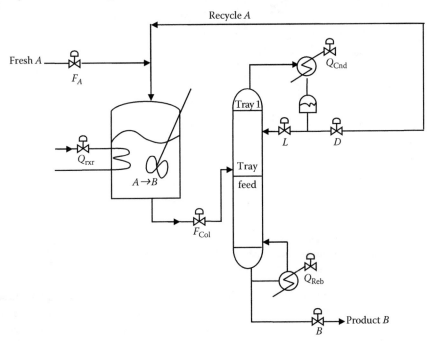

FIGURE 17.4 Single feed reactor–separator–recycle process.

(a)

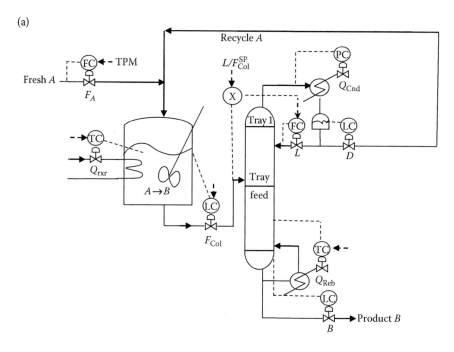

(b)

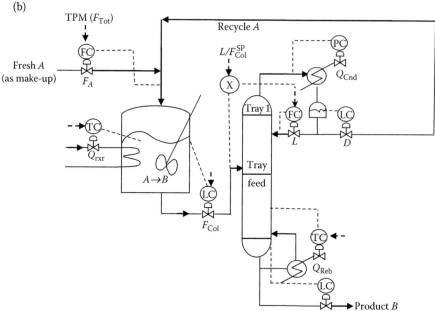

FIGURE 17.5 Alternative control structures for a single feed reactor–separator–recycle process. (a) Fresh feed as TPM (floating recycle). (b) Flow inside recycle loop as TPM (floating fresh feed).

Consider the control system in Figure 17.5a. The fresh feed rate is flow controlled and its set point determines the process throughput. In other words, the fresh feed flow rate acts as the throughput manipulator (TPM). Downstream, the reactor level is controlled by adjusting the reactor outflow. The reactor temperature is controlled by adjusting the reactor cooling duty. On the column, the reflux drum and sump levels are controlled by manipulating the distillate and bottoms flows, respectively, and the condenser pressure is regulated by adjusting condenser duty. The reflux rate is maintained in ratio with the column feed and the reboiler duty is manipulated to maintain the A impurity in the bottoms product. In this control structure, a small step perturbation in the fresh feed rate would result in a large swing in the recycle rate due to material snowballing. The perturbation may be deliberate, e.g., operator increasing the TPM set point to increase production; or an unmeasured disturbance such as a change in the fresh feed composition and a bias in the fresh feed flow sensor. The latter can be particularly confounding to operators with recycle rates exhibiting large swings for no apparent reason.

The problem with these large swings in the recycle rate is that all equipment in the material recycle loop must handle the large change in flow rate. Given that the turndown ratio of some of the unit operations such as distillation columns is not very large, one or more equipment in the recycle loop may hit maximum/minimum flow capacity constraints during the swings. A common example is a distillation column hitting its flooding/weeping limit requiring cumbersome operator intervention to bring it back to its normal hydraulic flow regime. Clearly, the control system with the TPM at a fresh feed makes smooth and steady process operation difficult due to material snowballing.

The problem is that the control system is designed such that the fresh feed rate (TPM) is independent so that the recycle rate must float to the appropriate value to fully consume the fresh A feed, and thus close the overall plant material balance. If we assume that the product B is pure, then by overall material balance, all fresh A leaves only as product B. At steady state, the fresh A rate must then equal the reactor B generation rate, i.e.,

$$F = kx_A V,$$

where k is the reaction rate constant, x_A is the reactor A molfraction, and V is the reactor volume. If F increases, x_A must increase in direct proportion. However, since $R \gg F$ due to the low conversion design, the recycle rate increases necessarily to effect the change in x_A, is disproportionately higher, which is the snowball effect.

Instead of fixing the fresh feed rate and letting the recycle rate float to close the overall plant material balance, we may fix a flow in the recycle loop and let the fresh feed float. Figure 17.5b shows one such control structure alternative that is often employed in the industry. The total (recycle + fresh) flow to the reactor (F_{rxr}) is held constant by manipulating the fresh feed. The fresh feed is thus fed as a make-up stream. In this control structure, the set point F_{rxr}^{SP} is the TPM. By tightly holding a flow inside the recycle loop, unpredictable swings in the recycle loop are avoided. This allows the operator to confidently steer the process to the desired steady state. In particular, one can steer the process to operate closer to the bottleneck equipment capacity constraint inside the recycle loop that limits the maximum throughput. One thus gets the higher maximum production from the same plant. We highlight that this alternative control structure follows our suggested guideline of "holding the recycle rate by manipulating what's going in or out of the recycle loop." We also note that other control structures that follow the above guideline can be devised (not shown).

17.4 Double Feed Reactor–Separator–Material Recycle Example

Consider a simple process for producing C via the irreversible addition of A to B. For simplicity, assume no side reactions. The overall process structure (Figure 17.6) then takes in fresh A and fresh B streams and outputs a C product stream with internal A and B component recycle streams. The steady-state overall component material balance dictates that

$$F_A = P(\hat{x}_C + \hat{x}_A)$$

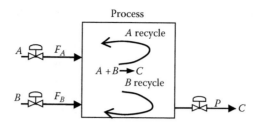

FIGURE 17.6 Input–output structure of an ideal $A + B \to C$ process.

$$F_B = P(\hat{x}_C + \hat{x}_B),$$

where the x terms denote product stream mol fractions.

Let us say that a conventional regulatory control system uses F_B as the TPM. If no constraints are imposed on $\hat{x}_A$ and $\hat{x}_B$, then even if F_A is independently specified, the overall reactant component balances would close with any unreacted A and B that is not recycled being taken out with the C product stream. If, however, a stringent product quality constraint requiring A and B impurity levels in the C product stream to be negligibly small is imposed (i.e., $\hat{x}_A \to 0$; $\hat{x}_B \to 0$; $\hat{x}_C \to 1$), then the overall component material balance requires that

$$F_A = F_B = P$$

In other words, if F_B is independently fixed (TPM), F_A cannot be set independently and the control system must ensure that F_A exactly matches F_B, else the overall component material balance cannot close without violating the product quality constraint. The imposition of a stringent product quality constraint thus implies all A and B fed to the process leave only as product C with no (negligible) leakage of A or B from the process. The two fresh feeds must then be exactly balanced as per the reaction stoichiometry (1 mol of fresh A for every mol of fresh B). Any imbalance implies that the reactant fed in excess necessarily builds up in the recycle loop. The recycle component inventories (or recycle component flow rates) thus behave as integrators with respect to any stoichiometric imbalance due to the stringent product quality constraint.

Assuming that F_B is fixed (TPM), the stoichiometric feed balancing requires that F_A be adjusted to maintain the A recycle inventory. This feedback adjustment usually is necessarily slow as the imbalance must pass through all the units in the forward path before its effect shows up in the recycle rate. Material integration thus introduces a subtle slow drifting mode in the recycle rate and hence plantwide dynamics.

Consider a particular flowsheet for our hypothetical $A + B \to C$ process, as in Figure 17.7 [4]. The process consists a cooled CSTR followed by a distillation column. Heavy C drops down the column and light A and B, recovered as distillate, being recycled back to the reactor. Two alternative control structures for this process are shown in Figure 17.8. The two control structures differ in the location of the TPM.

In Figure 17.8a, the TPM is on the fresh B feed, F_B, with F_A moving in ratio with F_B for stoichiometric feed balancing. Since flow measurements are never exact and the slightest excess of any reactant would necessarily build up in the recycle loop, the F_A/F_B ratio is adjusted to hold a reactant A composition in the reactor (x_A). This composition loop ensures exact stoichiometric feed balancing and thus closure of the steady-state overall plant component balances. The other control loops are similar to the single feed process studied previously and therefore not explained in detail. In Figure 17.8b control structure, the total flow to the reactor (F_{rxr}) is the TPM. F_B is manipulated to hold F_{rxr}

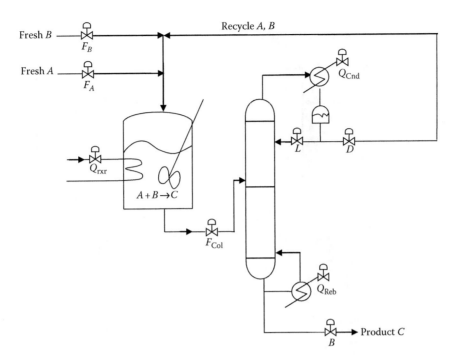

FIGURE 17.7 Double feed $A + B \rightarrow C$ reactor–separator recycle process.

with F_A moving in ratio with F_B. A reactor A composition controller adjusts the F_A/F_B ratio set point to ensure perfect stoichiometric feed balancing. The other loops are the same control structure, as shown in Figure 17.8a.

Even as both the control structures are consistent in that all independent material energy balances are closed by the control system, the control structure in Figure 17.8a is susceptible to snowballing. This is because one of the fresh feeds, F_B, is independently set. The action of the reactor composition loop ensures the other feed is exactly stoichiometrically balanced, i.e., $F_A = F_B$. Assuming pure C product, the recycle rate must then float such that the product C generation rate in the reactor equals F_B, i.e.,

$$F_B = k x_A x_B V.$$

Since k, V, and x_A are constant, x_A being held by the composition controller, similar to the previous example, if $R \gg F$ (i.e., small reactor), a small increase in F_B would require a large change in R to proportionately increase x_B so that the above equation holds for closure of the overall material balance. The control structure thus makes the plant susceptible to recycle loop snowballing.

In Figure 17.8b control structure, on the other hand, both the fresh feeds are fed as makeup streams. Holding F_{rxr} constant ensures the total flow going around the recycle loop is well regulated and does not swing unpredictably, while holding x_A constant ensures stoichiometric feed balancing. One may also think of these two loops as manipulating F_A and F_B to indirectly hold the component flow rate of A and B, respectively, going around the recycle loop. The recycle component flow rates are thus indirectly held while the production rate floats. Large unpredictable swings in the recycle rates due to the snowball effect are thus prevented. Note that this control structure again follows our suggested guideline of "maintain material/energy recycle rate by manipulating what's going in or out the recycle loop." The material recycle loop recycles A and B (two component recycles). The respective inlet rates into the recycle loop indirectly hold the A and B recycle flows.

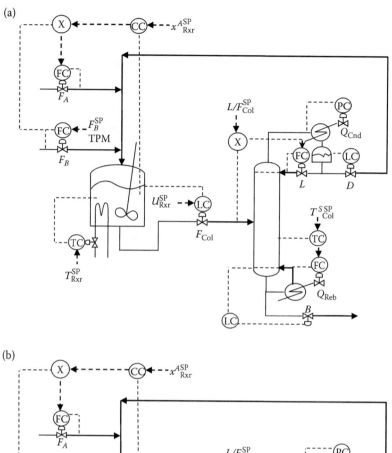

FIGURE 17.8 Alternative control structures for single feed reactor–separator–recycle process. (a) Fresh feed as TPM (floating recycle). (b) Flow inside recycle loop as TPM (floating fresh feed).

17.5 Considerations in Economic Process Operation

The preceding discussion suggests that the guideline of regulating the material/energy recycle rate by manipulating what is going in or out the loop provides an effective way of addressing the specific unique control issues due to recycle. With the recycle loops properly managed, "local" conventional level/pressure/temperature loops can be employed to close unit-specific material/energy balances. The regulatory control system thus effectively closes all the "local" and plantwide balances to drive the process to the steady state determined by the specific values of the regulatory control loop set points.

The next question then is what set point values should be provided to the regulatory controllers. These set point values determine the independent inventory amounts held in the plant and the consequent steady state at which it operates. For example, the reactor level set point determines the single pass reactor conversion and the consequent process steady state. Clearly, the inventory levels should be chosen for an economic objective such as minimizing the steady-energy consumption or waste generation or, alternatively, maximizing the production rate.

Typically, at the optimum steady state, multiple hard/soft constraints are active. Common active constraint examples include process operation at the minimum product quality guarantee to the customer, reactor operation at maximum holdup, hard equipment capacity constraints such as column operation at its flooding limit, etc. We should therefore adjust appropriate regulatory layer set points so as to drive the process operation as close as possible to the optimum steady state without violating any of the hard active constraints. This is illustrated in Figure 17.9 contrasting tight versus loose active constraint control with corresponding large versus small back-off from the hard active constraint limit. Of the remaining unconstrained regulatory set points, some would exhibit a sharp optimum with respect to the chosen set point value whilst others would be largely flat. To mitigate economic loss, it is imperative that the set point value for the former is regularly updated to maintain it near optimum. We now discuss approaches for tight active constraint control and managing the economically important unconstrained regulatory set points.

17.5.1 TPM Location Selection for Tight Dominant Active Constraint Control

One of the key decisions in designing the regulatory control system is the choice of the TPM. It dictates an outwardly radiating material balance control structure [5]. This is illustrated in Figure 17.10 for a four-units-in-series process. The orientation of the level (inventory) controllers upstream of the TPM is in the reverse direction of flow while the downstream level controllers are oriented in the direction of process flow. In this way, the upstream units supply the set flow while the downstream units process the set flow.

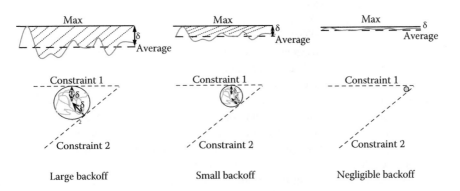

FIGURE 17.9   Tight control mitigates back-off from hard optimally active constraints.

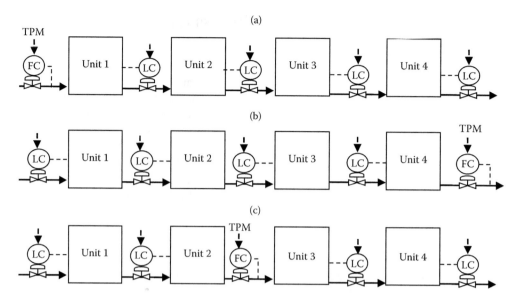

FIGURE 17.10 Outwardly radiating inventory control orientation from the TPM.

It is useful to qualitatively analyze the propagation of transients due to material imbalances for the process. Clearly, flow transients are absent on the unit feed that is the TPM. Due to disturbances in the various units, variability in the flows increases as one moves away from the TPM. Let us say Unit 3 is the bottleneck unit that first hits its maximum process limit as the throughput is ramped up. Now in case the fresh feed is used as the TPM, the average feed rate that Unit 3 processes is necessarily less than the maximum possible as sufficient back-off is necessary to prevent violation of the Unit 3 maximum processing limit. There is thus an unrecoverable production loss. On the other hand, if the TPM is located at the feed to Unit 3, there is negligible flow variability in the feed going to Unit 3 so that the necessary back-off is zero.

Typically, since recycle rates tend to increase disproportionately in response to an increase in the input flow into the recycle loop due to the snowball effect, some equipment in the recycle loop reaches its maximum capacity limit as production is ramped up. Usually, a back-off from this capacity limit has the highest economic penalty compared to the penalty due to back-off in the other active constraints. Clearly, by locating the TPM at the active bottleneck constraint, the back-off and hence economic penalty can be largely eliminated.

We note that the TPM can be located anywhere inside a material recycle loop and the material balance controllers upstream the TPM till the process fresh feed would ensure propagation of flow transients out of the recycle loop. Thus, by locating the TPM at the bottleneck constraint inside the material recycle loop and developing a consistent regulatory control system around the TPM, we achieve both tight active constraint control as well as mitigation of the snowball effect.

We highlight that of all the active constraints at the economic optimum, most are soft so that short-term deviations beyond the constraint limit is acceptable. Back-off is then not an issue. It is the hard constraints that require back-off since the constraint limit cannot be violated. Typically, these hard constraints are related to equipment capacity constraints beyond which damage to the equipment is likely. Usually, only the bottleneck equipment capacity constraint inside the recycle loop is the hard one. Also, for a given plant, the bottleneck equipment capacity constraint usually remains the same, i.e., the same particular constraint limits maximum production. Locating the TPM at the hard active constraint inside a material recycle loop thus achieves both back-off mitigation (good economics) and robust recycle management.

17.5.2 Managing Important Unconstrained Regulatory Set Points

Of the unconstrained regulatory layer set points, we need to ensure that the implemented set point value of those with a sharp optimum is close to optimum to avoid a large economic penalty. This may be achieved by driving the gradient of the economic objective with respect to the unconstrained set point to zero. Alternatively, the unconstrained regulatory set point is adjusted to control a self-optimizing process variable (SOPV). An SOPV, by definition, ensures near optimum operation regardless of process disturbances [6]. Designing an SOPV corresponding to an unconstrained regulatory set point can however be a nontrivial task.

17.6 Summary of Suggested Guidelines

The preceding long-winding discussion is summarized as a set of brief guidelines:

1. For robust control of recycle loops, a recycle rate inside the material/energy recycle loop should be held constant by manipulating either an inflow or outflow from the recycle loop.
2. For tight control of the bottleneck constraint inside a material recycle loop to maximize production, the TPM should be located at the bottleneck constraint with consistent outwardly radiating inventory control loops around the TPM.
3. An appropriate control strategy should be implemented for ensuring economically important unconstrained regulatory loop set points that remain close to their optimum. Alternatives include directly driving the gradient of the economic objective to zero or the indirect method of controlling a well-designed SOPV.

Example Case Studies

We are now ready to demonstrate the economic/sustainability benefit of the application of the above guidelines on two example processes. The first example is the hypothetical $A + B \rightarrow C$ process, while the second example is a ternary Petlyuk Column.

Example I

This example illustrates the economic impact of the TPM location for the hypothetical $A + B \rightarrow C$ process of Figure 17.7. For reasons of brevity, modeling details are omitted and the reader is referred to the article by Kanodia and Kaistha [7].As production is increased, flooding in the column limits maximum production. This is modeled as the column boil up, V, being constrained below V^{MAX} (bottleneck constraint). We evaluated five control structures, CS0–CS4, for their maximum achievable throughput. The control structures are shown in Figure 17.11 and differ from each other in the location of the TPM. In CS0, the column boil up set point V^{SP} is the TPM. The reboiler duty is manipulated to give extremely tight control of V so that the back-off from V^{MAX} is negligible. In CS1, the column feed is the TPM. In CS2, the feed to the reactor is the TPM. In CS3, the fresh A plus recycle rate is the TPM, while in CS4 the recycle stream is the TPM. Thus, the TPM location moves away from the bottleneck constraint in order from CS0 to CS4.

To systematically quantify the back-off in the alternative control structures, we considered a step increase in the B mol fraction in the fresh A feed (F_A) as the primary disturbance. For no disturbance, the TPM set point in all control structures can be chosen such that the steady column boilup is exactly V^{MAX}. Thus, for no disturbance, all control structures give the same maximum achievable throughput. However, as the magnitude of the disturbance is increased, swings occur in V for CS1–CS4 as the column feed composition and flow change during the ensuing transient. The TPM set point is then adjusted such that during the transient, V just touches V^{MAX} from below and does not violate the V^{MAX} constraint. Clearly, as the disturbance

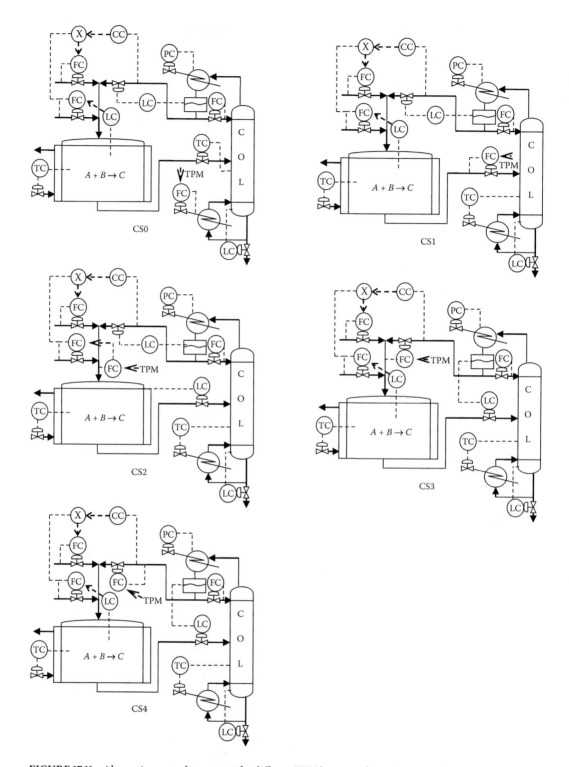

FIGURE 17.11 Alternative control structures for different TPM locations for $A + B \rightarrow C$ process.

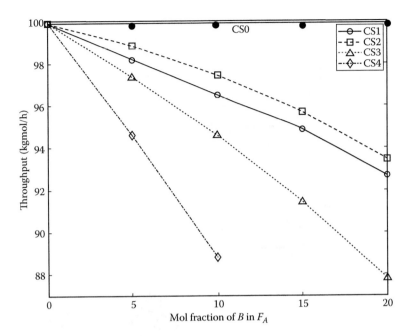

FIGURE 17.12   Variation in throughput derating due to back-off for alternative control structure.

magnitude increases, the TPM set point back-off would increase. The magnitude of the increase in TPM back-off however varies dramatically between the alternative control structures.

Figure 17.12 plots the variation in throughput (F_A) with the magnitude of the step disturbance for the alternative control structures. Expectedly, as the TPM location moves away from the bottleneck, the throughput derating necessary to avoid violating the V^{MAX} constraint for a given disturbance magnitude increases. For CS0 (TPM at bottleneck constraint), the throughput derating remains negligible regardless of the disturbance magnitude. For the worst case disturbance of a 20 mol% F_A composition change, the maximum achievable throughput using CS0 is more than 6% higher than the next best control structure. In other words, we get a significant 6% extra throughput merely via appropriate control structure design in light of the bottleneck constraint. This simple example illustrates the significant economic impact of TPM location choice in actual plant operation.

Example II

This second example is a ternary Petlyuk column, [8,9], separating benzene–toluene–xylene into its constituent near pure components and is based on a recent article by Kumari et al. [10]. The Petlyuk column, as shown in Figure 17.13, performs the easy benzene–xylene split in the prefractionator to mitigate the remixing of the intermediate boiler (toluene), compared to a conventional light-out-first or heavy-out-first two-column sequence. This reduces the inherent irreversibility of the process so that for the same product purities, the Petlyuk column design is significantly more energy efficient (literature claims 10%–40% savings).

The column has five steady-state operation DOF. Of the five DOF, three get used up to ensure on-target purity of the distillate $\left(x_{Bz}^D\right)$, side draw $\left(x_{Tol}^S\right)$, and bottoms $\left(x_{Xy}^B\right)$ streams. We then have two unconstrained DOF. Since, xylene is the principal side-draw impurity, we choose x_{Xy}^S as the fourth DOF leaving the prefractionator vapor rate and V_P as the fifth DOF. In the base-case operating conditions in Figure 17.13, V_P and x_{Xy}^S are adjusted to minimize Q_R. Figure 17.14 shows the variation in Q_R with respect to V_P for different values of x_{Xy}^S. Clearly, optimum is sharp with respect to both V_P and x_{Xy}^S.

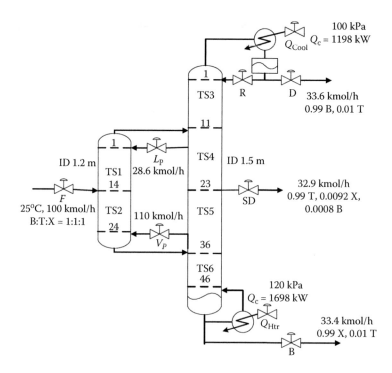

FIGURE 17.13 Ternary Petlyuk column with salient design and operating condition.

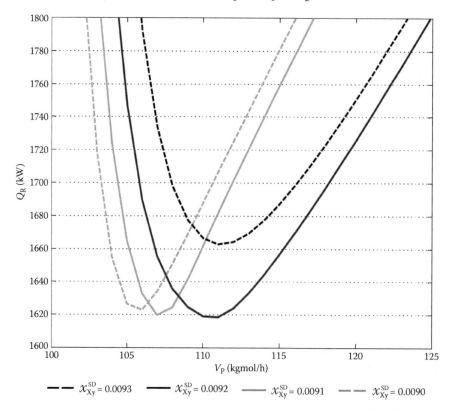

FIGURE 17.14 Variation in reboiler duty with unconstrained DOFs, V_P, and different values of x_{Xy}^{SD}.

For an operating process with on-target product purity for the three product streams, we would have x_{Bz}^D maintained by manipulating reflux (L), x_{Tol}^S maintained by manipulating side-draw rate (S), x_{Xy}^B maintained by manipulating reboiler duty (Q_R). With V_p and x_{Xy}^S as the other two DOFs, we have V_p under flow control and x_{Xy}^S regulated by adjusting the prefractionator rectifying section temperature controller set point (T_P^{SP}). The temperature controller manipulates the liquid to the prefractionator (L_p). The control structure thus obtained, CS^{CC}, is shown in Figure 17.15a.

Given that the optimum with respect to V_p and x_{Xy}^S is sharp, we expect large energy suboptimality for constant set point operation in the face of disturbances. This was indeed found to be true with the suboptimality being particularly severe for feed composition changes in particular directions. The base case feed is equimolar. For a ±6 mol% change in a feed component mol fraction (other two components remain equimolar) as the nominal disturbance, Q_R suboptimality for a benzene-rich, toluene-lean, or xylene-lean feed for constant set point operation was found to be particularly severe at >8%.

A deeper systematic investigation into column behavior reveals that the suboptimality manifests itself as noticeably larger curvature in either the prefractionator rectification section temperature profile or the tray section below the side-draw. To ensure near optimal operation, it is therefore suggested that V_p be adjusted to hold the prefractionator rectification section temperature profile curvature while the middle section temperature profile curvature is maintained by adjusting T_P^{SP}. The omitted technical details of the curvature variable are available in the article by Kumari et al. Suffice to say that the two temperature profile curvatures act as SOPVs, which ensures near optimal values of V_p and x_{Xy}^S. The control structure obtained, CS^{EC}, is shown in Figure 17.15b.

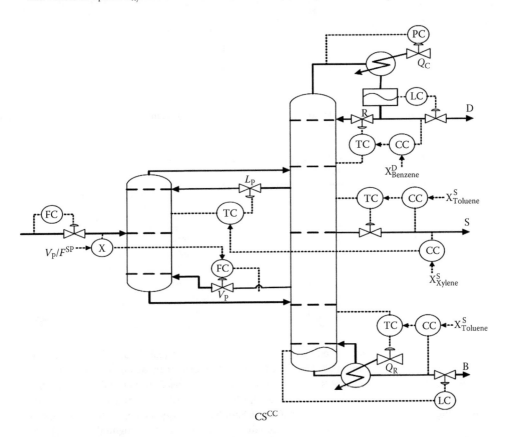

CS^{CC}

FIGURE 17.15 Petlyuk column control structures. (a) CS^{CC} for an on-target product purity operation (no curvature control).

(Continued)

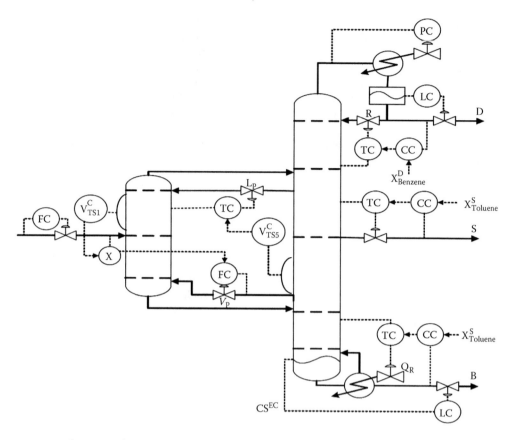

FIGURE 17.15 (Continued) (b) CS^{EC} for an on-target product purity with curvature control for energy-efficient operation.

TABLE 17.1 Steady-State Q_R Suboptimality Using CS^{CC} and CS^{EC} for Nominal Feed Composition Changes

Disturbance (mol%)	CS^{CC}		CS^{EC}	
	Q_R	% Suboptimality	Q_R	% Suboptimality
B 39	1859	16.2	1652	3.3
B 27	1740	3.0	1690	0.0
T 39	1736	1.9	1712	0.5
T 27	1758	11.5	1608	2.0
X 39	1651	2.2	1650	2.2
X 27	1798	8.1	1735	4.3

Table 17.1 quantitatively compares the steady-state Q_R suboptimality for column operation using CS^{CC} and CS^{EC} for the six different feed composition disturbances. Significant mitigation in the degree of suboptimality is accomplished when CS^{EC} is used. In particular, where CS^{CC} consumes 16.2% more reboiler duty than Q_R^{MIN}, CS^{EC} reduces the suboptimality to only 3.3% implying >10% energy savings compared to CS^{CC}. Significant energy savings for the toluene-lean or xylene-lean feed is also evident from the table. The case study thus highlights the significant improvement in energy-efficient process operation via proper management of the unconstrained DOFs using SOPVs.

17.7 Conclusions

In conclusion, this chapter has developed simple common sense heuristics/guidelines for sustainable operation of highly integrated chemical plants. We have developed the heuristic that the control system should be designed to directly/indirectly hold the recycle rate in a recycle loop by adjusting an inflow/outflow from the recycle loop. For maximizing production with the bottleneck constraint inside the recycle loop, this heuristic allows locating the TPM at the bottleneck constraint to eliminate back-off from the hard bottleneck constraint. We have argued that smart management of economically important unconstrained DOFs is necessary to ensure sustainable energy-efficient process operation. The two case studies demonstrate that the application of these heuristics can lead to significant sustainability benefit in the form of higher maximum production from a given plant as well as better energy efficiency.

References

1. Murthy Konda, N.V.S.N.; Rangaiah, G.P.; Krishnaswamy, P.R. Plantwide control of industrial processes. *Ind. Eng. Chem. Res.* 2005, 44(22), 8300–8313.
2. Luyben, W.L. Dynamics heat-exchanger bypass control. *Ind. Eng. Chem. Res.* 2011, 50(2), 965–973.
3. Luyben, W.L. Dynamics and control of recycle systems. 2. Comparison of alternative process designs. *Ind. Eng. Chem. Res.* 1993, 32(3), 476–486.
4. Luyben, W.L. Snowball effects in reactor/separator processes with recycle. *Ind. Eng. Chem. Res.* 1994, 33(2), 299–305.
5. Price, R.M.; Georgakis, C. Plantwide regulatory control design procedure using a tiered framework. *Ind. Eng. Chem. Res.*1993, 32, 2693–2705.
6. Skogestad, S. Plantwide control: The search for the self-optimizing control structure. *J. Proc. Control* 2000, 10(5), 487–507.
7. Kanodia, R.; Kaistha, N. Plant-wide control for throughput maximization: A case study. *Ind. Eng. Chem. Res.* 2010, 49, 210–221.
8. Wright, R.O. U.S. Patent 2,471,134, May 24, 1949.
9. Petlyuk, F.B.; Platonoy, V.M.; Slavinskii, D.M. Thermodynamically optimal method for separating multicomponent mixtures. *Int. Chem. Eng.* 1965, 5(3), 555–561.
10. Kumari, P.; Jagtap, R.; Kaistha, N. Control system design for energy efficient on-target product purity operation of a high purity Petlyuk Column. *Ind. Eng. Chem. Res.* 2014, 53(42), 16436–16452.

18

Evaluating Sustainability of Process, Supply Chain, and Enterprise: A Bio-Based Industry Case Study

Iskandar Halim and Arief Adhitya
Institute of Chemical and Engineering Sciences

R. Srinivasan
Indian Institute of Technology Gandhinagar

18.1 Introduction

The issues of sustainable production have gained significant attention and have become an important business factor in the chemical industry. Sustainable production can be defined as the creation of goods and services using processes and systems in a sustainable manner. This means that the activities involved should be nonpolluting; conserving energy and natural resources; economically viable; safe and healthful for workers, communities, and consumers; as well as socially and creatively rewarding for all working people [1]. In recent years, incidents affecting sustainability performances have hit many chemical companies hard and have appeared in the front news quite frequently [2]. Among the major incidents are environmental damages caused by emissions, presence of hazardous chemicals in products, unsafe plant operations, and labor disputes. Since most companies outsource far more than 50% of their incoming materials and services [2], there is thus a pressing need for them to evaluate the sustainability issues affecting their operations. This needs to be done beyond just the process/plant level. There is a need to include the entire supply chain of the enterprise, starting from material sourcing to manufacturing, storage, logistics, and distribution. Within the discipline of chemical engineering, such challenges might be best addressed through process systems engineering (PSE) approaches [3]. This is because, unlike the reductionist approach of other specialties, PSE adopts a holistic or system-level view of a plant or an enterprise. Such a system-level view is crucial for understanding the complex interactions between each component of sustainable production and the economic, environmental, and social dimensions of the plant or enterprise.

Setting up and managing a chemical enterprise involves a host of strategic, tactical, and operational decision-making activities related to raw material sourcing, product manufacturing, storage, logistics, and distribution [4]. Within the PSE field, simulation has played an important role in supporting such activities for achieving efficiency, cost reduction, and now the sustainability of the enterprise. Chemical process simulation, for example, has become a standard tool for analyzing the plant and evaluating various modifications efficiently, in a short time, without the need for extensive experimentation or pilot plant testing. It has also been successfully used for environmental studies, for instance, to evaluate the environmental impacts from various plant/process modifications [5–9]. Analogous to chemical process simulation is supply chain simulation, which is a powerful tool for analyzing the effects of changes in the supply chain design and configuration and operating policies along the supplier–manufacturer–distributor chain. Recently, it has also been applied to evaluate both the economic and environmental impacts of various supply chain decisions related to inventories, distribution network configuration, and ordering policy [10].

One essential element for assessing sustainability is indicator metrics—this will allow the enterprise to measure its progress toward sustainable production. In recent years, a number of indicator metrics have been proposed to enable a consistent comparison and identification of sustainable alternatives. To name a few, a methodology called Waste reduction (WAR) algorithm has been developed by the U.S. Environmental Protection Agency for evaluating the relative environmental impact of any given chemical process [11]. The basic principle of the WAR algorithm is environmental impact balance, which involves assigning an index value to each process material to indicate its potential impacts to different environmental categories such as toxicity, global warming, ozone depletion, and photochemical oxidation. Using the index, the total impact of a waste stream in the plant can be known by calculating the sum of each material index of the waste stream weighted by its flow rate. Another set of indicator metrics has been proposed by Azapagic and Perdan [12] based on life cycle thinking. In their approach, the sustainability of a plant or an enterprise is evaluated by considering the three aspects of sustainability: economic, environmental, and social. As in life cycle assessment (LCA), where the environmental impacts are expressed per functional unit of product or service delivered, in the life-cycle-based indicator metrics approach, each of the sustainability indicators is calculated based on the function that the system delivers. The environmental indicator includes the usual environmental categories used in the WAR algorithm such as global warming, ozone depletion and toxicity, as well as resource efficiencies and recyclability (material, energy, water, and land usage). The economic indicator measures the success of the enterprise economically—this is reflected by the company's profit economic value added. The social indicator is a measure of a company's attitude toward its own employees, suppliers, contractors, customers, and society at large. This is indicated by work satisfaction, income distribution, involvement in community projects, compliance to international standards of conduct, etc. Hence, unlike the WAR algorithm, which is only applicable to environmental aspect at the plant level, the life-cycle-based sustainability metrics can be used to evaluate the sustainability of the entire enterprise from raw material acquisition, product manufacturing, distribution, and disposal. IChemE in UK has also produced a set of metrics to enable process industry companies to measure the three components of enterprise sustainability [13]. Some of the indicator metrics used in this approach are similar to those of Azapagic and Perdan [12].

This chapter discusses the application of integrated simulation and indicator metrics calculation for holistically evaluating the sustainability of an enterprise at three different levels: manufacturing plant, supply chain, and multisite enterprise. The approach will be illustrated using several industrial case studies that utilize palm oil as the feedstock. The rest of the chapter is structured as follows: the next section describes sustainability issues in palm oil production. Section 18.3 presents our integrated simulation and metrics calculation approach to biodiesel plant. This is followed by application of the technique at the supply chain level of biodiesel production in Section 18.4 and multisite enterprise of palm-oil-based lubricant production in Section 18.5. Finally, Section 18.6 gives some concluding remarks.

18.2 Sustainability Issues of Palm Oil Production

Our modern society is heavily dependent on fossil fuels (oil, coal, and gas) to meet its energy needs. Several negative impacts due to this overreliance on fossil fuels have long been documented. The burning of fossil fuels releases greenhouse gases that trap heat that comes from the sun. As more greenhouse gases are emitted, more heat is trapped in the atmosphere causing an increase in the earth's surface temperature—the phenomenon known as global warming. Another major concern is the nonrenewability of fossil fuels. Fossil fuels are generally hydrocarbons that were formed and deposited as a result of decomposition of organic matters over millions of years ago. While new reserves are still being discovered and new technologies (e.g., hydraulic fracturing technique) that enable unlocking of fossil fuel deposits in unconventional places are being improved, our consumption of fossil fuels is also increasing at an ever faster rate. Hence, it is a matter of time that they will be completely depleted.

Palm oil is one important commodity that is widely used in our everyday life. In fact, about 50% of the goods that we use today contain palm oil as one of the ingredients [14]. In 2012, the global demand for palm oil was estimated to be 52.1 million tons with 85% of the supply coming from Southeast Asia (Indonesia and Malaysia) [15]. Besides being used as cooking oil and food ingredient, palm oil has other nonfood applications such as soaps, detergents, and cosmetics [16]. Another potential use of palm oil is as a substitute for mineral oil in the production of metalworking fluid (lubricant oil) [17].

Recent years have seen a surge in interest in palm oil biomass as an alternative to fossil fuels. In fact, palm oil biomass has become one of the largest biomass sources to replace the petroleum-based diesel oil. This is due to the fact that compared to other biomass, palm oil has the highest oil yield per unit land area [15]. It is also inherently renewable and carbon neutral in its life cycle. Further, the burning of its waste biomass produces less ash and sulfur dioxide emissions compared to coal or oil [18]. However, there are some downsides to utilizing palm oil biomass as an alternative fuel. While palm oil has become one of the most viable feedstocks for the production of biodiesel, the sustainability of palm oil production has been brought to question due to the following reasons [19]. The cultivation of palm oil has been linked with deforestation and destruction of peatland, leading to loss of wildlife and biodiversity. Further, palm oil production has been associated with air pollution due to the burning of forests and peatlands for the purpose of land clearing and also water pollution due to the heavy usage of pesticides and dumping of mill effluent to the water bodies. It has also sparked conflicts within the local communities who depend on forests for their livelihoods. Another concern associated with palm-oil-based biodiesel is its direct competition with food crops for agricultural land. Such competition will lead to escalation in food prices in the long run. All these issues, thus, highlight the complex and intricate connections between the economic, environmental, and social dimensions of palm oil production and use. Sustainable palm oil production should hence strive to balance between economic growth, social development, and environmental protection. We demonstrate the use of simulation-based methods for this purpose.

18.3 Plant-Level Sustainability Metrics and Simulation-Optimization

One way of improving the sustainability of palm-oil-based biodiesel production is by utilizing waste oil as the feedstock—this will eliminate direct competition with food crops. One common method of manufacturing such biodiesel is acid-based transesterification process, which involves the following steps:

- Transesterification: Initially, a stream of fresh methanol and sulfuric acid is mixed with the recycled stream for reaction with triolein, which is a major triglyceride component in the palm oil. The reaction, which is carried out at 80°C and 400 kPa pressure, converts 95% (by weight) of oil into biodiesel (methyl oleate) and glycerol by-products according to the following reaction:

$$\text{Triolein} + 3\text{ methanol} \rightarrow 3\text{ methyl oleate} + \text{glycerol}.$$

- Methanol recovery: The reaction product is cooled prior to distillation under a vacuum condition to recover the excess methanol. In this process, 94% (by weight) of methanol is recovered and recycled back to the reactor. The bottom stream is cooled and sent to the acid removal unit.

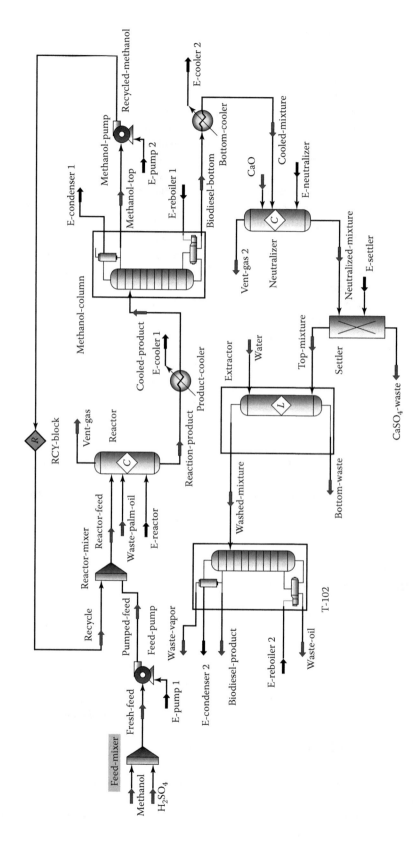

FIGURE 18.1 Flow sheet of acid-catalyzed biodiesel production process.

TABLE 18.1 Environmental and Economic Data for Biodiesel Process

Description	Value
Environmental Impact Index (Per kg)	
Methanol	0.495
Triolein	0.020
Methyloleate	0.056
Glycerol	1.838
H_2SO_4	0.659
Water	0
CaO	0.511
$CaSO_4$	0.249
Cost of Raw Material ($/kg)	
Methanol	0.18
H_2SO_4	0.06
Triolein (waste oil)	0.20
Water	0.01
CaO	0.04
Price of Product ($/kg)	
Methyloleate (biodiesel)	0.60
Cost of Waste Stream ($/kg)	
Waste vapor	0.05
Waste water	0.12
Cost of Energy ($/kWh)	
Pump	0.062
Cooler	0.003
Reactor	0.003
Condenser (distillation)	0.003
Reboiler 1	0.01
Reboiler 2	0.15
Neutralizer	0.03

- Acid removal: This step involves neutralizing the sulfuric acid component through calcium oxide (CaO) addition to produce $CaSO_4$ salt and water. This is followed by a settling process to remove $CaSO_4$. Complete removal of the acid component is assumed in this step.
- Water washing: This operation involves separating the biodiesel product from the glycerol, methanol, and acid catalyst stream. This is done in an extraction column by washing with water.
- Biodiesel purification: In the final stage, the biodiesel product is distilled further under vacuum condition to obtain the desired purity (99.6% by weight).

Figure 18.1 shows the flow sheet of the process, which has been adopted from Zhang et al. [20]. Table 18.1 shows the unit cost and the environmental impact index (WAR index) of each material in this process. In WAR, each process material is assigned an index value to indicate its potential impacts. Hence, a total environmental impact index ψ of a chemical k can be calculated as the sum of each environmental impact category of that chemical weighted by a factor α, as described by the following expression:

$$\psi_k = \sum_l \alpha_l \psi_{k,l}^s, \tag{18.1}$$

where α_l is a relative weighting factor for impact category type l independent of chemical k and $\psi_{k,l}^s$ is the specific potential environmental impact of chemical k for an environmental impact type l, which includes the following categories: global warming, ozone depletion, acid rain, smog formation, human toxicity, aquatic toxicity, and terrestrial toxicity. The relative weighting factor α_l allows the environmental impact of a chemical k, ψ_k, to be customized to specific or local conditions with the recommended value between 0 and 10 according to local needs and policies—for simplicity, we have set α_l to 1. Based on this index, the total impact of a waste stream in the plant $\dot{I}_{\text{waste}}$ can be calculated as

$$\dot{I}^{\text{waste}} = \dot{M}_m \times x_{km}^{\text{waste}} \times \psi k, \tag{18.2}$$

where $\dot{M}_m$ is the mass flow rate of waste stream m, x_{km}^{waste} is the mass fraction of chemical k in the stream m.

The process currently generates four waste streams: salt mixture of the settling column, waste water of the extraction column, vent gas of the purifier column, and waste oil of the biodiesel purification column. The total environmental impact of these waste streams, as calculated using WAR is 294. The process exhibits a $60/h profit, which is the difference between the sales revenue and the raw material, energy, and waste treatment costs. This profit value is quite low, thereby highlighting the need for subsidy for this process. The process can be made more sustainable through better utilization of raw material and energy—this can be investigated by optimizing the process conditions. We have used the HYSYS process simulator for this purpose, although other process simulators (such as Aspen Plus, PRO/II, and gPROMS) can be equally applied. We have identified the following variables as the key decision variables for optimization: feed streams of waste oil, methanol, H_2SO_4, the energy of the two coolers, and water flow rate. These will be optimized for the objectives of maximizing profit and at the same time minimizing environmental impact. While many techniques have been developed to facilitate multiobjective optimization, we have used a multiobjective simulated annealing algorithm [21] for this purpose. The basic principle of simulated annealing is modeled after the statistical mechanics of annealing of metals. In contrast to greedy search that moves in the direction of improving objective function values and is hence susceptible to local optima, simulated annealing seeks to reach the global minimum energy state by temporarily accepting even poorer solutions [22]. The interested reader is referred to Halim and Srinivasan [9] for detailed information about the simulation–optimization procedure. Table 18.2 shows the bounds of decision variables for simulation–optimization—these have been decided to ensure that flow sheet convergence (especially in the distillation column operation) can be attained during the simulation–optimization runs.

Figure 18.2 shows the Pareto optimal plot after 2000 runs. It clearly highlights the trade-off between the two objectives [9]. Compared to the base case, the maximum profit can be increased threefold to $189/h (216%), but at a cost of 3% increase in the environmental impact. The Pareto plot also shows the possibility of improving both the impact and profit. In this case, the maximum reduction in the environmental impact and improvement in the profit from the base values is found to be 4% and 36%,

TABLE 18.2 Initial Values and Ranges for Decision Variables in Biodiesel Process

Decision Variables	Base Value	Minimum Value	Maximum Value
Oil (kg/h)	1000	975	1100
Methanol (kg/h)	210	208	218
H_2SO_4 (kg/h)	150	138	166
Water (kg/h)	110	103	117
Energy of cooler 1 (kWh)	18.0	17.7	18.6
Energy of cooler 2 (kWh)	5.0	4.7	5.3

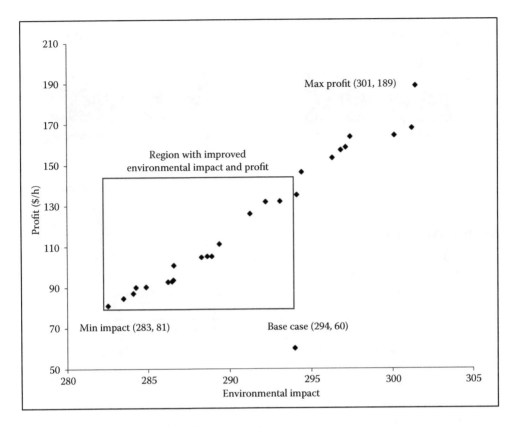

FIGURE 18.2   Pareto results for biodiesel production plant.

respectively. Besides this process condition optimization, some other measures to improve both the profit and environmental impact can also be investigated. These include recovery and recycling of glycerol and methanol from the waste streams and heat integration. However, these will require structural changes to the process.

18.4 Supply-Chain-Level Sustainability Metrics and Assessment

Beyond the process plant, it is also necessary to evaluate the sustainability of a production process at the supply chain level. Figure 18.3 shows the schematic flow of a typical palm oil biodiesel supply chain together with the inputs, outputs, and emissions. A similar metrics-based approach can be applied to measure the economic, environmental, and social performances along the supply chain, from plantation to manufacturing plant. Economic performance can be indicated by various costs incurred along the supply chain. Environmental sustainability can be evaluated through indicators such as energy consumption, CO_2 emission, eutrophication, and acidification [23]. For social impacts, the stakeholders concerned are the local community and workers along the supply chain. Hence, the social sustainability metrics can be measured through various socioenvironmental factors, such as deforestation, soil erosion, freshwater depletion, nutrient loss, pollution from chemicals, and biodiversity loss, as well as socioeconomic factors, such as employment, wages, child labor, healthcare, education, and human rights issues like land conflicts and ownership [24].

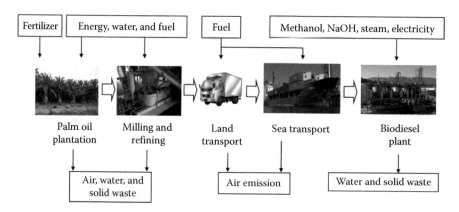

FIGURE 18.3   Palm oil biodiesel supply chain.

TABLE 18.3   Environmental Impacts (Per Ton Biodiesel) of Three Biodiesel Supply Chains

Indicators	Malaysia	Indonesia	Thailand
Energy consumption (GJ)	20.14	18.48	32.26
CO_2 emission (kg CO_2-eq)	6457	5865	8146
Eutrophication (kg O_2-eq)	117.96	13.92	1029.9
Acidification (kg SO_2-eq)	3.746	13.13	62.26

   To illustrate the sustainability assessment at the supply chain level, we compare three palm oil biodiesel case studies from literature: a typical biodiesel supply chain in Malaysia [25], a specific biodiesel supply chain in North Sumatra, Indonesia [26], and a specific biodiesel supply chain in Southern Thailand [27]. Here, we focus on the environmental and social aspects only. The environmental impacts that are calculated following the LCA-based approach of Achten et al. [23] comprise impacts from plantation, milling, and biodiesel production. Table 18.3 summarizes the environmental assessment results. This was obtained from simulation using Excel. As shown in the table, majority of the impacts originate from the use of fertilizers during the plantation stage. The results also show that among the three biodiesel supply chains, the one in Indonesia has the lowest environmental impacts except for acidification. This is due to the fact that this specific plantation in North Sumatra does not use man-made fertilizers and herbicides, but instead recycles its own wastes as fertilizers (empty fruit bunches and mill effluents) and fuel sources (palm oil fibers and shells), thereby reducing the environmental impacts [26].

   For social indicators, since no detailed information is available on the social impacts of these three biodiesel supply chains, we perform a general assessment on the social implications of each supply chain during the plantation stage using information that has been gathered from various literature sources [28]. The results are summarized in Table 18.4. While palm oil is a major employment sector in Southeast Asia, it comes with significant socioenvironmental costs including deforestation, soil erosion, depletion of freshwater resources, nutrient losses and soil nutrient depletion, pollution, and loss of biodiversity. Among the three countries, Indonesia has the highest number of land conflict cases. Alongside the land conflict are low wages, child labor, education, and health that still need to be addressed, especially in Indonesia and Malaysia, in order to achieve sustainability.

TABLE 18.4 Social Impacts of Palm Oil Biodiesel in Malaysia, Indonesia, and Thailand

Indicators	Malaysia	Indonesia	Thailand
Socioenvironmental			
Deforestation (%)	33	66	66
Soil erosions	14.9–79 tons/ha/year	57–1500 tons/year	
Depletion of freshwater resources	1.5–2.0 m³/tons fresh fruit bunch (FFB)	0.8 m³/tons FFB	1.0–1.3 m³/tons FFB
Nutrient losses and soil nutrient depletion	132 ton N, 43 tons P/year/ plantation	386,000 tons/year	
Pollution from chemicals	30 kg biological oxygen demand (BOD)/tons FFB	30 kg BOD/tons FFB	30 kg BOD/tons FFB
Biodiversity	85% drop in species types	50% drop in number of orangutans	53% drop in species types
Socioeconomic			
Land conflicts	40 cases in Sarawak	3500 cases in Indonesia	
Land rights and ownership	41% smallholders	44% smallholders	76% smallholders
Employment	400,000	1,000,000	650,000
Wages	US$92	US$32.8	US$150
Child labor	72,000	150,000	
Education	60% plantations without schools	Similar to Malaysia	Children go to school
Health	50% paraquat pesticide sprayers suffer skin or eye injuries, industrial accidents	Similar to Malaysia	Similar to Malaysia

Source: Tan, Y.N., Sustainability analysis of palm oil suppliers for biodiesel production using triple bottom line approach, Final Year Project Report. National University of Singapore, Singapore, 2011.

18.5 Enterprise-Level Sustainability Assessment through Simulation

The enterprise level includes multiple production plants within the same company and their supply chains. In this section, we consider a multisite palm-oil-based lubricant manufacturing enterprise as shown in Figure 18.4. The enterprise has a central sales department and three production plants in different locations. Each local plant has different departments performing specific functions: commercial, scheduling, operations, packaging, storage, procurement, and logistics. The customers and suppliers are external parties that constitute the supply chain. Material flows are denoted by solid arrows and information flows by dotted arrows. We can observe a parallel between a plant (with its multiple unit operations) and an enterprise supply chain (with its multiple entities). Like in a process plant, there are complex supply chain dynamics due to the numerous interactions among the entities. The impact of an entity's action on the overall indicator metrics may not be obvious. Like a process simulator that is used to study the impacts of different decision variables in a plant, a supply chain simulator could capture the behavior of the entities, their interactions, the resulting dynamics, and the impacts on the overall performance of the enterprise. Here, we use the dynamic simulation model of Adhitya and Srinivasan [29] integrated with environmental indicators for enterprise-level sustainability assessment [10]. The simulation model has been developed using MATLAB®/Simulink®, where flows of material, information, and finance are depicted by various mathematical, logical, and algorithmic operation blocks.

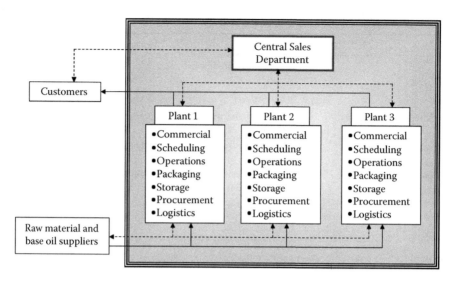

FIGURE 18.4   Multisite lubricant enterprise.

The enterprise supply chain operates in pull mode, i.e., triggered by customer demand. When the central sales department receives an order from a customer, it has to decide which plant to assign the order following an *order assignment policy*. For example, if assignment is based on earliest completion date, the central sales department asks the commercial department of each plant for projected completion date of that order and assigns the order to the plant that can deliver at the earliest date. This new order is then inserted into the job schedule by the scheduling department of the selected plant, based on a *job scheduling policy*, for example, first-come-first-serve and shortest processing time.

Production is carried out by the operations department following the job schedule and the product recipe, which defines the type and amount of raw materials needed to make the product. The required raw materials are transferred from storage to operations and after a certain processing time, products are collected. The products are then sent to the packaging department for packaging. The scheduling department removes the completed job from the schedule and proceeds to the next job. After packaging, the packaged products are immediately delivered to the customers by the logistics department; no product inventory is kept. The delivery time is calculated based on the distance between the plant and the customer, whose locations are represented through a pair of coordinates (x, y). The procurement department manages raw materials replenishment following a *procurement policy*. For example, in the reorder point policy, raw material is purchased from suppliers when its inventory falls below a reorder point to bring it back to the top-up level.

Economic performance is measured through profit and customer satisfaction of the overall enterprise. Profit is calculated as revenue from product sales minus costs:

$$\text{Profit} = \text{Revenue} - \begin{bmatrix} \text{Raw material cost} + \text{Variable operating cost} \\ + \text{Fixed operating cost} + \text{Packaging cost} \\ + \text{Transportation cost} + \text{Inventory cost} \\ + \text{Late penalty cost} \end{bmatrix} \qquad (18.3)$$

Customer satisfaction is calculated as the percentage of nonlate orders out of the total number of customer orders:

$$CS = \left(1 - \frac{\text{Number of late orders}}{\text{Number of orders}}\right) \times 100\%. \qquad (18.4)$$

Environmental performance is measured through eight indicators: acidification, global warming potential over 100 years (GWP 100), solid waste, water use, land use, ecotoxicity, nonrenewable energy consumption, and transportation. The stages included in the environmental impacts calculation are raw material acquisition, processing, packaging, and transportation from the plants to the customers.

Table 18.5 lists the environmental indicators per kg lubricant product [30]. In this case study, we compare two-order assignment policies. Under the *earliest completion date* policy, an order is assigned to the plant that can deliver the product to the customer at the earliest. On the other hand, the *nearest customer location* policy assigns the order to the plant nearest to the customer, which should minimize transportation cost and its associated environmental impact. Simulation is performed to compare these two order assignment policies and the results are shown in Table 18.6. The earliest completion date policy has a higher profit ($29.25 M vs. $28.08 M) due to more jobs completed, since orders are assigned to the plant that can deliver at the earliest, which is the least busy plant. Consequently, this also leads to higher customer satisfaction (100% vs. 94%). The nearest customer location policy results in a 17% reduction in transportation impact at the expense of profit and customer satisfaction. Other environmental

TABLE 18.5 Environmental Impacts (Per kg Lubricant) for Enterprise-Level Case Study

Indicators	Production	Packaging
Acidification (g SO$_2$)	6.55	0.43
GWP 100 (kg CO$_2$-eq)	0.855	0.2
Solid waste (g)	23.93	4.21
Water use (kg)	10.96	0.97
Land use (m^2)	3.89	0
Ecotoxicity (cg Pb-eq)	14.2	0
Nonrenewable energy (MJ)	26.39	1.37
Transportation (millipoints/m^3 km)	8	8

Source: Adhitya, A. et al., *Environ. Sci. Technol.*, 45, 10178–10185, 2011.

TABLE 18.6 Simulation Results for Two Order Assignment Policies

Indicators	Earliest Completion Date	Nearest Customer Location
Profit (M$)	29.25	28.08
Customer satisfaction (%)	100	94
Number of jobs completed	97	93
Acidification (g SO$_2$/kg product)	11.12	10.79
GWP 100 (kg CO$_2$-eq/kg product)	3.75	3.53
Solid waste (g/kg product)	31.32	31.07
Water use (kg/kg product)	15.21	14.95
Land use (m^2/kg product)	1.83	1.99
Ecotoxicity (cg Pb-eq/kg product)	6.83	7.42
Nonrenewable energy (MJv)	31.59	31.28
Transportation (millipoints/kg product)	16.95	14.12

indicators are comparable. We can see that simulation reveals and quantifies the trade-off between environmental and economic performance resulting from a policy decision at the enterprise level.

18.6 Conclusions

The current push toward sustainability has pressurized the process industries to embrace the concept of sustainable production as part of their core business values. This chapter presented a simulation and indicator metrics-based approach for sustainability evaluation that is done at three levels: manufacturing plant, supply chain, and multisite enterprise. Economic performance was measured through indicators such as profit, cost, or customer satisfaction. For environmental indicators, metrics such as the WAR algorithm and life-cycle-based indicator metrics were applied. Social impacts were measured through various socioenvironmental and socioeconomic factors such as deforestation, soil erosion, freshwater depletion, wages, child labor, healthcare, etc. The approach has been successfully tested using several interconnected case studies involving palm oil feedstock.

References

1. Lowell Center for Sustainable Production. 2016. What is sustainable production? http://www.sustainableproduction.org/abou.what.php (Accessed February 11, 2016).
2. Cheverton, P. and van der Velde, J.P. 2011. *Understanding the Professional Buyer What Every Sales Professional Should Know about How the Modern Buyer Thinks and Behaves.* London: Kogan Page Ltd.
3. Bakshi, B.R. and Fiksel, J. 2003. The quest for sustainability: Challenges for process systems engineering. *AIChE Journal* 49(6): 1350–1358.
4. Naraharisetti, P.K., Adhitya, A., Karimi, I.A. and Srinivasan, R. 2009. From PSE to PSE²—Decision support for resilient enterprises. *Computers and Chemical Engineering* 33(12): 1939–1949.
5. Dantus, M.M. and High, K.A. 1996. Economic evaluation for the retrofit of chemical processes through waste minimization and process integration. *Industrial Engineering Chemistry and Research* 35: 4566–4578.
6. Fu, Y., Diwekar, U.M., Young, D. and Cabezas, H. 2000. Process design for the environment: A multi-objective framework under uncertainty. *Clean Products and Processes* 2: 92–107.
7. Mata, T.M., Smith, R.L., Young, D.M. and Costa, C.A.V. 2003. Evaluating the environmental friendliness, economics and energy efficiency of chemical processes: Heat integration. *Clean Technologies and Environmental Policy* 5: 302–309.
8. Othman, M.R., Repke, J.U., Wozny, G. and Huang, Y. 2010. A modular approach to sustainability assessment and decision support in chemical process design. *Industrial and Engineering Chemistry Research* 49(17): 7870–7881.
9. Halim, I. and Srinivasan, R. 2011. A knowledge-based simulation-optimization framework and system for sustainable process operations. *Computers and Chemical Engineering* 35(1): 92–105.
10. Adhitya, A., Halim, I. and Srinivasan, R. 2011. Decision support for green supply chain operations by integrating dynamic simulation and LCA indicators: Diaper case study. *Environmental Science and Technology* 45: 10178–10185.
11. Cabezas, H., Bare, J.C. and Mallick, S.K. 1999, Pollution prevention with chemical process simulators: The generalized waste reduction (WAR) algorithm—Full version. *Computers and Chemical Engineering* 23(4–5): 623–634.
12. Azapagic, A. and Perdan, S. 2000. Indicators of sustainability development for industry: A general framework. *Process Safety and Environmental Protection* 78(4): 243–261.
13. Tallis, B. 2002. *Sustainable Development Progress Metrics.* IChemE Sustainable Development Working Group, Institution of Chemical Engineers, Rugby, UK.

14. Rainforest Rescue. 2016. PALM OIL: Facts about the ingredient that destroys the rainforest. https://www.rainforest-rescue.org/files/en/palm-oil-download.pdf (Accessed February 11, 2016).

15. Palm Oil Research. 2016. Untangling the great palm oil debate. http://www.palmoilresearch.org/statistics.html (Accessed February 11, 2016).

16. Soyatech. 2016. Palm oil facts. http://www.soyatech.com/Palm_Oil_Facts.htm (Accessed February 11, 2016).

17. Chang, T., Yunus, R., Rashid, U., Choong, T.S.Y., Biak, D.R.A. and Syam, A.M. 2015. Palm oil derived trimethylolpropane triesters synthetic lubricants and usage in industrial metalworking fluid. *Journal of Oleo Science* 64(2):143–151.

18. LCAworks. 2011. The availability of sustainable biomass for use in UK power generation. http://www.lcaworks.com/Low%20Carbon%20Bioelectricity%20in%20the%20UK.pdf. (Accessed February 11, 2016).

19. Centre for Science in the Public Interest. 2005. Cruel oil: How palm oil harms health, rainforest and wildlife. http://www.cspinet.org/palmoilreport/PalmOilReport.pdf (Accessed February 11, 2016).

20. Zhang, Y., Dube, M.A., McLean, D.D. and Kates, M. 2003. Biodiesel production from waste cooking oil: 1. Process design and technological assessment. *Bioresource Technology* 89: 1–16.

21. Sankararao, B. and Gupta, S.K. 2007. Multi-objective optimization of an industrial fluidized-bed catalytic cracking unit (FCCU) using two jumping gene adaptations of simulated annealing. *Computers and Chemical Engineering* 31: 1496–1515.

22. Kirkpatrick, S., Gelatt, C.D. and Vecchi, M.P. 1983. Optimization by simulated annealing. *Science* 220: 671–680.

23. Achten, W.M.J., Vandenbempt, P., Almeida, J., Mathijs, E. and Muys, B. 2010. Life cycle assessment of a palm oil system with simultaneous production of biodiesel and cooking oil in Cameroon. *Environmental Science and Technology* 44(12): 4809–4815.

24. Smeets, E., Faaij, A. and Lewandowski, I. 2005. The impact of sustainability criteria on the costs and potentials of bioenergy production. Report NWS-E-2005-6. Copernicus Institute, Utrecht University, Utrecht, The Netherlands.

25. Lam, M.K., Lee, K.T. and Mohamed, A.R. 2009. Life cycle assessment for the production of biodiesel: A case study in Malaysia for palm oil versus jatropha oil. *Biofuels, Bioproducts and Biorefining* 3(6): 601–612.

26. Hayashi, K. 2007. Environmental impact of palm oil industry in Indonesia. *Proceedings of International Symposium on EcoTopia Science 2007*, Nagoya University, Nagoya, Japan, 646–651.

27. Pleanjai, S., Gheewala, S.H. and Garivait, S. 2007. Environmental evaluation of biodiesel production from palm oil in a life cycle perspective. *Asian Journal on Energy and Environment* 8(1–2): 15–32.

28. Tan, Y.N. 2011. Sustainability analysis of palm oil suppliers for biodiesel production using triple bottom line approach. Final Year Project Report. National University of Singapore, Singapore.

29. Adhitya, A. and Srinivasan, R. 2010. Dynamic simulation and decision support for multisite specialty chemicals supply chain. *Industrial and Engineering Chemistry Research* 49: 9917–9931.

30. Nwe, E.S., Adhitya, A., Halim, I. and Srinivasan, R. 2010. Green supply chain design and operation by integrating LCA and dynamic simulation. *Computer-Aided Chemical Engineering* 28: 109–114.

Index